Structural Steel Design

A PRACTICE-ORIENTED APPROACH

Structural Steel Design

A PRACTICE-ORIENTED APPROACH

ABI AGHAYERE

Rochester Institute of Technology, Rochester, NY

JASON VIGIL

Consulting Engineer, Rochester, NY

Prentice Hall

Upper Saddle River, New Jersey
Columbus, Ohio

Library of Congress Cataloging-in-Publication Data

Aghayere, Abi O.
    Structural steel design: a practice-oriented approach / Abi Aghayere, Jason Vigil. – 1st ed.
      p. cm.
    Includes bibliographical references.
    ISBN-13: 978-0-13-234018-2
    ISBN-10: 0-13-234018-6
  1. Building, Iron and steel—Textbooks.   2. Steel, Structural—Textbooks.   I. Vigil, Jason, 1974   II. Title.
    TA684.A267 2009
    693′.71—dc22

                      2008038447

Editor in Chief: Vernon Anthony
Acquisitions Editor: Eric Krassow
Editorial Assistant: Sonya Kottcamp
Production Coordination: Aptara®, Inc.
Project Manager: Louise Sette
AV Project Manager: Janet Portisch
Operations Specialist: Laura Weaver
Art Director: Diane Ernsberger
Cover Designer: Michael Fruhbeis
Cover Image: Stockbyte
Director of Marketing: David Gesell
Marketing Manager: Derril Trakalo
Senior Marketing Coordinator: Alicia Wozniak

Table and Figure Credits:

Copyright © American Institute of Steel Construction, Inc. Reprinted with permission. All rights reserved. Pages 135, 172, 180, 190, 191, 193, 197, 409, 412, 414, 461, 559, 578, 583, and 584.

Portions of this publication reproduce sections from the 2003 *International Building Code*, International Code Council, Inc., Country Club Hills, IL. Reproduced with permission. All rights reserved. Pages 48, 98, 118.

This book was set in Times Roman by Aptara®, Inc. and was printed and bound by Hamilton Printing. The cover was printed by Phoenix Color Corp.

Pearson Education Ltd., London
Pearson Education Singapore Pte. Ltd.
Pearson Education Canada, Inc.
Pearson Education—Japan

Pearson Education Australia Pty. Limited
Pearson Education North Asia Ltd., Hong Kong
Pearson Educación de Mexico, S.A. de C.V.
Pearson Education Malaysia Pte. Ltd.

Prentice Hall
is an imprint of

www.pearsonhighered.com

10 9 8 7 6 5 4 3 2
ISBN-13: 978-0-13-234018-2
ISBN-10:   0-13-234018-6

To my wife, Josie, the love of my life and my greatest earthly blessing, and to my precious children, Osarhieme, Itohan, Odosa, and Eghosa, for their patience, encouragement and unflinching support; to my mother for instilling in me the virtue of excellence and hard work; and finally, and most importantly, to God Almighty, the utmost Structural Engineer, for the grace, inspiration, strength, and wisdom to complete this project.

— Abi Aghayere, *Rochester, NY*

I wish to express gratitude for the instruction and guidance over the years from my teachers and colleagues and the role that they played in my professional life. I am also thankful for my family—Michele, Adele, and Ivy—whose patience has made this endeavor possible. And finally, I wish to give thanks to my Lord and Savior Jesus Christ, who gives me the strength and wisdom to bring praise to His name in all that I do.

— Jason Vigil, *Rochester, NY*

PREFACE

The knowledge and expertise required to design steel-framed structures are essential for any architectural or structural designer, as well as students intending to pursue careers in the field of building design and construction. This textbook provides the essentials of structural steel design required for typical projects from a practical perspective so that students understand each topic *and* understand how to combine each of these topics into a project resulting in a fully designed steel structure. The American Institute for Steel Construction is actively encouraging educators to "expose their students to the design of steel building elements in a realistic building context." This text will help bridge the gap between the design of specific building components and the complete design of a steel structure. We provide details and examples that not only provide the reader with an essential background on structural steel design, but also provide subject material that closely mirrors details and examples that occur in practice.

INTENDED AUDIENCE

This text is ideal for students in a typical undergraduate course in structural steel design, and will sufficiently prepare students to apply the fundamentals of structural steel design to a typical project that they might find in practice. It is suitable for students in civil engineering, civil engineering technology, architectural engineering, and construction management and construction engineering technology programs. This text also serves as a good resource for practitioners of structural steel design because the approach taken is intended to be practical and easily applicable to typical, everyday projects. It is also a helpful reference and study guide for the Fundamentals of Engineering (FE) and the Professional Engineering (PE) exams. This text covers the course content for the Steel Design I course and a majority of the course content for the Steel Design II course in the curriculum for basic education of a structural engineer proposed by the National Council for Structural Engineering Associations (NCSEA).

UNIQUE FEATURES OF THIS TEXT

One of the focal points of this text is to help the reader learn the basics of steel design and how to practically apply that learning to real-world projects by combining the building code and material code requirements into the analysis and design process. This is essential for any practicing engineer or any student who wants to work in this field.

We use numerous details and diagrams to help illustrate the design process. We also introduce a student design project as part of the end-of-the-chapter problems to expose students to the important aspects of a real-world steel building design project. Several other unique features of this text are listed below.

1. The use of realistic structural drawings and practical real-world examples, including practical information on structural drawings.

2. An introduction to techniques for laying out floor and roof framing, and for sizing floor and roof decks. General rules of thumb for choosing beam spacing versus deck size and bay sizes, and beam and girder directions are discussed.

3. A discussion of other rules of thumb for sizing steel to allow for the quick design of common structural members (e.g., open-web steel joist, beams, and columns).

4. The calculation of gravity and lateral loads in accordance with the ASCE 7 provisions are included.

5. The design of column base plates and anchor rods for axial loads, uplift, and moments.

6. Step-by-step design of moment frames with a design example, as well as the design of moment connections.

7. An introduction to floor vibration analysis and design based on *AISC Design Guide No. 11*.

8. A chapter on practical considerations gives a holistic design view and helps reinforce the connection between structural element and member design and building design in practice.

9. A discussion of practical details showing transfer of lateral loads from roof and floor diaphragms to the lateral load resisting system.

10. An introduction to the analysis of torsion in steel members from a practical perspective.

11. An introduction to the strengthening and rehabilitation of steel structures.

12. Coverage of other topics, including beam copes and their reinforcing; X-braces using tension rods, clevises, and turnbuckles; stability bracing of beams and columns; beam design for uplift loads; ponding considerations; and introduction to coatings for structural steel.

INSTRUCTOR'S RESOURCES

An online Instructor's Manual is available to qualified instructors for downloading. To access supplementary materials online, instructors need to request an instructor access code. Go to **www.pearsonhighered.com/irc**, where you can register for an instructor access code. Within 48 hours after registering, you will receive a confirming e-mail, including an instructor access code. Once you have received your code, go to the site and log on for full instructions on downloading the materials you wish to use.

ACKNOWLEDGMENTS

We would like to thank the reviewers of this text: Paresh S. Shettigar, Hawkeye Community College; James Kipton Ping, Miami University; Jerald W. Kunkel, University of Texas–Arlington; Louis F. Geschwindner, Penn State University; Larry Bowne, Kansas State University; Robert Hamilton, Boise State University; and William M. Bulleit, Michigan Technological University.

In closing, the authors would welcome and appreciate any comments or questions regarding the contents of this book.

CONTENTS

CHAPTER 9   Bolted Connections   407

CHAPTER 10   Welded Connections   454

CHAPTER 14 **Practical Considerations in the Design of Steel Buildings   616**

Structural Steel Design

A PRACTICE-ORIENTED APPROACH

Introduction to Steel Structures

1.1 INTRODUCTION

The primary purpose of this book is to present the design procedures for steel buildings using a limit states or strength design approach in a practical, concise, and easy-to-follow format that is thorough enough to include the design of major structural steel elements found in steel buildings. In the United States, both the load and resistance factor design method (LRFD) and the allowable strength design method (ASD) are prescribed in the latest standard for the design of steel buildings from the American Institute for Steel Construction, AISC 360-05. Although the latest AISC manual follows a dual format, with the LRFD requirements placed side by side with the ASD requirements, only one of these two methods is usually taught in detail at most colleges. For the design of steel bridges in the United States, the mandated design method for bridges receiving federal funding is the LRFD method, and most civil engineering, architectural engineering, and civil engineering technology undergraduate programs in the United States offer at least one course in structural steel design using the LRFD method [1], [2]. In addition, many countries, including Australia, Canada, France, New Zealand, and the United Kingdom (among many others) have long adopted a limit states design approach similar to the LRFD method for the design of steel buildings, and most of the research in steel structures uses the limit states design approach. A comparative study of the cost differences between allowable stress design and LRFD methods for steel high-rise building structures indicated a cost savings of up to 6.9% in favor of the LRFD method [20]. In this era of global competitiveness, we believe that the trend is toward the LRFD method, and in view of the foregoing, we have adopted the LRFD method in this text.

The hallmark feature of this text is the holistic approach that includes the use of realistic structural plans and details in the examples, the discussion of structural loads and structural steel component design within the context of the entire structural building system, and the discussion of other pertinent topics that are essential to the design of real-world structural steel building projects in practice. A structural steel design project problem is introduced in Chapter 1, and subsequent end-of-chapter problems include some structural design project questions. The intent is that by the time the student works through all the chapters and the corresponding end-of-chapter problems pertaining to this design project problem, the student will have completed the design of an entire building, thus reinforcing the connection between the designs of the individual components within the context of an entire building. The importance of a practice-oriented approach in civil engineering education has also been highlighted and advocated for by Roessett and Yao [3]. It is instructive to note that the American Institute for Steel Construction recently developed a Web-based teaching tool for structural steel design that incorporates realistic structural drawings, the calculation of structural loads, and structural steel component design within the context of an entire building design case study, and the design is carried out using the LRFD method [4].

The intended audience for this book is students taking a first or second course in structural steel design, structural engineers, architects, and other design and construction professionals seeking a simple, practical, and concise guide for the design of steel buildings. The book will be well suited for a course involving a design project. The reader is assumed to have a working knowledge of statics, strength of materials, mechanics of materials or applied mechanics, and some structural analysis. We recommend that the reader have the *Steel Construction Manual,* 13th edition (AISC 325-05), available.

1.2 SUSTAINABLE DESIGN AND THE MANUFACTURE OF STRUCTURAL STEEL

There has been a trend in the United States toward sustainable building design and construction where the minimizing of the negative environmental impact is a major consideration in the design and construction of buildings. The most common and popular rating system for the design of "green" buildings is the United States Green Building Council's (USGBC) Leadership in Energy and Environmental Design (LEED) certification system introduced in 1998. This is a point-based building evaluation system that involves a checklist of the "quantifiable aspects of a project" [17]. In the LEED system, the following levels of certification are possible using third-party verification: LEED Silver, LEED Gold, and LEED Platinum. Several structural, as well as nonstructural, issues are considered in calculating the LEED points for buildings. In many cases, the nonstructural issues, such as natural lighting, the type of paint used, the type of heating, ventilation, and air-conditioning (HVAC) systems, and the type of roofing membrane and system, play a greater role in the calculation of the LEED points than do the structural components. Locally fabricated steel (i.e., within a 500-mi. radius) also scores higher on the LEED rating system because of the reduction in the environmental impact from reduced transportation distances.

In the past, steel was primarily manufactured by the refining of virgin iron ore, but today, only about 30% is made through this process because steel is a highly recyclable material. Most of the steel used for building construction in the United States today is made from some form of recycled scrap steel and about 95% of the steel used in structural shapes in the United States is from recycled steel scrap material. In the modern manufacture of new structural steel from old steel, steel scraps are fed into an electric arc furnace (EAF) where they are heated up to 3000°F. As the scraps are melted into liquid or molten steel, the resulting slag by-product floats to the surface, and this can be skimmed off to be used as aggregate in road construction [14]. The carbon content is continuously monitored during this heating process and the process is continued until the desired carbon content of the molten steel is achieved. Various chemical elements, such as manganese, vanadium, copper, nickel, and others, can be added to produce the desired chemical composition of the molten steel. After the "chemical fine tuning," the molten steel is ready to be cast and is poured into a mold to form the crude shape of the section desired [14]. The still red-hot rough shapes are cut to manageable lengths after which they undergo the finishing touches by passing them through machines that press the rough shapes into the desired sizes. After this process, the structural shapes are cooled and cut to lengths of between 30 ft. and 80 ft. to prepare them for transport to the steel fabricators, where the steel is further cut to lengths specific to a given project and modified to receive other connecting members.

1.3 STRUCTURAL STEEL AS A BUILDING MATERIAL

The forerunners to structural steel were cast iron and wrought iron and these were used widely in building and bridge structures until the mid-nineteenth century. In 1856, steel was first manufactured in the United States and since then it has been used in the construction of many buildings and bridge structures. Some notable examples of buildings constructed mainly of structural steel include the 1450-ft.-tall, 110-story Sears Tower in Chicago and the 1474-ft.-tall Taipei 101 building in Taiwan, with 101 floors. Structural steel is an alloy of iron and carbon and is manufactured in various standard shapes and sizes by steel rolling mills, and has a unit weight of 490 lb./ft.^{3}, a modulus of elasticity of 29,000 ksi, and a Poisson's ratio of approximately 0.30. The carbon content of commonly used structural steel varies from about 0.15% to about 0.30% by weight, with the iron content as high as 95% [5]. The higher the carbon content, the higher the yield stress, and the lower the ductility and weldability. Higher carbon steels are also more brittle. Structural steel is widely used in the United States for the construction of different types of building structures, from low-rise industrial buildings to high-rise office and residential buildings. Steel offers competitive advantages when a high strength-to-weight ratio is desired. Some of the advantages of structural steel as a building material include the following:

1. Steel has a high strength-to-weight ratio.
2. The properties of structural steel are uniform and homogeneous, and highly predictable.
3. It has high ductility, thus providing adequate warning of any impending collapse.
4. It can easily be recycled. In fact, some buildings have a majority of their components made of recycled steel.
5. Steel structures are easier and quicker to fabricate and erect, compared with concrete structures.
6. The erection of steel structures is not as affected by weather as is the use of other building materials, enabling steel erection to take place even in the coldest of climates.
7. It is relatively easier to make additions to existing steel structures because of the relative ease of connecting to the existing steel members.

Some of the disadvantages of steel as a building material include the following:

1. Steel is susceptible to corrosion and has to be protected by galvanizing or by coating with zinc-rich paint, especially structures exposed to weather or moisture, although corrosion-resistant steels are also available. Consequently, maintenance costs could be high compared to other structural materials.
2. Steel is adversely affected by high temperatures and therefore often needs to be protected from fire.
3. Depending on the types of structural details used, structural steel may be susceptible to brittle fracture due to the presence of stress concentrations, and to fatigue due to cyclic or repeated loadings causing reversals of stresses in the members and connections.

1.4 THE AISC MANUAL

The premier technical specifying and trade organization in the United States for the fabricated structural steel construction industry is the American Institute for Steel Construction (AISC). This nonprofit organization publishes and produces a number of technical manuals, design guides, and specifications related to the design and construction of steel buildings, such as the *AISC Manual* (AISC 325-05)—hereafter referred to as *"AISCM."* The *AISCM* includes the specification for the design of steel buildings (AISC 360-05) and the properties of standard steel shapes and sizes [6]. This manual, first published in 1923 and now in its 13th edition, consists of 17 chapters as listed below and provides the dimensions and properties of several standardized structural shapes, as well as several design aids, some of which will be used later in this text. The *AISCM* chapters are as follows:

Part 1: Dimensions and Properties

Part 2: General Design Considerations

Part 3: Design of Flexural Members

Part 4: Design of Compression Members

Part 5: Design of Tension Members

Part 6: Design of Members Subject to Combined Loading

Part 7: Design Considerations for Bolts

Part 8: Design Considerations for Welds

Part 9: Design of Connecting Elements

Part 10: Design of Simple Shear Connections

Part 11: Design of Flexible Moment Connections

Part 12: Design of Fully Restrained (FR) Moment Connections

Part 13: Design of Bracing Connections and Truss Connections

Part 14: Design of Beam Bearing Plates, Column Base Plates, Anchor Rods, and Column Splices

Part 15: Design of Hanger Connections, Bracket Plates, and Crane–Rail Connections

Part 16: Specifications and Codes

Part 17: Miscellaneous Data and Mathematical Information

The AISC Web site at www.aisc.org contains much information and resources related to steel design and construction, including *Modern Steel Construction* magazine and the Steel Solutions Center, among others. *Modern Steel Construction* regularly publishes useful and interesting articles related to the practical design and construction of steel structures.

1.5 PROPERTIES OF STRUCTURAL STEEL

The two most important properties of structural steel used in structural design are the tensile and ultimate strengths. These are determined by a tensile test that involves subjecting a steel specimen to tensile loading and measuring the load and axial elongation of the specimen until failure. The stress is computed as the applied load divided by the original cross-sectional area of the specimen, and the strain is the elongation divided by the original length of the specimen. A typical stress–strain curve for structural steel is similar to that shown in Figure 1-1a; it consists of a linear elastic region with a maximum stress that is equivalent to the yield strength, a plastic region in which the stress remains relatively constant at the

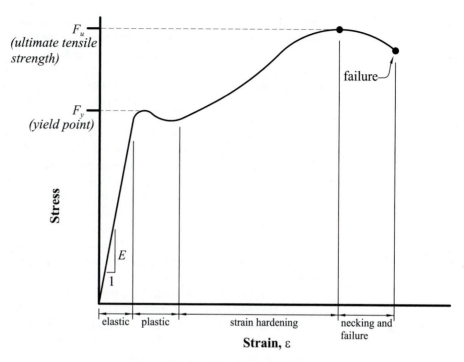

a. Steel with a defined yield point

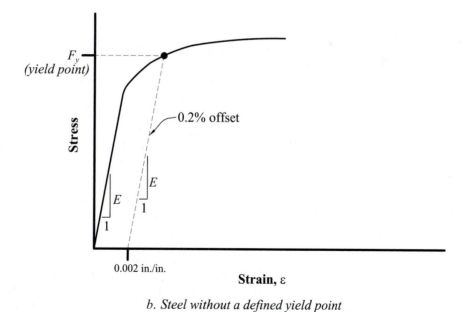

b. Steel without a defined yield point

Figure 1-1 Typical stress–strain diagram for structural steel.

yield stress as the strain increases, and a strain hardening region, the peak of which determines the tensile strength, F_u. Young's modulus, E, is the slope of the linear elastic or straight-line region of the stress–strain curve. The longer the flat horizontal or plastic region of the stress–strain curve, the more ductile the steel is. The ability of structural steel to sustain large deformations under constant load without fracture is called ductility; it is an important structural property that distinguishes structural steel from other commonly used building materials such as concrete and wood.

Where the steel stress–strain curve has no defined yield point, as is the case for high-strength steels, the yield strength is determined using the 0.2% offset method (see Figure 1-1b). The yield strength for this case is defined as the point where a line with a slope E passing through the 0.2% elongation value on the horizontal axis intersects the stress–strain curve. It should be noted that high-strength steels have much less ductility than mild steel. For practical design, the stress–strain diagram for structural steel is usually idealized as shown in Figure 1-2.

The behavior of structural steel discussed above occurs at normal temperatures, usually taken as between $-30°F$ and $+120°F$ [7]. Steel loses strength when subjected to elevated temperatures. At a temperature of approximately 1300°F, the strength and stiffness of steel is about 20% of its strength and stiffness at normal temperatures [22]. As a result of the adverse effect of high temperatures on steel strength and behavior, structural steel used in building construction is often fireproofed by spray-applying cementitious materials or fibers directly onto the steel member or by enclosing the steel members within plaster, concrete, gypsum board, or masonry enclosures. Intumescent coatings and ceramic wool wraps are also used as fire protection for steel members. Fire ratings, specified in terms of the time in hours it takes a structural assembly to completely lose its strength, are specified in the International Building Code (IBC) for various building occupancies. The topic of fire protection will be discussed in further detail in Chapter 14 of this text.

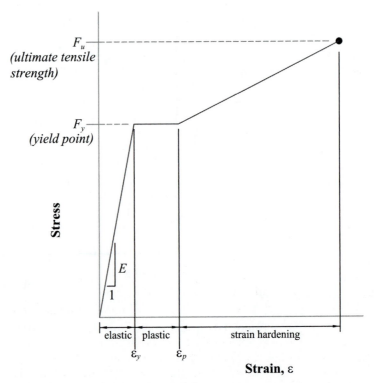

Figure 1-2 Idealized stress–strain diagram for structural steel.

Table 1-1 Structural steel material properties

Steel Type	ASTM Designation or Grade of Structural Steel	F_y (ksi)	F_u (ksi)
Carbon Steel	A36	36	58–80
	A53 Grade B	35	60
	A500 Grade B	42 or 46	58
	A500 Grade C	46 or 50	62
High-Strength, Low-Alloy	A913	50–70	60–90
	A992	50–65	65
	A572 Grade 50	50	65
Corrosion-Resistant, High-Strength, Low-Alloy	A242	50	70
	A588	50	70

where

F_u = Tensile strength, ksi, and

F_y = Yield strength, ksi.

Structural steel is specified using the American Society for Testing and Materials (ASTM) designation. Prior to the 1960s, steel used in building construction in the United States was made mainly from ASTM A7 grade, with a minimum specified yield stress of 33 ksi. The *AISCM* lists the different types of structural steel used in steel building construction, the applicable ASTM designation, and tensile and ultimate strength properties; these are shown in Table 1-1 for commonly used structural steels. ASTM A992 and A572 are the primary high-strength steels used for the main structural members in building construction in the United States, while ASTM A36 steels are typically used for smaller members such as angles, channels, and plates. Where resistance to corrosion is desired, as in the case of exposed steel members, ASTM A588 steel, which has essentially replaced ASTM A242 steel, could be used. This steel provides protection from corrosion through the formation over a period of time of a thin oxide coating on the surface of the structural member when exposed to the atmosphere. The use of this steel obviates the need for painting and it is frequently used in bridge structures. However, for many building structures where corrosion resistance is required, this protection is more commonly provided by coating the structural steel members with zinc-rich paint, or by hot-dip galvanizing the structural member, a process where the structural member is coated with zinc by dipping the entire member into a molten zinc bath at a temperature of approximately 850°F. The length and shape of the member to be galvanized is often limited by the size of the zinc bath used in the galvanizing process

Another property of steel that is of interest to the structural engineer is the coefficient of thermal expansion, which has an average value of 6.5×10^{-6} in./in. per °F for buildings with variations in temperatures of up to 100°F [6]. This property is used to calculate the expected expansion and contraction of a steel member or structure and is useful in determining the size of expansion joints in building structures or the magnitude of forces that will be induced in the structure if the movement is restrained. For enclosed heated and air-conditioned buildings, it is common practice by many engineers to use a temperature change of 50°F to 70°F. However, because buildings are usually unheated and unenclosed during construction, the temperature change may actually exceed these values, and the structural engineer would be

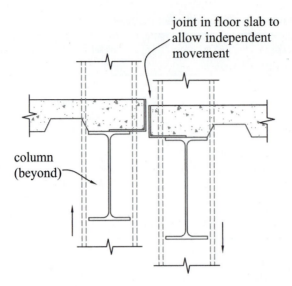

joint in floor slab to
allow independent
movement

column
(beyond)

a. double column connection

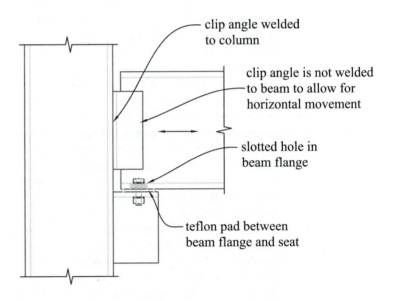

clip angle welded
to column

clip angle is not welded
to beam to allow for
horizontal movement

slotted hole in
beam flange

teflon pad between
beam flange and seat

b. beam connection

Figure 1-3 Expansion joint details.

wise to consider these increased temperature changes, which would vary depending on the location of the building [8]. Structurally, expansion joints in buildings are usually detailed either by using a line of double columns, that is, a column line on both sides of the expansion joint, or by using low-friction sliding bearings that are supported off of a bracket on columns on one side of the expansion joint (see Figures 1-3a and 1-3b). Great care should be taken to ensure that sliding bearing details allow for the anticipated movement because faulty details that do not allow expansion/contraction, or sliding bearings that are unintentionally restrained because of the buildup of debris could result in large unintended forces being transmitted to the structure, thereby causing structural failure as occurred in 2007 with the loading dock slab collapse at the Pittsburgh Convention Center [9].

Residual Stresses

Due to the different rates of cooling in a structural steel member during the final stages of the manufacturing process, initial stresses will exist in the member prior to any loads being applied. These preexisting stresses that are caused by the different cooling rates of the different fibers of the steel section are called residual stresses. The fibers that are the first to cool will be subjected to compressive stresses, whereas the fibers that are the last to cool will be subjected to tension stresses. Other processes that could result in residual stresses include cold bending or straightening, flame cutting of structural members, and the heat generated from the welding of structural members. The residual stresses are usually in internal equilibrium in a structural steel section and therefore have no impact on the plastic moment or tension capacity of a steel member. It does, however, affect the load deformation relationship of a structural member. The impact of residual stresses is most significant for axially loaded members, such as columns, because it causes a reduction in the modulus of elasticity, which decreases from the elasticity modulus (E) to the tangent modulus (E_T), thus resulting in a reduction in the Euler buckling load [11, 15, 16].

Brittle Fracture and Fatigue

Brittle fracture is the sudden failure of a structural steel member due to tensile stresses that cause a cleavage of the member, and it occurs without prior warning. Brittle fracture results from low ductility and poor fracture toughness of the structural member or connection. Other factors affecting brittle fracture include the presence of geometric discontinuities in a steel member, such as notches; rate of application of load; and temperature. The lower the temperature, the lower the ductility and toughness of the steel member.

Whereas *brittle* fracture of a steel member results from a few applications, or even a single application, of loading, *fatigue* failure occurs due to repeated applications of loading to a structure and it occurs over time, starting with a small fatigue crack. Members repeatedly loaded, primarily in tension, are more susceptible to developing fatigue cracks. In typical building structures, the number of cycles of loading is usually less than 100,000 cycles, whereas steel bridges can have more than 2 million cycles of loading during the life of the bridge [15]. The fatigue strength of a steel structure is usually determined at service load levels; it is a function of the stress category which, in turn, is greatly dependent on the connection details used in the structure (see *AISCM*, Table A-3.1). It is also a function of the stress range in the member, which is the algebraic difference between the maximum and minimum stress in a member or connection during one load cycle [15, 18]. The lower the stress range, the higher the fatigue strength of the member or connection. The calculated stress range should be less than the design or allowable stress range calculated from the equations in *AISCM*, Appendix 3.3 and Table A-3.1. It should be noted that since, in most buildings, the cycle of loading is less than 100,000 cycles, fatigue is typically not considered in the design of structural members in building structures, except for crane runway girders and members supporting machinery. However, fatigue is a major consideration in the design of steel bridges.

1.6 STRUCTURAL STEEL SHAPES AND ASTM DESIGNATION

The general requirements for the mechanical properties and chemical composition of rolled structural steel shapes, bars, and plates are given in the ASTM A6 specification. For hollow structural steel (HSS) sections and structural steel pipes, the ASTM A500 and ASTM A53 specifications, respectively, apply. Table 1-2 shows standard structural steel shapes and the corresponding ASTM designations or structural steel grades. The ASTM A6 specification prescribes the permissible maximum percentages of alloy elements such as carbon, manganese, chromium, nickel, copper, molybdenum, vanadium, and so forth in structural steel to ensure adequate weldability and resistance to corrosion and brittle fracture. In the specification, the percentage by weight of each of these chemical elements is combined to produce an equivalent percentage carbon content that is called the carbon equivalent (CE) [10]. Table 1-3 shows the major chemical elements in structural steel and their advantages and disadvantages [10, 15, 23]. The carbon equivalent is useful in determining the weldability of older steels in the repair or rehabilitation of existing or historical structures where the structural drawings and specifications are no longer available, and determining what, if any, special precautions are necessary for welding to these steels in order to prevent brittle fractures. To ensure good weldability already established above, the carbon equivalent, as calculated from equation (1-1), should be no greater than 0.5% [5, 11, 15]. Precautionary measures for steels with higher carbon equivalents include preheating the steel and using low-hydrogen welding electrodes. Alternatively, bolted connections could be used in lieu of welding.

Table 1-2 Structural steel shapes and corresponding ASTM designation

Structural Steel Shapes	ASTM Designation	Min F_y (ksi)	Min F_u (ksi)
W-shape	A913** A992*	50–70 50–65	60–90 65
M- and S-shapes	A36	36	58–80
Channels (C- and MC-shapes)	A36* A572 Grade 50	36 50	58–80 65
Angles and plates	A36	36	58–80
Steel Pipe	A53 Grade B	35	60
Round HSS	A500 Grade B* A500 Grade C	42 46	58 62
Square and Rectangular HSS	A500 Grade B* A500 Grade C	46 50	58 62

* Preferred material specification for the different shapes.

** A913 is a low-alloy, high-strength steel.

The equivalent carbon content or carbon equivalent is given as

$$CE = C + (Cu + Ni)/15 + (Cr + Mo + V)/5 + (Mn + Si)/6 \leq 0.5, \qquad (1\text{-}1)$$

Table 1-3 Alloy chemical elements used in structural steel

Chemical Element	Major Advantages	Disadvantages
Carbon (C)	Increases the strength of steel.	Too much carbon reduces the ductility and weldability of steel.
Copper (Cu)	When added in small quantities, it increases the corrosion resistance of carbon steel, as well as the strength of steel.	Too much copper reduces the weldability of steel.
Vanadium (V)	Increases the strength and fracture toughness of steel and does not negatively impact the notch toughness and weldability of steel.	
Nickel (Ni)	Increases the strength and the corrosion resistance of steel. Increases fracture toughness.	Reduces weldability
Molybdenum (Mo)	Increases the strength of steel. Increases corrosion resistance.	Decreases the notch toughness of steel.
Chromium (Cr)	Increases the corrosion resistance of steel when combined with copper, and also increases the strength of steel. It is a major alloy chemical used in stainless steel.	
Columbium (Cb)	Increases the strength of steel when used in small quantities.	Greatly reduces the notch toughness of steel.
Manganese (Mn)	Increases the strength and notch toughness of steel.	Reduces weldability.
Silicon (Si)	Used for deoxidizing of hot steel during the steelmaking process and helps to improve the toughness of the steel.	Reduces weldability.
Other alloy elements found in very small quantities include nitrogen; those elements permitted only in very small quantities include phosphorus and sulfur.		

Source: Refs [10], [15] and [23]

where

C = Percentage carbon content by weight,

Cr = Percentage chromium content by weight,

Cu = Percentage copper content by weight,

Mn = Percentage manganese content by weight,

Mo = Percentage molybdenum by weight,

Ni = Percentage nickel content by weight, and

V = Percentage vanadium content by weight.

Si = Percentage silicon by weight.

EXAMPLE 1-1

Carbon Equivalent and Weldability of Steel

A steel floor girder in an existing building needs to be strengthened by welding a structural member to its bottom flange. The steel grade is unknown and to determine its weldability, material testing has revealed the following percentages by weight of the alloy chemicals in the girder:

$$C = 0.25\%$$
$$Cr = 0.15\%$$
$$Cu = 0.25\%$$
$$Mn = 0.45\%$$
$$Mo = 0.12\%$$
$$Ni = 0.30\%$$
$$V = 0.12\%$$
$$Si = 0.20\%$$

Calculate the carbon equivalent (CE) and determine if this steel is weldable.

SOLUTION

Using equation (1-1), the carbon equivalent is calculated as

$$CE = 0.25\% + (0.25\% + 0.30\%)/15 + (0.15\% + 0.12\% + 0.12\%)/5$$
$$+ (0.45\% + 0.20\%)/6$$
$$= 0.47\% < 0.5\%.$$

Therefore, the steel is weldable.

However, because of the high carbon equivalent, precautionary measures such as low-hydrogen welding electrodes and preheating of the member are recommended. Since this is an existing structure, the effect of preheating the member should be thoroughly investigated so as not to create a fire hazard. If preheating of the member is not feasible, alternative strengthening approaches that preclude welding may need to be investigated.

EXAMPLE 1-2

Expansion Joints in Steel Buildings

A steel building is 600 ft. long with expansion joints provided every 200 ft. If the maximum anticipated temperature change is 50°F, determine the size of the expansion joint that should be provided.

SOLUTION

The coefficient of linear expansion for steel = 6.5×10^{-6} in./in. per °F

The anticipated expansion or contraction = $(6.5 \times 10^{-6}$ in./in.) $(200$ ft. $\times 12$ in./ft.) $(50°F) = 0.78$ in.

Since the portion of the building on both sides of the expansion joints can expand at the same rate and time, the minimum width of the expansion joint = $2 (0.78$ in.$) = 1.56$ in.

Therefore, use a 2-in.-wide expansion joint.

It should be noted that the nonstructural elements in the building, such as the exterior cladding (e.g., brick wall) and backup walls (e.g., block wall), as well as the interior partition walls, must be adequately detailed to accommodate the anticipated expansion and contraction.

1.7 STRUCTURAL STEEL SHAPES

There are two types of steel shapes available:

- Rolled steel shapes—These are standardized rolled shapes with dimensions and properties obtained from part 1 of the *AISCM* [6].
- Built-up shapes—Where standardized structural shapes cannot be used (e.g., where the load to be supported exceeds the capacity of the sections listed in the *AISCM*), built-up shapes could be made from plate stock. Examples include plate girders and box girders.

Rolled shapes are most commonly used for building construction, while built-up shapes are used in bridge construction. It should be noted that built-up shapes such as plate girders may also be used in building construction as transfer girders to carry heavy concentrated loads as may occur where multistory columns are discontinued at an atrium level to create a large column-free area, resulting in large concentrated loads that need to be supported.

Examples of the rolled standard shapes listed in part 1 of the *AISCM* include the following:

Wide-flanged: W-shapes and M-shapes

W-shapes are wide-flanged shapes that are commonly used as beams or columns in steel buildings. They are also sometimes used as the top and bottom chord members for trusses, and as diagonal braces in braced frames. The inner and outer flange surfaces of W-shapes are parallel, and M-shapes are similar to the W-shape, but they are not as readily available or widely used as W-shapes and their sizes are also limited. The listed M-shapes in *AISCM* have a maximum depth of 12.5 in. and a maximum flange width of 5 in. A W14 × 90, for example, implies a member with a nominal depth of 14 in. and a self-weight of 90 lb./ft. Similarly, an M12 × 10 indicates a miscellaneous shape with a nominal depth of 12 in. and a self-weight of 10 lb./ft. It should be noted that because of the variations from mill to mill in the fillet sizes used in the production of W-shapes, and also the wear and tear on the rollers during the steel shape production process, the decimal k-dimensions (k_{des}) specified for these shapes in part 1 of the *AISCM* should be used for design, while the fractional k-dimensions (k_{det}) are to be used for detailing.

S-shapes

S-shapes, also known as American Standard beams, are similar to W-shapes except that the inside flange surfaces are sloped. The inside face of the flanges usually have a slope

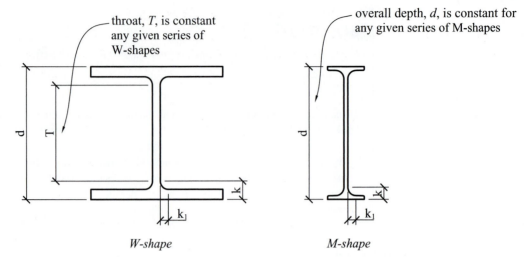

W-shape   *M-shape*

Figure 1-4 W- and M-shapes.

of 2:12, with the larger flange thickness closest to the web of the beam. These sections are commonly used as hoist beams for the support of monorails. An S12 × 35 implies a member with a 12-in. actual depth and a self-weight of 35 lb./ft. length of the member.

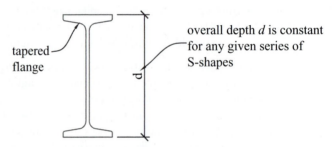

Figure 1-5 S-shape.

HP-shapes

HP-shapes are similar to W-shapes and are commonly used in bearing pile foundations. They have thicker flanges and webs, and the nominal depth of these sections is usually approximately equal to the flange width, with the flange and web thicknesses approximately equal.

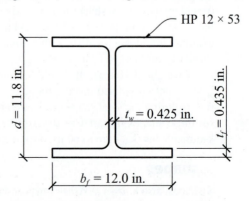

Figure 1-6 HP-shape.

Channel or C- and MC-shapes

Channels are C-shaped members with the inside faces of the channel flanges sloped. They are commonly used as beams to support light loads, such as in catwalks or as stair stringers, and they are also used to frame the edges of roof openings. C-shapes are American Standard channels, while MC-shapes are miscellaneous channels. A C12 × 30 member implies a C-shape with an actual depth of 12 in. and a weight of 30 lb./ft., while an MC 12 × 35 member implies a miscellaneous channel with an actual depth of 12 in. and a self-weight of 35 lb./ft.

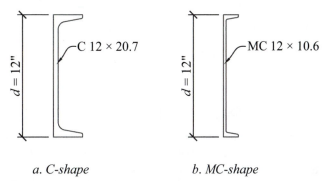

a. C-shape   b. MC-shape

Figure 1-7 C- and MC-shapes.

Angle (L) shapes

Angles are L-shaped members with equal or unequal length legs, and they are used as lintels to support brick cladding and block wall cladding, and as web members in trusses. They are also used as X-braces, chevron braces, or knee-braces in braced frames, and could be used as single angles or as double angles placed back-to-back. An angle with the designation L4 × 3 × ¼ implies a member with a long leg length of 4 in., a short leg length of 3 in., and a thickness of ¼ in. While all other rolled sections have two orthogonal axes (x–x and y–y) of bending, single angles have three axes (x–x, y–y, and z–z) about which the member could bend or buckle.

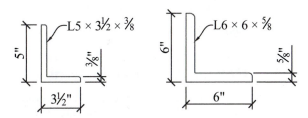

Figure 1-8 Angle shapes.

Structural Tees—WT-, MT-, and ST-shapes

Structural tees are made by cutting a W-shape, M-shape, or S-shape in half. For example, if a W14 × 90 is cut in half, the resulting shapes will be WT 7 × 45, where the nominal depth is 7 in. and the self-weight of each piece is 45 lb./ft. WT-shapes are commonly used as brace members and as top and bottom chords of trusses. They are also used to strengthen existing steel beams where a greater moment capacity is required. Similarly, ST- and MT-shapes are made from S-shapes and M-shapes, respectively.

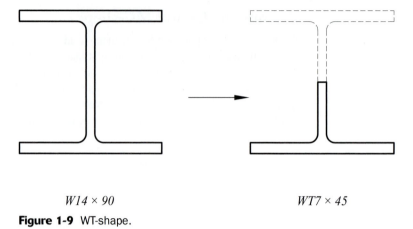

W14 × 90

WT7 × 45

Figure 1-9 WT-shape.

Plates and Bars

Plates and bars are flat stock members that are used as stiffeners, gusset plates, and X-braced members. They are also used to strengthen existing steel beams and as supporting members in built-up steel lintels. There is very little structural difference between bars and plates, and although historically, flat stock with widths not exceeding 8 in. were generally referred to as bars, while flat stock with widths greater than 8 in. were referred to as plates, it is now common practice to refer to flat stock universally as plates. As an example, a PL 6 × ½ implies a 6-in.-wide by ½-in.-thick plate. In practice, plate widths are usually specified in ½-in. increments, while thicknesses are specified in ⅛-in. increments. The practical minimum thickness is ¼-in., with a practical minimum width of 3 in. to accommodate required bolt edge distances but smaller sizes can be used for special conditions.

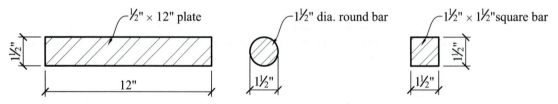

Figure 1-10 Plates and bars.

Hollow Structural Sections

Hollow structural section (HSS) members are rectangular, square, or round tubular members that are commonly used as columns, hangers, and braced-frame members. HSS members are not as susceptible to lateral torsional buckling and torsion as W-shape or other open sections. Therefore, they are frequently used as lintels spanning large openings, especially where eccentricity of the gravity loads may result in large torsional forces. Examples of HSS members are indicated below:

- HSS 6 × 4 × ¼ implies a rectangular hollow structural steel with outside wall dimensions of 6 in. in one direction and 4 in. in the orthogonal direction, and a wall thickness of ¼ in., except at the rounded corners.
- HSS 4 × 0.375 implies a round hollow structural steel with an outside wall diameter of 4 in. and a uniform wall thickness of ⅜ in.

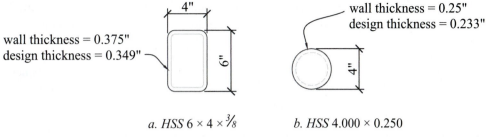

a. HSS $6 \times 4 \times \frac{3}{8}$ *b. HSS* 4.000×0.250

Figure 1-11 HSS and structural pipes.

Structural Pipes

Structural pipes are round structural tubes similar to HSS members (see Figure 1-11) that are sometimes used as columns. They are available in three strength categories: standard (Std), extra strong (X-strong), and double-extra strong (XX-strong). The bending moment capacity and the axial compression load capacity of these sections are tabulated in Tables 3-15 and 4-6, respectively, of the *AISCM*. Steel pipes are designated with the letter P, followed by the nominal diameter, and then the letter X for extra strong or XX for double-extra strong. For example, the designation P3 represents a nominal 3-in. standard pipe, P3X represents a 3-in. extra-strong pipe, and P3XX represents a 3-in. double-extra-strong pipe.

Built-up Sections

Built-up sections include welded plate girders and plates welded to the top or bottom flanges of W-sections. Plate girders are used to support heavy loads where the listed

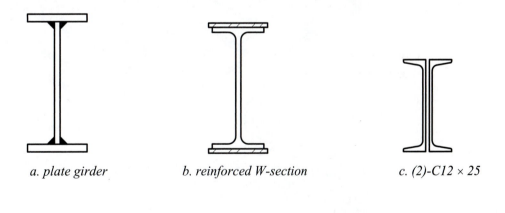

a. plate girder   *b. reinforced W-section*   *c. (2)-C12 × 25*

d. (2)-L6 × 4 × $\frac{3}{8}$ *e. S12 × 31.8 with*
                                                  C10 × 15.3 cap channel

Figure 1-12 Built-up sections.

standard steel sections are inadequate to support the loads. Built-up sections can also be used as lintels and as reinforcement for existing beams and columns. Other built-up shapes include double angles (e.g., 2L 5 × 5 × ½) and double channels (e.g., 2C 12 × 25) placed back-to-back in contact with each other or separated by spacers, and W- and M-shapes with cap channels that are used to increase the bending capacity of W- and S-shapes about their weaker (y–y) axis.

1.8 BASIC STRUCTURAL STEEL ELEMENTS

The basic structural steel elements and members that are used to resist gravity loads in steel-framed buildings as shown in Figures 1-13 and 1-14 will now be discussed.

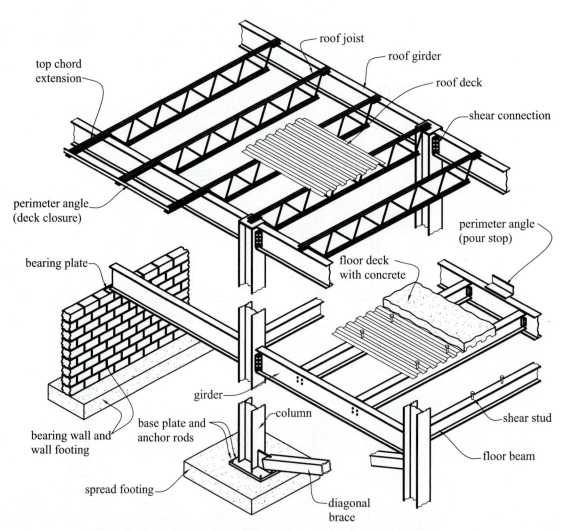

Figure 1-13 Typical steel building — basic structural elements (3-D).

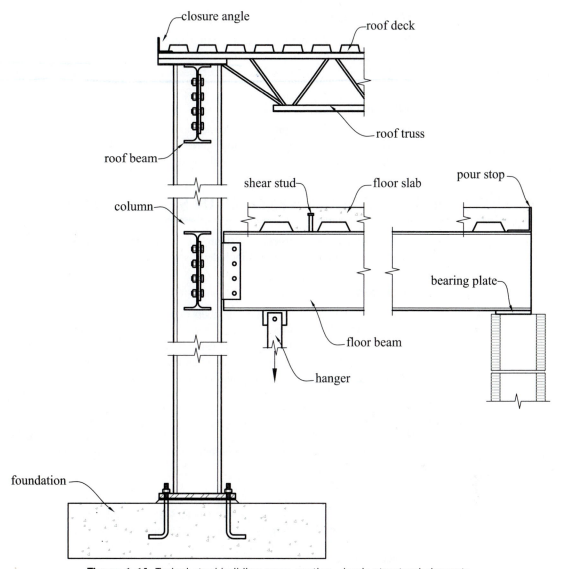

Figure 1-14 Typical steel building cross section—basic structural elements.

Beams and Girders

- The infill beams or joists support the floor or roof deck directly and spans between the girders. The roof or floor deck usually spans in one direction between the roof or floor infill beams.
- The girders support the infill beams and span between the columns. While beams are usually connected to the web of the columns, girders are typically connected to the column flanges.

Columns

These are vertical members that support axial compression loads only. They are sometimes referred to as struts. In practice, structural members are rarely subjected to pure compression loads alone.

Beam-columns

Beam-columns are members that support axial tension or axial compression loads in addition to bending loads. In practice, typical building columns usually act as beam-columns due to the eccentricity of the beam and girder reactions relative to the column centerline.

Hangers

Hangers are vertical members that support axial tension force only.

The reader should refer back to Figures 1-13 and 1-14 as the other structural elements are discussed in greater detail later in the text.

1.9 TYPES OF STRUCTURAL SYSTEMS IN STEEL BUILDINGS

The common types of structural systems (i.e., a combination of several structural members) used in steel building structures include trusses, moment frames, and braced frames.

Trusses

Trusses may occur as roof framing members over large spans or as transfer trusses used to support gravity loads from discontinuous columns above. The typical truss profile shown in Figure 1-15 consists of top and bottom chord members. The vertical and diagonal members are called web members. While the top and bottom chords are usually continuous members, the web members are connected to the top and bottom chords using bolted or welded connections.

Frames

Frames are structural steel systems used to resist lateral wind or seismic loads in buildings. The two main types of frames are moment frames and braced frames.

Moment Frames

Moment frames resist lateral loads through the bending rigidity of the beams/girders and columns. The connections between the beams/girders and the columns are designed and detailed as shown in Figure 1-16 to resist moments due to gravity and lateral loads.

Braced Frames

Braced frames (see Figure 1-17) resist lateral loads through axial compression and/or tension in the diagonal members. Examples include X-braced, chevron- or K-braced, and knee-braced frames. These frames are usually more rigid than a typical moment frame and exhibit smaller lateral deflections.

1.10 BUILDING CODES AND DESIGN SPECIFICATIONS

Building construction in the United States and in many parts of the world is regulated through the use of building codes that prescribe a consensus set of minimum requirements that will ensure public safety. A code consists of standards and specifications (or recommended practice), and covers all aspects of design, construction, and function of buildings, including occupancy and fire-related issues, but it only becomes a legal document within any jurisdiction after it is adopted by the legislative body in that jurisdiction. Once adopted by a jurisdiction, the code becomes the legal binding document for building construction in that

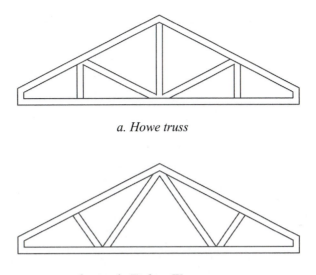

a. Howe truss

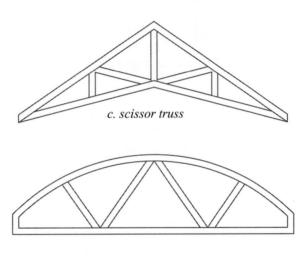

b. simple Fink or Warren truss

c. scissor truss

d. bowstring truss

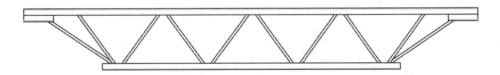

e. pre-engineered roof truss

Figure 1-15 Typical truss profiles.

locality and the design and construction professional is bound by the minimum set of requirements specified in the code. The International Building Code (IBC 2006) [12], published by the International Code Council (ICC), has replaced the former model codes—the Uniform Building Code (UBC), Building Officials and Code Administrators (BOCA), National Building Code (NBC), and the Standard Building Code (SBC)—and is fast becoming the most

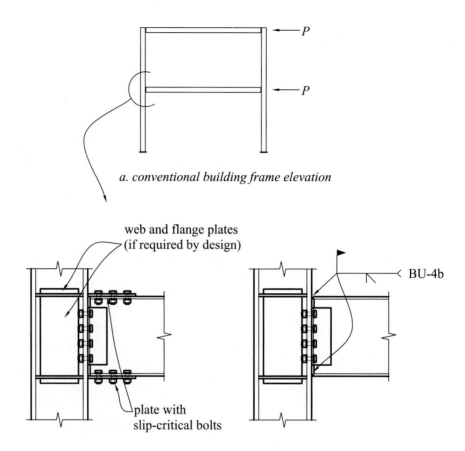

a. conventional building frame elevation

web and flange plates
(if required by design)

plate with
slip-critical bolts

BU-4b

b. bolted moment connection   *c. welded moment connection*

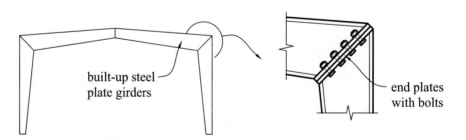

built-up steel
plate girders

end plates
with bolts

d. pre-engineered building frame elevation   *e. bolted moment connection*

Figure 1-16 Typical moment frames.

widely used building code in the United States for the design of building structures. The current edition of the IBC 2006 now references the ASCE 7 load standard [13] for calculation of all structural loads, including snow, wind, and seismic loads. The steel material section of the IBC references the AISC specifications as the applicable specification for the design of steel members in building structures.

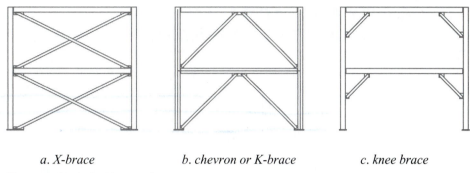

a. X-brace   b. chevron or K-brace   c. knee brace

Figure 1-17 Typical braced frames.

1.11 THE STRUCTURAL STEEL DESIGN AND CONSTRUCTION PROCESS

The design process for a structural steel building is iterative in nature and usually starts out with some schematic drawings developed by the architect for the owner of a building. Using these schematic drawings, the structural engineer carries out a preliminary design to determine the preliminary sizes of the members for each structural material and structural system (gravity and lateral) considered. This information is used to determine the most economical structural material and structural system for the building. After the structural material and systems are determined, then comes the final design phase where the roof and floor framing members and the lateral load systems are laid out and all the member sizes are proportioned to resist the applied loads with an adequate margin of safety. This results in a set of construction documents that include structural plans, sections, details, and specifications for each of the materials used in the project. After the final design phase comes the shop drawing and the construction phases during which the building is actually fabricated and erected. During the shop drawing phase, the steel fabricator's detailer uses the structural engineer's drawings to prepare a set of erection drawings and detail drawings that are sent to the structural engineer for review and approval. The shop drawing review process provides an opportunity for the design engineer to ensure that the fabrication drawings and details meet the design intent of the construction documents. Once the shop drawings are approved, steel fabrication and erection can commence. The importance of proper fabrication and erection procedures and constructible details to the successful construction of a steel project cannot be overemphasized.

In the United States and Canada, the design of simple connections (i.e., simple shear connections) is sometimes delegated by the structural engineer of record (EOR) to the steel fabricator, who then hires a structural engineer to design these connections using the loads and reactions provided on the structural drawings and/or specifications. The connection designs and the detail drawings of the connections are also submitted to the structural engineer of record for review and approval. In other cases, especially for the more complicated connections such as moment connections, the EOR may provide schematic or full connection designs directly to the fabricator. During the construction phase, although the structural engineer of record may visit the construction site occasionally, it is common practice for the owner to retain the services of a materials inspection firm to periodically inspect the fabrication and erection of the structural steel in order to ensure that the construction is being done in accordance with the structural drawings and specifications.

1.12 GRAVITY AND LATERAL LOAD PATHS AND STRUCTURAL REDUNDANCY

The load path is the trail that a load travels from its point of application on the structure until it gets to the foundation. Any structural deficiency in the integrity of the load path could lead to collapse; these commonly result from inadequate connections between adjoining structural elements rather than the failure of a structural member. The typical path that a gravity load travels as it goes from its point of application on the structure to the foundation is as follows: The load applied to the roof or floor deck or slab is transmitted horizontally to the beams, which in turn transfer the load horizontally to the girders. The girders and the beams along the column lines transfer the load as vertical reactions to the columns, which then transmit the load safely to the foundation and to the ground (i.e., the load travels from the Slab or Roof Deck $\rightarrow$ Beams $\rightarrow$ Girders $\rightarrow$ Columns $\rightarrow$ Foundations). This is illustrated in Figure 1-18.

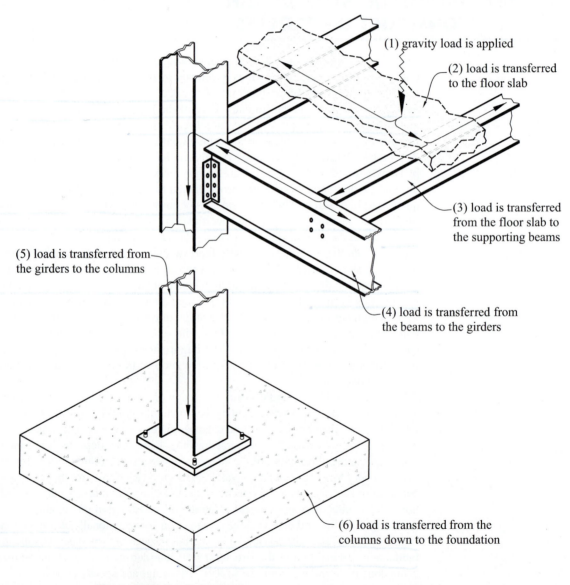

(1) gravity load is applied

(2) load is transferred to the floor slab

(3) load is transferred from the floor slab to the supporting beams

(5) load is transferred from the girders to the columns

(4) load is transferred from the beams to the girders

(6) load is transferred from the columns down to the foundation

Figure 1-18 Gravity load path.

For the lateral wind load path, the wind load is applied to the vertical wall surface, which then transfers the horizontal reactions to the horizontal roof or floor diaphragms. The horizontal diaphragm then transfers the lateral load to the lateral force resisting system (LFRS) parallel to the lateral load (e.g., moment frame, braced frame, or shear wall), and these lateral force resisting systems then transmit the lateral loads to the foundation and the ground (i.e., the lateral load travels from the Roof or Floor Diaphragm → Lateral Force Resisting System → Foundations. This is illustrated in Figure 1-19.

The seismic load path starts with ground shaking from a seismic event, which results in inertial forces being applied to the building structure, and these forces are assumed to be concentrated at the floor and roof levels. The lateral forces are then carried by the floor and roof diaphragms, which, in turn, transmit the lateral load to the lateral force resisting systems that are parallel to the lateral load, and the LFRS transmits the load to the foundation and then to the ground.

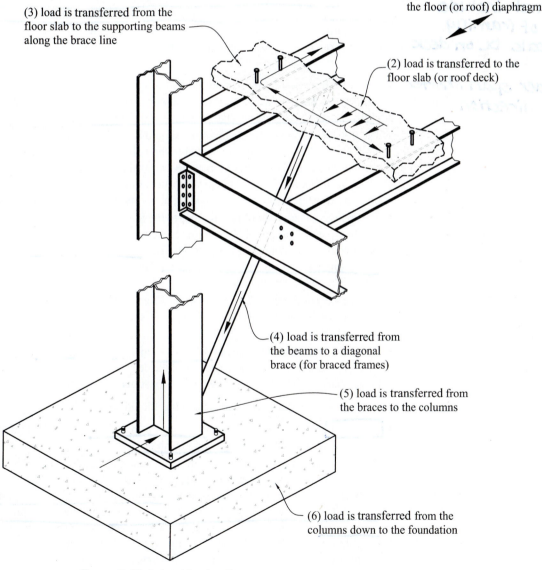

(1) lateral load is applied at the floor (or roof) diaphragm

(3) load is transferred from the floor slab to the supporting beams along the brace line

(2) load is transferred to the floor slab (or roof deck)

(4) load is transferred from the beams to a diagonal brace (for braced frames)

(5) load is transferred from the braces to the columns

(6) load is transferred from the columns down to the foundation

Figure 1-19 Lateral load path.

Structural redundancy—which is highly desirable in structural systems—is the ability of a structure to support loads through more than one load path, thus preventing progressive collapse. In a redundant structure, there are alternate load paths available so that failure of one member does not lead to failure of the entire structure; thus, the structure is able to safely transmit the load to the foundation through the alternate load paths.

1.13 ROOF AND FLOOR FRAMING LAYOUT

Once the gravity loads acting on a building are determined, the next step in the design process, before any of the structural members or elements can be designed, is the layout of the roof and floor framing and the lateral force-resisting systems. In this section, the criteria for the economical layout of roof and floor framing are presented. The self-weight of roof and floor framing (i.e., beams and girders) varies from approximately 5 psf to 10 psf. In calculating the allowable loads of roof and floor decks using proprietary deck load tables, the self-weight of the framing should be subtracted from the total roof or floor loads since the beams and girders support the deck and the loads acting on it.

Do not include SW of framing to calc DL on deck

In laying out roof or floor framing members, the following criteria should be noted for constructibility and economy:

Girder span shorter direction

- The filler beams or joists (supported by the girders) should be framed in the longer direction of the bay, while the girders should span in the shorter direction. Thus, the girder length should be less than or equal to the length of the filler beam or joist.
- The filler beams along the column lines should be connected to the web of the columns, while the girders should be connected to the column flanges because the girders support heavier loads and therefore have reactions that are, in general, greater than those of the infill beams. This arrangement ensures that the moments from the girder reactions are resisted by the column bending about its stronger axis (see Chapter 8).
- The span of the deck should be as close to the Steel Deck Institute (SDI) maximum allowable span [21] as possible to minimize the required number of filler beams or joists, and therefore the number of connections. The maximum deck span required to satisfy the Factory Mutual fire-rating requirements may be more critical than the SDI maximum allowable span, and should be checked.
- As much as possible, use 22-ga decks and 20-ga decks because these are the most commonly available deck sizes. Other deck sizes, such as the 18-ga deck can be obtained, but at a higher or premium cost.
- Decks are available in lengths of 30 to 42 ft. and widths of 2 to 3 ft. and these parameters should be considered in the design of the deck. It should be noted that deck Lengths longer than 30 ft will be too heavy for two construction workers to safely handle on site.

In selecting roof and floor decks, the following should be noted:

- Roof deck is readily available in 1½-in. and 3-in. depths, with gages ranging from 16 through 22-ga; however, the 22-ga wide-rib deck is more commonly used in practice.
- Floor deck is readily available in 1½-in., 2-in., and 3-in. depths, with gages ranging from 16 through 22-ga; however, the 20-ga wide-rib deck is more commonly used. Floor decks can be composite or noncomposite. Composite floor decks have protrusions inside the deck ribs that engage the hardened concrete within the ribs to provide the composite action. Noncomposite floor decks do not have these protrusions and are called form decks. The form deck does not act compositely with the concrete, but acts only as a form to support the wet weight of the concrete during construction. Therefore, the concrete slab has to be reinforced (usually with welded wire fabric) to support the applied loads.

- Noncomposite floor deck is available in depths ranging from ⁹⁄₁₆ in. to 3 in. and in gages ranging from 16 to 28-ga.
- Although 3-in.-deep deck costs more than 1½-in. deck, they are able to span longer distances and thus minimize the number of filler beams and connections used.
- To protect against corrosion, the roof and floor decks could be painted or galvanized (G60 galvanized or G90 galvanized), but where spray-applied fire protection is to be applied to the deck, care should be taken in choosing a classified paint product to ensure adequate bonding of the fireproofing to the metal deck.
- Preferably, the deck should be selected to span over at least four beams (i.e., the so-called 3-span deck). This 3-span deck configuration indicates that the minimum length of deck that the contractor can use on site will be three times the spacing between the beams or joists, and since the number of beam or joist spacings may not necessarily be a multiple of 3, the deck sheets will have to be overlapped.

In designing building structures, the reader should keep in mind the need for simplicity in the structural layout and details, and not just the weight of the material, because labor costs, which consist of steel fabrication and erection, are about 67% of the total construction costs, with material costs being only about 33% [19]. Thus, the least weight may not necessarily always result in the least cost. The sizing of roof and floor decks, and the layout of roof and floor framing members are illustrated in the following two examples.

EXAMPLE 1-3

Roof Framing Layout

A roof framing plan consists of 30-ft. by 40-ft. bays, supporting a total dead load of 30 psf and a roof snow load of 50 psf. Determine the layout of the roof deck and the size of the deck.

Assume a self-weight of 5 psf for the framing.

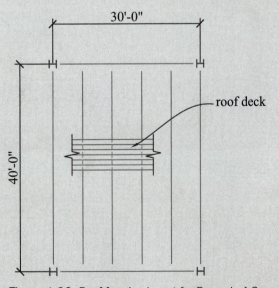

Figure 1-20 Roof framing layout for Example 1-3.

(continued)

SOLUTION

Dead load on the roof deck alone = 30 psf − 5 psf = 25 psf

Snow load on the roof = 50 psf

Total load on the roof deck = 25 psf + 50 psf = 75 psf

For our example, let us try a 1½-in. by 20-ga galvanized wide-rib deck (the reader should refer to the *Vulcraft Steel Roof and Floor Deck Manual* [21]) or similar deck manufacturers' catalogs.

Maximum SDI allowable deck span without shoring during construction = 7 ft. 9 in.

Preferably, the deck should span continuous over at least **four** beams (i.e., 3-span deck). It should be noted that the deck span selected must be less than or equal to 7 ft. 9 in. (i.e., the maximum span), and, in addition, the selected deck span must be a multiple of the shorter bay dimension.

Try a 7-ft. 6-in. span (a multiple of the 30-ft bay dimension) < 7 ft. 9 in. OK

Allowable load = 72 psf < 75 psf   Not Good.

The next lower multiple of 30 ft is 6 ft. 0 in. Therefore, try a deck span of 6 ft. 0 in., and 22-ga. deck resulting in five equal spaces per bay.

For this span, the allowable deck load = 89 psf > 75 psf   OK

Therefore, use **1.5-in. by 22-ga wide-rib galvanized metal deck.**

EXAMPLE 1-4

Floor Framing Layout

A typical floor framing plan consists of 30-ft by 40-ft bays and supports a total floor dead load of 80 psf and a floor live load of 150 psf. Determine the layout of the floor framing and the size of a composite floor deck assuming normal-weight concrete.

   Assume a self-weight of 7 psf for the framing.

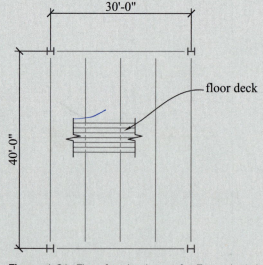

Figure 1-21 Floor framing layout for Example 1-4.

SOLUTION

Dead load on the floor deck alone = 80 psf − 7 psf = 73 psf,

Live load = 150 psf, and

The total floor load = 73 psf + 150 psf = 223 psf.

The reader should refer to the *Vulcraft Steel Roof and Floor Deck Manual* [21] or similar manufacturers' deck catalogs.

Try 2½-in. concrete slab on 3-in. by 20-ga galvanized composite metal deck.

Total Slab Depth = 2½ in. + 3 in. = 5½ in.

For the **3-span** condition, we find from the deck load tables that

Maximum allowable deck span without shoring during construction = 11 ft. 9 in.

Self-weight of concrete	≅ 50 psf	(see vulcraft deck load tables)
Self-weight of deck	≅ 2 psf	*Deck wt in manual includes CN wt*
	52 psf	

Total floor deck and concrete slab self-weight = 52 psf

Applied superimposed load on deck/slab = 223 − 52 = 171 psf

It should be noted that the deck span selected must be less than or equal to 11 ft. 9 in., which is the maximum allowable span for this particular deck without shoring when it spans over a minimum of **three spans.** In addition, the selected deck span must be a multiple of the shorter bay dimension.

  Try a 10-ft. 0-in. span (a multiple of the 30-ft. bay dimension) < 11 ft. 9 in. OK
  Allowable superimposed load ≅ 127 psf < 171 psf   N.G.
  The next lower multiple of 30 ft is 7 ft. 6 in. Therefore, try a deck span of 7 ft. 6 in., resulting in four equal spaces per bay.
  For this span, the **allowable superimposed** load = 247 psf (from the Vulcraft Deck Load tables) > 171 psf.   OK
  Therefore, use **2½-in. concrete slab on 3-in. by 20-ga galvanized composite metal deck.** *w/ framing @ 7'-6" spacing*

1.14 STUDENT DESIGN PROJECT PROBLEM

In this section, we introduce a structural steel building design project for the student to work on, and in the subsequent chapters, some of the end-of-chapter problems will be devoted to designing a component of this building using the concepts learned in each chapter. The design brief for this project is as follows:

1. **Office Building:** A steel-framed, two-story office building with plan dimensions of **72 ft. by 108 ft.** The floor-to-floor height is **15 ft** (see Figure 1-22).

   - Building is located in Buffalo, New York.
   - Main structural members are of structural steel.
   - The AISC LRFD specification and the ASCE 7 load standard should be used.

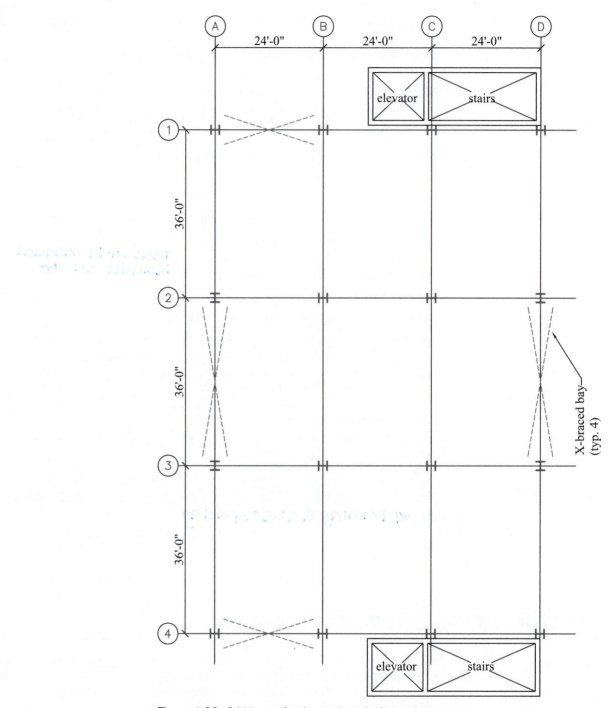

Figure 1-22 Grid layout for the student design project.

- The floors should be designed both as noncomposite and composite construction, with concrete slab on composite metal deck supported on steel infill beams and girders.

- Roofing is 5-ply plus gravel supported on a metal deck on noncomposite steel framing (open-web steel joist or steel infill beams on steel girders).

- Assume that the perimeter cladding is supported on the foundation wall at the ground floor level and bypasses the floor and the roof.
- Assume a 1-ft. edge distance at each floor level around the perimeter of the building.
- Assume that the stair and elevator are located outside of the 72-ft. by 108-ft. footprint on the east and west side of the building, and that they will be designed by others.
- Determine the critical lateral loading—seismic or wind—and use this to design the lateral force resisting systems.

2. **Drawings:** May be either large size, foldout size, 8½ in. by 11 in. and should include the following:

   Roof Plan that shows framing members and roof deck pattern. Indicate camber and end reactions.

   Floor Plan that shows framing members and floor deck structure. Indicate number of studs, camber, and end reactions.

   Foundation Plan that shows column sizes or marks.

   Elevations that show the ×-braces and connection configurations.

   Connection Details that show typical beam and girder connections, ×-braced connections, column and base plate schedule with *service* loads (for foundation design), and typical column base details.

3. **Loads:**

   Calculate the dead and live loads (floor live loads and roof live loads).

   Calculate the snow load, allowing for any snow load reduction.

   Follow the ASCE load standard to calculate the wind loads and seismic loads.

4. **Checklist of Design Items:**

   Gravity and Lateral Loads: Provide a load summation table for the floor and roof loads (gravity), and the wind and seismic loads (lateral). Provide a column load summation table.

   Roof members: Select open-web steel joists from a manufacturer's catalog. As an alternate option, also select wide flange roof beams. Design the roof girders as wide-flange sections.

   Floor Framing: Design the floor beams and girders both as noncomposite beams and as composite beams for comparison purposes.

   Columns: Design for maximum loads using the appropriate load combinations. Design columns for axial load and moments due to girder/beam connection eccentricities. Design the column base plates and anchor rods assuming a concrete strength (f_c') of 3000 psi and a 1-in. grout thickness.

   Floor and Roof Deck: Assume a 1½-in. roof deck and a 3-in. composite floor deck. Determine the exact gage and material weight from a deck manufacturer's catalog. Use normal-weight concrete and a concrete strength of $f_c' = 3000$ psi.

Lateral Force Resisting System

LFRS Option 1:

×-braces: Analyze the ×-brace frames manually or using a structural analysis software program. Design the brace connections and check all connection failure modes.

LFRS Option 2:

Moment frames: Assume that moment frames are used along grid lines 1 and 4 in lieu of the ×-braces, and reanalyze the frame and design the moment frame. Design the beam and girder connections and check all connection failure modes.

1.15 REFERENCES

1. Albano, Leonard D. Classroom assessment and redesign of an undergraduate steel design course: A case study. *ASCE Journal of Professional Issues in Engineering Education and Practice*: (October 2006) 306–311.

2. Gomez-Rivas, Alberto, and George Pincus. Structural analysis and design: A distinctive engineering technology program. *Proceedings of the 2002 American Society for Engineering Education Annual Conference and Exposition*. Montreal, Canada.

3. Roesset, Jose M., and James T. P. Yao. Suggested topics for a civil engineering curriculum. *Proceedings of the 2001 American Society for Engineering Education Annual Conference and Exposition*. Albuquerque, NM.

4. Estrada, Hector, Using the AISC steel building case study in a structural engineering course sequence. *Proceedings of the 2007 American Society for Engineering Education Annual Conference and Exposition*. Honolulu, HI.

5. Tamboli, Akbar, R. 1997. *Steel design handbook—LRFD method*. New York, NY: McGraw–Hill.

6. AISC. 2006. *Steel Construction Manual*, 13th ed. Chicago.

7. Spiegel, L., and G. Limbrunner. 2002. *Applied structural steel design*, 4th ed. Upper Saddle River, NJ: Prentice Hall.

8. Fisher, James M. Expansion joints: Where, when, and how. *Modern Steel Construction* (April 2005): 25–28.

9. Rosenblum, Charles L. Probers eye expansion joint in Pittsburgh slab mishap. *Engineering News–Record (ENR)* (February 8, 2007), http://enr.construction.com/news/safety/archives/070208a.asp (accessed May 26, 2008).

10. Brockenbrough, Roger L., and Frederick Merritt. 1999. Structural steel designer's handbook, 3rd ed. New York, NY: McGraw–Hill.

11. Lay, M. G. 1982. Structural steel fundamentals—An engineering and metallurgical primer. Australian Road Research Board.

12. International Codes Council. (ICC) 2006. *International Building Code—2006*. Falls Church, VA.

13. American Society of Civil Engineers. 2005. *ASCE-7, Minimum design loads for buildings and other structures*. Reston, VA.

14. Mckee, Bradford, and Timothy Hursely. Structural steel—How it's done. *Modern Steel Construction* (August 2007): 22–29.

15. Geschwindner, Louis F., Robert O. Disque, and Reidar Bjorhovde. 1994. *Load and resistance factor design of steel structures*. Prentice Hall.

16. Louis F. Geschwindner. 2008. *Unified design of steel structures*. John Wiley.

17. Farneth, Stephen. Sustaining the past. *Green Source—The Magazine of Sustainable Design* (October 2007): 25–27.

18. Galambos, Theodore V., F. J. Lin, and Bruce G. Johnston. 1996. *Basic steel design with LRFD*.: Prentice Hall.

19. Carter, C. J., T. M. Murray, and W. A. Thornton. Economy in steel. *Modern Steel Construction* (April 2002).

20. Sarma, Kamal C., and Hojjat Adeli. Comparative study of optimum designs of steel high-rise building structures using allowable stress design and load and resistance factor design codes. *Practice Periodical on Structural Design and Construction* (February 2005): 12–17.

21. Vulcraft. 2001. Vulcraft steel and roof deck manual, http://www.vulcraft.com/downlds/catalogs/deckcat.pdf (accessed May 26, 2008).

22. Gewain, Richard G., Nester R. Iwankiw, and Farid, Alfawakhiri. 2003. *Facts for steel buildings—Fire*, American Institute of Steel Constriction, Chicago, IL.

23. Mamlouk, Michael S., and John P. Zaniewski. 2006. *Materials for Civil and Construction Engineers*. Upper Saddle River, NJ: Prentice Hall.

1.16 PROBLEMS

1-1. List three advantages and disadvantages of steel as a building material, and research the Internet for the three tallest steel building structures in the world, indicating the types of gravity and lateral load resisting systems used in these buildings.

1-2. List the various types of standard shapes available in the *AISCM*.

1-3. What are the smallest and the largest wide flange or W-shapes listed in the *AISCM*?

1-4. Determine the self-weight, moment of inertia (I_x), and cross-sectional areas for the following hot-rolled standard sections:

   W14 × 22

   W21 × 44

   HSS 6 × 6 × 0.5

L6 × 4 × ½

C12 × 30

WT 18 × 128

1-5. Determine the weight, area, and moment of inertia (I_x) of the following built-up sections:

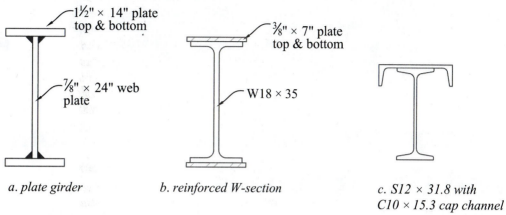

a. plate girder   b. reinforced W-section   c. S12 × 31.8 with
                                                             C10 × 15.3 cap channel

Figure 1-23 Compound shapes for problem 1-5.

1-6. List the basic structural elements used in a steel building.

1-7. Plot the idealized stress–strain diagram for a 6-in.-wide by ½-in.-thick plate and a 6-in.-wide by 1-in.-thick plate of ASTM A36 steel. Assume that the original length between two points on the specimen over which the elongation will be measured (i.e., the gage length) is 2 in.

1-8. Determine the most economical layout of the roof framing (joists and girders) and the gage (thickness) of the roof deck for a building with a 25-ft. by 35-ft. typical bay size. The total roof dead load is 25 psf and the snow load is 35 psf. Assume a 1½-in.-deep galvanized wide-rib deck and an estimated weight of roof framing of 6 psf.

1-9. Repeat problem 1-8 using a 3-in.-deep galvanized wide-rib roof deck.

1-10. Determine the most economical layout of the floor framing (beams and girders), the total depth of the floor slab, and the gage (thickness) of the floor deck for a building with a 30-ft. by 47-ft. typical bay size. The total floor dead load is 110 psf and the floor live load is 250 psf. Assume normal weight concrete, a 1½-in.-deep galvanized composite wide-rib deck, and an estimated weight of floor framing of 10 psf.

1-11. Repeat problem 1-9 using a 3-in.-deep galvanized composite wide-rib roof deck.

1-12. A steel floor girder in an existing building needs to be strengthened by welding a structural member to its bottom flange. The steel grade is unknown, but materials testing has revealed the following percentages by weight of the following alloy chemical elements in the girder:

C = 0.16%

Cr = 0.10%

Cu = 0.20%

Mn = 0.8%

Mo = 0.15%

Ni = 0.25%

V = 0.06%

Si = 0.20%

Calculate the carbon equivalent (CE) and determine the weldability of the structural steel.

1-13. A steel building is 900 ft. long, and it has been decided to provide expansion joints every 300 ft. If the maximum anticipated temperature change is 70°F, determine the size of the expansion joint.

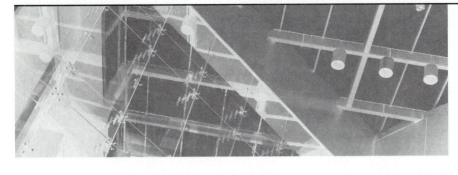

Design Methods, Load Combinations, and Gravity Loads

2.1 INTRODUCTION

The intent of structural design is to select member sizes and connections whose strength is higher than the effect of the applied loads and whose deflections and vibrations are within the prescribed limits. There are two main methods prescribed in the AISC specification [1] for the design of steel structures: the allowable strength design (ASD) method and the load and resistance factor design (LRFD) method; however, appendix 1 of the specification also allows inelastic methods of design such as the plastic design (PLD) method [1].

The LRFD requirements presented in the AISC 2005 specification are similar to the previous three LRFD specifications. The allowable *strength* design (ASD) method in the AISC 2005 specification is similar to the allowable *stress* design in previous specifications in the sense that both are carried out at the service load level. The difference between the two methods is that the provisions for the allowable strength design method are given in terms of forces in the AISC 2005 specification, while the provisions for the allowable stress design method were given in terms of stresses in previous specifications. It should be noted that in the current AISC specification, the design provisions for both the ASD and LRFD methods are based on limit state or strength design principles. In fact, the current *AISCM* presents a dual approach—ASD and LRFD—for all the design aids and tables, with the nominal strength being the same for both design methods. The three design methods for steel structures—LRFD, ASD, and the plastic design method—are discussed below, but for the reasons already discussed in Chapter 1, the focus of this text is on the LRFD method.

Load and Resistance Factor Design Method

In the load and resistance factor design (LRFD) method, the safety margin is realized by using load factors and resistance factors that are determined from probabilistic analysis based on a survey of the reliability indices inherent in existing buildings [2, 3] and a pre-selected reliability index. The load factors vary depending on the type of load because of the different degrees of certainty in predicting each load type, and the resistance factors pre-scribed in the AISC specification also vary depending on the load effects. For example, dead loads are more easily predicted than live or wind loads; therefore, the load factor for a dead load is generally less than that for a live load or a wind load. The load factors account for the possibility of overload in the structure.

In the ASD method, the safety margin is realized by reducing the nominal resistance by a factor of safety, and a single load factor is generally used for all loads. Since the LRFD method accounts for the variability of each load by using different load factors and the ASD assumes the same degree of variability for all loads, the LRFD method provides more uni-form reliability and level of safety for all members in the structure, even for different load-ing conditions. In the case of the ASD method, the level of safety is not uniform throughout the structure. For a comprehensive discussion of the reliability-based design approach, the reader is referred to reference [3].

As previously stated, the LRFD method uses a limit states design method (a limit state is the point at which a structure or structural member reaches its limit of usefulness). The basic LRFD limit state design equation requires that the design strength, ϕR_n, be greater than or equal to the sum of the factored loads or load effects. Mathematically, this can be written as

$$\phi R_n \geq R_u \tag{2-1}$$

where

$R_n =$ Theoretical or nominal strength or resistance of the member determined using the AISC specifications,

$R_u =$ Required strength or sum of the factored loads or load effects using the LRFD load combinations $= \Sigma Q_i \gamma_i$,

$Q_i =$ Service load or load effect,

$\gamma_i =$ Load factor (usually greater than 1.0), and

$\phi =$ Resistance or strength reduction factor (usually less than 1.0).

Note that the service load (i.e., the unfactored or working load), Q_i is the load applied to the structure or member during normal service conditions, while the factored or ultimate load, R_u, is the load applied on the structure at the point of failure or at the ultimate limit state.

Allowable Strength Design Method

In the allowable strength design (ASD) method, a member is selected so that the allowable strength is greater than or equal to the applied service load or load effect, or the required strength, R_a. The allowable strength is the nominal strength divided by a safety factor that is dependent on the limit state being considered, that is,

$$R_n / \Omega \geq R_a, \tag{2-2}$$

where

$R_n / \Omega =$ Allowable strength,

$R_a =$ Required *allowable* strength, or applied service load or load effect deter-mined using the ASD load combinations, and

$\Omega =$ Safety factor.

Plastic Design Method

Plastic design is an optional method in the AISC specification (see Appendix 1) that can be used in the design of continuous beams and girders. In the plastic design method, the structure is assumed to fail after formation of a plastic collapse mechanism due to the presence of plastic hinges. The load at which a collapse mechanism forms in a structure is called the collapse, or ultimate, load, and the load and resistance factors used for plastic design are the same as those used in the LRFD method. The plastic analysis and design of continuous beams is presented in Appendix B of this text.

2.2 STRENGTH REDUCTION OR RESISTANCE FACTORS

The strength reduction or resistance factors (ϕ) account for the variability of the material and section properties and are, in general, usually less than 1.0. These factors are specified for various limit states in the AISC specification, and are shown in Table 2-1.

Table 2-1 Resistance factors ϕ

Limit State	Resistance Factor (ϕ)
Shear	1.0 or 0.9
Flexure	0.90
Compression	0.90
Tension (yielding)	0.90
Tension (rupture)	0.75

2.3 LOAD COMBINATIONS AND LOAD FACTORS (IBC SECTION 1605 OR ASCE 7, SECTIONS 2.3 AND 2.4)

The individual structural loads acting on a building structure do not act in isolation, but may act simultaneously with other loads on the structure. Load combinations are the possible permutations and intensity of different types of loads that can occur together on a structure at the same time. The building codes recognize that all structural loads do not occur at the same time and their maximum values may not happen at the same time. The load combinations or critical combination of loads to be used for design are prescribed in the ASCE 7 load standard [2]. These load combinations include the overload factors, which are usually greater than 1.0, and account for the possibility of overload of the structure. For load combinations, including flood loads, F_a, and atmospheric ice loads, the reader should refer to Sections 2.3.3 and 2.3.4 of the ASCE 7 standard. The basic load combinations for LRFD are

1. $1.4 (D + F)$
2. $1.2 (D + F + T) + 1.6 (L + H) + 0.5 (L_r \text{ or } S \text{ or } R)$
3. $1.2D + 1.6 (L_r \text{ or } S \text{ or } R) + (L \text{ or } 0.8W)$
4. $1.2D + 1.6W + L + 0.5 (L_r \text{ or } S \text{ or } R)$

5. $1.2D + 1.0E + L + 0.2S$

6. $0.9D + (1.6W + 1.6H)$ (D always opposes W and H)

7. $0.9D + (1.0E + 1.6H)$ (D always opposes E and H)

For occupancies with tabulated floor live loads, L_o, not greater than 100 psf (except areas of public assembly and parking garages), it is permissible to multiply the live load, L, in load combinations 3, 4, and 5 by a factor of 0.50. In load combinations 6 and 7, the load factor of H should be set equal to zero when the structural action of H counteracts that due to W or E.

When designing for strength under service load conditions, the ASD load combinations (equations 8 through 15) below should be used. The basic load combinations for allowable strength design (ASD) are

8. $D + F$ *USUALLY SNOW GOVERNS OVER ROOF LL & RAIN*

9. $D + H + F + L + T$

10. $D + H + F + (L_r \text{ or } S \text{ or } R)$

11. $D + H + F + 0.75 (L + T) + 0.75 (L_r \text{ or } S \text{ or } R)$

12. $D + H + F + (W \text{ or } 0.7E)$ *↓D + ↓W/E*

13. $D + H + F + 0.75 (W \text{ or } 0.7E) + 0.75L + 0.75 (L_r \text{ or } S \text{ or } R)$

14. $0.6D + W + H$ (D always opposes W and H)   *↓D ↑W ⎫ LOWER DL*

15. $0.6D + 0.7E + H$ (D always opposes E and H) *↓D ↑E ⎬ IS MORE CONSERVATIVE*

GRAVITY & ↓ LATERAL LOADS

UPLIFT & OVERTURNING

The ASD load combinations (equations 8 through 15) are also used when designing for serviceability limit states such as deflections and vibrations. In load combinations 14 and 15, the load factor of H should be set equal to zero if the direction of H counteracts that due to W or E. It should be noted that for most building structures, the loads H, F, and T will be zero, resulting in more simplified load combination equations. In the above load combinations, downward loads have a positive ($+$) sign, while upward loads have a negative ($-$) sign. Load combinations 1 through 5, and 8 through 13 are used to maximize the downward acting loads, while load combinations 6 and 7, and 14 and 15 are used to maximize the uplift load or overturning effects. Therefore, in load combinations 6, 7, 14, and 15, the wind load, W, and the seismic load, E, take on only negative or zero values, while in all the other load combinations, W and E take on positive values. The notations used in the above load combinations are defined below.

E = Load effect due to *horizontal* and *vertical* earthquake-induced forces

     = $\rho Q_E + 0.2 S_{DS} D$ in load combinations 5, 12, and 13

     = $\rho Q_E - 0.2 S_{DS} D$ in load combinations 7 and 15

D = Dead load

Q_E = Horizontal earthquake load effect due to the base shear, V (i.e., forces, reactions, moments, shears, etc. caused by the horizontal seismic force)

$0.2 S_{DS} D$ = Vertical component of the earthquake force (affects mostly columns and footings)

S_{DS} = Design spectral response acceleration at short period

F = Fluid loads

H = Lateral soil pressure, hydrostatic pressures, and pressure of bulk materials

T = Self-straining force (e.g., temperature)

L = Floor live load

L_r = Roof live load

$$W = \text{Wind load}$$
$$S = \text{Snow load}$$
$$R = \text{Rain load}$$
$$\rho = \text{Redundancy coefficient}$$

A redundancy coefficient, ρ, must be assigned to the seismic lateral force resisting system in both orthogonal directions of the building. The value of the redundancy factor from ASCE 7, Sections 12.3.4.1 and 12.3.4.2 is as follows:

- For seismic design category (SDC) B or C: $\rho = 1.0$, and
- For SDC D, E, or F, it is conservative to assume: $\rho = 1.3$.

Applicable Load Combinations for Design

All structural elements must be designed for the most critical of the load combinations presented in the previous section. Since most floor beams are usually only subjected to dead load, D, and floor live load, L, the most likely controlling load combinations for floor beams and girders will be load combinations 1 or 2 for limit states design, and load combinations 8 and 9 for serviceability (deflections and vibrations) design. Roof beams and columns have to be designed or checked for all load combinations, but load combinations 2 and 9 are more likely to control the design of most floor beams and girders. For example problems 2-1 through 2-5, the loads H, F, and T are assumed to be zero as is the case in most building structures.

Special Seismic Load Combinations and the Overstrength Factor

For certain special structures and elements, the maximum seismic load effect, E_m, and the special load combinations specified in ASCE 7, Section 12.4.3 should be used. Some examples of special structures and elements for which the seismic force is amplified by the overstrength factor, Ω_o, include **drag strut** or **collector elements** (see ASCE 7, Section 12.10.2.1) in building structures, with the exception of light-frame buildings, and structural elements supporting discontinuous systems such as columns supporting discontinuous shear walls (see ASCE 7, Section 12.3.3.3). The seismic load effect, E_m, for these special elements is given as

$$E_m = \Omega_o Q_E \pm 0.2\, S_{\text{DS}}\, D,$$

where Ω_o is the overstrength factor from ASCE 7, Table 12.2-1.

EXAMPLE 2-1

Load Combinations, Factored Loads, and Load Effects

A simple supported floor beam 20 ft. long is used to support service or working loads as follows:

$$w_D = 2.5 \text{ kips/ft.},$$
$$w_L = 1.25 \text{ kips/ft.}$$

[Handwritten margin notes:]

GOVERNING LCs

FLOOR BEAMS/GIRDERS:
DL + LL

ROOF BEAMS & COLUMNS:
CHECK ALL COMBOS

APPLY $\dfrac{\Omega_o}{\rho}$ TO RISA GENERATED LOAD. LC MUST INCLUDE ρ. DO NOT FACTOR ENTIRE MODEL W/ Ω_o

a. Calculate the required shear strength or factored shear, V_u.

b. Calculate the required moment capacity or factored moment, M_u.

c. Determine the required *nominal* moment strength.

d. Determine the required *nominal* shear strength.

SOLUTION

For floor beams, load combinations 1 and 2 with dead and live loads only need to be considered for factored loads.

a. Factored loads:

The corresponding factored loads, w_u, are calculated as follows:

1. $w_u = 1.4D = 1.4\,w_D = 1.4(2.50 \text{ kip/ft.}) = 3.5 \text{ kip/ft.}$

2. $w_u = 1.2D + 1.6L = 1.2\,w_D + 1.6\,w_L = 1.2(2.5) + 1.6(1.25)$
$$= 5.0 \text{ kips/ft. (governs)}$$

$$\text{Maximum factored shear, } V_{u_{max}} = \frac{w_u L}{2} = \frac{(5)(20 \text{ ft.})}{2} = 50 \text{ kips}$$

b. Maximum factored moment, $M_{u_{max}} = \dfrac{w_u L^2}{8} = \dfrac{(5)(20 \text{ ft.})^2}{8} = 250 \text{ ft.-kips}$

Using the limit state design equation yields the following:

c. Required nominal moment strength, $M_n = M_u/\phi = 250/0.90 = 278 \text{ ft.-kips.}$

d. Required nominal shear strength, $V_n = V_u/\phi_v = 50/1.0 = 50 \text{ kips.}$

Note that $\phi = 0.9$ for shear for some steel sections. (see Chapter 6).

EXAMPLE 2-2

Load Combinations, Factored Loads, and Load Effects

Determine the required moment capacity, or factored moment, M_u, acting on a floor beam if the calculated service load moments acting on the beam are as follows: $M_D = 55 \text{ ft.-kip}$, $M_L = 30 \text{ ft.-kip}$.

SOLUTION

1. $M_u = 1.4\,M_D = 1.4(55 \text{ ft.-kip}) = 77 \text{ ft.-kips}$

2. $M_u = 1.2\,M_D + 1.6\,M_L = 1.2\,(55 \text{ ft.-kips}) + 1.6\,(30 \text{ ft.-kips})$
$$= 114 \text{ ft.-kips (governs)}$$

$M_u = 114 \text{ ft.-kips}$

EXAMPLE 2-3

Load Combinations, Factored Loads, and Load Effects

Determine the ultimate or factored load for a roof beam subjected to the following service loads:

Dead load = 35 psf
Snow load = 25 psf
Wind load = 15 psf upwards
10 psf downwards

SOLUTION

D = 35 psf
S = 25 psf
W = −15 psf or 10 psf

The values for service loads not given above are assumed to be zero; therefore,

Floor live load, L = 0
Rain live load, R = 0
Seismic load, E = 0
Roof live load, L_r = 0

Using the LRFD load combinations, and noting that only downward-acting loads should be substituted in load combinations 1 through 5, and upward wind or seismic loads in load combinations 6 and 7, the controlling factored load is calculated as follows:

1. $w_u = 1.4D = 1.4\,(35) = 49$ psf

2. $w_u = 1.2D + 1.6L + 0.5S$
   $= 1.2\,(35) + 1.6\,(0) + 0.5\,(25) = 55$ psf

3a. $w_u = 1.2D + 1.6S + 0.8W$
   $= 1.2\,(35) + 1.6\,(25) + 0.8\,(10) =$ **90 psf** (governs)

3b. $w_u = 1.2D + 1.6S + 0.5L$
   $= 1.2\,(35) + 1.6\,(25) + 0.5\,(0) = 82$ psf

4. $w_u = 1.2D + 1.6W + L + 0.5S$
   $= 1.2\,(35) + 1.6\,(10) + (0) + 0.5\,(25) = 71$ psf

5. $w_u = 1.2D + 1.0E + 0.5L + 0.2S$
   $= 1.2\,(35) + 1.0\,(0) + 0.5(0) + 0.2\,(25) = 47$ psf

6. $w_u = 0.9D + 1.6W$ (*D must* always oppose *W* in load combination 6)
   $= 0.9\,(35) + 1.6\,(−15)$ (*upward wind load is taken as negative*)
   $= 7.5$ psf

7. $w_u = 0.9D + 1.0E$ (*D must* always oppose *E* in load combination 7)
$$= 0.9\,(35) + 1.0\,(0)$$
$$= 31.5\ \text{psf}$$

In this example, load combinations 6 and 7 resulted in net positive (or downward) load. However, load combinations 6 and 7 may sometimes result in net negative (or upward) factored loads that would also have to be considered in the design of the structural member.

- For strength calculations, the controlling factored load is $w_u = $ **90 psf**.

If controlling service loads are required, these would be calculated using ASD load combinations 8 through 15.

EXAMPLE 2-4

Load Combinations, Factored Loads, and Load Effects

a. Determine the factored axial load or the required axial strength for a column in an office building with the given service loads below.

b. Calculate the required *nominal* axial compression strength of the column.

Given service axial loads on the column:

$P_D = 75$ kips (dead load)

$P_L = 150$ kips (floor live load)

$P_S = 50$ kips (snow load)

$P_W = \pm 100$ kips (wind load)

$P_E = \pm 50$ kips (seismic load)

SOLUTION

a. Note that downward loads take on positive values, while upward loads take on negative values. The factored loads are calculated as

1. $P_u = 1.4\,P_D = 1.4\,(75\ \text{kips}) = 105\ \text{kips}$

2. $P_u = 1.2\,P_D + 1.6\,P_L + 0.5\,P_S$
$$= 1.2\,(75) + 1.6\,(150) + 0.5\,(50) = \textbf{355 kips}\ (\text{governs})$$

3a. $P_u = 1.2\,P_D + 1.6\,P_S + 0.5\,P_L$
$$= 1.2\,(75) + 1.6\,(50) + 0.5\,(150) = 245\ \text{kips}$$

3b. $P_u = 1.2\,P_D + 1.6\,P_S + 0.8\,P_W$
$$= 1.2\,(75) + 1.6\,(50) + 0.8\,(100) = 240\ \text{kips}$$

4. $P_u = 1.2\,P_D + 1.6\,P_W + 0.5\,P_L + 0.5\,P_S$
$$= 1.2\,(75) + 1.6\,(100) + 0.5\,(150) + 0.5\,(50) = 350\ \text{kips}$$

(continued)

5. $P_u = 1.2\,P_D + 1.0\,P_E + 0.5\,P_L + 0.2\,P_S$
$= 1.2\,(75) + 1.0\,(50) + 0.5\,(150) + 0.2\,(50) = 225$ kips

Note that P_D must always oppose P_W and P_E in load combinations 6 and 7:

6. $P_u = 0.9\,P_D + 1.6\,P_W$
$= 0.9\,(75) + 1.6\,(-100) = \mathbf{-92.5\ kips}$ (governs)

7. $P_u = 0.9\,P_D + 1.0\,P_E$
$= 0.9\,(75) + 1.0\,(-50) = 17.5$ kips

- The factored axial *compression* load on the column is 355 kips.

- The factored axial *tension* force on the column is 92.5 kips.

  The column, base plate, anchor bolts, and foundation will need to be designed for both the downward factored load of 355 kips and the factored net uplift, or tension, load of 92.5 kips.

b. Nominal axial compression strength of the column, $P_n = P_u/\phi = 355/0.90 = 394$ kips.

EXAMPLE 2-5

Load Combinations—Factored Loads and Load Effects

Repeat Example 2-4 assuming that the structure is to be used as a parking garage.

SOLUTION

a. For parking garages or areas used for public assembly or areas with a floor live load, L, greater than 100 psf, the multiplier of the floor live load in load combinations 3, 4, and 5 is 1.0. The factored loads are calculated as follows:

1. $P_u = 1.4\,P_D = 1.4\,(75) = 105$ kips

2. $P_u = 1.2\,P_D + 1.6\,P_L + 0.5\,P_S$
$= 1.2\,(75) + 1.6\,(150) + 0.5\,(50) = 355$ kips

3a. $P_u = 1.2\,P_D + 1.6\,P_S + 1.0\,P_L$
$= 1.2\,(75) + 1.6\,(50) + 1.0\,(150) = 320$ kips

3b. $P_u = 1.2\,P_D + 1.6\,P_S + 0.8\,P_W$
$= 1.2\,(75) + 1.6\,(50) + 0.8\,(100) = 250$ kips

4. $P_u = 1.2\,P_D + 1.6\,P_W + 1.0\,P_L + 0.5\,P_S$
$= 1.2\,(75) + 1.6\,(100) + 1.0\,(150) + 0.5\,(50) = \mathbf{425\ kips}$ (governs)

5. $P_u = 1.2 P_D + 1.0 P_E + 1.0 P_L + 0.2 P_S$
   $= 1.2 (75) + 1.0 (50) + 1.0 (150) + 0.2 (50) = 300$ kips

Note that P_D must always oppose P_W and P_E in load combinations 6 and 7:

6. $P_u = 0.9 P_D + 1.6 P_W$
   $= 0.9 (75) + 1.6 (-100) = $ **−92.5 kips** (governs)

7. $P_u = 0.9 P_D + 1.0 P_E$
   $= 0.9 (75) + 1.0 (-50) = 17.5$ kips

- The factored axial *compression* load on the column is 425 kips.

- The factored axial *tension* force on the column is 92.5 kips.

The column, base plate, anchor bolts, and foundation will need to be designed for both the downward factored load of 425 kips and the factored net uplift, or tension, load of 92.5 kips.

b. Nominal axial compression strength of the column, $P_n = P_u/\phi = 425/0.90 = 473$ kips.

2.4 INTRODUCTION TO DESIGN LOADS

Structural loads are the forces that are applied on a structure (e.g., dead load, floor live load, roof live load, snow load, wind load, earthquake or seismic load, earth and hydrostatic pressure). The magnitude of these loads are specified in the ASCE 7 load standard; in this standard, buildings are grouped into different occupancy types, which are used to determine the importance factor, I, for snow, wind, seismic, or ice load calculations [2]. The importance factor is a measure of the consequence of failure of a building to public safety; the higher the importance factor, the higher the snow, wind, or seismic loads on the structure. ASCE 7, Table 1-1 should be used with ASCE 7, Tables 6-1, 7-4, 10-1, and 11.5-1 to determine the importance factors for wind loads, snow loads, ice loads, and seismic loads, respectively.

Gravity Load Resisting Systems

The two main types of floor systems used to resist gravity loads in building structures are *one-way* and *two-way* load distribution systems. For steel structures, the one-way load distribution system occurs much more frequently than the two-way system; consequently, only one-way systems are discussed further. There are several *one-way* load distribution systems that are used in steel buildings. These systems support loads in one-way action by virtue of their construction and because the bending strength in one direction is several times greater than the strength in the orthogonal direction. These types of one-way systems span in the stronger direction of the slab panel regardless of the aspect ratio of the panel. Examples of one-way systems used in steel buildings include

- Metal roof decks (used predominantly for roofs in steel buildings),
- Composite metal floor decks (used predominantly for floors in steel buildings), and
- Pre-cast concrete planks.

2.5 GRAVITY LOADS IN BUILDING STRUCTURES

The common types of gravity loads that act on building structures—*roof dead load, floor dead load, roof live load, snow load, and floor live load*—are discussed in the following sections.

Dead Loads

Dead loads are permanent or nonmoveable loads that act on a structure and include the weight of all materials that are permanently attached to the structure, including the self-weight of the structure. The dead loads can be determined with greater accuracy than any other type of load, and are not as variable as live loads. Examples of items that would be classified as dead loads include floor finishes, partitions, mechanical and electrical (M&E) components, glazing, and cladding. The floor and roof dead loads are typically uniform loads expressed in units of pounds or kips per square foot (psf or ksf) of the horizontal projected plan area; the dead load of sloped members, which is in units of psf of sloped area, must be converted to units of psf of the horizontal projected plan area. In certain cases, concentrated dead loads, such as a heavy safe with a small footprint, may also have to be considered. Typical checklists for roof and floor dead load components in steel buildings are shown below.

Ⓜ 17.26

Common Roof Dead Loads in Steel Buildings

Framing	5 psf to 8 psf
Fireproofing	2 psf
Metal deck	2 psf (1½″ deck)
	3 psf (3″ deck)
	6 psf (7½″ deck)
5 ply with gravel	6.5 psf
Rigid insulation	1.5 psf per inch of thickness
Plywood sheathing	0.4 psf per ⅛″ thickness
¼″ asphalt shingles	2.0 psf
Suspended ceiling	2.0 psf
Mechanical/electrical	5 psf to 10 psf

Common Floor Dead Loads in Steel Buildings

Framing	6 psf to 12 psf
Fireproofing	2 psf
Metal deck	1 psf (⁹⁄₁₆″ deck)
	2 psf (1½″ deck)
	3 psf (3″ deck)
Concrete	
lightweight	10 psf per inch of thickness
normal weight	12.5 psf per inch of thickness
w/metal deck	see deck manufacturers catalog
floor finishes	
¼″ ceramic tile	10 psf
1″ slate	15 psf
gypsum fill	6.0 psf per inch of thickness
⅞″ hardwood	4.0 psf
· partitions	15 psf
Suspended ceiling	2.0 psf
Mechanical/electrical	5 psf to 10 psf

The ASCE 7 load standard and the International Building Code (IBC 2006) [4] specify a minimum partition load of 15 psf, and partition loads need not be considered when the tabulated unreduced floor live load, L_o, is greater than 80 psf because of the low probability that partitions will be present in occupancies with these higher live loads. It should be noted that partition load is classified in ASCE 7 as a live load, but it is common in design practice to treat this load as a dead load, and this approach is adopted in this book.

Tributary Widths and Tributary Areas

In this section, the concepts of tributary widths and tributary areas are discussed. These concepts are used to determine the distribution of floor and roof loads to the individual structural members. The tributary width (TW) of a beam or girder is defined as the width of the floor or roof supported by the beam or girder, and is equal to the sum of one-half the distance to the adjacent beams to the right and left of the beam whose tributary width is being determined. The tributary width is calculated as

$$TW = \frac{1}{2}(\text{Distance to adjacent beam on the right}) + \frac{1}{2}(\text{Distance to adjacent beam on the left}).$$

The tributary area of a beam, girder, or column is the floor or roof area supported by the structural member. The tributary area of a beam is obtained by multiplying the span of the beam by its tributary width. The tributary area of a column is the plan area bound by lines located at one-half the distance to the adjacent columns surrounding the column whose tributary area is being calculated. The following should be noted:

- Beams are usually subjected to uniformly distributed loads (UDL) from the roof or floor slab or deck.
- Girders are usually subjected to concentrated, or point, loads due to the reactions from the beams. These concentrated loads or reactions from the beams have their own tributary areas.
- The tributary area of a girder is the sum of the tributary areas of all the concentrated loads acting on the girder.
- Perimeter beams and girders support an additional uniform load due to the loads acting on the floor or roof area extending from the centerline of the beam or girder to the edge of the roof or floor.

EXAMPLE 2-6

Calculation of Tributary Width and Tributary Area

Using the floor framing plan shown in Figure 2-1, determine the following:

  a. Tributary width and tributary area of a typical interior beam,
  b. Tributary width and tributary area of a typical spandrel or perimeter beam,
  c. Tributary area of a typical interior girder,
  d. Tributary area of a typical spandrel girder,
  e. Tributary area of a typical interior column,
  f. Tributary area of a typical corner column, and
  g. Tributary area of a typical exterior column.

(continued)

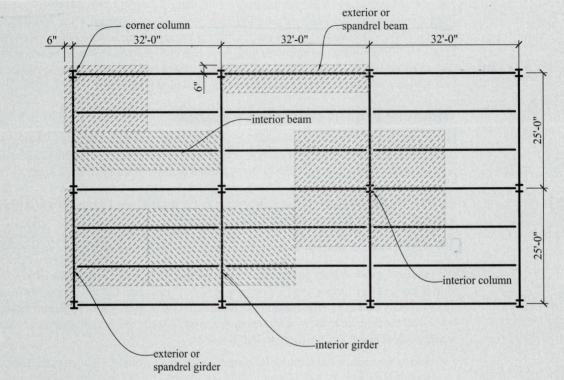

Figure 2-1 Tributary width and tributary areas.

SOLUTION

The tributary widths and areas are calculated in Table 2-2.

Table 2-2 Tributary widths and areas

Member	Tributary Width (TW)	Tributary Area (A_T)
Typical Interior Beam	25 ft./3 spaces = 8.33 ft.	(8.33 ft.) (32 ft.) = 267 ft.^{2}
Typical Spandrel Beam	(25 ft./3 spaces)/2 + 0.5 [ft.] edge distance = 4.67 ft.	(4.67 ft.) (32 ft.) = 150 ft.^{2}
Typical Interior Girder	—	$\left(\dfrac{267}{2}\text{ ft.}^2\right)(4\text{ beams}) = 534\text{ ft.}^2$
Typical Spandrel Girder	—	(0.5 ft. edge dist) (25 ft.) $+ \left(\dfrac{267}{2}\text{ ft.}^2\right)(2\text{ beams})$ $= 280\text{ ft.}^2$
Typical Interior Column	—	(32 ft.) (25 ft.) = 800 ft.^{2}
Typical Corner Column	—	(32 ft./2 + 0.5 ft. edge distance) $\times$ (25 ft./2 + 0.5 ft. edge distance) = 215 ft.^{2}
Typical Exterior Column (long side of building)	—	(32 ft./2 + 0.5 ft. edge distance) (25 ft.) = 413 ft.^{2}
Typical Exterior Column (short side of building)	—	(25 ft./2 + 0.5 ft. edge distance) (32 ft.) = 416 ft.^{2}

2.6   LIVE LOADS

In general, any load that is not permanently attached to the structure can be considered a live load. The three main types of live loads that act on building structures are floor live load, roof live load, and snow load. Floor live loads, L, are occupancy loads that are specified in ASCE 7, Table 4-1 or IBC, Table 1607.1; the magnitude depends on the use of the structure and the tributary area (TA). Floor live loads are usually expressed as uniform loads in units of pounds per square foot (psf) of horizontal plan area. In certain cases, the code also specifies alternate concentrated floor live loads (in lb. or kip) that need to be considered in the design; but in most cases, the uniform loads govern the design of structural members.

The two types of live loads that act on roof members are roof live load, L_r, and snow load, **S**. From the load combinations presented in Section 2.3, it becomes apparent that roof live loads and snow loads **do not** act together at the same time. Roof live loads rarely govern the design of structural members in the higher snow regions of the United States, except where the roof is a special purpose roof used for promenades or as a roof garden.

2.7   FLOOR LIVE LOADS (ASCE 7, TABLE 4-1 OR IBC 1607.1)

Floor live loads are occupancy loads that depend on the use of the structure. These loads are assumed to be uniform loads expressed in units of pounds per square foot (psf) and the values are specified in building codes such as the IBC and the ASCE 7 load standard. The live loads are determined from statistical analyses of a large number of load surveys. The codes also specify alternate concentrated live loads for some occupancies because for certain design situations, live load concentrations—as opposed to uniform live loads—may be more critical for design. The uniform floor live loads are most commonly used in practice; the concentrated live loads are used only in rare situations where, for example, punching shear may be an issue, such as in thin slabs or in the design of stair treads where the concentrated loads instead of the uniform loads control the design of the member. For a more complete listing of the recommended minimum live loads, the reader should refer to Table 2-3 [2-2, 2-4]. In general, the actual floor live loads are usually smaller than the live loads prescribed in the building codes or Table 2-3, but there are some unusual situations where the actual load may be greater than the live loads prescribed in the codes. For such cases, the actual live load should be used for design.

2.8   FLOOR LIVE LOAD REDUCTION

To account for the low probability that floor structural elements with large tributary areas will have their entire tributary area loaded with the live load at one time, the IBC and ASCE 7 load specifications allow for floor live loads to be reduced provided that certain conditions are satisfied. The reduced design live load of a floor, L, in psf, is given as

$$L = L_o \left[0.25 + \frac{15}{\sqrt{K_{LL}A_T}} \right]$$

(2.2)

Table 2-3 Minimum uniformly distributed and concentrated floor live loads

Minimum Uniformly Distributed and Concentrated Live Loads		
Occupancy	Uniform Load (psf)	Concentrated Load (lb.)
Balconies		
Exterior	100	—
One- and two-family residences only, and not exceeding 100 ft.²	60	—
Dining Rooms and Restaurants	100	—
Office Buildings		
Lobbies and first-floor corridors	100	2,000
Offices	50	2,000
Corridors above first floor	80	2,000
Residential (one- and two-family dwellings)	40	—
Hotels and Multifamily Houses		
Private rooms and corridors serving them	40	—
Public rooms and corridors serving them	100	
Roofs		
Ordinary flat, pitched, and curved roofs	20	
Promenades	60	
Gardens or assembly	100	
Single panel point of truss bottom chord or at any point along a beam		2000
Schools		
Classrooms	40	1,000
Corridors above first floor	80	1,000
First-floor corridors	100	1,000
Stairs and exit ways	100	300 lb. over an area of 4 in.²
One- and two-family residences only	40	
Storage		
Light	125	
Heavy	250	
Stores		
Retail		
First floor	100	1,000
Upper floors	75	1,000
Wholesale	125	1,000

Adapted from IBC, Table 1607.1 (ref. 4).

$\geq 0.50\,L_o$ for members supporting *one floor* (e.g., slabs, beams, and girders)

$\geq 0.40\,L_o$ for members supporting *two or more floors* (e.g., columns)

L_o = Unreduced design live load from Table 2-3 (IBC, Table 1607.1 or ASCE 7, Table 4-1)

K_{LL} = Live load element factor (see ASCE 7, Table 4-2)

$= 4$ (interior columns and exterior columns without cantilever slabs)

$= 3$ (edge columns with cantilever slab)

$= 2$ (corner columns with cantilever slabs, edge beams without cantilever slabs, interior beams)

$= 1$ (all other conditions)

$A_T =$ Summation of the floor tributary area, in ft.^{2}, supported by the member, excluding the roof area

- For beams and girders (including continuous beams or girders), A_T as defined in Section 2.5.
- For **one-way slabs**, A_T must be less than or equal to $1.5s^2$ where s is the slabspan.
- For a member supporting more than one floor area in multistory buildings, A_T will be the summation of all the applicable floor areas supported by that member.

The ASCE 7 load specification *does not* permit floor live load reductions for floors satisfying any one of the following conditions:

- $K_{LL} A_T \le 400$ ft.^{2};
- Floor live load, $L_o > 100$ psf; may be reduced 20% for members supporting 2 or more floors in nonassembly occupancies.
- Floors with occupancies used for assembly purposes, such as auditoriums, stadiums, etc., because of the high probability of overloading; or
- For passenger car garage floors, except that the live load is allowed to be reduced by 20% for members supporting two or more floors.

The following should be noted regarding the tributary area, A_T, used in calculating the reduced floor live load:

1. The infill beams are usually supported by girders which, in turn, are supported by columns, as indicated in our previous discussions on load paths. The tributary area, A_T, for beams is usually smaller than those for girders and columns and, therefore, beams will have smaller floor live load reductions than girders or columns. The question arises as to which A_T to use for calculating the loads on the girders.

2. For the design of the beams, use the A_T of the beam to calculate the reduced live load that is used to calculate the moments, shears, and reactions. These load effects are used for the design of the beam and the beam-to-girder or beam-to-column connections. Note that, in practice, because of the relatively small tributary areas for beams, it is common practice to neglect live load reduction for beams.

3. For the girders, recalculate the beam reactions using the A_T of the girder. These smaller beam reactions are used for the design of the girders only.

4. For columns, A_T is the summation of the tributary areas of all the floors with *reducible live loads* above the level at which the column load is being determined, excluding the roof areas.

2.9 ROOF LIVE LOAD

Roof live loads, L_r are the weight of equipment and personnel on a roof during maintenance of the roof or the weight of moveable nonstructural elements such as planters or other decorative elements, or the use of the roof for assembly purposes. Like floor live loads, the unreduced roof live loads are also tabulated in the ASCE 7 load standard. However, only live loads on ordinary flat, pitched, or curved roofs can be reduced as described in the following section.

Roof Live Load Reduction for Ordinary Flat, Pitched, and Curved Roofs

For ordinary flat, pitched, and curved roofs, the ASCE 7 load standard allows the roof live load, L_o, to be reduced according to the following formulas (*note that for all other types of roofs, $L_r = L_o$*):

$$\text{Design roof live load, in psf, } L_r = L_o R_1 R_2, \tag{2-3}$$

where

12 psf $\leq L_r \leq$ 20 psf, and

L_o = Roof live load from ASCE 7, Table 4-1.

The reduction factors, R_1 and R_2, are calculated as follows:

$R_1 = 1.0$ for $A_T \leq 200$ ft.^{2}

$R_1 = 1.2 - 0.001A_T$ for 200 ft.$^2 < A_T < 600$ ft.^{2}

$R_1 = 0.6$ for $A_T \geq 600$ ft.^{2}

$R_2 = 1.0$ for F ≤ 4

$R_2 = 1.2 - 0.05F$ for $4 <$ F < 12

$R_2 = 0.6$ for F ≥ 12

F = Number of inches of rise per foot for a pitched or sloped roof
    (e.g., F = 3 for a roof with a 3-in-12 pitch)
  = Rise-to-span ratio multiplied by 32 for an arch or dome roof

A_T = Tributary area in square feet (ft.^{2})

For landscaped roofs, it should be noted that the weight of the landscaped material should be included in the dead load calculations and should be computed assuming that the soil is fully saturated.

EXAMPLE 2-7

Roof Live Load

For the framing of the ordinary flat roof shown in Figure 2-2, determine the design roof live load, L_r, for the following structural members:

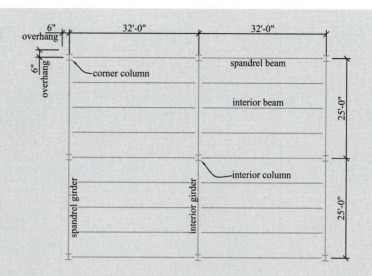

Figure 2-2 Roof framing for example 2-7.

a. Typical interior beam,

b. Typical spandrel or perimeter beam,

c. Typical interior girder,

d. Typical spandrel girder, and

e. Typical interior column.

SOLUTION

The unreduced roof live load, L_o, is 20 psf from ASCE-7, Table 4-1, and the tributary width and tributary areas are calculated in Table 2-4.

Table 2-4 Tributary widths and areas

Member	Tributary Width (TW)	Tributary Area (A_T)
Typical Interior Beam	25 ft./4 spaces = 6.25 ft.	(6.25 ft.) (32 ft.) = 200 ft.^{2}
Typical Spandrel Beam	(25 ft./4 spaces)/2 + 0.5 ft. edge distance = 3.63 ft.	(3.63 ft.) (32 ft.) = 116 ft.^{2}
Typical Interior Girder		$\left(\dfrac{200}{2}\,\text{ft.}^2\right)(6\text{ beams}) = 600\text{ ft.}^2$
Typical Spandrel Girder		$(0.5\text{ ft. edge dist})(25\text{ ft.})$ $+\left(\dfrac{200}{2}\,\text{ft.}^2\right)(3\text{ beams}) = 313\text{ ft.}^2$
Typical Interior Column		(32 ft.) (25 ft.) = 800 ft.^{2}

(continued)

a. Tributary area, $A_T = 200$ ft.^{2}, therefore, $R_1 = 1.0$
   For a flat roof, $F = 0$, therefore, $R_2 = 1.0$
   Using equation (2-3), the design roof live load, $L_r = 20$ (1.0) (1.0) $= 20$ psf

b. Tributary area, $A_T = 116$ ft.^{2}, therefore, $R_1 = 1.0$
   For a flat roof, $F = 0$, therefore, $R_2 = 1.0$
   Design roof live load, $L_r = 20$ (1.0) (1.0) $= 20$ psf

c. Tributary area, $A_T = 600$ ft.^{2}, therefore, $R_1 = 0.6$
   For a flat roof, $F = 0$; therefore, $R_2 = 1.0$
   Design roof live load, $L_r = 20$ (0.6) (1.0) $= 12$ psf

d. Tributary area, $A_T = 313$ ft.^{2}, therefore, $R_1 = 1.2 - 0.001$ (313) $= 0.89$
   For a flat roof, $F = 0$, therefore, $R_2 = 1.0$
   Design roof live load, $L_r = 20$ (0.89) (1.0) $= 17.8$ psf

e. Tributary area, $A_T = 800$ ft.^{2}, therefore, $R_1 = 0.6$
   For a flat roof, $F = 0$, therefore, $R_2 = 1.0$
   Design roof live load, $L_r = 20$ (0.6) (1.0) $= 12$ psf

To determine the total design load for the roof members, the calculated design roof live load, L_r, will need to be combined with the dead load and other applicable loads using the load combinations from Section 2.3.

EXAMPLE 2-8

Column Load With and Without Floor Live Load Reduction

A three-story building, with columns that are spaced at 20 ft. in both orthogonal directions, is subjected to the roof and floor loads shown below. Using a tabular format, calculate the cumulative **factored** and unfactored axial loads on a typical interior column **with and without live load reduction**. Assume a roof slope of ¼ in./ft. for drainage.

Roof Loads:
Dead load, $D_{roof} = 30$ psf
Snow load, $S = 30$ psf
Roof Live load, $L_r = $ per code
Second and Third Floor Loads:
Dead load, $D_{floor} = 110$ psf
Floor Live Load, $L = 40$ psf

SOLUTION

At each level, the tributary area, A_T, supported by a typical interior column is 20 ft. $\times$ 20 ft. = 400 ft.^{2}

Roof Live Load, L_r:
For an ordinary flat roof, L_o = 20 psf (ASCE 7, Table 4-1).
From Section 2.9, the roof slope factor, F, is 0.25; therefore, R_2 = 1.0.
Since the tributary area, A_T, of the column = 400 ft.^{2}, R_1 = 1.2 $-$ 0.001 (400) = 0.8
Using equation (2-3), the design roof live load is

$$L_r = L_o R_1 R_2 = 20\ R_1 R_2 = 20\ (0.8)\ (1.0) = 16\ \text{psf},$$
Since 12 psf $<$ L_r $<$ 20 psf; therefore, L_r = 16 psf.

L_r is smaller than the snow load, S = 30 psf; therefore, the snow load, S, is more critical than the roof live load, L_r, in the applicable load combinations. All other loads, such as W, H, T, F, R, and E, are zero for the roof or floors. The applicable *LRFD* load combinations from Section 2.3 that will be used to calculate the factored column axial loads are

1. $1.4D$

2. $1.2D + 1.6L + 0.5S$

3. $1.2D + 1.6S + 0.5L$

4. $1.2D + 0.5L + 0.5S$

5. $1.2D + 0.5L + 0.2S$

6. $0.9D$

7. $0.9D$

The corresponding ASD load combinations for calculating the unfactored column axial loads are

8. D

9. $D + L$

10. $D + S$

11. $D + 0.75L + 0.75S$

12. D

13. $D + 0.75L + 0.75S$

14. $0.6D$

15. $0.6D$

(continued)

> A close examination of these load combinations will confirm that load combinations 1 and 4 through 7 are not the most critical combinations for the factored axial load on the column. The most critical factored load combinations are load combinations 2 ($1.2D + 1.6L + 0.5S$) and 3 ($1.2D + 1.6S + 0.5L$). A similar examination of the ASD load combinations reveals that load combinations 9 ($D + L$) and 11 ($D + 0.75L + 0.75S$) are the two most critical load combinations for the calculation of the unfactored axial load on the typical interior column for this building.

The reduced or design floor live loads for the second and third floors are calculated using Table 2-5.

Using both load LRFD combinations 2 (i.e., $1.2D + 1.6L + 0.5S$) and 3 ($1.2D + 1.6S + 0.5L$), the maximum factored column axial loads with and without floor live load reductions are calculated in Table 2-6. The corresponding values for the unfactored column axial loads with and without floor live load reductions are calculated in Table 2-7.

Therefore, the ground floor column will be designed for a cumulative reduced factored axial compression load of 156 kips; the corresponding factored axial load without floor live load reduction is 177 kips. The reduction in factored column axial load due to floor live load reduction is only 12% for this three-story building. Thus, the effect of floor live load reduction on columns and column footings is not critical for low-rise buildings.

Although in the previous discussions and example we used the tributary area method in calculating the column loads, the column loads can also be determined by summing the reactions of all the beams and girders that frame into the column. This method will prove to be useful later when the columns have to be designed for combined axial loads and bending moments due to the eccentricity of the beam and girder connections.

Table 2-5 Reduced or design floor live load calculation table

Member	Levels Supported	A_T (summation of floor tributary area)	K_{LL}	Unreduced Floor Live Load, L_o (psf)	Live Load Reduction Factor, $0.25 + 15/\sqrt{K_{LL} A_T}$	Design Live Load, (L or S)
Third-floor column (i.e., column below roof)	Roof only	Floor live load reduction NOT applicable to roofs	—	—	—	30 psf (snow load)
Second-floor column (i.e., column below third floor)	1 Floor + Roof	(1 Floor) (400 ft.^{2}) = 400 ft.^{2}	$K_{LL} = 4$ $K_{LL} A_T = 1600 > 400$ ft.^{2} ∴ Live load reduction allowed	40 psf	$\left[0.25 + \dfrac{15}{\sqrt{1600}}\right]$ $= 0.625$	0.625 (40) = 25 psf ≥ 0.50 L_o
Ground or first-floor column (i.e., column below second floor)	2 Floors + Roof	(2 Floors) (400 ft.^{2}) = 800 ft.^{2}	$K_{LL} = 4$ $K_{LL} A_T = 3200 > 400$ ft.^{2} ∴ Live load reduction allowed	40 psf	$\left[0.25 + \dfrac{15}{\sqrt{3200}}\right]$ $= 0.52$	0.52 (40) = 21 psf ≥ 0.40 L_o

Table 2-6 Factored column load

Level	Tributary Area, A_T (ft.²)	Dead Load, D (psf)	Live Load, L_o or R (S or L_r or R on the roof) (psf)	Design Live Load Floor: S L Roof: S or L_r or R (psf)	Factored Uniform Load at Each Level, w_{u1} Roof: 1.2D + 0.5S Floor: 1.2D + 1.6L (psf)	Factored Uniform Load at Each Level, w_{u2} Roof: 1.2D + 1.6S Floor: 1.2D + 0.5L (psf)	Factored Column Axial Load, P, at Each Level, $(A_T)(w_{u1})$ or $(A_T)(w_{u2})$ (kips)	Cumulative Factored Axial Load, ΣP LC 2 (kips)	Cumulative Factored Axial Load, ΣP LC 3 (kips)	Maximum Cumulative Factored Axial Load, ΣP (kips)
With Floor Live Load Reduction										
Roof	400	30	30	30	51	84	20.4 or 33.6	20.4	33.6	**33.6**
Third Floor	400	110	40	25	172	145	68.8 or 58	89.2	91.6	**91.6**
Second Floor	400	110	40	21	166	143	66.4 or 57.2	156	149	**156**
Without Floor Live Load Reduction										
Roof	400	30	30	30	51	84	20.4 or 33.6	20.4	33.6	**33.6**
Third Floor	400	110	40	40	196	152	78.4 or 60.8	98.8	94.4	**98.8**
Second Floor	400	110	40	40	196	152	78.4 or 60.8	177	156	**177**

The maximum *factored* column loads (*with* floor live load reduction) are Third-story column (i.e., column below roof level) = 33.6 kips.
Second-story column (i.e., column below the third floor) = 91.6 kips.
First-story column (i.e., column below the second floor) = 156 kips.

The maximum factored axial column loads (*without* floor live load reduction) are Third-story column (i.e., column below roof level) = 33.6 kips.
Second-story column (i.e., column below the third floor) = 98.8 kips.
First-story column (i.e., column below the second floor) = 177 kips.

Table 2-7 Unfactored column load

Level	Tributary Area, A_T (ft.²)	Dead Load, D (psf)	Live Load, L_o or L_r or R (S or L_r or R on the roof) (psf)	Design Live Load Roof: S or L_r or R Floor: L (psf)	Unfactored Total Load at Each Level, w_{s1} Roof: D Floor: $D + L$ (psf)	Unfactored Total Load at Each Level, w_{s2} Roof: $D + 0.75S$ Floor: $D + 0.75L$ (psf)	Unfactored Column Axial Load at Each Level, $P = (A_T)(w_{s1})$ or $(A_T)(w_{s2})$ (kips)	Cumulative Unfactored Axial Load, ΣP_{D+L} (kips)	Cumulative Unfactored Axial Load, $\Sigma P_{D+0.75L+0.75S}$ (kips)	Maximum Cumulative Unfactored Axial Load, ΣP (kips)
With Floor Live Load Reduction										
Roof	400	30	30	30	30	52.5	12 or 21	12	21	**21**
Third Floor	400	110	40	25	135	128.8	54 or 51.5	66	72.5	**72.5**
Second Floor	400	110	40	21	131	125.8	52.4 or 50.3	118.4	122.8	**122.8**
Without Floor Live Load Reduction										
Roof	400	30	30	30	30	52.5	12 or 21	12	21	**21**
Third Floor	400	110	40	40	150	140	60 or 56	72	77	**77**
Second Floor	400	110	40	40	150	140	60 or 56	132	133	**133**

The maximum *unfactored* column loads (with floor live load reduction) are

Third-story column (i.e., column below roof level) = 21 kips.
Second-story column (i.e., column below the third floor) = 72.5 kips.
First-story column (i.e., column below the second floor) = 122.8 kips.

The maximum factored axial column loads (*without* floor live load reduction) are

Third-story column (i.e., column below roof level) = 21 kips.
Second-story column (i.e., column below the third floor) = 77 kips.
First-story column (i.e., column below the second floor) = 133 kips.

2.10   SNOW LOAD

A ground snow map of the United States showing the 50-yr. ground snow loads, p_g, is found in ASCE 7, Figure 7-1; however, for certain areas, specific snow load studies are required in order to establish the ground snow loads. The ground snow loads are specified in greater detail in the local building codes, and because relatively large variations in snow loads can occur even over small geographic areas, the local building codes appear to have more ac-curate snow load data for the different localities within their jurisdiction when compared to the snow load map given in ASCE 7. The value of the roof snow load depends on, but is usu-ally lower than, the ground snow load, p_g, because of the increased effect of wind at the higher levels. The roof snow load is also a function of the roof exposure, the roof slope, the use of the building, the temperature of the roof—whether it is heated or not—and the ter-rain conditions at the building site; however, it is unaffected by the tributary area of a struc-tural member. The steeper the roof slope, the smaller the snow load, because steep roofs are less likely to retain snow, and conversely, the flatter the roof slope, the larger the snow load. Depending on the type of roof, the snow load can either be a uniform balanced load or an unbalanced load. A balanced snow load is a uniform snow load over the entire roof surface, while an unbalanced snow load is a partial uniform or nonuniform distribution of snow load over the roof surface or a portion of the roof surface.

The procedure for calculating snow loads on a roof surface is as follows:

1. Determine the ground snow load from Figure 7-1 of ASCE 7 or from the local snow map of the area. In mountainous regions, local snow maps take on even greater im-portance as the ASCE 7 values are often low for these areas or are not given, and local building codes can override ASCE 7.

2. Determine the snow exposure factor, C_e, from ASCE 7, Table 7-2.

3. Determine the thermal factor, C_t, from ASCE 7, Table 7-3.

4. Determine the snow load importance factor, I_s, from ASCE 7, Tables 1-1 and 7-4.

5. Calculate the flat roof snow load, p_f, from ASCE 7, equation 7-1:

$$\text{Flat roof snow load, } p_f = 0.7 \, C_e \, C_t \, I_s \, p_g \text{ (psf)} \tag{2-4}$$

6. Determine the minimum flat roof snow load (ASCE 7, sections 7.3 and 7.3.4):

$$\text{If } p_g > 20 \text{ psf, } p_f \text{ (minimum)} = 20 I_s \text{ (psf)} \tag{2-5}$$
$$\text{If } p_g \leq 20 \text{ psf, } p_f \text{ (minimum)} = p_g I_s \text{ (psf)} \tag{2-6}$$

7. Determine the design snow load accounting for the sloped roof factor (ASCE 7, Figure 7-2):

$$\text{Design sloped roof snow load, } p_s = C_s \, p_f \text{ (psf)} \tag{2-7}$$

↖ SEE MIN.

8. Consider partial loading for continuous beams per ASCE 7, Section 7.5, where alter-nate span loading might create maximum loading conditions.

9. The unbalanced loading from the effects of wind must be considered per ASCE 7, Section 7.6.

10. Determine snow drift loads on lower roofs per ASCE 7, Section 7.7.

11. Determine snow drift loads at roof projections and parapet walls per ASCE 7, Section 7.8.

12. Determine the effects of sliding snow from higher sloped roofs onto lower roofs per ASCE 7, Section 7.9.

13. Check the requirements for rain-on-snow surcharge per ASCE 7, Section 7.10.

    Roofs with a slope of less than $W/50$ (where W is the horizontal distance in feet from eave to ridge) in areas where $p_g \leq 20$ psf, must be designed for an additional rain-on-snow surcharge load of 5 psf. It should be noted that this additional load only applies to the balanced load case, and does not need to be used for snow drift, sliding snow, and unbalanced or partial snow load calculations.

14. Where the roof slope is less than ¼ in./ft., the requirements for ponding instability should be checked per ASCE 7, Section 7.11. Ponding is the additional load that results from rain-on-snow or melted snow water acting on the deflected shape of a flat roof, which, in turn, leads to increased deflection and therefore increased roof loading.

The following should be noted with respect to using the "slippery surface" values in ASCE 7, Figure 7-2, to determine the roof slope factor, C_s:

- Slippery surface values can only be used where the roof surface is free of obstruction and sufficient space is available below the eaves to accept all of the sliding snow. Examples of slippery surfaces include metal; slate; glass; and bituminous, rubber, and plastic membranes with a smooth surface.

- Membranes with imbedded aggregates, asphalt shingles, and wood shingles should not be considered slippery.

EXAMPLE 2-9

Balanced Snow Load

An office building located in an area with a ground snow load of 85 psf, has an essentially flat roof with a slope of ¼ in. per foot for drainage. Assuming a partially exposed heated roof and terrain category 'C', calculate the design roof snow load using the ASCE 7 load standard.

SOLUTION

The ground snow load, $p_g = 85$ psf.

Roof slope, θ (for ¼ in./ft. of run for drainage) = 1.2 degrees

From ASCE 7, Table 7-2, exposure factor, $C_e = 1.0$ (partially exposed roof and terrain category C)

From ASCE 7, Table 7-3, thermal factor, $C_t = 1.0$ (for heated roof)

From IBC, Table 1604.5 (or ASCE 7, Tables 1-1 and 7-4), importance factor, $I_s = 1.0$

From ASCE 7, Figure 7-2 ($\theta = 1.2$ degrees), and assuming a *heated roof*, $C_s = 1.0$

From equation (2-4), the flat roof snow load, $p_f = 0.7\,C_e\,C_t\,I_s\,p_g$

$$= (0.7)\,(1.0)\,(1.0)\,(1.0)\,(85) = 59.5 \text{ psf}$$
$$> p_{f\,minimum} = 20I_s = 20(1.0) = 20 \text{ psf}$$

Using equation (2-7), the design roof snow load, $p_s = C_s\,p_f = (1.0)\,(59.5) = 59.5$ psf

The calculated snow load above will need to be combined with the dead load and all the other applied loads using the load combinations given in Section 2.3.

Windward and Leeward Snowdrift

Snowdrift loads on the lower levels of multilevel roofs are caused by wind transporting snow from the higher roof and depositing it onto the lower roof or to balconies or canopies. It can also occur where the wind encounters roof obstructions, such as high roof walls, penthouses, high parapet walls, and mechanical rooftop units, and the snow is deposited on the windward side of the obstruction. The snowdrift loads are additional snow loads on the lower roof that are superimposed on the balanced snow load. The distribution of the snowdrift load is assumed to be triangular in shape (see Figure 2-3). The two kinds of snowdrift are windward and leeward drift. *Windward* snowdrift occurs when the wind moves the snow from one area of a roof to another area on the same roof against a wall or other sufficiently high obstruction. *Leeward* snowdrift occurs as the wind moves the snow from an upper roof and deposits the snow on a leeward low roof adjacent to the wall of the higher roof building. It should be noted that the snowdrift load is an additional load and has to be superimposed on the balanced flat roof snow load, p_{SL}.

The procedure for calculating the maximum height of the triangular snowdrift load is as follows (ASCE 7, Sections 7.7, 7.8, and 7.9):

1. Calculate the density of snow, $\gamma(\text{pcf}) = 0.13\,p_g + 14 \leq 30$ pcf, $\qquad$ (2-8)

   where p_g = Ground snow load in pounds per square foot obtained from the snow map in the governing building code.

2. Calculate the height of balanced flat roof snow load, $h_b = p_f/\gamma$, where p_f is the flat roof snow load.

3. Calculate the difference in height, h (in feet), between the high and low roof, and calculate the additional wall height, h_c, available to accommodate the drifting snow, where $h_c = h - h_b$.

4. Where $h \leq h_b$ or $h_c/h_b < 0.2$, the snowdrift is neglected and the low roof is designed for the uniform balanced snow load, p_f.

5. The maximum drifting snow height in feet, h_d, is the higher of the values calculated using equations 2-9a and 2-9b in Table 2-8.

Table 2-8 Snowdrift heights

Type of Snowdrift	Height of Snowdrift, h_d
Windward snowdrift (i.e., snowdrift on the low roof on the *windward* side of the building	$h_d = 0.75(0.43[L]^{1/3}[p_g + 10]^{1/4} - 1.5)$ (2-9a) $L = L_L$ = Length of low roof $\geq$ 25 ft.
Leeward snowdrift (i.e., snow drift on the low roof on the *leeward* side of the building	$h_d = (0.43[L]^{1/3}[p_g + 10]^{1/4} - 1.5)$ (2-9b) $L = L_U$ = Total length of upper roof $\geq$ 25 ft.

6. h_d is the *larger* of the two values calculated from the previous step.
   If $h_d \leq h_c$, use h_d in steps 7 and 8, but if $h_d \geq h_c$, set $h_d = h_c$ in steps 7 and 8.

7. The maximum value of the triangular snowdrift load in pounds per square foot (psf) is given as

   $$p_{SD} = \gamma h_d \; (\text{psf}) \qquad\qquad (2\text{-}10)$$

8. This load has to be superimposed on the uniform balanced flat roof snow load, p_f.

9. The length of the triangular snowdrift load, w, is calculated as follows:

If $h_d \leq h_c$, $w = 4\,h_d$ (feet) (2-11a)

If $h_d > h_c$, $w = 4\,h_d^2/h_c \leq 8\,h_c$ (and use $h_d = h_c$) (2-11b)

Where the drift width, w, exceeds the length of the lower roof, L_L, the snowdrift load distribution should be truncated, but not reduced to zero, at the edge of the low roof (see Figure 2-3).

Sliding Snow Load

Where a higher pitched or gable roof is adjacent to a lower flat roof, there is a tendency for snow to slide from the higher roof onto the lower roof. The ASCE 7 load standard assumes that only 40% of the snow load on the pitched roof will slide onto the lower roof because of the low probability that the snow load on both the pitched roof and the low roof will be at their maximum values at the same time when sliding occurs. The magnitude of the snow that slides from a pitched roof (slippery roof with slopes greater than ¼ in./ft. or nonslippery roof with slopes greater than 2:12) onto a lower flat roof is assumed to be a uniform load, p_{SL}, distributed over a length of 15 ft. on the lower roof. This snow load is in addition to the flat roof balanced snow load, $p_{f\,lower}$, on the lower roof.

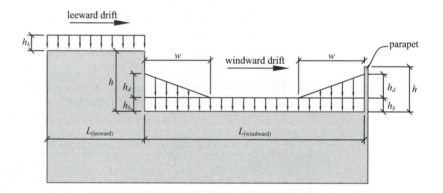

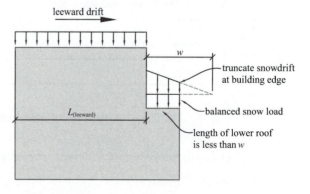

Figure 2-3 Snowdrift diagrams.

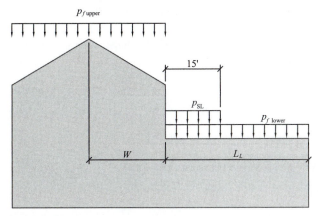

Figure 2-4 Sliding snow diagram.

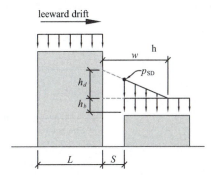

Figure 2-5 Effect of separation between adjacent buildings on snowdrift load.

Therefore, the maximum uniform snow load on the lower roof due to balanced and sliding snow is $p_{SL} + p_{f\,lower}$,

where

$$p_{SL} = \frac{0.4\,p_{f\,upper}\,W}{15\text{ ft.}},\qquad(2\text{-}12)$$

$W =$ Horizontal distance from the eave to the ridge of the higher roof as shown in Figure 2-4, and

$L_L =$ Length of the lower roof.

Where the length of the lower roof is less than 15 ft., the total sliding snow load on the lower roof will be proportionally smaller; however, the uniformly distributed sliding snow load in pounds per square foot is still calculated using equation (2-12).

Horizontal Separation Between Multilevel Adjacent Roofs

Where two adjacent buildings with different roof heights are separated by a horizontal distance S (in feet), as shown in Figure 2-5, the maximum snowdrift load, $\gamma_s h_d$, is modified by a factor of $(20 - S)/20$. Therefore, the maximum snowdrift load on the lower roof is given as

$$p_{SD} = \frac{(20 - S)}{20}\gamma_s h_d\qquad(2\text{-}13)$$

When $S \geq 20$ ft., no snowdrift load is assumed to occur on the lower roof.

Windward Snowdrift at Roof Projections

When drifting snow transported by wind is obstructed by a roof projection such as high parapets, signs, and rooftop units (RTU), snow will tend to accumulate on the windward side of the projection. This windward snowdrift load can be neglected if the width of the roof projection perpendicular to the snowdrift is less than 15 ft. [2] For all other cases, the snowdrift load distribution is assumed to be triangular (see Figure 2-6) in shape, with a maximum magnitude of

$$p_{SD} = \gamma_s h_{dp} \tag{2-14}$$

where

γ = Density of snow,

h_{dp} = Windward snowdrift height = $0.75(0.43\,[L]^{1/3}\,[p_g + 10]^{1/4} - 1.5)$ (2-15)

L = Length of roof on the windward side of the roof projection $\geq$ 25 ft.

The length of the triangular snowdrift load, w, is determined using equations (2-9a) and (2-9).

Partial Loading for Continuous and Cantilevered Roof Beams

In addition to designing continuous or cantilevered flat roof beams for the balanced design roof snow load, p_s, Section 7.5 of ASCE 7 prescribes a pattern of full and partial loading for the design of continuous beams. It involves applying one-half of the design snow load and the full design snow load in a checkered pattern that creates maximum load effects. The three different load cases specified in ASCE 7 are illustrated in Figure 2-7.

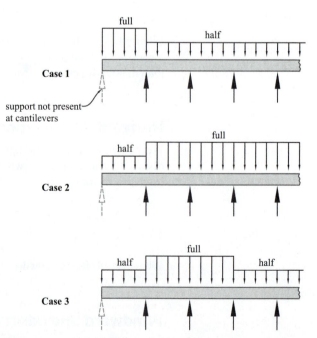

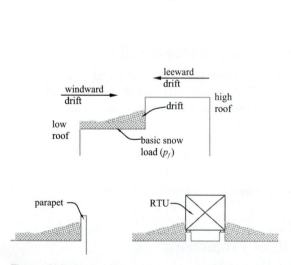

Figure 2-6 Snowdrift due to roof projections.

Figure 2-7 Partial loading diagram for continuous roof beams.

EXAMPLE 2-10

Balanced, Snowdrift, and Sliding Snow Load

A building located in an area with a ground snow load of 85 psf, has a lower flat roof adjacent to a pitched higher roof with a 30-degree slope as shown in Figure 2-8. Assume a fully exposed roof and terrain category C, and a warm roof with $C_t = 1.0$.

- Calculate the design snow loads for the upper roof using ASCE 7,
- Calculate the design snow load for the lower roof, considering snowdrift and sliding snow, and
- Determine the most critical average snow loads on beams A and B, assuming a typical beam spacing of 4 ft.

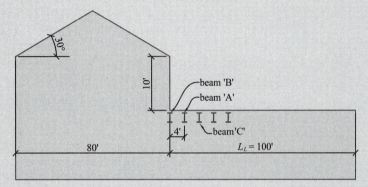

Figure 2-8 Building section for example 2-10.

SOLUTION

High-pitched Roof

The ground snow load, $p_g = 85$ psf

High roof slope, $\theta = 30$ degrees

From ASCE 7, Table 7-2, $C_e = 0.9$ (fully exposed roof and terrain category C)
From ASCE 7, Table 7-3, $C_t = 1.0$
From ASCE 7, Tables 1-1 and 7-4, importance factor, $I_s = 1.0$
From ASCE-7, Figure 7-2 (for $\theta = 30$ degrees), and with a *warm roof*, $C_s = 1.0$

Flat roof snow load, $\begin{aligned} p_{f\,upper} &= 0.7\, C_e\, C_t\, I_s\, p_g \\ &= (0.7)(0.9)(1.0)(1.0)(85) = 54 \text{ psf} \\ &> p_f\,(\text{minimum}) = 20 I_s = 20(1.0) = 20 \text{ psf} \end{aligned}$

Design roof snow load for the *higher* roof, $p_s = C_s\, p_f = (1.0)(54) = 54$ psf

Lower flat roof (adjacent to higher pitched roof)
Two load cases will be considered for the lower flat roof: drifting and sliding snow.

Case 1: Balanced Snow Load + Triangular Snowdrift Load

The ground snow load $p_g = 85$ psf

Lower roof slope, $\theta = 1.2$ degrees (i.e., ¼ in. per foot of slope for drainage)

(continued)

The roof is assumed to be fully exposed and a terrain category C is assumed. (*See ASCE 7, Section 6.5.6 for definitions of terrain categories.*)

From ASCE 7, Table 7-2, exposure factor, $C_e = 0.9$ (fully exposed roof and terrain category C)
From ASCE 7, Table 7-3, thermal factor, $C_t = 1.0$
From ASCE 7, Tables 1-1 and 7-4, importance factor, $I_s = 1.0$
From ASCE 7, Figure 7-2 (for $\theta = 1.2$ **degrees**), and assuming a *warm roof*, $C_s = 1.0$

$$\begin{aligned} \text{Flat roof snow load, } p_{f\,\text{upper}} &= 0.7\,C_e\,C_t\,I_s\,p_g \\ &= (0.7)(0.9)(1.0)(1.0)(85) = 54 \text{ psf} \\ &> p_f(\text{minimum}) = 20I_s = 20(1.0) = 20 \text{ psf} \end{aligned}$$

Design roof snow load for the *lower* roof, $p_s = C_s\,p_f = (1.0)(54) = 54$ psf

Snowdrift on lower roof:

1. Ground snow load, $p_g = 85$ psf
   Density of snow, $\gamma\,(\text{pcf}) = 0.13\,p_g + 14 \le 30$ pcf
   $\quad \gamma\,(\text{pcf}) = 0.13\,(85) + 14 = \textbf{25 pcf} \le 30$ pcf

2. Height of balanced flat roof snow load, $h_b = p_f/\gamma = 54 \text{ psf}/25 \text{ pcf} = 2.2$ ft.

3. Height difference between the high and low roof, $h = 10$ ft.
   Additional wall height available for drifting snow, $h_c = h - h_b = 10 \text{ ft.} - 2.2 \text{ ft.} = 7.8$ ft.

4. $h = 10 \text{ ft.} > h_b = 2.2 \text{ ft.}$, and $h_c/h_b = 7.8 \text{ ft.}/2.2 \text{ ft.} = 3.55 > 0.2$; therefore, snowdrift must be considered.

5. The maximum height in feet, h_d, of the **drifting snow** is calculated as shown in Table 2-9:

Table 2-9 Snowdrift heights

Type of Snowdrift	Height of Snowdrift, h_d
Windward snowdrift (i.e., snowdrift on the low roof on the *windward* side of the building	$h_d = 0.75(0.43[100]^{1/3}[85+10]^{1/4} - 1.5) = 3.55$ ft. $L = L_L = $ Length of low roof $= 100 \text{ ft} \ge 25$ ft.
Leeward snowdrift (i.e., snowdrift on the low roof on the *leeward* side of the building	$h_d = (0.43[80]^{1/3}[85+10]^{1/4} - 1.5) = 4.29$ ft. $L = L_U = $ Total length of upper roof $= 80 \text{ ft} \ge 25$ ft.

6. $h_d = $ **Larger** of the two values calculated from the previous step $= \textbf{4.29 ft.}$

   Since $h_d = 4.29 \text{ ft.} \le h_c = 7.8 \text{ ft.}$, use h_d in steps 7 and 8.

7. The maximum value of the triangular snowdrift load in pounds per square feet (psf) is given as

   $$p_{SD} = \gamma h_d = (25 \text{ psf})(4.29 \text{ ft.}) = \textbf{107 psf.}$$

   This load must be superimposed on the uniform balanced flat roof snow load.

8. The length of the triangular snowdrift load, w, is calculated as follows:

Since $h_d \leq h_c$, $w = 4\,h_d = (4)(4.29\text{ ft.}) = \textbf{17 ft.}$ (governs)
$\leq 8\,h_c = 8\,(7.8\text{ ft.}) = 63\text{ ft.}$

The resulting snowdrift load diagram is shown in Figure 2-9a.

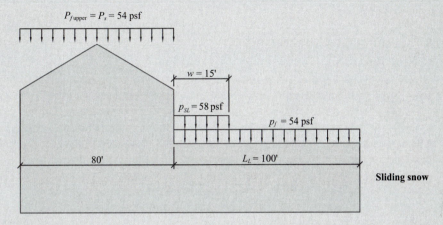

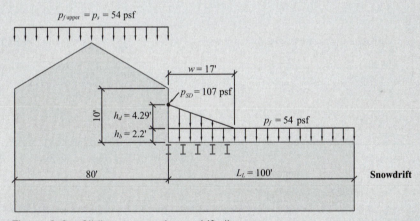

Figure 2-9a Sliding snow and snowdrift diagrams.

Case 2: Balanced Snow Load + Uniform *Sliding* Snow Load

From Case 1, $p_f = 54$ psf and $p_s = 54$ psf.

Sliding snow load on lower roof:
The ASCE 7 load standard prescribes the snow that slides from a pitched roof onto a lower flat roof as equal to a uniform load, p_{SL}, distributed over a length of **15 ft.** on the lower roof. This load is superimposed on the flat roof balanced snow load, $p_{f\,\text{lower}}$, for the lower roof.

W = Horizontal distance from eave to ridge of the higher roof = 40 ft. > 15 ft.

$p_{SL} = 0.4\,p_{f\,\text{upper}}\,W/15\text{ ft.} = (0.4)(54\text{ psf})(40\text{ ft.})/(15\text{ ft.}) = \textbf{58 psf}$ (uniformly distributed over 15 ft. length)

(continued)

Therefore, the maximum *uniform* snow load on the lower roof due to sliding snow is

$$p_{SL} + p_{f\,lower} = 58 + 54 = \mathbf{112\ psf.}$$

The resulting sliding snow diagram is shown in Figure 2-9a. The low roof structure must be designed and analyzed for *combined balanced snow plus snowdrift* and *combined balanced plus sliding snow*, to determine the worst case loading on the roof members.

Loads on Beams A and B Due to Snowdrift

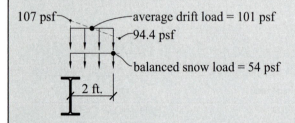

beam 'B' drift load

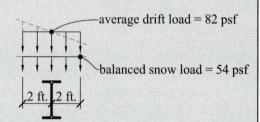

beam 'A' drift load

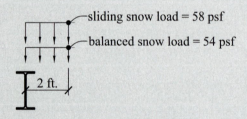

beam 'B' sliding snow load

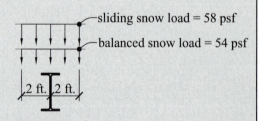

beam 'A' sliding snow load

Figure 2-9b Loads on Beam "A" and Beam "B."

Beam A (see Figure 2-9b)
From similar triangles, the average snowdrift load on beam A is obtained from Figure 2-9b

$$p_{AVG} = (17\ ft. - 4\ ft.)(107\ psf/17\ ft.) = 82\ psf,\ and$$

Snowdrift + Balanced snow loads, $\mathbf{S} = p_{AVG} + p_f = 82\ psf + 54\ psf = 136\ psf.$

Beam B (see Figure 2-9b)
From similar triangles, the snowdrift load midway between beams A and B is obtained from Figure 2-9b

$$y_{AB} = (17\ ft. - 2\ ft.)(107\ psf/17\ ft.) = 94.4\ psf,$$

Average snowdrift load on beam B $= (107 + 94.4)/2 = 101\ psf,\ and$

Snowdrift + Balanced snow load, $S =$ Average snowdrift $+ p_f = 101\ psf + 54\ psf = 155\ psf.$

Loads on Beams A and B Due to Sliding Snow

Beam A

Sliding snow + Balanced snow loads, $S = 58$ psf $+ p_f = 58$ psf $+ 54$ psf $= 112$ psf

This is less than the Snowdrift + Balanced snow load of 136 psf calculated previously for beam A; therefore, the most critical design snow load for beam A is $S_A = $ **136 psf.**

Beam B

Sliding snow + Balanced snow loads, $S = 58$ psf $+ p_f = 58$ psf $+ 54$ psf $= 112$ psf

This is less than the Snowdrift + Balanced snow load of 155 psf calculated previously for beam B; therefore, the most critical design snow load for beam B is $S_B = $ **155 psf.**

The design snow loads will need to be combined with the dead load and all other applicable loads using the load combinations in Section 2-3 to determine the most critical design loads for the beams. As a practice exercise, the reader should determine the most critical design snow load for beam C. Will snowdrift or sliding snow control the design for beam C?

EXAMPLE 2-11

Snowdrift Due to Rooftop Units and Parapets

A flat roof warehouse building located in an area with a ground snow load of 50 psf, has 3-ft.-high parapets around the perimeter of the roof and supports a 12-ft. by 22-ft. by 13-ft.-high cooling tower located symmetrically on the roof (see Figure 2-10). Assuming a partially exposed roof in a heated building in terrain category C,

- Calculate the flat roof and design snow loads for the warehouse roof using ASCE 7.
- Determine the design snowdrift load around the parapet, and
- Determine the design snowdrift load around the cooling tower.

(continued)

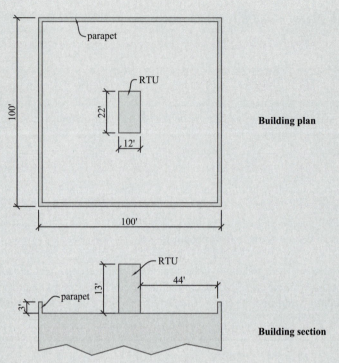

Building plan

Building section

Figure 2-10 Building section for example 2-11.

SOLUTION

Flat Roof

Case 1: Balanced Snow Load + Triangular Snowdrift Load

The ground snow load, $p_g = 50$ psf

Roof slope, $\theta = 1.2$ degrees (for ¼ in. per foot of slope for drainage)

The roof is assumed to be partially exposed and terrain category C is assumed. (*See ASCE 7, Section 6.5.6 for definitions of terrain categories.*)

From ASCE 7, Table 7-2, exposure factor, $C_e = 1.0$
From ASCE 7, Table 7-3, thermal factor, $C_t = 1.0$
From ASCE 7, Tables 1-1 and 7-4, importance factor, $I_s = 1.0$
From ASCE 7, Figure 7-2 (for $\theta = 1.2$ degrees), and assuming a *heated roof*, $C_s = 1.0$

Flat roof snow load, $p_f = 0.7\ C_e C_t I_s p_g = 0.7 \times 1.0 \times 1.0 \times 1.0 \times 50 = 35$ psf
$$> p_f(\text{minimum}) = 20(1.0) = 20 \text{ psf}$$

Design roof snow load for the flat roof, $p_s = C_s p_f = 1.0 \times 35 = $ **35 psf**

Snowdrift Around the Roof Parapet:

  1. Ground snow load, $p_g = 50$ psf

      Density of snow, $\gamma\,(\text{pcf}) = 0.13\,p_g + 14 \le 30$ pcf

      $\gamma = 0.13\,(50) + 14 = $ **20.5 pcf** ≤ 30 pcf

2. Height of balanced flat roof snow load, $h_b = p_f/\gamma = 35$ psf/20.5 pcf $= 1.71$ ft.

3. Height of parapet, $h = 3$ ft.

   Additional parapet height available for drifting snow, $h_c = h - h_b = 3$ ft. $-$ 1.71 ft. $= 1.29$ ft.

   Height of cooling tower, $h = 13$ ft.

   Additional cooling tower height available for drifting snow, $h_c = h - h_b = $ 13 ft. $- 1.71$ ft. $= 11.29$ ft.

4. *Parapet:*

   $h = 3$ ft. $> h_b = 1.71$ ft., and $h_c/h_b = 1.29$ ft./1.71 ft. $= 0.75 > 0.2$; therefore, snowdrift must be considered around this parapet.

   Cooling Tower:
   $h = 13$ ft. $> h_b = 1.71$ ft., and $h_c/h_b = 11.29$ ft./1.71 ft. $= 6.6$ ft. > 0.2 ft.; therefore, snowdrift must be considered around this cooling tower.

5. The maximum height in feet, h_d, of the **drifting snow around parapets and cooling tower** is calculated as follows:

   Parapet:

   $$L = \text{Length of the roof windward of the parapet} = 100 \text{ ft.} \geq 25 \text{ ft.}$$

   $$h_{dp} = \text{Snowdrift height around a parapet}$$

   $$= 0.75 \, (0.43 \, [L]^{1/3}[p_g + 10]^{1/4} - 1.5)$$

   $$= 0.75 \, (0.43 \, [100]^{1/3} \, [50 + 10]^{1/4} - 1.5) = 3.03 \text{ ft.}$$

   Cooling Tower:
   Since the cooling tower is symmetrically located on the roof, the length of the low roof on the windward side of the cooling tower is

   $$L = (100 \text{ ft.} - 12 \text{ ft. width of tower})/2 = 44 \text{ ft.} \geq 25 \text{ ft.};$$
   $$\text{therefore, use } L = 44 \text{ ft.}$$

   $$h_{dp} = \text{Snowdrift height around cooling tower}$$

   $$= 0.75 \, (0.43 \, [L]^{1/3}[p_g + 10]^{1/4} - 1.5)$$

   $$= 0.75 \, (0.43 \, [44]^{1/3} \, [50 + 10]^{1/4} - 1.5) = 2 \text{ ft.}$$

6. *Parapet:*

   $h_{dp} = $ **3 ft.**

   Since $h_{dp} = 3$ ft. $> h_c = 1.29$ ft.; therefore, use $h_{dp} = h_c = 1.29$ ft. for calculating p_{SD}.

   Cooling Tower:

   $h_{dp} = $ **2 ft.**

   Since $h_{dp} = 2$ ft. $< h_c = 11.29$ ft.; therefore, use $h_{dp} = 2$ ft. for calculating p_{SD}.

(continued)

7. The maximum value of the triangular snowdrift load in pounds per square feet (psf) is determined as follows:

Parapet:

$$p_{SD} = \gamma h_{dp} = (20.5 \text{ psf})(1.29 \text{ ft.}) = \textbf{26.4 psf}$$

This load must be superimposed on the uniform balanced flat roof snow load.

Cooling Tower:

$$p_{SD} = \gamma h_{dp} = (20.5 \text{ psf})(2 \text{ ft.}) = \textbf{41 psf}$$

This load must be superimposed on the uniform balanced flat roof snow load.

8. The length of the triangular snowdrift load, w, is calculated as follows:

Parapet:

$$\text{Since } h_{dp} > h_c, w = 4\, h_{dp}^2/h_c = 4\,(3.03 \text{ ft.})^2/1.29 \text{ ft.} = 28.5 \text{ ft.}$$
$$\leq 8\, h_c = 8(1.29 \text{ ft.}) = \textbf{10.3 ft.}$$
$$\therefore w = \textbf{10.3 ft.}$$

Cooling Tower:

$$\text{Since } h_{dp} < h_c, w = 4\, h_{dp} = 4\,(2 \text{ ft.}) = 8 \text{ ft.}$$
$$\leq 8\, h_c = 8(11.29 \text{ ft.}) = 90.3 \text{ ft.}$$
$$\therefore w = \textbf{8 ft.}$$

The resulting diagrams of snowdrift near the parapet and the cooling tower are shown in Figure 2-11.

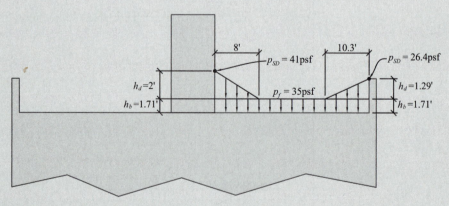

Figure 2-11 Diagrams of snowdrift for Example 2-11.

2.11 RAIN LOADS (ASCE 7, CHAPTER 8)

Rain loads are applicable only to flat roofs with parapets since accumulation of rain will generally not occur on roofs without parapets or on pitched roofs. The higher the roof parapet, the greater the rain load. Building [10] codes require that roofs with parapets have two independent drainage systems—primary and secondary (or overflow) drains at each drain location [2]. The secondary drain must be located at least 2 in. above the main roof level where the primary drain is located; the secondary drawn takes care of roof drainage in the event that the primary drain is blocked. The design rain load, R, is calculated based on the assumption that the primary drainage system is blocked (see Figure 2-12). Thus, the total depth of water to be considered is the depth from the roof surface to the inlet of the secondary drainage plus the depth of water that rises above the inlet of the secondary drainage due to the hydraulic head of the flowing water.

Roof drainage is a structural engineering, architectural, and mechanical engineering or plumbing issue; therefore, proper coordination is required among these disciplines to ensure adequate design. For flat roofs with slopes less than ¼ in. per foot for drainage, ponding or the additional rain load due to the deflection of the flat roof framing must also be considered in the design. Some roof structures have failed because of the additional rain load accumulated as a result of the deflection of the flat roof framing.

Assuming that the primary drainage is blocked, the rain load, R is given as

$$R \ (\text{psf}) \ = \ 5.2 \ (d_s \ + \ d_h), \tag{2-16}$$

where

d_s = Depth in inches from the undeflected roof surface to the inlet of the secondary drainage system (i.e., the static head of water),

d_h = Depth of water in inches above the inlet of the secondary drainage (i.e., the hydraulic head) obtained from ASCE 7, Table C8-1. The hydraulic head is a function of the roof area, A, drained by each drain, the size of the drainage system, and the flow rate, Q,

Q = Flow rate in gallons per minute = $0.0104 \ Ai$,

A = Roof area drained by the drainage system, ft^2, and

i = 100-yr., 1-hr. rainfall intensity (in inches per hour) for the building location specified in the plumbing code.

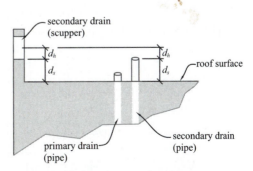

Figure 2-12 Rain drainage types.

EXAMPLE 2-12

Rain Loads

The roof plan for a building located in an area with a 100-yr., 1-hr. rainfall intensity of 4 in./hr. is shown in Figure 2-13 below.

    a. Assuming a 4-in.-diameter secondary drainage pipe that is set at 3 in. above the roof surface, determine the design rain load, R.

    b. Assuming a 6-in.-wide channel scupper secondary drainage system that is set at 3 in. above the roof surface, determine the design rain load, R.

    c. Assuming a 24-in.-wide, 6-in.-high closed scupper secondary drainage system that is set at 3 in. above the roof surface, determine the design rain load, R.

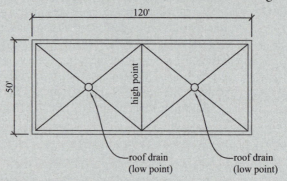

Figure 2-13 Roof drainage plan.

SOLUTION

    a. **4-in.-diameter secondary drainage**

        d_s = Depth in inches from the undeflected roof surface to the inlet of the secondary drainage system (i.e., the static head of water) = 3 in.

$$A = \text{Area drained by one secondary drainage} = \frac{120 \text{ ft.}}{2} (50 \text{ ft.}) = 3000 \text{ ft.}^2$$

        i = 100-yr., 1-hr. rainfall intensity = 4 in./hr. (given)

        $Q = 0.0104 \, Ai = (0.0104) (3000 \text{ ft.}^2) (4 \text{ in.}) = 125 \text{ gal./min.}$

Using ASCE 7, Table C8-1 for a **4-in.-diameter secondary drainage** system with a flow rate, $Q = 125$ gal./min., we find by linear interpolation that

    d_h = Depth of water in inches above the inlet of the secondary drainage system

$$= \frac{(2 \text{ in.} - 1 \text{ in.})}{(170 - 80)} (125 - 80) + 1 \text{ in.} = 1.5 \text{ in.}$$

Using equation (2-16), the rain load is

$$R \text{ (psf)} = 5.2 \, (d_s + d_h) = 5.2 \, (3 \text{ in.} + 1.5 \text{ in.}) = \textbf{24 psf}$$
$$\text{(assuming primary drainage is blocked).}$$

b. **6-in.-wide channel scupper**

d_s = Depth in inches from the undeflected roof surface to the inlet of the secondary drainage system (i.e., the static head of water) = 3 in.

A = Area drained by one secondary drainage = 3000 ft².

i = 100-yr., 1-hr. rainfall intensity = 4 in./hr.

Q = 0.0104 Ai = (0.0104) (3000 ft.²) (4 in.) = 125 gal./min.

Using ASCE 7, Table C8-1 for a **6-in.-wide channel scupper** secondary drainage with a flow rate, Q = 125 gal./min., we find by linear interpolation that

d_h = Depth of water in inches above the inlet of the secondary drainage system

$$= \frac{(4 \text{ in.} - 3 \text{ in.})}{(140 - 90)} (125 - 90) + 3 \text{ in.} = 3.7 \text{ in.}$$

Using equation (2-16), the rain load is

$$R \text{ (psf)} = 5.2 \, (d_s + d_h) = 5.2 \, (3 \text{ in.} + 3.7 \text{ in.}) = \textbf{35 psf}$$
$$\text{(assuming primary drainage is blocked)}$$

c. **24-in.-wide, 6-in.-high closed scupper**

d_s = Depth in inches from the undeflected roof surface to the inlet of the secondary drainage system (i.e., the static head of water) = 3 in.

A = Area drained by one secondary drainage = 3000 ft.²

i = 100-yr., 1-hr. rainfall intensity = 4 in./hr.

Q = 0.0104 Ai = (0.0104) (3000 ft.²) (4 in.) = 125 gal./min.

Using ASCE 7, Table C8-1 for a **24-in.-wide, 6-in.-high closed scupper** system with a flow rate, Q = 125 gal./min., we find by linear interpolation that

d_h = Depth of water in inches above the inlet of the secondary drainage

$$= \frac{(2 \text{ in.} - 1 \text{ in.})}{(200 - 72)} (125 - 72) + 1 \text{ in.} = 1.42 \text{ in.}$$

Using equation (2-16), the rain load is

$$R \text{ (psf)} = 5.2 \, (d_s + d_h) = 5.2 \, (3 \text{ in.} + 1.42 \text{ in.}) = \textbf{23 psf}$$
$$\text{(assuming primary drainage is blocked)}$$

2.12 ICE LOADS DUE TO FREEZING RAIN (ASCE-7, CHAPTER 10)

The weight of ice formed on exposed structures such as towers, cable systems, and pipes in the northern parts of the United States due to freezing rain must be accounted for in the design of these structures; it is quite common in these areas to see downed power lines and tree limbs due to the weight of accumulated ice loads. Figure 2-14 shows ice accumulation on

Figure 2-14 Ice accumulation on exposed tree limbs.

exposed tree limbs in Rochester, New York, during an ice storm. Similar ice accumulations occur on exposed structural steel members. The weight of ice on these structures is usually added to the snow load on the structure.

The procedure for calculating the ice load on a structural element is as follows:

1. Determine the occupancy category of the structure from ASCE 7, Table 1-1.
2. Determine the 50-yr. mean recurrence interval uniform nominal ice thickness, t, and the concurrent wind speed, V_c from ASCE-7, Figures 10-2a or 10-2b.
3. Determine the topographical factor, K_{zt}:

$$K_{zt} = (1 + K_1 K_2 K_3)^2 = 1.0 \text{ for flat land (per ASCE 7, Section 6.5.7.2),} \quad (2\text{-}17)$$

where the multipliers K_1, K_2, and K_3 account for wind speedup effect for buildings located on a hill or escarpment and are determined from ASCE 7, Figure 6-4.

4. Using the structure category from step 1, determine the importance factor, I_i, from ASCE 7, Table 10-1.
5. Determine the height factor, f_z:

$$f_z = \left(\frac{z}{33}\right)^{0.10} \leq 1.4 \quad (2\text{-}18)$$

where $z = $ Height of the structural member in feet above the ground.

6. The design uniform radial ice thickness in inches due to freezing rain is given as

$$t_d = 2.0 \, t \, I_i f_z \, (K_{zt})^{0.35} \quad (2\text{-}19)$$

7. The weight of ice on exposed surfaces of structural shapes and prismatic members is as follows:

Ice load, $D_i = \gamma_{ice}\pi t_d(D_c + t_d)$, in pounds per foot, $\qquad$ (2-20)

where

D_c = Characteristic dimension, shown for various shapes in ASCE 7-05, Figure 10-1, and

γ_{ice} = Density of ice = 56 pcf (minimum value per ASCE 7, Section 10.4.1).

For large 3-D objects, the volume of ice is calculated as follows:

$$\text{Flat plates: } V_i = \lambda\pi t_d \times \text{Area of one side of the plate} \qquad (2\text{-}21)$$
$$\lambda = 0.8 \text{ for vertical plates}$$
$$= 0.6 \text{ for horizontal plates}$$
$$\text{Domes or spheres: } V_i = \pi t_d \times (\pi r^2) \qquad (2\text{-}22)$$

The ice load, D_i, must now be combined with the dead load and all other applicable loads in the load combinations given in Section 2-3. It should be noted that the load combinations have to be modified according to ASCE 7, Sections 2.3.4 and 2.4.3 when ice loads are considered.

8. Determine the dead load, D, of the structural member in pounds per foot.
9. Determine the snow load, S on the ice-coated structural member.
10. Determine the live load, L, of the pipe due to the liquid carried in the pipe.
11. Determine the wind-on-ice load, W_i, using the methods discussed in Chapter 3 and taking into account the increased projected surface area of the ice-coated structural member that is exposed to wind (see Section 10 of ASCE 7).
12. Calculate the maximum total load using the modified load combinations that includes ice loading. The load combinations from Section 2.3 are modified per ASCE 7, Section 2.3.4 as follows:

2: $\quad 1.2 (D + F + T) + 1.6 (L + H) + 0.2D_i + 0.5S$

4: $\quad 1.2D + L + D_i + W_i + 0.5S$

6: $\quad 0.9D + D_i + W_i + 1.6H$

9: $\quad D + H + F + L + T + 0.7D_i$

10: $\quad D + H + F + 0.7D_i + 0.7W_i + S$

14: $\quad 0.6D + 0.7D_i + 0.7W_i + H$

EXAMPLE 2-13

Ice Loads

Determine the ice load, dead load, and live load on exposed 50-in.-diameter horizontal steel pipes for a chemical plant. The pipes have a wall thickness of 1 in. and carry a liquid with a density of 64 pcf. The top of the pipe is at an elevation of 100 ft. above the ground and the site is flat. Assume that the snow load is 35 psf, the density of the ice is 56 pcf, the density of the steel is 490 pcf, and the wind load on the pipe has been calculated to be 20 psf. Assume that the 50-yr uniform ice thickness, t, due to freezing rain is 1 in.

(continued)

SOLUTION

1. Determine the category of the structure (from ASCE 7, Table 1-1):

   Chemical plant = Category IV building (from ASCE 7, Table 1-1).

2. Determine the nominal ice thickness, t, and the concurrent wind speed, V_c, from ASCE 7, Figures 10-2a or 10-2b:

   $t = 1$ in. (50-yr uniform ice thickness due to freezing rain)

3. Determine the topographical factor, K_{zt}:

   $$K_{zt} = (1 + K_1 K_2 K_3)^2 = \mathbf{1.0} \text{ for } \textbf{flat land,}$$

4. Using the structure category from step 1, determine the importance factor, I_i, from ASCE 7, Table 10-1:

   $I_i = 1.25.$

5. Determine the height factor, f_z:

   $$f_z = \left(\frac{z}{33}\right)^{0.10} = \left(\frac{100}{33}\right)^{0.10}$$

   $$= 1.12 < 1.4$$

   $$\therefore f_z = 1.12$$

   where z = Height in feet above the ground = 100 ft.

6. The design uniform radial ice thickness in inches due to freezing rain is given as

   $$t_d = 2.0\, t\, I_i f_z\, (K_{zt})^{0.35} = (2.0)(1 \text{ in.})(1.25)(1.12)(1.0)^{0.35} = 2.8 \text{ in.}$$

7. The weight of ice on exposed surfaces of structural shapes and prismatic members is as follows:

   Ice load, $D_i = \gamma_{ice} \pi t_d (D_c + t_d)$
   $= (56 \text{ pcf})(2.8 \text{ in.}/12)\pi(50 \text{ in.} + 2.8 \text{ in.})/12 = 181 \text{ lb./ft.},$

   where

   D_c = Characteristic dimension, shown for various shapes in
   ASCE 7, Figure 10-1 = 50 in. for 50-in.-diameter pipe,
   γ_{ice} = Density of ice = 56 pcf (minimum value per ASCE 7, Section 10.4.1), and

8. Determine the dead load, D, of the structural member in pounds per foot:

   Dead load of pipe, $D = \left[\dfrac{\pi(50 \text{ in.})^2}{4} - \dfrac{\pi(50 \text{ in.} - 1 \text{ in.} - 1 \text{ in.})^2}{4} \right](490 \text{ pcf}/144)$

9. The snow load on the ice-coated pipe, $S = 35 \text{ psf } (D_c + t_d)$

   $$= 35 \text{ psf} \left(\frac{50 \text{ in.}}{12} + \frac{2.8 \text{ in.}}{12} \right)$$

   $$= 154 \text{ lb/ft.}$$

10. Determine the live load, L, of the pipe due to the liquid carried by the pipe:

$$\text{Live load in pipe, } L = \frac{\pi(50 \text{ in.} - 1 \text{ in.} - 1 \text{ in.})^2}{(4)(144)} \ (64 \text{ pcf}) \ = \ 805 \text{ lb./ft.}$$

11. Calculate the wind-on-ice load, W_i:

$$W_i = \text{wind pressure} \times (D_c + t_d) = (20 \text{ psf}) \ (50 + 2.8 \text{ in.})/12 = 88 \text{ lb./ft.}$$

12. The ice load, D_i, must now be combined with the dead load and other applicable loads to determine the maximum total load on the pipe. The most critical limit states load combinations for downward-acting loads are calculated below using modified load combinations, and recognizing that the loads F, T, and H are equal to zero, we get:

1. $1.4D = 1.4 \ (524) = 734 \text{ lb./ft.}$

2. $1.2D + 1.6L + 0.2D_i + 0.5S$
   $= 1.2 \ (524) + 1.6 \ (805) + 0.2 \ (181) + 0.5 \ (154) = \textbf{2030 lb./ft. (governs)}$

3. $1.2D + 1.6 \ (L_r \text{ or } S \text{ or } R) + (0.5L \text{ or } 0.8W)$
   $= 1.2 \ (524) + 1.6 \ (154) + 0.5 \ (805) = 1278 \text{ lb/ft.}$

4. $1.2D + L + D_i + W_i + 0.5S$
   $= 1.2 \ (524) + 805 + 181 + 88 + 0.5 \ (154) = 1780 \text{ lb./ft.}$

Therefore, the controlling factored-downward load on the pipe and for which the pipe will be designed is **2030 lb./ft.**

2.13 MISCELLANEOUS LOADS

The following miscellaneous loads will be discussed in this section of the text:

- Fluid loads,
- Flood loads,
- Self-straining loads (e.g., temperature),
- Lateral pressure due to soil, water, and bulk materials,
- Impact loads (see Table 2-10),
- Live loads from miscellaneous structural elements such as handrails, balustrades, and vehicle barriers (see Table 2-11), and
- Construction loads (see Table 2-12).

Fluid Loads, *F*

The ASCE 7 load standard uses the symbol F to denote loads due to fluids with well-defined pressures and maximum heights. This load is separate and distinct from the soil or hydrostatic pressure load, H, or the flood load, F_a. Not much guidance is given in ASCE 7 regarding this load, but since it has the same load factor as the dead load, D, in the LRFD load combinations, that would indicate that this load pertains to the weight of fluids that may be stored in a building or structure.

Table 2-10 Impact factors

Type of Load or Equipment	Impact Factor*
Elevator loads	2.0
Elevator machinery	2.0
Light machinery or motor-driven units	1.2
Reciprocating machinery or power-driven units	1.5
Hangers supporting floors and balconies	1.33
Monorail cranes (powered)	1.25
Cab-operated or remotely operated bridge cranes (powered)	1.25
Pendant-operated bridge cranes (powered)	1.10
Bridge cranes or monorail cranes with hand-geared bridge, trolley, and hoist	1.0

*All equipment impact factors shall be increased where larger values are specified by the equipment manufacturer.

Table 2-11 Live loads on miscellaneous structural elements (ASCE-7, Section 4.4)

Structural Element	Live Load
Handrails and ballustrades	50 lb./ft. applied at the top along the length of the handrail or ballustrade in any direction or Single concentrated load of 250 lb. applied in any direction at any point
Grab bar	Single concentrated load of 250 lb. applied in any direction at any point
Vehicle barriers for *passenger cars*	6000 lb. of horizontal load applied in any direction to the barrier system, acting at 18 in. above the floor or ramp surface over an area of 1 ft.^{2}
Fixed ladders with rungs	Single concentrated load of 300 lb. applied at any point to produce maximum load effect plus additional 300 lb. of concentrated load for every 10 ft. of ladder height
Rails of fixed ladders extending above floor or platform	Concentrated load of 100 lb. in any direction, at any height
Ship ladders with treads instead of rungs	Live load similar to stairs (typically 100 psf); see ASCE 7-05, Table 4-1
Anchorage for attachment of fall arrest equipment	Required factored load (per OSHA CFR 1926.502(d)[15]) = 5000 lb

Table 2-12 Construction live loads (from ASCE 37-05, Table 2 [5])

Class	Uniform Construction Live Load (psf)
Very Light Duty: sparsely populated with personnel; hand tools and small amounts of construction materials	20 psf
Light Duty: sparsely populated with personnel; hand-operated equipment	25 psf
Medium Duty: concentrations of personnel and equipment/materials	50 psf
Heavy Duty: heavy construction and placement of materials using motorized vehicles	75 psf

Flood Loads, F_a

This pertains to flood loads acting against a structure. The procedure for calculating flood loads is covered in Chapter 5 of ASCE 7. Most building structures are not subjected to flood loads, except for buildings in coastal regions. For load combinations, including flood load, refer to ASCE 7, Section 2.3.3.

Self-Straining Force, T

Self-straining loads (e.g., temperature) arise due to the restraining of movement in a structure caused by expansion or contraction from temperature or moisture change, creep, or differential settlement. If these movements are unrestrained, the temperature force is practically zero. This is the case for most building structures, except for posttensioned members, where restrained shrinkage could lead to self-straining loads in the member. Some cladding systems, when subjected to expansion due to temperature effects, could develop sizable self-straining forces, but these could be alleviated by proper detailing of the cladding connections.

Lateral Pressure Due to Soil, Water, and Bulk Materials, H

Lateral and hydrostatic pressures of soil on retaining walls and the lateral pressures exerted by bulk solids against the walls of bins and silos are denoted in ASCE 7 by the symbol H. This notation is also use to denote the upward hydrostatic pressures on base slabs and foundation mats of buildings since these upward forces usually act simultaneously with the lateral hydrostatic or soil pressures, depending on the elevation of the water table.

Impact Loads (ASCE 7, Section 4.7)

Impact loads are dynamic loads that are caused by the sudden application of a load on a structure, resulting in an amplification of the static live load by the so-called impact factor. Only live loads can cause impact; therefore, only the live load is amplified by the impact factors. Thus, the service live load for the structure is the static **live** load multiplied by an impact factor, which may range from 1.25 to greater than 2.0, depending on the cause of the impact and the elevation of the object relative to the structure. For structures subjected to impact loads from falling objects, the impact factor may be much larger than 2.0, depending on the height of the falling object above the structure. Examples of minimum impact factors specified in ASCE 7 are as shown in Table 2.10.

2.14 VERTICAL AND LATERAL DEFLECTION CRITERIA

The limits on vertical deflections due to gravity loads are intended to ensure user comfort and to prevent excessive cracking of plaster ceilings and architectural partitions. These deflection limits are usually specified in terms of the joist, beam, or girder span, and the deflections are calculated based on elastic analysis of the structural member under service or unfactored loads. The service loads, instead of the factored loads, are used in the deflection calculations because under the ultimate limit state, when failure is imminent, deflection of the structure is no longer important. However, under normal service conditions, the deflections are limited so that the occupants of the building do not become alarmed by any appreciable deflection, thinking the structure or member is about to collapse. The maximum allowable deflections recommended in IBC, Table 1604.3 are as follows [4]:

Maximum allowable *floor* deflection due to service live load $\leq \dfrac{L}{360}$,

Maximum allowable *floor* deflection due to service total dead plus live load $\leq \dfrac{L}{240}$,

Maximum allowable *roof* deflection due to service live load $\leq \dfrac{L}{180}$, and

Maximum allowable *roof* deflection due to service total dead plus live load $\leq \dfrac{L}{240}$,

where

L = Simple span length of flexural member.

For members that support masonry wall partition or cladding, and glazing, the allowable total deflection due to the weight of the cladding or partition wall, the superimposed dead load, and the live load should be limited to $\frac{L}{600}$ or 0.3 in., whichever is smaller, to reduce the likelihood of cracking of the cladding or partition wall [7].

It should be noted that it is the deflection that occurs after the cladding is in place (i.e., live load deflection) that would be most critical because as the masonry cladding is being installed, the beam or girder deflects, and the masonry will tend to fit the shape of the deflected member and any curvature in the wall can be corrected at the mortar joints by the mason. For prefabricated members or composite steel beams and girders, a camber is sometimes specified to help control the total deflection. To avoid too much camber, it is common practice to make the camber approximately equal to the dead load deflection of the structural member. Support restraints should also be considered and this is discussed in Chapters 6 and 7.

For operable partition walls, the deflection under superimposed loads should be no greater than ⅛ in. per 12 ft of wall length, yielding a deflection limit of $L/1152$ [6].

There are also limits placed on the **lateral** deflection of steel buildings. These lateral deflections are typically caused by wind or seismic loads. The 20-yr. return wind—which is approximately 70% of the 50-yr. return wind—is recommended in ASCE 7 to be used for wind drift calculations. The maximum total drift due to wind loads should be limited to 1/400 of the total building height and the interstory drift should be limited to 1/500 of the floor-to-floor height for buildings with brick cladding and 1/400 for all others [8]. Similar recommendations are provided in Ref. [9]. The lateral deflection limits for seismic loads are much higher than those for wind loads and are calculated using strength-level seismic forces because the design philosophy for earthquakes focuses on life safety and not serviceability conditions. These seismic lateral deflection limits are given in Table 12.12-1 of ASCE 7 [2].

2.15 REFERENCES

1. American Institute of Steel Construction. 2006. *Steel Construction Manual*, 13th ed. Reston, VA: AISC.

2. American Society of Civil Engineers. 2005. *Minimum Design Loads for Buildings and Other Structures.* Reston, VA: ASCE.

3. Ellingwood, B., T. V. Galambos, J. G. MacGregor, and C.A. Cornell. June 2008. *Development of a probability-based load criterion for American National Standard A58*, NBS Special Publication 577. Washington, DC: U.S. Department of Commerce, National Bureau of Standards.

4. International Codes Council. 2006. *International Building Code—2006,* Falls Church, VA: ICC.

5. American Society of Civil Engineers. 2002. ASCE 37-02, "Design Loads on Structures during Construction."

6. ASTM E 557-00, 2001 "Standard Guide for the Installation of operable Partitions," July.

7. ACI 530-05, "Building Code Requirement for Masonry Structures," American Concrete Institute.

8. CSA 2006. "CAN/CSA-S16-01," Canadian Standards Association.

9. Griffis, Lawrence G. 1993. "Serviceability Limit States Under Wind Loads," *Engineering Journal*, First Quarter.

10. International Codes Council, 2006. International Plumbing Code—2006, Falls Church, VA: ICC.

2.16 PROBLEMS

2-1. Define the term "limit state." What is the difference between the allowable strength design method (ASD) and the load and resistance factor design method (LRFD)?

2-2. What are the reasons for using resistance factors in the LRFD method? List the resistance factors for shear, bending, tension yielding, and tension fracture.

2-3. a. Determine the factored axial load or the required axial strength, P_u, of a column in an office building with a regular roof configuration. The service axial loads on the column are as follows:

$$P_D = 200 \text{ kips (dead load)}$$

$$P_L = 300 \text{ kips (floor live load)}$$

$$P_S = 150 \text{ kips (snow load)}$$

$$P_W = \pm 60 \text{ kips (wind load)}$$

$$P_E = \pm 40 \text{ kips (seismic load)}$$

b. Calculate the required *nominal* axial compression strength, P_n, of the column.

2-4. a. Determine the ultimate or factored load for a roof beam subjected to the following service loads:

$$\text{Dead load} = 29 \text{ psf (dead load)}$$

$$\text{Snow load} = 35 \text{ psf (snow load)}$$

$$\text{Roof live load} = 20 \text{ psf}$$

$$\text{Wind load} = 25 \text{ psf upwards}$$
$$15 \text{ psf downwards}$$

b. Assuming a roof beam span of 30 ft. and a tributary width of 6 ft., determine the factored moment and shear.

2-5. List the floor live loads for the following occupancies:

- Library stack rooms,
- Classrooms,
- Heavy storage,
- Light manufacturing, and
- Offices.

2-6. Determine the tributary widths and tributary areas of the joists, beams, girders, and columns in the roof framing plan shown below. Assuming a roof dead load of 30 psf and an essentially flat roof with a roof slope of ¼ in./ft. for drainage, determine the following loads using the ASCE 7 load combinations. Neglect the rain load, R, and assume the snow load, S, is zero: *psf*

a. Uniform dead and roof live loads on the typical roof beam in pounds per foot.

b. Concentrated dead and roof live loads on the typical roof girder in pounds per foot.

c. Total factored axial load on the typical interior column in pounds.

d. Total factored axial load on the typical corner column in pounds.

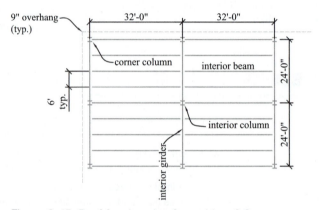

Figure 2-15 Roof framing plan for problem 2-6.

2-7. A three-story building has columns spaced at 18 ft. in both orthogonal directions, and is subjected to the roof and floor loads shown below. Using a column load summation table, calculate the cumulative axial loads on a typical interior column with and without live load reduction. Assume a roof slope of ¼ in./ft. per foot for drainage.

Roof Loads:

Dead load, $D_{roof} = 20$ psf

Snow load, $S = 40$ psf

Second- and Third-Floor Loads:

Dead load, $D_{floor} = 40$ psf

Floor live load, $L = 50$ psf

2-8. Determine the **dead load** (*with and without partitions*) in pounds per square foot of floor area for a steel building floor system with W24 × 55 beams (weighs 55 lb./ft.) spaced at 6 ft. 0 in. o.c. and W30 × 116 girders (weighs 116 lb./ft.) spaced at 35 ft. on centers. The floor deck is 3.5-in. normal weight concrete on 1.5 in. × 20 ga. composite steel deck.

- Include the weights of 1-in. light-weight floor finish, suspended acoustical tile ceiling, mechanical and electrical (assume an industrial building), and *partitions*.

- Since the beam and girder sizes are known, you must calculate the actual weight, in pounds per square foot, of the beam and girder by dividing their weights in in. pounds per foot by their tributary widths.

b. Determine the dead loads in kips/ft. for a typical interior beam and a typical interior girder. Assume that the girder load is uniformly distributed.

c. If the floor system in problem a is to be used as a heavy manufacturing plant, determine the controlling factored loads in kips/ft. for the design of the *typical beam* and the *typical girder*.

- Use the LRFD load combinations.

- Note that *partition* loads need not be included in the dead load calculations when the floor live load is greater than 80 psf.

d. Determine the factored shear, V_u, and the factored moment, M_u, for a typical beam and a typical girder.

- Assume that the beams and girders are simply supported.

- The span of the beam is 35 ft. (i.e., the girder spacing).

- The span of the girder is 30 ft.

2-9. The building with the steel roof framing shown in Figure 2-16 is located in Rochester, New York. Assuming terrain category C and a partially exposed roof, determine the following:

a. Balanced snow load on the *lower* roof, p_f.

b. Balanced snow load on the *upper* roof, p_f.

c. Design snow load on the *upper* roof, p_s.

d. Snow load distribution on the *lower* roof, considering sliding snow from the upper pitched roof.

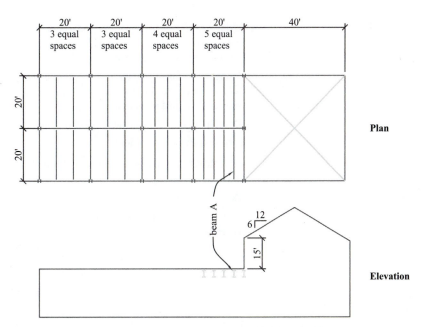

Figure 2-16 Roof plan and building elevation for problem 2-9.

e. Snow load distribution on the lower roof considering drifting snow.

f. Factored dead plus snow load in pounds per foot for the low roof beam A shown on the plan. *Assume a steel framed roof and a typical dead load of 29 psf for the steel roof.*

g. Factored moment, M_u, and factored shear, V_u, for beam A.

   Note that the beam is simply supported.

h. For the typical interior roof girder nearest the taller building (i.e., the interior girder supporting beam A, in addition to other beams), *draw* the dead load and snow load diagrams, showing all the numerical values of the loads in pounds per foot for:

   (1) Dead load and snowdrift loads, and

   (2) Dead load and sliding snow load.

   Assume that for the girder, the dead load, flat roof snow load, and sliding snow load will be uniformly distributed, and the snow drift load will be a linearly varying (trapezoidal) load.

i. For each of the two cases in problem h, determine the unfactored reactions at both supports of the simply supported interior girder due to dead load, snow load, and the factored reactions. Indicate which of the two snow loads (snowdrift or sliding snow) will control the design of this girder.

2-10. An eight-story office building consists of columns located 30 ft. apart in both orthogonal directions. The roof and typical floor gravity loads are given below:

   Roof Loads:

   Dead load $= 80$ psf

   Snow load $= 40$ psf

   Floor Loads:

   Floor dead load $= 120$ psf

   Floor live load $= 50$ psf

   a. Using the column tributary area and a column load summation table, determine the total unfactored and factored vertical loads in a typical interior column in the first story, neglecting live load reduction.

   b. Using the column tributary area and a column load summation table, determine the total unfactored and factored vertical loads in a typical interior column in the first story, considering live load reduction.

   c. Develop an Excel spreadsheet to solve problems a and b, and verify your results.

2-11. **Student Design Project Problem:**

   a. Calculate the dead, snow, roof live, and floor live loads on the roof and floor framing for the design project problem introduced in Chapter 1.

   b. Determine the most economical layout for the roof framing (joists, or infill beams and girders) and the gage (thickness) of the roof deck.

   c. Determine the most economical layout for the floor framing (infill beams and girders), the total depth of the floor slab, and the gage (thickness) of the floor deck.

Lateral Loads and Systems

3.1 LATERAL LOADS ON BUILDINGS

The types of lateral loads that may act on a building structure include wind loads, seismic loads, earth pressure, and hydrostatic pressures. These loads produce overturning, sliding, and uplift forces in the structure. In this text, only the two main types of lateral loads on steel structures (wind and seismic loads) will be discussed.

Wind Loads—Cause and Effect

All exposed structures are acted on by wind forces (see Figure 3-1); the surface nearest to the wind direction (i.e., the windward face) is acted on by a positive wind pressure; the surface opposite the wind direction (i.e., the leeward face) is acted on by a negative wind pressure (i.e., suction). The roof of the building is also subjected to negative (i.e., suction) and/or positive pressures. These wind pressures, expressed in pounds per square foot, act perpendicular to the building surfaces. The minimum exterior wind pressure is 10 psf per ASCE 7 and the minimum interior wind pressure for the design of interior elements is 5 psf.

Wind is actually a dynamic force because its velocity varies with time, but in the ASCE 7 load standard, the effects of wind forces are determined based on an equivalent static approach [1]. Wind force is a function of the wind speed or velocity (ASCE 7 uses the 3-sec. gust velocity), topography, building height and exposure, use of the building, building stiffness, and percentage of wall openings in the building. The 50-yr. return period wind load is typically used for the strength design of building structures for strength, while the 20-yr. and 10-yr. return period wind loads are used for lateral deflection or drift calculations. The ASCE 7 load standard recommends the 20-yr. return period wind loads for lateral drift calculations; the 20-yr. and 10-yr. return period wind loads to be used are approximately 70% and 55%, respectively, of the 50-yr. return period wind load. A total lateral drift and interstory drift limit of H/500 to H/400 is commonly specified in practice, where H is the total height of the building or the difference in height between adjacent floor levels.

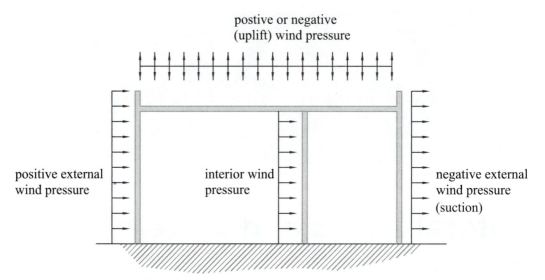

positive or negative
(uplift) wind pressure

positive external
wind pressure

interior wind
pressure

negative external
wind pressure
(suction)

Figure 3-1 Wind pressures on building surfaces.

Special attention should be paid to canopy or open structures (e.g., gas station structures) because these structures are susceptible to large upward wind pressures that may lift the roof off the building, and the wind pressures on canopies or overhangs of buildings are usually much higher than at other parts of the building.

Seismic or Earthquake Loads—Cause and Effect

Earthquakes are caused by the relative movement of the tectonic plates in the earth's crust, and these movements, which occur suddenly, originate at planes of weaknesses in the earth's crust called faults (e.g., the San Andreas fault), causing a release of stress that has built up, resulting in a release of massive amounts of energy [2]. This energy causes ground motion, which results in the vibration of buildings and other structures. Athough earthquake forces cause motion in all directions, only the horizontal and vertical motions are of the most significance.

The point at which the earthquake originates within the earth's crust is called the hypocenter, and the point on the earth's surface directly above the hypocenter is called the epicenter. The magnitude of earthquakes is measured by the Richter scale, which is a logarithmic measure of the maximum amplitude of the earthquake-induced ground vibration as recorded by a seismograph. The *theoretical elastic* dynamic force exerted on a structure by an earthquake is obtained from Newton's second law of motion:

$$F = Ma = (W/g)a = W(a/g), \tag{3-1}$$

where

M = Mass of structure,

a = Acceleration of the structure induced by the earthquake,

g = acceleration due to gravity,

a/g = Seismic coefficient, and

W = Weight of the structure.

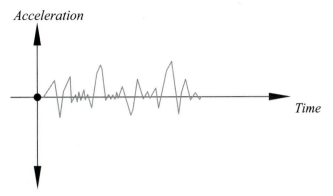

Figure 3-2 Acceleration–Time plot.

The ASCE 7 load standard uses a modified version of equation (3-1) to determine the equation for the seismic base shear on a structure during an earthquake. The code equation takes into account damping (or internal friction of the material), structural, and foundation properties. Seismic design in the United States is based on a 2,500-yr. earthquake, or an earthquake with a 2% probability of being exceeded in 50 years. The base shear is then converted to some "approximated" code equivalent lateral force at each floor level of the building. It should be noted that the lateral forces measured in buildings during actual earthquake events are usually greater than the code equivalent lateral forces. However, experience indicates that buildings that have been designed elastically to these code equivalent forces have always performed well during actual earthquakes. The reason for this is the ductility of building structures or the ability of structures to dissipate seismic energy, without failure, through inelastic action, such as cracking and yielding. During an earthquake event, the induced acceleration of the structure varies in an erratic manner, having low and high points as shown in Figure 3-2. A plot of the absolute maximum accelerations of buildings with different periods, T, yields a response spectrum similar to that shown in Figure 3-3. The ASCE 7 load standard uses a modified form of equation (3-1), together with a design response spectrum, to calculate the design seismic base shear on a structure.

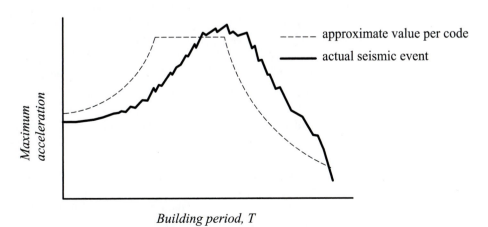

Figure 3-3 Response spectrum.

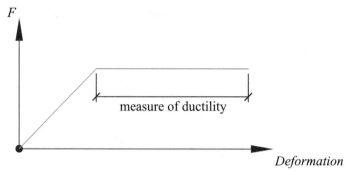

Figure 3-4 Ductility and the load-deformation relationship.

Ductility

Ductility is the ability of a structure or element to sustain large deformations and thus some structural damage under constant load without collapse. The length of the flat portion of the load-deformation plot of a structure (see Figure 3-4) is a measure of the ductility of the structure or member, and the longer the flat portion of the load-deformation plot, the more ductile the structure is. The more ductile that a structure is, the better the seismic resistance and behavior of the structure. Ductility is usually achieved in practice by proper *detailing* of the structure and its connections as prescribed in the materials sections of ASCE 7. In ASCE 7, ductility is accounted for by using the system response factor, R. This will be discussed later in this chapter.

Similarities and Differences Between Wind and Seismic Forces

The similarities and differences between wind and seismic forces affect the design provisions for these forces in the ASCE 7 load standard. These are summarized below:

- Both wind and seismic loads are dynamic in nature, but earthquakes are even more so than wind.
- Seismic forces on structures arise from ground motion and the inertial resistance of the structure to this motion, whereas wind forces on a building structure arise from the impact of the wind pressure on the exposed surfaces of the structure.
- Seismic forces on a structure depend on structural and foundation properties, and the dynamic properties of the earthquake. The softer the soil, the higher the earthquake forces are on the structure. Wind forces, however, depend mainly on the shape and surface area of the structure that is exposed to wind, but also on the period of the building.
- Because of the highly dynamic nature of earthquakes, compared with wind, safety is not necessarily ensured by using a stiffer structure for seismic resistance. In fact, the stiffer a structure is, the higher the seismic forces that the structure attracts. Therefore, in designing for seismic forces, the structural stiffness and the ductility of the structure are both equally important.
- The code-specified seismic forces are smaller than the actual elastic inertial forces induced by a seismic event; however, buildings have been known to perform well in earthquakes because of the ductility of these structures, which allows the structure to dissipate seismic energy through controlled structural damage, but without collapse [3]. Therefore, when designing for earthquake effects, it is not enough to design just for the code seismic forces;

the prescribed seismic detailing requirements in the materials sections of the International Building Code (IBC) also must be satisfied in order to ensure adequate ductility. On the other hand, when designing for wind forces, stiffness is a more important criterion, and ductility is not as important because of the lesser dynamic nature of wind.

- To calculate seismic forces, the base shear is first calculated, and then this base shear is converted into equivalent lateral forces at each floor level of the building using a linear or parabolic distribution based on the modal response of the structure. For wind forces, the design wind pressures are first calculated, followed by the calculation of the lateral forces at each level based on the vertical surface tributary area of each level, and then the wind base shear is calculated.

- For wind loads, two sets of lateral forces are required: the lateral wind forces on the main wind force resisting system (MWFRS) and the wind forces on smaller elements known as the components and cladding (C&C). For seismic design, two sets of lateral forces are required—the lateral forces, F_x, on the vertical lateral force resisting system (LFRS) and the lateral forces, F_P, on the horizontal diaphragms (i.e., the roof and floors)—because of the different dynamic behavior of the horizontal diaphragms compared with that of the vertical LFRS during an earthquake event. In addition, the seismic lateral forces on the parts and components (structural and nonstructural) of the building also need to be calculated.

3.2 LATERAL FORCE RESISTING SYSTEMS IN STEEL BUILDINGS

The different types of LFRS that are commonly used in steel buildings are discussed in this section. Each of these LFRS may be used solely to resist the lateral force in both orthogonal directions in a building, or a mixed LFRS or a combination of these systems may also be used. In taller buildings (30 stories or higher), a mixed LFRS of moment frames and shear walls is an efficient system for resisting lateral forces in the same direction [3]. However, for low- and mid-rise buildings, it is typical in design practice to use only one type of LFRS to resist the lateral force in any one direction. Depending on architectural considerations, the LFRS may be located internally within the building or on the exterior face of the building.

The lateral force distributed to each LFRS is a function of the in-plane rigidity of the roof and floor diaphragms (diaphragms can be classified as either flexible or rigid). The definitions of flexible and rigid diaphragms are given in ASCE 7, Sections 12.3.1.1 and 12.3.1.2.

If a roof or floor diaphragm is classified as rigid, the lateral wind or seismic forces are distributed to each LFRS in proportion to the lateral rigidities or stiffness of the LFRS. If a roof or floor diaphragm is classified as flexible, the lateral wind force along each line of LFRS is proportional to the tributary vertical surface area of the wall that receives the wind pressure. Where more than one LFRS exists along the same line or vertical plane, the lateral wind force parallel to and along that plane will be distributed to the LFRS along that line or vertical plane in proportion to the stiffness of the LFRS along that plane. For seismic forces in buildings with flexible diaphragms, the lateral seismic force on each LFRS is proportional to the tributary roof or floor plan area of each LFRS.

Fully Restrained Moment Connections (Rigid Frames)

For frames with fully restrained moment connections, sometimes called rigid frames (see Figure 3-5), the beams and girders are connected to the columns with moment-resisting connections, and the lateral load is resisted by the bending strength of the beams and columns. For maximum efficiency in steel buildings, the columns in the moment frames

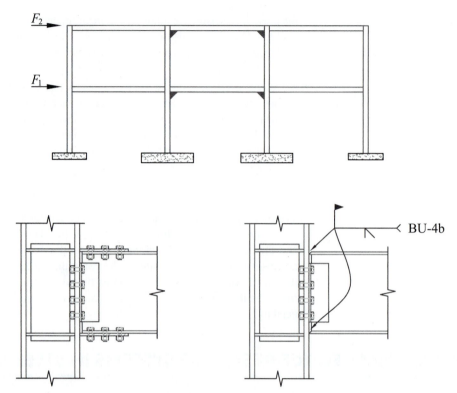

Figure 3-5 Steel moment frames.

in both orthogonal directions should be oriented so that they are subjected to bending about their strong axis, and the moment-resisting connection can be achieved by welding steel plates to the column flange and bolting or welding these plates to the beam/girder flanges. The steel beam or girder is connected to the column flange using shear connection plates or angles that may be welded or bolted to the beam web and column flange to support gravity loads.

Partially Restrained Moment Connections (Semi-rigid Frames)

For semi-rigid frames with partially restrained moment connections (see Figure 3-6), the rigidity of the beam-to-column connections is generally less than that of the fully restrained moment frame. In these connections, the flanges of the steel beam or girder are usually connected with angles to the column flange, and the beam web is connected to the column using shear connection plates or angles. However, at the roof level, the top flange of the beam or girder is typically connected to the column with a steel plate that also acts as a cap plate for the column. The deflection of a partially restrained moment frame is generally higher than that of a fully restrained moment frame.

Braced Frames

For braced frames, the lateral load is resisted through axial tension and/or compression forces in the diagonal bracing members. The beam-to-column connections in braced

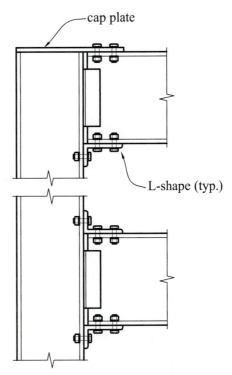

cap plate

L-shape (typ.)

Figure 3-6 Partially restrained moment connections.

frames are usually simple shear connections with no moment resisting capacity. Examples of braced frames are shown in Figure 3-7; these include X-bracing, chevron or K bracing, diagonal bracing, V-bracing, and knee bracing. The X-bracing and V-bracing offer the least flexibility for the location of doors, while chevron bracing, diagonal bracing, and knee bracing offer the most flexibility, and are usually the bracing systems preferred by architects.

Shear Walls

Shear walls (see Figure 3-8) are planar structural elements that act as vertical cantilevers fixed at their bases; they are usually constructed of concrete, masonry, plywood sheathing, or steel plates. These could be located internally within the building or on the exterior face of the building. The concrete or masonry walls around stair and elevator shafts may also be considered as shear walls. Shear walls are very efficient, lateral force resisting elements. Shear walls that are perforated by door or window openings are termed coupled shear walls, and these may be modeled approximately as moment frames. The actual strength of a coupled shear wall lies between the strength of the wall moment frame and an unperforated shear wall of the same overall dimensions.

Although moment frames are the least rigid of all the lateral force resisting systems, they provide the most architectural flexibility for the placement of windows and doors, while shear walls and X-brace frames provide the least architectural flexibility.

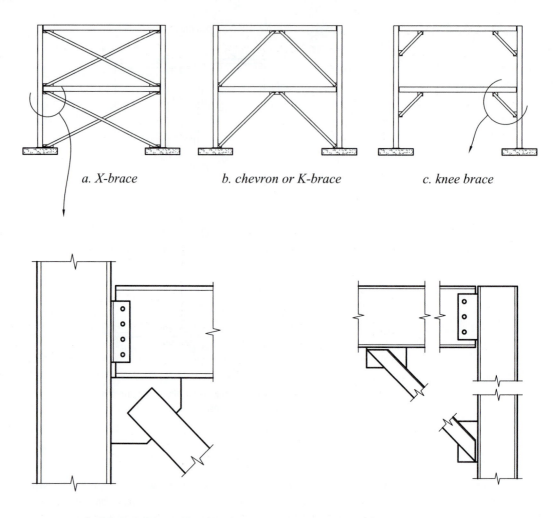

a. X-brace   *b. chevron or K-brace*   *c. knee brace*

d. X-brace connection detail   *e. knee brace connection detail*

Figure 3-7 Braced frames.

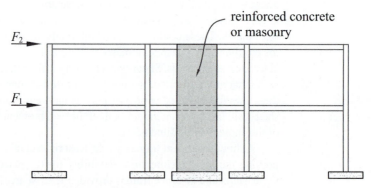

Figure 3-8 Shear walls.

3.3 IN-PLANE TORSIONAL FORCES IN HORIZONTAL DIAPHRAGMS

If the LFRS in a building with rigid diaphragms are not symmetrically placed, or the LFRS do not have equal stiffness, or if the center of mass (CM) or center of area (wind force acts only through the center of area of the wall surface) of the horizontal diaphragms does not coincide with the center of rigidity (CR) of the LFRSs, the building will be subjected to in-plane twisting, or rotational or torsional forces from the lateral wind or seismic forces. These in-plane torsional moments, when resolved, will result in an increase in the lateral force on some of the LFRS (i.e., positive torsion) and a decrease in the lateral force on other LFRS (i.e., negative torsion). Where in-plane torsional moments cause a decrease in the lateral force on an LFRS, the effect of in-plane torsion or twisting on that LFRS is usually ignored.

It should be noted that only buildings with rigid diaphragms can transmit in-plane torsional or rotational forces (see Figure 3-9); buildings with flexible diaphragms cannot resist rotational or torsional forces. The LFRS in buildings with flexible diaphragms can only resist direct lateral forces, and in proportion to their tributary areas; therefore, the LFRS must be placed in relation to the location and distribution of the masses supported. It is best to avoid structural system layouts that cause in-plane torsion or twisting of the roof or floor diaphragm. For example, a building with an open front with a rigid diaphragm would be subjected to large in-plane torsional forces, while a similar building with a flexible diaphragm will be unstable. Several buildings have collapsed during earthquakes due to the irregular layout of the LFRSs and the subsequent in-plane twisting of the horizontal diaphragms. One notable example is the JCPenney building in Anchorage that collapsed during the 1964 Alaskan earthquake [2]. It should be noted that for wind loads, the in-plane torsional moments must be calculated using the wind load cases shown in Figure 6.9 of the ASCE 7 load standard. The procedure for calculating the additional lateral forces caused by the in-plane torsional effects of seismic loads is beyond the scope of this text.

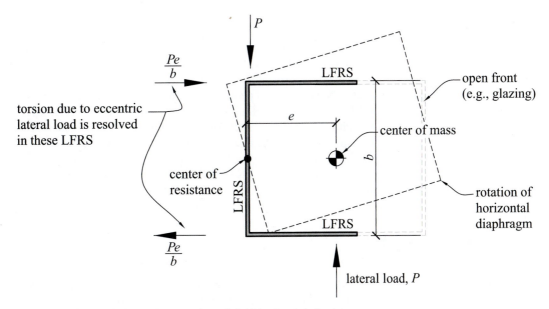

Figure 3-9 In-plane torsion of rigid horizontal diaphragms.

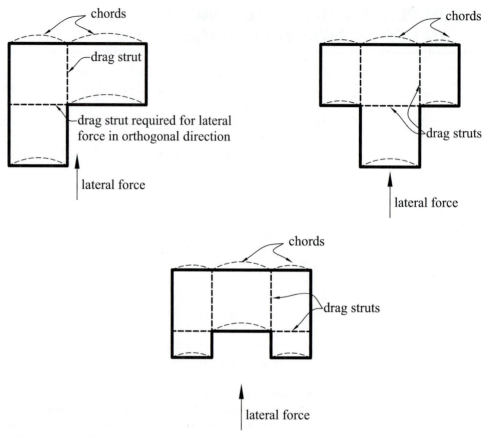

Figure 3-10 Diaphragm chords and drag struts.

Chords and Drag Struts (or Collectors)

Chords are structural elements located along the perimeter of the horizontal diaphragms, and they resist the tension and compression couple resulting from the in-plane bending of the diaphragm due to the lateral seismic or wind forces (see Figure 3-10). They are located perpendicular to the lateral load.

Drag struts or collectors are structural elements in the plane of the horizontal diaphragms; they are located parallel to the lateral forces and are in the same vertical plane as the LFRS. The drag strut transfers the lateral wind or seismic forces from the roof and floor diaphragms into the LFRS. They also help prevent differential or incompatible horizontal displacements of diaphragms in buildings with irregular shapes. Without drag struts or collectors, tearing forces would develop at the interface between the various diaphragm segments (see Figure 3-10). The ASCE 7 load standard requires that drag struts be designed for the special seismic force, E_m, which includes amplification of seismic forces by the overstrength factor, Ω_o.

3.4 WIND LOADS

The two conditions considered in the calculation of wind forces acting on building structures are the main wind force resisting system (MWFRS) and the components and cladding (C&C). The MWFRS consists of the roof and floor diaphragms, and shear walls, braced

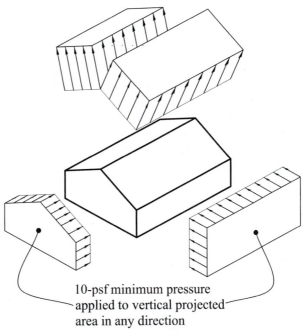

Figure 3-11 Minimum wind load diagram.

frames, and moment frames that are parallel to the wind force. The C&C are small individual structural components or members with the wind load acting perpendicular to the member. Examples of C&C include walls, stud wall, cladding, and uplift force on a roof deck fastener. The wind pressures on C&C are usually higher than the wind pressures on the MWFRS because of local spikes in wind pressure over small areas of the C&C. The C&C wind pressure is a function of the effective wind area, A_e, given as

$$A_e = \text{Span of member} \times \text{Tributary width} \geq (\text{Span of member})^2/3 \qquad (3\text{-}2)$$

For cladding and deck fasteners, the *effective wind area*, A_e, shall not exceed the area that is tributary to each fastener. In calculating wind pressure, positive pressures are indicated by a force "pushing" into the wall or roof surface, and negative pressures are shown "pulling" away the wall or roof surface. The minimum design wind pressure for MWFRS and C&C is 10 psf and is applied to the vertical projected area of the wall surfaces for the MWFRS, and normal to the wall or roof surface for C&C (see Figure 3-11 or ASCE 7, Figure C6-1).

3.5 CALCULATION OF WIND LOADS

The three methods available in the ASCE 7 load standard to calculate the design wind loads on buildings and other structures are as follows:

1. **The simplified method or method 1** uses projected areas with the net horizontal and vertical wind pressures assumed to act on the exterior projected area of the building as shown in Figure 3-12a. The simplified method is limited to buildings with mean roof heights not greater than 60 ft.

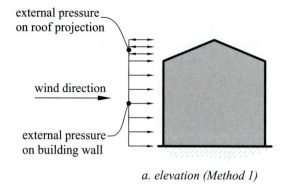

a. elevation (Method 1)

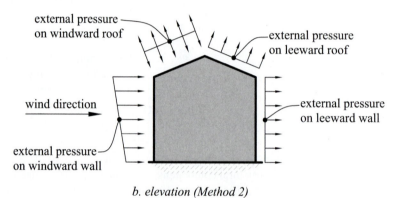

b. elevation (Method 2)

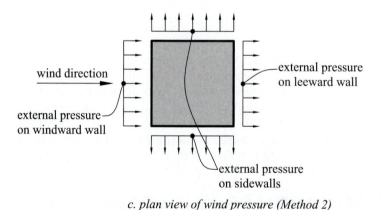

c. plan view of wind pressure (Method 2)

Figure 3-12 Wind pressure distribution (methods 1 and 2).

2. **The analytical method or method 2** assumes that wind pressures act perpendicular to the exterior wall and roof surfaces as shown in Figure 3-12b.

The analytical method is more cumbersome than the simplified method and is applicable to buildings with a regular shape (see ASCE 7, Section 6.2) that are not subjected to unusual wind forces, such as cross-wind loading, vortex shedding, galloping or flutter, and so forth. In the analytical method, positive pressures are applied to the windward walls (i.e., the walls receiving the wind pressure) and negative pressures or suction are applied to the leeward walls (i.e., the walls receiving the suction).

3. **The wind tunnel method or method 3** is used where methods 1 or 2 cannot be used. The wind tunnel procedure is used for very tall and wind-sensitive buildings, and slender buildings with a height-to-width ratio greater than 5.0 [4]. The wind pressures on buildings with unique topographic features should also be determined using the wind tunnel method.

Athough it might be expected that the simplified method (method 1) would yield more conservative (i.e., higher) values for the design wind pressures than the more cumbersome analytical method (method 2), comparisons of the two methods show that, in general, the analytical method yields design wind pressures, base shear, and overturning moment values that may be up to 50% higher than those obtained from the simplified method (method 1). It is instructive to note that the ASCE 7 simplified method was actually developed based on comprehensive wind tunnel testing done at the Boundary Layer Wind Tunnel Laboratory of the University of Western Ontario and may therefore be more accurate than is implied by the term "simplified" [5]. In design practice, the simplified method is more widely used for buildings with mean roof heights not exceeding 60 ft.

Exposure Categories

An exposure category is a measure of the terrain surface roughness and the degree of exposure or shielding of the building. The three main exposure categories are exposures B, C, and D. Exposure B is the most commonly occurring exposure category—approximately 80% of all buildings fall into this category—although many engineers tend to specify exposure C, for most buildings. The exposure category is determined from the ASCE 7 load standard by first establishing the type and extent of the ground surface roughness at the building site using Table 3-1, and then using Table 3-2 to determine the exposure category.

Basic Wind Speed

This is a 3-second gust wind speed in miles per hour based on a 50-yr. wind event. Wind speed increases as the height above the ground increases because of the reduced drag effect of terrain surface roughness at higher elevations.

The basic wind speeds at a height of 33 ft. above the ground for various locations in the United States are shown in ASCE-7, Figure 6-1.

Table 3-1 Ground surface roughness categories (ASCE 7, Section 6.5.6.2)

Ground Surface Roughness	Description
B	Urban, suburban, and mixed wooded areas with numerous spaced obstructions the size of a single-family dwelling or larger
C	Open terrain with scattered obstruction having heights generally less than 30 ft.; includes flat open country, grassland, and water surfaces in hurricane-prone regions
D	Flat, unobstructed areas and water surfaces outside hurricane-prone regions; includes smooth mudflats, salt flats, and unbroken ice

Table 3-2 Exposure categories (ASCE 7, Section 6.5.6.3)

Exposure Category	Description
B	Surface roughness B prevails in the *upwind** direction for a distance of $20h \geq 2600$ ft., where h is the height of the building.
C	Where exposure category B or D does not apply
D	Occurs where surface roughness D prevails in the *upwind** direction (over smooth water surfaces) for a distance of $20h \geq 5000$ ft., and exposure category D extends into *downwind** areas with a category B or C surface roughness for a distance of $20h \geq 600$ ft., where h is the height of the building.

* *Upwind* is the direction opposite the direction where the wind is coming from. For example, if wind is acting on a building from west to the east, then the upwind direction is west of the building while the downwind direction is east of the building.

3.6 SIMPLIFIED WIND LOAD CALCULATION METHOD (METHOD 1)

In the simplified method, which is only applicable to low-rise buildings, the wind forces are applied perpendicular to the *vertical and the horizontal projected areas* of the building. The wind pressure diagram for the MWFRS is shown in Figure 3-13. The horizontal pressures represent the combined windward and leeward pressures with the internal pressures

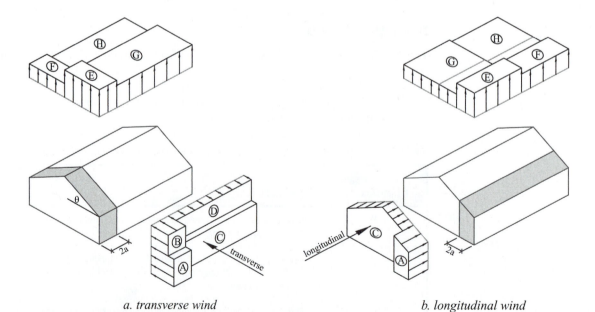

a. transverse wind   b. longitudinal wind

Figure 3-13 Wind pressure diagram for main wind force resisting system.
Adapted from Ref. [6]

Table 3-3 Definition of wind pressure zones—MWFRS

Zone	Definition
A	End-zone *horizontal* wind pressure on the *vertical* projected *wall* surface
B	End-zone *horizontal* wind pressure on the *vertical* projected *roof* surface
C	Interior-zone *horizontal* wind pressure on the *vertical* projected *wall* surface
D	Interior-zone *horizontal* wind pressure on the *vertical* projected *roof* surface
E	End-zone *vertical* wind pressure on the *windward* side of the *horizontal* projected *roof* surface
F	End-zone *vertical* wind pressure on the *leeward* side of the *horizontal* projected *roof* surface
G	Interior-zone *vertical* wind pressure on the *windward* side of the *horizontal* projected *roof* surface
H	Interior-zone *vertical* wind pressure on the *leeward* side of the *horizontal* projected *roof* surface
E_{OH}	End-zone *vertical* wind pressure on the *windward* side of the *horizontal* projected *roof overhang* surface
G_{OH}	Interior-zone *vertical* wind pressure on the *windward* side of the *horizontal* projected *roof overhang* surface

Where zone E or G falls on a roof overhang, the windward roof overhang wind pressures from ASCE 7, Figure 6-2 should be used.

canceling each other out; the vertical pressures include the combined effect of the external and internal pressures. The different wind pressure zones in Figure 3-13 are defined in Table 3-3 and it should be noted that for buildings with flat roofs, the simplified method yields a uniform horizontal wall pressure distribution over the entire height of the building.

The simplified procedure is applicable only if all of the following conditions are satisfied:

- Building has simple diaphragms (i.e., wind load is transferred through the roof and floor diaphragms to the vertical MWFRS),
- Building is enclosed (i.e., no large opening on any side of the building),
- Building has mean roof height that is less than or equal to the least horizontal dimension of the building,
- Building has mean roof height that is less than or equal to 60 ft.,
- Building is symmetrical,
- Building has an approximately symmetrical cross section in each direction with a roof slope ≤ 45°.

The simplified procedure (i.e., method 1) for calculating wind loads involves the following steps:

1. Determine the applicable wind speed for the building location from the ASCE 7 load standard.
2. Calculate the mean roof height and determine the wind exposure category.

3. Determine the applicable horizontal and vertical wind pressures as a function of the wind speed, roof slope, zones, and effective wind area by calculating

    a. P_{s30} for MWFRS from ASCE-7, Figure 6-2, and

    b. P_{net30} for C&C from ASCE-7, Figure 6-3.

Note the following:

- The tabulated values are based on an assumed exposure category of B, a mean roof height of 30 ft., and an importance factor of 1.0. These tabulated wind pressures are to be applied to the horizontal and vertical projections of the buildings.
- The horizontal wind pressure on the projected vertical surface area of the building is the sum of the external windward and external leeward pressures because the internal pressures cancel out each other. The resultant wind pressures are applied to one side of the building for each wind direction.
- The wind pressures on roof overhangs are much higher than at other locations on the roof because of the external wind pressures acting on both the bottom and top exposed surfaces of the overhang.

4. Obtain the design wind pressures (p_{s30} for MWFRS and p_{net30} for C&C), as a function of the tabulated wind pressures obtained in the previous step, the applicable height and exposure adjustment factor (λ from ASCE 7, Figure 6-2 or 6-3), the topography factor (K_{zt} from ASCE 7, Section 6.5.7), and the importance factor (I_w from ASCE 7, Tables 1-1 and 6-1). (Note that for most site conditions, $K_{zt} = 1.0$.)

$$P_{s30} \ (\text{MWFRS}) \ = \ \lambda \ K_{zt} \ I_w \ p_{s30} \geq 10 \ \text{psf} \tag{3-3}$$

$$P_{net30} \ (\text{C\&C}) \ = \ \lambda \ K_{zt} \ I_w \ p_{net30} \geq 10 \ \text{psf} \tag{3-4}$$

5. Apply the calculated wind pressures to the building as shown in ASCE 7, Figure 6-2 for MWFRS (End-zone width $= 2a$).

where

$a \leq 0.1 \times$ Least horizontal dimension of building,

    $\leq 0.4 \times$ Mean roof height of building, and

    ≥ 3 ft.

For a roof slope of less than 10 degrees, the eave height should be used in lieu of the mean roof height for calculating the end-zone width, $2a$.

6. Apply the calculated wind pressures to the building walls and roof as shown in ASCE 7, Figure 6-3 for C&C.

Note that for C&C,
End-zone width $= a$.

EXAMPLE 3-1

Simplified Method for Wind Loads (MWFRS)

Given a one-story office building 60 ft. $\times$ 90 ft. in plan and laterally braced with X-braces on all four sides and having a story height of 18 ft. as shown in Figure 3-14, determine the unfactored design wind forces on the MWFRS assuming that the building is located in a wind zone with $V = 90$ mph and exposure category D. Assume a flat roof with no overhang and assume that the building is enclosed.

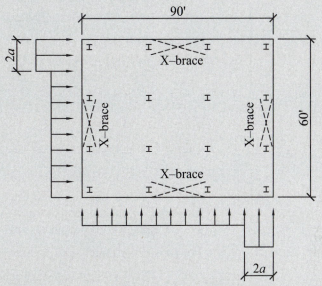

Figure 3-14 Wind load diagram for example 3-1.

SOLUTION

1. The 3-sec. gust wind speed = 90 mph (ASCE 7, Figure 6-1).

2. Mean roof height = 18 ft.
   Wind exposure category D

3. P_{s30} for MWFRS (from ASCE 7, Figure 6-2):
   With roof slope, $\theta = 0°$ and wind speed = 90 mph, the tabulated net horizontal wind pressures on a projected vertical surface area of the building are as follows:

   Net Horizontal Wind Pressures on MWFRS: Longitudinal Wind

   > End zone = 12.8 psf on wall
   >
   > Interior zone = 8.5 psf on wall

   Net Horizontal Wind Pressures on MWFRS: Transverse Wind

   > End zone = 12.8 psf on wall
   >
   > Interior zone = 8.5 psf on wall

   Note that the resultant wind pressure on the MWFRS is nonsymmetrical due to the nonsymmetrical location of the end zones and the higher wind pressures acting on the end zones as shown in ASCE 7, Figure 6-2. For simplicity, we will be using an average horizontal wind pressure in the examples in this text; the asymmetrical nature of the wind loading on the building and the torsional effect of such loading should always be considered.

(continued)

For the MWFRS, the end-zone width is $2a$ per ASCE 7, Figure 6-2, where

$a \leq 0.1 \times$ Least horizontal dimension of building,

$\quad \leq 0.4 \times$ Mean roof height of building, and

$\quad \geq 3$ ft.

4. For a category II building (see ASCE 7, Table 1-1), the importance factor, $I_w = 1.0$. (office building)

For a **longitudinal wind**, end-zone width is $2a$ per ASCE 7, Figure 6-2, where

$a \leq 0.1 \times 60$ ft. $= 6.0$ ft. *(governs)*,

$\quad \leq 0.4 \times 18$ ft. $= 7.2$ ft., and

$\quad \geq 3$ ft.

For simplicity, in our calculations, we will use the average horizontal wind pressure acting on the building as a whole, calculated as follows:

The average horizontal wind pressure is

$$P_{s30} = \frac{(12.8 \text{ psf})(2)(6 \text{ ft.}) + (8.5 \text{ psf})[90 \text{ ft.} - (2)(6 \text{ ft.})]}{90 \text{ ft.}}$$

$$= 9.07 \text{ psf } (\textit{transverse}), \text{ and}$$

$$P_{s30} = \frac{(12.8 \text{ psf})(2)(6 \text{ ft.}) + (8.5 \text{ psf})[60 \text{ ft.} - (2)(6 \text{ ft.})]}{60 \text{ ft.}}$$

$$= 9.36 \text{ psf } (\textit{longitudinal}),$$

Adjusting for height and exposure (equation (3-3)),

$\lambda = 1.52$ (by interpolation, ASCE 7, Table 6-2), and

$I_w = 1.0$.

The uniform horizontal design wind pressure, $P = \lambda K_{zt} I_w p_{s30} \geq 10$ psf, is

$P_{\text{transverse}} = (1.52)(1.0)(1.0)(9.07 \text{ psf}) = 13.8 \text{ psf, and}$

$P_{\text{longitudinal}} = (1.52)(1.0)(1.0)(9.36 \text{ psf}) = 14.2 \text{ psf.}$

The total unfactored load at the roof level for the building as a whole is

$F_{\text{transverse}} = (13.8 \text{ psf})(90 \text{ ft.})(18 \text{ ft./2}) = 11,178 \text{ lb., and}$

$F_{\text{longitudinal}} = (14.2 \text{ psf})(60 \text{ ft.})(18 \text{ ft./2}) = 7668 \text{ lb.}$

The total lateral force at the roof level for each X-brace is calculated below and shown in Figure 3-15:

$F_{\text{transverse}} = (11,178 \text{ lb./2}) = 5590 \text{ lb.}$

$F_{\text{longitudinal}} = (7668 \text{ lb./2}) = 3834 \text{ lb.}$

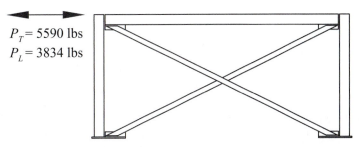

$P_T = 5590$ lbs
$P_L = 3834$ lbs

Figure 3-15 Wind loads on each X-brace frame.

EXAMPLE 3-2

Simplified method for wind loads (MWFRS)

The typical floor plan and elevation of a six-story office building measuring 100 ft. × 100 ft. in plan and laterally braced with ordinary moment resisting frames is shown in Figure 3-16. The building is located in an area with a 3-second wind speed of 90 mph. Determine the factored design wind forces on the MWFRS and C&C. Assume that the building is enclosed and that the roof is flat (except for the minimum roof slope for drainage), with no overhang.

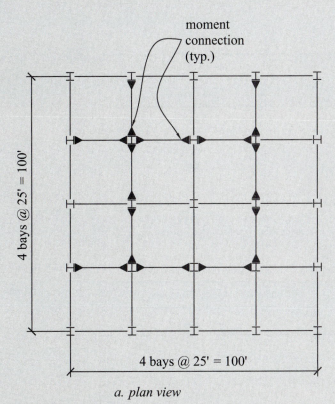

a. plan view

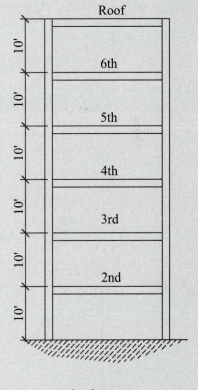

b. elevation

Figure 3-16 Building plan and elevation for example 3-2.

(continued)

SOLUTION

1. The 3-sec. gust wind speed is 90 mph.

2. • Building is enclosed,
   • Roof slope, $\theta \approx 0$,
   • Mean roof height = 60 ft.,
   • Importance factor, $I_w = 1.0$,
   • Effective wind area, $A_e = 10$ ft.^{2} (assumed) Note: This is conservative for C&C, and
   • Wind exposure category is C.

Note that wind exposure category C is the most commonly assumed in design practice, although most buildings are in category B.

3. Obtain P_{s30} for MWFRS from ASCE 7, Figure 6-2 and P_{net30} for C&C from ASCE 7, Figure 6-3.
   For the MWFRS, the wind pressures act on *the projected* vertical and horizontal surfaces of the building.
   For C&C, we have assumed an effective wind area, $A_e \leq 10$ ft.^{2}, which is conservative. For larger effective wind areas, recalculate the wind pressures by selecting the appropriate tabulated wind pressures corresponding to the effective wind area from ASCE 7, Figure 6-3.

   P_{s30} for MWFRS from ASCE 7, Figure 6-2:

   The MWFRS tabulated wind pressures obtained from ASCE 7, Figure 6-2 for a building with a roof slope, $\theta = 0$ and a 3-sec. gust wind speed = 90 mph are shown in Tables 3-4, 3-5, and 3-6.

Table 3-4 Tabulated horizontal wind pressures*

Zone	Tabulated Horizontal Pressures, psf
	Load Case 1
End Zone A	12.8
End Zone B**	
Interior Zone C	8.5
Interior Zone D**	

 * Horizontal wind pressures for longitudinal, as well as transverse, wind.
** Note that zones B and D do not exist for buildings with flat roofs.

Table 3-5 Tabulated vertical wind pressures on roofs*

Zone	Tabulated Vertical Pressures, psf
	Load Case 1
End Zone E	−15.4
End Zone F	−8.8
Interior Zone G	−10.7
Interior Zone H	−6.8

* Vertical wind pressures for longitudinal, as well as transverse, wind.

Table 3-6 Tabulated vertical wind pressures at overhangs*

Zone	Tabulated Vertical Pressures, psf
	Load Case 1
End-Zone E_{OH}	−21.6
End-Zone G_{OH}	−16.9

* Vertical wind pressures for longitudinal, as well as transverse, winds.

Since there are no overhangs in this building, the vertical overhang pressures will be neglected.

P_{net30} for C&C from ASCE 7, Figure 6-3:

The C&C tabulated wind pressures obtained from ASCE 7, Figure 6-3 for a building with the following design parameters are given in Tables 3-7 and 3-8:

- Flat roof (i.e., roof slope, $\theta = 0$),
- Wind speed = 90 mph, and
- Assuming an effective wind area, $A_e \leq 10$ ft.^{2}, *which is conservative* in most cases.

Table 3-7 Tabulated horizontal wind pressures on wall* (C&C)

Zone	Tabulated Horizontal Pressures, psf	
	+ve Pressure	−ve Pressure or Suction
Wall *End* Zone **5**	14.6	−19.5
Wall *Interior* Zone **4**	14.6	−15.8

*Horizontal wind pressures for longitudinal, as well as transverse, winds.

Table 3-8 Tabulated vertical wind pressures on roofs

Zone	Tabulated Vertical Pressures, psf	
	+ve Pressure*	−ve Pressure or Suction*
Roof *Interior* Zone 1	5.9	−14.6
Roof *End* Zone 2	5.9	−24.4
Roof *Corner* Zone 3	5.9	−36.8

* Positive pressure indicates downward wind loads and negative pressure indicates upward wind loads that causes uplift.

4. For a category II building, the wind importance factor, $I_w = 1.0$.

For longitudinal wind, the end-zone width is $2a$ per ASCE 7, Figure 6-2,

(continued)

where

$a \leq 0.1 \times 100$ ft. $= 10$ ft. (governs),

$\leq 0.4 \times 60$ ft. $= 24$ ft., and

≥ 3 ft.

For simplicity in our calculations, as discussed previously, we will use the average horizontal wind pressure acting on the building calculated as follows:

$$P_{s30} = \frac{(12.8 \text{ psf})(2)(10 \text{ ft.}) + (8.5 \text{ psf})[100 \text{ ft.} - (2)(10 \text{ ft.})]}{100 \text{ ft.}} = 9.36 \text{ psf } (\textit{transverse and longitudinal}).$$

5. Using the location of the building and assuming exposure category C, the design horizontal forces on the MWFRS from ASCE 7 Section 6.5.6 are given in Table 3-9. In the simplified wind load calculation method in ASCE 7, the design horizontal wind pressures are assumed to be uniform for the full height of the building as shown in Figure 3-17. The equivalent lateral forces at each floor level due to the design wind pressure are shown in Figure 3-17. Since the building is square in plan, the longitudinal lateral wind loads will be equal to the transverse lateral wind loads.

Table 3-9 MWFRS—Longitudinal Wind

Level	Height	Exposure/ Height Coefficient, λ at mean roof height	Average Horizontal Wind Pressure, P_{s30}, psf	Design Horizontal Wind Pressure, $P_s = \lambda I_w P_{s30}$ ≥ 10 psf	TOTAL Unfactored Wind Load at each level of the building, kip	Unfactored Lateral Load at Each Level on each MWFRS (i.e., each moment frame), kips
Roof	60	1.62	9.36 psf	(1.62)(1.0)(9.36) = 15.2 psf	(15.2 psf)(100 ft.)(10 ft./2) = 7.6 kips	7.6 kip/2 = 3.8 kips
Sixth Floor	50	1.62	9.36 psf	15.2 psf	(15.2 psf)(100 ft.)(10 ft./2) + (15.2 psf)(100 ft.)(10 ft./2) = 15.2 kips	15.2 kip/2 = 7.6 kips
Fifth Floor	40	1.62	9.36 psf	15.2 psf	(15.2 psf)(100 ft.) (10 ft./2) + (15.2 psf)(100 ft.)(10 ft./2) = 15.2 kips	7.6 kips
Fourth Floor	30	1.62	9.36 psf	15.2 psf	(15.2 psf)(100 ft.) (10 ft./2) + (15.2 psf)(100 ft.)(10 ft./2) = 15.2 kips	7.6 kips
Third Floor	20	1.62	9.36 psf	15.2 psf	(15.2 psf)(100 ft.) (10 ft./2) + (15.2 psf)(100 ft.)(10 ft./2) = 15.2 kips	7.6 kips
Second Floor	10	1.62	9.36 psf	15.2 psf	(15.2 psf)(100 ft.) (10 ft./2) + (15.2 psf)(100 ft.)(10 ft./2) = 15.2 kips	7.6 kips
Base Shear					83.6 kip	41.8 kips

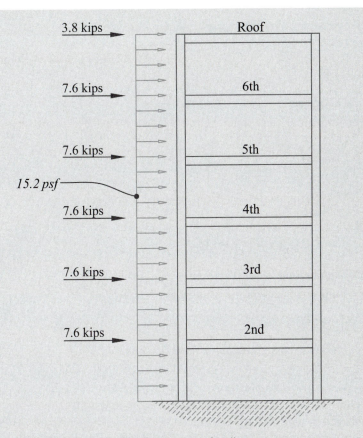

Figure 3-17 Horizontal distribution of wall pressures.

To calculate the factored wind load at each level, the unfactored wind load at each level is multiplied by the maximum wind load factor of 1.6 given in the LRFD load combinations in Chapter 2.

Wind Base Shear (Unfactored):

The unfactored wind base shear on the building as a whole is obtained from column 6 in Table 3-9.

Longitudinal: $V = 83.6$ kips

Transverse: $V = 83.6$ kips

The unfactored wind base shear on the each moment frame is obtained from column 7 in Table 3-9.

Longitudinal: $V = 41.8$ kips

Transverse: $V = 41.8$ kips

Longitudinal Wind (Unfactored):

The unfactored overturning moment for the building as a whole is

$(7.6)(60 \text{ ft.}) + (15.2)(50 \text{ ft.}) + (15.2)(40 \text{ ft.}) + (15.2)(30 \text{ ft.}) + (15.2)(20 \text{ ft.}) + (15.2)(10 \text{ ft.})$
$= 2736$ ft.-kips.

(continued)

The **unfactored** overturning moment for each moment frame is

$$(3.8)(60 \text{ ft.}) + (7.6)(50 \text{ ft.}) + (7.6)(40 \text{ ft.}) + (7.6)(30 \text{ ft.}) + (7.6)(20 \text{ ft.}) + (7.6)(10 \text{ ft.})$$
$$= 1368 \text{ ft.-kips.}$$

Wind Base Shear (Factored):

The factored wind base shear on the building as a whole is obtained as follows:

Longitudinal: $V_u = 1.6(83.6 \text{ kips}) = 134 \text{ kips}$

Transverse: $V_u = 1.6(83.6 \text{ kips}) = 134 \text{ kips}$

The factored wind base shear on the each moment frame is obtained as follows:

Longitudinal: $V_u = 1.6(41.8 \text{ kips}) = 67 \text{ kips}$

Transverse: $V_u = 1.6(41.8 \text{ kips}) = 67 \text{ kips}$

Transverse and Longitudinal Wind (Factored):

The factored overturning moment for the building as a whole is

$$1.6(2736 \text{ ft.-kips}) = 4378 \text{ ft.-kips}$$

The factored overturning moment for each moment frame is

$$1.6(1368 \text{ ft.-kips}) = 2189 \text{ ft.-kips}$$

Calculate the design vertical uplift wind pressures on the roof (MWFRS):

These are obtained by multiplying the tabulated vertical wind pressures for MWFRS from step 2 by the height and exposure adjustment factor and the importance factor. The design parameters are

- Exposure category C,
- Mean roof height = 60 ft.,
- Height/Exposure adjustment coefficient from ASCE 7, Figure 6-2, $\lambda = 1.62$, and
- Importance factor, $I_w = 1.0$.

The design uplift pressures are shown in Tables 3-10 and 3-11.

Table 3-10 Design vertical wind pressures on roofs* (MWFRS)

Zone	Design Vertical Pressures, $\lambda \, I_w \, P_{s30}$**, psf
	Load Case 1
End Zone E	$(1.62)(1.0)(-15.4) = -25$
End Zone F	$(1.62)(1.0)(-8.8) = -14.3$
Interior Zone G	$(1.62)(1.0)(-10.7) = -17.3$
Interior Zone H	$(1.62)(1.0)(-6.8) = -11$

* Vertical wind pressures for longitudinal, as well as transverse, winds.

**See Table 3-5 for P_{s30}.

Table 3-11 Tabulated vertical wind pressures at overhangs* (MWFRS)

Zone	Design Vertical Pressures, $\lambda\, I_w\, P_{s30}$**, psf
	Load Case 1
End-Zone E_{OH}	$(1.62)(1.0)(-21.6) = -35$
End-Zone G_{OH}	$(1.62)(1.0)(-16.9) = -27.4$

* Vertical wind pressures for longitudinal, as well as transverse, winds; the overhang pressures will be neglected for this building since it has no roof overhangs.

** See Table 3-6 for P_{s30}.

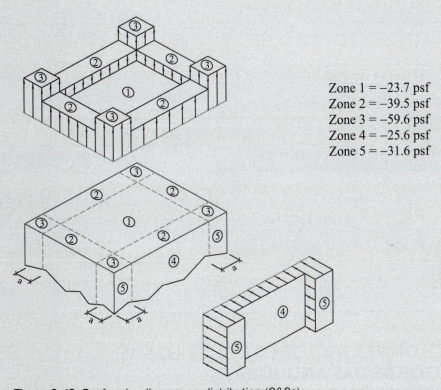

Zone 1 = –23.7 psf
Zone 2 = –39.5 psf
Zone 3 = –59.6 psf
Zone 4 = –25.6 psf
Zone 5 = –31.6 psf

Figure 3-18 Roof and wall pressure distribution (C&Cs).

6. Calculate the design wind pressures for C&C—wall and roof.

   The design wind pressures are obtained by multiplying the tabulated wind pressures for C&C from step 2 by the height/exposure adjustment factor and the importance factor. The design parameters are

   - Exposure category C,
   - Mean roof height = 60 ft.,
   - Height/Exposure adjustment coefficient from ASCE 7, Figure 6-3, $\lambda = 1.62$, and
   - Importance factor, $I_w = 1.0$.

The design pressures for C&C are shown in Figure 3-18 and Tables 3-12 and 3-13.

(continued)

For C&C, the end-zone width is a per ASCE 7, Figure 6-3,

where

$a = 10$ ft. (see step 4)

Table 3-12 Design horizontal wind pressures on wall* (C&C)

Zone	Design Horizontal Pressures, $\lambda I_w P_{net30}$**, psf	
	+ve Pressure	−ve Pressure or Suction
Wall *Interior* Zone **4**	$(1.62)(1.0)(14.6) = 23.7$	$(1.62)(1.0)(-15.8) = -25.6$
Wall *End* Zone **5**	$(1.62)(1.0)(14.6) = 23.7$	$(1.62)(1.0)(-19.5) = -31.6$

* Horizontal wind pressures for longitudinal, as well as transverse, winds.
** See Table 3-7 for P_{net30}.

Table 3-13 Design vertical wind pressures on roof (C&C)

Zone	Design Vertical Pressures, $\lambda I_w P_{net30}$*, psf	
	+ve Pressure**	−ve Pressure or Suction**
Roof *Interior* Zone **1**	$(1.62)(1.0)(5.9) = 9.6$ (use 10 psf minimum)	$(1.62)(1.0)(-14.6) = -23.7$
Roof *End* Zone **2**	$(1.62)(1.0)(5.9) = 9.6$ (use 10 psf minimum)	$(1.62)(1.0)(-24.4) = -39.5$
Roof *Corner* Zone **3**	$(1.62)(1.0)(5.9) = 9.6$ (use 10 psf minimum)	$(1.62)(1.0)(-36.8) = -59.6$

* See Table 3-8 for P_{net30}.

** +ve Pressure indicates downward wind loads and −ve pressure indicates upward wind loads that cause uplift.

3.7 EFFECT OF NET FACTORED UPLIFT LOADS ON ROOF BEAMS AND JOISTS

The net factored uplift roof loads due to wind are normally calculated using, the C&C wind pressures. Note that the C&C roof pressures calculated for example 3-2 were obtained assuming the smallest effective wind area, A_e, of 10 ft.^{2}, which is conservative for most structural members, except for members with very small tributary areas. For illustration purposes, assume that the dead, snow, and roof live loads on the roof framing have already been determined previously as follows for a roof beam in Example 3-2:

Dead load, $D = 25$ psf

Snow load, $S = 35$ psf

Roof live load, $L_r = 20$ psf (actual value depends on the tributary area of the member under consideration)

The design vertical wind pressures on the roof can now be calculated. Using the wind pressures for C&C, and assuming an effective area, A_e, of at least 100 ft.^{2} (because most roof beams and girders typically will have at least this much tributary area), we use ASCE 7, Figure 6-3 to determine the design wind load as follows:

From equation (3-2), the effective area, A_e = Tributary area $\geq$ (Span of member)^2/3.

The design vertical wind pressure, W, psf = P_{net} (C&C) = $\lambda I_w P_{net30} \geq 10$ psf,

where

λ = Height/Exposure adjustment coefficient at the mean roof height,

I_w = Importance factor for wind (ASCE 7, Tables 1-1 and 6-1), and

P_{net30} = Tabulated wind pressure, psf (based on wind speed and effective wind area, A_e).

At a mean roof height of 60 ft. and exposure category C, the height/exposure adjustment factor, C_e =1.62. The design wind pressures on the roof are calculated in Table 3-14 as follows:

Table 3-14 Design vertical wind pressures on roof framing (C&C)

Zone	Design Vertical Pressures, $\lambda\,I_w\,P_{net30}$, psf	
	+ve Pressure*	−ve Pressure or Suction*
Roof *Interior* Zone **1**	(1.62)(1.0)(4.7)= 7.7 (use 10 psf minimum)	(1.62)(1.0)(−13.3) = −21.6
Roof *End* Zone **2**	(1.62)(1.0)(4.7)= 7.7 (use 10 psf minimum)	(1.62)(1.0)(−15.8) = −25.6
Roof *Corner* Zone **3**	(1.62)(1.0)(4.7)= 7.7 (use 10 psf minimum)	(1.62)(1.0)(−15.8) = −25.6

*Positive pressure indicates downward wind loads and negative pressure indicates upward wind loads that cause uplift. Assume an effective area, $A_e \geq$ 100 ft.^{2}. Note the difference in vertical wind pressures because of the A_e value used in this table, compared with an A_e of 10 ft.^{2} used in Table 3-13.

Using the largest wind pressure values (conservative, but it saves time!), we obtain the governing downward and upward wind loads from Table 3-14 as follows:

W = +10.0 psf (this +*ve* wind load will be used in load combinations 1 through 5)

W = −25.6 psf (this −*ve* wind load value will be used in load combinations 6 and 7)

Summary of Roof Loads

The following is a summary of the loads acting on the roof of the building:

Dead load, D = 25 psf

Snow load, S = 35 psf

Maximum roof live load, L_r = 20 psf

Wind load, W = +10.0 psf and −25.6 psf

The LRFD load combinations from Chapter 2 will now be used to determine the governing or controlling load case for this steel-framed roof; for illustrative purposes, we will assume that the roof beams have a span of 20 ft. and a tributary width of 5 ft.

LFRD Load Combinations

1: 1.4 (25 psf) = 35 psf
2: 1.2 (25 psf) + 1.6 (0) + 0.5 (35 psf) = 48 psf
3: 1.2 (25 psf) + 1.6 (35 psf) + 0.8 (10.0 psf) = **94 psf** (governs downward)
4: 1.2 (25 psf) + 1.6 (10.0 psf) + 0 + 0.5 (35 psf) = 63.5 psf
5: 1.2 (25 psf) + 0 + 0 + 0.2 (35 psf) = 37 psf
6: 0.9 (25 psf*) + 1.6 (−25.6 psf) = **−18.5 psf** (governs upward)

*This is the dead load assumed to be present when the wind acts on the building. The actual value may be less than the dead load used for calculating the maximum factored downward load in load combinations 1 through 5. This dead load value should be carefully determined so as not to create an unconservative design for uplift wind forces.

Load Effects in Roof Beams with Net Uplift Loads

The roof beams in the previous example will have to be designed for a downward-acting load of 94 psf and a net factored uplift load of −18.5 psf. The girders will have to be designed for the corresponding beam reactions. Note that this beam must be checked for both downward and uplift loads. Most roof beams have their top edges fully braced by roof decking or framing, but in designing beams for moments due to net uplift loads, the unbraced length of the compression edge of the beam (which is the bottom edge for uplift loads) will, in most cases, be equal to the full span of the beam. For steel beams, this could lead to a substantial reduction in strength that could make the uplift moments more critical than the moments caused by the factored downward loads.

Assuming that the roof beam span, L, is 20 ft. with a tributary width of 5 ft., the moments and reactions are calculated as follows:

Factored uniform downward load, w_u (downwards) = (94 psf)(5 ft.) = 470 lb./ft.

Maximum +ve moment, M_u^{+ve} = (470 lb./ft.)(20 ft.)^2/8 = 23,500 ft.-lb.
(unbraced length, L_u = 0 ft.)

Maximum downward load reaction, R_u^{+ve} = (470 lb./ft.)(20 ft.)/2 = 4700 lb.

Net factored uniform uplift load, w_u (upwards) = (−18.5 psf)(5 ft.) = −92.5 lb./ft.

Maximum −ve moment, M_u^{-ve} = (−92.5 lb./ft.)(20 ft.)^2/8 = −4625 ft.-lb.
(unbraced length, L_u = 20 ft.)

Maximum upward load reaction, R_u^{-ve} = (−92.5 lb./ft.)(20 ft.)/2 = −925 lb.

If open-web steel joists were used for the roof framing instead of the infill steel beams, a net uplift wind pressure could lead to the collapse of these joists if the uplift load is not adequately taken into account in the design of the joists. When subjected to a net uplift wind load, the bottom chord of the open-web steel joists and the long diagonal members, which are typically in tension under downward gravity loads (and, in many cases, may have slenderness ratios of between 200 and 300), will be in compression due to the uplift loads. Under this condition, these members may be inadequate to resist the resulting compression loads unless they have been designed for this load reversal. The combination of a light roof system and an inaccurately calculated wind uplift load led to the collapse of a commercial warehouse roof in the Dallas area in 2001 [7]. In this warehouse structure, the net uplift due to wind loads caused stress reversals in the roof joist end web and bottom chord members, which had only been designed to resist axial tension forces.

3.8 CALCULATION OF SEISMIC LOADS

Seismic loads are calculated differently for the primary system (i.e., the lateral force resisting system) than for parts and components such as architectural, mechanical, and electrical fixtures. In the United States, seismic load calculations are based on a 2% probability of exceeding the design earthquake in 50 years or the so-called 2500-yr. earthquake. The seismic load calculations for the primary system are covered in ASCE 7, Chapter 12, while the seismic load calculations for parts and components are covered in ASCE 7, Chapter 13.

The intent of the ASCE 7 seismic design provisions is to allow limited structural damage without collapse during an earthquake. The inelastic action of structures during a seismic event, resulting from cracking and yielding of the members, causes an increase in the damping ratio and in the fundamental period of vibration of the structure; thus, seismic forces are reduced due to inelastic action.

The seismic design category (SDC) determines the applicable seismic analysis procedure, the seismic detailing requirements, quality assurance plans, and height limitations for building structures. The SDC is a function of

- Building location,
- Building use and occupancy category (see ASCE 7, Tables 11.5-1, 11.6-1, and 11.6-2), and
- Soil type.

The six SDCs identified in ASCE 7 are given in Table 3-15, and the four occupancy categories and their corresponding seismic importance factors are tabulated in Table 3-16 (ASCE 7, Table 1-1). The step-by-step procedure for determining the seismic design category is given in Table 3-17.

Table 3-15 Seismic design categories

Seismic Design Category (SDC)	Application
A	• Applies to structures (regardless of use) in regions where ground motions are minor, even for very long periods.
B	• Applies to occupancy category I and II structures in regions where moderately destructive ground shaking is anticipated.
C	• Applies to occupancy category III structures in regions where moderately destructive ground shaking is anticipated. • Applies to occupancy category I and II, structures in regions where somewhat more severe ground shaking is anticipated.
D	• Applies to occupancy category I, II, and III structures in regions where destructive ground shaking is anticipated, BUT not located close to major active faults.
E	• Applies to occupancy category I and II structures in regions located close to major active faults.
F	• Applies to occupancy category III structures in regions located close to major active faults.

Table 3-16 Occupancy category and seismic importance factor

Occupancy Category	Type of Occupancy	Seismic Importance Factor, I_E
I	Buildings that represent a low hazard to human life in the event of failure (e.g., buildings that are not always occupied)	1.0
II	Standard occupancy buildings	1.0
III	Assembly buildings: Buildings that represent a substantial hazard to human life in the event of failure	1.25
IV	Essential and hazardous facilities e.g., police and fire stations, hospitals, aviation control towers, power-generating stations, water treatment plants, and national defense facilities	1.50

Table 3-17 Determination of the seismic design category

Step	Short-period ground motion, S_s	Long-period ground motion, S_1
1. Determine spectral response accelerations for the building location from ASCE 7, Figures 22-1 through 22-14, or from other sources.	At short (0.2-sec.) period, S_s (site class B), given as a fraction or percentage of g.	At long (1-sec.) period, S_1 (site class B), given as a fraction or percentage of g. Check if notes in step 8 are applicable.
2. Determine site class (usually specified by the geotechnical engineer) or ASCE 7, Chapter 20.		
• If site class is F	Do site-specific design.	Do site-specific design.
• If data available for shear wave velocity, standard penetration resistance (SPT), and undrained shear strength	Choose from site class A through E.	Choose from site class A through E.
• If no soil data available	Use site class D.	Use site class D.
3. Determine site coefficient for acceleration or velocity (percentage of g).	Determine F_a from ASCE 7, Table 11.4-1.	Determine F_v from ASCE 7, Table 11.4-2.
4. Determine soil-modified spectral response acceleration (percentage of g).	$S_{MS} = F_a S_s$ (ASCE 7, equation 11.4-1)	$S_{M1} = F_v S_1$ (ASCE 7, equation 11.4-2)
5. Calculate the design spectral response acceleration (percentage of g).	$S_{DS} = 2/3\ S_{MS}$ (ASCE 7, equation 11.4-3)	$S_{D1} = 2/3\ S_{M1}$ (ASCE 7, equation 11.4-4)
6. Determine occupancy category of the structure from ASCE 7, Table 1-1. (see Table 3-16)		
7. Determine seismic design category (SDC).	Use ASCE 7, Table 11.6-1.	Use ASCE 7, Table 11.6-2.
8. Select the most severe SDC (see ASCE 7, Section 11.6) from step 7.	Compare columns 2 and 3 from step 7 and select the more severe SDC value. In addition, note the following: • For occupancy categories I, II, or III (see ASCE 7, Table 1-1), with mapped $S_1 \geq 0.75g$, SDC = E. • For occupancy category IV, with mapped $S_1 \geq 0.75g$, SDC = F.	

It is recommended, whenever possible, to endeavor to be in SDC A, B, or C, but it should be noted that the SDC value for any building will depend largely on the soil conditions at the site and the structural properties of the building. Note that the site coefficients F_a and F_v increase as the soil becomes softer. If SDC D, E, or F is obtained, this will trigger special detailing requirements, and the reader should refer to ASCE 7 and the materials sections of the International Building Code (IBC) for several additional requirements.

3.9 SEISMIC ANALYSIS OF BUILDINGS USING ASCE 7

There are several methods available for seismic analysis in the ASCE 7 load specification. The appropriate method of analysis to be used will depend on the SDC obtained in Table 3-17. The permitted seismic analysis methods for SDC B through F is described in ASCE 7 Table 12.6-1. The seismic analysis procedure for structures in SDC A is described below.

Seismic Analysis Procedure for SDC A (ASCE 7, Sections 11.7.1 and 11.7.2)

The minimum lateral force procedure is permitted for buildings in SDC A. The lateral force is calculated as follows:

$$F_x = 0.01 \ W_x, \tag{3-5}$$

where

W_x = Portion of the total seismic dead load (see Section 3-10) tributary to or assigned to level x, and

F_x = Seismic lateral force at level x.

Simplified Analysis Procedure for Simple Bearing Wall or Building Frame Systems (ASCE 7, Section 12.14)

If all the conditions listed in ASCE 7, Section 12.14.1.1 are met, the simplified procedure described below can be used. Some of the more common conditions include the following:

Building must be in occupancy category I and II in site class A through D and not exceed three stories in height above grade. The seismic force resisting system shall be a bearing wall system or a building frame system (see ASCE 7, Table 12.14-1) and no irregularities are permitted.

For the simplified method, the seismic base shear and the lateral force at each level are given in equations (3-6) and (3-7), respectively.

Seismic Base Shear:

$$V = \frac{F \ S_{DS}}{R} \ W, \tag{3-6}$$

Seismic Lateral Force at Level x:

$$F_x = \left(\frac{w_x}{W}\right)\left(\frac{F \ S_{DS}}{R} W\right) = \frac{F \ S_{DS}}{R} w_x, \tag{3-7}$$

where

R = Structural system response modification factor (ASCE 7, Table 12.2-1),

W_x = Portion of the total seismic dead load (see Section 3-10) tributary to or assigned to level x, and

W = Total seismic dead load,

F = 1.0 for one-story buildings,

= 1.1 for two-story buildings, and

= 1.2 for three-story buildings.

3.10 EQUIVALENT LATERAL FORCE METHOD (ASCE 7, SECTION 12.8)

The equivalent lateral force method is the most widely used of all the seismic analysis methods. The equivalent lateral force method can be used for most structures in SDC A through F, except for certain structures with period greater than 3.5 seconds and structures with geometric irregularities. (See ASCE 7 Table 12.6-1.) Using this method, the factored seismic base shear (applied separately in each of the two orthogonal directions) is calculated as

$$V = C_s W, \qquad (3\text{-}8)$$

where

W is the total seismic dead load (including cladding loads) plus other loads listed below:

- 25% of floor live load for warehouses and structures used for storage of goods, wares, or merchandise. (public garages and open parking structures are excepted.)
- Partition load or 10 psf, whichever is greater. (Note: This only applies when an allowance for a partition load was included in the floor load calculations.)
- Total operating weight of permanent equipment. (*For practical purposes, use 50% of the mechanical room live load for schools and residential buildings, and 75% for mechanical rooms in industrial buildings.*)
- 20% of the flat roof or balanced snow load, if the flat roof snow load, P_f, exceeds 30 psf. *A higher snow load results in a tendency for the bottom part of the accumulated snow to adhere to the structure and thus contribute to the seismic load* [1].

$$W = W_{2nd} + W_{3rd} + W_{4th} + \ldots + W_{roof} = \sum_{x\,=\,2nd}^{roof} W_x \qquad (3\text{-}9)$$

where

W_x = Seismic dead load tributary to level x.

The seismic coefficient is given as

$$C_s = S_{DS}/(R/I_E) \qquad (3\text{-}10)$$

$$\leq S_{D1}/(TR/I_E) \text{ for } T \leq T_L$$

$$\leq S_{D1} T_L/(T^2 R/I_E) \text{ for } T > T_L$$

$$\geq 0.5 S_1/(R/I_E) \text{ for buildings and structures with } S_1 \geq 0.6g,$$

$$\geq 0.01$$

where

I_E = Importance factor from ASCE 7, Table 1-1

S_{DS} = Short-period design spectral response acceleration,

S_{D1} = 1-sec. design spectral response acceleration,

S_1 = Mapped 1-sec. spectral acceleration,

T = Period of vibration of the structure,

T_L = Long-period transition period (see ASCE 7, Section 11.4.5 and Figures 22-15 through 22-20, [1]), and

R = Structural system response modification factor (see ASCE 7, Table 12.2-1).

The structural system response modification factor, R is a measure of the ductility of the seismic lateral force resisting system; it accounts for the inelastic behavior of the lateral force resisting system. An R-value of 1.0 corresponds to a purely elastic structure, but it would be highly uneconomical to design structures assuming elastic behavior. The use of higher R-values that would reduce the base shear values to well below the elastic values is borne out of experience from previous earthquakes in which buildings have been known to resist earthquake forces significantly higher than the design base shear values because of inelastic behavior and inherent redundancy in the building structure.

The seven basic structural systems prescribed in ASCE 7, Table 12.2-1 are

1. **Bearing Wall Systems:** Lateral force resisting system (LFRS) that supports both gravity and lateral loads. Therefore, R-values are smaller because the performance is not as good as that of the building frame system because it supports dual loading (e.g., concrete or masonry shear walls that support both gravity and lateral loads).

2. **Building Frame Systems:** LFRS that supports only lateral load. It has better structural performance because it supports a single load; therefore, the R-values are higher than those for bearing wall systems. Thus, building frame systems are more economical than bearing wall systems. Examples of building frame systems include braced frames, masonry shear walls in steel-framed buildings, and steel plate shear walls.

   It should be noted that concrete or masonry shear walls that do not support gravity loads may also be classified as building frame systems, and it is not necessary that the walls be isolated from the building frame. If a shear wall that is built integral with a building frame has confined columns at the ends of the wall and/or confined columns within the wall length, and has a beam or girder immediately above and in the plane of the wall spanning between these columns to support the gravity loads, such a shear wall can be classified as a building frame system; the reason is that if the shear wall panel itself were removed, the gravity loads can still be supported by the end columns and/or the confined columns, and the beam or girder spanning between these columns [5].

3. **Moment Resisting Frame Systems**

4. **Dual Systems:** Combined shear wall and moment resisting frame.

5. **Shear Wall–Frame Interactive System:** Ordinary reinforced concrete moment frames and ordinary reinforced concrete shear walls.

6. **Inverted Pendulum and Cantilevered Column Systems:** Structures where a large proportion of their total weight is concentrated at the top of the structure (e.g., water storage towers).

7. ***Steel Systems not specifically detailed for seismic resistance (R = 3),*** excluding cantilevered column systems: This system is only applicable for SDC A, B, or C. Using this system, which is frequently chosen by design professionals in practice, obviates the need for special steel detailing requirements. Using R-values greater than 3 triggers the special detailing requirements in the AISC Seismic Detailing Provisions for Steel Buildings (AISC 341-05) [8]. The authors recommend that an R-value of 3 or less be used in design whenever possible to avoid the increased costs associated with the seismic detailing requirements that are triggered when R exceeds 3.

In addition to the R-values provided in ASCE 7, Table 12.2-1, the following parameters are also provided in the table:

- Deflection amplification factor, C_d
- Overstrength factor, Ω_o,
- Structural system limitations and building height limits as a function of the SDC, and
- ASCE 7 sections where material-specific design and detailing requirements are specified.

A plot of the IBC design response spectra that gives the maximum acceleration versus period, T, for various buildings is shown in Figure 3-19.

Fundamental Period, *T* (all types of buildings)

The most commonly used equation for calculating the approximate fundamental period for all types of buildings is given as

$$T = Approximate \text{ fundamental period of building} = T_a = C_t(h_n^x), \qquad (3\text{-}11)$$

where

C_t and x are obtained from Table 3-18 (ASCE 7, Table 12.8-2), and
h_n = Height (in feet) from the base to the highest level (i.e., roof) of building.

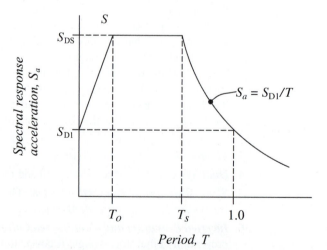

Figure 3-19 Design response spectra. Adopted from Ref. [6]

Table 3-18 C_t values for various structural systems

Structural System	C_t	x
Steel Moment Resisting Frames	0.028	0.8
Eccentrically Braced Frames (EBF)	0.03	0.75
All other structural systems	0.02	0.75

Fundamental Period, *T* (moment frames only)

An alternate equation for the fundamental period that is applicable only to concrete or steel moment frames, and is valid only for structures not exceeding 12 stories with a story height of at least 10 ft., is given as

$$T = T_a = 0.1\,N, \tag{3-12}$$

where N is the number of stories in the building.

The higher the fundamental period, T_a, of the structure, the smaller the seismic force on the structure is. To minimize the effect of the error in calculating the fundamental period, the ASCE 7 load specification sets an upper limit on the period using a factor C_u. If a dynamic structural analysis, including structural properties and deformational characteristics, is used to determine the fundamental period of lateral vibration of the structure, the calculated period is limited to a maximum value determined as follows:

$$T_{\max} = C_u T_a, \tag{3-13}$$

where

C_u = Factor that depends on S_{D1} and is obtained from ASCE 7, Table 12.8-1, and

T_a = Approximate period of vibration as determined previously.

Note that equation (3-11) is more commonly used in practice, and equation (3-13) is used only when the natural period of the structure is determined from a dynamic structural analysis.

3.11 VERTICAL DISTRIBUTION OF SEISMIC BASE SHEAR, *V*

Since most structures are multiple degrees-of-freedom systems with several modes of vibration, the distribution of the seismic lateral force is a combination of the contributions from all the significant modes of vibration of the structure.

The force distribution to each level is a function of the seismic weight, W_x, at that level, the height or stiffness of the structure, and the predominant mode of vibration. The exponent k in equation 3-15 is an attempt to capture the contributions from the higher modes of vibration.

Table 3-19 k Values

Building Period, T, in seconds	k
≤ 0.5	1 (*no whiplash effect*)
$0.5 < T < 2.5$	$1 + 0.5\,(T - 0.5)$
≥ 2.5	2

The factored seismic lateral force at any level of the vertical LFRS is given as

$$F_x = C_{vx}\,V, \tag{3-14}$$

where

$$C_{vx} = \frac{W_x h_x^k}{\displaystyle\sum_{i=1}^{n} W_i h_i^k}, \tag{3-15}$$

W_x = Portion of the total gravity load of the building, W, that is tributary to level x (includes weight of floor or roof, plus weight of perimeter or interior walls tributary to that level),

W_i = Portion of the total gravity load of the building, W, that is a tributary to level i (includes weight of floor or roof, plus weight of perimeter or interior walls tributary to that level),

h_i and h_x = Height (in feet) from the base to level i or x,

k = Exponent related to the building period (refer to Table 3-19),

Note: The height of a vertical wall that is tributary to a particular level, x, is the distance from a point midway between level x and level $x + 1$ to a point midway between level x and level $x - 1$.

Level i = Any level in the building ($i = 1$ for first level above the base),

Level x = That level which is under design consideration, and

Level n = Uppermost level of the building.

3.12 STRUCTURAL DETAILING REQUIREMENTS

After the seismic forces on a structure have been determined and the lateral load resisting systems have been designed for these forces, the structure must also be detailed to conform to the structural system requirements that are required by the seismic design category (SDC). Buildings in SDC A, B, or C do not generally require stringent detailing requirements. However, buildings in SDC D, E, or F require stringent detailing requirements for the seismic force resisting system and other components of the building. The reader should refer to ASCE 7, Chapter 14 for the specified design and detailing requirements as a function of the SDC and the construction material [1].

EXAMPLE 3-3

Seismic Lateral Forces in a Two-story Building

Calculate the seismic forces at each level for the office building shown in Figure 3-20. The seismic accelerations are $S_S = 0.31$ and $S_1 = 0.10$. The dead load on the second floor is 80 psf and the dead load on the roof is 30 psf; ignore the weight of the cladding. The flat roof snow load, P_f, is 42 psf. The lateral force resisting system is a structural steel system not specifically detailed for seismic resistance. Soil conditions are unknown. Use the minimum lateral force procedure and the simplified procedure.

Calculate the lateral force on each X-braced frame assuming there is one X-brace on each of the four exterior walls.

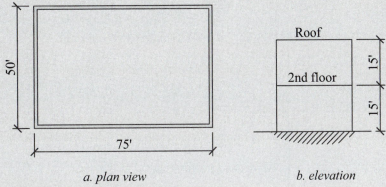

a. plan view   b. elevation

Figure 3-20 Building plan for example 3-3.

SOLUTION

1. Determine the seismic design category (SDC) using Table 3-17.
    The SDC is not applicable to the simplified or minimum lateral force calculation methods.

2. Determine the method of seismic analysis to be used from ASCE 7, Table 12.6-1.
    The minimum lateral force procedure and the simplified procedure are specified for this example.

3. Calculate the dead load at each level, W_i, and the total dead load, W (see Table 3-20).
    The flat roof snow load for this building is $P_f = 42$ psf > 30 psf; therefore, 20% of the snow load must be included in the calculation of the seismic dead load, W.

Table 3-20 Assigned seismic weights at each level of building

Level	Height from Base, h_i	Weight, W_i
Roof	30 ft.	$W_{roof} = (30 \text{ psf})(75 \text{ ft.})(50 \text{ ft.}) +$ $(0.20)(42 \text{ psf})(75 \text{ ft.})(50 \text{ ft.}) = $ **144 kip**
Second	15 ft.	$W_2 = (80 \text{ psf})(75 \text{ ft.})(50 \text{ ft.}) = $ **300 kip**

Note: $W = \Sigma W_i = 444$ kip

(continued)

4. Determine the seismic coefficient, C_s.

For the minimum lateral force method, the seismic response coefficient is essentially 0.01.

For the simplified method, the coefficient is obtained as follows:
Soil site class = D (since site conditions are unknown)

$$S_S = 0.31$$
$$S_1 = 0.10$$
$$R = 3 \text{ (ASCE 7, Table 12.2–1; system not specifically detailed}$$
$$\text{for seismic resistance)}$$
$$F_a = 1.4 \text{ (ASCE 7, Table 11.4.1)}$$
$$S_{MS} = F_a S_S = (1.4)(0.31) = 0.434$$
$$S_{DS} = 2/3 \, S_{MS} = (2/3)(0.434) = 0.289$$
$$F = 1.1 \text{ (two-story building)}$$

The seismic coefficient is calculated from equation (3-6) as

$$C_s = \frac{FS_{DS}}{R} = \frac{(1.1)(0.289)}{3.0} = 0.106.$$

5. Calculate the seismic base shear, V.

Minimum Lateral Force:

$$V = 0.01W = (0.01)(444) = 4.44 \text{ kips}$$
$$F_R = 0.01W_R = (0.01)(144) = 1.44 \text{ kips}$$
$$F_2 = 0.01W_2 = (0.01)(300) = 3.0 \text{ kips}$$

Simplified Procedure:

$$V = \frac{FS_{DS}W}{R} = \frac{(1.1)(0.289)(444)}{3.0} = 47.1 \text{ kips}$$

$$F_R = \frac{FS_{DS}W_R}{R} = \frac{(1.1)(0.289)(144)}{3.0} = 15.3 \text{ kips}$$

$$F_2 = \frac{FS_{DS}W_2}{R} = \frac{(1.1)(0.289)(300)}{3.0} = 31.8 \text{ kips}$$

Total lateral force on each X-brace at each level is calculated as follows:

$$F_{roof} = (1.44/2) = 0.72 \text{ kip (minimum lateral force)};$$
$$15.3/2 = 7.65 \text{ kips (simplified)}$$
$$F_{2nd} = (3.0/2) = 1.5 \text{ kips (minimum lateral force)};$$
$$31.8/2 = 15.9 \text{ kips (simplified)}$$

The lateral forces are shown in Figure 3-21.

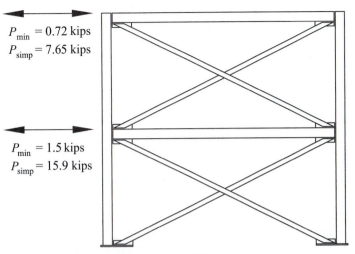

$P_{min} = 0.72$ kips
$P_{simp} = 7.65$ kips

$P_{min} = 1.5$ kips
$P_{simp} = 15.9$ kips

Figure 3-21 Seismic loads on each X-brace.

EXAMPLE 3-4

Seismic Lateral Forces in a Multistory Building

The typical floor plan and elevation of a six-story office building measuring 100 ft. by 100 ft. in plan and laterally braced with ordinary moment resisting frames is shown in Figure 3-22. Determine the seismic forces on the build-

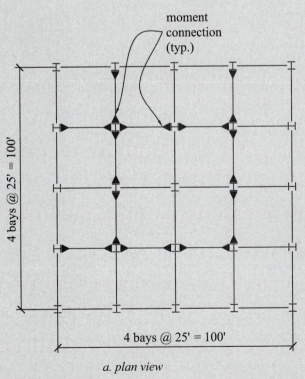

moment
connection
(typ.)

4 bays @ 25' = 100'

4 bays @ 25' = 100'

a. plan view

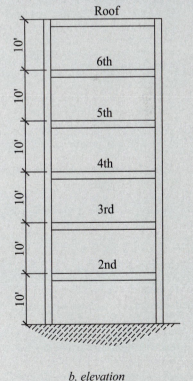

Roof

6th

10'

5th

10'

4th

10'

3rd

10'

2nd

10'

10'

b. elevation

Figure 3-22 Building plan and elevation for example 3-4.

(continued)

ing and on each moment frame assuming the following design parameters:

$$\text{Roof dead load} = 25 \text{ psf}$$
$$\text{Snow load} = 31.5 \text{ psf}$$
$$\text{Floor dead load} = 75 \text{ psf (includes partition load)}$$
$$\text{Cladding (glazing)} = 20 \text{ psf}$$
$$S_S = 0.25g$$
$$S_1 = 0.072g$$

SOLUTION

1. Determine the seismic design category (SDC) (see Table 3-21.)

Table 3-21 Determining the seismic design category for Example 3-4

Step	SHORT-PERIOD ground motion, S_s	LONG-PERIOD ground motion, S_1
1. Determine spectral response accelerations for the building location from ASCE 7, Figures 22-1 through 22-14 (2% probability of exceedance (PE) in 50 years).	$S_s = 0.25g$ Use the fraction of g (i.e., 0.25) in the calculations.	$S_1 = 0.072g < 0.75g$ Use the fraction of g (i.e., 0.072) in the calculations.
2. Determine site class (usually specified by the geotechnical engineer.		
• If site class is F	Do site-specific design.	Do site-specific design.
• If data available for shear wave velocity, standard penetration resistance (SPT), and undrained shear strength	Choose from site class A through E.	Choose from site class A through E.
• If no soil data available	**Use site class D.**	**Use site class D.**
3. Determine site coefficient for acceleration or velocity (percentage of g).	$F_a = 1.6$ (ASCE 7, Table 11.4-1)	$F_v = 2.4$ (ASCE 7, Table 11.4-2)
4. Determine soil-modified spectral response acceleration (percentage of g).	$S_{MS} = F_a S_s = (1.6)(0.25) = 0.40$ (ASCE 7, equation 11.4-1)	$S_{M1} = F_v S_1 = (2.4)(0.072) = 0.17$ (ASCE 7, equation 11.4-2)
5. Calculate the design spectral response acceleration (percentage of g).	$S_{DS} = \frac{2}{3} S_{MS} = (\frac{2}{3})(0.405) = \mathbf{0.27}$ (ASCE 7, equation 11.4-3)	$S_{D1} = \frac{2}{3} S_{M1} = (\frac{2}{3})(0.17) = \mathbf{0.12}$ (ASCE 7, equation 11.4-4)
6. Determine occupancy category of the structure from ASCE 7, Table 1-1.	Standard occupancy building $\Rightarrow I_E = 1.0$	Standard occupancy building $\Rightarrow I_E = 1.0$
7. Determine seismic design category (SDC).	**SDC = B** (ASCE 7, Table 11.6-1)	**SDC = B** (ASCE 7, Table 11.6-2)
8. Choose most severe SDC (i.e., the higher SDC value).	Compare the second and third columns from the previous step $\Rightarrow$ **USE SDC = B**	

2. Determine the method of seismic analysis to be used.

   From Table 3-21, the SDC is found to be B. Therefore, from ASCE 7, Table 12.6-1, we find that the equivalent lateral force method is one of the permitted methods of seismic analysis for this building.

3. Calculate the dead load at each level, W_i, and the total dead load, W (see Table 3-22).

   The flat roof snow load given for this building is $P_f = 31.5$ psf > 30 psf. Therefore, 20% of the snow load must be included in the seismic dead load calculations.

Table 3-22 Assigned seismic weights at each level of building

Level	Height from Base, h_i	Weight, W_i
Roof	60 ft.	$W_{roof} = (25$ psf$)(100$ ft.$)(100$ ft.$) + (20\%)(31.5$ psf$)(100$ ft.$)(100$ ft.$) +$ $(20$ psf$)(2)(100$ ft. $+ 100$ ft.$)(10$ ft.$/2) = $ **353 kips**
Sixth Floor	50 ft.	$W_6 = (75$ psf$)(100$ ft.$)(100$ ft.$) + (20$ psf$)(2)(100$ ft. $+ 100$ ft.$)[(10$ ft. $+ 10$ ft.$)/2]$ $= $ **830 kips**
Fifth Floor	40 ft.	$W_5 = (75$ psf$)(100$ ft.$)(100$ ft.$) + (20$ psf$)(2)(100$ ft. $+ 100$ ft.$)[(10$ ft. $+ 10$ ft.$)/2]$ $= $ **830 kips**
Fourth Floor	30 ft.	$W_4 = (75$ psf$)(100$ ft.$)(100$ ft.$)+(20$ psf$)(2)(100$ ft. $+ 100$ ft.$)[(10$ ft. $+ 10$ ft.$)/2]$ $= $ **830 kips**
Third Floor	20 ft.	$W_3 = (75$ psf$)(100$ ft.$)(100$ ft.$) + (20$ psf$)(2)(100$ ft. $+ 100$ ft.$)[(10$ ft. $+ 10$ ft.$)/2]$ $= $ **830 kips**
Second Floor	10 ft.	$W_2 = (75$ psf$)(100$ ft.$)(100$ ft.$) + (20$ psf$)(2)(100$ ft. $+ 100$ ft.$)[(10$ ft. $+ 10$ ft.$)/2]$ $= $ **830 kips**

Note: $W = \Sigma W_i = 4503$ kips.

4. Determine the seismic coefficient, C_s.

   For this building, the lateral loads are resisted solely by the moment frames. Therefore, from ASCE 7, Table 12.2-1, we could select ordinary steel moment frames (system C-4) for which we obtain the following parameters: $R = 3.5$, $C_d = 3$, and $\Omega_o = 3$.

   The equivalent lateral force method is permitted in SDC A through C.

   Note that in order to use system C-4, certain detailing requirements must be met, and the AISC 341 specification for seismic detailing of steel buildings [8] must be used (see ASCE 7, Sections 12.2.5 and 14.1).

   However, if a "steel system not specifically detailed for seismic resistance" or system H is used (*whenever possible and for economic reasons, this is a highly recommended system for steel buildings.*) the less stringent AISC 360 specification [9] for structural steel buildings is allowed to be used. From a cost point of view for this building, we will adopt system H, and the following parameters are obtained from ASCE 7, Table 12.2-1: $R = 3$, $C_d = 3$, and $\Omega_o = 3$. (*Note that system H is only permitted in SDC A, B, or C*).

   The equivalent lateral force method is permitted for this building with SDC B.

   $C_t = 0.028$ and $x = 0.8$ (from Table 3-18 for steel moment resisting frames)

   $h_n = $ Roof height $= 60$ ft.

   (continued)

$T = T_a$ = Approximate period of the building = $C_t \, (h_n^{0.8})$
$= 0.028 \times (60^{0.8}) = 0.74 < T_L = 6$ s (see ASCE 7, Figure 22-15)

C_s = Seismic response coefficient = $S_{DS}/[R/I_E] = 0.27/(3/1.0) = 0.09$
$\leq S_{D1}/[TR/I_E] = 0.12/(0.74)(3)(1.0) = 0.054$
$\Rightarrow C_s = 0.054 > 0.01$

5. Calculate the seismic base shear, V.

$V = C_s W = (0.054)(4503 \text{ kips}) = 243 \text{ kips}$

(This is the seismic force in both the N–S and E–W directions.)

6. Determine the vertical distribution of the seismic base shear (i.e., determine F_x at each level).

For $T = 0.74$ sec., and from Table 3-19, k = $1 + 0.5(0.74 - 0.5) = 1.12$

The seismic lateral forces at each level of the building are calculated in Table 3-23.

Table 3-23 Seismic lateral force

Level	Height from Base, h_i	Dead Load at each Level, W_i	$W_i(h_i)^k$	$C_{vx} = W_i(h_i)^k/\Sigma W_i(h_i)^k$	$F_x = C_{vx}V$
Roof	60 ft.	353 kips	34618	0.154	37 kips
Sixth floor	50 ft.	830 kips	66363	0.295	72 kips
Fifth floor	40 ft.	830 kips	51687	0.230	56 kips
Fourth floor	30 ft.	830 kips	37450	0.167	41 kips
Third floor	20 ft.	830 kips	23781	0.106	26 kips
Second floor	10 ft.	830 kips	10942	0.049	12 kips
			$\Sigma W_i(h_i)^k = 224{,}841$	$\Sigma F_i = 244$ kips ≈ 243 kips	

- The F_x forces calculated in Table 3-23 are the factored seismic forces acting at each level of the building in both the N–S and the E–W directions.
- If the building has *rigid diaphragms* (as most steel buildings do), the F_x forces will be distributed to the LFRS in each direction in proportion to the relative stiffness of the moment frames.
- If the building has a *flexible diaphragms,* the forces are distributed in proportion to the plan area of the building tributary to each LFRS.

3.13   REFERENCES

1. American Society of Civil Engineers. 2005. ASCE 7, *Minimum Design Loads for Buildings and Other Structures.* Reston, VA.

2. Green, Norman B. 1981. *Earthquake Resistant Building Design and Construction.* New York: Van Nostrand Reinhold.

3. McNamara, Robert J. 2005. Some current trends in high rise structural design. *Structure* (September): 19–23.

4. Gamble, Scott. 2003. Wind tunnel testing, a breeze through. *Structure* (November).

5. Ghosh, S. K., and Susan Dowty. 2007. Code simple. *Structural Engineer* (February): 18.

6. International Codes Council. 2003. *International Building Code—2003*. Falls Church, VA.

7. Nelson, Erik L., D. Ahuja, Stewart M. Verhulst, and Erin Criste. 2007. The source of the problem. *Civil Engineering* (January): 50–55.

8. AISC. 2006. Seismic Design Manual. ANSI/AISC 341-05 and ANSI-AISC 358-05, American Institute of Steel Construction, Chicago.

9. American Institute of Steel Construction. 2006. *Steel Construction Manual,* 13th ed. Chicago.

3.14   PROBLEMS

3-1.   For a two-story office building 140 ft. by 140 ft. in plan and with a floor-to-floor height of 13 ft. located in your city, calculate the following wind loads assuming an X-brace is located on each exterior wall:

   a.   Average horizontal wind pressure in the transverse and longitudinal directions.

   b.   Total wind base shear in the transverse and longitudinal directions.

   c.   Force to each X-brace frame in the transverse and longitudinal directions.

Assume the building is enclosed and exposure category D.

3-2.   For the building in Problem 3-1, calculate the following seismic loads:

   a.   Seismic base shear and force at each level, assuming the minimum lateral force procedure.

   b.   Seismic base shear and force at each level, assuming the simplified procedure.

Assume a roof dead load of 25 psf and a flat roof snow load of 35 psf. Include the weight of the cladding around the perimeter of the building in the weight of the roof and floor levels. Use $S_{DS} = 0.27$, $S_{D1} = 0.12$, and $R = 3.0$.

3-3.   A five-story office building, 80 ft. by 80 ft. in plan, with a floor-to-floor height of 12 ft. and an essentially flat roof, is laterally supported by 10-ft-long shear walls on each of the four faces of the building. The building is located in New York City (assume a 120-mph basic wind speed, exposure category C, and a category I building).

   a.   For the MWFRS, determine the unfactored wind loads at each floor, the base shear, and the overturning moments at the base of the building using ASCE 7, method 1 (simplified method).

   b.   Assuming two shear walls in each direction, determine the unfactored wind lateral force at each floor level for each shear wall, the base shear, and the overturning moment.

   c.   Repeat part b using factored wind loads (i.e., using the load factor for wind).

3-4.   Find the following for the building described in problem 3-3. Assume that the rigid diaphragms and the ordinary reinforced concrete shear walls support gravity, as well as lateral, loads.

   a.   Determine the factored seismic lateral force at each level of the building in the N–S and E–W directions. (Neglect torsion.)

   b.   Calculate the factored seismic force at each level for a typical shear wall in the N–S and E–W directions.

   c.   Calculate the factored total seismic base shear for a typical shear wall in the N–S and E–W directions.

   d.   Calculate the factored seismic overturning moment at the base of a typical shear wall in the N–S and E–W directions.

e. If instead of having shear walls on each of the four faces of the building (i.e., two shear walls in each direction), the building has five ordinary concentric steel X-brace frames in a building frame system (located 20 ft. apart) in both the N–S and E–W directions, recalculate the factored seismic force at each level of a typical interior X-brace frame, the factored base shear, and the overturning moment. Assume that the structural steel system is not specifically detailed for seismic resistance.

f. Recalculate the forces and moments in problem e, assuming that the building has flexible diaphragms.

Assume the following design parameters:

- Average dead load for each floor is 150 psf.
- Average dead load for roof is 30 psf; the balanced roof snow load, P_f, is 35 psf; and the ground snow load, P_g, is 50 psf.
- Average weight of perimeter cladding is 60 psf of vertical plane.
- Building is a Non-essential facility.
- Floor and roof diaphragms are rigid (parts a-e).
- Shear walls are bearing wall systems with ordinary reinforced concrete shear walls.
- Short-term spectral acceleration, $S_S = 0.25g$.
- 1-sec. spectral acceleration, $S_1 = 0.07g$.
- No geotechnical report is available.
- Neglect torsion.

3-5. The roof of a one-story, 100-ft. by 120-ft. warehouse, with a story height of 20 ft. is framed with open-web steel joists and girders as shown in Figure 3-23. Assuming a roof dead load of 15 psf, determine the net factored wind uplift load on a typical interior joist. The building is located in Dallas, Texas.

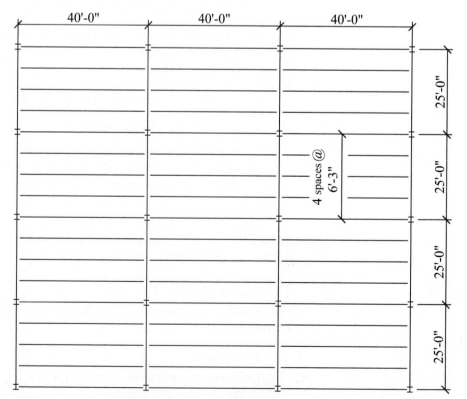

Figure 3-23 Warehouse roof framing plan for problem 3-5.

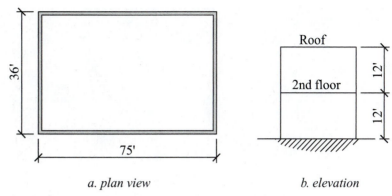

a. plan view   *b. elevation*

Figure 3-24 Building plan and elevation for problem 3-6.

3-6. A two-story steel structure, 36 ft. by 75 ft. in plan, is shown below (see Figure 3-24) with the following given information. The floor-to-floor height is 13 ft., and the building is enclosed and located in Rochester, New York, on a site with a category C exposure. Assuming the following additional design parameters, calculate the following:

Floor dead load $=$ 100 psf

Roof dead load $=$ 30 psf

Exterior walls $=$ 10 psf

Snow load, P_f $=$ 40 psf

Site class $=$ D

Importance, I_e $=$ 1.0

S_S $=$ 0.25%

S_1 $=$ 0.07%

R $=$ 3.0

a. The total horizontal wind force on the MWFRS in both the transverse and longitudinal directions.

b. The gross vertical wind uplift pressures and the net vertical wind uplift pressures on the roof (MWFRS) in both the transverse and longitudinal directions.

c. The seismic base shear, V, in kips.

d. The lateral seismic load at each level in kips.

3-7. **Student Design Project Problem**

Calculate the wind loads and the seismic loads for the design project introduced in Chapter 1. Determine the design lateral forces at each level of the building for the two LFRS options given in the design project brief in Chapter 1.

Tension Members

4.1 INTRODUCTION

Tension members are axially loaded members stressed in tension and are used in steel structures in various forms. They are used in trusses as web and chord members, hanger and sag rods, diagonal bracing for lateral stability, and lap splices such as in a moment connection (see Figure 4-1 for examples of tension members).

Beams and columns are subjected to compression buckling (such as lateral-torsional Buckling, Euler Buckling, and Local Buckling) and must be checked for this failure mode, but tension members are not subjected to the same lateral instability since compression stresses do not exist. The exception to this is the special case when the applied tension load is eccentric to the member in question, inducing an applied moment and therefore creating the possibility of lateral instability.

The basic design check for a tension member is to provide enough cross-sectional area to resist the applied tensile force. In practice, however, pure tension members do not typically exist in this form and several additional factors must be considered. One common example is tension members with nonuniform cross sections, such as the case when a tension member is connected with bolts at the ends. Eccentric loading must also be considered, such as a single steel angle with a concentric load connected to a gusset plate. Even though slenderness is not a direct design concern, the AISC specification does recommend an upper limit on the slenderness ratio L/r for tension members. This upper L/r limit is equal to 300 for tension members and 200 for compression members, where L is the length of the member and r is the radius of gyration. The recommendation does not apply to rods or hangers in tension and is not absolutely required for tension members.

4.2 ANALYSIS OF TENSION MEMBERS

For members subjected to tension, the two basic modes of failure are tensile yielding and tensile rupture. Tensile yielding occurs when the stress on the *gross area* of the section is large enough to cause excessive deformation. Tensile rupture occurs when the stress on the

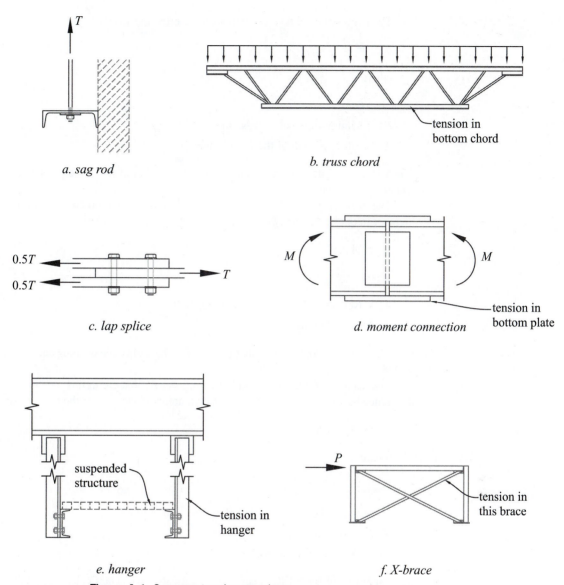

a. sag rod

b. truss chord

tension in bottom chord

c. lap splice

d. moment connection

tension in bottom plate

e. hanger

f. X-brace

Figure 4-1 Common tension members.

effective area of the section is large enough to cause the member to fracture, which usually occurs across a line of bolts where the tension member is weakest.

The expression for tensile yielding on the gross area is

$$\phi P_n = \phi F_y A_g, \tag{4-1}$$

where

$\phi = 0.90,$

F_y = Minimum yield stress, and

A_g = Gross area of the tension member.

The expression for tensile rupture on the effective area is

$$\phi P_n = \phi F_u A_e, \tag{4-2}$$

where

$\phi = 0.75,$

$F_u = $ Minimum tensile stress, and

$A_e = $ Effective area of the tension member.

The design strength of a tension member is the smaller of the two expressions indicated in equations (4-1) and (4-2).

The *gross area*, A_g, of a tension member is simply the total cross-sectional area of the member in question. The *effective area*, A_e, of a tension member is described as follows:

$$A_e = A_n U, \tag{4-3}$$

where

$A_n = $ Net area of the tension member, and

$U = $ Shear lag factor.

Note that for a tension member that is connected by welds, the net area equals the gross area (i.e., $A_n = A_g$).

The net area of a tension member with fasteners that are in line (see Figure 4-2) is the difference between the gross cross-sectional area and the area of the bolt holes:

$$A_n = A_g - A_{\text{holes}} \tag{4-4}$$

where

$A_{\text{holes}} = n(d_b + \frac{1}{8})t$

$n = $ number of bolt holes along the failure plane,

$d_b = $ bolt diameter,

$t = $ material tickness.

Section B3.13 of the AISC specification indicates that when calculating the net area for shear and tension, an additional $\frac{1}{16}$ in. should be added to the hole size to account for the roughened edges that result from the punching or drilling process. For standard holes, the

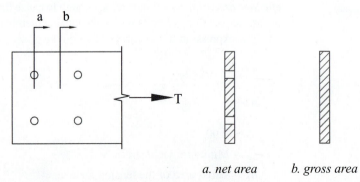

a. net area   *b. gross area*

Figure 4-2 Tension member with in-line fasteners.

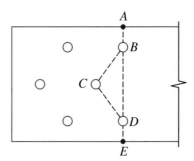

Figure 4-3 Tension member with diagonal fasteners.

hole size used for strength calculations would be the value from the *AISCM*, Table J3.3, which is the nominal hole dimension plus $\frac{1}{16}$ in. Since the nominal hole size is $\frac{1}{16}$ in. larger than the fastener for standard (STD) holes, the actual hole size used in the design calculations will be $\frac{1}{16}$ in. + $\frac{1}{16}$ in. = $\frac{1}{8}$ in. for most bolted connections in tension.

For tension members with a series of holes in a diagonal or zigzag pattern, which might be used when bolt spacing is limited there may exist several possible planes of failure that need to be investigated. When the failure plane crosses straight through a line of bolts (line *ABCE* in Figure 4-3), then the net area is as noted in equation (4-4). For a failure plane where one or more of the failure planes is at an angle (line *ABCE* in Figure 4-3), then the following term is added to the net width of the member for each diagonal portion that is present along the failure plane:

$$\frac{s^2}{4g},\tag{4-5}$$

where

s = Longitudinal center-to-center spacing or pitch between two consecutive holes, and

g = Transverse center-to-center spacing or gage between two consecutive holes.

This modification accounts for the increase in strength due to the added cross-sectional area at an angle in the failure plane. In Figure 4-3, note that failure plane *ABCE* has two diagonal failure planes: *BC* and *CD*. The expression for the net width then becomes

$$w_n = w_g - \sum d_h + \sum \frac{s^2}{4g},\tag{4-6}$$

where

w_n = Net width,

w_g = Gross width, and

d_h = Hole diameter.

Multiplying Eq. (4-6) by the thickness of the member yields

$$w_n t = w_g t - \sum d_h t + \sum \frac{s^2}{4g} t.\tag{4-7}$$

Since $A_n = w_n t$ and $A_g = w_g t$, equation (4-7) can be simplified as follows:

$$A_n = A_g - \sum d_h t + \sum \frac{s^2}{4g} t\tag{4-8}$$

shaded area not
directly connected;
has lower stress

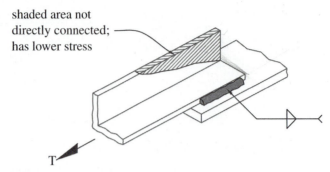

T

Figure 4-4 Shear lag effect.

The shear lag factor (U in equation (4-3)) accounts for the nonuniform stress distribution when some of the elements of a tension member are not directly connected, such as a single angle or WT member (see Figure 4-4).

Table D3.1 of the *AISCM* gives the value for the shear lag factor, U, for various connection configurations. With the exception of plates and round hollow structural sections (HSS) members with a single concentric gusset plate and longitudinal welds, the shear lag factor is

$$U = 1 - \frac{\bar{x}}{\ell},$$ (4-9)

where

$\bar{x}$ = Distance from the centroid of the connected part to the connection plane, and

ℓ = Connection length.

The variables $\bar{x}$ and ℓ are illustrated in Figure 4-5.

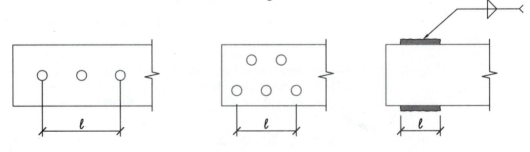

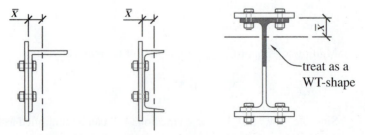

Figure 4-5 Determination of $\bar{x}$ and ℓ.

The shear lag factor for plates and round HSS members with a single concentric gusset plate are included in Table 4-1. Also shown in Table 4-1 are alternate values of U for single angles and W, M, S, and HP shapes that may be used in lieu of equation (4-9). Note that the calculated value of U should be greater than 0.60 for all cases unless eccentricity effects are accounted for (see Sections H1.2 or H2 of the AISC specification).

Table 4-1 Shear lag factor for common tension member connections

Tension Member Type	Description	Shear Lag Factor, U	Example
Plate	All bolted*	$U = 1.0$	
	All welded	$U = 1.0$	
	Transverse weld	$U = 1.0$	
	Longitudinal welds	$\ell \geq 2w, U = 1.0$ $1.5w \leq \ell < 2w, U = 0.87$ $w \leq \ell < 1.5w, U = 0.75$	
Round HSS	Single concentric gusset plate	$\ell \geq 1.3D, U = 1.0$ $D \leq \ell < 1.3D, U = 1 - \dfrac{\bar{x}}{\ell}$ $\bar{x} = \dfrac{D}{\pi}$	
Rectangular HSS	Single concentric gusset plate	$\ell \geq H, U = 1 - \dfrac{\bar{x}}{\ell}$ $\bar{x} = \dfrac{B^2 + 2BH}{4(B + H)}$	

(continued)

Table 4-1 (continued)

Tension Member Type	Description	Shear Lag Factor, U	Example
Rectangular HSS	Two-sided gusset plate	$\ell \geq H,\ U = 1 - \dfrac{\bar{x}}{\ell}$ $\bar{x} = \dfrac{B^2}{4(B + H)}$	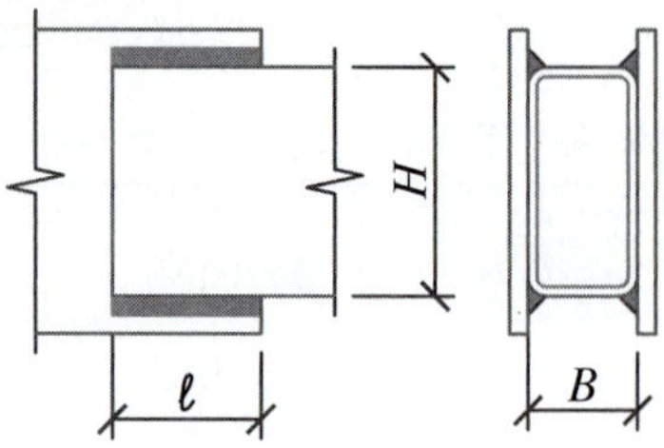
W, M, S, or HP, or Tees Cut from These Shapes	Flange connected with three or more fasteners per line in the direction of the load	$b_f \geq \dfrac{2}{3}d,\ U = 0.90$ $b_f < \dfrac{2}{3}d,\ U = 0.85$	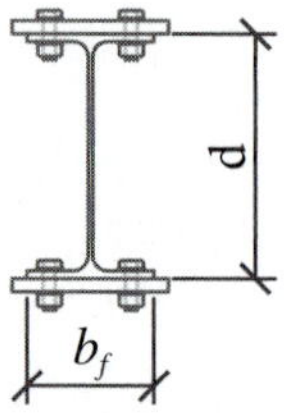
	Web connected with four or more fasteners per line in the direction of the load	$U = 0.70$	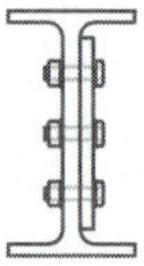
Single Angle	Four or more fasteners per line in the direction of the load	$U = 0.80$	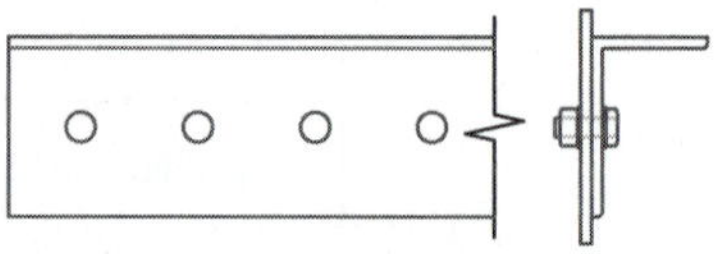
	Two or three fasteners per line in the direction of the load	$U = 0.60$	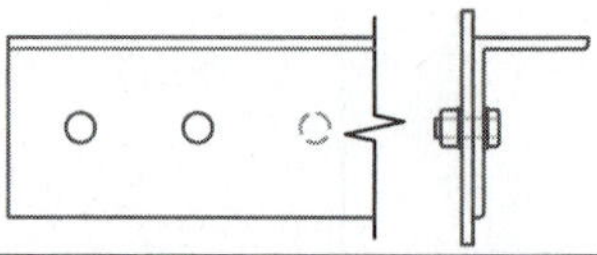

*For bolted splice plates, $A_e = A_n \leq 0.85\,A_g$ ($U = 1.0$).

Adapted from AISCM, Table D3.1.

EXAMPLE 4-1

U-Value for a Bolted Connection

For the bolted tension member shown in Figure 4-6, determine the shear lag factor, U; the net area, A_n; and the effective area, A_e.

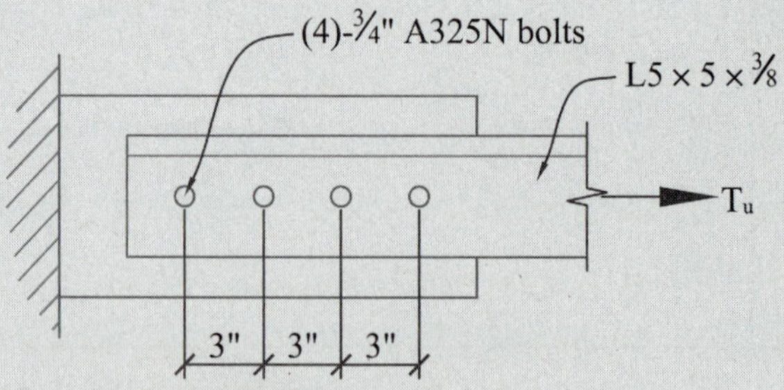

Figure 4-6 Details for Example 4-1.

SOLUTION

From the section property tables in part 1 of the *AISCM*, we find that for an $L5 \times 5 \times \frac{3}{8}$,

$$\bar{x} = 1.37 \text{ in.}$$
$$A_g = 3.61 \text{ in.}^2$$

Shear Lag Factor:

$$U = 1 - \frac{\bar{x}}{\ell}$$

$$= 1 - \frac{1.37 \text{ in.}}{9 \text{ in.}} = 0.848$$

Alternatively, $U = 0.80$ from Table 4-1. The larger value of $U = 0.848$ can be used.

Net Area of the Angle:

$$A_n = A_g - A_{\text{holes}}$$

$$= (3.61) - \left(\frac{3}{4} + \frac{1}{8}\right)(0.375) = 3.28 \text{ in.}^2$$

Effective Area:

$$A_e = A_n U$$

$$= (3.28)(0.848) = 2.78 \text{ in.}^2$$

EXAMPLE 4-2

U-Value for a Welded Connection

For the welded tension member shown in Figure 4-7, determine the shear lag factor, U; the net area, A_n; and the effective area, A_e.

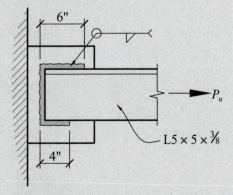

Figure 4-7 Detail for Example 4-2.

(continued)

SOLUTION

From the section property tables in part 1 of the *AISCM*, we find that for an L5 × 5 × ⅜,

$$\bar{x} = 1.37 \text{ in., and}$$
$$A_g = 3.61 \text{ in.}^2$$

Shear Lag Factor:

$$U = 1 - \frac{\bar{x}}{\ell}$$

$$= 1 - \frac{1.37 \text{ in.}}{4 \text{ in.}} = 0.657$$

ℓ = Smaller of the longitudinal weld lengths of 4 in. and 6 in ∴ ℓ = 4 in. There is not an alternate value to use from Table 4-1, so $U = 0.657$.

Since there are no holes, $A_n = A_g = 3.61 \text{ in.}^2$.

Effective Area:

$$A_e = A_n U$$
$$= (3.61)(0.657) = 2.37 \text{ in.}^2$$

EXAMPLE 4-3

Maximum Factored Load in a Tension Member

Determine the maximum factored load that can be applied in tension to the plate shown in Figure 4-8. The material is ASTM A36; it is welded on three sides to the gusset plate.

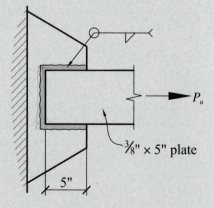

Figure 4-8 Detail for Example 4-3.

SOLUTION

From the *AISCM*, Table 2-4:

$F_y = 36$ ksi

$F_u = 58$ ksi to 80 ksi (use $F_u = 58$ ksi)

Gross and Effective Area:

$A_g = (0.375 \text{ in.})(5 \text{ in.}) = 1.88 \text{ in.}^2$

$A_n = A_g$ (no bolt holes)

$U = 1.0$ (Table 4-1, all-welded plate)

$A_e = A_n U$

$\quad = (1.88 \text{ in.}^2)(1.0) = 1.88 \text{ in.}^2$

From equation (4-1), the strength based on gross area is

$\phi P_n = \phi F_y A_g$

$\quad = (0.9)(36)(1.88 \text{ in.}^2) = 60.8$ kips

From equation (4-2), the strength based on effective area is

$\phi P_n = \phi F_u A_e$

$\quad = (0.75)(58)(1.88 \text{ in.}^2) = 81.6$ kips

The smaller value controls, so $P_u = 60.8$ kips

EXAMPLE 4-4

Tension Member Analysis

Determine if the channel is adequate for the applied tension load shown in Figure 4-9. The channel is ASTM A36; it is connected with four ⅝-in. diameter bolts. Neglect block shear.

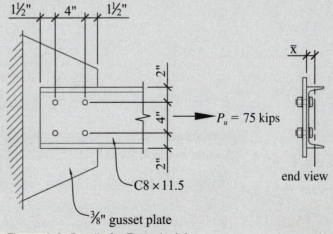

Figure 4-9 Details for Example 4-4.   *(continued)*

SOLUTION

From the *AISCM*, Table 1-5:

$$A_g = 3.37 \text{ in.}^2$$
$$\bar{x} = 0.572$$
$$t_w = 0.220 \text{ in.}$$

From the *AISCM* Table 2-3:

$$F_y = 36 \text{ ksi}$$
$$F_u = 58 \text{ ksi to } 80 \text{ ksi} \left(\text{use } F_u = 58 \text{ ksi} \right)$$

Net Area of the Channel:

$$A_n = A_g - A_{\text{holes}}$$
$$= (3.37) - \left[(2)\left(\frac{5}{8} + \frac{1}{8} \right)(0.220) \right] = 3.04 \text{ in.}^2$$

Effective Area of the Channel:

$$U = 1 - \frac{\bar{x}}{\ell}$$
$$= 1 - \frac{0.572 \text{ in.}}{4 \text{ in.}} = 0.857$$
$$A_e = A_n U$$
$$= (3.04)(0.857) = 2.61 \text{ in.}^2$$

From equation (4-1), the strength based on gross area is

$$\phi P_n = \phi F_y A_g$$
$$= (0.9)(36)(3.37) = 109 \text{ kips} > P_u = 75 \text{ kips. OK}$$

From equation (4-2), the strength based on effective area is

$$\phi P_n = \phi F_u A_e$$
$$= (0.75)(58)(2.61 \text{ in.}^2) = 113 \text{ kips} > P_u = 75 \text{ kips. OK}$$

(Note: Block shear should also be checked, this is covered in Section 4-3.)

EXAMPLE 4-5

Tension Member Analysis with Staggered Bolts

Determine the maximum factored load that can be applied in tension to the angle shown in Figure 4-10. The angle is ASTM A36; it is connected with four ¾-in. diameter bolts. Neglect block shear.

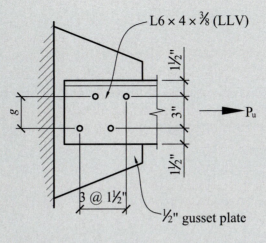
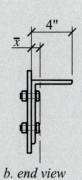

b. end view

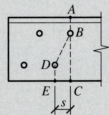

a. failure planes

Figure 4-10 Details for Example 4-5.

SOLUTION

From the *AISCM* Table 1-7:

$A_g = 3.61$ in.^{2}
$\bar{x} = 0.933$ in.
$t = 0.375$ in.

From the *AISCM*, Table 2-3:

$F_y = 36$ ksi
$F_u = 58$ ksi to 80 ksi (use $F_u = 58$ ksi)

Net Area of the Angle:

$$A_n = A_g - \sum d_h t + \sum \frac{s^2}{4g} t$$

Failure Plane *ABC*:

$$A_n = 3.61 - \left[\left(\frac{3}{4} + \frac{1}{8} \right)(0.375) \right] + 0 = 3.28 \text{ in.}^2$$

(continued)

Failure Plane *ABDE*:

$$A_n = 3.61 - \left[(2)\left(\frac{3}{4} + \frac{1}{8}\right)(0.375) \right] + \left[\frac{(1.5)^2}{(4)(3)}(0.375) \right] = 3.02 \text{ in.}^2$$

The failure plane along *ABDE* controls, since it has a smaller net area.

Effective Area of the Angle:

$$U = 1 - \frac{\bar{x}}{\ell}$$

$$= 1 - \frac{0.933}{(3)(1.5)} = 0.792$$

Alternatively, $U = 0.60$ from Table 4-1. The larger value is permitted to be used, so $U = 0.792$.

$$A_e = A_n U$$

$$= (3.02)(0.792) = 2.39 \text{ in.}^2$$

From equation (4-1), the strength based on gross area is

$$\phi P_n = \phi F_y A_g$$

$$= (0.9)(36)(3.61) = 116 \text{ kips}$$

From equation (4-2), the strength based on effective area is

$$\phi P_n = \phi F_u A_e$$

$$= (0.75)(58)(2.39 \text{ in.}^2) = 104 \text{ kips}$$

The smaller value controls, so $P_u = 104$ kips. (Note: Block shear should also be checked, this is covered in Section 4-3.)

4.3 BLOCK SHEAR

In the previous sections, we discussed the strength of members in pure tension only. In addition to checking the connected ends of tension members for tensile failure, there exist certain connection configurations where tensile failure could be accompanied by shear failure such that a block of the tension member tears away (see Figure 4-11). This failure plane usually occurs along the path of the centerlines of the bolt holes for bolted connections. This type of failure could also occur along the perimeter of welded connections.

For this mode of failure, it is assumed that the tension member ruptures in both shear and tension. Therefore, both the shear and tension failure planes contribute to the strength of the connection.

The nominal strength based on shear yielding is

$$R_n = 0.6 \, F_y A_{gv}, \tag{4-10a}$$

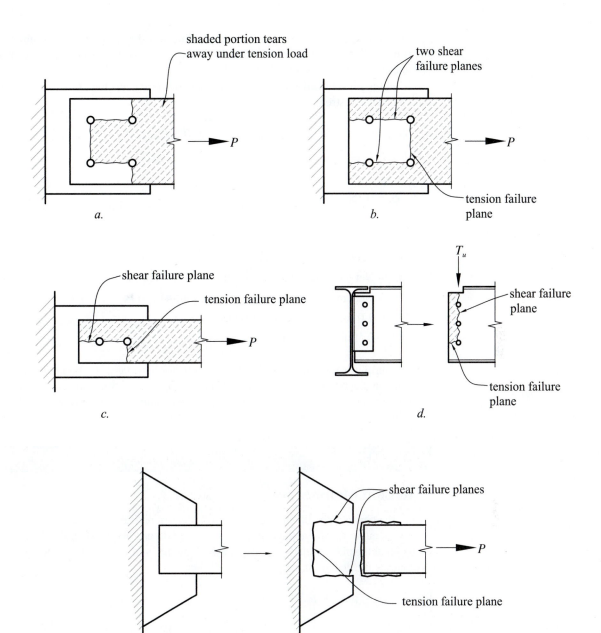

Figure 4-11 Block shear failure.

and the nominal strength based on shear rupture is

$$R_n = 0.6\,F_u A_{nv}, \qquad (4\text{-}10\text{b})$$

where

A_{gv} = gross area subject to shear, and

A_{nv} = Net area subject to shear (see eq. (4-4)).

To determine the design strength in shear yielding and shear rupture, the nominal strength R_n, is multiplied by a ϕ-factor of 1.0 and 0.75, respectively, when the shear does not occur simultaneously with tension stress.

Combining the available tension and shear strength yields the expression for the available block shear strength:

$$\phi P_n = \phi(0.60 F_u A_{nv} + U_{bs} F_u A_{nt}) \le \phi(0.60 F_y A_{gv} + U_{bs} F_u A_{nt}), \qquad (4\text{-}11)$$

where

$\phi = 0.75,$

$F_u =$ Minimum tensile stress,

$F_y =$ Minimum yield stress,

$A_{gv} =$ Gross area subjected to shear,

$A_{nt} =$ Net area subjected to tension (see eq. (4-4)),

$A_{nv} =$ Net area subjected to shear (see eq. (4-4)), and

$U_{bs} = 1.0$ for uniform tension stress

$\quad\ = 0.50$ for nonuniform tension stress.

The U_{bs} term in equation (4-11) is a reduction factor that accounts for a nonuniform stress distribution. Section C-J4.3 of the *AISCM* gives examples of connections with uniform and nonuniform tension stress distribution, but the most common case is to have a uniform stress distribution and, therefore, $U_{bs} = 1.0$ for most cases.

EXAMPLE 4-6

Tension Member with Block Shear

For the connection shown in Example 4-4, determine if the channel and gusset plate are adequate for the applied tension load considering block shear. Assume that the width of the plate is such that block shear along the failure plane shown in Figure 4-12 controls the design of the plate.

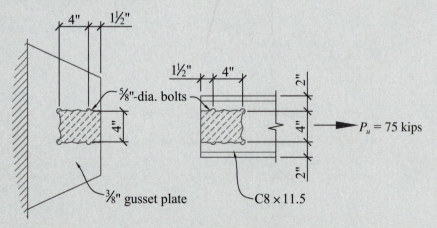

a. block shear in plate   *b. block shear in channel*

Figure 4-12 Details for Example 4-6.

From the *AISCM* Table 2-3:

$F_y = 36$ ksi

$F_u = 58$ ksi to 80 ksi (use $F_u = 58$ ksi)

Plate Dimensions:

$A_{gv} = (2)(4 + 1.5)(0.375) = 4.12$ in.^{2}

$A_{nv} = A_{gv} - A_{\text{holes}}$

$$= 4.12 - \left[(2\ \text{holes})(1.5)\left(\frac{5}{8} + \frac{1}{8} \right)(0.375) \right] = 3.28\ \text{in.}^2$$

GUSSET PLATE WIDTH

$A_{nt} = A_{gt} - A_{\text{holes}}$

$$= [(4)(0.375)] - \left[\left(\frac{5}{8} + \frac{1}{8} \right)(0.375) \right] = 1.21\ \text{in.}^2$$

Channel Dimensions:

$A_{gv} = (2)(4 + 1.5)(0.220) = 2.42$ in.^{2}

$A_{nv} = A_{gv} - A_{\text{holes}}$

$$= 2.42 - \left[(2\ \text{holes})(1.5)\left(\frac{5}{8} + \frac{1}{8} \right)(0.220) \right] = 1.92\ \text{in.}^2$$

$A_{nt} = A_{gt} - A_{\text{holes}}$

$$= [(4)(0.220)] - \left[\left(\frac{5}{8} + \frac{1}{8} \right)(0.220) \right] = 0.715\ \text{in.}^2$$

$U_{bs} = 1.0$ (tension stress is uniform)

The available block shear strength is found from equation (4-11):

$$\phi P_n = \phi(0.60 F_u A_{nv} + U_{bs} F_u A_{nt}) \leq \phi(0.60 F_y A_{gv} + U_{bs} F_u A_{nt}).$$

For the plate,

$$\phi P_n = 0.75[(0.60)(58)(3.28) + (1.0)(58)(1.21)]$$
$$\leq 0.75[(0.60)(36)(4.12) + (1.0)(58)(1.21)]$$
$$= 138\ \text{kips} > 119\ \text{kips}.$$

The smaller value controls, so the available strength of the plate in block shear is 119 kips, which is greater than the applied load of $P_u = 75$ kips.

For the channel,

$$\phi P_n = 0.75[(0.60)(58)(1.92) + (1.0)(58)(0.715)]$$
$$\leq 0.75[(0.60)(36)(2.42) + (1.0)(58)(0.715)]$$
$$= 81.2\ \text{kips} > 70.3\ \text{kips}.$$

The smaller value controls, so the available strength of the channel in block shear is 70.3 kips, which is less than the applied load of $P_u = 75$ kips, so the channel is not adequate in block shear. *(Note: The bolts also should be checked for shear and bearing, but bolt strength is covered later in this text.)*

EXAMPLE 4-7

Block Shear in a Gusset Plate

For the HSS-to-gusset plate connection shown in Figure 4-13, determine the required length, ℓ, required to support the applied tension load considering the strength of the gusset plate only. The plate is ASTM A529, grade 50.

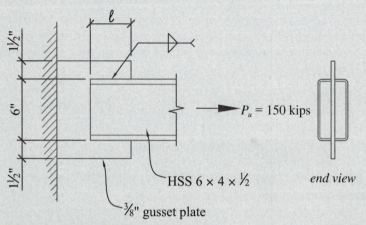

Figure 4-13 Details for Example 4-7.

SOLUTION

From the *AISCM*, Table 2-4:

$F_y = 50$ ksi

$F_u = 70$ ksi to 100 ksi (use $F_u = 70$ ksi)

From equation (4-1), the strength based on the gross area of the plate is

$$\phi P_n = \phi F_y A_g$$
$$= (0.9)(50)(1.5 + 6 + 1.5)(0.375) = 151 \text{ kips} > 150 \text{ kips. OK}$$

From equation (4-2), the strength based on effective area is

$$\phi P_n = \phi F_u A_e \ (A_e = A_g, \text{ load is concentric on the plate and bolts are not used})$$
$$= (0.75)(70)(1.5 + 6 + 1.5)(0.375) = 177 \text{ kips} > 150 \text{ kips. OK}$$

The available block shear strength is found from equation (4-11):

$$\phi P_n = \phi(0.60 F_u A_{nv} + U_{bs} F_u A_{nt}) \le \phi(0.60 F_y A_{gv} + U_{bs} F_u A_{nt})$$
$$U_{bs} = 1.0 \text{ (tension stress is uniform)}$$

Since $A_{nv} = A_{gv}$, the right-hand side of the equation will control. Solving for A_{gv},

$$\phi P_n = \phi(0.60 F_y A_{gv} + U_{bs} F_u A_{nt})$$
$$150 = (0.75)[(0.60)(50)(A_{gv}) + (1.0)(70)(6)(0.375)]$$

$A_{gv} = 1.42$ in.^{2}

$\ell_{min} = \dfrac{1.42}{(2)(0.375)} = 1.89$ in.

The minimum length of engagement is 1.89 in. in order to develop adequate strength in the gusset plate. Note that from Table 4-1, the minimum length, ℓ, is the height of the connecting HSS member, or 6 in. ($\ell > H$). Therefore, the minimum length, ℓ, is actually 6 in. From Table 4-1, the shear lag factor for the HSS member is

$$\bar{x} = \frac{B^2 + 2BH}{4(B + H)}$$

$$= \frac{(4)^2 + (2)(4)(6)}{4(4 + 6)} = 1.6$$

$$U = 1 - \frac{\bar{x}}{\ell}$$

$$= 1 - \frac{1.6}{6} = 0.733.$$

This value would then be used to determine the strength of the HSS member in tension.

EXAMPLE 4-8

Hanger in Tension with Block Shear

The W12 × 53 tension member shown in Figure 4-14 has two rows of three 1-in.-diameter A325N bolts in each flange. Assuming ASTM A572 steel and considering the strength of the W12 × 53 only,

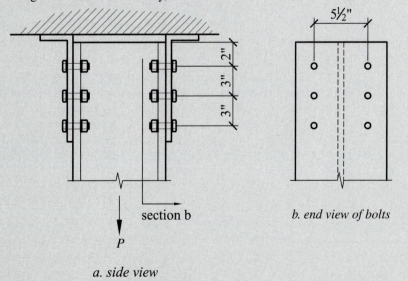

a. side view

b. end view of bolts

Figure 4-14 Details for Example 4-8.

(continued)

1. Determine the design tension strength of the W12 × 53,
2. Determine the service dead load that can be supported if there is no live load, and
3. If a service dead load, $P_D = 100$ kips is applied, what is the maximum service live load, P_L, that can be supported?

SOLUTION

1. For A572 steel, $F_y = 50$ ksi, $F_u = 65$ ksi (*AISCM* Table 2-3).

   For 1-in.-diameter bolts, $d_{hole} = 1$ in. + ⅛ in. = 1.125 in.

   From part 1 of the AISCM, for W12 × 53,

   $A_g = 15.6$ in.^{2}

   $b_f = 10$ in.

   $t_f = 0.575$ in.

   $d = 12.1$ in.

   $\bar{x} = 1.02$ in. (*AISCM* Table 1-8, use value for WT shape)

   $A_n = A_g - \Sigma A$ holes

   $\quad = 15.6 - (4\,\text{holes})(1.125)(0.575) = 13.0$ in.^{2}

   From Table 4-1, we obtain $U = 0.90$, since $b_f > \dfrac{2}{3}d$; 10 in. $> \dfrac{2}{3}(12.1) = 8.07$ in.

   Using the calculation method,

   $$U = 1 - \frac{\bar{x}}{\ell} = 1 - \frac{1.02}{6} = 0.83 \rightarrow \text{The larger value of } U = 0.90 \text{ may be used.}$$

   ℓ = centerline distance between outer bolts = 3 in. + 3 in. = 6 in.

   $A_e = UA_n = (0.9)(13.0) = 11.7$ in.^{2}

   a. Yielding failure mode: $\phi P_n = 0.9 A_g F_y = (0.9)(50)(15.6) = 702$ kips

   b. Fracture failure mode: $\phi P_n = 0.75 F_u A_e = (0.75)(65)(11.7) = 570$ kips

   c. Block shear failure mode:

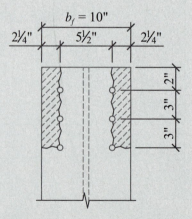

Figure 4-15 Block shear in W12 × 53.

$$A_{gv} = (4)(2 \text{ in.} + 3 \text{ in.} + 3 \text{ in.})(0.575 \text{ in.}) = 18.4 \text{ in.}^2$$

$$A_{gt} = (4)(2.25 \text{ in.})(0.575 \text{ in.}) = 5.18 \text{ in.}^2$$

$$A_{nv} = A_{gv} - \Sigma A_{\text{holes}}$$

$$= 18.4 - (4)(2.5)(1.125 \text{ in.})(0.575 \text{ in.}) = 11.9 \text{ in.}^2$$

$$A_{nt} = A_{gt} - \Sigma A_{\text{holes}}$$

$$= 5.18 - (4)\left(\frac{1}{2}\right)(1.125 \text{ in.})(0.575 \text{ in.}) = 3.88 \text{ in.}^2$$

The available block shear strength is found from equation (4-11) ($U_{bs} = 1.0$):

$$\phi P_n = \phi(0.60 F_u A_{nv} + U_{bs} F_u A_{nt}) \leq \phi(0.60 F_y A_{gv} + U_{bs} F_u A_{nt})$$

$$\phi P_n = (0.75)[(0.60)(65)(11.9) + (1.0)(65)(3.88)]$$

$$\leq (0.75)[(0.60)(50)(18.4) + (1.0)(65)(3.88)]$$

$$537 \text{ kips} < 603 \text{ kips}$$

$$\phi P_n = 537 \text{ kips (block shear capacity)}$$

Summary:

$$\begin{aligned}
\text{Yielding failure mode:} &\quad \phi P_n = 702 \text{ kips} \\
\text{Fracture failure mode:} &\quad \phi P_n = 570 \text{ kips} \\
\text{Block shear failure mode:} &\quad \phi P_n = 537 \text{ kips}
\end{aligned}$$

The smallest of the three values governs $\therefore$ *Design strength,* $\phi P_n = 537$ kips

2. $P_L = 0$; the two load combinations to be considered will be

$$P_u = 1.4 P_D \qquad \text{Governs}$$

$$P_u = 1.2 P_D + 1.6 P_L = 1.2 P_D$$

Setting the design strength equal to the applied loads yields

$$P_u = 1.4 P_D \leq \phi P_n = 537 \text{ kips}$$

$$1.4 P_D \leq 537 \text{ kips}$$

Solving for $P_D = 383$ kips (maximum unfactored dead load that can be supported)

3. Given $P_D = 100$ kips, determine the live load, P_L, that can be safely supported:

$$P_u = 1.2 P_D + 1.6 P_L$$

$$P_u = 1.2 P_D + 1.6 P_L \leq \phi P_n = 537 \text{ kips}$$

$$(1.2)(100) + 1.6 P_L \leq 537 \text{ kips}$$

Solving for $P_L = 260$ kips (maximum unfactored live load that can be supported)

4.4 DESIGN OF TENSION MEMBERS

The design of tension members will require consideration of failure modes not specifically addressed in this chapter. For tension members with welded connections, the design strength of the welds in shear and tension must be considered. For tension members with bolted connections, the design strength of the bolts in shear, tension, and bearing must be considered. Load eccentricity effects at the connection points also must be considered. In this section, we will consider the design strength in tension of the actual member only. The reader is referred to Chapters 9 and 10 for the design of the connections for tension members.

Tension members need to have enough gross cross-sectional area for strength in yielding and enough effective area for strength in fracture. Note that the effective area accounts for shear lag effects.

In addition to having enough design strength in yielding and fracture, block shear at the connected ends needs to be checked. In some cases, there might be more than one mode of failure in block shear. The design strength of the tension member is the smallest of the strength in yielding, fracture, and block shear.

Slenderness effects should also be considered. The AISC specification recommends a slenderness limit L/r of 300 to prevent flapping, flutter, or sag of the member, but this is not a mandatory requirement. If this slenderness limit cannot be met, the member can be pretensioned to reduce the amount of sag. The amount of this pretension force can vary between 5% and 10%, and the pretension force will reduce the design strength of the member accordingly by increasing the forces used for design. The AISC specification suggests the following pretension or "draw" values:

Table 4-2 Recommended pretension values for slender tension members

Length of Tension Member, L	Pretension or "Draw" Member Length Deduction
$L \le 10$ ft.	0 in.
10 ft. $< L \le 20$ ft.	$\frac{1}{16}$ in.
20 ft. $< L \le 35$ ft.	$\frac{1}{8}$ in.
$L > 35$ ft.	$\frac{3}{16}$ in.

When members are fabricated shorter in accordance with the above table, they will be drawn up in the field in order to reduce the amount of flap or sag in the member. However, it can be shown that the amount of load to the tension member and connections will increase between 20% to 50% when using the values in Table 4-2, which is not desirable. The authors suggest that an increase of 5% to 10% has proven to be effective in practice and they would recommend using these lower values. The authors further recommend that the designer consult with the steel fabricator on any given project to determine the proper amount of draw that might be appropriate based on the actual fabrication and erection procedures.

Figure 4-16 shows a possible pretension detail where the tension member is intentionally fabricated shorter than the actual length so that once the connection is tightened to its final position, the amount of sag is reduced.

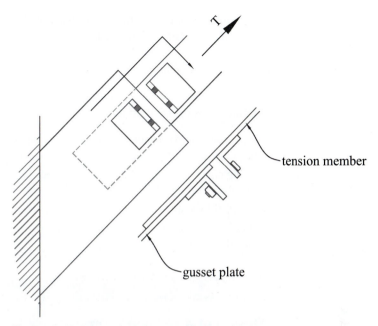

Figure 4-16 Pretensioned connection.

The design of a tension member can be summarized as follows:

1. Determine the minimum gross area from the tensile yielding failure mode equation:

$$A_g \geq \frac{P_u}{0.9F_y} \qquad (4\text{-}12)$$

2. Determine the minimum net area from the tensile fracture failure mode equation:

$$A_n \geq \frac{P_u}{0.75F_uU}, \qquad (4\text{-}13)$$

where the net area is found from equation (4-4):

$$A_n = A_g - \Sigma A_{\text{holes}} \geq \frac{P_u}{0.75F_uU}$$

$$A_g \geq \frac{P_u}{0.75F_uU} + \Sigma A_{\text{holes}} \qquad (4\text{-}14)$$

3. Use the larger A_g value from equations (4-13) and (4-14), and select a trial member size based on the larger value of A_g.
4. For tension members, AISC specification Section D1 *suggests* that the slenderness ratio KL/r_{min} should be ≤ 300 to prevent flapping or flutter of the member,

where

K = Effective length factor (usually assumed to be 1.0 for tension members),

L = Unbraced length of the tension member, and

r_{min} = *Smallest* radius of gyration of the member.

The smallest radius of gyration for rolled sections can be obtained from part 1 of the *AISCM*. For other sections, such as plates, the radius of gyration can be calculated from

$$r_{min} = \sqrt{\frac{I_{min}}{A_g}} > \frac{L}{300}, \tag{4-15}$$

where I_{min} is the smallest moment of inertia.

If equation (4-15) cannot be satisfied (i.e., the member is too slender), the member should be pretensioned. Allow for 5% to 10% pretension force in the design of the member.

5. Using equation (4-11), determine the block shear capacity of the selected tension member.

If ϕP_n (block shear) is greater than P_u, the member is adequate.

If ϕP_n (block shear) is less than P_u, increase the member size and repeat step 5 until ϕP_n (block shear) $\geq P_u$.

EXAMPLE 4-9

Design of a Tension Member

Design the X-brace in the first story of the building shown in Figure 4-17, which is subjected to wind loads. Use a steel plate that conforms to ASTM A36.

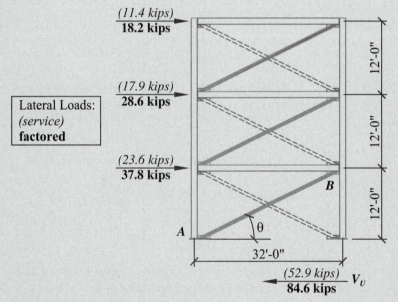

Figure 4-17 Detail for Example 4-9.

Some X-brace configurations have slender members such that they can only support loads in tension. In this figure, all of the members are assumed to be too slender to support compression loads. Only the shaded members support lateral loads in the assigned direction of the lateral loads.

The lateral loads shown below are the loads acting on each X-braced frame. The wind loads acting on the entire building must be distributed to the various braced frames in the building in the direction of the lateral load. If the diaphragm is assumed to be *rigid*, the lateral load is distributed in proportion to the stiffness of each braced frame. If the diaphragm is *flexible*, the lateral load is distributed in proportion to the tributary area of each braced frame.

SOLUTION

The maximum load factor for wind is 1.6 (from the ASCE 7 load combinations).

Loads to Each Level:

Service Loads:

$P_r = 11.4$ kips

$P_3 = 17.9$ kips

$P_2 = 23.6$ kips

Factored Loads:

$(1.6)(11.4) = 18.2$ kips

$(1.6)(17.9) = 28.6$ kips

$(1.6)(23.6) = 37.7$ kips

V_u (base shear) $= 18.2 + 28.6 + 37.8 = 84.6$ kips

$$\theta = \tan^{-1}\left(\frac{12 \text{ ft.}}{32 \text{ ft.}}\right) = 20.6°$$

$$T_{AB} = \frac{V_u}{\cos\theta} = \frac{84.6}{\cos 20.6} = 90.4 \text{ kips}$$

We will cover bolts and welds in Chapters 9 and 10, but for now, assume the shear strength of A325N bolts in single shear to be as follows:

$\phi R_n = 15.9$ kips for ¾-in.-diameter bolt

$\phi R_n = 21.6$ kips for ⅞-in.-diameter bolt

1. $A_g \geq \dfrac{P_u}{0.9F_y}$ (allow for added 5% to 10% pretension ∴ use 7.5% pretension)

$$A_g \geq \frac{(90.4)(1.075)}{0.9(36)} = 2.99 \text{ in.}^2$$

Try ½-in. × 6-in. plate, $A_g = (0.5)(6) = 3.0$ in.^{2}

2. Number of bolts:

Five ¾-in.-diameter bolt → $\phi R_n = (5)(15.9) = 79$ kips < 90.4 kips

Five ⅞-in.-diameter bolt → $\phi R_n = (5)(21.6) = $ **108 kips** > 90.4 kips

Six ¾-in.-diameter bolt → $\phi R_n = (6)(15.9) = 95$ kips > 90.4 kips

Use five ⅞-in. diameter bolts in a single line.

The shear lag factor, U, is 1.0 for plates connected with bolts (from Table 4-1).

$$A_g \geq \frac{P_u}{0.75F_uU} + \Sigma A_{holes}$$

(continued)

$$A_g \geq \frac{(90.4)(1.075)}{(0.75)(58)(1.0)} + (1 \text{ hole})\left(\frac{7}{8} + \frac{1}{8}\right)(0.5) = 2.73 \text{ in.}^2 < 3.0 \text{ in.}^2 \text{ OK}$$

3. From step 1, $A_g = 2.99$ in.^{2} and from step 2, $A_g = 2.73$ in.^{2}; both are less than the gross area of the trial member size for 3.0 in.^{2}.

4. Check slenderness ratio:

$$L = \sqrt{(12)^2 + (32)^2} = 34.2 \text{ ft.}$$

$$I_{min} = \frac{bh^3}{12} = \frac{(6)(0.5)^3}{12} = 0.0625 \text{ in.}^2$$

$$r_{min} = \sqrt{\frac{I_{min}}{A_g}} > \frac{L}{300}$$

$$r_{min} = \sqrt{\frac{0.0625}{3}} > \frac{(34.2)(12)}{300}; 0.144 < 1.37 \rightarrow \text{Slenderness limit is exceeded}$$

The assumption to pretension the X-brace is justified.

5. Check block shear strength.

The bolt spacing and configuration shown in Figure 4-18 will be used.

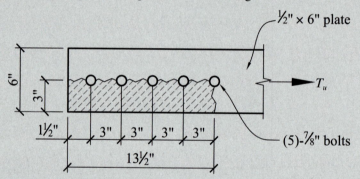

Figure 4-18 Block shear in the ½-in. by 6-in. plate.

$$A_{gv} = (13.5)(0.5) = 6.75 \text{ in.}^2$$

$$A_{nv} = A_{gv} - \Sigma A_{holes}$$

$$= 6.75 - (4.5 \text{ holes})\left(\frac{7}{8} + \frac{1}{8}\right)(0.5) = 4.5 \text{ in.}^2$$

$$A_{gt} = (3)(0.5) = 1.5 \text{ in.}^2$$

$$A_{nt} = A_{gt} - \Sigma A_{holes}$$

$$= 1.5 - (0.5 \text{ holes})\left(\frac{7}{8} + \frac{1}{8}\right)(0.5) = 1.25 \text{ in.}^2$$

$$\phi P_n = \phi(0.60F_u A_{nv} + U_{bs}F_u A_{nt}) \leq \phi(0.60F_y A_{gv} + U_{bs}F_u A_{nt})$$

$$\phi P_n = 0.75[(0.60)(58)(4.5) + (1.0)(58)(1.25)]$$

$$\leq 0.75[(0.60)(36)(6.75) + (1.0)(58)(1.25)]$$

$$171 \text{ kips} > 163 \text{ kips}$$

$$\phi P_n = 163 \text{ kips (block shear capacity)} > T_u = 90.4 \text{ kips OK}$$

Use a 6-in. by ½-in. plate with five ⅞-in.-diameter A325N bolts.

EXAMPLE 4-10

Design of a Single-Angle Tension Member

Design a tension member given the following:

- Service loads: $P_D = 40$ kips, $P_L = 66$ kips;
- Single angle required;
- Unbraced length, $L = 20$ ft.;
- ASTM A36 steel; and
- Two lines of four ¾-in.-diameter bolts.

SOLUTION

$P_u = 1.4 \, P_D = (1.4)(40) = 56$ kips

$P_u = 1.2 \, P_{DL} + 1.6 \, P_{LL} = [(1.2)(40)] + [(1.6)(66)] = 154$ kips

1. $A_g \geq \dfrac{P_u}{0.9 F_y}$ (assume slenderness ratio $< L/300$)

    $A_g \geq \dfrac{(154)}{(0.9)(36)} = 4.75$ in.^{2}

2. Shear lag factor, U, is 0.80 for single angles (from Table 4-1). Alternatively, U may be calculated using $\ell = (3)(3 \text{ in.}) = 9$ in. (three spaces at 3 in.), but an angle size would have to be assumed.

    $A_g \geq \dfrac{P_u}{0.75 F_u U} + \Sigma A_{\text{holes}}$

    $A_g \geq \dfrac{(154)}{(0.75)(58)(0.80)} + (2 \text{ holes})\left(\dfrac{3}{4} + \dfrac{1}{8}\right)(t)$

    $A_{g \, \text{required}} = 4.43 + 1.75t$, where t is the thickness of the angle

    $r_{\min} = \dfrac{L}{300} = \dfrac{(20)(12)}{300} = 0.80$ in.

Summary of angle selection

t	A_g Required		Selected Angle	r_z
	A_g (Step 1)	A_g (Step 2)		
¼ in.	4.75 in.^{2}	4.86 in.^{2}	None worked	–
5/16 in.	4.75 in.^{2}	4.97 in.^{2}	None worked	–
⅜ in.	4.75 in.^{2}	5.08 in.^{2}	None worked	–
7/16 in.	4.75 in.^{2}	5.19 in.^{2}	L8 × 6 × 7/16 (wt. = 20.2 lb./ft.)	1.31 in.
½ in.	4.75 in.^{2}	5.30 in.^{2}	L6 × 6 × ½ (wt. = 19.6 lb./ft.)	1.18 in.
			L8 × 4 × ½ (wt. = 19.6 lb./ft.)	0.863 in.

(continued)

3. Select L8 × 4 × ½ because of its lighter weight; also, it would have greater block shear capacity than the L6 × 6 × ½ (same weight).

4. Slenderness ratio checked in step 2.

5. Check block shear capacity. The spacing of the bolts will have to be assumed. See Figure 4-19 for the assumed bolt layout.

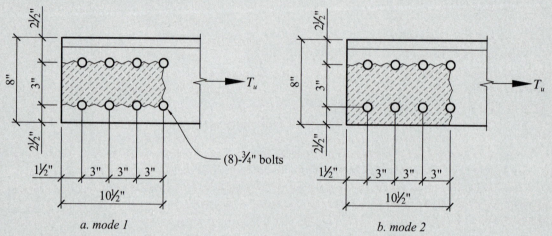

a. mode 1   *b. mode 2*

Figure 4-19 Block shear in L8 × 4 × ½.

Mode 1 Block Shear:

$$A_{gv} = (2)(10.5)(0.5) = 10.5 \text{ in.}^2$$
$$A_{nv} = A_{gv} - \Sigma A_{\text{holes}}$$
$$= 10.5 - (2)(3.5 \text{ holes})\left(\frac{3}{4} + \frac{1}{8}\right)(0.5) = 7.43 \text{ in.}^2$$

$$A_{gt} = (3)(½) = 1.5 \text{ in.}^2$$
$$A_{nt} = A_{gt} - \Sigma A_{\text{holes}}$$
$$= 1.5 - (2)(0.5 \text{ holes})\left(\frac{3}{4} + \frac{1}{8}\right)(0.5) = 1.06 \text{ in.}^2$$

$$\phi P_n = \phi(0.60F_u A_{nv} + U_{bs}F_u A_{nt}) \leq \phi(0.60F_y A_{gv} + U_{bs}F_u A_{nt})$$
$$\phi P_n = 0.75[(0.60)(58)(7.43) + (1.0)(58)(1.06)]$$
$$\leq 0.75[(0.60)(36)(10.5) + (1.0)(58)(1.06)]$$
$$240 \text{ kips} > 216 \text{ kips}$$

$$\phi P_n = 216 \text{ kips (mode 1 block shear capacity)} > T_u = 154 \text{ kips OK}$$

Mode 2 Block Shear:

$$A_{gv} = (10.5)(0.5) = 5.25 \text{ in.}^2$$
$$A_{nv} = A_{gv} - \Sigma A_{\text{holes}}$$

$$= 5.25 - (3.5 \text{ holes})\left(\frac{3}{4} + \frac{1}{8}\right)(0.5) = 3.72 \text{ in.}^2$$

$$A_{gt} = (3 + 2.5)(\tfrac{1}{2}) = 2.75 \text{ in.}^2$$

$$A_{nt} = A_{gt} - \Sigma A_{\text{holes}}$$

$$= 2.75 - (1.5 \text{ holes})\left(\frac{3}{4} + \frac{1}{8}\right)(0.5) = 2.09 \text{ in.}^2$$

$$\phi P_n = \phi(0.60 F_u A_{nv} + U_{bs} F_u A_{nt}) \leq \phi(0.60 F_y A_{gv} + U_{bs} F_u A_{nt})$$

$$\phi P_n = 0.75[(0.60)(58)(3.72) + (1.0)(58)(2.09)]$$

$$\leq 0.75[(0.60)(36)(5.25) + (1.0)(58)(2.09)]$$

$$188 \text{ kips} > 175 \text{ kips}$$

$$\phi P_n = 175 \text{ kips } (\textit{mode 2 block shear capacity}) > T_u = 154 \text{ kips OK}$$

Select an $L8 \times 4 \times \frac{1}{2}$ with two lines of four ¾-in.-diameter bolts.

4.5 TENSION RODS

Rods with a circular cross section are commonly used in a variety of structural applications. Depending on the structural application, tension rods might be referred to as hanger rods or sag rods. Hangers are tension members that are hung from one member to support other members. Sag rods are often provided to prevent a member from deflecting (or sagging) under its own self-weight, as is the case with girts on the exterior of a building (see Figure 4-1a). Tension rods are also commonly used as diagonal bracing in combination with a clevis and turnbuckle to support lateral loads.

There are two basic types of threaded rods. The more commonly used type is a rod where the nominal diameter is greater than the root diameter. The tensile capacity is based on the available cross-sectional area at the root where the threaded portion of the rod is the thinnest (see Figure 4-20). The other type of threaded rod is one with an upset end. The threaded end of an upset rod is such that the root diameter equals the nominal diameter. Upset rods are not commonly used because the fabrication process can be cost-prohibitive.

As stated previously, the slenderness ratio (L/r) of tension members should be less than 300, but this requirement does not apply to rods or hangers in tension. The AISC specification does not limit the size of tension rods, but the practical minimum diameter of the rod should not be less than ⅝ in. since smaller diameter rods are more susceptible to damage during construction.

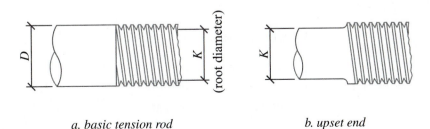

a. basic tension rod   b. upset end

Figure 4-20 Basic tension rod and upset-end tension rod.

The design strength of tension rods is the same as for bolts in tension (see Chapter 9). The design strength of a tension rod is given in Section J3.6 of the AISC specification as

$$\phi R_n = \phi F_n A_b,\qquad(4\text{-}16)$$

where

$\phi = 0.75,$

$F_n =$ Nominal tension stress from *AISCM* Table J3.2, and

$A_b =$ Nominal unthreaded body area.

From the *AISCM* Table J3.2,

$$F_n = 0.75 F_u.\qquad(4\text{-}17)$$

The 0.75 factor in equation (4-17) accounts for the difference between the nominal un-threaded body diameter and the diameter of the threaded rod at the root where the stress is critical. Combining equations (4-16) and (4-17) yields

$$\phi R_n = \phi 0.75 F_u A_b,\qquad(4\text{-}18)$$

The F_u term in the above equations is the minimum tensile stress of the threaded rod. There are several acceptable grades of threaded rods that are available (*AISCM* Section A3.4 and *AISCM* Table 2-5), the most common of which are summarized in Table 4-3.

When tension rods are used as diagonal bracing (see Figure 4-21), they are commonly used in combination with a clevis at the ends and possibly a turnbuckle to act as a splice for the tension rod. Clevises and turnbuckles are generally manufactured in accordance with ASTM A29, grade 1035, but the load capacities are based on testing done by the various manufacturers. However, there has been enough independently published test data that AISC has developed load tables for standard clevises and turnbuckles (see *AISCM* Tables 15-3 and 15-5). It should be noted that the factor of safety for clevises and turnbuckles is higher in the AISC load tables since these connectors are commonly used in hoisting and rigging where the loads are cyclical and are therefore subjected to fatigue failure, which is more critical.

Table 4-3 Grades of threaded rods

Material Specification		Diameter Range, in.	F_y, ksi	F_u, ksi
ASTM A36		Up to 10	36	'58–80
ASTM A193 Gr. B7 *(corrosion-resistant)*		4 to 7	–	100
		2.5 to 4	–	115
		2.5 and under	–	125
ASTM F1554	Grade 36	0.25 to 4	36	58–80
	Grade 55	0.25 to 4	55	75–95
	Grade 105	0.25 to 3	105	125–150

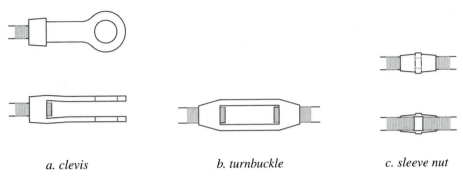

a. clevis   *b. turnbuckle*   *c. sleeve nut*

Figure 4-21 Connectors for tension rods.

Clevises are designated by a number (2 through 8) and have a corresponding maximum diameter associated with each number designation. Each clevis also has a corresponding maximum pin diameter associated with each number designation. The pins are also proprietary and are generally designed to have a capacity equal to or greater than that of the clevis, provided that the diameter of the pin is 125% greater than the diameter of the threaded rod. The pin diameters are given in *AISCM* Table 15-4 and do not need to be designed; however, the gusset plate that the pins connect to need to be checked for shear, tension, and bearing.

Turnbuckles are designated by the diameter of the connecting threaded rod. Alternatively, sleeve nuts (*AISCM* Table 15-6) can be used in lieu of turnbuckles. Sleeve nuts develop the full capacity of the tensile strength of the threaded rod, provided that the threaded rod has the proper thread engagement. Sleeve nuts are generally manufactured to conform to ASTM A29, grade 1018. Turnbuckles are usually preferred over sleeve nuts from a cost standpoint. Figure 4-21 shows the above-mentioned connecting elements for tension rods.

Tension members connected with a single pin, as is the case with tension rods used as diagonal bracing, are subject to the failure modes covered in Section D5 of the AISC specification. Pin-connected members differ from a single bolt connection in that deformation is not permitted in the gusset plate for pins so that the pins can rotate freely. There are three main failure modes that need to be checked for pin-connected members: tensile rupture on the net area (Figure 4-22a), shear rupture on the effective area (Figure 4-22b), and bearing on the projected area of the pin (Figure 4-22c).

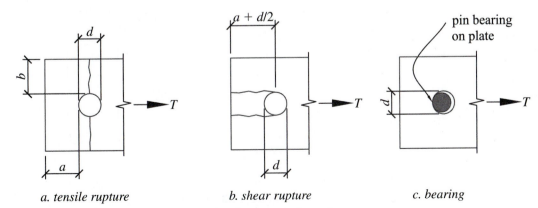

a. tensile rupture   *b. shear rupture*   *c. bearing*

Figure 4-22 Failure modes for pin-connected members.

The expression for tensile rupture on the net effective area is given as

$$\phi P_n = \phi 2 t b_{\text{eff}} F_u;$$ (4-19)

the expression for shear rupture on the effective area is given as

$$\phi P_n = \phi 0.6 F_u A_{sf};$$ (4-20)

and the expression for bearing on the projected area of the pin is

$$\phi P_n = \phi 1.8 F_y A_{pb},$$ (4-21)

where

$$\phi = 0.75,$$

$$A_{sf} = 2t\left(a + \frac{d}{2}\right) \text{ in.}^2,$$

a = Shortest distance from the edge of the hole to the plate edge parallel to the direction of the force,

b_{eff} = $2t + 0.63$ in. < b,

b = Distance from the edge of the hole to the plate edge perpendicular to the direction of the force,

d = Pin diameter (noted as p in the *AISCM* Table 15-4),

t = Plate thickness,

A_{pb} = Projected bearing area ($A_{pb} = dt$),

F_y = Minimum yield stress, and

F_u = Minimum tensile stress.

For pin-connected members, there are also dimensional requirements for the gusset plate that must be satisfied, and these are indicated below and are shown in Figure 4-23. Note that the edges of the gusset plate can be cut 45°, provided that the distance from the edge of the hole to the cut is not less than the primary edge distance.

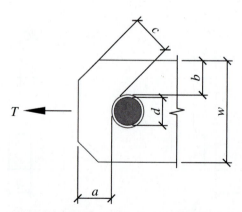

Figure 4-23 Dimensional requirements for pin-connected members.

$$a \geq 1.33b_{\text{eff}} \tag{4-22}$$

$$w \geq 2b_{\text{eff}} + d \tag{4-23}$$

$$c \geq a \tag{4-24}$$

EXAMPLE 4-11

Tension Rod Design

1. For the braced frame shown in Figure 4-24, design the threaded rod, clevis, and turnbuckle for the applied lateral load shown. The threaded rod conforms to ASTM A36 and the clevises and turnbuckles conform to ASTM A29, grade 1035.

2. Determine if the gusset plate connection is adequate. The plate is ASTM A36.

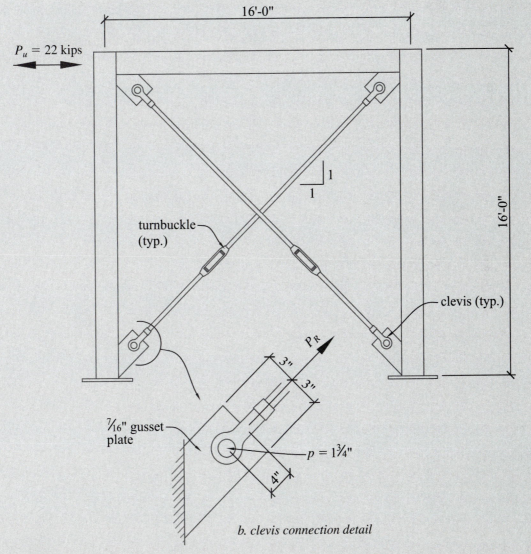

16'-0"

$P_u = 22$ kips

turnbuckle (typ.)

clevis (typ.)

16'-0"

P_R

3"

3"

$\frac{7}{16}$" gusset plate

$p = 1\frac{3}{4}$"

4"

b. clevis connection detail

Figure 4-24 Details for Example 4-11.

(continued)

SOLUTION

Load to each tension rod: *(Note: Only one tension rod is engaged when the lateral load is applied since the threaded rod is too slender to support compression loads.)* We will assume that the slenderness ratio, *L/r* is greater than 300 and account for a pretension force of 10%.

$$P_R = \frac{(22k)(1.10)}{\cos 45} = 34.3 \text{ kips (factored load on tension rod and connectors)}$$

From the *AISCM*, Table 2-5,

$F_y = 36$ ksi

$F_u = 58$ ksi to 80 ksi (use $F_u = 58$ ksi)

1. From *AISCM* Table 15-3, a No. 3 clevis is required ($\phi R_n = 37.5$ kips > 34.3 kips). The maximum threaded rod diameter allowed is 1-3/8 in. and the maximum pin diameter is p = 1¾-in.
   From *AISCM* Table 15-4, a No. 3 clevis can be used with a pin that varies in diameter from 1 in. to 1¾ in.
   From *AISCM* Table 15-5, a turnbuckle with a rod diameter of 1¼-in. is required ($\phi R_n = 38$ kips > 34.3 kips).
   Recall that the pin diameter must be at least 125% of the threaded rod diameter.

   $$D_{\text{pin}}(\text{required}) = 1.25D_{\text{rod}} = (1.25)(1.25) = 1.57 \text{ in. (use a 1¾-in. pin)}$$

   Check the 1¼-in. threaded rod:

   $A_b = 1.23$ in.^{2} (*AISCM* Table 7-2)

   $\phi R_n = \phi 0.75 F_u A_b$

   $\quad = (0.75)(0.75)(58)(1.23) = 40.1 \text{ kips} > 34.3 \text{ kips}$

   Use a 1¼-in. threaded rod with a No. 3 clevis and a turnbuckle.

2. Check tensile rupture on the net effective area:

   $$b = \frac{(3 + 3) - \left(1.75 + \dfrac{1}{16}\right)}{2} = 2.09 \text{ in.}$$

   $b_{\text{eff}} = 2t + 0.63 \text{ in.} < b$

   $\quad = (2)(7/16) + 0.63 = 1.5 \text{ in.} < b = 2.09 \text{ in. (use } b_{\text{eff}} = 1.5 \text{ in.)}$

   $\phi P_n = \phi 2t b_{\text{eff}} F_u$

   $\quad = (0.75)(2)(7/16)(1.5)(58) = 57.1 \text{ kips} > 34.3 \text{ kips OK}$

   Check shear rupture on the effective area:

   $$a = 4 - \left(\frac{1.75 + \dfrac{1}{16}}{2}\right) = 3.09 \text{ in.}$$

$$A_{sf} = 2t\left(a + \frac{d}{2}\right)$$

$$= (2)(7/16)\left(3.09 + \frac{1.75}{2}\right) = 3.46 \text{ in.}^2$$

$$\phi P_n = \phi 0.6 F_u A_{sf}$$

$$= (0.75)(0.6)(58)(3.46) = 90.5 \text{ kips} > 34.3 \text{ kips OK}$$

Check bearing on the projected area of the pin:

$$\phi P_n = \phi 1.8 F_y A_{pb}$$

$$= (0.75)(1.8)(36)(1.75)(7/16) = 37.2 \text{ kips} > 34.3 \text{ kips OK}$$

Check dimensional requirements:

$$a \geq 1.33 b_{eff}$$
$$3.09 \text{ in.} \geq 1.33(1.5) = 2 \text{ in. OK}$$

$$w \geq 2 b_{eff} + d$$
$$(3 + 3) = 6 \text{ in.} \geq (2)(1.5) + 1.75 = 4.75 \text{ in. OK}$$

$c \geq a$, not applicable

The 7/16-in. gusset plate is adequate.

4.6 REFERENCES

1. American Institute of Steel Construction. 2006. *Steel construction manual*, 13th ed. Chicago IL: AISC.

2. American Institute of Steel Construction. 2002. *Steel design guide series 17: High-strength bolts—A primer for structural engineers*. Chicago, IL Kulak, Geoffrey

3. American Institute of Steel Construction. 2006. *Steel design guide series 21: Welded connections—A primer for structural engineers*. Chicago, IL Muller, Duane

4. McCormac, Jack. 1981. *Structural steel design,* 3rd ed. New York: Harper and Row.

5. Salmon, Charles G., and John E. Johnson. 1980. *Steel structures: Design and behavior*, 2nd ed. New York: Harper and Row.

6. Smith, J. C. 1988. *Structural steel design: LRFD fundamentals*. New York: Wiley.

7. Segui, William. 2006. *Steel design*, 4th ed. Toronto, ON Thomson Engineering.

8. Limbrunner, George F., and Leonard Spiegel. 2001. *Applied structural steel design*, 4th ed. New York: Prentice Hall.

4.7 PROBLEMS

4-1. Determine the tensile capacity of the ¼-in. by 6-in. plate shown in Figure 4-25. The connection is made with ⅝-in.-diameter bolts, the plate is ASTM A529, grade 50.

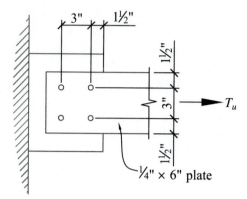

Figure 4-25 Detail for problem 4-1.

4-2.   Determine the following for the connection shown in Figure 4-26, assuming the bolts are ¾-in. diameter and the angle and plate is ASTM A36

1.   Tensile capacity of the angle, and
2.   Required gusset plate thickness to develop the tensile capacity of the angle determined in part 1 above.

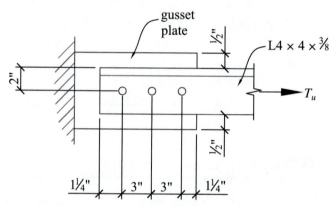

Figure 4-26 Detail for problem 4-2.

4-3.   Determine the design strength of the ⅜″ × 4″ plate shown in Figure 4-27, assuming the steel is ASTM A36.

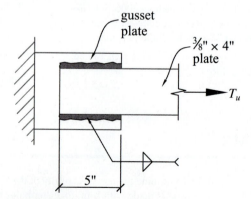

Figure 4-27 Detail for problem 4-3.

4-4. Determine the design strength of the connection shown in Figure 4-28, assuming the steel is ASTM A529, grade 50. The plates are ⅜ in. thick. Neglect the strength of the bolts.

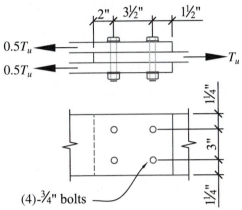

(4)-¾" bolts

Figure 4-28 Details for problem 4-4.

4-5. Determine the net area of the members shown in Figure 4-29.

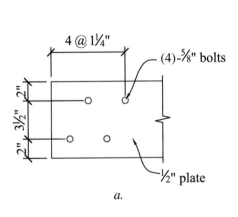

a.

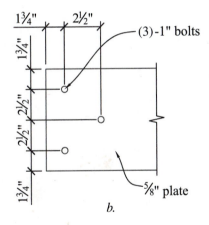

b.

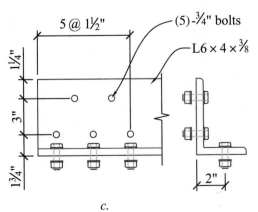

c.

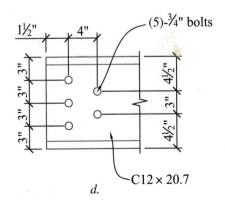

d.

Figure 4-29 Details for problem 4-5.

4-6. For the two-story braced frame shown, design the following, assuming the threaded rod and gusset plate conforms to ASTM A36 and the clevises and turnbuckles conform to ASTM A29, grade 1035:

1. Clevis, turnbuckle, and threaded rod at each level, and
2. Gusset plate, assuming a ⅜-in. thickness.

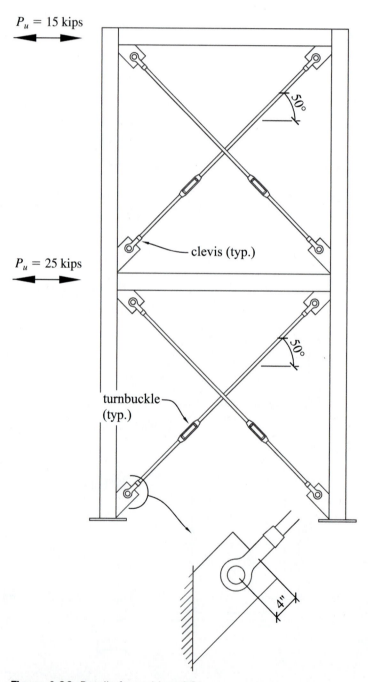

Figure 4-30 Details for problem 4-6.

4-7. Using the lateral loads and geometry shown in problem 4-6, complete the following takes, assuming the tension member is an L4 × 4 × ⅜, ASTM A36 and connected as shown in Figure 4-31:

1. Determine if the L4 × 4 × ⅜ tension member is adequate.
2. Design the gusset plate for the full tension capacity of the angle.

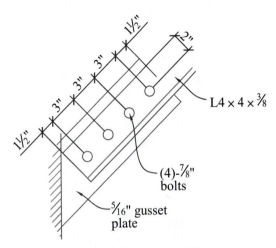

Figure 4-31 Details for problem 4-7.

4-8. Determine the maximum factored tensile force that can be applied as shown in Figure 4-32 and based on the following:

1. Capacity of the angle in tension only, and
2. Capacity of the angle in block shear.

The angle is ASTM A572, grade 50.

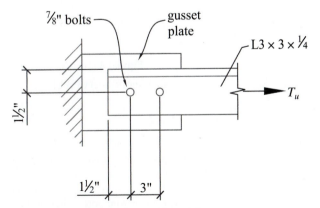

Figure 4-32 Details for problem 4-8.

4-9. For the truss shown in Figure 4-33, determine if member *CD* is adequate for the service loads shown. The steel is ASTM A36. Ignore the strength of the gusset plate.

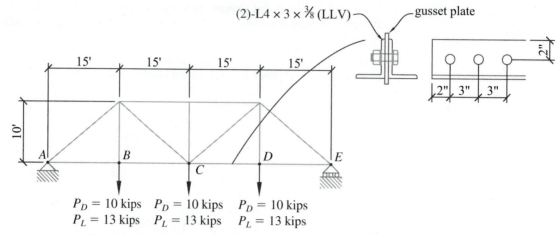

Figure 4-33 Details for problem 4-9.

4-10. For the canopy support details shown in Figure 4-34.

1. Determine the required threaded rod size for member *AB,* assuming ASTM A307, grade C steel for the fastener; and
2. Determine an appropriate clevis and gusset plate size and thickness, assuming the plate at point *A* is 6 in. wide. Use ASTM A36 steel for the plate. The clevis conforms to ASTM A29, grade 1035.

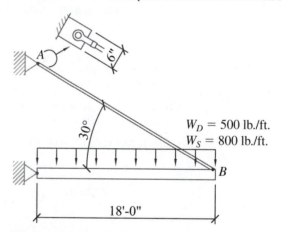

Figure 4-34 Details for problem 4-10.

Student Design Project Problems

4-11. For the X-braces in the student design project (Figure 1-22), design the following, assuming ASTM A36 steel:

1. Using the lateral wind forces previously determined, design the X-braces at the upper and lower levels for the transverse direction. Use double-angles and assume a single line of three ⅞-in.-diameter bolts at the ends. Assume the X-braces are tension-only members and check the limit states for tension on gross and net area, and block shear. Use ASTM A36 steel.
2. Using the lateral wind forces previously determined, design the gusset plates at the upper and lower levels for the transverse direction that connect the X-braces to the intersecting beams and columns. Assume a single line of three ⅞-in.-diameter bolts at the ends
3. Repeat part 1 for the seismic forces.
4. Repeat part 2 for the seismic forces.

4-12. Repeat problem 4-11 for the longitudinal direction.

5

Compression Members Under Concentric Axial Loads

5.1 INTRODUCTION

There are few situations where structural steel elements are subjected to concentric compressive axial forces without any accompanying bending moment. Examples include truss web members, compression chords of some trusses, and some columns in buildings. Smaller compression members are sometimes called posts or struts. In this chapter, we will cover the analysis and design of structural members subject to axial compression with no accompanying bending moment. In Chapter 8, we will discuss beam–columns, that is, structural steel elements subjected to combined axial loads and bending moments, which may occur due to eccentrically applied axial load or to bending loads acting within the length of the member. In structural steel, the common shapes used for columns are wide flange shapes, round and square hollow structural sections (HSS), and built-up sections. For truss members, double- or single-angle shapes are used, as well as round and square HSS and WT-shapes (see Figure 5-1).

5.2 COLUMN CRITICAL BUCKLING LOAD

Consider the two axially loaded members shown in Figure 5-2. In Figure 5-2a, the column is short enough that the failure mode is by crushing compression. This is called a short column. For the longer column shown in Figure 5-2b, the failure mode is buckling at the midspan of the member. This is called a slender, or long, column. Intermediate columns fail by a combination of buckling and compression.

For a pure compression member, the axial load at which the column begins to bow outward is called the Euler critical buckling load. Assuming a perfectly straight member without any initial crookedness and no residual stresses, the Euler critical buckling load for

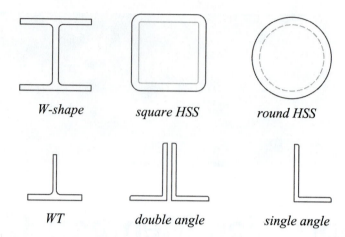

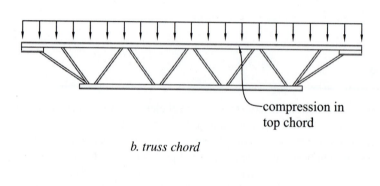

a. compression member types

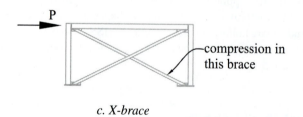

b. truss chord

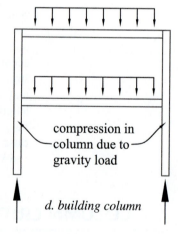

Figure 5-1 Basic compression member types.

a column with pinned ends is

$$P_e = \frac{\pi^2 EI}{L^2},$$

(5-1)

where

P_e = Elastic critical buckling load, lb.,

E = Modulus of elasticity, 29×10^6 psi,

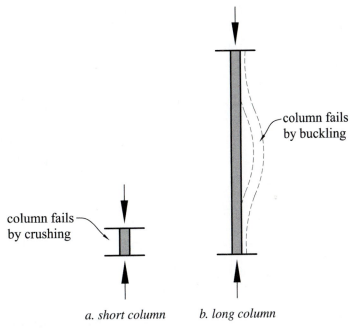

a. short column   b. long column

Figure 5-2 Column failure modes.

I = Moment of inertia, in.^{4}, and

L = Length of the column between brace points, in.

Knowing that $I = Ar^2$ and that the compression stress on any member is $f_c = P/A$, we can express the Euler critical buckling load in terms of stress as

$$F_e = \frac{\pi^2 E}{(L/r)^2},$$ *BETTER EQN than P_c* (5-2)

where

F_e = Euler elastic critical buckling stress, psi,

A = Cross-sectional area, in.^{2}, and

r = Radius of gyration, in.

Equations (5-1) and (5-2) assume that the ends of the column are pinned. For other end conditions, an adjustment or effective length factor, K, is applied to the column length. The effective length of a column is defined as KL, where K is usually determined by one of two methods:

1. *AISCM*, Table C-C2.2—The recommended design values from this table are commonly used in design practice to determine the effective lengths of columns because the theoretical values assume idealized end support conditions. This table, reproduced in Figure 5-3, is especially useful for preliminary design when the size of the beams, girders, and columns are still unknown. In Figure 5-3, *a* through *c* represent columns in braced frames, while *d* through *f* represent columns in unbraced frames. It should be noted that for building columns supported at the top and bottom ends, it is common design practice to assume that the column is pinned at both ends,

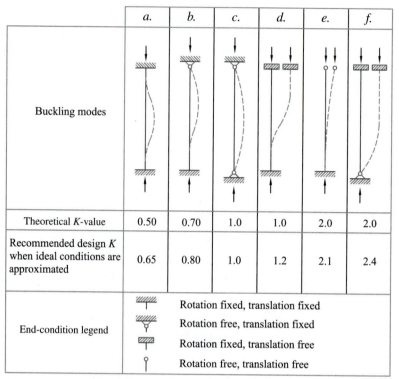

	a.	b.	c.	d.	e.	f.
Buckling modes						
Theoretical K-value	0.50	0.70	1.0	1.0	2.0	2.0
Recommended design K when ideal conditions are approximated	0.65	0.80	1.0	1.2	2.1	2.4

End-condition legend:

Rotation fixed, translation fixed

Rotation free, translation fixed

Rotation fixed, translation free

Rotation free, translation free

Adapted from Table C-C2.2[1]

Figure 5-3 Buckling length coefficients, K.

resulting in a practical effective length factor, K, of 1.0. For columns fixed at both ends, the recommended design value is $K = 0.65$.

2. Nomographs or alignment charts (*AISCM*, Tables C-C2.3 and C-C2.4)—The alignment charts use the actual restraints at the girder-to-column connections to determine the effective length factor, K. They provide more accurate values for the effective length factor than *AISCM*, Table C-C2.2 (see Figure 5-3), but the process of obtaining these values is more tedious than the first method, and the alignment charts can only be used if the initial sizes of the columns and girders are known. This method will be discussed later in this chapter.

When the column end conditions are other than pinned, equations (5-1) and (5-2) are modified as follows:

$$P_e = \frac{\pi^2 EI}{(KL)^2} \tag{5-3}$$

$$F_e = \frac{\pi^2 E}{(KL/r)^2} \tag{5-4}$$

The term KL/r is called the slenderness ratio, and the AISC specification recommends limiting the column slenderness ratio such that

$$\frac{KL}{r} \leq 200 \text{ for compression members.} \tag{5-5}$$

Although the above limit is not mandatory, it should be noted that this is the cutoff point for the *AISCM* design tables for compression members.

Braced Versus Unbraced Frames

In using the alignment charts or Table C-C2.2 of the *AISCM*, it is necessary to distinguish between *braced* and *unbraced frames*. *Braced frames* exist in buildings where the lateral loads are resisted by diagonal bracing or shearwalls as shown in Figure 5-4*a*. The beams and girders in braced frames are usually connected to the columns with simple shear connections,

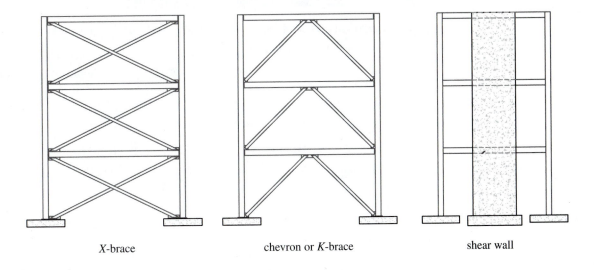

| *X*-brace | chevron or *K*-brace | shear wall |

a. braced frames

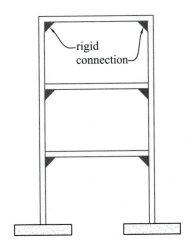

b. unbraced frames

Figure 5-4 Braced and unbraced frames.

and thus there is very little moment restraint at these connections. The ends of columns in braced frames are assumed to have no appreciable relative lateral sway; therefore, the term *nonsway* or *sidesway-inhibited* is used to describe these frames. The effective length factor for columns in braced frames is taken as 1.0. In *unbraced* or *moment frames*, (Figure 5-4b) the lateral loads are resisted through bending of the beams, girders, and columns, and thus the girder-to-column and beam-to-column connections are designed as moment connections. The ends of columns in unbraced frames undergo relatively appreciable sidesway; therefore, the term *sway* or *sway-uninhibited* is used to describe these frames. The effective length of columns in moment frames is usually greater than 1.0.

EXAMPLE 5-1

Determination of Effective Length Factor, *K*, using the *AISCM* Table C-C2.2

Determine the effective length factor for the ground floor columns in the following frames (see Figure 5-5).

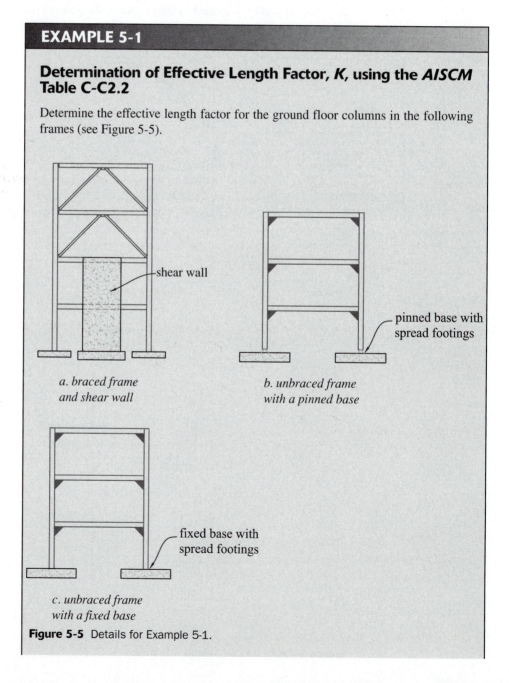

a. braced frame and shear wall

b. unbraced frame with a pinned base

c. unbraced frame with a fixed base

Figure 5-5 Details for Example 5-1.

SOLUTION

a. **Braced Frame**

Since the building is braced by diagonal braces and shear walls, the K-value for all columns in the building is assumed to be 1.0.

b. **Unbraced Frames (Moment Frame with Pinned Column Bases)**

Since the bottom ends of the ground floor columns are pinned, the effective length factor, K, for each column at this level in the moment frame is 2.4.

c. **Unbraced Frames (Moment Frame with Fixed Column Bases)**

Since the bottom ends of the ground floor columns are fixed, the effective length factor, K, for each column at this level in the moment frame is 1.2.

5.3 COLUMN STRENGTH

The assumptions used in the derivation of equations (5-1) through (5-4) assume idealized support conditions that cannot be achieved in real-life structural members. To account for initial crookedness, residual stresses, and end restraints in the compression member, the AISC specification defines the design compressive strength of a column as follows:

$$\phi_c P_n = \phi_c F_{cr} A_g, \tag{5-6}$$

where

$\phi_c = 0.90$,

$P_n = $ Nominal compressive strength, kips,

$F_{cr} = $ Flexural buckling stress (see below), ksi, and

$A_g = $ Gross cross-sectional area of the column, in.^{2}.

The flexural buckling stress, F_{cr}, is determined as follows:

When $\dfrac{KL}{r} \le 4.71\sqrt{\dfrac{E}{F_y}}$ (or when $F_e \ge 0.44F_y$),

$$F_{cr} = \left[0.658^{\frac{F_y}{F_e}}\right]F_y; \tag{5-7}$$

when $\dfrac{KL}{r} > 4.71\sqrt{\dfrac{E}{F_y}}$ (or when $F_e < 0.44F_y$),

$$F_{cr} = 0.877F_e. \tag{5-8}$$

Equation (5-7) accounts for the case where inelastic buckling dominates the column behavior because of the presence of residual stresses in the member, while equation (5-8) accounts for elastic buckling in long or slender columns.

5.4 LOCAL STABILITY OF COLUMNS

The preceding section was based on the global strength and buckling of the column member as a whole. In this section, we will look at the local stability of the individual elements that make up the column section. Local buckling (see Figure 5-6) leads to a reduction in the strength of a compression member and prevents the member from reaching its overall compression capacity. To avoid or prevent local buckling, the AISC specification prescribes limits to the width-to-thickness ratios of the plate components that make up the structural member. These limits are given in section B4 of the *AISCM*.

In Section B4 of the *AISCM*, three possible local stability parameters are defined: *compact*, *noncompact*, or *slender*. A *compact* section reaches its cross-sectional material strength, or capacity, before local buckling occurs. In a *noncompact* section, only a portion of the cross-section reaches its yield strength before local buckling occurs. In a slender section, the cross-section does not yield and the strength of the member is governed by local buckling. The use of slender sections as compression members is not efficient or economical; therefore, the authors do not recommend their use in design practice.

There are also two type of elements of a column section that are defined in the *AISCM*: *stiffened* and *unstiffened*. Stiffened elements are supported along both edges parallel to the applied axial load. An example of this is the web of an I-shaped column where the flanges are connected on either end of the web. An unstiffened element has only one unsupported edge parallel to the axial load—for example, the outstanding flange of an I-shaped column that is connected to the web on one edge and free along the other edge (see Figure 5-7).

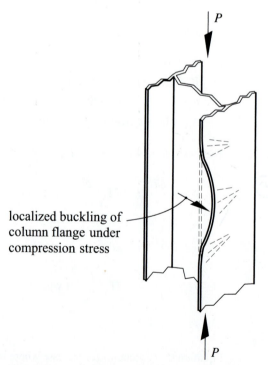

localized buckling of column flange under compression stress

Figure 5-6 Local buckling of column under axial compression load.

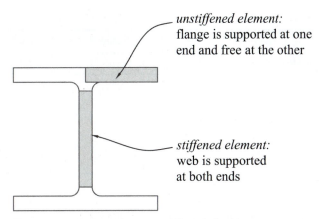

unstiffened element:
flange is supported at one
end and free at the other

stiffened element:
web is supported
at both ends

Figure 5-7 Stiffened and unstiffened elements.

The limiting criteria for compact, noncompact, and slender elements as a function of the width-to-thickness ratio is shown in Table 5-1. When elements of a compression member exceed the limits for noncompact shapes, such an element is said to be slender and a reduction is applied to the flexural buckling stress, F_{cr}, in equations (5-7) and (5-8). For elements that are compact or noncompact, equations (5-7) and (5-8) can be used directly.

Table 5-1 Limiting width–thickness ratios for compression elements

	Description	λ_p (compact)	λ_r (noncompact)	Details
Unstiffened	Flanges of I-shaped sections			
	Outstanding legs of double angles in continuous contact	N/A	$\dfrac{b}{t} \le 0.56\sqrt{\dfrac{E}{F_y}}$	
	Flanges of C-shapes			

(*continued*)

Table 5-1 (continued)

	Description	Limiting Width–Thickness Ratio		Details
		λ_p *(compact)*	λ_r *(noncompact)*	
Unstiffened	Flanges of WT-shapes	N/A	$\dfrac{b}{t} \le 0.56\sqrt{\dfrac{E}{F_y}}$	
	Stems of WT-shapes	N/A	$\dfrac{d}{t} \le 0.75\sqrt{\dfrac{E}{F_y}}$	
	Outstanding legs of double angles not in continuous contact	N/A	$\dfrac{b}{t} \le 0.45\sqrt{\dfrac{E}{F_y}}$	
Stiffened	Webs of I-shaped sections	N/A	$\dfrac{h}{t_w} \le 1.49\sqrt{\dfrac{E}{F_y}}$	
	Webs of C-shapes			
	Square or rectangular HSS	$\dfrac{b}{t} \le 1.12\sqrt{\dfrac{E}{F_y}}$	$\dfrac{b}{t} \le 1.40\sqrt{\dfrac{E}{F_y}}$	*use longer dimension for b*
	Round HSS or pipes	N/A	$\dfrac{D}{t} \le 0.11\left(\dfrac{E}{F_y}\right)$	

Note: N/A = not applicable.

For column shapes with slender elements, the following reduction factors apply to the yield stress, F_y:

When $\dfrac{KL}{r} \leq 4.71\sqrt{\dfrac{E}{QF_y}}$ (or when $F_e \geq 0.44QF_y$),

$$F_{cr} = Q\left[0.658^{\frac{QF_y}{F_e}}\right]F_y; \qquad\qquad\qquad (5\text{-}9)$$

when $\dfrac{KL}{r} > 4.71\sqrt{\dfrac{E}{QF_y}}$ (or when $F_e < 0.44QF_y$),

$$F_{cr} = 0.877F_e, \qquad\qquad\qquad (5\text{-}10)$$

where

$Q = Q_sQ_a$ for members with slender elements $\qquad\qquad (5\text{-}11)$

$\quad = 1.0$ for compact and noncompact shapes,

Q_s = Reduction factor for unstiffened elements (see *AISCM*, Section E7.1), and

Q_a = Reduction factor for stiffened elements (see *AISCM*, Section E7.2).

Most wide flange shapes that are listed in the *AISCM* do not have slender elements; therefore, the reduction factor, Q, is 1.0 for most cases. There are, in fact, very few sections listed in the *AISCM* that have slender elements and these are usually indicated by a footnote. However, some HSS (round and square), double-angle shapes, and WT-shapes are made up of slender elements.

5.5 ANALYSIS PROCEDURE FOR COMPRESSION MEMBERS

It is sometimes necessary to determine the strength of an existing structural member for which the size is known; this process is called analysis, as opposed to design, where the size of the member is unknown and has to be determined. There are several methods available for the analysis of compression members and these are discussed below. The first step is to determine the effective length, KL, and the slenderness ratio, KL/r, for each axis of the column. For many shapes, both KL and r are different for each axis (see Figure 5-8).

Method 1: Use equations (5-6) through (5-8), using the larger of $\dfrac{K_xL_x}{r_x}$ and $\dfrac{K_yL_y}{r_y}$.

Method 2: AISC Available Critical Stress Tables (*AISCM*, Table 4-22)

This gives the critical buckling stress, ϕF_{cr}, as a function of $\dfrac{KL}{r}$ for various values of F_y.

For a given $\dfrac{KL}{r}$, determine ϕF_{cr} from the table using the larger of $\dfrac{K_xL_x}{r_x}$ and $\dfrac{K_yL_y}{r_y}$

(e.g., when $\dfrac{KL}{r} = 97$ and $F_y = 36$ ksi, *AISCM*, Table 4-22 gives $\phi F_{cr} = 19.7$ ksi).

Knowing the critical buckling stress, the axial design capacity can be calculated from the equation

$$\phi P_{cr} = \phi F_{cr}A_g,$$

where A_g is the gross cross-sectional area of the compression member.

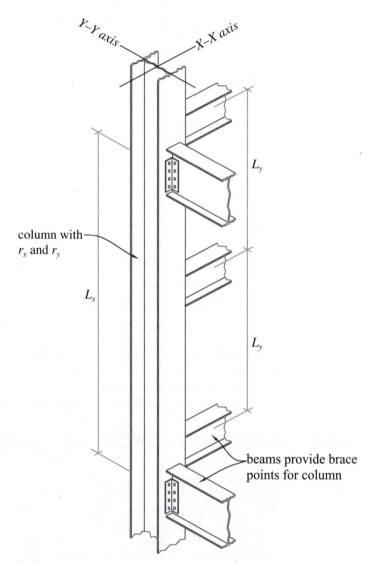

Figure 5-8 Effective length and slenderness ratio.

Method 3: AISCM Available Compression Strength Tables (*AICSM*, Tables 4-1 through 4-12)

These tables give the design strength, $\phi_c P_n$, of selected shapes for various effective lengths, *KL*, and for selected values of F_y. Go to the appropriate table with *KL*, using the larger of

$$\frac{K_x L_x}{\left(\dfrac{r_x}{r_y}\right)} \text{ and } K_y L_y.$$

Notes:

- Ensure that the slenderness ratio for the member is not greater than 200, that is,

$$\frac{KL}{r} \leq 200.$$

- Check that local buckling will not occur, and if local buckling limits are not satisfied, modify the critical buckling stress, ϕF_{cr}, using equations (5-9) through (5-11).
- Use column load tables (i.e., the available compression strength tables) whenever possible. (*Note:* Only a few selected sections are listed in these tables, but these are typically the most commonly used for building construction.)
- Equations (5-6) through (5-8) can be used in all cases for column shapes that have no *slender* elements.

EXAMPLE 5-2

Column Analysis Using the AISC Equations

Calculate the design compressive strength of a W12 × 65 column, 20 ft. long, and pinned at both ends. Use ASTM A572 steel.

$$L_x = L_y = 20 \text{ ft.}$$
$$K = 1.0 \text{ (Figure 5-3)}$$
$$F_y = 50 \text{ ksi}$$
$$A_g = 19.1 \text{ in.}^2$$

Obtain the smallest radius of gyration, r, for W12 × 65 from *AISCM*, part 1. For a W12 × 65,

$$r_x = 5.28 \text{ in., and}$$
$$r_y = 3.02 \text{ in.} \leftarrow \text{Use the smaller value, since } KL \text{ is the same for both axes.}$$
$$\frac{KL}{r} = \frac{(1.0)(20 \text{ ft.})(12)}{3.02} = 79.5 < 200 \text{ OK}$$

Check the slenderness criteria for compression elements:

$$b_f = 12 \text{ in. } (b = 12/2 = 6 \text{ in.})$$
$$t_f = 0.605 \text{ in.}$$
$$t_w = 0.39 \text{ in.}$$
$$h = d - 2k_{\text{des}} = 12.1 - (2)(1.20) = 9.7 \text{ in.}$$

(*Note:* k_{des}, used for design, is smaller than k_{det}, used for detailing. The difference in these values is due to the variation in the fabrication processes.)

$$\frac{b}{t} \le 0.56 \sqrt{\frac{E}{F_y}}; \frac{6}{0.605} = 9.92 < 0.56 \sqrt{\frac{29,000}{50}} = 13.48 \text{ OK}$$

$$\frac{h}{t_w} \le 1.49 \sqrt{\frac{E}{F_y}}; \frac{9.7}{0.39} = 24.88 < 1.49 \sqrt{\frac{29,000}{50}} = 35.88 \text{ OK}$$

Determine the flexural buckling stress, F_{cr}:

$$4.71 \sqrt{\frac{E}{F_y}} = 4.71 \sqrt{\frac{29,000}{50}} = 113.4$$

(continued)

Since $\dfrac{KL}{r} = 79.5 < 113.4$, use equation (5-7) to determine F_{cr}.

$$F_e = \frac{\pi^2 E}{(KL/r)^2} = \frac{\pi^2 29{,}000}{(79.5)^2} = 45.3 \text{ ksi}$$

$$F_{cr} = \left[0.658^{\frac{F_y}{F_e}}\right] F_y = \left[0.658^{\frac{50}{45.3}}\right](50) = 31.5 \text{ ksi}$$

$$\phi F_{cr} = (0.90)(31.5) = 28.4 \text{ ksi}$$

The design strength of the column is then determined from equation (5-6):

$$\phi_c P_n = \phi_c F_{cr} A_g$$
$$= (0.90)(31.5)(19.1) = 541 \text{ kips}$$

From *AISCM*, Table 4-22, ϕF_{cr} could be obtained directly by entering the table with $KL/r = 79.5$ and $F_y = 50$ ksi. A value of about $\phi F_{cr} = 28.4$ is obtained, which confirms the calculation above. Alternatively, the design strength could be obtained directly from AISC, Table 4-1 (i.e., the column load tables). Go to the table with $KL = 20$ ft. and obtain $\phi_c P_n = 541$ ksi.

EXAMPLE 5-3

Analysis of Columns Using the *AISCM* Tables

Determine the design compressive strength for a pin-ended HSS $8 \times 8 \times \frac{3}{8}$ column of ASTM A500, grade B steel with an unbraced length of 35 ft.

SOLUTION

Unbraced column length, $L = 35$ ft.

Pin-ended column: $K = 1.0$, $KL = (1.0)(35 \text{ ft.}) = 35$ ft.

ASTM A500 steel: $F_y = 46$ ksi

For an HSS $8 \times 8 \times \frac{3}{8}$ from part 1 of the *AISCM*, we find that

$$A_g = 10.4 \text{ in}^2.$$
$$r_{(x)} = r_y = 3.10 \text{ in.}$$
$$\frac{KL}{r} = \frac{(1.0)(35 \text{ ft.})(12)}{3.10} = 135.5 < 200 \quad \text{OK}$$

From *AISCM* Table 4-22, ϕF_{cr} is obtained directly by entering the table with $KL/r = 135.5$ and $F_y = 46$ ksi. A value of about $\phi F_{cr} = 12.3$ ksi is obtained; therefore, the design strength is

$$\phi_c P_n = \phi_c F_{cr} A_g = (12.3)(10.4) = 128 \text{ kips.}$$

Alternatively, the design strength could be obtained directly from the *AISCM* column load tables (Table 4-4). Enter the table with $KL = 35$ ft. and obtain $\phi_c P_n = 128$ ksi.

Alternate Check:
Check the slenderness criteria for compression elements:

$$\frac{b}{t} \leq 1.40\sqrt{\frac{E}{F_y}};\ 19.9 < 1.40\sqrt{\frac{29{,}000}{46}} = 35.1 \text{ OK}$$

$$\left(\frac{b}{t} = 19.9 \text{ from part 1 of the } AISCM\right)$$

Determine the flexural buckling stress, F_{cr}:

$$4.71\sqrt{\frac{E}{F_y}} = 4.71\sqrt{\frac{29{,}000}{46}} = 118.2$$

Since $\dfrac{KL}{r} = 135.5 > 118.2$, use equation (5-8) to determine F_{cr}:

$$F_e = \frac{\pi^2 E}{(KL/r)^2} = \frac{\pi^2 29{,}000}{(135.5)^2} = 15.6 \text{ ksi}$$

$$F_{cr} = 0.877 F_e = (0.877)(15.6) = 13.7 \text{ ksi}$$

The design strength of the column is then determined from equation (5-6):

$$\phi_c P_n = \phi_c F_{cr} A_g$$
$$= (0.90)(13.7)(10.4) = 128 \text{ kips}$$

5.6 DESIGN PROCEDURES FOR COMPRESSION MEMBERS

The design procedures for compression members are presented in this section, starting with the design procedure for sections not listed in the *AISCM* column design tables.

1 For Members Not Listed in the *AISCM* Column Tables:

a. Calculate the factored axial compression load or the required axial strength, P_u, and assume a value for the critical buckling stress, $\phi_c F_{cr}$, that is less than the yield stress, F_y.
b. Determine the required gross area, $A_{g\ required}$, which should be greater than or equal to $\dfrac{P_u}{\phi_c F_{cr}}$.
c. Select a section from part 1 of the *AISCM* with $A_g > A_{g\ required}$.
   - Check that $KL/r \leq 200$ for each axis.
d. For the section selected in step 1c, compute the actual $\phi_c F_{cr}$, using either *AISCM*, Table 4-22, or equation (5-7) or (5-8).

e. Compute the design strength, $\phi_c P_n = (\phi_c F_{cr})A_g$, of the selected shape from step c.
   - If $\phi_c P_n \geq P_u$, the section selected is adequate; go to step g.
   - If $\phi_c P_n < P_u$, the column is **inadequate**; go to step f.

f. Use $\phi_c F_{cr}$, obtained in step d, to repeat steps a through e until $\phi_c P_n$ is just greater than or equal to the factored load, P_u (approximately 5% is a suggested value).

g. Check local buckling (see Table 5-1).

2 For Members Listed in the *AISCM* Column Tables:

In designing columns using the column load tables, follow these steps:

a. Calculate P_u (i.e., the factored load on the column).

b. Obtain the recommended effective length factor, K, from Figure 5-3 and calculate the effective length, KL, for each axis.

c. Enter the column load tables (*AISCM*, Tables 4-1 through 4-21) with a KL value that is the larger of $\dfrac{K_x L_x}{\left(\dfrac{r_x}{r_y}\right)}$ and $K_y L_y$, and move horizontally until the **lightest** column section is found with a design strength, $\phi_c P_n >$ the factored load, P_u.

It is recommended to use the column load tables whenever possible because they are the easiest to use.

EXAMPLE 5-4

Design of Axially Loaded Columns Using the *AISCM* Tables

Select a W14 column of ASTM A572, grade 50 steel, 14 ft. long, pinned at both ends, and subjected to the following service loads:

$$P_D = 160 \text{ kips}$$
$$P_L = 330 \text{ kips}$$

SOLUTION

- A572, grade 50 steel: $F_y = 50$ ksi
- Pinned at both ends, $K = 1.0$
- $L = 14$ ft: $KL = (1.0)(14) = 14$ ft.

The factored load, $P_u = 1.2P_D + 1.6P_L = 1.2(160) + 1.6(330) = 720$ kips. From the column load tables in part 4 of the *AISCM*, find the W14 tables. Enter these tables with $KL = 14$ ft. and find the lightest W14 with $\phi_c P_n > P_u$. We obtain a **W14 × 82** with $\phi_c P_n = 774$ kips $> P_u = 720$ kips (*Note*: $\phi_c P_n = 701$ kips for W14 × 74.)

EXAMPLE 5-5

Column Design Using Sections Listed in the *AISCM* Column Tables

Using the *AISCM* column design tables, select the lightest column for a factored compression load, $P_u = 194$ kips, and a column length, $L = 24$ ft. Use ASTM A572, grade 50 steel and assume that the column is pinned at both ends.

SOLUTION

$$K = 1.0$$
$$KL = (1.0)(24 \text{ ft.}) = 24 \text{ ft.}$$

For $KL = 24$ ft., we obtain the following $\phi_c P_n$ from the column load tables (*AISCM*, Table 4-1):

	Selected Size	$\phi_c P_n$, kips
W8 × ??	W8 × 58	205
W10 × ??	W10 × 49	254
W12 × ??	W12 × 53	261
W14 × ??	W14 × 61	293

Always select the lightest column section if other considerations (such as architectural restrictions on the maximum column size) do not restrict the size.

Therefore, use a **W10 × 49** column.

EXAMPLE 5-6

Column Design for Sections Not Listed in the *AISCM* Column Load Tables

Select a W18 column of ASTM A36 steel, 26 ft. long, and subjected to a factored axial load of 500 kips. Assume that the column is pinned at both ends.

SOLUTION

Since W18 shapes are not listed in the *AISCM* column load tables, we cannot use these tables to design this column. Procedure 1 in Section 5.6 will be followed.

$$P_u = 500 \text{ kips,}$$
$$KL = 1.0 \times 26 \text{ ft.} = 26 \text{ ft.}$$

(continued)

Cycle 1:

1. Assume $\phi_c F_{cr} = 20$ ksi $< F_y$ (36 ksi).

2. $A_{g \text{ required}} \geq \dfrac{P_u}{\phi_c F_{cr}} = \dfrac{500}{20} = 25$ in.^{2}

3. Select W18 section from part 1 of the *AISCM* with $A_g \geq 25$ in.^{2}.

   Try **W18 × 86** with $A_g = 25.3$ in.^{2},

   $r_x = 7.77$ in.^{2}, and

   $r_y = 2.63$ in.^{2}.

   $\dfrac{KL}{r_{\min}} = \dfrac{(1.0)(26 \text{ ft.})(12)}{2.63} = 118.6 < 200$ OK

4. Go to *AISCM*, Table 4-22, with the KL/r value from step 3 and obtain $\phi_c F_{cr} = 15.5$ ksi.

5. $\phi_c P_n = (\phi_c F_{cr})(A_g) = (15.5 \text{ ksi})(25.3 \text{ in.}^2) = 392$ kips $< P_u = 500$ kips

The selected column is not adequate. Therefore, proceed to cycle 2.

Cycle 2:

1. Assume $\phi_c F_{cr} = 15.5$ ksi (from step 4 of the previous cycle).

2. $A_{g \text{ required}} \geq \dfrac{P_u}{\phi_c F_{cr}} = \dfrac{500}{15.5} = 32.2$ in.^{2}

3. Try **W18 × 119** with $A_g = 35.1$ in.$^2 > A_{g \text{ required}}$,

   $r_x = 7.90$ in., and

   $r_y = 2.69$ in.

   $\dfrac{KL}{r_{\min}} = \dfrac{(1.0)(26 \text{ ft.})(12)}{2.69} = 116 < 200$ OK

4. Go to AISC, Table 4-22 with $KL/r = 116$ and obtain $\phi_c F_{cr} = 16.0$ ksi.

5. The compression design strength is

   $\phi_c P_n = (\phi_c F_{cr})(A_g) = (16.0 \text{ ksi})(35.1 \text{ in.}^2) = 561$ kips $> P_u = 500$ kips OK

   Therefore, the **W18 × 119** column is adequate.

6. Check the slenderness criteria for compression elements:

   $b_f = 11.3$ in.$(b = 11.3/2 = 5.65$ in.)

   $t_f = 1.06$ in.

   $t_w = 0.655$ in.

   $h = d - 2k_{\text{des}} = 19.0 - (2)(1.46) = 16.08$ in.

(*Note:* k_{des} is smaller than k_{det} and should be used for the design. The difference in these values is due to the variation in the fabrication processes.)

$$\frac{b}{t} \le 0.56\sqrt{\frac{E}{F_y}}; \frac{5.65}{1.06} = 5.33 < 0.56\sqrt{\frac{29,000}{36}} = 15.9 \quad \text{OK}$$

$$\frac{h}{t_w} \le 1.49\sqrt{\frac{E}{F_y}}; \frac{16.08}{0.655} = 24.5 < 1.49\sqrt{\frac{29,000}{36}} = 42.2 \quad \text{OK}$$

This implies that the column section is not slender; therefore, the design strength calculated above does not have to be reduced for slenderness effects.

EXAMPLE 5-7

Analysis of Columns with Unequal, Unbraced Lengths Using the *AISCM* Tables

Determine the design compressive strength of the following column:

- W14 × 82 column
- A572, grade 50 steel
- Unbraced length for strong (*X–X*) axis bending = 25 ft.
- Unbraced length for weak (*Y–Y*) axis bending = 12.5 ft.

SOLUTION

$$K_x L_x = (1.0)(25) = 25 \text{ ft.}$$
$$K_y L_y = (1.0)(12.5) = 12.5 \text{ ft.}$$

From part 1 of the *AISCM*, we find the following properties for W14 × 82:

$$A_g = 24.0 \text{ in.}^2$$
$$r_x = 6.05 \text{ in.}$$
$$r_y = 2.48 \text{ in.}$$

1. Using *AISCM*, Table 4-22,

$$\frac{K_x L_x}{r_x} = \frac{(25 \text{ ft.})(12)}{6.05} = 49.6 < 200 \quad \text{OK, and}$$

$$\frac{K_y L_y}{r_y} = \frac{(12.5 \text{ ft.})(12)}{2.48} = 60.5 < 200 \quad \text{OK, the larger } KL/r \text{ value governs.}$$

Going into *AISCM*, Table 4-22, with a *KL/r* value of 60.5, we obtain $\phi_c F_{cr} = 34.4$ ksi.

Column design strength, $\phi_c P_n = \phi_c F_{cr} A_g = (34.4)(24.0 \text{ in.}^2) = $ **825 kip**

(continued)

2. Using the column load tables,

$$\frac{K_x L_x}{\left(\dfrac{r_x}{r_y}\right)} = \frac{25 \text{ ft.}}{\left(\dfrac{6.05}{2.48}\right)} = 10.25 \text{ ft.}$$

(*Note:* r_x/r_y is also listed at the bottom of the column load tables.)

$K_y L_y = 12.5 \leftarrow$ The larger value governs.

Going into the column load table for W14 $\times$ 82(F_y = 50 ksi) in part 4 of the *AISCM* with KL = 12.5 ft., we obtain

KL	$\phi_c P_n$, kips
12.0	846
12.5	828
13.0	810

The compression design strength, $\phi_c P_n$, is 828 kips. Note that the column load tables also indicate whether or not the member is slender. For members that are slender, the column load tables account for this in the tabulated design strength; therefore, the local stability criteria does not need to be checked.

EXAMPLE 5-8

Design of Columns with Unequal, Unbraced Lengths Using the *AISCM* Tables

Select an ASTM A572, grade 50 steel column to resist a factored compression load, P_u = 780 kips. The unbraced lengths are L_x = 25 ft. and L_y = 12.5 ft. and the column is pinned at each end.

SOLUTION

1. P_u = 780 kip

2. $K_x L_x$ = (1.0)(25 ft.) = 25 ft. (strong axis)

   $K_y L_y$ = (1.0)(12.5 ft.) = 12.5 ft. (weak axis)

3. Initially, assume that the *weak* axis governs $\Rightarrow KL = K_y L_y$. Go to the column load table with $KL = K_y L_y$ = 12.5 ft.

   *W12 $\times$ 79 $\phi_c P_n$ = 874 kips
    W14 $\times$ 82   = 828 kips

   *Try this section because it is the lightest.

4. For W12 × 79,

$r_x = 5.34$ in.,

$r_y = 3.05$ in., and

$r_x/r_y = 1.75$ (from *AISCM* Table 4-1);

thus,

$$\frac{K_x L_x}{\left(\dfrac{r_x}{r_y}\right)} = \frac{25 \text{ ft.}}{1.75} = 14.3 \text{ ft.} \leftarrow \text{The larger value governs, and}$$

$K_y L_y = 12.5.$

Therefore, the original assumption in step 3—that the weak axis governs—was incorrect; the strong axis actually governs.

5. Go into the W12 × 79 column load tables with $\dfrac{K_x L_x}{\left(\dfrac{r_x}{r_y}\right)} = 14.3$ ft. and obtain the

column axial load capacity by linear interpolation:

KL	$\phi_c P_n$, kips
14.0	836
14.3	827
15.0	809

For W12 × 79, $\phi_c P_n = 827$ kips $> P_u = 780$ kips　OK

Use a W12 × 79 column.

5.7 ALIGNMENT CHARTS OR NOMOGRAPHS (SEE *AISCM*, TABLES C-C2.3 AND C-C2.4)

As discussed previously in this chapter, the alignment charts, or nomographs, are an alternate method for determining the effective length factor, K. These nomographs take into account the restraints provided at the ends of the column by the beams or girders framing into the columns. They provide more accurate K-values, but require knowledge of the sizes of the beams, girders, and columns, and are more cumbersome to use. Two charts are presented in the *AISCM*: sidesway inhibited (i.e., buildings with braced frames or shearwalls), reproduced in Figure 5-9, and sidesway uninhibited (i.e., buildings with moment frames), reproduced in Figure 5-10. The following assumptions have been used in deriving these alignment charts, or nomographs [1]:

1. Behavior is purely elastic.
2. All members have a constant cross section.
3. All joints are rigid.

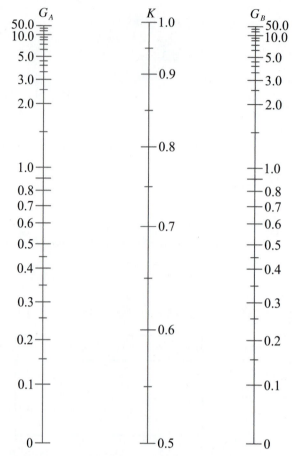

Figure 5-9 Alignment chart: Sidesway inhibited (i.e., braced frames). [1]

4. For columns in sidesway-inhibited frames (i.e., braced frames), rotations at opposite ends of the restraint beams or girders are equal in magnitude and opposite in direction, producing single-curvature bending.

5. For columns in sidesway-uninhibited frames, rotations at opposite ends of the restraining beams or girders are equal in magnitude and direction, producing double- or reverse-curvature bending.

6. The stiffness parameters, $L\sqrt{P/EI}$, of all columns are equal.

7. Joint restraint is distributed to the column above and below the joint in proportion to EI/L for the two columns.

8. All columns buckle simultaneously.

9. No significant axial compression force exists in the beams or girders.

To use these charts, the relative stiffness, G, of the columns, compared to the girders (or beams) meeting at the joint at both ends of each column, is calculated using equation (5-12):

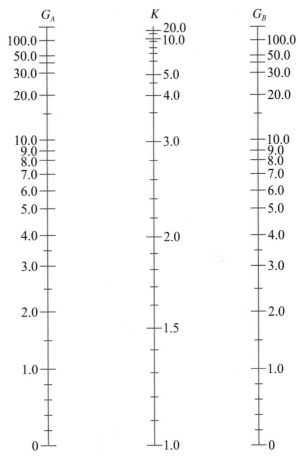

Figure 5-10 Alignment chart: Sidesway uninhibited
(i.e., moment frames). [1]

$$G = \frac{\text{Total column stiffness at the joint}}{\text{Total girder stiffness at the joint}}$$

$$= \frac{\sum\left[\dfrac{\tau_a E_c I_c}{L_c}\right]}{\sum\left[\dfrac{\tau_g E_g I_g}{L_g}\right]} = \frac{\left[\dfrac{\tau_a E_c I_c}{L_c}\right]_{\text{bottom}} + \left[\dfrac{\tau_a E_c I_c}{L_c}\right]_{\text{top}}}{\left[\dfrac{\tau_g E_g I_g}{L_g}\right]_{\text{left}} + \left[\dfrac{\tau_g E_g I_g}{L_g}\right]_{\text{right}}}, \tag{5-12}$$

where

E_c, E_g = Modulus of elasticity for columns and girders, respectively;

$\quad I_c$ = Moment of inertia of column in the plane of bending of the frame;

$\quad I_g$ = Moment of inertia of the girders in the plane of bending of the column;

L_c, L_g = Unsupported or unbraced length of the columns and girders, respectively;

$\quad \tau_a$ = Column stiffness modification factor for inelasticity from Table 5-2

$\quad\quad$ = 1.0 if the assumptions on pages 189 and 190 are satisfied; and

Table 5-2 Column reduction factors,*τ_a

$\frac{P_u}{A_g}$, ksi	F_y			
	36 ksi	42 ksi	46 ksi	50 ksi
45	–	–	–	–
44	–	–	–	0.0599
43	–	–	–	0.118
42	–	–	–	0.175
41	–	–	0.0262	0.231
40	–	–	0.0905	0.285
39	–	–	0.153	0.338
38	–	–	0.214	0.389
37	–	0.057	0.274	0.438
36	–	0.127	0.331	0.486
35	–	0.194	0.387	0.532
34	–	0.260	0.441	0.577
33	–	0.323	0.492	0.620
32	0.0334	0.384	0.542	0.660
31	0.115	0.443	0.590	0.699
30	0.194	0.500	0.636	0.736
29	0.270	0.554	0.679	0.771
28	0.344	0.606	0.720	0.804
27	0.414	0.655	0.759	0.835
26	0.481	0.701	0.796	0.863
25	0.545	0.745	0.830	0.890
24	0.606	0.786	0.861	0.913
23	0.663	0.823	0.890	0.934
22	0.716	0.858	0.915	0.953
21	0.766	0.890	0.938	0.969
20	0.811	0.917	0.957	0.982
19	0.853	0.942	0.974	0.992
18	0.890	0.962	0.986	0.998
17	0.922	0.979	0.996	1.0
16	0.949	0.991	1.0	1.0
15	0.971	0.999	1.0	1.0
14	0.988	1.0	1.0	1.0
13	0.998	1.0	1.0	1.0
12	1.0	1.0	1.0	1.0
11	1.0	1.0	1.0	1.0
10	1.0	1.0	1.0	1.0
9	1.0	1.0	1.0	1.0
8	1.0	1.0	1.0	1.0
7	1.0	1.0	1.0	1.0
6	1.0	1.0	1.0	1.0
5	1.0	1.0	1.0	1.0

*Adapted from *AISCM*, Table 4-21 [1].

Table 5-3 Girder stiffness modification factors, τ_g

	Girder Far-End Condition	Girder Stiffness Modification Factor, τ_g
Sidesway Uninhibited (i.e., unbraced or moment frames)	Far end is fixed	⅔
	Far end is pinned	0.5
Sidesway Inhibited (i.e., braced frames)	Far end is fixed	2.0
	Far end is pinned	1.5

τ_g = Girder stiffness modification factor from Table 5-3

= 1.0 if the assumptions on pages 189 and 190 are satisfied.

For steel structures, the modulus of elasticity of the column, E_c, is the same value as the modulus of elasticity of the girder or beam, E_g; thus, E_c and E_g will cancel out of equation (5-12). In real-life structures, the assumptions listed above are usually approximately satisfied. Where assumption 1 is not satisfied—implying inelastic behavior in the column—the column elastic modulus of elasticity, E_c, in equation (5-12) needs to be replaced by the lower tangent modulus of elasticity, E_t. Thus, the column elastic stiffness in equation (5-12) will be reduced by the ratio E_t/E_c or the stiffness reduction factor, which is denoted by τ_a. The stiffness reduction factors are tabulated in Table 5-2 for different values of the yield strength, F_y. The use of the stiffness reduction factor yields a lower G-factor from equation (5-12), and therefore a lower effective length factor, K, and hence a higher column capacity, resulting in a more economical design.

In the calculation of G, fully restrained (FR) moment connections are assumed at the girder-to-column connections at both ends of the girders. For other situations (i.e., where assumptions 4 and 5 or page 190 are not satisfied), the stiffness of the beams or girders are modified by the adjustment factors given in Table 5-3 for the various far-end support conditions of the girders. Note that when the near end of a girder is pinned (i.e., simple shear connection), the girder or beam stiffness reduction factor is zero. The use of these modification factors yields conservative values of the effective length factor, K.

Although the theoretical G-value for a column with a pinned base (e.g., column supported on spread footing) is infinity, a practical value of 10 is recommended in the *AISCM* because there is no perfect pinned condition. Similarly, the G-value for a column with a fixed base is theoretically zero, but a value of 1.0 is recommended for practical purposes.

The alignment charts assume FR moment connections at the girder-to-column connections (i.e., rigid joints) and, as such, they are mostly useful for columns in unbraced or moment frames (see Figure 5-11 or *AISCM*, Table C-C2.4). The girder-to-column connections in braced frames are usually simple shear connections with no moment restraints; therefore, for a braced frame with pinned column bases, the G-factor will be 10 for both the top and bottom ends of a typical column in the frame. Using these G-factors in the sway-inhibited alignment chart (Figure 5-10 or *AISCM*, Table C-C2.3) yields an effective length factor, K, of 0.96. This value is not much different from the effective length factor of 1.0 that is obtained for a similar pin-ended column using Figure 5-3. It should be noted that the effective length factor, K, of 1.0 is the recommended practical value for columns that are supported at both ends in building structures. Consequently, the alignment charts will only be used in this text for unbraced or moment frames (i.e., sidesway uninhibited).

The procedure for using the alignment charts to determine the effective length factor, K, for a column is as follows:

1. Calculate the factored axial load, P_u, on the column. It is assumed that at this stage, the girder and beam sizes, and the preliminary column sizes have already been determined.

2. Calculate the stiffness of the girders and columns, and, where necessary, modify the girder stiffness using the adjustment factors in Table 5-3 based on the support conditions at the far ends of the girders.

   Note: Where a girder is pinned at the joint under consideration (i.e., connected to the column with a simple shear connection at the near joint), that girder stiffness (i.e., EI/L of the girder) will be taken as zero in calculating the G-factor at that joint.

3. Where necessary, modify the column stiffness by the inelasticity reduction factor, τ_a, from Table 5-2.

4. Calculate the G-factors at both ends of the column. Assume that G_A is the G-factor calculated at the *bottom* of the column and G_B is the G-factor calculated at the *top* of the column.

5. For a pinned column base, use $G_A = 10$; for a fixed column base, use $G_A = 1.0$.

6. Plot the G_A and G_B factors on the corresponding vertical axes of the applicable alignment chart. For unbraced or moment frames, use the alignment chart shown in Figure 5-11.

7. Join the two plotted points (i.e., G_A and G_B) with a straight line; the point at which the vertical K-axis on the alignment chart is intercepted gives the value of the effective length factor, K.

The alignment charts (or nomographs) yield more accurate values for the effective length factor, K, than Figure 5-3; however, they can only be used if the preliminary sizes of the beams, girders, and columns are known, unlike Figure 5-3, which is independent of member size. The recommended design values for the effective length factor, K, from Figure 5-3 can be used for preliminary, as well as final, design.

EXAMPLE 5-9

Effective Length Factor of Columns Using Alignment Charts

For the two-story moment frame shown in Figure 5-11, the preliminary column and girder sizes have been determined as shown. Assume in-plane bending about the strong axes for the columns and girders, and assume columns supported by spread footings. The factored axial loads on columns *BF* and *FJ* are 590 kip and 140 kip, respectively, and $F_y = 50$ ksi.

1. Determine the effective length factor, K, for columns *BF* and *FJ* using the alignment charts, assuming elastic behavior.

2. Determine the effective length factor, K, for columns *BF* and *FJ* using the alignment charts, assuming inelastic behavior.

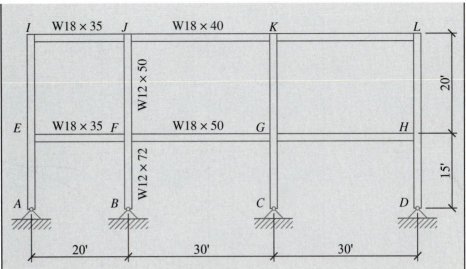

Figure 5-11 Moment frame for Example 5-9.

SOLUTION

The moments of inertia for the given column and girder sections are as follows:

Member	Section	I_{xx} (in.⁴)	Length (ft.)	I_{xx}/L
FJ	W12 × 50	391	20	19.6
BF	W12 × 72	597	15	39.8
IJ, EF	W18 × 35	510	20	25.5
JK	W18 × 40	612	30	20.4
FG	W18 × 50	800	30	26.7

1. Elastic Behavior
Column *BF*:

Joint *B*: The bottom of column *BF* is supported by a spread footing that provides little or no moment restraint to the column; therefore, it is assumed to be pinned. Thus, $G_A = 10$ (This is the practical value recommended in the *AISCM* as discussed earlier.)

Joint *F*:

$$\tau_a = 1.0 \text{ and } \tau_g = 1.0$$

$$G_B = \frac{\sum \left(\dfrac{\tau_a E_c I_c}{L_c} \right)}{\sum \left(\dfrac{\tau_g E_g I_g}{L_g} \right)} = \frac{E[\overset{BF}{(1.0)(39.8)} + \overset{FJ}{(1.0)(19.6)}]}{E[\underset{EF}{(1.0)(25.5)} + \underset{FG}{(1.0)(26.7)}]} = 1.14 \text{ (top of column } BF)$$

(continued)

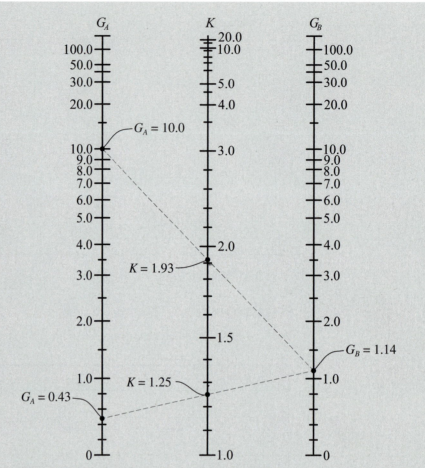

Figure 5-12 Alignment chart for elastic behavior.

Entering the alignment chart for unbraced frames (see Figure 5-11) with a G_A of 10 at the bottom of column BF and a G_B of 1.14 at the top of the column, and joining these two points with a straight line, yields a K-value of **1.93**, as shown in Figure 5-12.

Column FJ:

$\tau_a = 1.0$ and $\tau_g = 1.0$

Joint F: $G_A = \dfrac{\sum\left(\tau_a \dfrac{E_c I_c}{L_c}\right)}{\sum\left(\tau_g \dfrac{E_g I_g}{L_g}\right)} = \dfrac{E[(1.0)(39.8) + (1.0)(19.6)]}{E[(1.0)(25.5) + (1.0)(26.7)]} = 1.14$ (bottom of column FJ)

Joint J: $G_B = \dfrac{\sum\left(\dfrac{E_c I_c}{L_c}\right)}{\sum\left(\dfrac{E_g I_g}{L_g}\right)} = \dfrac{E[(1.0)(19.6)]}{E[(1.0)(25.5) + (1.0)(20.4)]} = 0.43$ (top of column FJ)

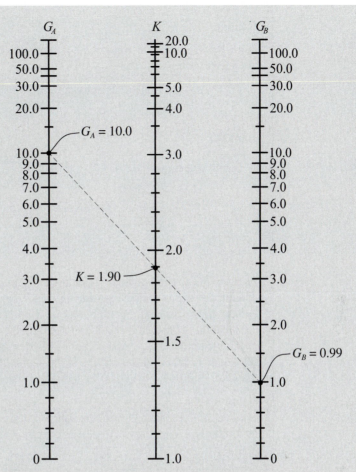

Figure 5-13   Alignment chart for inelastic behavior.

Going to the alignment chart for moment frames (see Figure 5-11) with a G_A of 1.14 at the bottom of column *FJ* and a G_B of 0.43 at the top of the column, and joining these two points with a straight line, yields a *K*-value of 1.25, as shown in Figure 5-13.

2. Inelastic Behavior

Column	Size	Area, A_g, in.^{2}	Factored Axial Load, P_u, kips	P_u/A_g	Stiffness Reduction Factor, τ_a
BF	W12 × 72	21.1	590	28.0	0.804
FJ	W12 × 50	14.6	140	9.6	1.0

(continued)

Column *BF*:

Joint *B*: The bottom of column *BF* is supported by a spread footing; therefore, it is assumed to be pinned. Thus,

$G_A = 10$ (This is the practical value recommended in the *AISCM* as discussed earlier.)

Joint *F*:

$\tau_a = 0.804$ and $\tau_g = 1.0$

$$G_B = \frac{\sum \left(\dfrac{\tau_a E_c I_c}{L_c} \right)}{\sum \left(\dfrac{\tau_g E_g I_g}{L_g} \right)} = \frac{E[(0.804)(39.8) + (1.0)(19.6)]}{E[(1.0)(25.5) + (1.0)(26.7)]} = 0.99 \;\;\text{(top of column *BF*)}$$

Entering the alignment chart for unbraced frames with a G_A of 10 at the bottom of column *BF* and a G_B of 0.99 at the top of the column, and joining these two points with a straight line, yields a *K*-value of **1.90**, as shown in Figure 5-14. It can be seen that the difference in the effective length factor and the effect of inelasticity are negligible for this example. It is common in design practice to conservatively assume the elastic behavior of the column because the effective length factor obtained in that case will be higher than for inelastic behavior.

Column *FJ*:
Column *FJ* is unchanged from the solution in part 1 because the stiffness reduction factor for this column is 1.0; therefore, the effective length factor will be as calculated in part 1.

5.8 REFERENCES

1. American Institute of Steel Construction. 2006. *Steel construction manual*, 13th ed. Chicago.

2. International Codes Council. 2006. *International building code—2006*. Falls Church, VA.

3. American Society of Civil Engineers. 2005. *ASCE-7, Minimum design loads for buildings and other structures*. Reston, VA.

4. American Concrete Institute. 2005. *ACI 318, Building code requirements for structural concrete and commentary*. Farmington Hills, MI.

5.9 PROBLEMS

5-1. Determine the design strength of the column shown in Figure 5-14 using the following methods:

1. Design equations (i.e., equations (5-6) through (5-8)); check slenderness.

2. Confirm the results from part 1 using *AISCM*, Table 4-22.

3. Confirm the results from part 1 using *AISCM*, Table 4-4.

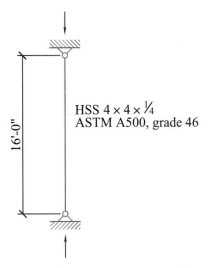

Figure 5-14 Details for problem 5-1.

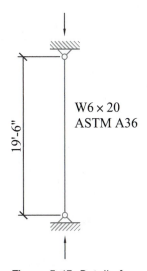

Figure 5-15 Details for problem 5-2.

5-2. Determine the design strength of the column shown in Figure 5-15 using the following methods:

1. Design equations (i.e., equations (5-6) through (5-8)); check slenderness.

2. Confirm the results from part 1 using *AISCM* Table 4-22.

5-3. A column with an unbraced length of 18 ft. must resist a factored load of $P_u = 200$ kips. Select the lightest W-shape to support this load, consider W8, W10, W12, and W14 shapes. The steel is ASTM A572, grade 50.

5-4. A pipe column with a factored load of 80 k has an unbraced length of 10 ft. Select the lightest shape, using the following:

1. Standard pipe (STD),

2. Extra strong pipe (XS), and

3. Double extra strong pipe (XXS).

Steel is ASTM A53, grade B.

What is the maximum unbraced length permitted for a 4-in extra strong (XS) pipe column to support this load?

5-5. A W-shape column must support service loads of $D = 200$ kips and $L = 300$ kips. The unbraced length in the x-direction is 30 ft., 15 ft. in the y-direction. Select the lightest *W*-shape to support this load.

5-6. Determine if a W8 × 28 column is adequate to support a factored axial load of $P_u = 175$ kips with unbraced lengths, $L_x = 24$ ft. and $L_y = 16$ ft., and pinned-end conditions. The steel is ASTM A992, grade 50.

5-7. The preliminary column and girder sizes for a two-story moment frame are shown in Figure 5-16. Assuming in-plane bending about the strong axes for the columns and girders, determine the effective length factor, K, for columns CG and GK using the alignment chart and assuming elastic behavior.

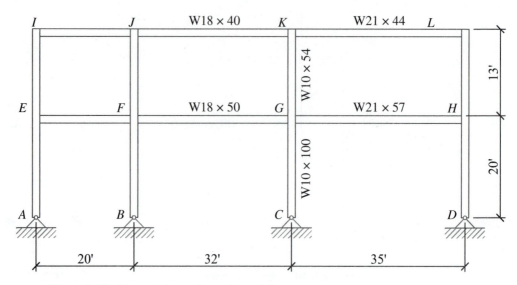

Figure 5-16   Moment frame for problem 5-7.

Student Design Project Problem

Neglecting the Beam-to-column and girder-to-column connection eccentricities (and therefore the moments caused by those eccentricities), calculate the cumulative factored axial loads on the typical interior, exterior, and corner columns for each level of the building. Determine the required column size for each level of the building, assuming an effective length factor, K, of 1.0.

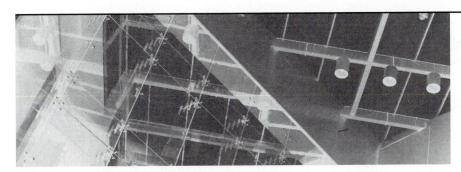

Noncomposite Beams

6.1 INTRODUCTION

Beams are the most common members found in a typical steel structure. Beams are primarily loaded in bending about a primary axis of the member. Beams with axial loads are called beam-columns, and these will be covered in Chapter 8. Common types of beam are illustrated in Figure 6-1.

Beams can be further classified by the function that they serve. A girder is a member that is generally larger in section and supports other beams or framing members. A joist is typically a lighter section than a beam—such as an open-web steel joist. A stringer is a diagonal member that is the main support beam for a stair. A lintel (or loose lintel) is usually a smaller section that frames over a wall opening. A girt is a horizontal member that supports exterior cladding or siding for lateral wind loads.

The basic design checks for beams includes checking bending, shear, and deflection. The loading conditions and beam configuration will dictate which of the preceding design parameters controls the size of the beam.

We will now review some of the basic principles of bending mechanics. When a beam is subjected to bending loads, the bending stress in the extreme fiber is defined as

$$f_b = \frac{Mc}{I} = \frac{M}{S}.$$

(6-1)

And the yield moment is defined as

$$M_y = F_y S,$$

(6-2)

where

f_b = Maximum bending stress,

M_y = Yield moment,

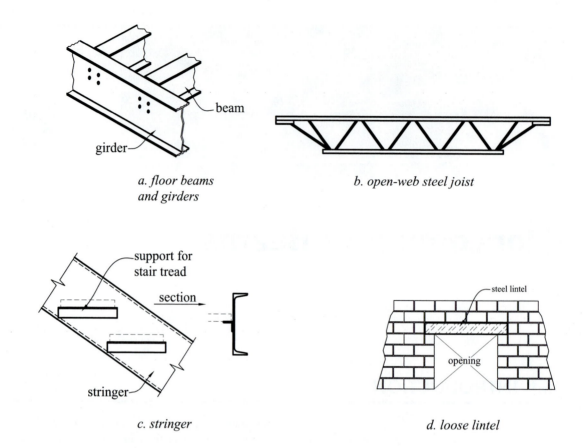

a. floor beams
and girders

b. open-web steel joist

c. stringer

d. loose lintel

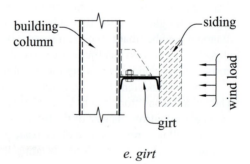

e. girt

Figure 6-1 Common beam members.

F_y = Yield stress,
M = Bending moment,
c = Distance from the neutral axis to the extreme fiber,
I = Moment of inertia, and
S = Section modulus.

The above formulation is based on the elastic behavior of the beam. However, if we assume that the extreme fiber of the steel sections yields and that any additional moment and bending

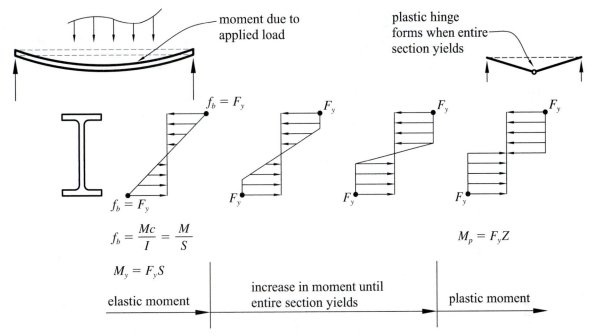

Figure 6-2 Stress distribution for bending members.

stress is distributed to the remaining steel section toward the centroid of the beam, such that the remaining portions of the beam are also brought to the yield limit, a plastic hinge will form in the beam. A plastic hinge occurs when the entire cross section of the beam is at its yield point, not just the extreme fiber. The moment at which a plastic hinge is developed in a beam is called the plastic moment and is defined as

$$M_p = F_y Z. \tag{6-3}$$

where

M_p = plastic moment, and

Z = plastic section modulus.

The plastic moment is the maximum moment, or nominal bending strength of a beam with full lateral stability. For standard wide flange shapes, the ratio of the plastic moment, M_p, to the yield moment, M_y, usually varies from 1.10 to 1.25 for strong axis bending (Z_x, S_x), and 1.50 to 1.60 for weak axis bending. The stress distribution for the above discussion is illustrated in Figure 6-2.

The design parameters for shear and deflection will be discussed later. A summary of the basic load effects for common beam loading conditions is shown below in Table 6-1. *AISCM*, Table 3-23 has several other loading conditions beyond what is shown below.

6.2 CLASSIFICATION OF BEAMS

All flexural members are classified as either compact, noncompact, or slender, depending on the width-to-thickness ratios of the individual elements that form the beam section. There are also two type of elements that are defined in the AISC specification *stiffened* and *unstiffened*

Table 6-1 Summary of shear, moment, and deflection formulas

Loading	Loading Diagram	Maximum Shear	Maximum Moment	Maximum Deflection
Uniformly loaded simple span	 *a. uniformly loaded*	$V = \dfrac{wL}{2}$	$M = \dfrac{wL^2}{8}$	$\Delta = \dfrac{5wL^4}{384EI}$
Concentrated load at midspan	 *b. concentrated load at midspan*	$V = \dfrac{P}{2}$	$M = \dfrac{PL}{4}$	$\Delta = \dfrac{PL^3}{48EI}$
Concentrated loads at ⅓ points	 *c. concentrated loads at ⅓ points*	$V = P$	$M = \dfrac{PL}{3}$	$\Delta = \dfrac{PL^3}{28EI}$

Loading	Loading Diagram	Maximum Shear	Maximum Moment	Maximum Deflection
Uniformly loaded, cantilever	w L V M *d. uniformly loaded, cantilever*	$V = wL$	$M = \dfrac{wL^2}{2}$	$\Delta = \dfrac{wL^4}{4EI}$
Concentrated load at end of cantilever	P L V M *e. concentrated load at end of cantilever*	$V = P$	$M = PL$	$\Delta = \dfrac{PL^3}{3EI}$

elements. Stiffened elements are supported along both edges parallel to the load. An example of this is the web of an I-shaped beam because it is connected to flanges on either end of the web. An unstiffened element has only one unsupported edge parallel to the load; an example of this is the outstanding flange of an I-shaped beam that is connected to the web on one side and free on the other end.

Table 6-2 gives the upper limits for the width-to-thickness ratios for the individual elements of a beam section. These ratios provide the basis for the beam section. When the width-to-thickness ratio is less than λ_p, then the section is *compact*. When the ratio is greater than λ_p but less than λ_r, then the shape is *noncompact*. When the ratio is greater than λ_r, the section is classified as *slender*.

The classification of a beam is necessary since the design strength of the beam is a function of its classification.

The width-to-thickness ratios are also given in part 1 of the *AISCM* for structural shapes, so there is usually no need to calculate this ratio.

Table 6-2 Limiting width–thickness ratios for flexural elements

	Decsription	Limiting Width–Thickness Ratio		Details
		λ_p λ_{pf} (flange) λ_{dw} (web) (compact)	λ_r λ_{rf} (flange) λ_{rw} (web) (noncompact)	
Unstiffened	Flanges of I-shaped sections			
	Flanges of C-shapes	$\dfrac{b}{t} \leq 0.38\sqrt{\dfrac{E}{F_y}}$	$\dfrac{b}{t} \leq 1.0\sqrt{\dfrac{E}{F_y}}$	
	Flanges of WT-shapes			
	Outstanding legs of single angles	$\dfrac{b}{t} \leq 0.54\sqrt{\dfrac{E}{F_y}}$	$\dfrac{b}{t} \leq 0.91\sqrt{\dfrac{E}{F_y}}$	

	Description	Limiting Width-Thickness Ratio		Details
		λ_p λ_{pf} (flange) λ_{dw} (web) (compact)	λ_r λ_{rf} (flange) λ_{rw} (web) (noncompact)	
Stiffened	Webs of I-shaped sections	$\dfrac{h}{t_w} \leq 3.76\sqrt{\dfrac{E}{F_y}}$	$\dfrac{h}{t_w} \leq 5.70\sqrt{\dfrac{E}{F_y}}$	
	Webs of C-shapes			
	Square or rectangular HSS	$\dfrac{h}{t} \leq 2.42\sqrt{\dfrac{E}{F_y}}$	$\dfrac{h}{t} \leq 5.70\sqrt{\dfrac{E}{F_y}}$	use longer dimension for 'h'
	Round HSS or pipes	$\dfrac{D}{t} \leq 0.07\dfrac{E}{F_y}$	$\dfrac{D}{t} \leq 0.31\dfrac{E}{F_y}$	

Adapted from Table B4.1 of the *AISCM*.

EXAMPLE 6-1

Classification of a W-Shape

Determine the classification of a W18 × 35 and a W21 × 48 for F_y = 50 ksi. Check both the flange and the web.

From part 1 of the *AISCM*,

W18 × 35	W21 × 48
$\dfrac{b_f}{2t_f} = 7.06$	$\dfrac{b_f}{2t_f} = 9.47$
$\dfrac{h}{t_w} = 53.5$	$\dfrac{h}{t_w} = 53.6$

Flange:

$$\lambda_{pf} = 0.38\sqrt{\frac{E}{F_y}} = 0.38\sqrt{\frac{29,000}{50}} = 9.15 > 7.06;$$

$$\therefore \text{ the W18} \times 35 \text{ flange is compact.}$$

$$\lambda_{rf} = 1.0\sqrt{\frac{E}{F_y}} = 1.0\sqrt{\frac{29,000}{50}} = 24.0 > 9.47 > 9.15;$$

$$\therefore \text{ the W21} \times 48 \text{ flange is noncompact.}$$

Web:

$$\lambda_{pw} = 3.76\sqrt{\frac{E}{F_y}} = 3.76\sqrt{\frac{29,000}{50}} = 90.5 > 53.6;$$

$$\therefore \text{ the W18} \times 35 \text{ and W21} \times 48 \text{ webs are compact.}$$

Note that shapes that are noncompact for bending are noted by subscript f in Part 1 of the *AISCM*.

6.3 DESIGN STRENGTH IN BENDING FOR COMPACT SHAPES

The basic design strength equation for beams in bending is

$$M_u \leq \phi_b M_n \tag{6-4}$$

where

M_u = Factored moment,
ϕ_b = 0.9,
M_n = Nominal bending strength, and
$\phi_b M_n$ = Design bending strength.

The nominal bending strength, M_n, is a function of the following:

1. Lateral–torsional buckling (LTB),
2. Flange local buckling (FLB), and
3. Web local buckling (WLB).

Flange local buckling and web local buckling are localized failure modes and are only of concern with shapes that have noncompact webs or flanges, which will be discussed in further detail later.

Lateral–torsional buckling occurs when the distance between lateral brace points is large enough that the beam fails by lateral, outward movement in combination with a twisting action (Δ and θ, respectively, in Figure 6-3). Beams with wider flanges are less susceptible to lateral–torsional buckling because the wider flanges provide more resistance to lateral displacement. In general, adequate restraint against lateral–torsional buckling is accomplished by the addition of a brace or similar restraint somewhere between the centroid of the member and the compression flange (see Figure 6-3b). For simple-span beams supporting normal gravity loads, the top flange is the compression flange, but the bottom flange could be in compression for continuous beams or beams in moment frames.

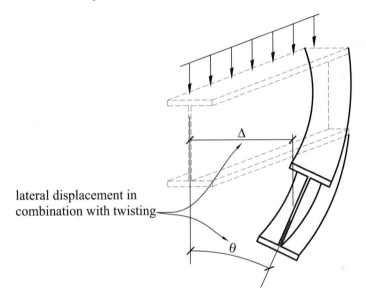

lateral displacement in combination with twisting

a. lateral torsional buckling behavior

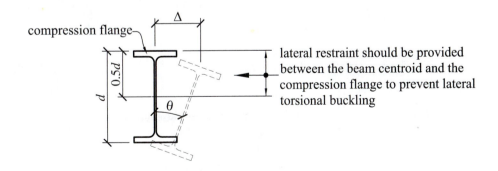

b. lateral torsional buckling restraint

Figure 6-3 Lateral–Torsional buckling.

Lateral–torsional buckling can be controlled in several ways, but it is usually dependant on the actual construction details used. Beams with a metal deck oriented perpendicular to the beam span are considered fully braced (Figure 6-4a), whereas the girder in Figure 6-4b is not considered braced by the deck because the deck has very little stiffness to prevent lateral displacement of the girder. The girder in Figure 6-4c would be considered braced by the intermediate framing members and would have an unbraced length L_b.

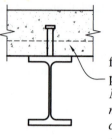

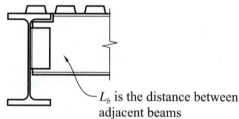

full lateral stability provided ($L_b = 0$)
Note: Full lateral stability may not be provided until concrete cures in some cases.

b. beam section

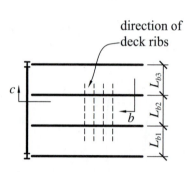

direction of deck ribs

a. typical floor plan

deck ribs are weak in this direction

L_b is the distance between adjacent beams

c. girder section

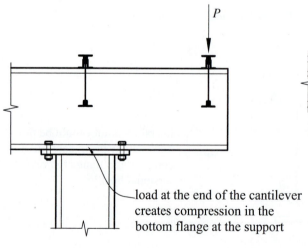

P

load at the end of the cantilever creates compression in the bottom flange at the support

d. compression flange in a cantilever beam

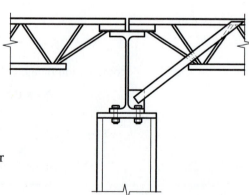

e. lateral bracing of the compression flange in a cantilever beam

Figure 6-4 Lateral bracing details.

When full lateral stability is provided for a beam, the nominal moment strength is the plastic moment capacity of the beam ($M_p = F_y Z_x$ see Section 6-1). Once the unbraced length reaches a certain upper limit, lateral–torsional buckling will occur and therefore the nominal bending strength will likewise decrease. The failure mode for lateral–torsional buckling can be either inelastic or elastic. The AISC specification defines the unbraced length at which inelastic lateral–torsional buckling occurs as

$$L_p = 1.76 r_y \sqrt{\frac{E}{F_y}}. \tag{6-5}$$

L_p is also the maximum unbraced length at which the nominal bending strength equals the plastic moment capacity. The unbraced length at which elastic lateral–torsional buckling occurs is

$$L_r = 1.95 r_{ts} \frac{E}{0.7 F_y} \sqrt{\frac{Jc}{S_x h_o}} \sqrt{1 + \sqrt{1 + 6.76 \left(\frac{0.7 F_y}{E} \frac{S_x h_o}{Jc} \right)^2}}, \tag{6-6}$$

where

$$r_{ts} = \left(\frac{\sqrt{I_y C_w}}{S_x} \right)^{1/2}, \tag{6-7}$$

$$c = \frac{h_o}{2} \sqrt{\frac{I_y}{C_w}} \text{ (for channel shapes)}, \tag{6-8}$$

$c = 1.0$ (for I-shapes),

F_y = Yield strength,

E = Modulus of elasticity,

J = Torsional constant,

S_x = Section modulus (x-axis),

I_y = Moment of inertia (y-axis),

C_w = Warping constant, and

h_o = Distance between flange centroids.

When lateral–torsional buckling is not a concern (i.e., when the unbraced length, $L_b \le L_p$), the failure mode is flexural yielding. The nominal bending strength for flexural yielding is

$$M_n = M_p = F_y Z_x \quad \text{(when } L_b \le L_p\text{)}, \tag{6-9}$$

where

F_y = Yield strength, and

Z_x = Plastic section modulus (from part 1 of the *AISCM*).

For compact I-shapes and C-shapes when $L_p < L_b < L_r$, the nominal flexural strength is

$$M_n = C_b \left[M_p - (M_p - 0.7 F_y S_x) \left(\frac{L_b - L_p}{L_r - L_p} \right) \right] \le M_p. \tag{6-10}$$

In the above equation, the term $0.7F_y S_x$ is also referred to as M_r, which corresponds to the limiting buckling moment when $L_b = L_r$ and is the transition point between inelastic and elastic lateral–torsional buckling (see location of M_r in Figure 6-5).

For compact I-shapes and C-shapes, when $L_b > L_r$, the nominal flexural strength is

$$M_n = F_{cr}S_x \leq M_p, \tag{6-11}$$

where

$$F_{ct} = \frac{C_b \pi^2 E}{\left(\dfrac{L_b}{r_{ts}}\right)^2}\sqrt{1 + 0.078\frac{Jc}{S_x h_o}\left(\frac{L_b}{r_{ts}}\right)^2}, \tag{6-12}$$

C_b = Moment gradient factor

$$= \frac{12.5M_{max}}{2.5M_{max} + 3M_A + 4M_B + 3M_C}R_m \leq 3.0, \tag{6-13}$$

M_{max} = Absolute value of the maximum moment in the unbraced segment,

M_A = Absolute value of the moment at the ¼ point of the unbraced segment,

M_B = Absolute value of the moment at the centerline of the unbraced segment,

M_C = Absolute value of the moment at the ¾ point of the unbraced segment,

R_m = Section symmetry factor

= 1.0 for doubly symmetric members (I-shapes)

= 1.0 for singly symmetric sections in single-curvature bending

$$= 0.5 + 2\left(\frac{I_{yc}}{I_y}\right)^2 \text{ for singly symmetric shapes subjected to reverse curvature}$$

bending, and (6-14)

I_{yc} = Moment of inertia of the compression flange about the y-axis.

For doubly symmetric shapes, I_{yc} is approximately equal to $I_y/2$. For reverse-curvature bending, I_{yc} is the moment of inertia of the smaller flange.

The moment gradient factor, C_b, accounts for the possibility that the entire beam will not be subject to the maximum moment for the entire length of the beam when lateral–torsional buckling controls. It is conservative to assume that $C_b = 1.0$ for any loading condition, which implies that the applied moment is constant across the entire beam. In lieu of equation (6-13), values of C_b can also be obtained from Table 6-3, which is based on equation (6-13).

The variation in bending strength with respect to the unbraced length is shown in Figure 6-5, which summarizes the preceding discussion of bending strength for beams. This figure shows the bending strength for both compact and noncompact shapes. The bending strength of noncompact shapes will be discussed in a later section, but is indicated here for completeness.

There are three distinct zones shown in the figure, the first being where lateral–torsional buckling does not occur and the bending strength is a constant value of M_p. The second and third zones show how the bending strength decreases due to inelastic and elastic lateral–torsional buckling as the unbraced length increases.

The point on the curve at which the bending strength starts to decrease (i.e., when the unbraced length, L_b, equals L_p) is indicated with a darkened circle. The point at which the bending strength undergoes a transition from inelastic to elastic lateral–torsional buckling (i.e., when the unbraced length, L_b, equals L_r and when the nominal bending strength, M_n, equals M_r) is indicated with an open circle. The use of this curve and the curves in the *AISCM* will be discussed in a later section.

Table 6-3 Values of C_b for simple-span beams

Load Description	Lateral Bracing	C_b	
Concentrated load at midspan	None		a. concentrated load at midspan, no lateral bracing
	At load point		b. concentrated load at midspan, lateral brace at midspan
Concentrated load at ⅓ points	None		c. concentrated load at ⅓ points, no lateral bracing
	At all load points		d. concentrated load at ⅓ points, lateral bracing at ⅓ points
Concentrated load at ¼ points	None		e. concentrated load at ¼ points, no lateral bracing
	At all load points		f. concentrated load at ¼ points, lateral bracing at ¼ points
Uniformly loaded	None		g. uniformly loaded, no lateral bracing
	At midspan		h. uniformly loaded, lateral brace at midspan
	At ⅓ points		i. uniformly loaded, lateral bracing at ⅓ points
	At ¼ points		j. uniformly loaded, lateral bracing at ¼ points

Adapted from Table 3-1 of the *AISCM*.

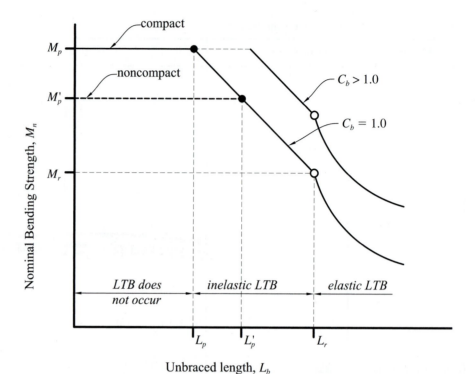

Figure 6-5 Bending strength with respect to unbraced length.

EXAMPLE 6-2

Moment Gradient Factor

Determine the moment gradient factor, C_b, for the beam shown in Figure 6-6.

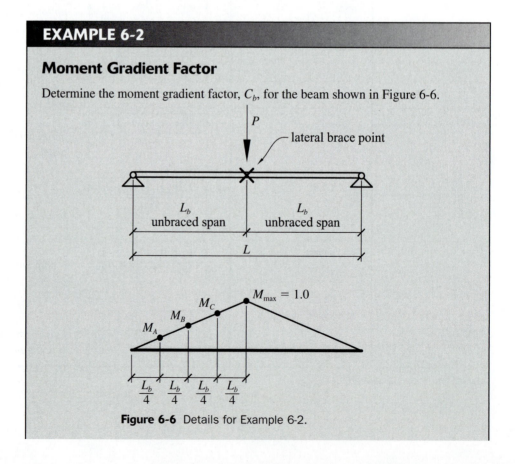

Figure 6-6 Details for Example 6-2.

Assuming a maximum unit moment of $M_{max} = 1.0$, the moments at various locations can be determined by linear interpolation.

From equation (6-13),

$$C_b = \frac{12.5 M_{max}}{2.5 M_{max} + 3 M_A + 4 M_B + 3 M_C} R_m \leq 3.0$$

$$= \frac{12.5(1.0)}{2.5(1.0) + 3(0.25) + 4(0.5) + 3(0.75)} (1.0) \leq 3.0$$

$$= 1.67 \text{ (agrees with Table 6-3).}$$

EXAMPLE 6-3

Bending Strength of a Wide Flange Beam

Determine the design bending strength, or moment capacity, $\phi_b M_n$, for a W14 × 74 flexural member of ASTM A572, grade 50 steel, assuming

1. Continuous lateral support;
2. Unbraced length = 15 ft., $C_b = 1.0$; and
3. Unbraced length = 15 ft., $C_b = 1.30$.

Check the compact section criteria:

From part 1 of the *AISCM*,

$$\frac{b_f}{2t_f} = 6.41, \text{ and}$$

$$\frac{h}{t_w} = 25.4.$$

Flange:

$$\lambda_{pf} = 0.38 \sqrt{\frac{E}{F_y}} = 0.38 \sqrt{\frac{29,000}{50}} = 9.15 > 6.41,$$

∴ the flange is compact.

Web:

$$\lambda_{pw} = 3.76 \sqrt{\frac{E}{F_y}} = 3.76 \sqrt{\frac{29,000}{50}} = 90.5 > 25.4,$$

∴ the web is compact.

1. Continuous lateral support, $L_b = 0$ (upper linear zone in Figure 6-5)

$$M_n = M_p = F_y Z_x$$

$$= (50)(126) = 6300 \text{ in. kips or } 525 \text{ ft.–kips}$$

$$\phi_b M_n = (0.9)(525) = \textbf{473 ft. kips}$$

(continued)

2. $L_b = 15$ ft. and $C_b = 1.0$

$$L_p = 1.76r_y\sqrt{\frac{E}{F_y}} = 1.76(2.48)\sqrt{\frac{29,000}{50}} = 105.1 \text{ in.} = 8.76 \text{ ft.}$$

$$L_r = 1.95r_{ts}\frac{E}{0.7F_y}\sqrt{\frac{Jc}{S_xh_o}}\sqrt{1+\sqrt{1+6.76\left(\frac{0.7F_yS_xh_o}{E}\frac{}{Jc}\right)^2}}$$

$$= 1.95(2.82)\frac{(29,000)}{0.7(50)}\sqrt{\frac{(3.87)(1.0)}{(112)(13.4)}}\sqrt{1+\sqrt{1+6.76\left(\frac{(0.7)(50)(112)(13.4)}{(29,000)(3.87)(1.0)}\right)^2}}$$

$$= 371.2 \text{ in.} = 31.0 \text{ ft.}$$

Since $L_p < L_b < L_r$, the nominal moment strength is found from equation (6-10):

$$M_n = C_b\left[M_p - (M_p - 0.7F_yS_x)\left(\frac{L_b - L_p}{L_r - L_p}\right)\right] \le M_p$$

$$M_n = (1.0)\left[6300 - (6300 - (0.7)(50)(112))\left(\frac{15 - 8.76}{31.0 - 8.76}\right)\right] \le 6300$$

$M_n = 5632$ in.-kips or 469 ft.-kips

$\phi_bM_n = (0.9)(469) = \textbf{422 ft.-kips}$

3. $L_b = 15$ ft. and $C_b = 1.3$

$$M_n = C_b\left[M_p - (M_p - 0.7F_yS_x)\left(\frac{L_b - L_p}{L_r - L_p}\right)\right] \le M_p$$

$$M_n = 1.3\left[6300 - (6300 - (0.7)(50)(112))\left(\frac{15 - 8.76}{31.0 - 8.76}\right)\right] \le 6300$$

$M_n = 7321$ in.-kips > 6300 in.-kips, $\therefore$ flexural yielding controls.

$M_n = M_p = 6300$ in.-kips $= 525$ ft.-kips

$\phi_bM_n = (0.9)(525) = \textbf{473 ft.-kips}$

6.4 DESIGN STRENGTH IN BENDING FOR NONCOMPACT AND SLENDER SHAPES

In the previous section, we considered the flexural strength of compact shapes. In this section, we will consider the strength of noncompact shapes. There are a few noncompact shapes that are available, but there are no standard shapes that are considered slender. Furthermore, all of the available sections in the *AISCM* have compact webs, so this limit state does not have to be considered. Built-up plate girders can have slender flanges and webs, but the design strength of these sections will be considered in a later chapter. The following list indicates the available sections that have noncompact flanges for $F_y = 50$ ksi (also noted with a footnote f in the *AISCM*): M4 × 6, W6 × 8.5, W6 × 9, W6 × 15, W8 × 10, W8 × 31, W10 × 12, W12 × 65, W14 × 90, W14 × 99, W21 × 48.

For compression flange local buckling of noncompact shapes, the nominal flexural strength is

$$M_n = M_p' = \left[M_p - (M_p - 0.7F_yS_x)\left(\frac{\lambda - \lambda_{pf}}{\lambda_{rf} - \lambda_{pf}}\right)\right], \tag{6-15}$$

where

$$\lambda = \frac{b_f}{2t_f} = \text{Width-to-thickness ratio of flange (see Table 6-2)},$$

$\lambda_{pf} = \lambda_p = $ Limiting slenderness ratio for a compact flange (see Table 6-2), and

$\lambda_{rf} = \lambda_r = $ Limiting slenderness ratio for a noncompact flange (see Table 6-2).

Referring back to Figure 6-5, the limiting unbraced length at which lateral–torsional buckling occurs is noted as $L_p{}'$, which is greater than L_p, but less than L_r:

$$L_p{}' = L_p + (L_r - L_p)\left(\frac{M_p - M_p{}'}{M_p - M_r}\right), \tag{6-16}$$

where

$$M_r = 0.7 F_y S_x. \tag{6-17}$$

EXAMPLE 6-4

Bending Strength of a Noncompact Shape

Determine the design moment for a W10 × 12 with (1) $L_b = 0$, and (2) $L_b = 10$ ft. The yield strength is $F_y = 50$ ksi and $C_b = 1.0$.

1. A W10 × 12 is noncompact ∴ Equation (6-15) is used to determine the nominal moment capacity.
   From Table 6-2,

$$\lambda_{pf} = 0.38\sqrt{\frac{E}{F_y}} = 0.38\sqrt{\frac{29,000}{50}} = 9.15$$

$$\lambda_{rf} = 1.0\sqrt{\frac{E}{F_y}} = 1.0\sqrt{\frac{29,000}{50}} = 24.0$$

From *AISCM* Table 1-1, $\lambda = \dfrac{b_f}{2t_f} = 9.43$,

$\lambda_{pf} < \lambda < \lambda_{rf}$ ∴ flange is non-compact

$$M_p = F_y Z_x = (50)(12.6) = 630 \text{ in.-kips}$$

$$M_n = M_p{}' = \left[M_p - (M_p - 0.7F_y S_x)\left(\frac{\lambda - \lambda_{pf}}{\lambda_{rf} - \lambda_{pf}}\right)\right]$$

$$M_n = M_p{}' = \left[630 - (630 - 0.7(50)(10.9))\left(\frac{9.43 - 9.15}{24.0 - 9.15}\right)\right]$$

$$= 625 \text{ in.-kips} = 52.1 \text{ ft.-kips}$$

$$\phi_b M_n = (0.9)(52.1) = \textbf{46.9 ft.-kips}$$

(continued)

2. Determine L_p' and L_r:

$$L_p = 1.76 r_y \sqrt{\frac{E}{F_y}} = 1.76(0.785)\sqrt{\frac{29,000}{50}} = 33.3 \text{ in.} = 2.77 \text{ ft.}$$

$$L_r = 1.95 r_{ts} \frac{E}{0.7F_y} \sqrt{\frac{Jc}{S_x h_o}} \sqrt{1 + \sqrt{1 + 6.76\left(\frac{0.7F_y}{E}\frac{S_x h_o}{Jc}\right)^2}}$$

$$= 1.95(0.983)\frac{(29,000)}{0.7(50)} \sqrt{\frac{(0.0547)(1.0)}{(10.9)(9.66)}} \sqrt{1 + \sqrt{1 + 6.76\left(\frac{0.7(50)}{29,000}\frac{(10.9)(9.66)}{(0.0547)(1.0)}\right)^2}}$$

$$= 96.6 \text{ in.} = 8.05 \text{ ft.}$$

$$M_r = 0.7 F_y S_x = (0.7)(50)(10.9) = 382 \text{ in.-kips}$$

$$L_p' = L_p + (L_r - L_p)\left(\frac{M_p - M_p'}{M_p - M_r}\right)$$

$$L_p' = 2.77 + (8.05 - 2.77)\left(\frac{630 - 625}{630 - 382}\right) = 2.87 \text{ ft.}$$

Since $L_b > L_r$, lateral–torsional buckling will control and equation (6-11) will be used to determine the bending strength:

$$F_{cr} = \frac{C_b \pi^2 E}{\left(\frac{L_b}{r_{ts}}\right)^2} \sqrt{1 + 0.078\frac{Jc}{S_x h_o}\left(\frac{L_b}{r_{ts}}\right)^2}$$

$$F_{cr} = \frac{(1.0)\pi^2(29,000)}{\left(\frac{(10)(12)}{0.983}\right)^2} \sqrt{1 + 0.078\frac{(0.0547)(1.0)}{(10.9)(9.66)}\left(\frac{(10)(12)}{0.983}\right)^2} = 24.3 \text{ ksi}$$

$$M_n = F_{cr} S_x \leq M_p = 630 \text{ in.-kips}$$

$$= (24.3)(10.9) = 265.1 \text{ in.-kips} = \textbf{22.1 ft.-kips}$$

$$\phi_b M_n = (0.9)(22.1) = \textbf{19.9 ft.-kips}$$

6.5 DESIGN FOR SHEAR

In the design process for steel beams, shear rarely controls the design; therefore, most beams need to be designed only for bending and deflection. Special loading conditions, such as heavy concentrated loads or heavy loads on a short span beam, might cause shear to control the design of beams.

From mechanics of materials, the general formula for shear stress in a beam is

$$f_y = \frac{VQ}{Ib} \tag{6-18}$$

where

f_v = shear stress at the point under consideration,

V = vertical shear at a point along the beam under consideration,

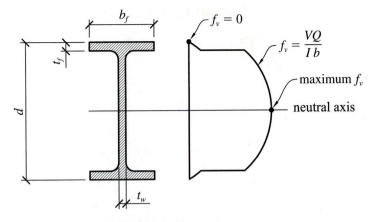

a. actual shear stress

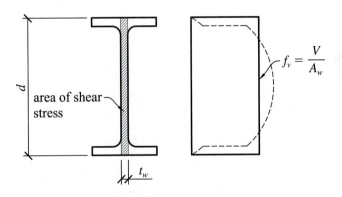

b. approximated shear stress

Figure 6-7 Shear in a beam.

I = moment of inertia about the neutral axis, and

b = thickness of the section at the point under consideration.

The variation in shear stress across the section of a W-shape is shown in Figure 6-7a. Note here that the shear stress in the flange is much smaller that the stress in the web because the variable, b, in equation 6-18 would be the flange width or the web thickness when calculating the shear stress in the beam flange and beam web respectively. For common W-shapes, the flange width can range between 10 to 20 times the thickness of the beam web. Additionally, the distribution of shear stress in the beam flange does not occur as indicated in equation 6-18, because of the low aspect ratio between the flange thickness and flange width (Ref. 9). Equation 6-18 is more directly applicable to steel sections with a high aspect ratio with respect to the direction of the load. For this reason, the AISC specification allows the design for shear to be based on an approximate or average shear stress distribution as shown in Figure 6-7b, where the shear stress is concentrated only in the vertical section of the beam, for which the aspect ratio between the beam depth, d, and the web thickness, t_w, is generally high. In the AISC specification, the shear yield stress is taken as 60% of the yield stress, F_y.

The design shear strength is defined as

$$\phi_v V_n = \phi_v 0.6 F_y A_w C_v, \tag{6-19}$$

where

F_y = Yield stress,
A_w = Area of the web = dt_w,
C_v = Web shear coefficient (see below), and
ϕ_v = 0.9 or 1.0 (see below).

Since the shear stress is concentrated in the beam web, localized buckling of the web needs to be checked. A web slenderness limit for local web buckling if there are I-shaped members is defined as

$$\frac{h}{t_w} \leq 2.24 \sqrt{\frac{E}{F_y}}. \tag{6-20}$$

When this limit is satisfied, local web buckling does not occur and $C_v = 1.0$ and $\phi_v = 1.0$. Most I-shaped members meet the criteria in equation (6-20), except for the following shapes for $F_y = 50$ ksi: W12 × 14, W16 × 26, W24 × 55, W30 × 90, W33 × 118, W36 × 135, W40 × 149, and W44 × 230.

In Part 1 of the *AISCM*, shapes that do not meet the web slenderness criteria are marked with a superscript v.

For the I-shaped members listed above and for all other doubly and singly symmetric shapes and channels (excluding round HSS), $\phi_v = 0.9$ and the web shear coefficient, C_v, is as follows:

For $\dfrac{h}{t_w} \leq 1.10 \sqrt{\dfrac{k_v E}{F_y}}$, $C_v = 1.0$.

For $1.10 \sqrt{\dfrac{k_v E}{F_y}} < \dfrac{h}{t_w} \leq 1.37 \sqrt{\dfrac{k_v E}{F_y}}$,

$$C_v = \frac{1.10 \sqrt{\dfrac{k_v E}{F_y}}}{\dfrac{h}{t_w}}. \tag{6-21}$$

For $\dfrac{h}{t_w} > 1.37 \sqrt{\dfrac{k_v E}{F_y}}$,

$$C_v = \frac{1.51 E k_v}{\left(\dfrac{h}{t_w}\right)^2 F_y}, \tag{6-22}$$

where $k_v = 5$ for unstiffened webs with $h/t_w < 260$, except that $k_v = 1.2$ for the stem of T-shapes.

For all steel shapes, $C_v = 1.0$, except for the following for $F_y = 50$ ksi: M10 × 7.5, M10 × 8, M12 × 10, M12 × 10.8, M12 × 11.8, M12.5 × 11.6, and M12.5 × 12.4.

EXAMPLE 6-5

Shear Strength of a Steel Shape

Determine the design shear strength of the following, using $F_y = 50$ ksi for the W-shapes and $F_y = 36$ ksi for the C-shape:

1. W16 $\times$ 26, and
2. W18 $\times$ 50.
3. C12 $\times$ 20.7

1. W16 $\times$ 26

   From *AISCM*, Table 1-1,

   $$t_w = 0.25 \text{ in.}$$
   $$d = 15.7 \text{ in.}$$
   $$\frac{h}{t_w} = 56.8$$

   $$\frac{h}{t_w} \leq 2.24\sqrt{\frac{E}{F_y}}$$

   $$2.24\sqrt{\frac{29,000}{50}} = 54.0 < h/t_w = 56.8, \therefore \phi_v = 0.9.$$

   Determine C_v:

   $$\frac{h}{t_w} \leq 1.10\sqrt{\frac{k_v E}{F_y}}$$

   $$1.10\sqrt{\frac{(5)(29,000)}{50}} = 59.2 > h/t_w = 56.8, \therefore C_v = 1.0.$$

   From equation (6-19),

   $$\phi_v V_n = \phi_v 0.6 F_y A_w C_v = (0.9)(0.6)(50)(0.25)(15.7)(1.0) = \textbf{106 kips.}$$

2. W18 $\times$ 50

   From *AISCM*, Table 1-1,

   $$t_w = 0.355 \text{ in.}$$
   $$d = 18.0 \text{ in.}$$

   For a W18 $\times$ 50 $C_v = 1.0$ and $\phi_v = 1.0$.

   From equation (6-19),

   $$\phi_v V_n = \phi_v 0.6 F_y A_w C_v = (1.0)(0.6)(50)(0.355)(18.0)(1.0) = \textbf{192 kips.}$$

3. C12 $\times$ 20.7

   From *AISCM*, Table 1-5,

   $$t_w = 0.282 \text{ in.}$$
   $$d = 12 \text{ in.}$$

   For a C12 $\times$ 20.7, $C_v = 1.0$ and $\phi_v = 0.9$.

   From equation (6-19),

   $$\phi_v V_n = \phi_v 0.6 F_y A_w C_v = (0.9)(0.6)(36)(12)(0.282)(1.0) = \textbf{65.8 kips.}$$

6.6 BEAM DESIGN TABLES

The design bending strength of W-shapes and C-shapes with respect to the unbraced length is given in AISC, Tables 3-10 and 3-11, respectively. These tables assume a moment gradient factor of $C_b = 1.0$, which is conservative for all cases, and yield strengths of $F_y = 50$ ksi for W-shapes and $F_y = 36$ ksi for C-shapes. These curves are similar to the curve shown in Figure 6-5, except that the AISC tables have the ϕ-factor incorporated into the design strength.

In the beam design tables, the sections that appear in bold font are the lightest and, therefore, the most economical sections available for a given group of section shapes; these sections should be used especially for small unbraced lengths where possible.

For beams with C_b greater than 1.0, multiply the moment capacity calculated using these tables by the C_b value to obtain the actual design moment capacity of the beam for design moments that correspond to unbraced lengths greater than L_p. Note that $C_b \phi M_n$ must always be less than ϕM_p.

AISCM, Tables 3-2 through 3-5 can be used to select the most economical beam based on section properties. AISCM, Table 3-2 lists the plastic section modulus, Z_x, for a given series of shapes, with the most economical in one series at the top of the list in bold font. The most economical shapes for I_x, Z_y, and I_y are provided in AISCM, Tables 3-3, 3-4, and 3-5, respectively.

AISCM, Table 3-6 provides a useful summary of the beam design parameters for W-shapes. The lower part of the table provides values for ϕM_p, ϕM_r, ϕV_n, L_p, and L_r for any given shape. The upper portion of the table provides the maximum possible load that a beam may support based on either shear or bending strength.

AISCM, Table 3-6 can also be used to determine the design bending strength for a given beam if the unbraced length is between L_p and L_r. When the unbraced length is within this range, the design bending strength is

$$\phi_b M_n = \phi_b M_p - BF(L_b - L_p), \tag{6-23}$$

where BF is a constant found from AISCM, Table 3-6. Note that equation (6-23) is a simpler version of equation (6-10).

The following examples will illustrate the use of the AISC beam design tables.

EXAMPLE 6-6

Design Bending Strength Using the *AISCM* Tables (Compact Shape)

Confirm the results from Example 6-3 using AISCM, Table 3-10.

W14 × 74, $F_y = 50$ ksi

1. $L_b = 0$

From AISCM, Table 3-10, $\phi_b M_n = 473$ ft.-kips (same as Example 6-3).

2. $L_b = 15$ ft., $C_b = 1.0$

From AISCM, Table 3-10, $\phi_b M_n = 422$ ft.-kips (same as Example 6-3).

3. $L_b = 15$ ft., $C_b = 1.3$

From AISCM, Table 3-10, $\phi_b M_n = 422$ ft.-kips Multiplying by C_b yields

$$(1.3)(422) = 548.6 \text{ ft.-kips}$$

Since this is greater than $\phi_b M_p$, the yield strength controls; therefore, $\phi_b M_n = 473$ ft.-kips

Alternatively, from *AISCM*, Table 3-6,

$$\phi_b M_p = 473 \text{ ft.-kips} \quad L_p = 8.76 \text{ ft.}$$
$$\phi_b M_r = 294 \text{ ft.-kips} \quad L_r = 33.1 \text{ ft.}$$
$$\text{BF} = 8.03$$
$$\phi_b M_n = \phi_b M_p - \text{BF}(L_b - L_p) = 473 - (8.03)(15 - 8.76)$$
$$= 422 \text{ ft.-kips (equivalent to Part 2)}$$

EXAMPLE 6-7

Design Bending Strength Using the *AISCM* Tables (Noncompact Shape)

Confirm the results from Example 6-4 using *AISCM*, Table 3-10:

$$W10 \times 12, \, F_y = 50 \text{ ksi}, \, C_b = 1.0$$

1. $L_b = 0$

From Table 3-10, $\phi_b M_n = 47$ ft.-kips (same as Example 6-4).

From Table 3-10, values of $L_p' = 2.8$ ft. and $L_r = 8$ ft. are obtained (solid, dark circle, and open circle, on the curves).

2. $L_b = 10$ ft.

From Table 3-10, $\phi_b M_n = 20$ ft.-kips (same as Example 6-4).

Alternatively, from *AISCM*, Table 3-6,

$$\phi_b M_p = 46.9 \text{ ft.-kips} \quad L_p = 2.87 \text{ ft.}$$
$$\phi_b M_r = 28.6 \text{ ft.-kips} \quad L_r = 8.05 \text{ ft.}$$

EXAMPLE 6-8

Design Shear Strength Using the *AISCM* Tables

Confirm the results from Example 6-5 using the *AISCM*.

1. W16 × 26

From *AISCM*, Table 3-6, $\phi V_n = 106$ kips (same as Example 6-5).

2. W18 × 50

From *AISCM*, Table 3-6, $\phi V_n = 192$ kips (same as Example 6-5).

3. C12 × 20.7

From *AISCM*, Table 3-8, $\phi V_n = 65.8$ kips (same as Example 6-5).

6.7 SERVICEABILITY

In addition to designing for bending and shear, beams also need to be checked for serviceability. There are two main serviceability requirements: deflection and floor vibrations. Floor vibrations are covered in Chapter 12. For beams, deflections must be limited such that the occupants of the structure perceive that the structure is safe. Excessive deflections will often lead to vibration problems.

The deflection equations for common loading conditions were given in Table 6-1. The basic deflection limits are found in Section 1604 of the International Building Code (IBC) and are summarized in Table 6-4 below. Note that only service level loads are used for serviceability considerations.

The deflection limits in Table 6-4 do not consider the effects of ponding, which is discussed further in Chapter 14. In some cases, a beam can be cambered upward to counteract the dead load such that the beam will be in a somewhat flat position prior to the application of live or other loads. The amount of camber varies between 75% and 85% of the actual dead load deflection in order to prevent the possibility of over eambering of the beam and because the end connections provide more end restraint than what is typically assumed in design.

When designing members that support masonry, ACI 530 (ref. 24) requires a deflection limit of $L/600$ or a 0.3-in. maximum, where L is the beam span. When designing members that support cranes, the vertical deflection limit varies from $L/600$ for light cranes to $L/1000$ for heavy cranes (see ref. 16), where the applied load is the crane lifting capacity. For lateral loads on cranes, the deflection limit is $L/400$, where the lateral load is taken as 20% of the crane lifting capacity.

For cantilever beams, the length, L, used in the deflection limit equations is twice the span of the cantilever, since the deflection at the end of a cantilever beam is equivalent to the midspan deflection of a simple span beam (see ref. 2).

In a design situation, it is common to develop approximate deflection equations in order to allow for quicker selection of a member based on deflection limitations.

The deflection for a uniformly loaded, simple-span beam is

$$\Delta = \frac{5wL^4}{384EI},$$

and the maximum moment is

$$M = \frac{wL^2}{8}.$$

Table 6-4 Deflection limits for beams

Member Description	Live Load	Snow Load or Wind Load	Dead Plus Live Load
Roof Members			
Supporting plaster ceiling	L/360	L/360	L/240
Supporting nonplaster ceiling	L/240	L/240	L/180
Not supporting ceiling	L/180	L/180	L/120
Floor Members	L/360	N/A	L/240

Converting the units such that the moment is in ft.-kips and the beam length is in feet yields

$$8M = wL^2$$

$$\Delta = \frac{5(8M)L^2}{384EI} = \frac{(5)(8)(M)(L^2)(1728)}{(384)(29,000)I} = \frac{ML^2}{(161.1)I}$$

$$\Delta = \frac{ML^2}{(161.1)I}, \tag{6-24}$$

where

Δ is in inches,

M is in ft.-kips,

L is in feet, and

I is in in.[4].

Knowing that the two basic deflection limits are L/240 and L/360, equation (6-24) can be modified such that the moment of inertia is calculated as follows:

$$\frac{(L)(12)}{(240)} = \frac{ML^2}{(161.1)I}$$

$$I_{required} = \frac{ML}{8.056} \quad \text{(required moment of inertia for } L/240) \tag{6-25}$$

Similarly, for the L/360 case,

$$I_{required} = \frac{ML}{5.37} \quad \text{(required moment of inertia for } L/360), \tag{6-26}$$

where

M is in ft.-kips,

L is in feet, and

I is in in.[4].

Equations (6-24), (6-25), and (6-26) allow for quick selection of a shape based on a known moment and beam span. These equations can also be used for the approximate sizing of a beam with nonuniform loads by using the maximum moment due to the nonuniform loads as M in the preceding equations. Note that by inspection, it can be seen that when the live load is more than twice the dead load, live load (or L/360) deflections will control. When the dead load is more than half of the live load, then total load deflection limit (or L/240) will control.

For the case of a concentrated load at midspan of a simple-span beam, similar equations can be developed for quicker selection of a steel shape:

$$\Delta = \frac{PL^3}{48EI}$$

$$\frac{(L)(12)}{(240)} = \frac{PL^3(1728)}{(48)(29,000)I}$$

Solving for I,

$$I_{required} = \frac{PL^2}{40.28} \quad \text{(required moment of inertia for } L/240\text{)}. \tag{6-27}$$

Similarly, for the $L/360$ case,

$$I_{required} = \frac{PL^2}{26.85} \quad \text{(required moment of inertia for } L/360\text{)}. \tag{6-28}$$

For the uniformly loaded beam case,

$$\frac{(L)(12)}{(240)} = \frac{5wL^4}{384EI}$$

$$\frac{(L)(12)}{(240)} = \frac{5wL^4(1728)}{384(29,000)I}.$$

Solving for I,

$$I_{required} = \frac{wL^3}{64.44} \quad \text{(required moment of inertia for } L/240\text{)}. \tag{6-29}$$

Similarly, for the L/360 case,

$$I_{required} = \frac{wL^3}{42.96} \quad \text{(required moment of inertia for } L/360\text{)}, \tag{6-30}$$

where

P is in kips,

w is in ft.-kips,

L is in feet, and

I is in in.[4]

The above deflection and required moment of inertia equations are summarized in Table 6-5 below.

Table 6-5 Summary of common deflection equations[1]

Loading	Deflection	L/240	L/360
Variable moment[2]	$\Delta = \dfrac{ML^2}{(161.1)I}$	$I_{reqd} = \dfrac{ML}{8.056}$	$I_{reqd} = \dfrac{ML}{5.37}$
Concentrated load at midspan	$\Delta = \dfrac{PL^3}{806I}$	$I_{reqd} = \dfrac{PL^2}{40.28}$	$I_{reqd} = \dfrac{PL^2}{26.85}$
Uniformly loaded	$\Delta = \dfrac{wL^4}{1289I}$	$I_{reqd} = \dfrac{wL^3}{64.44}$	$I_{reqd} = \dfrac{wL^3}{42.96}$

[1] M is the maximum moment in ft.-kips, P is in kips, w is in kips/ft., L is in ft., and I is in in.[4]

[2] Loading is based on a uniformly distributed load on a simple-span beam

6.8 BEAM DESIGN PROCEDURE

The typical design procedure for beams involves selecting a member that has adequate strength in bending and adequate stiffness for serviceability. Shear typically does not control, but it should be checked as well. The design process is as follows:

1. Determine the service and factored loads on the beam. Service loads are used for deflection calculations and factored loads are used for strength design. The weight of the beam would be unknown at this stage, but the self-weight can be initially estimated and is usually comparatively small enough not to affect the design.

2. Determine the factored shear and moments on the beam.

3. Select a shape that satisfies strength and deflection criteria. One of the following methods can be used:

   a. For shapes listed in the AISC beam design tables, select the most economical beam to support the factored moment. Then check deflection and shear for the selected shape.

   b. Determine the required moment of inertia using Table 6-5. Select the most economical shape based on the moment of inertia calculated, and check this shape for bending and shear.

   c. For shapes not listed in the AISC beam design tables, an initial size must be assumed. An estimate of the available bending strength can be made for an initial beam selection; then check shear and deflection. A more accurate method might be to follow the procedure in step b above.

4. Check floor vibrations (see Chapter 12).

EXAMPLE 6-9

Floor Beam and Girder Design

For the floor plan shown below in Figure 6-8, design members B1 and G1 for bending, shear, and deflection. Compare deflections with $L/240$ for total loads and $L/360$ for live loads. The steel is ASTM A992, grade 50; assume that $C_b = 1.0$ for bending. The dead load (including the beam weight) is assumed to be 85 psf and the live load is 150 psf. Assume that the floor deck provides full lateral stability to the top flange of B1. Ignore live load reduction. Use the design tables in the *AISCM* where appropriate.

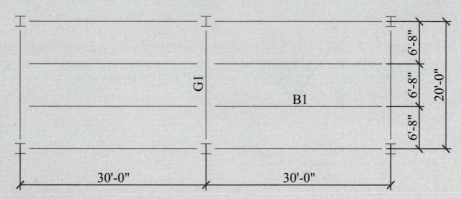

Figure 6-8 Floor plan for Example 6-9.

(continued)

SOLUTION

Since the dead load is more than half of the live load, the total load deflection of $L/240$ will control.

Summary of Loads (see Figure 6-9):

$p_s = 85 + 150 = 235$ psf $= 0.235$ ksf (total service load)

$p_u = (1.2)(85) + (1.6)(150) = 342$ psf $= 0.342$ ksf (total factored load)

B1	**G1**

B1

Tributary width $= 6'-8''$

$w_s = (6.67)(0.235) = 1.57$ kips/ft.

$w_u = (6.67)(0.342) = 2.28$ kips/ft.

$V_u = \dfrac{w_u L}{2} = \dfrac{(2.28)(30)}{2} = 34.2$ kips

$M_u = \dfrac{w_u L^2}{8} = \dfrac{(2.28)(30)^2}{8} = 257$ ft.-kips

G1

Tributary area $= (6'-8'')(30') = 200$ ft.^{2}

$P_s = (200)(0.235) = 47$ kips

$P_u = (200)(0.342) = 68.4$ kips

$V_u = P_u = 68.4$ kips

$M_u = \dfrac{P_u L}{3} = \dfrac{(68.4)(20)}{3} = 456$ ft.-kips

The loading diagrams for B-1 and G-1 are shown in Figure 6-9.

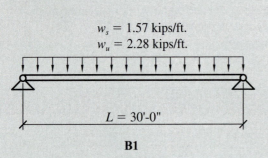

$w_s = 1.57$ kips/ft.
$w_u = 2.28$ kips/ft.

$L = 30'-0''$

B1

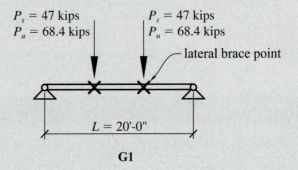

$P_s = 47$ kips
$P_u = 68.4$ kips

$P_s = 47$ kips
$P_u = 68.4$ kips

— lateral brace point

$L = 20'-0''$

G1

Figure 6-9 Loading for B1 and G1.

1. Design of Beam B1

We will use the beam charts and the required moment of inertia method to select a beam size.

$M_u = 257$ ft.-kips and $L_b = 0$

From *AISCM*, Table 3-10, a W16 $\times$ 40 is selected as the most economical size for bending, with $\phi_b M_n = 274$ ft.-kips However, note that a W18 $\times$ 40 has the same beam weight and, therefore, virtually the same cost, but provides more strength ($\phi_b M_n = 294$ ft.-kips) and more stiffness ($I = 612$ in.^{4} versus $I = 518$ in.^{4}). Therefore, a W18 $\times$ 40 is initially selected.

From Table 6-5 (total load controls deflection),

$$I_{required} = \frac{wL^3}{64.44} = \frac{(1.57)(30)^3}{64.44} = 658 \text{ in.}^4$$

The required moment of inertia is greater than the moment of inerita of the W18 $\times$ 40, which is 612 in.^{4}; therefore, a new size needs to be selected.

From *AISCM* Table 1-1,

> W16 × 50, $I = 659$ in.^{4}
>
> W18 × 46, $I = 712$ in.^{4}
>
> **W21 × 44, $I = 843$ in.^{4}** ← *Select*
>
> W24 × 55, $I = 1350$ in.^{4}

The W21 × 44 is the lightest, so this beam is selected. Alternatively, from *AISCM*, Table 3-3, a W21 × 44 is found to be the lightest section with a moment of inertia greater than 658 in.^{4} Checking the moment capacity with *AISCM*, Table 3-10, we find that $\phi_b M_n = 358$ ft.-kips $> M_u = 257$ ft.-kips.

Checking shear, note that a W21 × 44 does not have a slender web; therefore, the design shear strength is determined from equation (6-19), with $C_v = 1.0$ and $\phi_v = 1.0$:

$$\phi_v V_n = \phi_v 0.6 F_y A_w C_v = (1.0)(0.6)(50)(0.35)(20.7)(1.0) = \textbf{217 kips} > V_u = \textbf{34.2 kips} \text{ OK}$$

Alternatively, the shear strength can be found from *AISCM*, Table 3-6 ($\phi V_n = 217$ kips, same as above). *A W21 × 44 is selected for member B1.*

2. Design of Girder G1

We will use the beam charts and the required moment of inertia method to select a beam size.

$$M_u = 456 \text{ ft.-kips and } L_b = 6.67 \text{ ft.}$$

From *AISCM*, Table 3-10, a W24 × 55 is selected as the most economical size for bending, with $\phi_b M_n = 460$ ft.-kips

Checking deflection,

$$\Delta = \frac{PL^3}{28EI} = \frac{(47)[(20)(12)]^3}{(28)(29,000)(1350)} = 0.593 \text{ in.} < \frac{L}{240} = \frac{(20)(12)}{240} = 1 \text{ in. OK}$$

As an alternate approximate check, use equation (6-24) to determine the deflection:

$$M_s = \frac{P_s L}{3} = \frac{(47)(20)}{3} = 313.3 \text{ ft.-kips}$$

$$\Delta = \frac{ML^2}{(161.1)I} = \frac{(313.3)(20)^2}{(161.1)(1350)} = 0.589 \text{ in.}$$

Recall that equation (6-24) was developed for the uniform load case, but the results for the concentrated loads at ⅓ points were close to the actual value of $\Delta = 0.589$ in. (less than a 3% difference). Therefore, equations (6-25) and (6-26) could have reasonably been used for a quick size selection based on stiffness.

Check Shear:

Note that from Section 6-5, the design shear strength is determined from equation (6-19), with $C_v = 1.0$ and $\phi_v = 0.9$:

$$\phi_v V_n = \phi_v 0.6 F_y A_w C_v = (0.9)(0.6)(50)(0.395)(23.6)(1.0) = \textbf{251 kips} > V_u = \textbf{68.4 kips} \text{ OK}$$

Alternatively, the shear strength can be found from *AISCM*, Table 3-6 ($\phi V_n = 251$ kips, same as above). *A W24 × 55 is selected for member G1.*

6.9 BIAXIAL BENDING AND TORSION

Biaxial bending is the bending of the beam about both axes (the *x–x* and *y–y* axes). Pure biaxial bending occurs when the loads to each axis are applied directly through the shear center which is the point within a member such that when loads are applied through that point, twisting will not occur. When the applied loads do not pass through the shear center, as is often the case with singly symmetric shapes, torsion will occur. Examples of these beams are crane girders, purlins for roof framing, and unbraced beams providing lateral support to exterior cladding. Each of these examples has an applied load in the *x* and *y* directions, but since the applied loads in each case do not always pass through the shear center, torsional stresses will occur in addition to bending stresses (see Figures 6-10 and 6-11).

When torsion occurs in a steel section, the effect of warping must be considered. Warping occurs primarily in open sections such as W, C, and L shapes. Warping of a W shape is a condition in which the top and bottom flanges of the cross section have deflected in such

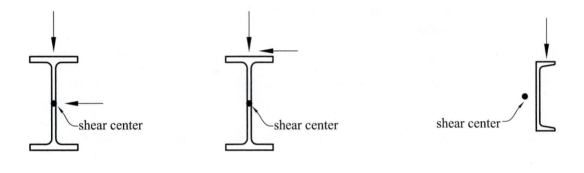

a. pure biaxial bending   b. biaxial bending with torsion   c. torsion in a singly symmetric shape

Figure 6-10 Biaxial bending and torsion loading.

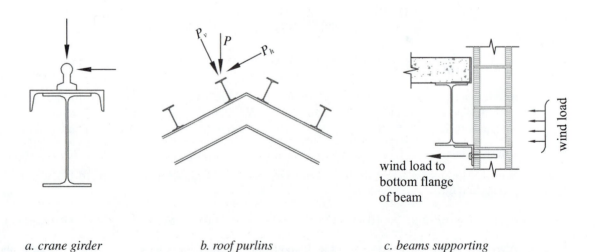

a. crane girder   b. roof purlins   c. beams supporting cladding for lateral loads

Figure 6-11 Biaxial bending and torsion examples.

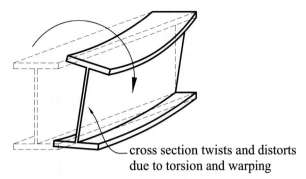

cross section twists and distorts
due to torsion and warping

Figure 6-12 Warping action.

a way that they are no longer parallel to each other (see Figure 6-12). The torsional resistance of an open section is the summation of the torsional stiffness of each of the elements
of the beam section. When a closed section such as a square or circular tube is subjected to
torsion, each element of the section rotates without warping; that is, the plane sections remain virtually plane after rotation. The torsional resistance of a closed section is much larger
than that of an open section, since the torsional stresses can be equally distributed in a closed
section. For this reason, closed sections are highly recommended when any significant torsion is to be resisted.

The general relationship between the torsional moment and the angle of twist is

$$\theta = \frac{TL}{GJ}, \tag{6-31}$$

where

θ = angle of twist, radians,

T = torsional moment,

L = unit length,

G = shear modulus of elasticity = 11,200 ksi for steel, and

J = torsional resistance constant.

The torsional resistance constant J, is equal to the polar moment for circular sections. For rectangular sections, J is slightly less than the polar moment of inertia; for open sections, J is
much less than the polar moment of inertia. The value of J for any section can be calculated
[19, 20] but tabulated values for standard sections are found in Part 1 of the *AISCM*. Since
closed sections are not subject to warping, the general relationship given in equation 6-31 can
be applied directly to various torsional loading conditions on closed sections. Table 6-6 provides the formulas for rotation for closed sections subject to various torsional loads. Note that,
for closed sections, the end conditions are usually assumed to be torsionally fixed, since
warping in closed sections is negligible.

To account for the effects of warping in open sections, tabulated values of a warping
constant, C_w, are given in Part 1 of the *AISCM*. When a W-shaped beam is subjected to torsional loading, the supported ends of the beam are generally restrained against rotation with
standard shear connections and the torsional stresses are concentrated in the beam flanges.
In the middle portion of the beam span, rotation is generally unrestrained and torsional
stresses are distributed to the beam web as well. A torsion bending constant, a, is suggested

Table 6-6 Torsion displacement equations for closed sections

Loading	θ (radians)	Detail
cantilever–concentrated torsion at end	$\theta = \dfrac{PeL}{GJ}$	
cantilever–uniform torsion	$\theta = \dfrac{wel^2}{2GJ}$	
simple span–concentrated torsion	$\theta = \dfrac{Peab}{LGJ}$ at $a = b$: $\theta = \dfrac{PeL}{4GJ}$	
simple span–uniform torsion	$\theta = \dfrac{weL^2}{8GJ}$	

T = concentrated torsion = Pe
t = uniformly distributed torsion = we

in [22] to determine the approximate location along the length of the beam where the effects of torsional restraint are negligible, and this constant is defined as

$$a = \sqrt{\frac{EC_w}{GJ}} \tag{6-32}$$

Table 6-7 [21] provides the approximate values of the flange moment and rotation in W-shapes. (The concept of flange moment will be discussed in the next section.) Note that the end conditions are assumed to be torsionally fixed for cantilever beams, a requirement for equilibrium, and torsionally pinned for simple span conditions. While rotation is virtually zero at the ends of a simple span beam with shear connections, it is conservative to assume torsionally pinned ends for W-shapes. For a W-shape to be torsionally fixed, the top and bottom flanges must have full restraint at the supported ends and the flanges should be connected with stiffener plates such that a tube section is created at the ends of the beam. These stiffeners should extend a length equal to the beam depth in order to be effective [23]. Figure 6-13 shows examples of torsionally pinned and torsionally fixed end conditions for W-shapes. AISC Design Guide 9 [11] provides a more detailed coverage of determining the torsional displacement for various loading and boundary conditions.

In this next section, we will discuss a conservative analysis procedure for beams with either biaxial bending or bending plus torsion. There are more exact methods of analysis that could be used [11, 12, 18] but it is often desirable to use a quicker, more conservative method to expedite the design process in practice. The approach taken here will be to resolve any torsional loading into an equivalent lading parallel to a primary axis of the member in

Table 6-7 Approximate torsion displacement and flange moment equations for W-shapes

Loading		Flange Moment, M_f	θ (radians)	Detail
cantilever—concentrated torsion at end	$L/a < 0.5$	$M_f = \dfrac{PeL}{h_o}$	$\theta = 0.32\left(\dfrac{Pea}{GJ}\right)\left(\dfrac{L}{a}\right)^3$	$T = Pe$
	$0.5 < L/a < 2.0$	$M_f = \dfrac{Pea}{h_o}\left[0.05 + 0.94\left(\dfrac{L}{a}\right) - 0.24\left(\dfrac{L}{a}\right)^2\right]$	$\theta = \dfrac{Pea}{GJ}\left[-0.029 + 0.266\left(\dfrac{L}{a}\right)^2\right]$	
	$2.0 < L/a$	$M_f = \dfrac{Pea}{h_o}$	$\theta = \left(\dfrac{Pe}{GJ}\right)(L - a)$	
cantilever—uniform torsion	$L/a < 0.5$	$M_f = \dfrac{weL^2}{2h_o}$	$\theta = 0.114\left(\dfrac{weLa}{GJ}\right)\left(\dfrac{L}{a}\right)^3$	$t = we$
	$0.5 < L/a < 3.0$	$M_f = \dfrac{weLa}{h_o}\left[0.041 + 0.423\left(\dfrac{L}{a}\right) - 0.068\left(\dfrac{L}{a}\right)^2\right]$	$\theta = \dfrac{weLa}{GJ}\left[-0.023 + 0.029\left(\dfrac{L}{a}\right) + 0.86\left(\dfrac{L}{a}\right)^2\right]$	
	$3.0 < L/a$	$M_f = \dfrac{weLa}{h_o}\left(1 - \dfrac{a}{L}\right)$	$\theta = \left(\dfrac{weLa}{GJ}\right)\left(\dfrac{L}{2a} - 1 + \dfrac{a}{L}\right)$	
simple span—concentrated torsion	$L/a < 1.0$	$M_f = \dfrac{PeL}{4h_o}$	$\theta = 0.32\left(\dfrac{Pea}{GJ}\right)\left(\dfrac{L}{2a}\right)^3$	$T = Pe$
	$1.0 < L/a < 4.0$	$M_f = \dfrac{Pea}{2h_o}\left[0.05 + 0.94\left(\dfrac{L}{2a}\right) - 0.24\left(\dfrac{L}{2a}\right)^2\right]$	$\theta = \dfrac{Pea}{GJ}\left[-0.029 + 0.266\left(\dfrac{L}{2a}\right)^2\right]$	
	$4.0 < L/a$	$M_f = \dfrac{Pea}{4h_o}$	$\theta = \left(\dfrac{Pe}{GJ}\right)\left(\dfrac{L}{2} - a\right)$	
simple span—uniform torsion	$L/a < 1$	$M_f = \dfrac{weL^2}{8h_o}$	$\theta = 0.094\left(\dfrac{weLa}{GJ}\right)\left(\dfrac{L}{2a}\right)^3$	$t = we$
	$1 < L/a < 6.0$	$M_f = \dfrac{weLa}{h_o}\left[0.097 + 0.094\left(\dfrac{L}{2a}\right) - 0.0255\left(\dfrac{L}{2a}\right)^2\right]$	$\theta = \dfrac{weLa}{GJ}\left[-0.032 + 0.062\left(\dfrac{L}{2a}\right) + 0.052\left(\dfrac{L}{2a}\right)^2\right]$	
	$6.0 < L/a$	$M_f = \dfrac{wea^2}{h_o}$	$\theta = \left(\dfrac{weLa}{GJ}\right)\left(\dfrac{L}{8a} - \dfrac{a}{L}\right)$	

Adapted from Ref. [21]

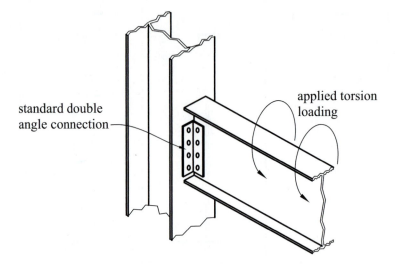

a. torsionally pinned

standard double
angle connection

applied torsion
loading

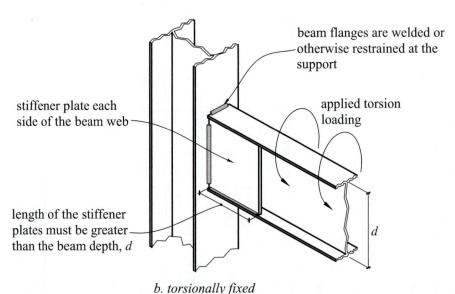

b. torsionally fixed

beam flanges are welded or
otherwise restrained at the
support

stiffener plate each
side of the beam web

applied torsion
loading

length of the stiffener
plates must be greater
than the beam depth, *d*

d

Figure 6-13 Torsionally pinned and torsionally fixed ends for W-shapes.

question such that the member is treated as being subjected to biaxial bending.
There are three cases of biaxial bending of beams that we will consider:

Case 1: Beams in which the load passes *through* the shear center (see Figure 6-14)

Case 2: Beams in which the load *does not* pass through the shear center, but the
vertical component does pass through the shear center. The *shear center*
(SC) is the point through which the load must act if there is to be no twist-

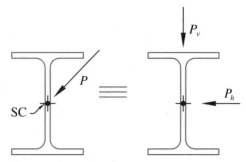

Figure 6-14 Inclined load passing through the shear center.

ing of the beam section. The location of the shear center is given in Part 1 of the *AISCM* for standard shapes. (See Figure 6-15 for the Case 2 loading diagram.)

Case 3: Vertical load that *does not* pass through the shear center (see Figure 6-16).

For Case 1, where the inclined load passes through the shear center, the load is resolved into a horizontal and a vertical component, each of which passes through the shear center. This will result in no twisting of the beam and will cause simple bending about both the *x*- and *y*-axes.

An interaction equation is used to determine whether the member is adequate for combined bending. This equation is given in the AISC specification as

$$\frac{M_{ux}}{\phi_b M_{nx}} + \frac{M_{uy}}{\phi_b M_{ny}} \leq 1.0, \tag{6-33}$$

where

M_{ux} = Factored moment about the *x*-axis,

M_{uy} = Factored moment about the *y*-axis,

M_{nx} = Nominal bending strength for the *x*-axis,

M_{ny} = Nominal bending strength for the *y*-axis, and

ϕ_b = 0.9.

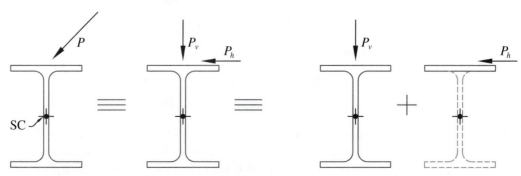

Figure 6-15 Inclined load *not* passing through the shear center.

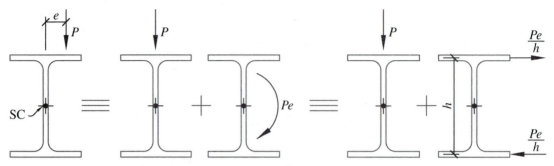

Figure 6-16 Vertical load eccentric to the shear center.

The nominal bending strength for the x-axis has been discussed previously in this chapter. For the y-axis, lateral-torsional buckling is not a limit state, since the member does not buckle about the strong axis when the weak axis is loaded. For shapes with compact flanges, the nominal bending strength about the y-axis is

$$M_{ny} = M_{py} = F_y Z_y \leq 1.6 F_y S_y, \tag{6-34}$$

where

M_{ny} = Nominal bending strength about the y-axis,
M_{py} = Plastic bending strength about the y-axis,
F_y = Yield stress,
Z_y = Plastic section modulus about the y-axis, and
S_y = Section modulus about the y-axis.

For shapes with noncompact flanges, the nominal bending strength about the y-axis is

$$M_{ny} = M_{py} - (M_{py} - 0.7 F_y S_y)\left(\frac{\lambda - \lambda_p}{\lambda_r - \lambda_p}\right) \tag{6-35}$$

For Case 2, where the inclined load does not pass through the shear center but the vertical component does, the load is resolved into a vertical component and a horizontal component located at the top flange (see Figure 6-15). The horizontal component is concentrated near the major half of the y-axis shape. The interaction equation for this case is

$$\frac{M_{ux}}{\phi_b M_{nx}} + \frac{M_{uy}}{0.5(\phi_b M_{ny})} \leq 1.0 \tag{6-36}$$

For Case 3, where there is a vertical load eccentric to the shear center, the load is resolved into a vertical load coincident with the shear center and horizontal forces located at the top and bottom flanges (see Figure 6-15).

The moment about the y-axis (i.e., the flange moment) could be taken as $M_{ny} = \dfrac{Pe}{h_o}$, and the interaction equation for Case 2. [Equation (6-36) could be used.] This moment is applied in such way that each flange displaces in the opposite direction. This is the warping behavior that was previously discussed, so Case 3 is equivalent to beams loaded in pure torsion. For W-shaped beams, the formulas for flange moment and rotation in Table 6-7 are recommended. Practical examples of Case 3 loading are illustrated in Figure 6-17.

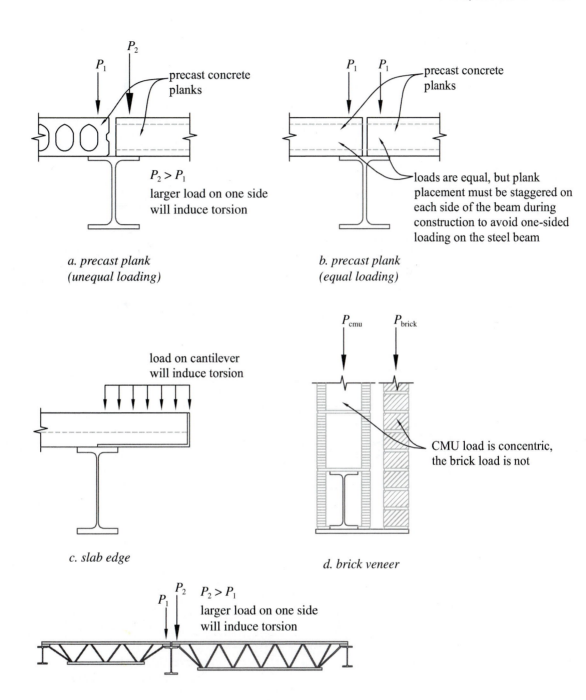

*a. precast plank
(unequal loading)*

*b. precast plank
(equal loading)*

c. slab edge

d. brick veneer

e. girder supporting unequal spans

Figure 6-17 Common torsion examples.

References [11, 12, 18] provide more detailed coverage of torsion in steel design, but a simplified approach (i.e., Case 3 loading) is often used in practice.

In practice, it is common to provide adequate detailing in lieu of allowing any significant torsion, since the analysis of members for torsion can be quite cumbersome. The two most common methods for controlling torsion in practice are to provide a steel section that is closed

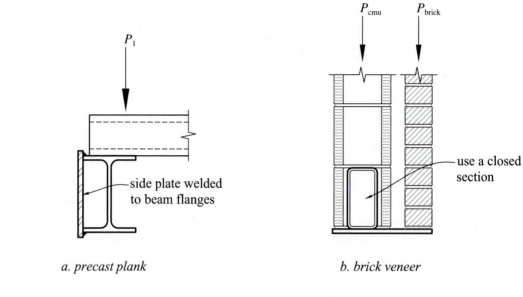

a. precast plank

b. brick veneer

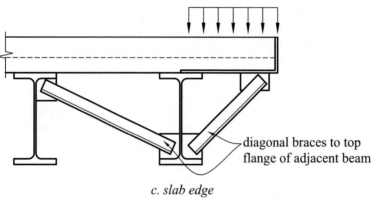

c. slab edge

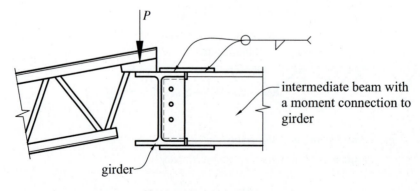

d. moment connection

Figure 6-18 Details used to control torsion.

(such as a hollow structural section) or to provide adequate lateral bracing. Closed sections are able to distribute torsional stresses effectively around the perimeter of the section, whereas other sections, such as W-shapes, rely on the stiffness of the individual components (web, flange) that make up the section to resist torsion. The primary measure of the torsional stiffness of a member is the torional constant, J, which is found for each shape in Part I of the *AISCM*. To provide a comparative example, the torsional stiffness of a W8 $\times$ 31 is found to be $J = 0.536$ in.[4]. An HSS member of equivalent size and weight would be an HSS 8 $\times$ 8 $\times$ 5/16, which has torsional stiffness of $J = 136$ in.[4], nearly 250 times the value obtained for the W8 $\times$ 31. The torsional stiffness constant, J, is directly proportional to the rotation, so the W8 $\times$ 31 would undergo a rotation that is much greater than the rotation of an equivalent closed section for the same loading.

Providing adequate lateral bracing will also help to control torsion in that the addition of a lateral brace will decrease the length, L, used in the analysis of torsion (see Tables 6-6 and 6-7). Figure 6-18 indicates common details used to control torsion.

EXAMPLE 6-10

Torsion in a Spandrel Beam

Determine whether the beam shown in Figure 6-19 is adequate for combined bending loads. The floor-to-floor height is 12 ft.; the beam span is 15ft. and is unbraced for this length on both the x- and y-axes. The loads shown are service loads.

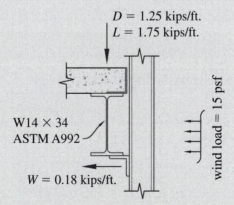

Figure 6-19 Details for Example 6-10.

SOLUTION

Since only the bottom flange of the beam is subjected to y–y axis bending, this corresponds to Case 2 biaxial bending. The beam is subjected to biaxial bending from the vertical gravity loads and horizontal wind loads. Thus,

Lateral wind load, $W = 15$ psf $\times$ 12 ft. tributary height $= 0.18$ kips/ft.

The critical load combination for these loads is $1.2D + 1.6W + 0.5L$

Vertical Load:

$$1.2(1.25) + 0.5(1.75) = 2.38 \text{ kips/ft.}; \quad M_{ux} = \frac{w_{ux}L^2}{8} = \frac{(2.38)(15)^2}{8} = \mathbf{67 \text{ ft.-kips}}$$

(continued)

Horizontal Load:

$$1.6(0.18) = 0.288 \text{ kips/ft.}; \quad M_{uy} = \frac{w_{uy}L^2}{8} = \frac{(0.288)(15)^2}{8} = \textbf{8.1 ft.-kips}$$

From *AISCM*, Table 3-10, $\phi_b M_{nx} = 132$ ft.-kips ($L_b = 15$ ft, W14 × 34)

A W14 × 34 has compact flanges, so the design bending strength in the y-axis is found from equation (6-34):

$$M_{ny} = M_{py} = F_y Z_y \leq 1.6 F_y S_y$$
$$= (50)(10.6) < (1.6)(50)(6.91) = 530 \text{ in-kips} < 553 \text{ in.-kips}$$
$$\phi_b M_{ny} = \frac{(0.9)(530)}{12} = 39.7 \text{ ft.-kips.}$$

Checking the interaction equation for Case 2,

$$\frac{M_{ux}}{\phi_b M_{nx}} + \frac{M_{uy}}{0.5(\phi_b M_{ny})} \leq 1.0; \quad \frac{67}{132} + \frac{8.1}{0.5(39.7)} = 0.92 < 1.0$$

The W14 × 34 beam is adequate for biaxial bending.

EXAMPLE 6-11

Torsion in a Spandrel Beam Supporting Brick Veneer

Determine whether the beam shown in Figure 6-20 is adequate for combined bending loads and torsion. The floor-to-floor height is 12 ft. and the beam span is 17 ft. The loads shown are service loads. Use a unit weight of 40 psf for the brick veneer, and assume that the top flange has continuous lateral support.

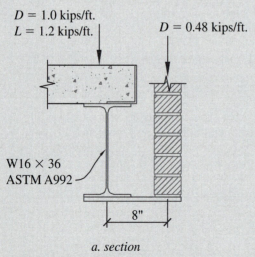
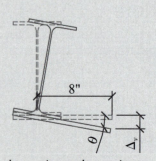

$D = 1.0$ kips/ft.
$L = 1.2$ kips/ft.

$D = 0.48$ kips/ft.

W16 × 36
ASTM A992

8"

8"

θ

Δ_v

a. section   *b. rotation under torsion*

Figure 6-20 Details for Example 6-11.

SOLUTION

The loading shown is equivalent to Case 3. We will use the both methods presented for Case 3 loading to compare the results.

Veneer weight, $w_v = (0.040\text{ ksf})(12\text{ ft.}) = 0.48\text{ kips/ft.}$

The critical load combination for these loads is $1.2D + 1.6L$.

Vertical Load

$$1.2(1.0 + 0.48) + 1.6(1.2) = 3.70\text{ kips/ft.}; M_{ux} = \frac{w_{ux}L^2}{8} = \frac{(3.70)(17)^2}{8} = \mathbf{134\text{ ft.-kips}}$$

Horizontal Load

$$M = Pe = (1.2)(0.48)(8\text{ in.}/12) = 0.384\text{ ft.-kips/ft}$$

$$h_o = d - t_f = 15.9\text{ in.} - 0.43\text{ in.} = 15.47\text{ in.} = 1.29\text{ ft.}$$

$$\frac{Pe}{h_o} = \frac{0.384}{1.29} = 0.298\text{ kips/ft.} \rightarrow M_{uy} = \frac{w_{uy}L^2}{8} = \frac{(0.298)(17)^2}{8} = \mathbf{10.8\text{ ft.-kips}}$$

From *AISCM* Table 3-6, $\phi_b M_{nx} = 240\text{ ft.-kips}$ $(L_b = 0\text{ ft.})$

$$M_{ny} = M_{py} = F_y Z_y \leq 1.6 F_y S_y$$
$$= (50)(10.8) < (1.6)(50)(7) = 540\text{ in-k} < 560\text{ in.-kips}$$
$$\phi_b M_{ny} = \frac{(0.9)(540)}{12} = \mathbf{40.5\text{ ft.-kips}}$$

Checking the interaction equation for Case 2:

$$\frac{M_{ux}}{\phi_b M_{nx}} + \frac{M_{uy}}{0.5(\phi_b M_{ny})} \leq 1.0; \quad \frac{134}{240} + \frac{10.8}{0.5(40.5)} = 1.09 > 1.0$$

The W16 × 36 is not adequate for biaxial bending.

As a comparison, we will use the approximate equation from Table 6-7 to see if the preceding method of analysis was too conservative.

From Part 1 of the *AISCM*,

$$C_w = 1460\text{ in.}^6$$

$$J = 0.545\text{ in.}^4$$

$$a = \sqrt{\frac{EC_w}{GJ}} = \sqrt{\frac{(29,000)(1460)}{(11,200)(0.545)}} = 83.3\text{ in.}$$

$$\frac{L}{a} = \frac{(17)(12)}{83.3} = 2.45$$

(continued)

From Table 6-7,

$$w = (1.2)(0.48) = 0.576 \text{ kips/ft.} = 0.048 \text{ kips/in.}$$

$$e = 8 \text{ in.}$$

$$L = (17)(12) = 204 \text{ in.}$$

$$M_r = \frac{weLa}{h_o}\left[0.097 + 0.094\left(\frac{L}{2a}\right) - 0.0255\left(\frac{L}{2a}\right)^2\right]$$

$$M_f = \frac{(0.048)(8)(204)(83.3)}{15.47}\left[0.097 + 0.094\left(\frac{204}{(2)(83.3)}\right) - 0.0255\left(\frac{204}{(2)(83.3)}\right)^2\right]$$

$$M_f = 73.3 \text{ in.-kips} = 6.11 \text{ ft. kips}$$

This value is less than the value previously calculated (10.8 ft.-kips). Checking the interaction equation for the new value gives

$$\frac{M_{ux}}{\phi_b M_{nx}} + \frac{M_{uy}}{0.5(\phi_b M_{ny})} \leq 1.0; \quad \frac{134}{240} + \frac{6.11}{0.5(40.5)} = 0.86 < 1.0$$

The W16 × 36 is adequate for biaxial bending.

Check Deflection:

For deflections, the combined weight of the veneer and the live load will be used and will be compared with a deflection limit of the smaller of $L/600$ and 0.3″. (A 0.3″ deflection controls only for $L > 15', 0''$). Further, the Brick Industry Association [5] recommends a maximum torsional rotation of 1/16 in., which will also be checked, Thus,

$$w = 1.2 + 0.48 = 1.68 \text{ kips/ft.}$$

$$\Delta = \frac{5wL^4}{384EI} = \frac{5(1.68/12)(17 \times 12)^4}{(384)(29 \times 10^6)(448)} = 0.243 \text{ in.} < \frac{L}{600} = \frac{(17)(12)}{600} = 0.34 \text{ in. or } 0.3 \text{ in., OK}$$

Check torsional rotation:

$$w = 0.480 \text{ kips/ft} = 0.04 \text{ kips/in.}$$

From Table 6-7,

$$\theta = \frac{weLa}{GJ} = \left[-0.032 + 0.062\left(\frac{L}{2a}\right) + 0.052\left(\frac{L}{2a}\right)^2\right]$$

$$\theta = \frac{(0.04)(8)(204)(83.3)}{(11{,}200)(0.545)}\left[-0.032 + 0.062\left(\frac{204}{(2)(83.3)}\right) + 0.052\left(\frac{204}{(2)(83.3)}\right)^2\right]$$

$$\theta = 0.109 \text{ radians}$$

$$(0.109)\left(\frac{180}{\pi}\right) = 6.22°$$

With reference to Figure 6-20b, the vertical displacement for this rotation is

$$\Delta_v = (\tan 6.22)(8'') = 0.872 \text{ in.}$$

This amount of twist far exceeds the deflection limit and would not be adequate. The reader should confirm that adding lateral support at ⅓ points would decrease the torsional rotation to about 0.11° with $\Delta_v = 0.015$ in.) and would be recommended here.

As a comparison, we will now consider the torsional rotation of an equivalent closed section. An HSS16 × 8 × ¼ will be selected. From Part 1 of the AISCM, $J = 300$ in.[4]. From Table 6-6, the torsional rotation is

$$\theta = \frac{weL^2}{8GJ} = \frac{(0.04)(8)(17 \times 12)^2}{(8)(11{,}200)(300)} = 0.0005 \text{ radians}$$

$$(0.0005)\left(\frac{180}{\pi}\right) = 0.028°$$

The vertical displacement for this rotation is

$$\Delta_v = (\tan 0.028)(8'') = 0.004 \text{ in.}$$

This displacement is more than 200 times less than that of the W16 × 36 for the same span. The maximum permissible deflection [5] is ⅟₁₆ in., or 0.0625 in., so the HSS16 × 8 × ¼ would be adequate for this loading.

6.10 BEAM BEARING

In typical steel structures, steel beams and girders are connected to other steel members by some combination of gusset plates, clip angles, welds, and bolts to transfer the end reactions (see Chapters 9 and 10). In some cases, the end reaction of a beam is transferred in direct bearing onto masonry, concrete, or another steel member. When steel beams are supported in this way, a steel bearing plate is used to spread the load out over a larger surface area. In the case of bearing on concrete or masonry, the bearing plate is large enough such that the bearing strength of the concrete or masonry is not exceeded. In the case of bearing on another steel section, the bearing plate is designed to be large enough such that local buckling does not occur in the supporting steel section. Figure 6-21 indicates these common beam bearing conditions.

For practical purposes, N is usually a minimum of 6 in. and B is usually greater than or equal to the beam flange width, b_f (see Figure 6-21). This allows for reasonable construction tolerances in placing the bearing plate and beam. Both the plate dimension B and N should be selected in increments of 1 in., and the plate thickness, t_p, is usually selected in increments of ¼ in.

The basic design checks for beam bearing are web yielding and web crippling in the beam, plate bearing and plate bending in the plate, and bearing stress in the concrete or masonry.

Web yielding is the crushing of a beam web subjected to compression stress due to a concentrated load. When a concentrated load occurs at or near the end of the beam, the compression stress distribution is less than if the load were placed on the interior portion of the beam (see Figure 6-21).

The compression stress is assumed to be distributed on a ratio of 1:2.5 through the beam flange and inner radius. Multiplying this distance by the web thickness and yield stress gives the following equations for web yielding:

For $x > d$,

$$\phi_{wy}R_n = \phi_{wy}(5k + N)F_y t_w, \tag{6-37}$$

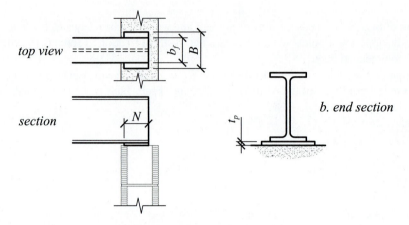

a. beam bearing on masonry

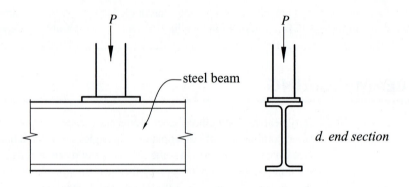

c. bearing on a steel beam

Figure 6-21 Bearing on masonry and steel.

For $x \leq d$,

$$\phi_{wy}R_n = \phi_{wy}(2.5\,k + N)F_y t_w, \tag{6-38}$$

where

$\phi_{wy} = 1.0$ (resistance factor for web yielding),

R_n = Nominal design strength,

t_w = Beam web thickness,

N = Bearing length,

d = Beam depth,

x = Distance from the end of the beam to the concentrated load,

k = Section property from *AISCM*, Part 1, and

F_y = yield strength, ksi.

Web crippling occurs when the concentrated load causes a local buckling of the web. The design strength for web crippling is

For $x \geq \dfrac{d}{2}$,

$$\phi_{wc}R_n = \phi_{wc}0.8t_w^2\left[1 + 3\left(\frac{N}{d}\right)\left(\frac{t_w}{t_f}\right)^{1.5}\right]\sqrt{\frac{EF_yt_f}{t_w}}. \tag{6-39}$$

For $x < \dfrac{d}{2}$ and $\dfrac{N}{d} \leq 0.2$,

$$\phi_{wc}R_n = \phi_{wc}0.4t_w^2\left[1 + 3\left(\frac{N}{d}\right)\left(\frac{t_w}{t_f}\right)^{1.5}\right]\sqrt{\frac{EF_yt_f}{t_w}}. \tag{6-40}$$

For $x < \dfrac{d}{2}$ and $\dfrac{N}{d} > 0.2$,

$$\phi_{wc}R_n = \phi_{wc}0.4t_w^2\left[1 + \left(\frac{4N}{d} - 0.2\right)\left(\frac{t_w}{t_f}\right)^{1.5}\right]\sqrt{\frac{EF_yt_f}{t_w}}, \tag{6-41}$$

where

ϕ_{wc} = 0.75 (resistance factor for web crippling),

R_n = Nominal design strength,

t_w = Beam web thickness,

t_f = Beam flange thickness,

N = Bearing length,

d = Beam depth,

x = Distance from the end of the beam to the concentrated load,

k = Section property from the AISCM, Part 1,

E = 29×10^6 psi, and

F_y = Yield strength, ksi.

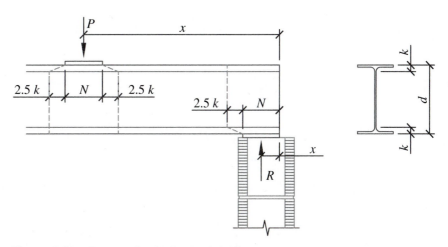

Figure 6-21e Beam web crippling and yielding parameters.

The bearing strength of the supporting concrete or masonry in crushing on the full support area is

$$\phi_{cb}P_p = \phi_{cb}0.85f'_cA_1. \qquad (6\text{-}42)$$

When the bearing is on less than the full area of concrete support, the bearing strength is

$$\phi_{cb}P_p = 0.85f'_cA_1\sqrt{\frac{A_2}{A_1}} \le 1.7f'_cA_1, \qquad (6\text{-}43)$$

where

$\phi_{cb} = 0.65$ (resistance factor for concrete bearing),

P_p = Nominal design strength,

f'_c = 28-day compressive strength of the concrete or masonry,

A_1 = Area of steel bearing = BN, and

A_2 = Maximum area of the support geometrically similar and concentric with the loaded area

$= (B + 2e)(N + 2e)$. Note that the dimension e is the minimum distance from the edge of the plate to the edge of the concrete support.

Note that the strength reduction factor given in the AISC specification for bearing on concrete is 0.60. However, ACI 318 [4] recommends a value of 0.65, which will be used here. The dimensional parameters for A, and A_2 are indicated in Figure 6-22.

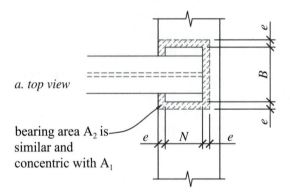

a. top view

bearing area A_2 is similar and concentric with A_1

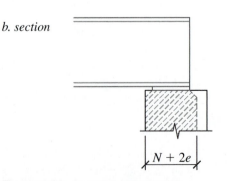

b. section

Figure 6-22 A_1 and A_2 parameters.

The bearing plate strength in bending also has to be checked. From equation (6-9), the design strength in bending for a plate is

$$\phi_b M_n = \phi_b M_p = \phi_b F_y Z_x,$$ (6-44)

where

$\phi_b = 0.9,$

$F_y = $ Yield stress, and

$$Z_x = \frac{N t_p^2}{4} \text{ (plastic section modulus for a plate).}$$ (6-45)

From Figure 6-23, the maximum factored moment is

$$M_u = \frac{R_u \ell^2}{2B}$$ (6-46)

Combining equations (6-44), (6-45), and (6-46) yields

$$\phi_b M_n > M_u$$

$$(0.9)(F_y)\left(\frac{N t_p^2}{4}\right) = \frac{R_u \ell^2}{2B}$$

Solving for the plate thickness, t_p, yields

$$t_p \geq \sqrt{\frac{2 R_u \ell^2}{0.9 B N F_y}},$$ (6-47)

where

$t_p = $ Plate thickness,

$B, N = $ Bearing plate dimensions,

$R_u = $ Factored reaction,

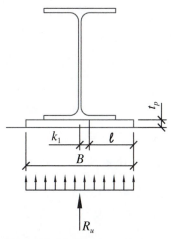

Figure 6-23 Bearing plate bending.

$$\ell = \text{Moment arm for plate bending} = \frac{B - 2k_1}{2}, \tag{6-48}$$

k_1 = Dimensional constant for beam (from Part 1 of the *AISCM*), and

F_y = Yield stress.

For the limit states of web yielding and web crippling, a pair of transverse stiffeners or a web doubler plate is added to reinforce the beam section when the design strength is less than the applied loads. This topic, including the design of the stiffener and doubler plates, is covered in Chapter 11.

The design procedure for bearing plates can be summarized as follows:

1. Determine the location of the load relative to the beam depth (dimension x in Figure 6-21).
2. Assume a value for the bearing plate length, N.
3. Check the beam for web yielding and web crippling for the assumed value of N; adjust the value of N as required.
4. Determine the bearing plate width, B, such that the bearing plate area, $A_1 = BN$, is sufficient to prevent crushing of the concrete or masonry support.
5. Determine the thickness, t_p, of the beam bearing plate so that the plate has adequate strength in bending.

EXAMPLE 6-12

Web Yielding and Web Crippling in a Beam

Check web yielding and web crippling for the beam shown in Figure 6-24. The steel is ASTM A572, grade 50.

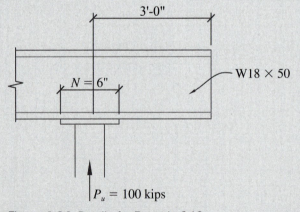

Figure 6-24 Details for Example 6-12.

SOLUTION

For W18 × 50, we obtain the following properties from *AISCM*, Table 1-1:

$k = 0.972$ in.

$t_w = 0.355$ in.

$d = 18$ in.

$t_f = 0.57$ in.

1. $x = 3'\text{-}0'' > d$ and $> d/2$

2. $N = 6$ in. (given)

3a. *Web Yielding*

$$\phi_{wy}R_n = \phi_{wy}(5\,k + N)F_y t_w = 1.0[(5)(0.972) + 6](50)(0.355)$$
$$= \mathbf{192\ kips} > P_u = \mathbf{100\ kips\ OK}$$

3b. *Web Crippling*

$$\phi_{wc}R_n = \phi_{wc}0.8t_w^2\left[1 + 3\left(\frac{N}{d}\right)\left(\frac{t_w}{t_f}\right)^{1.5}\right]\sqrt{\frac{EF_y t_f}{t_w}}$$

$$\phi_{wc}R_n = (0.75)(0.8)(0.355)^2\left[1 + 3\left(\frac{6}{18.0}\right)\left(\frac{0.355}{0.57}\right)^{1.5}\right]\sqrt{\frac{(29,000)(50)(0.57)}{0.355}}$$

$$\phi_{wc}R_n = 172\ \text{kips} > P_u = 100\ \text{kips} \quad \text{OK}$$

The W18 × 50 beam is adequate for web yielding and web crippling.

EXAMPLE 6-13

Beam Bearing on a Concrete Wall

A W18 × 50 beam is simply supported on 10-in.-thick concrete walls at both ends as shown in Figure 6-25. Design a beam bearing plate at the concrete wall supports assuming the following:

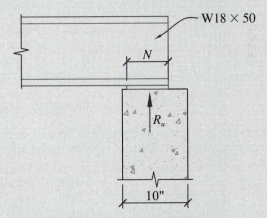

Figure 6-25 Details for Example 6-13.

– Beam span = 20 ft. center-to-center of support

– ASTM A36 steel for the beam and bearing plate

– $D = 1.5$ kips/ft

– $L = 2$ kips/ft

– $f'_c = 4000$ psi

(continued)

SOLUTION

For a W18 × 50, we obtain the following properties from *AISCM*, Table 1-1:

$k = 0.972$ in.

$t_w = 0.355$ in.

$d = 18$ in.

$t_f = 0.57$ in.

$k_1 = 13/16$

The reactions at each end are

$$w_u = 1.2D + 1.6L = (1.2)(1.5) + (1.6)(2) = 5.0 \text{ kips/ft}$$

$$R_u = \frac{w_u L}{2} = \frac{(5.0)(20)}{2} = 50 \text{ kips}$$

Design Steps:

1. $x < 5$ in., (half of the wall thickness) $\therefore x < d$ and $x < d/2$.

2. Assume $N = 6$ in. (recommended practical value)

3a. *Web Yielding*

$$\phi_{wy} R_n = \phi_{wy}(2.5\,k + N)F_y t_w = 1.0[(2.5)(0.972) + 6](36)(0.355)$$
$$= \mathbf{107\ kips} > R_u\ \mathbf{50\ kips\ OK}$$

3b. *Web Crippling*

$$\frac{N}{d} = \frac{6}{18.0} = 0.33 > 0.2;(\text{ equation (6-41) is used:}$$

$$\phi_{wc} R_n = \phi_{wc} 0.4 t_w^2 \left[1 + \left(\frac{4N}{d} - 0.2\right)\left(\frac{t_w}{t_f}\right)^{1.5}\right]\sqrt{\frac{E F_y t_f}{t_w}}$$

$$\phi_{wc} R_n = (0.75)(0.4)(0.355)^2\left[1 + \left(\frac{(4)(6)}{18} - 0.2\right)\left(\frac{0.355}{0.57}\right)^{1.5}\right]$$

$$\times \sqrt{\frac{(29,000)(36)(0.57)}{0.355}}$$

$$\phi_{wc} R_n = \mathbf{72.0\ kips} = R_u = \mathbf{50\ kips\ OK}$$

4. *Plate Bearing*

From Equation (6-42),

$$\phi_{cb} P_p = \phi_{cb} 0.85 f'_c A_1$$
$$50 = (0.65)(0.85)(4)(6 \text{ in.})B$$

Solving for B yields $B_{min} = 3.78$ in.

It is practical to use a value of B at least equal to or greater than the beam flange width:

$b_f = 7.495$ in. $\Rightarrow$ Try **B = 8 in.**

Therefore, the trial bearing plate size ($B \times N$) is 8 in. $\times$ 6 in.

5. Determine the plate thickness:

$$\ell = \frac{B - 2k_1}{2} = \frac{8 - (2)\left(\dfrac{13}{16}\right)}{2} = 3.19 \text{ in.}$$

$$t_p \geq \sqrt{\frac{2R_u\ell^2}{0.9BNF_y}} = \sqrt{\frac{(2)(50)(3.19)^2}{(0.9)(8)(6)(36)}} = 0.81 \text{ in.}$$

Select plate thickness in increments of ¼ in.; therefore,

Use a 6-in. $\times$ 8-in. $\times$ 1-in. bearing plate.

6.11 BEARING STIFFENERS

Bearing stiffeners are the plates used at the bearing points of a beam when the beam does not have sufficient strength in the web to support the end reaction or concentrated load. The limit states for this condition are web local yielding, web crippling, and web sidesway buckling. The design provisions for web local yielding and web crippling are covered in the previous section, and the design of the stiffener plates for these two limit states is covered in Chapter 11.

Web sidesway buckling can occur when a concentrated compressive force is applied to a beam and the relative lateral movement between the loaded compression flange and the tension flange is not restrained at the location of the concentrated force. When this happens, the flanges remain parallel while the web buckles. The concentrated compressive force could be applied at a point within the length of the beam, or the force could be the end reaction.

When the compression flange is restrained against rotation, the limit state of web sidesway buckling is as follows:

For $\dfrac{h/t_w}{\ell/b_f} \leq 2.3$,

$$\phi R_n = \frac{C_r t_w^3 t_f}{h^2}\left[1 + 0.4\left(\frac{h/t_w}{\ell/b_f}\right)^3\right]. \tag{6-49}$$

For $\dfrac{h/t_w}{\ell/b_f} > 2.3$, web sidesway buckling does not have to be checked.

When the compression flange is not restrained against rotation, the limit state of web sidesway buckling is as follows:

For $\dfrac{h/t_w}{\ell/b_f} \leq 1.7$,

$$\phi R_n = \frac{C_t t_w^3 t_f}{h^2}\left[0.4\left(\frac{h/t_w}{\ell/b_f}\right)^3\right]. \tag{6-50}$$

For $\dfrac{h/t_w}{\ell/b_f} > 1.7$, web sidesway buckling does not have to be checked.

In the preceding equations,

h = Clear distance between the flanges for built-up shapes

= Clear distance between the flanges less the fillets for rolled shapes,

t_w = Web thickness,

h/t_w (From Part 1 of the *AISCM*, for standard sections),

ℓ = Largest unbraced length along either the top or bottom flange at the load point,

b_f = Flange width,

t_f = Flange thickness,

C_r = 960,000 ksi when $M_u < M_y$

= 480,000 ksi when $M_u > M_y$,

$M_y = F_y S_x$,

F_y = Yield stress, and

S_x = Section modulus.

When the stress from the concentrated compressive force is greater than the design strength of the web, either bearing stiffeners or lateral bracing are required at the location of the force. When bearing stiffeners are provided to resist the full compressive force, the limit states of web local buckling, web crippling, and web sidesway buckling do not have to be checked.

Bearing stiffeners are designed as short columns when they are provided to reinforce the web of a beam subjected to concentrated loads or the web of a beam at an end reaction. The section properties of the stiffened beam section are as shown in Figure 6-26. The AISC specification allows a portion of the web to be included in the calculation of the design

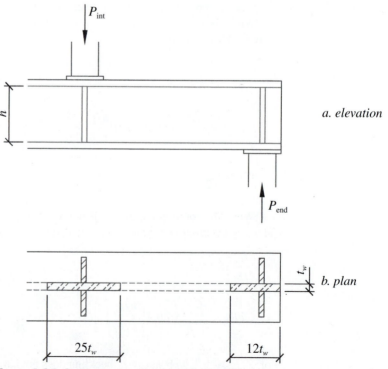

Figure 6-26 Section properties for bearing stiffeners.

compressive strength of the localized section. For bearing stiffeners at the end of a member, the section of web included has a maximum length of $12t_w$; for interior stiffeners, the maximum length is $25t_w$. The effective length factor for stiffeners is $K = 0.75$.

The calculated section properties are then used to determine the design strength of the stiffener in compression. For connection elements such as bearing stiffeners, AISC specification Section J4.4 permits the design compressive strength to be as follows:

$$\text{For } \frac{KL}{r} \leq 25, \quad \phi_c P_n = \phi F_y A_g, \tag{6-51}$$

Where

$\phi_c = 0.9$,

$F_y =$ Yield stress,

$A_g =$ Gross area of the bearing stiffener section,

$K = 0.75$ for bearing stiffeners,

$L = h$, and

$r =$ Least radius of gyration for the bearing stiffener section.

For $KL/r > 25$, the provisions from Chapter 5 apply, or *AISCM*, Table 4-22 can be used to determine $\phi_c F_{cr}$ for any value of KL/r.

The limit state of bearing strength also needs to be checked, but rarely controls the design of stiffeners. The design bearing strength for stiffeners is

$$\phi_{pb} R_n = \phi_{pb} 1.8 F_y A_{pb}, \tag{6-52}$$

where

$\phi_{pb} = 0.75$,

$F_y =$ Yield stress, and

$A_{pb} =$ Cross-sectional area of the bearing stiffeners.

The bearing stiffeners are usually welded to the flanges and the web of the beam. However, the stiffener is not required to be welded to the compression flange. For the weld to the web, the difference between the total concentrated force and the smallest design strength for web local yielding, web crippling, and web sidesway buckling can be used to determine the weld size.

EXAMPLE 6-14

Bearing Stiffeners

Determine the design bearing strength at the end of the W18 × 50 beam shown in Figure 6-27. The steel is grade 50.

Recall that when bearing stiffeners are provided, web local yielding, web crippling, and web sidesway buckling do not have to be checked. (Web local yielding and web crippling were checked in the previous examples.)

(continued)

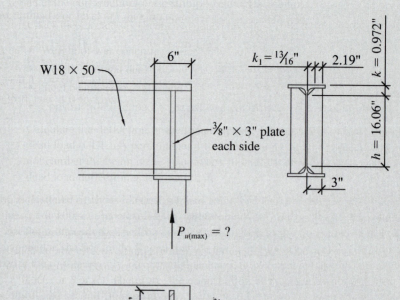

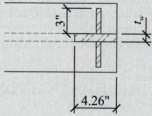

Figure 6-27 Details for Example 6-14.

SOLUTION

From *AISCM*, Table 1-1, we find the following properties:

$$k = 0.972 \text{ in.} \quad k_1 = 13/16 \quad t_w = 0.355 \text{ in.}$$
$$d = 18 \text{ in.} \quad t_f = 0.57 \text{ in.} \quad bf = 7.495 \text{ in.}$$

Section properties of the stiffened area about the web are

$$L_w = 12t_w = (12)(0.355) = 4.26 \text{ in.}$$
$$h = d - 2k = 18'' - (2)(0.972 \text{ in.}) = 16.06 \text{ in.}$$

Shape	A, in.²	I, in.⁴	d, in.	Ad², in.⁴	I + Ad², in.⁴
Stiff. plates	2.25	1.69	1.68	6.33	8.02
Web	1.51	0.016	0	0	0.016
Σ =	3.76				8.03

$$r = \sqrt{\frac{I}{A}} = \sqrt{\frac{8.03}{3.76}} = 1.46 \text{ in.}$$

Slenderness Ratio:

$$\frac{KL}{r} = \frac{(0.75)(16.06)}{1.46} = 8.24 < 25$$

Since the slenderness ratio is less than 25, we can use equation (6-51):

$$\phi_c P_n = \phi F_y A_g = (0.9)(50)(3.76) = 169 \text{ kips}$$

Bearing:
The bearing area will be the area of the end of the plate, excluding the fillets:

$$A_{pb} = (2)(3/8)(3 - 13/16) = 1.64 \text{ in.}^2$$
$$\phi_{pb} R_n = \phi_{pb} 1.8 F_y A_{pb} = (0.75)(1.8)(50)(1.64) = 110 \text{ kips.}$$

Bearing strength controls, so the maximum factored reaction is 110 kips.
 Weld strength is covered in Chapter 10, but the weld to the web will be covered here for completeness (see Section J2.2 of the AISC specification).

Minimum weld size = 3/16 in.

Maximum weld size = 5/16 in.

$$\phi R_n = 1.392 DL$$

where

ϕR_n = Design weld strength (kips)
D = Weld size in sixteenths of an inch (e.g., $D = 5$ for a ⁵⁄₁₆-inch weld)
L = Weld length, inches
$L = 2h = (2)(16.06 \text{ in.}) = 32.12 \text{ in.}$

$$110 \text{ kips} = (1.392)(D)(32.12)$$

Solving for D gives $D = 2.46$; therefore, use a 3/16 weld to beam web.

6.12 OPEN-WEB STEEL JOISTS

Open-web steel joists can be used in lieu of steel beams in either floor or roof framing, but are more commonly used for roof framing. These joists are sometimes called bar joists in practice, because at one time, many were fabricated with round bars as the web members. Most joists are manufactured with double-angle members used for the top and bottom chords, and either single or double-angle web members. Some joists are manufactured with proprietary, nonstandard steel sections, and the load-carrying capacity of these sections would have to be determined from the manufacturers' load tables.
 The main advantages of open-web steel joists are the following:

1. They are lighter in weight than rolled shapes for a given span.
2. An open web allows for easy passage of duct work and electrical conduits.
3. They may be more economical than rolled shapes, depending on the span length.

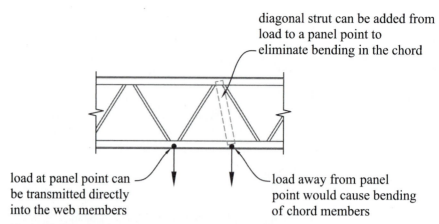

Figure 6-28 Loading at and away from joist panel points.

The main disadvantages of open-web steel joists are the following:

1. They cannot easily support concentrated loads away from panel points (see Figure 6-28).
2. The light weight could result in vibration problems if joists are used for floor framing (see Chapter 12 for further discussion).
3. They may not be economical for floor framing because of the closer joist spacing that is required due to the heavy floor loads.
4. Future structural modifications are not as easy to accomplish with joists.

There are a variety of manufacturers of open-web steel joists, but the most commonly used publication for the selection of these members is the *Catalog of Standard Specifications and Load Tables for Steel Joists and Joist Girders*, published by the Steel Joist Institute (SJI) [7].

The load tables in this catalog are identified by a designation that corresponds to a certain strength and stiffness. Many individual manufacturers will provide joists based on these load tables and the corresponding SJI specifications. The basic types of joists are summarized below and are included in the load tables in Appendix A.

K-Series Joists

These are the most common joists; they are used as the primary members for roof or floor framing. They are selected from the SJI catalog by a number designation (e.g., 14K1). The first number—14—represents the overall depth of the joist, and the last number—1—is the series. A14K3 would have more strength than the 14K1. They are listed in the catalog with a certain load-carrying capacity based on a certain span. The load-carrying capacity has two numbers listed in pounds per lineal foot. For the LRFD tables, the first, or upper, number is the total factored load-carrying capacity of the joist. The second, or lower, number (often in red font) is the service live load that will produce a deflection of L/360 for floors and L/240 for roof members. The standard joist seat depth for k-series joists is 2½ inches (see Figure 6-29a). Larger depths or inclined seats can be used for special conditions.

K-series joists are generally economical for spans of up to 50 ft. and vary in overall depth from 8 in. to 30 in.

KCS-Series Joists

KCS joists are K-series joists with the ability to support a constant shear across the span (hence the CS designation). KCS joists are designed to support a constant moment across all interior panel points and a constant shear across the entire length. KCS joists are used for nonuniform loading conditions such as equipment loads or trapezoidal snowdrift loads. The designer simply needs to calculate the maximum shear and moment for a given special joist loading and select a KCS joist with a corresponding shear and moment capacity. The joist seat depth for KCS joists is also 2½ inches.

Type S and R Extensions

Top chord extensions (often designated as TCX) are used at perimeter conditions where a cantilever is desired. These extensions are often designed for the same load as that for the main joist; however, two standard cantilevered sections are provided in the SJI specification— the S-type and the R-type. The S-type implies that only the upper seat angles are extended, whereas the R-type is a stronger section where the entire depth of the joist seat is extended (see Figures 6-29c and 6-29d).

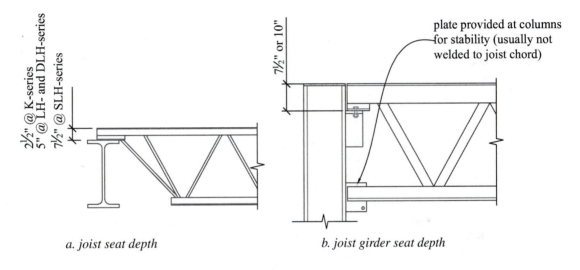

a. joist seat depth

b. joist girder seat depth

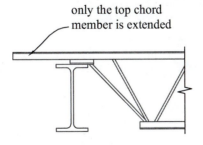

c. type S extension

d. type R extension

Figure 6-29 Joist seat types.

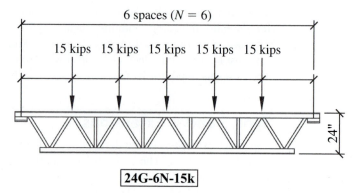

Figure 6-30 Joist girder designation.

LH-Series and DLH-Series

These are long-span joists that are used for spans of up to about 130 ft. The letter D in the DLH designation indicates a deeper section than the LH series. The designation is similar to that of K-series joists (e.g., 32LH06). The number 32 is the overall depth and the 06 is a series designation. LH- and DLH-series joists vary in depth from 18 in. to 72 in. The joist seat depth is 5 inches for LH and most DLH joists. Some of the larger DLH joists require a 7½-inch joist seat depth.

Joist Girders

Joist girders are of open-web construction and are members that usually support steel joists. Joist girders are designed as steel trusses that support concentrated loads from the joists that frame into them. They are designated by their overall depth, the number of panel points, and the loads at each panel point, (e.g., 24G-6N-15k) (see Figure 6-31). The 24 is the overall depth, 6N is the number of panel points, and 15k is the total load at each point (factored load for the LRFD tables and service loads for the ASD tables). The seat depth for joist girders is either 7½ inches or 10 inches depending on the loading and configuration of the joist girder (see Figure 6-29b).

When a required joist design does not fit one of the above categories, a special joist is often designated where the designer provides a special joist loading diagram for the joist manufacturer to use in designing the joist. Examples of this would include joists with concentrated loads, joists and joist girders with nonuniform loads, and joists and joist girders that have end moments such as would be the case of joists in moment frames. See [17] for more detailed coverage of the design of joists and joist girders.

EXAMPLE 6-15

Selection of a K-series Joist

Select a K-series using the SJI specifications to support the following loads for the framing shown in Figure 6-31.

Roof dead load $=$ 30 psf, snow load $=$ 35 psf

Joist tributary width $=$ 6 ft.

Joist span $=$ 25 ft.

Total load (factored) $= [(1.2)(30) + (1.6)(35)]6\text{ ft.} = 552\text{ lb./ft.}$

Live load (service) $= (35)(6\text{ ft.}) = 210\text{ lb./ft.}$

From the SJI load tables for K-series joists, the joist selections are as shown in Table 6-8.

Table 6-8 Joist selection

Joist Selection	Total Load Capacity, lb./ft.	Live Load Capacity, lb./ft.	Joist Weight lb./ft.	
16K6	576	238	8.1	
18K5	600	281	7.7	
20K4	594	312	7.6	← Select
22K4	657	381	8.0	

20K4 is the most economical joist for the given loads.

EXAMPLE 6-16

Selection of a Joist Girder

Select a joist girder using the SJI specifications for member JG1 to support a total roof dead load of 20 psf and a snow load of 40 psf.

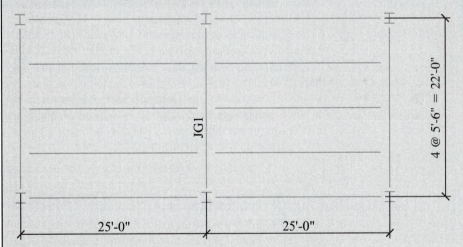

Figure 6-31 Roof framing plan for Example 6-16.

Total load (factored) $= [(1.2)(20) + (1.6)(40)] = 88$ psf

Concentrated load at each panel point:

$$P_u = (88 \text{ psf})(5.5 \text{ ft.})\left(\frac{25 \text{ ft.}}{2} + \frac{25 \text{ ft.}}{2}\right) = 12.1 \text{ kips}$$

The factored load used in Table 6-9 was $P_u = 12.0$ kips, which is close to the actual load of $P_u = 12.1$ kips. Generally speaking, larger joist girder depths will lead to a

(continued)

lighter overall section. The actual designation for this joist girder is **28G-4N-12.1k**. The largest depth joist girder was selected here, but architectural or other design constraints might dictate a shallower section.

Table 6-9 Joist girder selection

Joist Girder Span	Joist Spaces, N	Depth	Joist Weight, lb./ft.	
22 ft.	4N @ 5.5	20	19	
		24	17	
		28	16	← Select

6.13 FLOOR PLATES

Floor plates are a type of decking material used mainly in industrial applications as a floor deck for mezzanines or similar types of structures. There are several other types of decking materials, which are generally proprietary, but only the floor plate is addressed in the *AISCM* (see Figure 6-32). The selection of proprietary decking is done by using data provided by the specific manufacturer. The most common floor plate has a raised pattern and is often called a *diamond plate* because of the shape of the raised patterns.

A floor plate should conform to ASTM A786, which has a minimum yield stress of $F_y = 27$ ksi. Higher grades could be specified, such as A36, but availability should be considered. For deflection, a relatively low limit of $L/100$ is recommended by AISC.

AISCM, Tables 3-18a and 3-18b are selection tables for floor plates of various thicknesses and superimposed surface load capacities for spans from 18 in. to 7 ft. These tables are based conservatively on a simple-span condition for the floor plate. The plate is selected in 1/8-in. increments for thicknesses less than 1 in. and is selected in ¼ in. increments for thicknesses greater than 1 in.

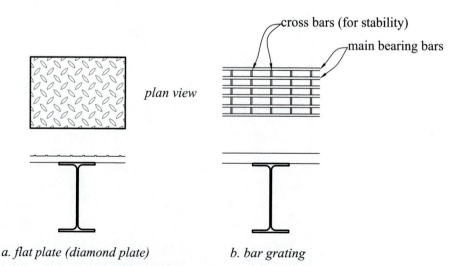

plan view

cross bars (for stability)

main bearing bars

a. flat plate (diamond plate)   *b. bar grating*

Figure 6-32 Various floor deck types for industrial applications.

EXAMPLE 6-17

Steel Floor Plate

Determine the required thickness of a steel plate floor deck to support a live load of 125 psf. The plate conforms to ASTM A786 and the span between the supports is $3'-6''$.

The dead load of the floor plate will have to be assumed. Assuming a $\frac{3}{8}$-in.-thick plate, the dead load is

$$\left(\frac{\frac{3}{8}}{12}\right)(490 \text{ lb/ft.}^3) = 16 \text{ psf.}$$

The factored load is

$$(1.2)(16) + (1.6)(125) = 220 \text{ psf} = 0.220 \text{ ksf}$$

Check the strength using equations (6-44) and (6-45):

$$\phi_b M_p = \phi_b F_y Z_x$$

Solving for Z_x,

$$Z_x = \frac{\dfrac{(12)(0.220)(3.5)^2}{8}}{(0.9)(27)} = 0.167 \text{ in.}^3/\text{ft.}$$

$$Z_x = \frac{bt^2}{4} = 0.167 = \frac{(12)t^2}{4}$$

$$t_{min} = 0.236 \text{ in.} = \textbf{¼-in.-thick plate}$$

Using *AISCM*, Table 3-18b, we find that a ¼-in.-thick plate has a factored load capacity of 245 psf, which also has less weight than the assumed value for a ⅜-in.-thick plate.

Check Deflection:

$$\Delta = \frac{5wL^4}{384EI}, \text{ where } \Delta = L/100.$$

Solving for *I* (use service load for deflection),

$$I = \frac{(100)(5)(0.150/12)[(3.5)(12)]^3}{(384)(29,000)} = 0.0346 \text{ in.}^4/\text{ft}$$

$$I = \frac{bt^3}{12} = 0.0346 = \frac{(12)t^3}{12}$$

$$t_{min} = 0.326 \text{ in.} \rightarrow \textbf{Select } \tfrac{3}{8}\textbf{-in. plate.}$$

(continued)

Using *AISCM*, Table 3-18a, we find that a 3/8-in.-thick plate has a service load capacity of 190 psf. Note that a ¼-in.-thick plate is adequate for bending, but has a service load capacity of only 56.4 psf for deflection considerations.

6.14 REFERENCES

1. American Institute of Steel Construction. 2006. *Steel construction manual*, 13th ed. Chicago. AISC.

2. International Codes Council. 2006. *International building code—2006*. Falls Church, VA: ICC (INTL Codes Council).

3. American Society of Civil Engineers. 2005. *Minimum design loads for buildings and other structures*. Reston, VA.

4. American Concrete Institute. 2008. *Building code requirements for structural concrete and commentary*, ACI 318. Farmington Hills, MI.

5. Brick Industry Association (BIA). 1987. *Technical notes on brick construction: Structural Steel untels #31B, BIA*. Reston, VA.

6. Vulcraft. 2001. *Steel roof and floor deck*. Florence, SC. Vulcraft/Nucor.

7. Steel Joist Institute. 2005. *Standard specifications—Load tables and weight tables for steel joists and joist girders*, 42nd ed. Steel Joist Institute. Myrtle Beach, SC.

8. Myrtic Beach, SL 2008. *Structural steel design*, 4th ed. Prentice Hall. Upper Saddle River, NJ.

9. Segui, William. 2006. *Steel design*, 4th ed. Toronto: Thomson Engineering.

10. Limbrunner, George F., and Leonard Spiegel. 2001. *Applied structural steel design*, 4th ed. Prentice Hall.

11. American Institute of Steel Construction. 2003. *Steel design guide series 9: Torsional analysis of structural steel members*. Chicago. AISC.

12. Lin, Philip H. Third Quarter: 1977. *Simplified design of torsional loading of rolled steel members. Engineering Journal.*

13. Abu-Saba Elias G. 1995. *Design of steel structures*. New York, Chapman & Hall.

14. Bhatt, P. and H. M. Nelson. 1990. *Marshall and Nelson's structures*, 3rd ed. Longman. London, UK.

15. Disque, R. O. *Applied Plastic Design in Steel*. New York, Van Nostrand Reinhold.

16. American Institute of Steel Construction. 2005. *Steel design guide series 7: Industrial buildings—Roofs to anchor rods*. Chicago, IL.

17. Fisher, James, Michael West, and Julius Van de Pas. 2002. *Designing with Vulcraft steel joists, joist girders, and steel deck*, 2nd ed. Milwaukee: Nucor.

18. Salmon, C. G., and J. E. Johnson. 1990. *Steel structures: Design and behavior*, 3rd ed. New York: Harper & Row.

19. Blodgett, Omer. Design of welded structures, Cleveland: The James F. Lincoln Arc Welding Foundation.

20. Tmoshenko, Stephen P., and Gere, James M. 1961. *Theory of elastic stability*, 2nd ed. New York: McGraw-Hill.

21. Glambos, Theodore V., 1996. F.J. L/N, and Bruce G. Johnston. *Basic steel design with LRFD*. Upper Saddle River, NJ: Prentice Hall.

22. Johnston, Bruce G. 1982. "Design of W-shapes for combined bending and torsion." *Engineering Journal, AISC*, 2nd Quarter 65–85.

23. Hotchkiss, John G. 1966. "Torsion of rolled sections in building structures." *Engineering Journal, AISC*, 19–45.

24. ACI 530. 2005. Building code requirements for masonry structures. Farmington Hills, MI: American Concrete Institute.

6.15 PROBLEMS

6-1. Draw a design moment, ϕM_n, versus unbraced length, L_b, curve for a W21 × 50 beam for ASTM A992 steel. Include the following points and calculations:

a. Web and flange slenderness ratios

b. L_p and L_r

c. Design moments for $L_b < L_p$, $L_p < L_b < L_r$, and $L_b = 15$ ft.

6-2. Determine the design moment for a W14 × 22 beam with (a) $L_b = 0$ and (b) $L_b = 14$ ft. The yield strength is $F_y = 36$ ksi and $C_b = 1.0$.

6-3. For the floor framing shown below in Figure 6-33, select the most economical W-shape for members B1 and G1. The floor dead load is 75 psf (including the weight of the framing) and the live load is 80 psf. Use ASTM A992, grade 50 steel. Check bending, shear, and deflection. Assume that B1 has full lateral stability and G1 is braced at the beam connections.

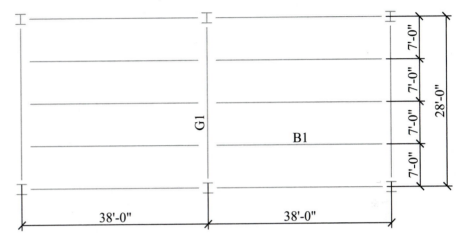

Figure 6-33 Details for Problem 6-3.

6-4. Determine the most economical size for the WF beam supporting the floor loads shown below in Figure 6-34 based on bending and deflection. The loads shown are service loads and the steel is ASTM A992, grade 50. Assume $C_b = 1.0$ and $L_b = 0$.

$w_D = 400$ lb./ft.
$w_L = 900$ lb./ft.

$L = 34'\text{-}6''$

Figure 6-34 Details for Problem 6-4.

6-5. Determine whether the beam shown below in Figure 6-35 is adequate for the given loads considering bending and shear only. The steel is ASTM A36.

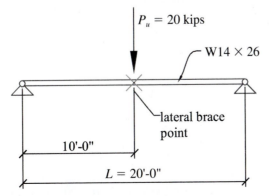

$P_u = 20$ kips

W14 × 26

lateral brace point

10'-0"

$L = 20'\text{-}0''$

Figure 6-35 Details for Problem 6-5.

6-6. Determine the maximum factored loads that can be applied to the beam shown below in Figure 6-36 based on web crippling and web yielding. The steel is ASTM A992, grade 50.

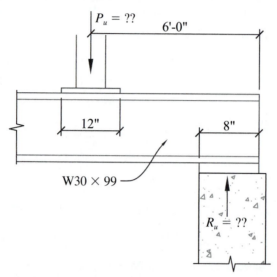

Figure 6-36 Details for Problem 6-6.

6-7. Determine whether the following is adequate for the connection shown below in Figure 6-37. The beam is ASTM A992, grade 50 and the bearing plate is ASTM A36. The concrete strength is $f'_c = 3500$ ksi.

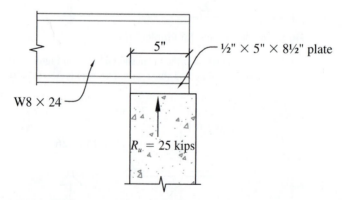

Figure 6-37 Details for Problem 6-7.

6-8. Design a bearing plate using ASTM A572, grade 50 steel for a factored reaction of $R_u = 65$ k. Check web crippling and web yielding in the beam. Use $f'_c = 3$ ksi.

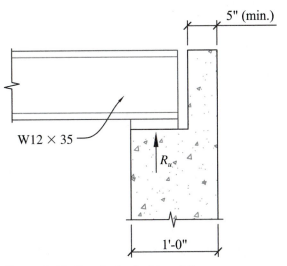

5" (min.)

W12 × 35

R_u

1'-0"

Figure 6-38 Details for Problem 6-8.

6-9. Select the most economical open-web steel joist J1 for the floor framing plan shown below in Figure 6-39 and select the most economical W-shape for member G1. The dead load is 65 psf and the floor live load is 80 psf. Consider bending, deflection, and shear. The steel is ASTM A992, grade 50. Assume that the unbraced length, L_b, is 3 ft. for member G1.

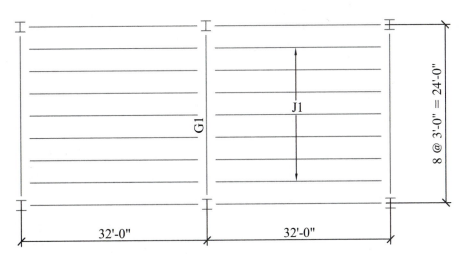

G1

J1

8 @ 3'-0" = 24'-0"

32'-0"   32'-0"

Figure 6-39 Details for Problem 6-9.

6-10. Select the most economical open-web steel joist J1 for the roof framing plan shown below in Figure 6-40 and specify a joist girder for member JG1. The dead load is 25 psf and the flat-roof snow load is 60 psf.

6-11. Determine the maximum span allowed for a ¼-in.-thick floor plate with a superimposed live load of 100 psf for strength and deflection. Compare the results with *AISCM*, Tables 3-18a and 3-18b. The plate conforms to ASTM A786.

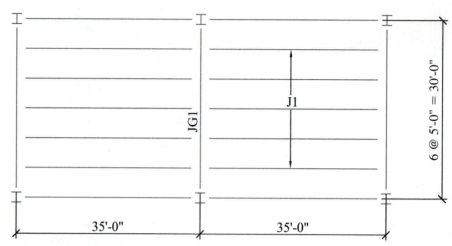

Figure 6-40 Details for Problem 6-10.

Student Design Project Problems:

6-12. For the floor framing in the student design project (see Figure 1-22), design the typical interior floor beams and girders as noncomposite.

6-13. Repeat Problem 6-12 for the typical perimeter beams.

Student Project Problems:

6-14. For the roof framing in the student design project, design the typical interior roof beams and girders as open-web steel joists and joist girders.

6-15. For the roof framing in the student design project, design the typical exterior roof beams and girders as W-shapes.

Composite Beams

7.1 INTRODUCTION

In steel-framed building construction, the floor deck system can be made of wood, steel, or concrete. With wood-framed floor decks, the steel beams are usually spaced farther apart, with the wood beams or trusses spanning between the steel beams. Steel floor decks can be either bar grating or a flat steel plate (see Chapter 6), with the supporting steel beams spaced at closer intervals. The most common floor system used with steel beams is a concrete slab with a metal deck.

The concrete floor deck can occur in various forms, the most common of which is shown in Figure 7-1. A corrugated metal deck (Figure 7-1a) is commonly used in steel building construction. The metal deck acts as a form for the wet concrete and also can provide strength to the floor deck system. A reinforced concrete slab without the corrugated metal deck (see Figure 7-1b) can also be used as the floor deck; this system is more commonly used in bridge construction. Another type of concrete floor deck system that is commonly used consists of steel beams with precast slab panels (see Figures 12-14 and 12-15). This type of system is not considered a composite beam system. In the past, steel floor beams and columns were commonly encased in concrete (Figure 7-1c). The steel framing was encased in concrete, with the concrete providing adequate fireproofing to the steel beams. Currently, it is generally more economical to spray the steel beams, and sometimes the corrugated metal deck, with a lightweight fireproofing product. Steel framing encased in concrete is not commonly used today in construction; therefore, our focus will be on steel framing with concrete on a composite metal floor.

The corrugated metal deck used in composite construction can serve several purposes. A *form* deck acts as a form for the wet concrete, but must have reinforcing in the concrete in order to provide adequate strength to span between supporting beams, since a form deck usually does not have enough strength to support more than the weight of the concrete (see Figure 7-1d). The reinforcement in the slab is usually a welded wire fabric (WWF). A *composite* deck is usually strong enough to support more than just the weight of the ✂

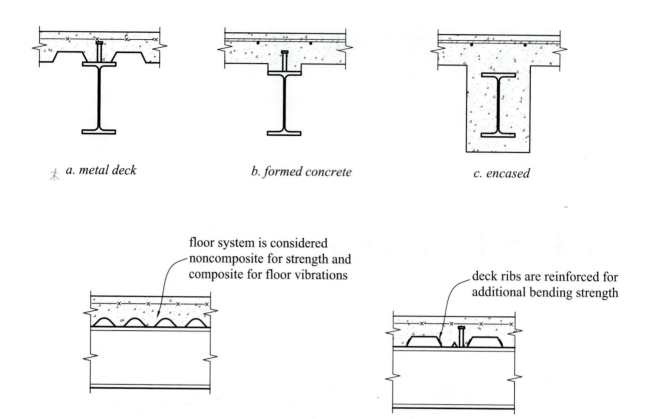

a. metal deck   *b. formed concrete*   *c. encased*

floor system is considered
noncomposite for strength and
composite for floor vibrations

deck ribs are reinforced for
additional bending strength

d. form deck   *e. composite deck*

Figure 7-1 Types of composite beams.

concrete and is often used in composite construction. The concrete slab supported by a composite deck often has a layer of WWF to control shrinkage cracking in the slab.

A slab system with a form deck does not usually have headed studs to engage the concrete with the steel beam (see Section 7.2 for a further discussion on headed studs); therefore, the beams are designed as noncomposite. By contrast, a composite slab system usually has headed studs and is designed as a composite system. For floor vibrations, a floor system with a metal deck and concrete is analyzed as if composite action occurs even if headed studs or other shear connectors are not used (see Chapter 12).

In a concrete and composite metal floor deck system supported by steel framing, greater economy can be achieved if the floor deck and steel framing can be made to act in concert to resist gravity loads. The combination of dissimilar materials to form an equivalent singular structural element is called *composite construction* and can occur in various forms. Our focus in this chapter will be on the combination of steel beams and a concrete in composite metal floor deck.

In order for steel beams and the floor deck to work together in resisting gravity loads, there needs to be an adequate horizontal force transfer mechanism at the interface where the two materials meet to prevent slippage between the surfaces (see Figure 7-2). This force transfer is accomplished by using shear connectors, which are commonly headed studs, but can also be channels or some other type of deformed connector. Headed shear studs are almost exclusively used in bridge and building construction due to their ease of installation, and so we will focus on these types of shear connectors in this chapter.

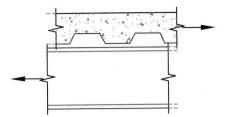

Figure 7-2 Slippage between a steel beam and a concrete deck.

7.2 SHEAR STUDS

Headed shear studs should conform to ASTM A108. These studs are welded to the top flange of a steel beam and spaced at regular intervals to adequately transfer the horizontal shear (see Figure 7-3). When used with a corrugated metal deck, the spacing of the studs coincides with the spacing of the ribs of the decking. More than one row of shear connectors can be provided, but there are some dimensional limitations that often limit the use of multiple rows. (We will discuss these limitations later.)

The number of shear connectors provided will determine how much of the concrete slab is engaged or acting in combination with the steel beam. Floor systems where a smaller number of shear connectors are provided are called *partially composite* because only a portion of the concrete slab is engaged. A fully composite system is one in which enough shear connectors are provided to completely engage the concrete slab. In this case, there is an upper limit to the number of studs that can be provided because adding more studs beyond this limit will not contribute to the strength of the floor system.

In a floor system with a corrugated metal deck, there are steel beams where the deck ribs will run perpendicular to the axis of the beam and steel beams where the deck ribs are parallel with the beams. In building construction, the members that are oriented perpendicular to the span of the slab system are usually refered to as beams, and the members that support the beams and are oriented parallel to the span of the slab system are usually called girders (see Figure 7-4).

BEAMS ⊥ DECK SPAN
GIRDERS ‖ DECK SPAN

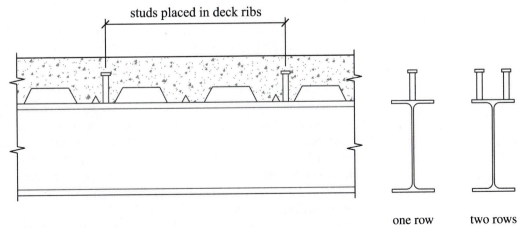

studs placed in deck ribs

one row   two rows

Figure 7-3 Headed studs.

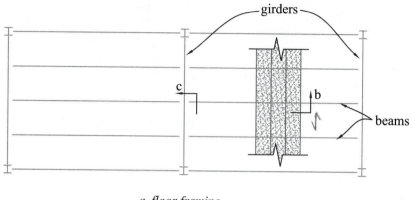

a. floor framing

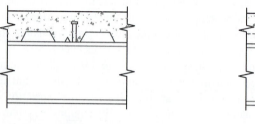

b. deck ribs parallel (beam) *c. deck ribs perpendicular (girder)*

Figure 7-4 Floor beams and floor girders.

There are several requirements for headed stud placement in the AISC specification; these are summarized below and are illustrated in Figure 7-5.

1. Except for corrugated metal deck, the minimum lateral cover around headed studs is 1 in.

2. The maximum stud diameter, D_s, is $2.5t_f$ (t_f is the beam flange thickness); if the studs are located directly over the beam web, then this provision does not apply. For studs placed in formed steel deck, the maximum stud diameter is limited to ¾ in.

3. The minimum stud spacing along the longitudinal axis of the beam is $6D_s$ (D_s is the stud diameter).

4. The maximum stud spacing along the longitudinal axis of the beam must be less than $8Y_{con}$, or 36 in. (Y_{con} is the total slab thickness).

5. The minimum stud spacing across the flange width is $4D_s$.

6. The minimum stud length is $4D_s$.

7. For formed steel deck, the rib height, h_r, must be less than or equal to 3 in.

8. For formed steel deck, the minimum rib width, w_r, must be greater than 2 in., but shall not be taken as less than the clear width across the top of the deck.

9. For formed steel deck, the studs should extend at least 1½ in. above the top of the deck, with at least ½ in. of concrete cover over the top of the stud.

10. For formed steel deck, the deck must be anchored to the supporting steel beams at intervals not exceeding 18 in. The anchorage can be provided by some combination of spot welds (also called puddle welds), mechanical connectors, or welding the headed stud through the deck.

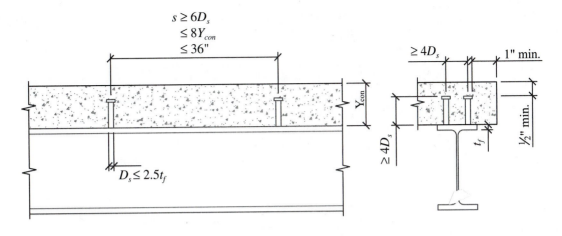

a. formed concrete

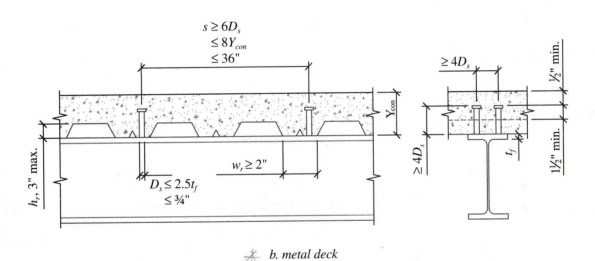

b. metal deck

Figure 7-5 Dimensional requirements for headed studs.

The number of shear studs that are provided is a function of the required strength of the composite section, which will be discussed later. When studs are placed in corrugated metal deck, it is ideal to place the stud in the middle of the deck rib. However, the deck ribs are usually reinforced in the center, thus forcing the stud to be offset within the deck rib (see Figure 7-6). Studs should be placed on the side of the deck rib closest to the end of the beam because more load can be transmitted to the stud through the concrete due to the additional concrete cover (see [5]). This is called the strong position. When studs are located in the deck rib in the weak position, the shear strength is decreased by about 25%.

When shear studs are required by design, the number of studs required between the point of maximum moment and the point of zero moment is denoted as N_s. For the case of a uniformly loaded beam (see Figure 7-7a), the maximum moment occurs at midspan and therefore a total of N_s studs are provided on each side of the beam centerline.

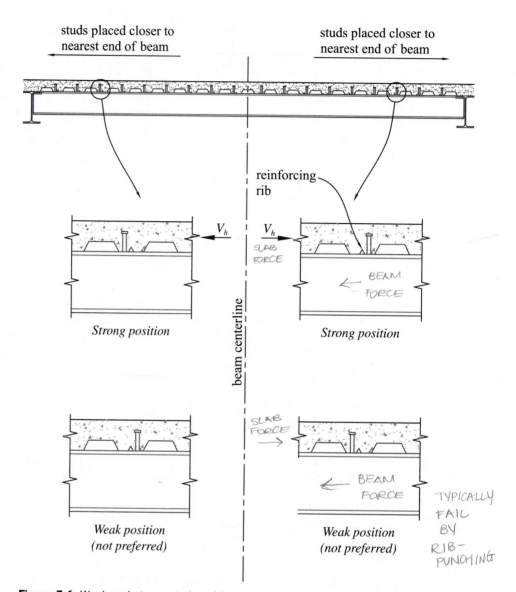

Figure 7-6 Weak and strong stud positions.

For a beam with two symmetrically placed concentrated loads (Figure 7-7b), a total of N_s studs are provided at each end of the beam up to the concentrated load. The middle third of the beam has a constant moment (i.e., zero shear) and therefore does not require shear studs; however, a nominal number of shear studs are commonly provided in practice at the discretion of the designer, to account for slight variations in loading that may create a small moment gradient in this region. A common layout for this condition would be to add studs at 24 in. on center. When three or more symmetrically placed concentrated loads are present, the moment diagram approaches the uniformly loaded case; thus, the number of shear studs provided is similar to what is shown in Figure 7-7a, where the stud spacing is uniform along the beam length.

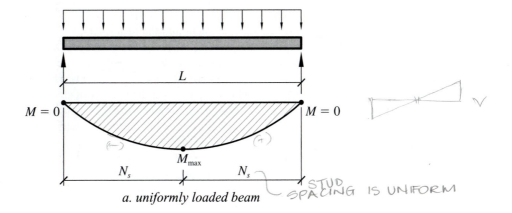

$M = 0$ $M = 0$

M_{max}

N_s N_s

a. uniformly loaded beam

STUD SPACING IS UNIFORM

ZERO SHEAR = ZERO STUDS REQ*

$M = 0$ $M = 0$

M_{max} M_{max}

N_s $N_{\text{required}} = 0*$ N_s

b. beam with symmetrically placed concentrated loads

* A nominal number of studs are provided to account for any induced shear due to unbalanced loads

*TYP STUDS @ 24" O.C.

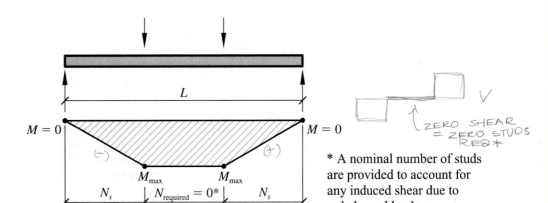

$M = 0$ $M = 0$

CONC LOAD

M_2

$M_1 = M_{\text{max}}$

N_2

N_1 N_1

c. beam with nonsymmetrical loads

Figure 7-7 Stud placement.

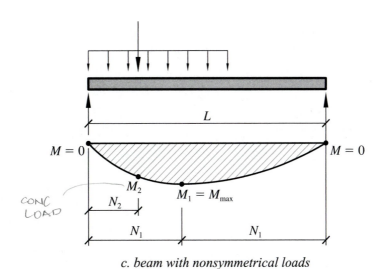

For beams with nonsymmetrical loads, a total of N_s studs are still provided between the point of maximum moment and the nearest point of zero moment (see Figure 7-7c). This creates a situation in which the required stud layout is not symmetrical about the beam centerline; thus, in some cases the designer will add more studs so that the spacing is uniform to avoid errors in the field placement of the studs. There is an additional provision for concentrated loads which requires that the number of studs placed between a concentrated load and the nearest point of zero moment shall be sufficient to develop the moment at the concentrated load. This is illustrated in Figure 7-7c, where a total of N_1 studs are provided between the point of maximum moment, M_1, and the nearest point of zero moment. A total of N_2 studs are required between the location of the concentrated load, where the moment is M_2, and the nearest point of zero moment. In this case, M_1 is the maximum moment in the beam and M_2 is the moment at the concentrated load.

7.3 COMPOSITE BEAM STRENGTH

In order to analyze a composite beam section made of different materials (i.e., steel and concrete), we need to develop an equivalent model to determine the behavior of the composite section under loads. Figure 7-8a shows the typical composite beam section, with a steel beam and a concrete deck. In order to analyze this composite section, we need to transform the concrete section into an equivalent steel section because the modulus of elasticity, and therefore the behavior of the two materials, is different under loading.

Once the concrete section is transformed into an equivalent steel section, then the section properties of the composite beam can be determined. The area of concrete is transformed by dividing the area of the concrete by the modular ratio, n. The modular ratio is defined as follows:

$$n = \frac{E_s}{E_c},\tag{7-1}$$

where

n = Modular ratio,

E_s = Modulus of elasticity for steel (29×10^6 psi),

E_c = Modulus of elasticity for concrete.

$$= w_c^{1.5}\sqrt{f'_c},\tag{7-2}$$

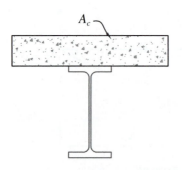

A_c

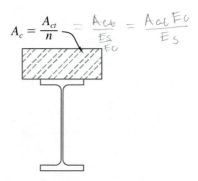

$A_c = \dfrac{A_{ct}}{n}$ $= \dfrac{A_{ct}}{E_s} = \dfrac{A_{ct}\,E_c}{E_s}$

a. composite beam section   *b. equivalent composite beam section*

Figure 7-8 Transformed composite beam section.

Note: ACI 318 defines the modulus of elasticity for concrete as $w_c^{1.5}33\sqrt{f'_c}$, with f'_c in pounds per square inch; however, AISC has adopted an approximate value that uses kips per square inch for f'_c. The AISC value will be used here for consistency.

w_c = Unit weight of the concrete, lb./ft.^{3}, and

f'_c = 28-day compressive strength of the concrete, ksi.

The area of the concrete section (Figure 7-8a) is then transformed into an equivalent steel section (Figure 7-8b) by dividing the area of the concrete by the modular ratio:

$$A_{ct} = \frac{A_c}{n},\qquad(7\text{-}3)$$

where

A_{ct} = Transformed concrete area, and

A_c = Concrete area

  = bt_c.

Once the concrete slab is transformed into an equivalent steel section, the section properties of the composite section can then be determined. There are three possible cases that must be considered, each corresponding to the location of the plastic neutral axis (PNA) of the composite section. The PNA is the axis of equal area; that is, the area above the PNA equals the area below the PNA. For positive bending, the composite section area below the PNA is in tension and the area above the PNA is in compression. We will first look at the horizontal strength of the shear connectors. The horizontal shear due to the compression force above the PNA is assumed to be resisted by the shear connectors. This horizontal shear is taken as the lowest of the following three failure modes:

Crushing of the concrete:

$$V' = 0.85f'_c A_c \qquad(7\text{-}4)$$

Tensile yielding of the steel beam:

$$V' = F_y A_s \qquad(7\text{-}5)$$

Strength of the shear connectors:

$$V' = \Sigma Q_n, \qquad(7\text{-}6)$$

where

V' = Horizontal force in the shear connectors,

f'_c = 28-day compressive strength of the concrete,

A_c = Concrete area

  = bt_c,

A_s = Area of the steel beam,

F_y = Minimum yield stress in the steel beam, and

Q_n = Nominal strength of the shear connectors between the points of maximum positive and zero bending moment.

(handwritten margin notes:) (+) BENDING   C   PNA   T   HORZ SHEAR DUE TO COMP. RESISTED BY HSA

The nominal strength of a single shear stud is

$$Q_n = 0.5 A_{sc} \sqrt{f'_c E_c} \leq R_g R_p A_{sc} F_u, \tag{7-7}$$

where

A_{sc} = Cross-sectional area of the shear stud,

R_g = Reduction coefficient for corrugated deck (see Table 7-1)

     = 1.0 for formed concrete slabs (no deck),

R_p = Reduction coefficient for corrugated deck (see Table 7-1)

     = 1.0 for formed concrete slabs (no deck), and

F_u = minimum tensile strength of the shear connector

     = 65 ksi for ASTM A108 (see *AISCM*, Table 2-5).

The shear stud strength, Q_n, can also be determined from *AISCM*, Table 3-21.

The number of shear connectors provided is a function of how much of the concrete slab needs to be engaged to provide the required design strength. In many cases, it is economical

Table 7-1 Reduction coefficients R_g and R_p

Framing Condition		R_g	R_p	Notes
No deck interference (or no deck for formed concrete slabs)		1.0	1.0	Shear connectors are welded directly to the beam flange; miscellaneous deck fillers can only be placed over less than 50% of the beam flange
Deck ribs oriented parallel to beam (i.e., girders)	$\frac{w_r}{h_r} \geq 1.5$	1.0	0.75	
	$\frac{w_r}{h_r} < 1.5$	0.85	0.75	Value for R_g = 0.85 only applies for a single stud
Deck ribs oriented perpendicular to beam	1 stud per rib	1.0	0.6	Values for R_p may be increased to 0.75 when $e_{mid\text{-}ht} \geq$ 2 in. (i.e., when studs are placed in the ("strong" position, see Figure 7-6)
	2 studs per rib	0.85	0.6	
	3 or more shear studs per rib	0.7	0.6	

Adapted from Ref [1]

Notes:

     w_r = Average width of the deck rib

     h_r = Deck rib height

     $e_{mid\text{-}ht}$ = Horizontal distance from the face of the shear stud to the mid-height of the adjacent deck rib in the direction of the load

to provide only enough shear studs for strength such that the concrete slab is partially engaged, which is called *partially composite action*. For any composite condition, the section properties can be determined from the parallel axis theorem. Recall from statics that the location of the neutral axis of a composite shape is

$$\bar{y} = \frac{\Sigma Ay}{\Sigma A} \tag{7-8}$$

The composite moment of inertia, which is called the transformed moment of inertia here since the concrete slab will be transformed into an equivalent steel section, is defined as

$$I_{tr} = \Sigma(I + Ad^2) \tag{7-9}$$

The transformed section properties calculated using the full slab depth assumes that there is full-composite action and that enough shear connectors are provided to achieve this condition. For this to be the case, the strength of the shear connectors, ΣQ_n, must be equal to or greater than the compression force in the slab. When the strength of the shear connectors is less than the maximum compression force in the slab, a partially composite section results and the section properties are reduced. The commentary in the *AISCM* gives the reduced section properties for the partially composite condition as

$$I_{eff} = I_s + \sqrt{\frac{\Sigma Q_n}{C_f}}(I_{tr} - I_s), \tag{7-10}$$

where

I_s = Moment of inertia of the steel section,

I_{tr} = Moment of inertia of the fully composite section,

C_f = Smaller of $0.85f'_c A_c$ and $F_y A_s$ (V' from eqs. (7-4) and (7-5)), and

$$\frac{\Sigma Q_n}{C_f} \geq 0.25.$$

The reduction term $\dfrac{\Sigma Q_n}{C_f}$ in equation (7-10) represents the degree of compositeness of a section. This term must be greater than 0.25 (i.e., composite sections are required to have at least 25% composite action, per the AISC specification).

The width of the concrete slab that is effective in the composite section is a function of the beam spacing and the length of the beam. The effective slab width on each side of the beam centerline (see Figure 7-9) is the smallest of

- $\frac{1}{8}$ of the beam span,
- $\frac{1}{2}$ of the distance to the adjacent beam, or
- The edge of the slab distance (for edge beams).

For formed-steel deck, the effective thickness of the concrete slab is reduced to account for the deck ribs. When the deck ribs are oriented perpendicular to the beam, the concrete below the top of the deck is neglected. When the deck ribs are oriented parallel to the beam,

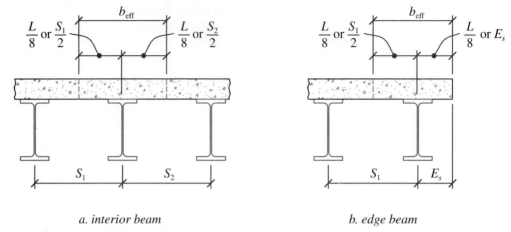

a. interior beam   *b. edge beam*

Figure 7-9 Effective slab width.

(i.e., girder) the concrete below the deck ribs may be included. The effective concrete slab thickness is usually taken as the average thickness for this case (see Figure 7-10).

We will now consider the bending strength of a composite section. There are three possible locations of the plastic neutral axis (PNA): within the concrete slab, within the beam flange, and within the beam web. With partial-composite action, the PNA is usually located within the steel section (in the flange or in the web). This is because a partially composite section engages a smaller amount of the concrete slab in compression; therefore, any additional cross-sectional area needed for compression is taken in the upper portion of the steel beam.

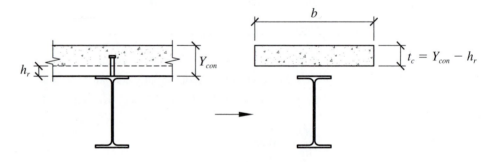

a. deck ribs perpendicular (beam)

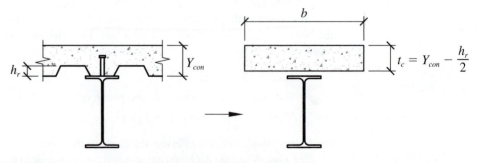

b. deck ribs parallel (girder)

Figure 7-10 Effective slab depth.

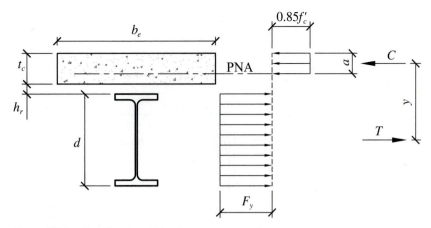

Figure 7-11 The PNA is within the concrete slab.

When the PNA is within the concrete slab (Figure 7-11), the available moment can be determined by taking a summation of moments as follows:

$$\phi M_n = \phi T y \text{ (or } \phi C y),\qquad(7\text{-}11)$$

where

T = Tension force in the steel,

C = Compression force in the concrete, and

y = Distance between T and C (see Figure 7-11).

This is because a partially composite section engages a smaller amount of the concrete slab in compression; Therefore, any additional cross-sectional area needed for compression is taken in the upper portion of the steel beam.

The depth of the compression stress block is

$$a = \frac{A_s F_y}{0.85 f'_c b_e},\qquad(7\text{-}11a)$$

where

A_s = Area of the steel section,

F_y = Yield stress,

f'_c = 28-day compressive strength of the concrete, and

b_e = Effective slab width.

The equation for the design moment is given as

$$\phi_b M_n = \phi A_s F_y \left(\frac{d}{2} + h_r + t_c - \frac{a}{2} \right),\qquad(7\text{-}11b)$$

where

h_r = Deck thickness (height of the deck ribs),

t_c = Concrete thickness above the deck, and

d = Beam depth.

Note that the total slab thickness, Y_{con}, equals $h_r + t_c$.

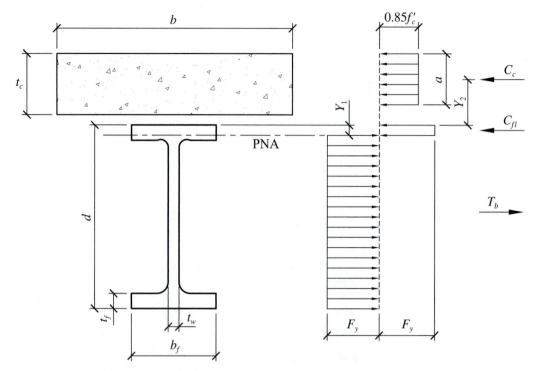

Figure 7-12 The PNA is within the beam flange.

Recall from Chapter 6 that for bending, $\phi = 0.9$.

When the PNA is within the top flange (Figure 7-12), the compression force in the concrete slab is at its maximum value. Therefore,

$$C_c = 0.85f'_c A_c, \tag{7-12}$$

where C_c is the compression force in the concrete slab and A_c is the area of concrete. The compression force in the top flange of the steel beam is

$$C_{fl} = b_f Y_1 F_y, \tag{7-13}$$

where

b_f = Beam flange width, and

Y_1 = Distance from the PNA to the top flange.

To determine the tension in the bottom portion of the beam, T_b, we assume that the entire beam has yielded in tension and we subtract the portion that is in compression:

$$T_b = A_s F_y - b_f Y_1 F_y. \tag{7-14}$$

The summation of horizontal forces yields

$$T_b = C_c + C_{fl}. \tag{7-15}$$

Combining equations (7-12) through (7-15) yields an expression for Y_1:

$$A_sF_y - b_fY_1F_y = 0.85f'_cA_c + b_fY_1F_y$$

$$Y_1 = \frac{A_sF_y - 0.85f'_cA_c}{2F_yb_f}, \tag{7-16}$$

where Y_1 is the distance from the top flange to the PNA.

The available moment can be determined by summing moments about the PNA, which yields

$$\phi M_n = \phi_b\left[\left(0.85f'_cA_c(Y_1 + Y_2) + 2F_yb_fY_1\left(\frac{Y_1}{2}\right) + A_sF_y\left(\frac{d}{2} - Y_1\right)\right)\right], \tag{7-17}$$

where Y_2 is the distance from the top of the beam flange to the centroid of the concrete compressive force (C_c) and $\phi_b = 0.9$.

When the PNA is located within the web of the steel beam, a similar analysis can be made to determine Y_1 (see Figure 7-13). The compression force in the slab, C_c, is found from equation (7-12). The compression force in the beam flange is

$$C_{fl} = b_ft_fF_y, \tag{7-18}$$

where t_f is the beam flange thickness.

The compression force in the upper part of the beam web is

$$C_w = t_wF_y(Y_1 - t_f), \tag{7-19}$$

where t_w is the beam web thickness.

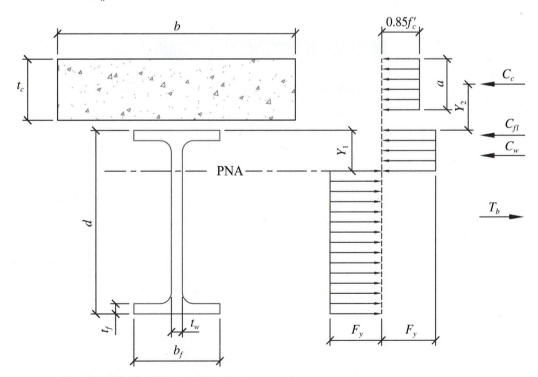

Figure 7-13 The PNA is within the beam web.

The tension force in the bottom portion of the beam is

$$T_b = A_s F_y - b_f t_f F_y - t_w F_y (Y_1 - t_f). \tag{7-20}$$

The summation of horizontal forces yields

$$T_b = C_c + C_{fl} + C_w. \tag{7-21}$$

Combining equations and solving for Y_1,

$$A_s F_y - b_f t_f F_y - t_w F_y (Y_1 - t_f) = 0.85 f'_c A_c + b_f t_f F_y + t_w F_y (Y_1 - t_f)$$

$$Y_1 = \frac{A_s F_y - 0.85 f'_c A_c - 2 b_f t_f F_y}{2 t_w F_y} + t_f. \tag{7-22}$$

The available moment can be determined by summing moments about the PNA, which yields

$$\phi M_n = \phi \left[0.85 f'_c A_c (Y_1 + Y_2) + 2 b_f t_f F_y \left(Y_1 - \frac{t_f}{2} \right) \right.$$

$$\left. + 2 t_w F_y (Y_1 - t_f) \left(\frac{Y_1 - t_f}{2} \right) + A_s F_y \left(\frac{d}{2} - Y_1 \right) \right]. \tag{7-23}$$

Several examples will follow to illustrate how to calculate the section properties and design bending strength for each possible PNA location.

EXAMPLE 7-1

Transformed Section Properties: Full-Composite Action

For the composite section shown in Figure 7-14, determine the transformed moment of inertia. Assume that the section is fully composite and that the concrete has a density of 145 pcf and a 28-day strength of 3.5 ksi.

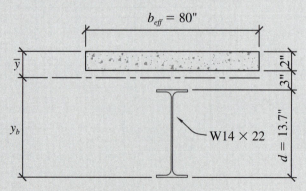

Figure 7-14 Details for Example 7-1.

SOLUTION

From *AISCM*, Table 1-1,

W14 × 22

$A = 6.49$ in.^{2}

$d = 13.7$ in.

$I = 199$ in.^{4}

Transforming the concrete section into an equivalent steel section,

$$E_c = w_c^{1.5}\sqrt{f_c'}$$
$$= (145)^{1.5}\sqrt{3.5} = 3266 \text{ ksi}$$

$$n = \frac{E_s}{E_c}$$

$$n = \frac{29,000}{3266} = 8.87$$

$$A_{ct} = \frac{A_c}{n}$$

$$= \frac{(80)(2)}{8.87} = 18.02 \text{ in.}^2$$

Using the top of the concrete slab as the datum, we develop Table 7-2.

Table 7-2 Transformed section properties

Element	A	y	Ay	I	$d = y - \bar{y}$	$I + Ad^2$
Slab	18.02	1	18.02	6.01	−2.87	154.4
W14 × 22	6.49	11.85	76.9	199	7.98	612.2
$\Sigma =$	**24.51**		**94.92**			$I_{tr} = $ **766.6**

$$I_{ct} = \frac{bt_3^c}{12n}$$

$$= \frac{(80)(2)^3}{(12)(8.87)} = 6.01 \text{ in.}^4$$

$$\bar{y} = \frac{\Sigma Ay}{\Sigma A}$$

$$\bar{y} = \frac{94.92}{24.51} = 3.87 \text{ in.}$$

$$I_{tr} = \Sigma(I + Ad^2)$$

$$I_{tr} = 766.6 \text{ in.}^4 \text{ (see Table 7-2)}$$

EXAMPLE 7-2

Design Bending Strength of a Composite Section

Determine the design bending strength of the composite section given in Example 7-1, assuming full-composite action and ASTM A992 steel.

SOLUTION

Since there is full-composite action, the compressive force in the concrete is the smaller of $0.85f'_cA_c$ (crushing of the concrete) and F_yA_s (tensile yielding of the steel beam).

$$C = 0.85f'_cA_c$$
$$= (0.85)(3.5)(2)(80) = 476 \text{ kips}$$
$$C = F_yA_s$$
$$= (50)(6.49) = 324.5 \text{ kips}$$

Tensile yielding in the steel controls, which means that only a portion of the concrete slab is required to develop the compressive force. Figure 7-15 shows the stress distribution in the slab.

From equation (7-4), the effective slab depth can be determined as follows:

$$C = 0.85f'_cA_c$$
$$C = 0.85f'_cab$$
$$324.5 = (0.85)(3.5)(a)(80)$$
$$a = 1.36 \text{ in.}$$

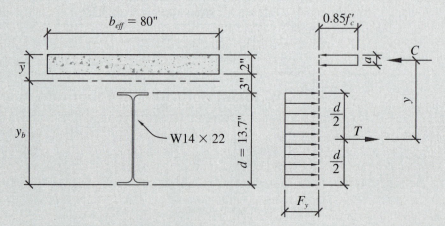

Figure 7-15 Stress distribution for Example 7-2.

The moment arm between the tensile and compressive forces is

$$y = \frac{d}{2} + 3 + 2 - \frac{a}{2}$$

$$y = \frac{13.7}{2} + 3 + 2 - \frac{1.36}{2} = 11.16 \text{ in.}$$

The design bending strength is found from equation (7-11):

$$\phi M_n = \phi Ty \text{ (or } \phi Cy) = (0.9)(324.5)(11.16) = 3260 \text{ in.-kips} = 271 \text{ ft.-kips.}$$

EXAMPLE 7-3

Transformed Section Properties: Partial-Composite Action

Determine the section properties and design moment capacity of the composite section given in Example 7-1 assuming that (8)-¾"ASTM A108 headed studs are provided between points of maximum and zero moments in the deck profile shown in Figure 7-16. Assume that studs are placed in the "strong" position.

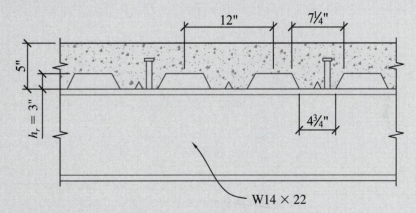

Figure 7-16 Metal deck profile for Example 7-3.

SOLUTION

From *AISCM*, Table 1-1,

W14 × 22
$A = 6.49 \text{ in.}^2$
$d = 13.7 \text{ in.}$
$I = 199 \text{ in.}^4$
$b_f = 5.00 \text{ in.}$

(continued)

$$t_f = 0.335 \text{ in.}$$

$$t_w = 0.23 \text{ in.}$$

The horizontal force in the shear connectors (V') is the smaller of the following:

$$V' = 0.85f'_c A_c$$

$$= (0.85)(3.5)(2)(80) = 476 \text{ kips, or}$$

$$V' = A_s F_y$$

$$= (50)(6.49) = 324.5 \text{ kips.}$$

From Table 7-1, for one row of shear studs, deck ribs perpendicular, to the beam,

$$R_g = 1.0$$

$$R_p = 0.75(\text{"strong" position of the stud in the deck rib})$$

$$A_{sc} = \frac{\pi d^2}{4} = \frac{\pi(0.75)^2}{4} = 0.441 \text{ in.}^2$$

$$Q_n = 0.5A_{sc}\sqrt{f'_c E_c} \le R_g R_p A_{sc} F_u$$

$$= (0.5)(0.441)\sqrt{(3.5)(3266)} \le (1.0)(0.75)(0.441)(65)$$

$$= 23.5 \text{ kips} > 21.5 \text{ kips}$$

$$Q_n = 21.5 \text{ kips (this value agrees with } AISCM, \text{ Table 3–21})$$

$$V' = \Sigma Q_n$$

$$= (8)(21.5) = 172 \text{ kips}$$

The degree of compositeness is found from the reduction term in equation (7-10), with C_f being 324.5 kips (the smaller of eqs. (7-4) and (7-5)):

$$\frac{\Sigma Q_n}{C_f} = \frac{172}{324.5} = 0.53, \text{ or about 53\% composite action.}$$

This section is at least 25% composite, so we can proceed to calculate the reduced section properties:

$$I_{eff} = I_s + \sqrt{\frac{\Sigma Q_n}{C_f}(I_{tr} - I_s)}$$

$$= 199 + \sqrt{\frac{172}{324.5}(766.6 - 199)} = 612 \text{ in.}^4$$

Since $\Sigma Q_n < A_s F_y$, there must be an additional compressive force within the beam section and therefore the PNA lies somewhere within the beam section. From equation (7-4), the effective slab depth can be determined as follows:

$$C = 0.85f'_c A_c$$

$$C = 0.85f'_c ab$$
$$172 = (0.85)(3.5)(a)(80)$$
$$a = 0.722 \text{ in.}$$

Assuming that the PNA is within the beam flange, Y_1 is determined from equation (7-16):

$$Y_1 = \frac{A_s F_y - 0.85f'_c A_c}{2F_y b_f}$$

$$Y_1 = \frac{(6.49)(50) - (0.85)(3.5)(0.722)(80)}{(2)(50)(5.00)} = 0.305 \text{ in.}$$

$Y_1 = 0.305 < t_f = 0.335$; therefore, the PNA is within the beam
    flange as assumed.

The design strength is then determined from equation (7-17):

$$Y_2 = 3 \text{ in. } + 2 \text{ in. } - \frac{0.722 \text{ in.}}{2} = 4.64 \text{ in.}$$

Note that Y_2 is the distance from the top of the beam flange to the centroid of the concrete compressive force.

$$\phi M_n = \phi \left[0.85f'_c A_c(Y_1 + Y_2) + 2F_y b_f Y_1 \left(\frac{Y_1}{2} \right) + A_s F_y \left(\frac{d}{2} - Y_1 \right) \right]$$

$$\phi M_n = 0.9 \Big[[(0.85)(3.5)(0.722)(80)(0.305 + 4.64)]$$

$$+ \left[(2)(50)(5.00)(0.305)\left(\frac{0.305}{2} \right) \right] + \left[(6.49)(50)\left(\frac{13.7}{2} - 0.305 \right) \right] \Big]$$

$$\phi M_n = 2697 \text{ in.-kips } = 225 \text{ ft.-kips}$$

EXAMPLE 7-4

PNA in the Slab

Determine the design moment strength for the beam shown in Figure 7-17. The concrete has a design strength of 4 ksi; use $F_y = 50$ ksi for the steel beam. Assume full-composite action.

(continued)

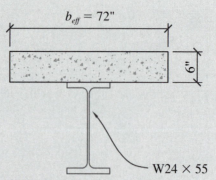

Figure 7-17 Details for Example 7-4.

SOLUTION

From *AISCM*, Table 1-1,

W24 × 55
$A = 16.2$ in.²
$d = 23.6$ in.
$I = 1350$ in.⁴

Determine the location of the PNA (start with eq. (7-11a)):

$$a = \frac{A_s F_y}{0.85 f'_c b_e} = \frac{(16.2)(50)}{(0.85)(4)(72)} = 3.31 \text{ in.} < 6 \text{ in.; therefore, the PNA is in the slab.}$$

The design bending strength is found from equation (7-11b):

$$\phi_b M_n = \phi A_s F_y \left(\frac{d}{2} + h_r + t_c - \frac{a}{2} \right) = (0.9)(16.2)(50)\left(\frac{23.6}{2} + 0 + 6 \text{ in.} - \frac{3.31}{2} \right)$$

$$= 11{,}769 \text{ in.-kips} = 980 \text{ ft.-kips.}$$

EXAMPLE 7-5

PNA in the Beam Web

Determine the design moment strength for the beam shown in Figure 7-18. The concrete has a 28-day strength of 4 ksi; use $F_y = 50$ ksi for the steel beam. Assume 25% composite action.

From *AISCM*, Table 1-1,

W18 × 35
$A = 10.3$ in.² $t_f = 0.425$ in.
$d = 17.7$ in. $b_f = 6.00$ in.
$I = 510$ in.⁴ $t_w = 0.300$ in.

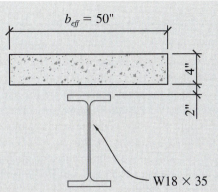

Figure 7-18 Details for Example 7-5.

C_c = Smaller of $0.85f'_cA_c$ or F_yA_s (V' from eqs. (7-4) and (7-5))

$0.85f'_cA_c = (0.85)(4)(50)(4) = 680$ kips

$F_yA_s = (50)(10.3) = 515$ kips ← Controls

Assuming 25% composite action, $C_c = (0.25)(515) = 129$ kips.

$C_c = 0.85f'_cab \rightarrow 129$ kips $= (0.85)(4)(a)(50$ in.$) \rightarrow a = 0.757$ in.

$Y_2 = h_r + t_c - \dfrac{a}{2} = 2 + 4 - \dfrac{0.757}{2} = 5.62$ in.

Use equations (7-22) and (7-23), to determine Y_1 as follows, (if Y_1 is greater than t_f, then PNA is, in fact, in the beam web):

$$Y_1 = \frac{A_sF_y - 0.85f'_cA_c - 2b_ft_fF_y}{2t_wF_y} + t_f$$

$$= \frac{(10.3)(50) - (0.85)(4)(50)(0.757) - (2)(6.0)(0.425)(50)}{(2)(0.300)(50)} + 0.425 = 4.80 \text{ in.}$$

Since Y_1 is greater than the beam flange thickness ($Y_1 = 4.80$ in. $> t_f = 0.425$ in.) the PNA must be in the beam web. The design bending strength is then found from equation (7-23) as follows:

$$\phi M_n = \phi\left[0.85f'_cA_c(Y_1 + Y_2) + 2b_ft_fF_y\left(Y_1 - \frac{t_f}{2}\right) + 2t_wF_y(Y_1 - t_f)\left(\frac{Y_1 - t_f}{2}\right) + A_sF_y\left(\frac{d}{2} - Y_1\right)\right]$$

$$\phi M_n = 0.9\left[(0.85)(4)(50)(0.757)(4.80 + 5.62) + (2)(6.0)(0.425)(50)\left(4.80 - \frac{0.425}{2}\right)\ldots\right]$$

$$\left[\ldots + (2)(0.300)(50)(4.80 - 0.425)\left(\frac{4.80 - 0.425}{2}\right) + (10.3)(50)\left(\frac{17.7}{2} - 4.80\right)\right] = 4396 \text{ in.-kips}$$

$\phi M_n = $ **367 ft.-kips**

7.4 SHORING

Before discussing the deflection of a composite section, we need to first look at the concept of shored versus unshored construction. Shoring can be provided under the beams and girders in order to allow the concrete floor to cure and reach its design strength prior to imposing any load on the steel beams. Once the shores are removed, the beam will have an instantaneous deflection due to the weight of the concrete floor slab. With unshored construction, the weight of the wet concrete is superimposed on the steel floor beams prior to the concrete curing, thus causing the beams to deflect under the weight of the concrete. The instantaneous deflection of the beams in the shored scheme will not be as much as it is in the unshored scheme since the beam in the former case will be in a composite state and thus have greater stiffness.

From a design standpoint, there is no conclusive data that favors either scheme. From a construction standpoint, the unshored scheme is usually preferred because it avoids labor required to install and remove the shores. However, the disadvantage of unshored construction is that because the beams deflect under the wet weight of the concrete additional concrete will be required to achieve a flat floor surface, resulting in concrete ponding. Ponding occurs when the deflected shape of a beam loaded with wet concrete allows additional concrete to accumulate, (see Figure 7-19b). This creates a situation where the builder must account for additional concrete that must be placed and the designer must account for the added dead load. For common floor framing systems, the additional concrete required due to ponding can range from 10% to 15%. One way to mitigate concrete ponding is to camber the beams. It would be ideal to camber the beams an amount equal to the deflection of the wet concrete, but this is not recommended because if the beam is cambered too much, then the slab might end up being too thin at mid span of the beam and there would not be enough concrete coverage for the headed studs (Figure 7-19c). For this reason, most designers provide camber that is equivalent to 75% to 85% of the deflection due to the dead load of the concrete. This reduction accounts for the possibility of overestimating the dead load, as well as the fact that the calculated deflection usually does not account for the actual support conditions of the beams (i.e., the typical deflection equations assume pinned ends, whereas the end conditions have some degree of fixity). Beams less than 25 ft. long should not be cambered and the minimum camber should be at least ¾ in. These limits ensure economy in the fabrication and cambering processes.

With unshored construction, the beams must be designed to support the concrete slab, as well as the temporary construction loads present while the slab is being placed. In floor systems with a formed-steel deck, the floor deck is usually adequate to brace the top flange of the beams against lateral–torsional buckling because the deck is oriented in the strong orthogonal direction, with the ribs perpendicular to the beam. However, the floor deck is usually not adequate to provide lateral stability for the girders because the deck is oriented in the weak orthogonal direction, with the deck ribs parallel to the girder. The design must therefore consider the unbraced length of the beams and girders during slab construction phase loading. Most designers use a construction live load of 20 psf for the construction phase design check [6]. (Recall that construction live loads were discussed in Chapter 2.)

For shored construction, the advantages are that all of the strength and deflection checks are based on the composite condition and the strength of the steel beam alone is not a factor when the concrete is still wet. Aside from the fact that added labor and materials will be required for shored construction, one key disadvantage is that cracks are likely to occur over the supporting girders and sometimes over the beams as well. One way to mitigate this occurrence is to add reinforcement over the beams and girders to control the cracking

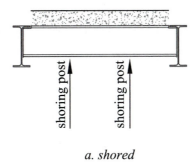

a. shored

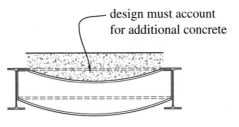

b. unshored (no camber)

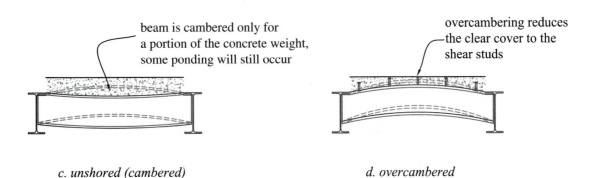

c. unshored (cambered)   *d. overcambered*

Figure 7-19 Shored and unshored beams.

(see Figure 7-20). Even with unshored construction, cracking is somewhat common over the supporting girders, so rebar is often added over the girders in either scheme.

Another way to mitigate cracking over the beams and girders in a shored scheme is to place the shoring such that some amount of deflection in the beams will occur, while minimizing the amount of ponding. Figure 7-21 shows one recommended shoring placement scheme that allows some deflection and minimizes ponding. In this scheme, shores are placed under the girders where a beam intersects and shores are placed at a distance of $L/5$ from the ends of the beams.

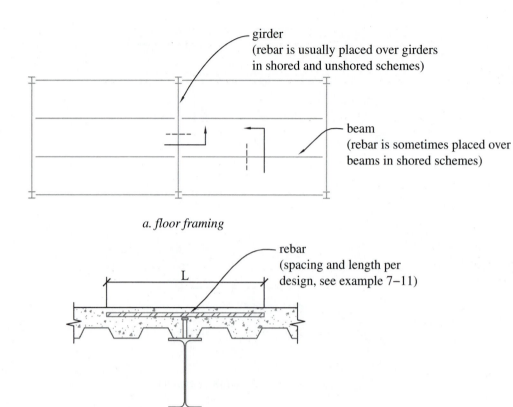

girder
(rebar is usually placed over girders
in shored and unshored schemes)

beam
(rebar is sometimes placed over
beams in shored schemes)

a. floor framing

L

rebar
(spacing and length per
design, see example 7–11)

b. section

Figure 7-20 Crack control over beams and girders.

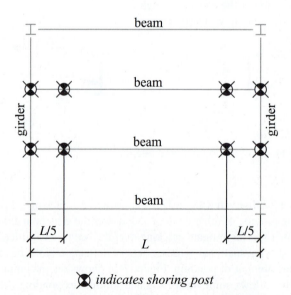

beam

beam

girder

beam

girder

beam

L/5

L/5

L

🞫 *indicates shoring post*

Figure 7-21 Shoring to allow some deflection and
minimize ponding.

EXAMPLE 7-6

Construction Phase Loading

For the floor framing shown in Figure 7-22, determine the following, assuming ASTM A992 steel and a slab weight of 75 psf (neglect the self-weight of the framing).

1. Required camber in the W18 × 35 beam for an unshored scheme,
2. Required camber in the W18 × 71 girder for an unshored scheme,
3. Adequacy of the W18 × 35 beam for construction loading, and
4. Adequacy of the W18 × 71 girder for construction loading.

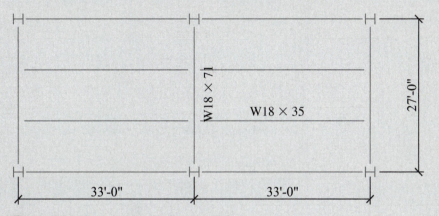

Figure 7-22 Floor framing for Example 7-6.

SOLUTION

The following load combinations are applicable (recall that the recommended construction live load is 20 psf from Chapter 2):

$1.4D$
$1.2D + 1.6L$
$(1.4)(75 \text{ psf}) = 105 \text{ psf}$
$(1.2)(75 \text{ psf}) + (1.6)(20 \text{ psf}) = 122 \text{ psf} = W_u \leftarrow$ Controls

Uniform load on beam:

Tributary width of the beam = 9 ft.

$w_D = (75 \text{ psf})(9 \text{ ft.}) = 675 \text{ lb./ft. } (56.25 \text{ lb./in.}) \text{ Unfactored dead load}$
$w_u = (122 \text{ psf})(9 \text{ ft.}) = 1098 \text{ lb./ft. } (1.098 \text{ kips/ft.}) \text{ Factored construction load}$

Concentrated loads on the girder:

Span of the beam = 33 ft.

$P_D = (75 \text{ psf})(33 \text{ ft.})(9 \text{ ft.}) = 22{,}275 \text{ lb. or } 22.3 \text{ kips}$
$P_L = (20 \text{ psf})(33 \text{ ft.})(9 \text{ ft.}) = 5940 \text{ lb. or } 5.94 \text{ kips}$
$P_u = (1.2)(22.3) + (1.6)(5.94) = 36.3 \text{ kips}$

(continued)

Camber

For the required camber, 75% of the slab weight will be assumed.

The camber for the beam is

$$\Delta_{cB} = (0.75)\frac{5wL^4}{384EI}$$

$$= (0.75)\frac{(5)(56.25)[(33)(12)]^4}{(384)(29 \times 10^6)(510)} = 0.913 \text{ in. Use a } \%\text{-in. camber.}$$

The camber for the girder is

$$\Delta_{cG} = (0.75)\frac{PL^3}{28EI}$$

$$= (0.75)\frac{(22.3)(27 \times 12)^3}{(28)(29 \times 10^6)(1170)} = 0.60 \text{ in. No camber.}$$

Recall that the minimum recommended camber is ¾ in. and the minimum recommended beam span is 25 ft. for cambered beams.

Construction Phase Strength

The factored bending moment due to construction loading for the beam is

$$M_{cB} = \frac{w_u L^2}{8} = \frac{(1.098)(33)^2}{8} = 150 \text{ ft.-kips}$$

The factored bending moment due to construction loading for the girder is

$$M_{cG} = \frac{P_u L}{3} = \frac{(36.3)(27)}{3} = 327 \text{ ft.-kips}$$

From *AISCM*, Table 3-10, the design bending strength in the beam and girder are

$$\phi M_n = 250 \text{ ft.-kips} > 150 \text{ ft.-kips (at } L_b = 0 \text{ ft. for W18} \times 35, \text{ OK), and}$$
$$\phi M_n = 500 \text{ ft.-kips} > 327 \text{ ft.-kips (at } L_b = 9 \text{ ft. for W18} \times 71, \text{ OK).}$$

7.5 DEFLECTION

Composite beams have a larger moment of inertia than noncomposite beams of the same size once the concrete slab cures and reaches its required strength. The designer must first consider whether or not the slab is shored (see Section 7.4). For shored construction, all of the deflection will occur "post-composite" and will be a function of the transformed moment of inertia. For unshored construction, the weight of the concrete will cause an initial deflection based on the moment of inertia of the steel beam alone; then any live load and superimposed and sustained dead load deflections will be based on the transformed moment of inertia.

When a composite beam is subjected to sustained loads, the concrete slab will be in a constant state of compression and subject to *creep*. Creep is deformation that occurs slowly over a period of time due to the constant presence of compression loads or stresses. It is difficult to quantify the amount of creep in a composite section, but designers will estimate this deflection by using a modular ratio of $2n$ to $3n$ instead of the n calculated in equation (7-1). AISC recommends a modular ratio of $2n$ for calculating long-term deflections [3]. The creep deflection due to sustained dead loads is not significant in typical steel buildings. Creep should be accounted for when there is a large amount of sustained live load. A more detailed coverage of this topic is found in [9].

In Section 7.3, we discussed the calculation of the transformed moment of inertia, I_{tr}, for a composite section, as well as the effective moment of inertia, I_{eff}, for a partially composite section. In many cases, the terms I_{tr} and I_{eff} are used interchangeably to describe the stiffness of a composite section. In this text, we will use the term I_{eff} to describe either a fully composite or partially composite section (in equation (7-10), note that $I_{eff} = I_{tr}$ for a fully composite section).

AISC specification C-I3.1 states that deflections calculated using I_{eff} are overestimated by 15% to 30%, based on testing. It is therefore recommended that the actual moment of inertia used for calculating deflections be taken as $0.75I_{eff}$. Therefore,

$$I_{actual} = 0.75I_{eff}, \tag{7-24}$$

where I_{actual} is the moment of inertia used for calculating deflections.

Alternatively, we can use the lower-bound moment of inertia tabulated in the lower-bound elastic moment of inertia table (Table 3-20 in the *AISCM*). The lower-bound moment of inertia (I_{LB}) is the moment of inertia at the ultimate limit state, which is less than the moment of inertia at the serviceability limit state, where deflections are calculated. For comparison purposes, I_{LB} is equivalent to I_{actual}, given in equation (7-24). To obtain I_{LB}, enter the table with a Y_2 value from the ultimate design stage and a Y_1 value corresponding to ΣQ_n, obtained in the ultimate design stage. The I_{LB} obtained is then used to calculate deflection.

The deflection limits for beams are found in the International Building Code (IBC), Section 1604.3. Composite beams are normally floor members and so, from the IBC, Table 1604.3, live load deflection is limited to $L/360$ and total load deflection is limited to $L/240$. For shored construction, the deflection limits are as follows:

$$\Delta_{LL} \leq \frac{L}{360}, \text{ and} \tag{7-25}$$

$$\Delta_{TL} \leq \frac{L}{240}, \tag{7-26}$$

where

Δ_{LL} = Live load deflection,

Δ_{TL} = Total load deflection (dead plus live load), and

L = Beam span.

For unshored construction, the live load deflection is as indicated in equation (7-25), but the total load deflection must also account for the construction phase deflection. Therefore, for the unshored condition,

$$\Delta_{TL} = \Delta_{SDL} + \Delta_{LL} + \Delta_{CDL}, \tag{7-27}$$

where

Δ_{SDL} = Superimposed dead load deflection (post-composite dead load),

Δ_{LL} = Live load deflection, and

Δ_{CDL} = Construction phase dead load deflection less any camber (pre-composite dead load).

Note that Δ_{SDL} and Δ_{LL} will be a function of the composite moment of inertia (from equation (7-24)) and Δ_{CDL} will be a function of the moment of inertia of the steel beam only (i.e., noncomposite moment of inertia). Note also that the construction dead load, Δ_{CDL}, can be minimized by specifying a camber.

Floor vibrations are an additional serviceability consideration for composite as well as noncomposite members and are covered in Chapter 12.

7.6 COMPOSITE BEAM ANALYSIS AND DESIGN USING THE AISC TABLES

The analysis process presented in the previous sections can be tedious when performing hand calculations, which is why most composite beam designs are carried out with the aid of a computer program. As an alternative, there are several design aids in the *AISCM* that can be used. *AISCM*, Table 3-19 is used to determine the design bending strength of a composite section and *AISCM*, Table 3-20 is used to determine the moment of inertia, I_{LB}. Both tables are a function of Y_1 (location of the PNA), Y_2 (location of the compression force in the slab), and ΣQ_n (magnitude of the compression force).

Referring to Figure 7-23, there are seven possible locations of the PNA, starting from the top flange of the steel beam to some point within the web of the beam. If it is found that

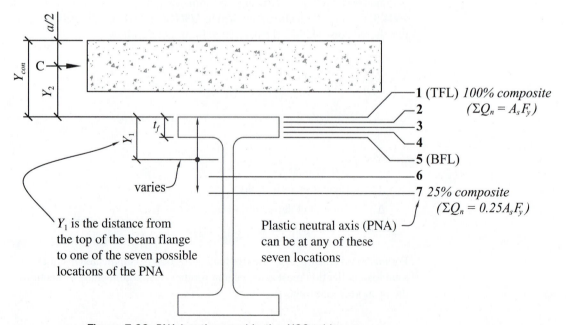

Figure 7-23 PNA location used in the AISC tables.

the PNA is above the top of the beam flange, then the values obtained from PNA location 1 are conservatively used. Full-composite action occurs at PNA location 1 (i.e., $\Sigma Q_n = A_s F_y$), whereas PNA location 7 represents 25% composite action, which is the lowest allowed by the AISC specification.

The examples that follow illustrate the use of composite design tables in the *AISCM*.

EXAMPLE 7-7

Composite Beam Deflections

Determine whether the composite beam in Example 7-3 is adequate for immediate and long-term deflections, assuming the following conditions:

Slab dead load = 50 psf (ignore self-weight of the beam)

Superimposed dead load = 25 psf (partitions, mechanical and electrical)

Live load = 50 psf (assume that 20% of this is sustained)

Beam span = 28' − 6"

Beam spacing = 8 ft.

SOLUTION

Loads to Beam:

$$w_{CDL} = (50 \text{ psf})(8 \text{ ft.}) = 400 \text{ plf } (33.3 \text{ lb./in.})$$
$$w_{SDL} = (25 \text{ psf})(8 \text{ ft.}) = 200 \text{ plf } (16.7 \text{ lb./in.})$$
$$w_{LL} = (50 \text{ psf})(8 \text{ ft.}) = 400 \text{ plf } (33.3 \text{ lb./in.})$$

From Example 7-3,

$$I = 199 \text{ in.}^4 \text{ (noncomposite moment of inertia)}$$
$$I_{eff} = 612 \text{ in.}^4 \text{ (the calculated effective moment of inertia)}$$

From equation (7-24), the moment of inertia used for long-term deflections is

$$I_{actual} = 0.75I_{eff} = (0.75)(612) = 459 \text{ in.}^4$$

As an alternate check, recall from Example 7-3 that $Y_1 = 0.305$ in., $Y_2 = 4.64$ in., and $\Sigma Q_n = 172$ kips. From *AISCM*, Table 3-20, we find that $I_{LB} = 496$ in.^{4} (by interpolation). Recall that research indicates that the actual moment of inertia is between 15% and 30% less than the calculated value of I_{eff}. This explains the variation between $I_{actual} = 459$ in.^{4} and $I_{LB} = 496$ in.^{4}

Since there is no specific guidance in the AISC specification as to which value to use, we will proceed with the more conservative value of $I_{actual} = 459$ in.^{4}.

Construction Phase Deflection:

$$\Delta_{CDL} = \frac{5w_{CDL}L^4}{384EI}$$

$$= \frac{(5)(33.3)[(28.5)(12)]^4}{(384)(29 \times 10^6)(199)} = 1.03 \text{ in.}$$

(continued)

If this beam were to be cambered, we would take 75% of the construction phase dead load deflection as the camber, as follows:

$$\Delta_c = (0.75)(1.03 \text{ in.}) = 0.771 \text{ in. Use } \tfrac{3}{4}\text{-in. camber.}$$

The construction phase deflection would then be reduced by ¾ in.:

$$\Delta_{CDL} = 1.03 \text{ in.} - 0.75 \text{ in.} = 0.28 \text{ in.}$$

A construction phase deflection of 1.03 in. would likely add more concrete in the floor slab due to ponding than would be desired, so a camber of ¾ in. will be specified. The post-composite deflection is

$$\Delta_{SDL} = \frac{5w_{SDL}L^4}{384EI}$$

$$= \frac{(5)(16.7)[(28.5)(12)]^4}{(384)(29 \times 10^6)(459)} = 0.223 \text{ in.}$$

$$\Delta_{LL} = \frac{5w_{LL}L^4}{384EI}$$

$$= \frac{(5)(33.3)[(28.5)(12)]^4}{(384)(29 \times 10^6)(459)} = 0.446 \text{ in.}$$

$$\Delta_{TL} = \Delta_{SDL} + \Delta_{LL} + \Delta_{CDL}$$

$$= 0.223 \text{ in.} + 0.446 \text{ in.} + 0.28 \text{ in.} = 0.949 \text{ in.}$$

The deflection limits are found from equations (7-25) and (7-26):

$$\Delta_{LL} \leq \frac{L}{360}$$

$$0.446 < \frac{(28.5)(12)}{360} = 0.95 \text{ in., OK for live load deflection}$$

$$\Delta_{TL} \leq \frac{L}{240}$$

$$0.949 < \frac{(28.5)(12)}{240} = 1.43 \text{ in., OK for total load deflection}$$

It can be seen by inspection that the ¾-in. camber is needed to meet the total load deflection limit.

We must also consider the long-term deflection to account for creep effects. To do this, the effective moment of inertia must be recalculated using a modular ratio of $2n$. From Example 7-1, $n = 8.87$; therefore,

$$2n = 17.74$$

$$A_{ct} = \frac{A_c}{n} = \frac{(80)(2)}{17.74} = 9.01 \text{ in.}^2$$

$$I_{ct} = \frac{bt_3^c}{12n} = \frac{(80)(2)^3}{(12)(17.74)} = 3.0 \text{ in.}^4$$

Using the top of the concrete slab as the datum, we develop Table 7-3.

Table 7-3 Composite beam section properties

Element	A	y	Ay	I	$d = y - \bar{y}$	$I + Ad^2$
Slab	9.01	1	9.01	3.0	−4.54	188.7
W14 × 22	6.49	11.85	76.9	199	6.31	457.4
$\Sigma =$	15.5		85.91			$I_{tr} = 646.1$

$$\bar{y} = \frac{\Sigma Ay}{\Sigma A} = \frac{85.91}{15.5} = 5.54 \text{ in.}$$

$$I_{tr} = \Sigma(I + Ad^2)$$

$$I_{tr} = 646.1 \text{ in.}^4 \text{ (see Table 7-3)}$$

Accounting for the partial-composite behavior (see Example 7-3),

$$I_{eff} = I_s + \sqrt{\frac{\Sigma Q_n}{C_f}}(I_{tr} - I_s) = 199 + \sqrt{\frac{172}{324.5}}(646.1 - 199) = 524.5 \text{ in.}^4, \text{ and}$$

$$I_{actual} = 0.75 I_{eff} = (0.75)(524.5) = 393 \text{ in.}^4.$$

The ratio of the actual moment of inertia used for total loads versus that used for sustained loads is

$$\frac{459 \text{ in.}^4}{393 \text{ in.}^4} = 1.168.$$

This ratio can be used to calculate the actual sustained load deflection. The amount of sustained live load specified for this example is 20% of the total live load, and the remaining 80% is assumed to be transient. The total sustained load deflection is then

$$\Delta_{TL(sust)} = 1.168(\Delta_{SDL} + 0.2\Delta_{LL}) + \Delta_{CDL} + 0.8\Delta_{LL}$$

$$= 1.168[0.223 + (0.2)(0.446)] + 0.28 + (0.8)(0.446) = 1.02 \text{ in.}$$

This is less than the $L/240 = 1.43$ in. calculated previously, so the section is adequate for sustained loads.

EXAMPLE 7-8

Composite Design Strength Using the *AISCM* Tables

Determine the design strength of the beam in Example 7-2 using *AISCM*, Table 3-19.

SOLUTION

From example 7-2,

C_c = 324.5 k, and

a = 1.36 in.;

therefore,

$$Y_2 = h_r + t_d - \frac{a}{2} = 3 \text{ in.} + 2 \text{ in.} - \frac{1.36}{2} = 4.32 \text{ in.}$$

From *AISCM*, Table 3-19, ϕM_n = 272 ft.-kips (by interpolation).

EXAMPLE 7-9

Composite Design Strength Using the *AISCM* Tables

Determine the design strength of the beam in Example 7-3 using *AISCM*, Table 3-19.

SOLUTION

From Example 7-3,

C = 172 kips

a = 0.722 in.

Y_1 = 0.305 in.

Y_2 = 4.64 in.

From *AISCM*, Table 3-19, ϕM_n = 227 ft.-kips (by linear interpolation).

EXAMPLE 7-10

Composite Design Strength Using the *AISCM* Tables

Determine the design strength of the beam in Example 7-5 using *AISCM*, Table 3-19.

SOLUTION

From Example 7-5,

C_c = 129 kips

Y_1 = 4.8 in.

Y_2 = 5.62 in.

From *AISCM*, Table 3-19, ϕM_n = 368 ft.-kips (by linear interpolation).

7.7 COMPOSITE BEAM DESIGN

There are several factors that must be considered in the design of composite floors. One key constraint is that the required floor structure depth is usually specified early in the design stage. In some cases, the depth of the steel beams will be limited if the floor-to-floor height of a building is limited. The beam spacing, another factor that should be considered early in the design stage, is generally a function of the slab strength, but tighter beam spacing could be required to minimize the steel beam depth. The slab design is mainly a function of the beam spacing (i.e., thinner slabs can be used with smaller beam spacing and thicker slabs are required for larger beam spacing). One must also consider the fire rating of the floor structure. Many occupancy categories will require that a floor structure have a certain fire rating, usually measured in hours. The Underwriters Laboratory (UL) has tested several floor assembly types and assigned ratings for each assembly type. (Fireproofing is covered in Chapter 14.) It is generally more economical and desirable to obtain the required fire rating by selecting a concrete slab with enough thickness so that the steel beams and steel deck do not have to be fireproofed.

In order to determine the most economical floor framing system for any building, the designer may have to perform several iterations of framing schemes, considering all of the variables noted above, to determine the best system. We will assume at this point that the slab spacing and the beam spacing have already been determined, so that we can proceed with the composite beam or girder design.

The following steps are given as a guide for the design of composite beams:

1. Tabulate the design loads (dead load, superimposed dead load, live load, and sustained live load). Tabulate service loads to be used for deflection and tabulate factored loads for strength checks.

2. Compute the maximum factored shear, V_u, and moment, M_u, for design, and the maximum factored moment during the construction phase, M_{uc}, for unshored construction.

3. Estimate the beam weight. Assuming that the PNA is in the concrete slab, we can use equation (7-11) to determine the moment capacity, $\phi M_n = \phi Ty$. Using $M_u = \phi M_n$,

   $T = A_s F_y$, and $y = \left(\dfrac{d}{2} + Y_{con} - \dfrac{a}{2} \right)$, we can solve for the required area of steel:

$$A_s = \frac{M_u}{\phi Fy \left(\dfrac{d}{2} + Y_{con} - \dfrac{a}{2} \right)}, \tag{7-28}$$

An assumed value of $a = 0.4t_c$ is recommended here (see Figure 7-10 for t_c).

In equation (7-28), Y_{con} is the distance from the top of the steel beam to the top of the slab. Knowing that the density of steel is 490 lb./ft.^{3}, we can solve for the beam weight:

$$w = \left(\frac{A_s}{144} \right) 490 = \frac{M_u}{\phi Fy \left(\dfrac{d}{2} + Y_{con} - \dfrac{a}{2} \right)}$$

$$w = \frac{3.4 M_u}{\phi Fy \left(\dfrac{d}{2} + Y_{con} - \dfrac{a}{2} \right)}, \tag{7-29}$$

where w is in pounds per linear foot. For various assumed values of the nominal beam depth, d, in inches, the required minimum beam weight, in pounds per linear foot, is calculated, and the lightest weight beam is selected.

4. Conduct the construction phase strength and deflection check (unshored beams only):

   a. Check the selected steel beam as a noncomposite section to support the following loads:

      - Weight of wet concrete,
      - Weight of metal deck,
      - Steel beam self-weight, and
      - Construction live load = 20 psf (weight of workers and equipment).

   b. Calculate the deflection of the noncomposite beam under construction phase dead load (CDL). A percentage of this deflection can be the specified camber. Do not camber the beam if the span is less than 25 ft. or if the construction phase dead load deflection is less than ¾ in.

5. Calculate $A_s F_y$ for the selected steel section, then calculate the degree of composite-ness. For 100% composite action (fully composite), $\Sigma Q_n = A_s F_y$. The minimum degree of compositeness is 25% (i.e., $\Sigma Q_n = 0.25 A_s F_y$).

6. Calculate the effective flange width, b, of the concrete slab. The effective width on each side of the beam centerline is the smallest of

   - ⅛ of the beam span,
   - ½ of the distance to the adjacent beam, or
   - The edge of the slab distance (for edge beams).

7. Calculate the actual depth of the effective concrete flange, a (this might be different from the value assumed in step 3):

$$a = \frac{\Sigma Q_n}{0.85 f'_c b}. \tag{7-30}$$

8. Compute the distance from the beam top flange to the centroid of the effective concrete flange:

$$Y_2 = Y_{con} - \frac{a}{2}. \tag{7-31}$$

9. Use the value of Y_2 from step 8 and the assumed value of ΣQ_n from step 5. Go to the composite beam selection table (Table 3-19 in the *AISCM*) corresponding to the beam chosen in step 3. Determine ϕM_n, the design bending strength of the composite section, linearly interpolating if necessary:

   a. If $\phi M_n > M_u$, the beam section is adequate.
   b. If $\phi M_n < M_u$, the beam is inadequate, and the following options should be considered:

      - Increase the degree of compositeness (i.e., increase ΣQ_n up to $\leq A_s F_y$).
      - Use larger a beam size.

10. Check the shear strength of the steel beam, $\phi V_n > V_u$, where $\phi_v V_n = \phi_v 0.6 F_y A_w C_v$.

    Recall that for webs of I-shaped members with $\dfrac{h}{t_w} \leq 2.24 \sqrt{\dfrac{E}{F_y}}$, $\phi_v = C_v = 1.0$

    (see Chapter 6 for other values of C_v). Alternatively, ϕV_n can be obtained for W-shapes from the maximum total uniform load table (Table 3-6 in the *AISCM*).

11. Select shear stud spacing (see Section 7.2 and Figure 7-7). Recall that $N_s = \dfrac{\Sigma Q_n}{Q_n}$ and N_s is the number of studs between the point of maximum moment and the point of zero moment.

12. Check deflections. There are two methods for computing the actual moment of inertia of the composite section, which is required in the deflection calculations:

    a. Use the lower-bound moment of inertia tabulated in the lower-bound elastic moment of inertia table (Table 3-20 in the *AISCM*), or

    b. Determine the transformed moment of inertia and calculate I_{eff} from equation (7-10). Then determine the actual moment of inertia ($I_{actual} = 0.75I_{eff}$) from equation (7-24), which is then used to calculate deflections.

    Recall that the modular ratio used for short-term deflections is n and for long-term (sustained load) deflection, the modular ratio is $2n$. Deflections are generally limited to $L/360$ for live loads and $L/240$ for total loads.

13. Check floor vibrations (see Chapter 12).

14. Compute the reinforcement required in the concrete slab and over the girders (see Section 7-8, Practical Considerations).

EXAMPLE 7-11

Composite Beam and Girder Design

Given the floor plan shown in Figure 7-24, design a typical filler beam B1 and girder G1. The floor consists of a 3.5-in. normal weight concrete slab on 1.5-in. × 20-ga. galvanized composite metal deck with $6 \times 6 - W2.9 \times 2.9$ welded wire fabric (WWF) to reinforce the slab. Assume ASTM A572, grade 50 steel and a concrete strength, $f'_c = 3.5$ ksi. Use a floor live load of 150 psf.

Floor Loads:

Weight of the concrete slab	= 51 psf
Weight of the steel deck	= 2 psf
Beam self-weight (35 lb./ft./6.67 ft.)	= 5 psf (assumed)
Girder self-weight (50 lb./ft./30 ft.)	= 2 psf (assumed)
Partitions	= 20 psf
Ceiling + Mechanical/Electrical	= 5 psf
Construction phase dead load (CDL)	= 51 + 2 + 5 + 2 = 60 psf
Superimposed dead load (SDL)	= 20 + 5 = 25 psf
Total floor dead load (FDL)	= 85 psf
Floor live load (FLL)	**= 150 psf**

For 1.5-in. metal deck, deck rib width, $w_r = 3.5$ in. > 2 in. OK, and Deck rib depth, $h_r = 1.5$ in.

Note: These values must be obtained from a deck manufacturer's catalog.

3.5-in. concrete slab on top of deck (i.e., $t_c = 3.5$ in. > 2-in. minimum) OK

(continued)

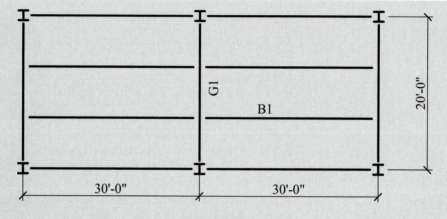

a. framing plan

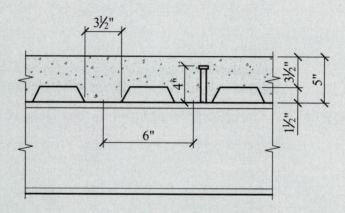

b. floor deck details

Figure 7-24 Details for Example 7-11.

Stud diameter = ¾ in. ≤ 2.5 t_f ∴ Minimum flange thickness = 0.75/2.5 = 0.3 in.

Total slab depth, $Y_{con} = t_c + h_r$ = 3.5 in. + 1.5 in. = 5 in.

Assuming a 1-in. clear concrete cover over the head of the stud, we have

Stud length, $H_s = Y_{con} - 1$ in. = 4 in.

Maximum longitudinal stud spacing = Smaller of $8Y_{con}$ = 40 in. or 36 in.
Minimum longitudinal stud spacing = $6D_s$ = 4.5 in.
Minimum transverse spacing of stud = $4D_s$ = 3 in.

Design of composite beam B1:

1. Composite design (loads and moments):

   Floor dead load   = 85 psf
   Floor live load   = 150 psf

Tributary width, TW $=$ 6.67 ft.

Service live load, $w_{LL} =$ (150 psf)(6.67 ft.) $=$ 1.0 kips/ft.
Superimposed dead load, $w_{SDL} =$ (25 psf)(6.67 ft.) $=$ 0.17 kips/ft. (0.014 kips/in.)
Ultimate factored total load, $w_u =$ $[(1.2)(85 \text{ psf}) + (1.6)(150 \text{ psf})](6.67 \text{ ft.})$
$$= 2.28 \text{ kips/ft.}$$

2. Factored moment, $M_u =$ (2.28 kips/ft.) (30 ft.)^2/8 $=$ 257 kips/ft

   Factored shear, $V_u = \dfrac{(2.28 \text{ kips/ft})(30 \text{ ft})}{2} = 34.2 \text{ kips}$

   Service live load reaction, $R_{LL} = \dfrac{(1 \text{ kip/ft})(30 \text{ ft})}{2} = 15 \text{ kips}$

   Superimposed dead load reaction, $R_{SDL} = \dfrac{(0.17 \text{ kips/ft})(30 \text{ ft})}{2} = 2.5 \text{ kips}$

3. Assume $a = 0.4t_c = (0.4)(3.5 \text{ in.}) = 1.4 \text{ in.}$

   $Y_{con} =$ 5 in.
   $F_y =$ 50 ksi
   $\phi =$ 0.90

   For $d =$ 10 in.,

   $$w = \frac{3.4M_u}{\phi F_y\left(\dfrac{d}{2} + Y_{con} - \dfrac{a}{2}\right)} = \frac{(3.4)(257 \times 12)}{(0.9)(50)\left(\dfrac{10}{2} + 5 - \dfrac{1.4}{2}\right)} = 26 \text{ lb./ft.}$$

   For $d =$ 12 in., beam weight $=$ 23 lb./ft.
   For $d =$ 14 in., beam weight $=$ **21 lb./ft.**

   Try a W14 $\times$ 22 beam; $A_s =$ 6.49 in.^{2} and moment of inertia, $I =$ 199 in.^{4}.

4. Construction phase strength and deflection check:

   Construction phase dead load (CDL) $=$ 60 psf
   Construction phase live load $\quad\quad=$ 20 psf
   Tributary width (TW) of beam $\quad=$ 6.67 ft.

   The construction phase factored total load is

   $w_u =$ 1.4DL $\times$ TW $=$ (1.4)(60 psf)(6.67 ft.) $=$ 0.56 kips/ft. OR
   $w_u =$ (1.2DL $+$ 1.6LL) $\times$ TW $=$ $[(1.2)(60 \text{ psf}) + (1.6)(20 \text{ psf})](6.67 \text{ ft})$
   $\quad\quad=$ 0.694 kip/ft. $\leftarrow$ Governs.

   The construction phase dead load is

   $w_{CDL} =$ (60 psf)(6.67 ft.) $=$ 0.4 kip/ft. (0.033 kip/in.).

   Construction phase dead load reaction, $R_{CDL} = \dfrac{(0.4 \text{ kips/ft.})(30 \text{ ft.})}{2} = 6 \text{ kips.}$

   The construction phase factored shear, $V_u = \dfrac{(0.7 \text{ kips/ft.})(30 \text{ ft.})}{2} = 10.4 \text{ kips.}$

The construction phase factored moment is

$$M_u = \frac{(0.694)(30)^2}{8} = 78 \text{ ft.-kips (top flange of beam is assumed to be fully}$$

$$\text{braced by the deck; therefore, } L_b = 0).$$

From the beam design selection table (*AISCM*, Table 3-6 or 3-10), we obtain the design moment capacity of the noncomposite beam:

$$\phi M_n \text{ for W14} \times 22 = 125 \text{ ft.-kips} > M_u = 78 \text{ ft.-kips. OK}$$

The construction phase dead load deflection is

$$\Delta = \frac{5wL^4}{384EI} = \frac{(5)(0.033)[(30)(12)]^4}{(384)(29,000)(199)} = 1.26 \text{ in.} > \frac{3}{4} \text{ in.} \therefore \text{ Camber required}$$

$$\Delta_c = (0.75)(1.25 \text{ in.}) = 0.94 \text{ in. Use 1–in. camber.}$$

$$\Delta_{CDL} = 1.25 \text{ in.} - 1 \text{ in. (Camber)} = 0.25 \text{ in.}$$

5. A_sF_y for W14 × 22 = (6.49 in.^{2})(50 ksi) = 325 kips

   ΣQ_n must be $\le A_sF_y$ (Choose a value between 25% A_sF_y and 100% A_sF_y.)
   Assume $\Sigma Q_n = 325$ kips $\Rightarrow$ 100% composite action

6. Effective concrete flange width, *b*, is the smaller of

   (⅛)(30 ft.) + (⅛)(30 ft.) = 7.5 ft., or
   (½)(6.67 ft.) + (½)(6.67 ft.) = 6.67 ft. (80 in.), governs.

7. Depth of the effective concrete flange is

$$a = \frac{\Sigma Q_n}{0.85f'_c b} = \frac{325}{(0.85)(3.5)(80)} = 1.37 \text{ in.}$$

8. Distance from the top of the steel beam top flange to the centroid of the effective concrete flange is

$$Y_2 = Y_{con} - 0.5a = 5 \text{ in.} - 0.5 \times 1.37 \text{ in.} = 4.32 \text{ in.}$$

9. Using the composite beam selection table (*AISCM*, Table 3-19) with $Y_2 = 4.32$ in. and the $\Sigma Q_n = 325$ kips assumed in step 5, we obtain, by linear interpolation,

$$\phi M_n = 273 \text{ ft.-kips} > M_u = 257 \text{ ft.-kips OK, and}$$

$$Y_1 = 0.0 \text{ in.} = \text{Distance from top of steel beam to the PNA.}$$

   If ϕM_n had been much less than M_u, we would have had to increase the beam size to W14 × 26 or W16 × 26, since we are already at 100% composite action.

10. Factored shear, $V_u = 35$ kips.

    The design shear strength is from Chapter 6 or *AISCM*, Table 3-6:

$$\phi_v V_n = \phi_v 0.6F_y A_w C_v = (1.0)(0.6)(50)(0.23)(13.7)(1.0)$$

$$= 94.5 \text{ k} > V_u = 34.2 \text{ kips. OK}$$

11. $N_s = \dfrac{\Sigma Q_n}{Q_n} =$ Number of studs between the point of maximum moment and the point of zero moment.

- Deck rib depth, $h_r = 1.5$ in.
- Stud length, $H_s = 4$ in. $< h_r + 3$ in. $= 4.5$ in.
- $N_r = 1$ (assuming a single row of studs)
- Deck rib width, $w_r = 3.5$ in.

For metal deck ribs perpendicular to the beam and placed in the strong position, the shear stud capacity reduction factors are from Table 7-1:

$$R_g = 1.0$$
$$R_p = 0.75$$
$$Q_n = 0.5 A_{sc} \sqrt{f'_c E_c} \le R_g R_p A_{sc} F_u$$
$$E_c = w_c^{1.5} \sqrt{f'_c} = (145)^{1.5} \sqrt{3.5} = 3266 \text{ ksi}$$
$$Q_n = 0.5\left(\frac{\pi(0.75)^2}{4}\right)\sqrt{(3.5)(3266)} \le (1.0)(0.75)\left(\frac{\pi(0.75)^2}{4}\right)65$$

23.6 kips $>$ 21.5 kips

Or, from *AISCM*, Table 3-21, $Q_n = 21.5$ kip

The number of studs between the point of maximum moment and the point of zero moment is

$$N_s = \frac{\Sigma Q_n}{Q_n} = \frac{325}{21.5} = 15.1 \therefore \text{ use 16 studs,}$$

where N_s is the number of studs between the point of maximum moment and the point of zero moment. Since the loading on the beam is symmetrical, the total number of studs on the beam is

$$N = 2N_s = (2)(16) = 32 \text{ studs}$$
$$\text{Stud spacing, } s = \frac{(30 \text{ ft.})(12)}{32} = 11.25 \text{ in.} < 36 \text{ in. OK}$$
$$> 4.5 \text{ in. OK}$$

Note: For composite beams the actual stud spacing will depend on the spacing of the deck flutes. This implies that the final spacing of the studs will need to be less than or equal to the value above, depending on the spacing of the deck flutes (i.e., the final stud spacing in a composite beam must be a multiple of the deck flute spacing). For example, if the deck flute spacing on this beam were 12 in., it would not be physically possible to place 32 studs in one row on the beam that is 30 feet long (i.e., the beam would have no more than 30 deck flutes available). We would then have to consider two rows of studs or an increase in the beam size, since we are already at 100% composite action. For 1½-in. composite metal deck, the usual rib spacing is about 6 in., so 32 studs could be placed in one row on this beam.

12. We will now check deflections. We need to calculate the lower-bound moment of inertia, I_{LB}, of the composite beam using the values of $Y_1 = 0.0$ in. and $Y_2 = 4.32$ in. obtained from step 9. Using the lower-bound moment of inertia table for a W14 × 22 (*AISCM*, Table 3-20), we obtain $I_{LB} = 606$ in.[4] by linear interpolation.

The live load deflection is

$$\Delta = \frac{5wL^4}{384EI} = \frac{(5)(0.084)[(30)(12)]^4}{(384)(29,000)(606)} = 1.04 \text{ in.}$$

$$\frac{L}{360} = \frac{(30)(12)}{360} = 1 \text{ in. (close to } 1.04 \text{ in. therefore, OK)}$$

Since the beam is cambered, the total deflection, Δ_{TL}, which is the sum of the deflections due to superimposed dead load ($w_{SDL} = 0.014$ kip/in.), live load ($w_{LL} = 0.084$ kip/in.) and the dead load not accounted for in the camber, is

$$\Delta_{TL} = \Delta_{SDL} + \Delta_{LL} + \Delta_{CDL}$$

$$\Delta_{TL} = \frac{5wL^4}{384EI} = \frac{(5)(0.084 + 0.014)[(30)(12)]^4}{(384)(29,000)(606)} + 0.25 \text{ in.} = 1.46 \text{ in.}$$

$$\frac{L}{240} = \frac{(30)(12)}{240} = 1.5 \text{ in.} > 1.46 \text{ in. OK}$$

Use a W14 × 22 beam ($N = 32$ and Camber = 1 in.)

Design of Composite Girder G1:

1. Loads (see Figure 7-25):

   Girder tributary width = 30 ft.
   Factored load, P_u = 2 beams × 34.2 kips = 68.4 kips
   Service live load, P_{LL} = 2 beams × 15 kips = 30 kips
   Service superimposed dead load P_{SDL} = 2 beams × 2.5 kips = 5.0 kips

2. Factored moment, M_u = (68.4 kips) (6.67 ft.) = 456 ft.-kips
   Factored shear, V_u = 70 kips

3. Assume $a = 0.4\, t_c$

$$t_c = Y_{con} - \frac{h_r}{2} \text{ (see Figure 7-10)}$$

$$= 5 - \frac{1.5}{2} = 4.25 \text{ in.}$$

$$a = (0.4)(4.25) = 1.7 \text{ in.}$$

$$Y_{con} = 5 \text{ in., } F_y = 50 \text{ ksi, } \phi = 0.90$$

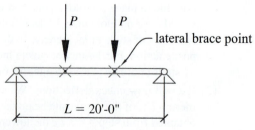

Figure 7-25 Loading for G1.

For $d = 16$ in.,

$$w = \frac{3.4 M_u}{\phi F_y \left(\dfrac{d}{2} + Y_{con} - \dfrac{a}{2}\right)} = \frac{(3.4)(456 \times 12)}{(0.9)(50)\left(\dfrac{16}{2} + 5 - \dfrac{1.7}{2}\right)} = 35 \text{ lb./ft.}$$

For $d = 18$ in., beam weight $= 32$ lb./ft.
For $d = 21$ in., beam weight $= 29$ lb./ft.

Try a W18 $\times$ 35 beam; $A_s = 10.3$ in.^{2} and moment of inertia, $I = 510$ in.^{4}.

4. Construction phase strength and deflection check:

Construction phase factored load, $P_u = 2$ beams $\times$ 10.4 kips $= 20.8$ kips
Construction phase dead load, $P_{CDL} = 2$ beams $\times$ 6 kips $= 12$ kips
Construction phase factored moment, $M_u = (20.8 \text{ kips})(6.67 \text{ ft.})$
$$= 139 \text{ ft.-kips} \ (L_b = 6.67 \text{ ft.})$$

From the beam design selection table (*AISCM*, Table 3-10), we obtain the design moment capacity of the noncomposite section:

$$\phi M_n \text{ for } W18 \times 35 = 220 \text{ ft.-kips} > M_u = 139 \text{ ft.-kips OK}$$

The construction phase dead load deflection of the noncomposite section is

$$\Delta_{CDL} = \frac{PL^3}{28EI} = \frac{(12)[(20)(12)]^3}{(28)(29,000)(510)} = 0.40 \text{ in.} < \tfrac{3}{4} \text{ in.}$$

$\therefore$ No camber is required.

Since no camber is required, the calculated construction phase dead load deflection will be added in step 12 to the superimposed dead load and live load deflections, to obtain the total load deflection, Δ_{TL}.

5. $A_s F_y$ for W18 $\times$ 35 $= (10.3 \text{ in.}^2)(50 \text{ ksi}) = 515 \text{ kips}$

ΣQ_n must be $\leq A_s F_y$ (Choose a value between 25% $A_s F_y$ and 100% $A_s F_y$.)
Assume $\Sigma Q_n = 515$ kips (i.e., 100% composite action).

6. Effective concrete flange width, b, is the smaller of

$(\tfrac{1}{8})(20 \text{ ft.}) + (\tfrac{1}{8})(20 \text{ ft.}) = 5 \text{ ft. (60 in.), or} \leftarrow \text{Controls}$
$(\tfrac{1}{2})(30 \text{ ft.}) + (\tfrac{1}{2})(30 \text{ ft.}) = 30 \text{ ft. (360 in.).}$

7. Depth of the effective concrete flange is

$$a = \frac{\Sigma Q_n}{0.85 f'_c b} = \frac{515}{(0.85)(3.5)(60)} = 2.89 \text{ in.}$$

8. Distance from the top of the steel beam to the centroid of the effective concrete flange is

$$Y_2 = Y_{con} - 0.5a = 5 - (0.5)(2.89) = 3.55 \text{ in.}$$

9. Using the composite beam selection table for W18 × 35 (*AISCM*, Table 3-19) with $Y_2 = 3.55$ in. and $\Sigma Q_n = 515$ kips, and we obtain

$$\phi M_n = 478 \text{ ft.-kips} > M_u = 456 \text{ ft.-kips OK, and}$$

$$Y_1 = 0.0 \text{ in.} = \text{Distance from top of steel beam to PNA.}$$

10. $V_u = 68.4$ kips. The design shear strength is from Chapter 6 or *AISCM*, Table 3-6:

$$\phi_v V_n = \phi_v 0.6 F_y A_w C_v = (1.0)(0.6)(50)(0.30)(17.7)(1.0)$$
$$= 159 \text{ kips} > V_u = 68.4 \text{ kips. OK}$$

11. $N_s = \dfrac{\Sigma Q_n}{Q_n} = $ Number of studs between the points of maximum and zero moments

$$w_r / h_r = 3.5/1.5 = 2.33$$

From Table 7-1,

$$R_g = 1.0$$
$$R_p = 0.75$$

$$Q_n = 0.5 A_{sc} \sqrt{f'_c E_c} \leq R_g R_p A_{sc} F_u$$
$$= w_c^{1.5} \sqrt{f'_c} = (145)^{1.5} \sqrt{3.5} = 3266 \text{ ksi}$$

$$Q_n = 0.5 \left(\frac{\pi(0.75)^2}{4} \right) \sqrt{(3.5)(3266)} \leq (1.0)(0.75) \left(\frac{\pi(0.75)^2}{4} \right) 65$$

23.6 kips > 21.5 kips

Or, from AISC, Table 3-21, $Q_n = 18.3$ kips.

The number of studs between the point of maximum moment and the nearest point of zero moment is

$$N_s = \frac{\Sigma Q_n}{Q_n} = \frac{515}{21.5} = 23.9 \therefore \text{use 24 studs.}$$

N_s is the number of studs between the point of maximum moment and the nearest point of zero moment (within 6′-8″ from each end of the girder).

$$\text{Stud spacing, } s = \frac{(6.67 \text{ ft.})(12)}{24} = 3.33 \text{ in.} < 36 \text{ in. OK}$$

$$< 4.5 \text{ in. Not good}$$

The required spacing is too close, so either the beam size needs to be increased, since we are at 100% composite action, or two rows of studs could be used, since that would effectively double the stud spacing from 3.33 in. to 6.67 in., which is greater than the minimum calculated value of 4.5 in. This particular design has a relatively high number of studs, so a larger beam size will be selected. Note that a rule of thumb is that one shear stud equates to 10 pounds of steel [10]. We will examine this relationship later in the design.

Use a W18 × 46.

$A_s F_y$ for W18 × 46 = 13.5 × 50 ksi = 675 kips

From Table 3-19, it can be seen that the design moment for a W18 × 46 is about $\phi M_n = 480$ ft.-kips ($> M_u = 456$ ft.-kips) at about 25% composite action. We will assume 40% composite action.

$$a = \frac{\Sigma Q_n}{0.85 f'_c b} = \frac{(0.4)(675)}{(0.85)(3.5)(60)} = 1.52 \text{ in.}$$

$$Y_2 = Y_{con} - 0.5a = 5 \text{ in.} - 0.5 \times 1.52 \text{ in.} = 4.24 \text{ in.}$$

$$Y_1 = 1.61 \text{ in. (by linear interpolation)}$$

$$N_s = \frac{\Sigma Q_n}{Q_n} = \frac{(0.4)(675)}{21.5} = 12.6 \therefore \text{ use 13 studs}$$

$$\text{Stud spacing, } s = \frac{(6.67 \text{ ft.})(12)}{13} = 6.15 \text{ in.} < 36 \text{ in. OK}$$

$$> 4.5 \text{ in. OK}$$

Since we have concentrated loads acting on the girder and the maximum moments occur at these concentrated loads, N_s is the number of studs on the beam from the concentrated load location to the point of zero moment (i.e., the ends of the girder). Only a nominal number of studs (e.g., 4 studs or studs at 24 in. on center) are typically provided between the concentrated loads on the girder, since the moment gradient or horizontal shear between these points is negligible. Therefore, the total number of studs provided on the girder is specified as $N = 13, 4, 13$.

12. We will now check the deflections. Using values of $Y_2 = 4.24$ in. and $Y_1 = 1.61$ in., in the lower-bound moment of inertia table for W18 × 46 (*AISCM*, Table 3-20) we obtain the lower-bound moment of inertia for the composite section, $I_{LB} = 1389$ in.[4]. Recall that the service loads on the girder are as follows:

$$P_{LL} = 30 \text{ kips} \qquad P_{SDL} = 5.0 \text{ kips} \qquad P_{CDL} = 12 \text{ kips}$$

Construction phase dead load deflection for the W18 × 46 is

$$\Delta_{CDL} = \frac{PL^3}{28EI} = \frac{(12)[(20)(12)]^3}{(28)(29,000)(712)} = 0.285 \text{ in.}$$

Live load deflection:

$$\Delta_{LL} = \frac{PL^3}{28EI} = \frac{(30)[(20)(12)]^3}{(28)(29,000)(1389)} = 0.365 \text{ in.}$$

$$\frac{L}{360} = \frac{(20)(12)}{360} = 0.67 \text{ in. OK}$$

Total load deflection:

$$\Delta_{TL} = \Delta_{SDL} + \Delta_{LL} + \Delta_{CDL}$$

$$\Delta_{TL} = \frac{PL^3}{28EI} = \frac{(30 + 5.0)(20 \times 12)^3}{(28)(29,000)(1389)} + 0.285 = 0.71 \text{ in.}$$

The total initial dead load deflection, Δ_{CDL}, is included here, since no camber was specified in step 3.

$$\frac{L}{240} = \frac{(20)(12)}{240} = 1.5 \text{ in.} > \Delta_{TL}, \text{ OK}$$

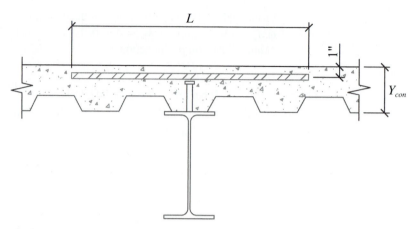

Figure 7-26 Rebar over G1.

Use a W18 × 46 girder (N = 13, 4, 13).

See Chapter 12 for floor vibrations.

As a comparison, we increased the beam size in lieu of using extra shear studs on a smaller beam size. The total increase in beam weight was (46 plf − 35 plf) (20 ft.) = 220 lb. The change in the number of required shear studs was 48 − 26 = 22 studs. Given that one shear stud is equivalent to 10 pounds of steel (see rule of thumb on page 315) we have realized some economy in increasing the beam size.

13. Determine additional rebar over girders.

Concrete strength, f'_c = 3.5 ksi

Yield strength of rebar, f_y = 60 ksi

Effective concrete flange width of composite girder, b = 60 in. (see step 6 of the girder design)

Factored load on slab, w_u = (1.2)(85 psf) + (1.6)(150 psf) = 0.342 kip/ft.^{2}

Maximum negative moment, M_u (neg.) = $\dfrac{(0.342 \text{ kips/ft.})(2.5 \text{ ft.})^2}{2}$

$\qquad\qquad\qquad$ = 1.07 ft.-kips/ft.

The approximate required area of rebar (in square inches per foot width of slab) can be derived from reinforced concrete design principles as

Area of reinforcing steel required, $A_s \approx M_u/4d > A_s$ min.,

where

d = Effective depth of concrete slab in inches,

M_u is the factored moment in ft-kips/ft width of slab,

Note: The inconsistency of the units in the equation for A_s has already been accounted for by the constant, 4, in the denominator.

$d = Y_{con} - 1$ in. = 5 in. − 1 in. = 4 in.,

$A_{s \min}$ = (0.0018)(12 in.)(Y_{con}) = (0.0018)(12 in.)(5 in.) = 0.11 in.^{2}/ft. width of slab, and

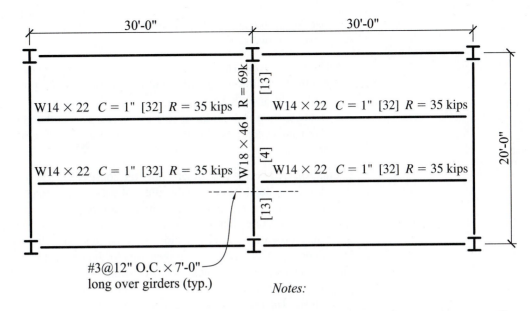

Figure 7-27 Shows how the preceding design would be indicated on a framing plan. Note that a typical framing plan should show the beam sizes, camber, number of studs, and end reaction.

$$A_{s\text{ required}} = \frac{1.07 \text{ ft.-kips}}{(4)(4 \text{ in.})} = 0.07 \text{ in.}^2/\text{ft. width of slab} < A_{s\text{ min}}.$$

Therefore, use $A_s = A_{s\text{ min}} = 0.11$ in.^{2}/ft. width of slab.

Use no. 3 top bars @ 12 in. on centers (o.c.) × 7 ft. long over G1.

The length of the rebar is a function of the required development length of the reinforcement (see [2], Chapter 12). The required development length of a #3 bar is less than 24 in. past the point of maximum moment, but is also 12 in. past any point of stress, so the total length of the #3 bar is the effective slab width, 60 in. plus 12 in. for each side, or 7 ft. long. Figure 7-27 summarizes the floor design for this example.

7.8 PRACTICAL CONSIDERATIONS

1. In the United States, composite beams and girders are usually unshored, with the deck and the bare steel beam supporting the dead and live loads during the construction phase. It is more expensive to shore the beams and girders, but shored construction results in smaller size beams and girders.

2. Camber implies the upward bowing of the beam or girder. Steel beams can be cambered in one of two ways:

   - Cold cambering: The unheated beam is forced into the cambered shape by passing the beam through guides that have been set at the predetermined radius to achieve the specified camber.
   - Heat Cambering: Similar to cold cambering except that the beam is heated before cambering; heat cambering is the more expensive of the two cambering processes.

3. For uncambered beams and girders, limit the construction phase dead load deflection, Δ_{CDL}, to $L/360$ to minimize the effect of concrete ponding.

   Because of concrete ponding, additional concrete is needed to achieve a level floor due to the deflection of the floor beams and girders. As concrete is first poured, the beams and girders deflect, resulting in more concrete being required to achieve a level floor, which in turn results in increased loading, which in turn leads to more deflections, and thus leads to more concrete being required to achieve a level floor. This process continues until the beam and girders reaches equilibrium and the floor becomes level. The more flexible the beams and girders are, the greater the effect of concrete ponding on the floor system.

4. For uncambered beams and girders, add 10% to 20% more concrete dead weight to allow for the additional weight that will result from concrete ponding due to the deflection of the steel beam or girder during the construction phase.

5. Calculation of the construction phase deflection or the required camber usually assumes pinned supports for the beams and girders, but some restraint against deflection will be provided by the simple shear connections at the ends of the beams and girders, and thus, the actual construction phase deflection may be less than calculated. This could lead to an overestimation of the required camber, which is not desirable.

6. Overestimating the required camber could lead to problems, resulting in a floor slab with less than adequate depth at the critical sections. This could lead to inadequate cover for the shear studs at these sections. In order to avoid this situation, and to account for the restraint provided by the beam or girder connections, it is advisable to specify 75% to 85% of the construction phase dead load Δ_{CDL}, as the required camber.

7. Metal decks usually come in widths of 2 to 3 ft. and lengths of up to 42 ft. Specify 3-span decks whenever possible (i.e., decks that are long enough to span over four or more beams in order to achieve the maximum strength of the deck. Avoid single-span decks whenever possible; it is more susceptible to ponding and it is not as strong as the 2-span or 3-span decks.

8. Stud diameters could be ½ in., ⅝ in., or ¾ in., but ¾-in.-diameter studs are the most commonly used.

9. Where studs cannot be placed at the center of the deck flute, offset the stud toward the nearest end support of the beam or girder.

10. If the number of studs required in the beam or girder exceeds the number that can be placed in each flute in a single row, lay out the balance of studs in double rows starting from both ends of the beam or girder.

11. If the number of studs required exceeds the number that can be placed in every second flute, place the studs in every second flute and add the remaining studs to the deck flutes in between, starting from both ends of the beam or girder. This is an alternative to uniformly spacing the studs in every flute throughout the beam or girder.

12. A rule of thumb to maintain economy in the balance between adding shear studs and increasing the beam size is that one shear stud is equivalent to 10 pounds of steel.

13. The reactions of composite beams or girders are usually higher than those of comparable noncomposite beams or girders. Steel fabricators usually design the end connections for one-half of the maximum total factored uniform load (see *AISCM*, Table 3-6), but this is often not adequate for composite beams or girders. To account for the higher end reactions in composite beams or girders, either specify on the plan the actual reactions at the ends of the composite beams or girders, or specify that the composite beam or girder connections be designed for three-quarters of the maximum total factored uniform load.

14. The effect of floor openings on the composite action of beams and girders is a function of the size and location of the openings. Reference[11] provides an analytical approach for calculating the effective width of composite beams with floor openings. A conservative but quick approach for considering the effect of floor openings on composite beams is as follows: If the floor opening is located only on one side of the beam, the beam is considered as an L-shaped beam for calculating the effective width at the location of the opening, but the beam is still assumed to support the full tributary width of the floor on both sides of the beam (less the reduction in load due to the floor opening). In the case where there are floor openings at the same location on both sides of the beam, the beam is considered to be non-composite at that location, and the load on the beam will include the load from the tributary width on both sides of the beam (less the reduction in load due to the floor openings). It is recommended that additional reinforcement be added at the edges of the floor openings to control cracking.

7.9 REFERENCES

1. American Institute of Steel Construction. 2006. *Steel construction manual*, 13th ed. Chicago AISC.

2. American Concrete Institute. 2008. *Building code requirements for structural concrete and commentary*. Farmington Hills, MI.

3. American Institute of Steel Construction. 2005. *Steel design guide series 5: Low- and medium-rise steel buildings*.

4. Tamboli, Akbar. 1997. *Steel design handbook—LRFD method*. New York: McGraw Hill.

5. Easterling, Samuel, David Gibbings, and Thomas Murray. Second Quarter, 1993. Strength of shear studs in steel deck on composite beams and joists. *AISC Engineering Journal*.

6. Hansell, W. C., T. V. Galambos, M. K. Ravindra, and I. M. Viest. 1978. Composite beam criteria in LRFD. *Journal of the Structural Division*, ASCE 104 (No. ST9).

7. Segui, William. 2006. *Steel Design*, 4th ed. Toronto: Thomson Engineering.

8. Limbrunner, George F. and Leonard Spiegel. 2001. *Applied structural steel design*, 4th ed. Upper Saddle River, NJ: Prentice Hall.

9. Vest, I. M., Colaco, J. P., Furlong, R. W., Griffis, L. G., Leon, R. T., and Wyllie, L. A. 1997. *Composite construction design for buildings*. New York: McGraw Hill.

10. Carter, Charles J., Murray, Thomas M., and Thornton, William A. 2000. "Economy in steel" *Modern Steel Construction*, April 2000.

11. Weisner, Kenneth B. 'Composite beams with slab openings', Modern Steel Construction, March 1996, pp. 26–30.

7.10 PROBLEMS

7-1. Determine the transformed moment of inertia for the sections shown in Figure 7-28. The concrete has a density of 115 pcf and a 28-day strength of 3 ksi. Assume full-composite action.

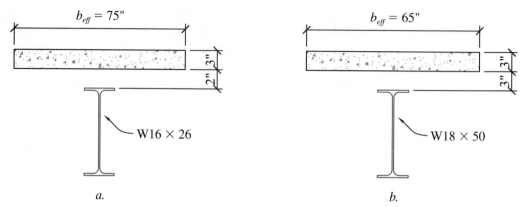

Figure 7-28 Details for Problem 7-1.

7-2. Determine the design strength of the composite section given in Problem 1, assuming full-composite action and ASTM A992 steel. Confirm the results with *AISCM*, Table 3-19.

7-3. Determine the following for the section shown in Figure 7-29. The concrete has a density of 145 pcf and a 28-day strength of 3.5 ksi. The steel is ASTM A992. Assume $Q_n = 21.5$ kips for one stud.

    a. Transformed moment of inertia

    b. Design moment strength, $\phi_b M_n$, assuming full-composite action

    c. Design moment strength, $\phi_b M_n$, assuming 40% action

    d. The number of ¾-in. ASTM A108 studs required between the points of maximum moment required for 100% and 40% composite action

Confirm the results from b, c and d using the *AISCM* tables.

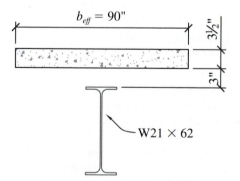

Figure 7-29 Details for Problem 7-3.

7-4. For the beam shown in Figure 7-30, determine the effective moment of inertia and the design moment capacity for 40% composite action and compare the results with *AISCM*, Tables 3-19 and 3-20. Use $f'_c = 3.5$ ksi and a concrete density of 145 pcf.

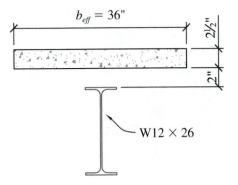

Figure 7-30 Details for Problem 7-4.

7-5. For the floor framing shown in Figure 7-31, design the composite members B1 and G1. The floor construction is 3-in. composite deck plus 3½-in. normal weight concrete (6½ in. total thickness). The steel is ASTM A992, grade 50 and the concrete has a 28-day strength of 3.5 ksi. Design for flexure, shear, and deflection, considering dead and live loads.

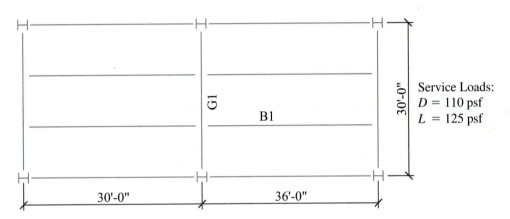

Figure 7-31 Details for Problem 7-5.

Student Design Project Problems

7-6. For the floor framing in the student design project (Figure 1-22), design the typical interior floor beams and girders as composite members.

7-7. Repeat Problem 7-6 for the typical perimeter beams.

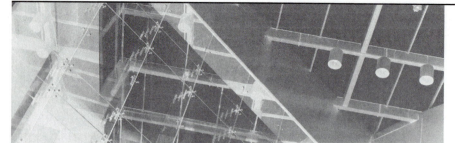

Compression Members Under Combined Axial and Bending Loads

8.1 INTRODUCTION TO BEAM–COLUMNS

Structural members that are subjected to combined axial and bending loads are called beam–columns. Beam–columns could be part of braced frames or unbraced frames (i.e., moment frames); the design of these columns will differ depending on whether the building frame is braced or unbraced. In this chapter, we will discuss beam–columns in a typical steel-framed building. In Chapter 5, we covered the design of columns in pure compression, which rarely exists in buildings. Generally speaking, most building columns are actually beam–columns because of how they are loaded, so the majority of this chapter will focus on building columns.

Braced Frames

In buildings with braced frames, the lateral loads are resisted by diagonal bracing or shear walls. Braced frames are also referred to as nonsway frames or sidesway-inhibited frames. In braced frames, the beams and girders are connected to the columns with simple shear connections that have practically little or no moment restraint. The moments in the columns, M_{nt} (i.e., **n**o **t**ranslation moments), are nonsway moments that result from the eccentricity of the beam and girder reactions. For these frames, the sway moment, M_{lt} (i.e., **l**ateral **t**ranslation moment), is zero.

Unbraced Frames

Unbraced, or moment, frames (also referred to as sway frames or sidesway-uninhibited frames) resist lateral loads through bending in the columns and girders, and the rigidity of the girder-to-column moment connections. The moments in these frames are a combination of no translation moments, M_{nt}, and lateral translation moments, M_{lt}.

Types of Beam–Columns

The different situations where beam–columns might occur in building structures are discussed in this section.

Type 1: Columns in buildings with braced frames

In this case, the moments result from the eccentricity of the girder and beam reactions. Therefore, the moment due to the reaction eccentricity is

$$M = Pe,$$

where e is the eccentricity of the girder or beam reactions as shown in Figure 8-1.

Type 2: Exterior columns and girts

For buildings with large story heights (e.g., > 12 ft.), there might not be a cladding system that can economically span from floor to floor to resist the wind load perpendicular to the face of the cladding; therefore, it may be necessary to use beams in the plane of the cladding to reduce the span of the metal cladding. These beams, known as girts, are subjected to bending in the horizontal plane due to wind loads perpendicular to the face of the cladding. They usually consist of channels, with their webs parallel to the horizontal plane and the toes pointing downward. They are often oriented this way so that debris or moisture does not accumulate on the member. The channel girts, because of their orientation, are also subjected to weak axis bending due to the self-weight of the girt. To minimize the vertical deflection due to self-weight, sag rods are used as shown in Figure 8-2. It should also be noted that the exterior columns in the plane of the cladding will also be subjected to bending loads from the wind pressure perpendicular to the face of the cladding, in addition to the axial loads on the column.

Type 3: Truss chords

Top and bottom chords of trusses (see Figure 8-3) where the members are subjected to combined axial loads and bending that could be caused by floor or roof loads applied to the top or bottom chord between the panel points of the truss, or moments induced due to the continuity of the top and bottom chords, are another type of beam–column.

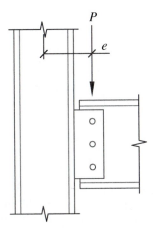
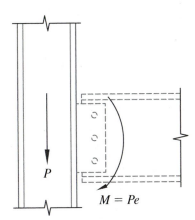

Figure 8-1 Type 1 columns in braced frames.

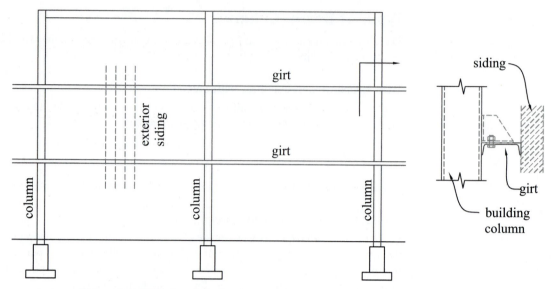

Figure 8-2 Wall elevation showing exterior columns, girts, and sag rods.

Type 4: Hangers with eccentric axial loads

Hangers with eccentric axial loads where the structural member is subjected to combined axial tension and bending are a type of beam–column that occurs in lighter structures, such as a catwalk or a mezzanine (see Figure 8-4).

Type 5: Moment, or unbraced, frames

Moment frames consist of columns and beams or girders that are subjected to bending moments due to lateral wind or seismic loads in addition to the axial loads on the beams and columns (see Figure 8-5).

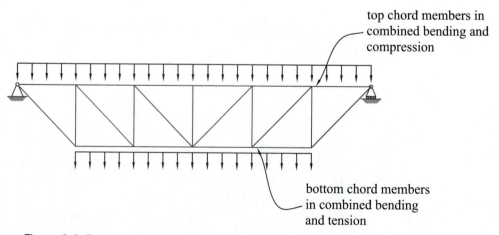

Figure 8-3 Top and bottom chords of trusses.

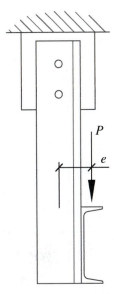

Figure 8-4 Hanger with eccentric axial load.

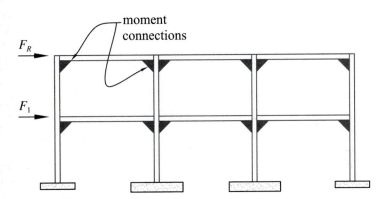

Figure 8-5 Moment, or unbraced, frames.

8.2 EXAMPLES OF TYPE 1 BEAM–COLUMNS

In this section, we present several examples of type 1 beam–columns in building structures. The connection eccentricities result in unbalanced moments in the columns due to unequal reactions from the girders or the beams on adjacent sides of a column.

Unbalanced moments occur primarily in columns in a building with braced frames (i.e., buildings that are braced with masonry shear walls, concrete shear walls, or steel-plate shear walls or diagonal bracing such as X-bracing, chevron bracing, or single diagonal bracing). The moments in the beam–columns occur due to the eccentricity of the reaction that is transferred to the column at the girder-to-column and beam-to-column connections. Thus, moments about two orthogonal axes will exist in a typical building column and this will be more critical for corner columns and slender columns. The different types of beam or girder-to-column connections and the resulting moments due to reaction eccentricities are shown in Figures 8-6 through 8-9.

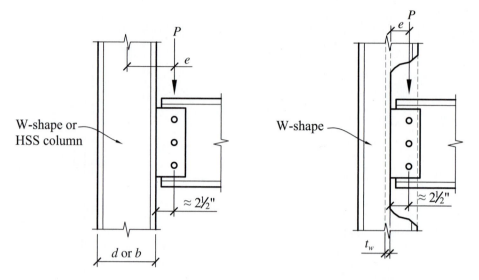

a. connection to column flange   *b. connection to column web*

$$e = \frac{d}{2} + 2.5" \quad \text{for beams framing into column flange}$$

$$e = \frac{t_w}{2} + 2.5" \quad \text{for beams framing into column web}$$

$$e = \frac{b}{2} + 2.5" \quad \text{for beams framing into face of HSS}$$

Figure 8-6 Simple shear connection eccentricity.

1. Simple shear connection eccentricity (see Figure 8-6)

    The majority of the connections in steel buildings are standard simple shear connections. The eccentricity, e, is the distance between the centerline of the column and the location of the bolt line on the beam or girder.

2. Seated connection eccentricity (see Figure 8-7)

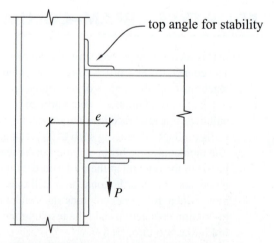

Figure 8-7 Seated connection eccentricity.

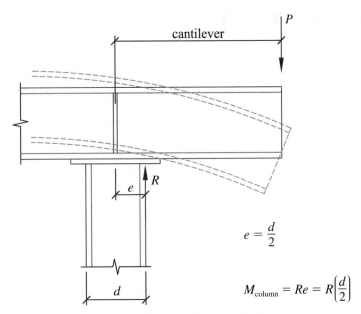

$$e = \frac{d}{2}$$

$$M_{\text{column}} = Re = R\left(\frac{d}{2}\right)$$

Figure 8-8 Top-connected connection eccentricity.

3. Top-connected connection eccentricity (see Figure 8-8)
4. End-plate connection eccentricity (see Figure 8-9)

    For end-plate connections, the connection eccentricity for strong axis bending is taken as one-half the distance from the face of the column flange to the centerline of the column (see Figure 8-9). The eccentricity for weak axis bending will be one-half the web thickness for wide flange columns; this is practically negligible for wide flange columns and therefore can be ignored in design.

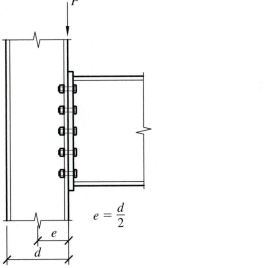

$$e = \frac{d}{2}$$

$$e \approx 0$$

a. connection to the column flange   *b. connection to the column web*

Figure 8-9 End-plate connection eccentricity.

8.3 COLUMN SCHEDULE

Before discussing the design of beam–columns, we introduce a tabular format for presenting column design information on structural drawings known as a *column schedule*. The column schedule (see Table 8-1) is an organized and efficient tabular format for presenting the design information for all columns in a building structure. The information presented typically includes column sizes, factored axial loads at each level, and the location of column splices, as well as anchor rod and base plate sizes. Different columns in the building with identical loadings are usually grouped together as shown in Table 8-1. The lower section of the column schedule show the distance from the floor datum to the underside of the column base plate, and the base plate and anchor bolt sizes.

In typical low- to mid-rise buildings, the steel columns are usually spliced every two or three floors since the maximum column length that can be transported safely is approximately 60 ft. and the practical column height that can be erected safely on the site with guy-wire bracing before the beams and girders are erected is also limited. For high-rise buildings, it is not uncommon to splice columns every four floors to achieve greater economy (see Figure 8-10). See Chapter 11 for further discussion of column splices.

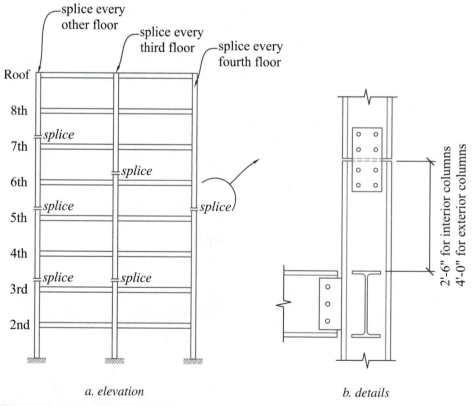

a. elevation   *b. details*

Figure 8-10 Column splice locations.

Table 8-1 Column schedule

Level *(elevations are top of steel or deck bearing)*	Grid Line			
	A-1 A-2	B-1 B-2	C-1 C-2	D-1 D-2
Roof +52'-0"	5" — W8 × 24 — 40 kips	11" — W8 × 24 — 40 kips	5" — W10 × 39 — 65 kips	7" — W10 × 39 — 65 kips
4th Flr. +39'-0"	W8 × 24 — 80 kips (4'-0")	W8 × 24 — 80 kips	W10 × 39 — 130 kips	W10 × 39 — 130 kips
splice location 3rd Flr. +26'-0"	W8 × 40 — 120 kips	W8 × 40 — 120 kips	W10 × 49 — 195 kips	W10 × 49 — 195 kips
2nd Flr. +13'-0"	W8 × 40 — 160 kips	W8 × 40 — 160 kips	W10 × 49 — 260 kips	W10 × 49 — 260 kips
Gnd. Flr. +0'-0"	8"	8"	1'-0"	1'-0"
underside of base plate elevation				
Base Plate $B \times N \times t$	14" × 14" × ¾"	14" × 14" × ¾"	16" × 18" × ¾"	16" × 18" × ¾"
Anchor rods	(4)-¾" diam. × 12" embedment with 4" hook	(4)-¾" diam. × 12" embedment with 4" hook	(4)-1" diam. × 24" embedment with heavy hex nut	(4)-1" diam. × 24" embedment with heavy hex nut

8.4 BEAM–COLUMN DESIGN

The beam–column design interaction equations from Chapter H of the AISC specification for *doubly and singly symmetric members subject to biaxial bending and axial load* are given as follows [1]:

AISC equation H1-1a is

$$\text{If } \frac{P_u}{\phi P_n} \geq 0.20, \quad \frac{P_u}{\phi P_n} + \frac{8}{9}\left(\frac{M_{ux}}{\phi_b M_{nx}} + \frac{M_{uy}}{\phi_b M_{ny}}\right) \leq 1.0; \tag{8-1}$$

AISC equation H1-1b is

$$\text{If } \frac{P_u}{\phi P_n} < 0.20, \quad \frac{P_u}{2\phi P_n} + \left(\frac{M_{ux}}{\phi_b M_{nx}} + \frac{M_{uy}}{\phi_b M_{ny}}\right) \leq 1.0, \tag{8-2}$$

where

P_u = Factored axial compression or tension load or the required axial strength,

ϕP_n = Compression design strength or tension design strength,

For *compression* members in *braced frames*, ϕP_n is calculated using an effective length factor, K, that is typically less than or equal to 1.0; $K = 1.0$ is commonly used in practice.

For *compression* members in *moment frames*, ϕP_n is calculated using an effective length factor, K, that is typically greater than 1.0; $K = 2.1$ is an approximate value that is commonly used for preliminary design. More accurate K-values can be determined using the nomograghs or alignment charts (see Chapter 5).

For *tension* members, ϕP_n is the smaller of $0.9A_g F_y$, $0.75A_e F_u$, or the block shear capacity.

M_{ux} = Factored bending moment about the *x*-axis (i.e., the strong axis) of the member,

M_{uy} = Factored bending moment about the *y*-axis (i.e., the weak axis) of the member,

$\phi_b M_{nx}$ = Design moment capacity for bending about the strong axis of the member (see beam design in Chapter 6), and

$\phi_b M_{ny}$ = Design moment capacity for bending about the weak axis of the member ($\phi_b Z_y F_y \leq 1.5\,\phi_b S_y F_y$, where $\phi_b = 0.9$).

Note that for the case of beam–columns with axial compression loads and bending moments, the factored moments about the *x–x* and *y–y* axes (i.e., M_{ux} and M_{uy}, respectively) must include the effect of the slenderness of the compression member (i.e., the so-called *P*-delta effects). This will be discussed in the following sections.

8.5 MOMENT MAGNIFICATION, OR *P*-DELTA, EFFECTS

When an axial *compression* load is applied to a beam–column that has some initial crookedness or deflection and is supported at both ends (i.e., nonsway), additional moments are produced. This is the first type of *P*-delta (*P*-δ) effect (see Figure 8-11a).

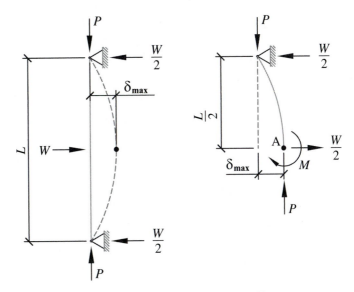

Figure 8-11a Moment amplification from P-δ effects.

Summing the moments of the free-body diagram in Figure 8-11a about point A yields

$\Sigma M_A = 0 \Rightarrow M - P\delta - (W/2)(L/2) = 0$, and it follows that

$$M = \frac{WL}{4} + P\delta. \tag{8-3}$$

The $WL/4$ term in equation (8-3) is the first-order moment, while the second term is known as the P-δ moment. The total second-order moment, M, in equation (8-3) can be rewritten as

$$M = B_1[M_{unt}], \tag{8-4}$$

where

B_1 = Moment amplification factor due to the column deflection between laterally supported ends of the column. This applies to individual beam–columns in nonsway frames or braced frames, and

M_{unt} = First-order nonsway moments.

The first-order moment, M_{unt}, may be caused by lateral loads applied between the supported ends of the column or due to the eccentricity of beam and girder reactions.

When an axial *compression* load is applied to a beam–column that is subjected to relative **lateral** sway at the ends of the member, additional moments are produced due to the destabilizing effect of the axial load as it undergoes the relative translation, Δ. This is the second type of P-delta (or P-Δ) effect and is applicable to beam–columns in moment frames.

Summing the moments of the free-body diagram in Figure 8-11b about point A yields

$\Sigma M_A = 0 \Rightarrow -M + P\Delta + WL = 0$, and

$$M = WL + P\Delta. \tag{8-5}$$

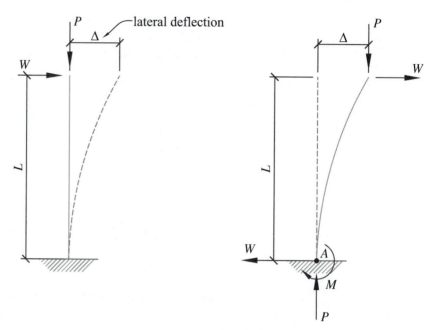

Figure 8-11b Moment amplification from P-Δ effects.

The WL term in equation (8-5) is the first-order moment, while the second term is the second-order, or P-Δ, moment.

Thus, $M = B_2 \left[M_{ult(\text{1st order})} \right],$ (8-6)

where

> $B_2 =$ Moment amplification factor due to **lateral** deflection of the top end of the column relative to the bottom end. This magnification factor is applicable to all columns in moment frames for the story under consideration; and

> $M_{ult\,(\text{1st order})} =$ First-order lateral sway moments. These moments are caused by wind or seismic loads or by unbalanced gravity loads.

8.6 STABILITY ANALYSIS AND CALCULATION OF THE REQUIRED STRENGTHS OF BEAM–COLUMNS

The AISC specification presents three methods for the stability analysis of building frames and the calculation of the required strengths or factored loads and moments in beam–columns. These methods are discussed below [1, 2].

Leaning Columns

Leaning columns are beam–columns that are laterally braced by moment frames or other lateral force-resisting systems in the plane of bending and do not participate in any way in resisting lateral loads. They possess no lateral stiffness, but depend on the moment frames for their lateral support. In designing leaning columns, an effective length factor, K, of 1.0 is usually assumed; therefore, the axial loads on the leaning columns must be considered in analyzing the moment frames that provide lateral bracing to these columns. Yura proposed a method for including the effect of leaning columns on moment frames that involves designing the moment frame columns for additional axial loads *in the plane of bending* for which the moment frame provides lateral support to the leaning columns [3]. The additional load on the restraining columns is the total axial load on all leaning columns distributed to the moment frame columns. In this text, the distribution of the leaning column axial load to the moment frame columns is assumed to be proportional to the plan tributary area of the moment frame columns.

First-Order Analysis (AISC Specification, Section C2.2b)

A first-order analysis is a structural analysis of a building frame where the effects of geometric nonlinearities (or P-delta effects) are not included. The method uses **unreduced stiffnesses and cross-sectional areas** for the columns and girders. The AISC specification allows the first-order analysis loads and moments to be used in the design of beam–columns only when the factored axial compression loads (or required compression strength) is not greater than 50% of the yield strength (i.e., $P_u \le 0.5\, P_y$) for **all** members whose flexural stiffnesses are considered to contribute to the lateral stiffness of the frame. If this condition is satisfied, an effective length factor, $K = 1.0$ is used for the design of the beam–columns, but the total moments must still be amplified by the nonsway moment magnification factor, B_1. For all load combinations, an additional notional lateral load must also be applied in both orthogonal directions; this is in addition to any applied lateral loads. This notional load is given as

$$N_i = 2.1\,(\Delta/L)P_i \ge 0.0042\,P_i,$$

where

$$P_i = \text{Cumulative factored gravity load applied at level } i,$$
$$\Delta/L = \text{Maximum ratio of } \Delta \text{ to } L \text{ for all stories in the building,}$$
$$\Delta = \text{First-order interstory drift due to factored (LRFD) load combinations, and}$$
$$L = \text{Story height.}$$

This method will not be discussed further in this text.

Amplified First-Order Analysis or the Effective Length Method (AISC Specification, Section C2.2a)

This is an indirect second-order analysis where the first-order moments are amplified by the B_1 and B_2 moment magnification factors as demonstrated in the previous section. The method uses **unreduced stiffnesses and cross-sectional areas** for the columns and girders, and the analysis is carried out at the factored load level. This method is limited to building frames where the sway moment magnification factor, B_2, does not exceed 1.50. Where B_2 exceeds 1.50, the AISC specification requires that the direct analysis method be used. For braced frames, the effective length factor, K, is taken as 1.0, and for sway or moment frames, the effective length factor is determined from Figure 5-3 or the alignment charts presented in Chapter 5. Also, when $B_2 \le 1.1$, the columns can be designed with an effective length factor, $K = 1.0$.

In addition, this method requires that all **gravity-only** load combinations include a minimum notional lateral load of $0.002\, P_i$ applied at each level in both orthogonal directions, where P_i is the **cumulative** factored gravity load on the column at the story under consideration. In many practical situations, for the ***gravity-only*** load combinations, the lateral translation will be small, even with the notional lateral load, except for highly asymmetrical frames, so it is practical to assume the nonsway case for the gravity-only load combination. Thus, B_2 can be assumed to be 1.0 and the sway, or translation, moment, M_{lt}, is assumed be negligible for the gravity-only load combinations.

Direct Second-Order Analysis Method (AISC Specification, Appendix 7)

In this method, the second-order moments and the axial loads in a **moment frame** are obtained directly and explicitly by performing a **second-order**, or ***P*-delta (*P*-Δ)**, computer-aided analysis of the moment frame, taking into account geometric nonlinearities, imperfections, and inelasticity [1, 2]. Since the design moments obtained from this method are actually the second-order moments, M_u, the design of the columns is carried out with the moment magnification factors, B_1 and B_2, taken as 1.0. This second-order analysis must include **all of the gravity loads** tributary to the moment frame being analyzed, including the axial loads on any ***leaning columns***. In the direct analysis method, geometric imperfections are accounted for by the application of a notional lateral load, which is usually a certain percentage of the gravity loads on the frame, and inelasticity can be taken into account by using reduced flexural and axial stiffness for the columns and girders. There are no limitations to the use of the direct analysis method.

Table 8-2 summarizes the requirements discussed above for the three methods of stability analysis in the AISC specification.

Table 8-2 Summary of AISC specification requirements for the stability analysis and design of moment frames*

	Amplified First-Order Analysis Method	Direct Analysis Method	First-Order Analysis Method
Limits of Applicability	Yes	No	Yes
Type of structural analysis	Approximate second-order analysis	Second-order analysis	First-order analysis
Member stiffness used in the structural analysis	Gross EI and EA	Reduced EI and EA to account for inelastic behavior	Gross EI and EA
Is a notional load required?	Yes (for the gravity-only load combinations)	Yes	Yes (as an additional lateral load)
Effective length factor, K, used in the analysis	Sway buckling effective length factor ($K \geq 1.0$). Use alignment charts or Figure 5-3.	$K = 1$	$K = 1$

*Adapted from ref. 2, Courtesy of Dr. Shankar Nair.

Note:
E = Modulus of elasticity EA = Axial stiffness
A = Cross-sectional area EI = Bending stiffness
I = Moment of inertia

8.7 MOMENT MAGNIFICATION FACTORS FOR AMPLIFIED FIRST-ORDER ANALYSIS

In this text, the amplified first-order analysis will be used for the stability analysis of beam–columns in moment frames. The equations for calculating the moment magnification factors, B_1 and B_2, according to the AISC specification will now be presented.

Nonsway Moment Magnification Factor, B_1

The nonsway moment magnification factor, B_1, is calculated as follows from the AISC specification:

$$B_1 = \frac{C_m}{1 - \dfrac{P_u}{P_{e1}}} \geq 1.0, \tag{8-7}$$

where the elastic buckling load of the column, P_{e1}, is calculated as

$$P_{e1} = \frac{\pi^2 EA}{\left(\dfrac{KL}{r}\right)^2}, \tag{8-8}$$

where KL/r is the slenderness ratio about the **axis of bending**.

Alternatively, $\boldsymbol{P_{e1x}}$ and $\boldsymbol{P_{e1y}}$ can also be obtained for **W-shapes** from the bottom rows of the column load tables in Section 4 of the *AISCM* (*AISCM*, Table 4-1) as follows:

$$P_{e1}, \text{ in kips } = \frac{10^4}{(KL)^2} \times \text{Corresponding value from the column load tables,} \tag{8-9}$$

where

KL is in inches in equation (8-9),

$K \leq 1.0$ (a practical value of K for columns in braced frames = 1.0), and

A = Gross cross-sectional area of the beam–column.

The moment reduction coefficient, C_m, used in equation (8-7) accounts for the effect of moment gradient in the column, and is obtained as follows:

1. For beam–columns with **no** transverse loads between the supports,

$$C_m = 0.6 - 0.4\frac{M_1}{M_2}, \tag{8-10}$$

where

$\dfrac{M_1}{M_2}$ = Absolute ratio of bending moment at the ends of the member (M_1 is the **smaller** end moment, M_2 is the **larger** end moment)

$\qquad$ = $-$ve for single-curvature bending (see Figure 8-12a), and

$\qquad$ = $+$ve for double-curvature bending (see Figure 8-12a).

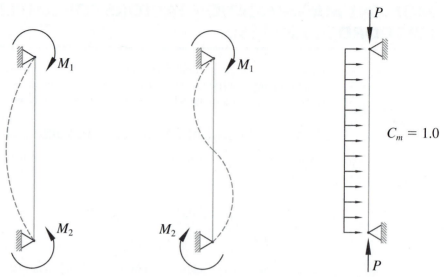

Figure 8-12a Single- and double-curvature bending.

Figure 8-12b Beam–columns with transverse loads.

Examples of columns that are subjected to double-curvature bending include exterior columns and columns in moment frames; for these columns, the maximum possible C_m value from equation (8-10) is 0.6. In addition, the exterior and interior columns at the lowest level of a building will also have a maximum possible C_m value of 0.6 if the bases of the columns are pinned. On the other hand, interior columns above the ground-floor level may be subjected to single-curvature or double-curvature bending, depending on the live load pattern on the beams and girders at the floor levels coinciding with the top and bottom of the column. The maximum possible C_m value for these interior columns is 1.0. For beam–columns with transverse loads between the supports, $C_m = 1.0$, as shown in Figure 8-12b. Single and double curvature bending is illustrated in Figure 8-12a.

Since several combinations of column end moments, M_1 and M_2, are possible, C_m can take on many different values for the same column, depending on the load combinations considered at floor levels at the top and bottom of the column. To simplify the determination of the C_m factor, and because B_1 is almost always 1.0 for many practical cases, the suggested approximate values for C_m, as depicted in Figure 8-13, can be used [4, 5].

Sway Moment Magnification factor, B_2, for Unbraced or Moment Frames

The sway moment magnifier, B_2, is calculated for all columns in **each story** of a moment frame using the approximate AISC specification equation in equations (8-11a) and (8-11b). Thus, all columns in a given story will have the same sway magnification factor.

$$B_2 = \frac{1}{1 - \alpha\dfrac{\sum P_u}{\sum P_{e2}}} \geq 1.0, \qquad (8\text{-}11a)$$

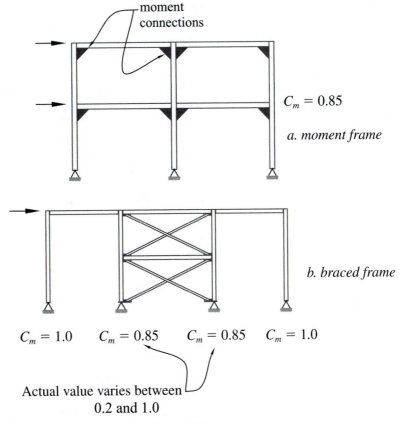

Figure 8-13 Approximate values of C_m.

or

$$B_2 = \frac{1}{1 - \dfrac{\Delta_{oh}}{L}\,\alpha\dfrac{\Sigma P_u}{\Sigma H}} \geq 1.0, \tag{8-11b}$$

where

ΣP_u = Sum of the factored loads for **all** columns in the **story** under consideration,

ΣP_{e2} = Sum of the buckling capacity for **all** columns in the **story** under consideration

$\qquad = \Sigma \dfrac{\pi^2 EA}{(KL/r)^2}$ for **all** columns in the story, \hfill (8-12)

$\quad \alpha$ = 1.0 (LRFD)

ΣH = Factored horizontal or lateral shear in the story under consideration,

$\quad L$ = Story or floor-to-floor height of the moment frame,

Δ_{oh} = Interstory drift caused by the factored lateral shear, ΣH, and

$\dfrac{\Delta_{oh}}{L}$ = Drift limit for factored loads (typical values range from 1/500 to 1/400).

For practical situations, the drift index, or limit, can be assumed to be 1/500 for lateral wind loads—which is an interstory drift limit commonly used in design practice to satisfy serviceability requirements under wind loads. Since factored gravity and lateral loads are used in equation (8-11b), the drift limit should be modified to the factored load level. Therefore, a drift limit of 1/(500/1.6), or 1/312, may be used at the factored load level for moment frames subjected to wind loads. For seismic loads, the drift limits given in Table 12.12-1 of the ASCE 7 load specification should be used [6]. Note that equation (8-11b) is more convenient to use than equation (8-11a) because the column and girder sizes do not have to be known to use this equation.

The effective length factor, K, for moment frames is typically greater than 1.0. A practical value of the effective length factor, K, for moment frames is **2.1** from Figure 5-3, but more accurate values of K can be obtained using the alignment charts presented in Chapter 5.

Total Factored Moment, M_u, in a Beam–Column

The total factored second-order moment or required moment strength is

$$M_u = B_1 M_{unt} + B_2 M_{ult},\qquad(8\text{-}13)$$

where

M_{unt} = Factored moments in the beam–column when no appreciable sidesway occurs (nt = No translation), and

*The M_{unt} moments are caused by **gravity loads** acting at the simple-shear beam-to-column connection eccentricities.*

M_{ult} = Factored moments in the beam–column caused by

- Wind or earthquake loads on the frame,
- The restraining force necessary to prevent sidesway in a symmetrical frame loaded with asymmetrically placed gravity loads, and
- The restraining force necessary to prevent sidesway in an asymmetrical frame loaded with symmetrically placed gravity loads.

(lt = Lateral translation)

For most reasonably symmetric moment frames, the M_{ult} moments are caused only by lateral wind or seismic loads.

We can rewrite the previous M_u equation for the x- and y-axes of bending as follows:

$$M_{ux} = B_{1x} M_{untx} + B_{2x} M_{ultx} \text{ and}\qquad(8\text{-}14a)$$
$$M_{uy} = B_{1y} M_{unty} + B_{2y} M_{ulty}, \text{ respectively.}\qquad(8\text{-}14b)$$

Note that for braced frames, there are no lateral translation moments; therefore,

$$M_{ultx} = 0 \quad \text{and} \quad M_{ulty} = 0.$$

8.8 UNBALANCED MOMENTS, M_{NT}, FOR COLUMNS IN BRACED FRAMES DUE TO THE ECCENTRICITY OF THE GIRDER AND BEAM REACTIONS

Unbalanced moments occur in columns due to differences in the reactions of adjacent beams and girder spans that frame into a column (see Figure 8-14). Differences in the girder and beam reactions may occur due to differences in the span and loading, or it may

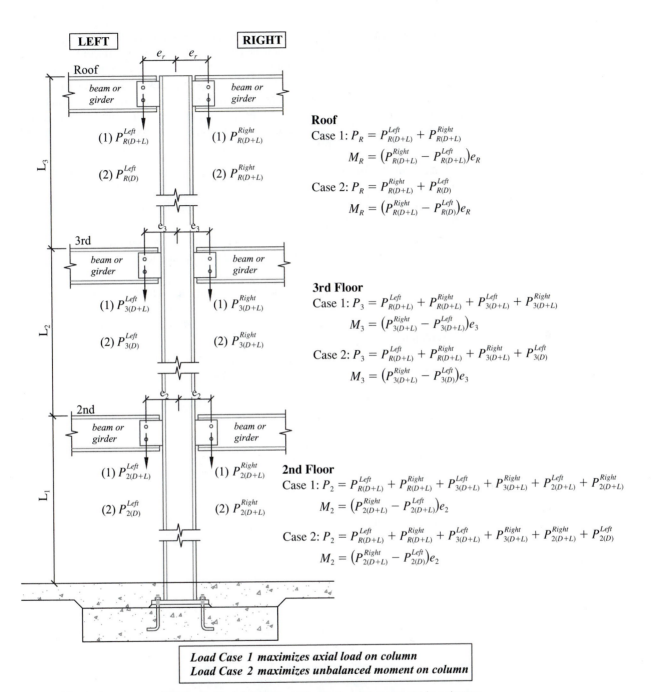

LEFT **RIGHT**

e_r e_r

Roof

beam or girder   beam or girder

(1) $P^{Left}_{R(D+L)}$ (1) $P^{Right}_{R(D+L)}$

(2) $P^{Left}_{R(D)}$ (2) $P^{Right}_{R(D+L)}$

Roof

Case 1: $P_R = P^{Left}_{R(D+L)} + P^{Right}_{R(D+L)}$

$\qquad M_R = \left(P^{Right}_{R(D+L)} - P^{Left}_{R(D+L)}\right)e_R$

Case 2: $P_R = P^{Right}_{R(D+L)} + P^{Left}_{R(D)}$

$\qquad M_R = \left(P^{Right}_{R(D+L)} - P^{Left}_{R(D)}\right)e_R$

e_3 e_3

3rd

beam or girder   beam or girder

(1) $P^{Left}_{3(D+L)}$ (1) $P^{Right}_{3(D+L)}$

(2) $P^{Left}_{3(D)}$ (2) $P^{Right}_{3(D+L)}$

3rd Floor

Case 1: $P_3 = P^{Left}_{R(D+L)} + P^{Right}_{R(D+L)} + P^{Left}_{3(D+L)} + P^{Right}_{3(D+L)}$

$\qquad M_3 = \left(P^{Right}_{3(D+L)} - P^{Left}_{3(D+L)}\right)e_3$

Case 2: $P_3 = P^{Left}_{R(D+L)} + P^{Right}_{R(D+L)} + P^{Right}_{3(D+L)} + P^{Left}_{3(D)}$

$\qquad M_3 = \left(P^{Right}_{3(D+L)} - P^{Left}_{3(D)}\right)e_3$

e_2 e_2

2nd

beam or girder   beam or girder

(1) $P^{Left}_{2(D+L)}$ (1) $P^{Right}_{2(D+L)}$

(2) $P^{Left}_{2(D)}$ (2) $P^{Right}_{2(D+L)}$

2nd Floor

Case 1: $P_2 = P^{Left}_{R(D+L)} + P^{Right}_{R(D+L)} + P^{Left}_{3(D+L)} + P^{Right}_{3(D+L)} + P^{Left}_{2(D+L)} + P^{Right}_{2(D+L)}$

$\qquad M_2 = \left(P^{Right}_{2(D+L)} - P^{Left}_{2(D+L)}\right)e_2$

Case 2: $P_2 = P^{Left}_{R(D+L)} + P^{Right}_{R(D+L)} + P^{Left}_{3(D+L)} + P^{Right}_{3(D+L)} + P^{Right}_{2(D+L)} + P^{Left}_{2(D)}$

$\qquad M_2 = \left(P^{Right}_{2(D+L)} - P^{Left}_{2(D)}\right)e_2$

L_3 L_2 L_1

Load Case 1 maximizes axial load on column
Load Case 2 maximizes unbalanced moment on column

Figure 8-14 Axial loads and unbalanced moments in columns.

occur due to the skipping of live loads. Ordinarily, live load skipping should be considered simultaneously at the top and bottom of each column; however, this results in many possible load cases.

To minimize the number of load cases, the authors have chosen in this text to consider two load cases that will give conservative results. Load case 1 maximizes the factored axial load on the column and load case 2 maximizes the unbalanced moments about both orthogonal axes of the column. This is accomplished by placing or removing the entire live load from the beams and girders framing into the column in order to achieve the maximum effect. It should be remembered that the dead load always remains on the beams and girders at all times, and only the live load may be skipped. It is also assumed that the unbalanced moment at one end of a column is not affected by the unbalanced moment at the other end.

In calculating the axial loads and unbalanced moments for columns in braced frames as shown in Figure 8-14, the following should be noted:

- All loads are **factored** and the factored load combinations from Chapter 2 should be used.

- The loading conditions shown are for a three-story building, but could be applied to a building of any height.

- In summing the column loads, the maximum factored dead plus live loads from the floors/roof above must be included using the appropriate load combinations as indicated in Table 8-3.

- The moments in Figure 8-14 are the M_{unt} moments at **each level**; it should be noted that the moments are **not** cumulative as are the axial loads. The M_{unt} moments at any level are caused by the **unbalanced** reactions from the beams and girders framing into the column at that level.

- The factored axial loads at **each** floor level are defined as follows (note that these are **not** the cumulative axial loads):

$$P_{roof\,D+L} = 1.2P_{roof\,DL} + 1.6(P_{Lr}\,or\,P_S\,or\,P_R)$$
$$P_{roof\,D} = 1.2P_{roof\,DL}$$

$$P_{3D+L} = 1.2P_{3DL} + 1.6P_{3L}$$
$$P_{3D} = 1.2P_{3DL}$$

$$P_{2D+L} = 1.2P_{2DL} + 1.6P_{2L}$$
$$P_{2D} = 1.2P_{2DL}$$

where

DL = Service or unfactored dead load,

L = Service or unfactored live load,

L_r = Unfactored roof live load,

Table 8-3 Applicable load combinations for cumulative axial load in columns

Level	Cumulative Axial load in Column	Applicable Load Combination for Cumulative Axial Load
Roof level	P_{roof}	$1.2D + 1.6\,(L_r\,or\,S\,or\,R)$
nth floor	P_n	$1.2D + 1.6L + 0.5\,(L_r\,or\,S\,or\,R)$
Third floor	P_3	$1.2D + 1.6L + 0.5\,(L_r\,or\,S\,or\,R)$
Second floor	P_2	$1.2D + 1.6L + 0.5\,(L_r\,or\,S\,or\,R)$

S = Unfactored snow load,

R = Unfactored rain load,

D = Factored dead load,

$D+L$ = Factored dead plus live load,

P_{roof} = Factored axial load at the roof level,

P_3 = Factored axial load at the third-floor level, and

P_2 = Factored axial load at the second-floor level.

Moment Distribution Between Columns

The distribution of no-translation moments (i.e., moment split) between the columns above and below a given floor level (see Figure 8-15) is a function of the following factors: the continuity of the column and the type of column splice (if any), the floor-to-floor heights, and the moments of inertia of the columns above and below the floor level. The column just below the roof level has to resist 100% of the unbalanced moment at the roof level because there is no column above the roof level with which to split the unbalanced moment.

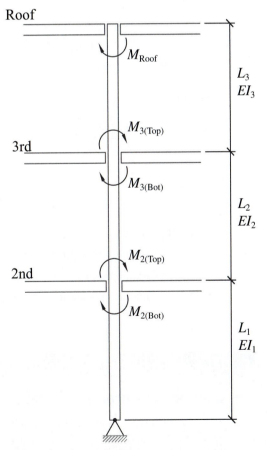

Figure 8-15 Distribution of M_{nt} moments in columns.

Roof:

$$M = M_{\text{roof}}$$

Third Floor: The total column moment is split between the columns above and below the third floor, based on the ratio of the column stiffnesses as follows:

$$\frac{M_{top}}{M_{bot}} = \frac{EI_{top}}{L_{top}} \bigg/ \frac{EI_{bot}}{L_{bot}} = \frac{EI_3/L_3}{EI_2/L_2}.$$

To simplify the analysis, it can be assumed that I_{bot} (or I_2) $= I_{top}$ (or I_3). For practical situations, there is not much loss in accuracy with this assumption. In fact, this assumption is widely used in practice in the design of steel buildings. Therefore, the moment split between the upper and lower columns at a floor level is inversely proportional to the length of the columns.

$$\Rightarrow \frac{M_{top}}{M_{bot}} = \frac{1/L_3}{1/L_2}$$

$$M_3 = M_{3top} + M_{3bot}$$

$$M_{3bot} = \left(\frac{1/L_2}{1/L_2 + 1/L_3}\right)M_3$$

$$M_{3top} = M_3 - M_{3bot}$$

Second Floor:

$$M_{2bot} = \left(\frac{1/L_1}{1/L_1 + 1/L_2}\right)M_2$$

$$M_{2top} = M_2 - M_{2bot},$$

where

M_{3bot} = Column moment just *below* the third-floor level,

M_{3top} = Column moment just *above* the third-floor level,

M_{2bot} = Column moment just *below* the second-floor level, and

M_{2top} = Column moment just *above* the second-floor level.

EXAMPLE 8-1

Design of Beam–Columns in a Braced Frame

The typical floor and roof plans for a three-story braced-frame building are shown in Figures 8-16 and 8-17, respectively. The beams and girders are connected to the columns with simple shear connections (e.g., double angles or shear plates). Design columns C1 and C2 for the gravity loads shown assuming a floor-to-floor height of 10 ft.

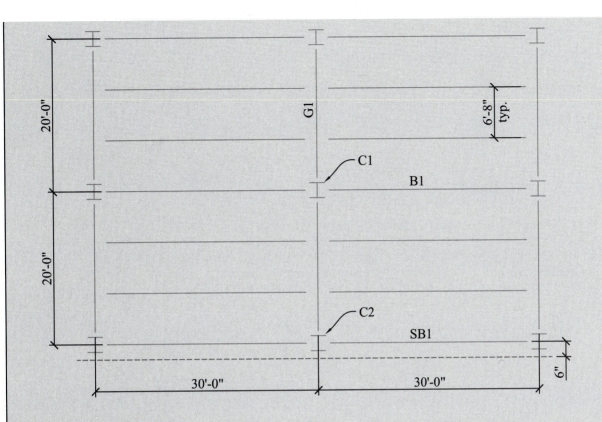

Figure 8-16 Typical floor plan.

Floor Loads

Dead load = 85 psf
Live load = 150 psf

Factored roof load:
$$w_u = (1.2)(85) + (1.6)(150) = 0.342 \text{ ksf}$$

Service roof load:
$$w_{s(D+L)} = 85 + 150 = 0.235 \text{ ksf (Dead + Live)}$$
$$w_{s(L)} = 150 = 0.150 \text{ ksf (Live)}$$

Factored roof dead load:
$$w_{u(D)} = (1.2)(85) = 0.10 \text{ ksf}$$

SOLUTION

The first step in the solution process is to determine the governing moments and the factored cumulative axial loads at each level of the column for the two load cases considered (i.e., load cases 1 and 2) in Figure 8-14.

Load Calculations for Column C1

All Loads are Factored

(continued)

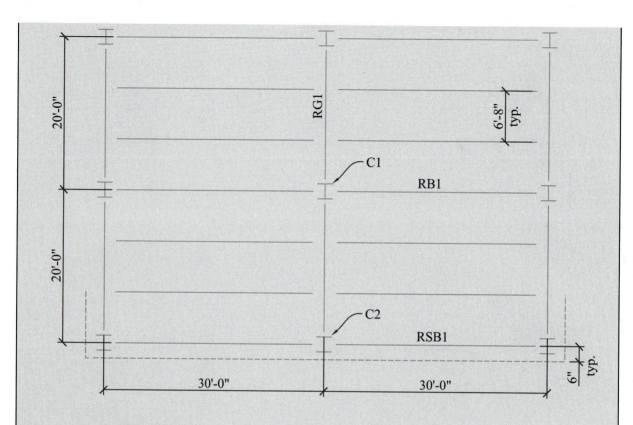

Roof Loads

Dead load = 30 psf
Live load = 35 psf (Snow)

Factored roof load:
$$w_u = (1.2)(30) + (1.6)(35) = 0.092 \text{ ksf}$$

Service roof load:
$$w_{s(D+L)} = 30 + 35 = 0.065 \text{ ksf (Dead + Snow)}$$
$$w_{s(L)} = 35 = 0.035 \text{ ksf (Snow)}$$

Factored roof dead load:
$$w_{u(D)} = (1.2)(30) = 0.036 \text{ ksf}$$

Figure 8-17 Roof plan.

Loads to Column C1 at the roof level (See Figures 8-18 and 8-19)

Girder and Beam Eccentricities:

e_r girders   = 2½ in. + ½ Column depth
= 2½ in. + ½ (8 in.) Assuming a minimum W8 column
= 6½ in. = **0.54 ft.**

e_r beams   = 2½ in. + ½ Column web thickness
= 2½ in. + ½ (0.5 in.) assumed
= 2.75 in. ≈ **0.25 ft.**

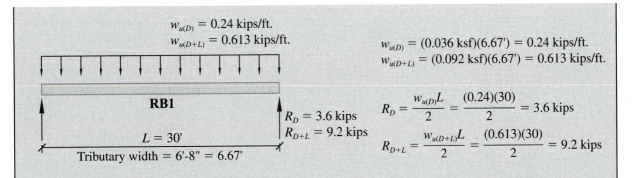

$w_{u(D)} = 0.24$ kips/ft.
$w_{u(D+L)} = 0.613$ kips/ft.

$w_{u(D)} = (0.036 \text{ ksf})(6.67') = 0.24$ kips/ft.
$w_{u(D+L)} = (0.092 \text{ ksf})(6.67') = 0.613$ kips/ft.

RB1

$R_D = 3.6$ kips
$R_{D+L} = 9.2$ kips

$L = 30'$
Tributary width = 6'-8" = 6.67'

$R_D = \dfrac{w_{u(D)}L}{2} = \dfrac{(0.24)(30)}{2} = 3.6$ kips

$R_{D+L} = \dfrac{w_{u(D+L)}L}{2} = \dfrac{(0.613)(30)}{2} = 9.2$ kips

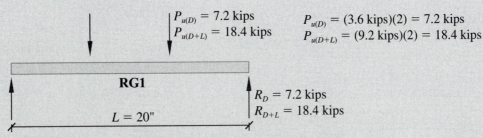

$P_{u(D)} = 7.2$ kips
$P_{u(D+L)} = 18.4$ kips

$P_{u(D)} = (3.6 \text{ kips})(2) = 7.2$ kips
$P_{u(D+L)} = (9.2 \text{ kips})(2) = 18.4$ kips

RG1

$R_D = 7.2$ kips
$R_{D+L} = 18.4$ kips

$L = 20"$

Figure 8-18 Calculation of roof beam and girder reactions to C1.

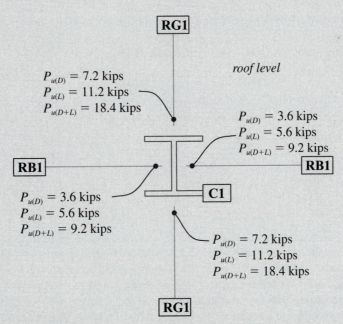

RG1

roof level

$P_{u(D)} = 7.2$ kips
$P_{u(L)} = 11.2$ kips
$P_{u(D+L)} = 18.4$ kips

$P_{u(D)} = 3.6$ kips
$P_{u(L)} = 5.6$ kips
$P_{u(D+L)} = 9.2$ kips

RB1

RB1

$P_{u(D)} = 3.6$ kips
$P_{u(L)} = 5.6$ kips
$P_{u(D+L)} = 9.2$ kips

C1

$P_{u(D)} = 7.2$ kips
$P_{u(L)} = 11.2$ kips
$P_{u(D+L)} = 18.4$ kips

RG1

Figure 8-19 Reactions of roof beams and girders framing into column C1.

(continued)

W8 is the minimum wide-flange column size typically used in design practice. Smaller column sizes are not frequently used because the flange width of these columns do not provide enough room to accommodate the double-angle connections used to connect the girders to the columns. Where the girders are connected to the columns using shear tabs or plates, a column size smaller than W8 × 31 may be used provided that it is adequate for resisting the applied loads and moments.

Factored Loads:

$$w_u = 0.092\,\text{ksf}(6.67\,\text{ft.}) = 0.613\,\text{kips/ft.}$$
$$w_{u(D)} = 0.036\,\text{ksf}(6.67\,\text{ft.}) = 0.24\,\text{kips/ft.}$$
$$P_D = 3.6\,\text{kips (2 beams)} = 7.2\,\text{kips}$$
$$P_{D+L} = 9.2\,\text{kips (2 beams)} = 18.4\,\text{kips}$$

Load Case 1: The axial load is maximized.

$$P_{roof} = \underset{(P^R_{DL+LL}\,beams)}{9.2} + \underset{(P^L_{DL+LL}\,beams)}{9.2} + \underset{(P^R_{DL+LL}\,girders)}{18.4} + \underset{(P^L_{DL+LL}\,girders)}{18.4} = 55\,\text{kips}$$

$$M_{rx-x} = (18.4 - 18.4)\underset{\uparrow\,(e_r\,\textbf{girder})}{(0.54\,\text{ft.})} = 0\,\text{ft.-kips}$$

$$M_{ry-y} = (9.2 - 9.2)\underset{\uparrow\,(e_r\,beams)}{(0.25\,\text{ft.})} = 0\,\text{ft.-kips}$$

$$P = 55\,\text{kips}$$
$$M_{rx-x} = 0\,\text{ft.-kips}$$
$$M_{ry-y} = 0\,\text{ft.-kips}$$

For this case, the column is designed for axial load only, since the moments are zero.

Load Case 2: The moments at this floor level are maximized.

$$P_{roof} = \underset{(P^R_{DL+LL}\,beams)}{9.2} + \underset{(P^L_{DL}\,beams)}{3.6} + \underset{(P^R_{DL+LL}\,girders)}{18.4} + \underset{(P^L_{DL}\,girders)}{7.2} = 38\,\text{kips}$$

$$M_{rx-x} = (18.4 - 7.2)(0.54\,\text{ft.}) = 6\,\text{ft.-kips}$$
$$M_{ry-y} = (9.2 - 3.6)(0.25\,\text{ft.}) = 1.4\,\text{ft.-kips}$$

$$P = 38\,\text{kips}$$
$$M_{rx-x} = 6\,\text{ft.-kips}$$
$$M_{ry-y} = 1.4\,\text{ft.-kips}$$

For this load case, the column is designed or checked for combined axial load plus bending.

The reader can observe that the column moments at the roof level are resisted solely by the column below the roof level since there is no column above this level. For the roof loads that will be cumulatively added to the floor loads below, the appropriate load factor must be applied to the live loads. Recall that the load combination $1.2D + 1.6\,(L_r$ or S or $R)$ applies only to the column load at the roof level only. For cumulative loads at the lower levels, the applicable load combination is $1.2D + 1.6L + 0.5\,(L_r$ or S or $R)$. Therefore, the contribution from the roof level to the cumulative loads at the lower levels (i.e., third- and second-floor levels) is

$$P_{roof} = 1.2D + 0.5\,(L_r\text{ or }S\text{ or }R) = 7.2 + 7.2 + 3.6 + 3.6 + 0.5\,(11.2 + 11.2 + 5.6 + 5.6) = 38.4\,\text{kips.}$$

Loads to Column C1 at the Third Floor: (See Figures 8-20 and 8-21)

Girder and Beam Eccentricities:

e_3 girders $= 2\frac{1}{2}$ in. $+ \frac{1}{2}$ Column depth

$= 2\frac{1}{2}$ in. $+ \frac{1}{2}$ (8 in.) assuming minimum W8 column

$= 6\frac{1}{2}$ in. $= 0.54$ ft.

e_3 beams $= 2\frac{1}{2}$ in. $+ \frac{1}{2}$-in. Column web thickness

$= 2\frac{1}{2}$ in. $+ \frac{1}{2}$ (0.5 in.) assumed

$= 2.75$ in. $\approx$ **0.25 ft.**

Factored Loads:

w_u $= 0.342$ ksf (6.67 ft.) $= 2.3$ kips/ft.

$w_{u(D)}$ $= 0.10$ ksf (6.67 ft.) $= 0.67$ kips/ft.

P_D $= 10$ kips (2 beams) $= 20$ kips

$P_{D+L} = 35$ kips (2 beams) $= 70$ kips

Moment Split in the Column:

$$M_{3bot} = \left(\frac{1/L_2}{1/L_2 \,+\, 1/L_3} \right) M_3; \; L_2 = 10 \text{ ft. and } L_3 = 10 \text{ ft.}$$

$$= \left(\frac{1/10}{1/10 \text{ ft.} \,+\, 1/10 \text{ ft.}} \right) M_3$$

$$= 0.5 \, M_3$$

Load Case 1: The axial load is maximized.

P_3 $= \underset{(\text{max } P_{roof})}{38.4} + \underset{(P_{DL+LL}^{R} \, beam)}{(35} + \underset{(P_{DL+LL}^{L} \, beam)}{35)} + \underset{(P_{DL+LL}^{R} \, girder)}{(70} + \underset{(P_{DL+LL}^{L} \, girder)}{70)} = 248$ kips

M_{3x-x} $= (70 - 70) \times e_3$ **girder** $= 0$ kips (0.54 ft.) $= 0$ ft.–kips

M_{3y-y} $= (35 - 35) \times e_3$ **beam** $= 0$ kips (0.25 ft.) $= 0$ ft.–kips

P_3 $= 248$ kips

$M_{3botx-x} = 0.5(0 \text{ ft.-kips}) = 0$ ft.-kips

$M_{3boty-y} = 0.5(0 \text{ ft.-kips}) = 0$ ft.-kips

For this load case, design the column for axial load only since we have zero moments.

Load Case 2: The moments at this floor level are maximized.

P_3 $= \underset{(\text{max } P_{roof})}{38.4} + \underset{(P_{DL+LL}^{R} \, beam)}{(35} + \underset{(P_{DL}^{L} \, beam)}{10)} + \underset{(P_{DL+LL}^{R} \, girder)}{(70} + \underset{(P_{DL}^{L} \, girder)}{20)} = 173$ kips

$M_{3x-x} = (70 - 20) \times e_3$ **girder** $= 50$ kips (0.54 ft.) $= 27$ ft.-kips

$M_{3y-y} = (35 - 10) \times e_3$ **beam** $= 25$ kips (0.25 ft.) $= 6.3$ ft.-kips

(continued)

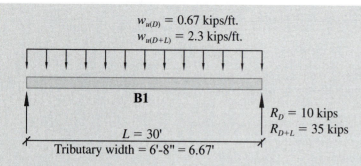

$$w_{u(D)} = (0.10 \text{ ksf})(6.67') = 0.67 \text{ kips/ft.}$$
$$w_{u(D+L)} = (0.342 \text{ ksf})(6.67') = 2.3 \text{ kips/ft.}$$

$$R_D = \frac{w_{u(D)}L}{2} = \frac{(0.67)(30)}{2} = 10 \text{ kips}$$

$$R_{D+L} = \frac{w_{u(D+L)}L}{2} = \frac{(2.3)(30)}{2} = 35 \text{ kips}$$

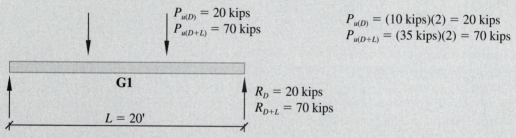

$$P_{u(D)} = (10 \text{ kips})(2) = 20 \text{ kips}$$
$$P_{u(D+L)} = (35 \text{ kips})(2) = 70 \text{ kips}$$

Figure 8-20 Calculation of floor beam and girder reactions to C1 at the third floor.

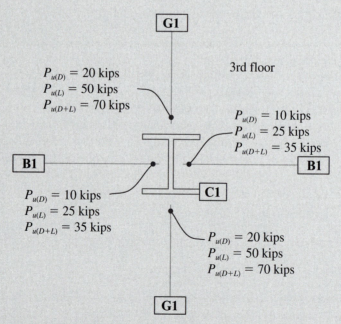

Figure 8-21 Reactions of floor beams and girders framing into the column C1 at the third floor.

The moment at the third-floor level will be distributed between the columns above and below that level in a ratio that is inversely proportional to the length of the columns. Therefore, the load and moments on the column just below the third-floor for **load case 2** are calculated as

$$P_3 = 173 \text{ kips}$$
$$M_{3botx-x} \cong 0.5(27 \text{ ft.-kips}) = 14 \text{ ft.-kips}$$
$$M_{3boty-y} \cong 0.5(6.3 \text{ ft.-kips}) = 3.2 \text{ ft.-kips}$$

Design or check the column for combined axial load plus bending.
    The 0.5 term in the above equations for moment is the moment distribution factor.

Loads to Column C1 at the Second Floor: (See Figures 8-22 and 8-23)

Girder and Beam Eccentricities:

e_2 girders   = 2½ in. + ½ (Column depth)
                      = 2½ in. + ½ (10 in.) assuming W10 column
                      = 7.5 in. = **0.63 ft.**

e_2 beams   = 2½ in. + ½ (Column web)
                      = 2½ in. + ½ (0.5 in.) assumed
                      = 2.75 in. ≈ **0.25 ft.**

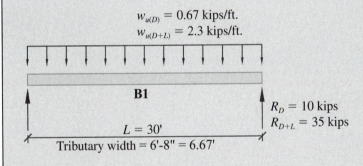

$$w_{u(D)} = 0.67 \text{ kips/ft.}$$
$$w_{u(D+L)} = 2.3 \text{ kips/ft.}$$

B1

$$R_D = 10 \text{ kips}$$
$$R_{D+L} = 35 \text{ kips}$$

$$L = 30'$$
Tributary width = 6'-8" = 6.67'

$$w_{u(D)} = (0.10 \text{ ksf})(6.67') = 0.67 \text{ kips/ft.}$$
$$w_{u(D+L)} = (0.342 \text{ ksf})(6.67') = 2.3 \text{ kips/ft}$$

$$R_D = \frac{w_{u(D)}L}{2} = \frac{(0.67)(30)}{2} = 10 \text{ kips}$$

$$R_{D+L} = \frac{w_{u(D+L)}L}{2} = \frac{(2.3)(30)}{2} = 35 \text{ kips}$$

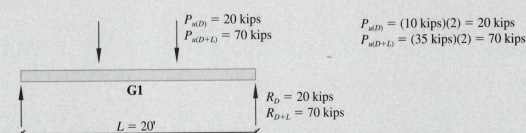

$$P_{u(D)} = 20 \text{ kips}$$
$$P_{u(D+L)} = 70 \text{ kips}$$

G1

$$R_D = 20 \text{ kips}$$
$$R_{D+L} = 70 \text{ kips}$$

$$L = 20'$$

$$P_{u(D)} = (10 \text{ kips})(2) = 20 \text{ kips}$$
$$P_{u(D+L)} = (35 \text{ kips})(2) = 70 \text{ kips}$$

Figure 8-22 Calculation of floor beam and girder reactions to C1 at the second floor.

(continued)

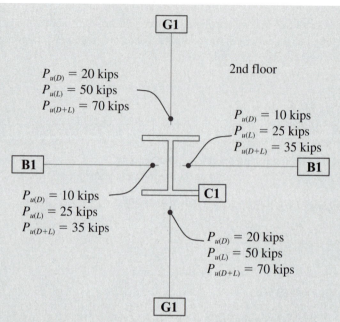

Figure 8-23 Reactions of floor beams and girders framing into column C1 at the second floor.

Factored Loads:

$$w_u \quad = 0.342\,\text{ksf}(6.67\,\text{ft.}) = 2.3\,\text{kips/ft.}$$
$$w_{u(D)} = 0.10\,\text{ksf}(6.67\,\text{ft.}) = 0.67\,\text{kips/ft.}$$

$$P_{DL} \quad = 10\,\text{kips}\,(2\,\text{beams}) = 20\,\text{kips}$$
$$P_{DL+LL} = 35\,\text{kips}\,(2\,\text{beams}) = 70\,\text{kips}$$

Moment Split in Columns:

$$M_{2bot} = \left(\frac{1/L_1}{1/L_1 + 1/L_2}\right)M_2, \text{ where } L_1 = 10\,\text{ft. and } L_2 = 10\,\text{ft.}$$

$$= \left(\frac{1/10\,\text{ft.}}{1/10\,\text{ft.} + 1/10\,\text{ft.}}\right)M_2$$

$$= 0.5M_2$$

Load Case 1: The axial load is maximized.

$$P_2 \quad = 248 + (35 \quad + \quad 35) \quad + \quad (70 \quad + \quad 70) = 458\,\text{kips}$$
$$\quad\quad (\max P_3)\ (P^R_{DL+LL}\,beam)\ (P^L_{DL+LL}\,beam)\ (P^R_{DL+LL}\,girder)\ (P^L_{DL+LL}\,girder)$$
$$M_{2x-x} = (70 - 70) \times e_2\,\textbf{girder} = (0\,\text{kips})(0.63\,\text{ft.}) = 0\,\text{ft.-kips}$$
$$M_{2y-y} = (35 - 35) \times e_2\,\textbf{beam} = (0\,\text{kips})(0.25\,\text{ft.}) = 0\,\text{ft.-kips}$$

$P_2 = 458$ kips

$M_{2botx-x} = 0.5(0 \text{ ft.-kips}) = 0 \text{ ft.-kips}$

$M_{2boty-y} = 0.5(0 \text{ ft.-kips}) = 0 \text{ ft.-kips}$

For this load case, design the column for axial load only.

Load Case 2: The moments at this floor level are maximized.

$P_2 = 248 + (35 + 10) + (70 + 20) = 383$ kips
      (max P_3) (P^R_{DL+LL} beam) (P^L_{DL} beam) (P^R_{DL+LL} girder) (P^L_{DL} girder)

$M_{2x-x} = (70 - 20) \times e_2 \textbf{ girder} = (50 \text{ kips})(0.63 \text{ ft.}) = 32 \text{ ft.-kips}$

$M_{2y-y} = (35 - 10) \times e_2 \textbf{ beam} = (25 \text{ kips})(0.25 \text{ ft.}) = 6.3 \text{ ft.-kips}$

The moment at the second-floor level will be distributed between the columns above and below that level in a ratio that is inversely proportional to the length of the columns. Therefore, the load and moments on the column just below the second-floor for **load case 2** are calculated as

$P_2 = 383$ kips

$M_{2botx-x} = 0.5(32 \text{ ft.-kips}) = 16 \text{ ft.-kips}$

$M_{2boty-y} = 0.5(6.3 \text{ ft.-kips}) = 3.2 \text{ ft.-kips}$

Design or check column for combined axial load plus bending.

The 0.5 term in the above equations for moment is the moment distribution factor. The summary of the factored loads and moments for column C1 is shown in Table 8-4.

Table 8-4   Summary of **factored loads and moments** for column **C1**

Level	Load Case 1	Load Case 2
Roof	$P = 55$ kips	$P = 38$ kips
	$M_{r-xx} = 0$ ft.-kips	$M_{r-xx} = 6$ ft.-kips
	$M_{r-yy} = 0$ ft.-kips	$M_{r-yy} = 1.4$ ft.-kips
Third Floor	$P = 248$ kips	$P = 173$ kips
	$M_{3botxx} = 0$ ft.-kips	$M_{3botxx} = 14$ ft.-kips
	$M_{3botyy} = 0$ ft.-kips	$M_{3botyy} = 3.2$ ft.-kips
Second Floor	$P = 458$ kips	$P = 383$ kips
	$M_{2botxx} = 0$ ft.-kips	$M_{2botxx} = 16$ ft.-kips
	$M_{2botyy} = 0$ ft.-kips	$M_{2botyy} = 3.2$ ft.-kips

(continued)

Load Calculations for Column C2

All loads are factored.

Loads to Column C2 at the Roof Level: (See Figures 8-24 and 8-25)

Girder and Beam Eccentricities:

e_r **girders** $= 2\frac{1}{2}$ in. $+ \frac{1}{2}$ (Column depth)

$= 2\frac{1}{2}$ in. $+ \frac{1}{2}$ (8 in.) assuming minimum W8 column

$= 6\frac{1}{2}$ in. $= $ **0.54 ft.**

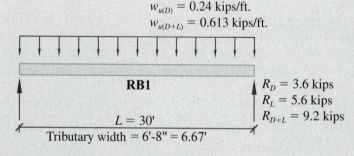

$w_{u(D)} = 0.24$ kips/ft.
$w_{u(D+L)} = 0.613$ kips/ft.

RB1

$L = 30'$
Tributary width $= 6'\text{-}8" = 6.67'$

$R_D = 3.6$ kips
$R_L = 5.6$ kips
$R_{D+L} = 9.2$ kips

$w_{u(D)} = (0.036 \text{ ksf})(6.67') = 0.24$ kips/ft.
$w_{u(D+L)} = (0.092 \text{ ksf})(6.67') = 0.613$ kips/ft.

$R_D = \dfrac{w_{u(D)}L}{2} = \dfrac{(0.24)(30)}{2} = 3.6$ kips

$R_{D+L} = \dfrac{w_{u(D+L)}L}{2} = \dfrac{(0.613)(30)}{2} = 9.2$ kips

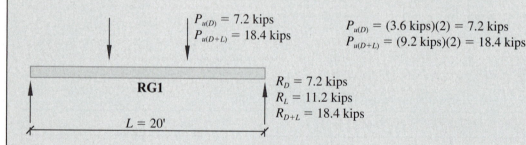

$P_{u(D)} = 7.2$ kips
$P_{u(D+L)} = 18.4$ kips

$P_{u(D)} = (3.6 \text{ kips})(2) = 7.2$ kips
$P_{u(D+L)} = (9.2 \text{ kips})(2) = 18.4$ kips

RG1

$L = 20'$

$R_D = 7.2$ kips
$R_L = 11.2$ kips
$R_{D+L} = 18.4$ kips

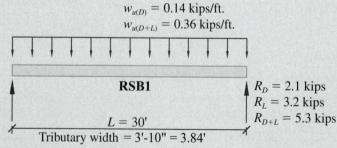

$w_{u(D)} = 0.14$ kips/ft.
$w_{u(D+L)} = 0.36$ kips/ft.

RSB1

$L = 30'$
Tributary width $= 3'\text{-}10" = 3.84'$

$R_D = 2.1$ kips
$R_L = 3.2$ kips
$R_{D+L} = 5.3$ kips

$w_{u(D)} = (0.036 \text{ ksf})(3.84') = 0.14$ kips/ft.
$w_{u(D+L)} = (0.092 \text{ ksf})(3.84') = 0.36$ kips/ft.

$R_D = \dfrac{w_{u(D)}L}{2} = \dfrac{(0.14)(30)}{2} = 2.1$ kips

$R_{D+L} = \dfrac{w_{u(D+L)}L}{2} = \dfrac{(0.36)(30)}{2} = 5.3$ kips

Figure 8-24 Calculation of roof beam and girder reactions to C2.

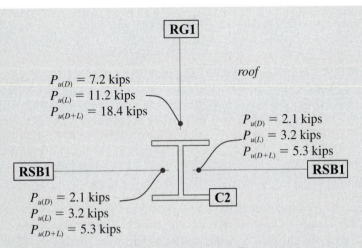

Figure 8-25 Reactions of roof beams and girders framing into column C2.

e_r **beams** $= 2\frac{1}{2}$ in. $+ \frac{1}{2}$ (Column web thickness)

$= 2\frac{1}{2}$ in. $+ \frac{1}{2}$ (0.5 in.) assumed

$= 2.75$ in. $\approx$ **0.25 ft.**

Factored Loads:

$w_u \quad = 0.092\,\text{ksf}\,(6.67\,\text{ft.}) = 0.613\,\text{kips/ft.}$

$w_{u(D)} = 0.036\,\text{ksf}\,(6.67\,\text{ft.}) = 0.24\,\text{kips/ft.}$

$P_D \quad = 3.6\,\text{kips}\,(2\,\text{beams}) = 7.2\,\text{kips}$

$P_{D+L} = 9.2\,\text{kips}\,(2\,\text{beams}) = 18.4\,\text{kips}$

For **perimeter or spandrel beams**, the tributary width, TW $= (6.67\,\text{ft.}/2) + 0.5$-ft. edge distance $= 3.84$ ft. Therefore, the factored loads on the spandrel beam are

$w_u \quad = 0.092\,\text{ksf}\,(3.84\,\text{ft.}) = 0.36\,\text{kips/ft.}$

$w_{u(D)} = 0.036\,\text{ksf}\,(3.84\,\text{ft.}) = 0.14\,\text{kips/ft.}$

Load Case 1: The axial load is maximized.

$P_{roof} = \quad 5.3 \quad + \quad 5.3 \quad + \quad 18.4 \quad + \quad 0 = 29\,\text{kips}$
$\qquad\quad (P^R_{DL+LL}\,beams)\,(P^L_{DL+LL}\,beams)\,(P^R_{DL+LL}\,girders)\,(P^L_{DL+LL}\,girders)$

$M_{rx-x} = (18.4 - 0)(0.54\,\text{ft.}) = 10\,\text{ft.-kips}$
$\qquad\qquad\qquad \uparrow (e_r,\,girder)$

$M_{ry-y} = (5.3 - 5.3)(0.25\,\text{ft.}) = 0\,\text{ft.-kips}$
$\qquad\qquad\qquad \uparrow (e_r\,beams)$

$P \quad\;\; = 29\,\text{kips}$

$M_{rx-x} = 10\,\text{ft.-kips}$

$M_{ry-y} = 0\,\text{ft.-kips}$

For this case, the column is designed for axial load plus bending.

(continued)

Load Case 2: The moments at this floor level are maximized.

$$P_{roof} = \underset{(P^R_{DL+LL}\ beams)}{5.3} + \underset{(P^L_{DL}\ beams)}{2.1} + \underset{(P^R_{DL+LL}\ girders)}{18.4} + \underset{(P^L_{DL}\ girders)}{0} = 26\ kips$$

$$M_{rx-x} = (18.4 - 0)(0.54\ ft.) = 10\ ft.\text{-}kips$$
$$M_{ry-y} = (5.3 - 2.1)(0.25\ ft.) = 1.0\ ft.\text{-}kips$$

$$P = 26\ kips$$
$$M_{rx-x} = 10\ ft.\text{-}kips$$
$$M_{ry-y} = 1.0\ ft.\text{-}kips$$

For this load case, the column is designed or checked for combined axial load plus bending.

The column moments at the roof level are resisted solely by the column below the roof level because there is no column above that level. For the roof loads that will be cumulatively added to the floor loads below, the appropriate load factor must be applied to the live loads. Recall that the load combination $1.2D + 1.6(L_r$ or S or $R)$ applies only to the column load at the roof level only. For cumulative loads at the lower levels, the applicable load combination is $1.2D + 1.6L + 0.5(L_r$ or S or $R)$. Therefore, the contribution of the roof level to the cumulative loads at the lower levels (i.e., third- and second-floor levels) is

$$P_{roof} = 1.2D + 0.5(L_r\ or\ S\ or\ R) = 7.2 + 2.1 + 2.1 + 0.5(11.2 + 3.2 + 3.2)$$
$$= 20.2\ kips$$

Loads to Column C2 at the Third Floor: (See Figures 8-26 and 8-27)

Girder and Beam Eccentricities:

e_3 **girders** $= 2\frac{1}{2}$ in. $+ \frac{1}{2}$ (Column depth)

                $= 2\frac{1}{2}$ in. $+ \frac{1}{2}$ (8 in.) assuming minimum W8 column

                $= 6\frac{1}{2}$ in. $= $ **0.54 ft.**

e_3 **beams** $= 2\frac{1}{2}$ in. $+ \frac{1}{2}$ (Column web thickness)

                $= 2\frac{1}{2}$ in. $+ \frac{1}{2}$ (0.5 in.) assumed

                $= 2.75$ in. $\approx$ **0.25 ft.**

Factored Loads:

$$w_u = 0.342\ ksf\ (6.67\ ft.) = 2.3\ kips/ft.$$
$$w_{u(D)} = 0.10\ ksf\ (6.67\ ft.) = 0.67\ kips/ft.$$

For the reactions for girder G1, refer to the load calculations for column C1.

For **perimeter or spandrel beams**, the tributary width,

$$TW = (6.67\ ft./2) + 0.5\text{-}ft.\ edge\ distance = 3.84\ ft.$$

Therefore, the loads on the spandrel beams are

$$w_u = 0.342\ ksf\ (3.84\ ft.) = 1.32\ kips/ft.$$
$$w_{u(D)} = 0.10\ ksf\ (3.84\ ft.) = 0.39\ kips/ft.$$

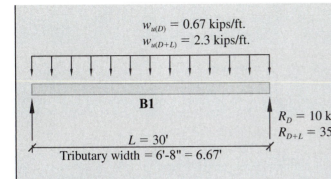

$w_{u(D)} = 0.67$ kips/ft.
$w_{u(D+L)} = 2.3$ kips/ft.

$w_{u(D)} = (0.10 \text{ ksf})(6.67') = 0.67$ kips/ft.
$w_{u(D+L)} = (0.342 \text{ ksf})(6.67') = 2.3$ kips/ft.

$$R_D = \frac{w_{u(D)}L}{2} = \frac{(0.67)(30)}{2} = 10 \text{ kips}$$

$R_D = 10$ kips
$R_{D+L} = 35$ kips

$$R_{D+L} = \frac{w_{u(D+L)}L}{2} = \frac{(2.3)(30)}{2} = 35 \text{ kips}$$

B1

$L = 30'$
Tributary width $= 6'\text{-}8" = 6.67'$

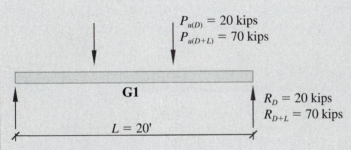

$P_{u(D)} = 20$ kips
$P_{u(D+L)} = 70$ kips

$P_{u(D)} = (10 \text{ kips})(2) = 20$ kips
$P_{u(D+L)} = (35 \text{ kips})(2) = 70$ kips

G1

$R_D = 20$ kips
$R_{D+L} = 70$ kips

$L = 20'$

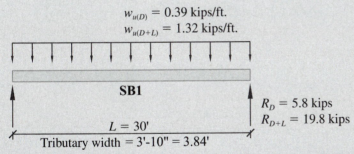

$w_{u(D)} = (0.10 \text{ ksf})(3.84') = 0.39$ kips/ft.
$w_{u(D+L)} = (0.342 \text{ ksf})(3.84') = 1.32$ kips/ft.

$w_{u(D)} = 0.39$ kips/ft.
$w_{u(D+L)} = 1.32$ kips/ft.

$$R_D = \frac{w_{u(D)}L}{2} = \frac{(0.39)(30)}{2} = 5.8 \text{ kips}$$

SB1

$R_D = 5.8$ kips
$R_{D+L} = 19.8$ kips

$$R_{D+L} = \frac{w_{u(D+L)}L}{2} = \frac{(1.32)(30)}{2} = 19.8 \text{ kips}$$

$L = 30'$
Tributary width $= 3'\text{-}10" = 3.84'$

Figure 8-26 Calculation of floor beam and girder reactions to C2 at the third floor.

Moment Split in the Column:

$$M_{3bot} = \left(\frac{1/L_2}{1/L_2 + 1/L_3}\right)M_3; \quad L_2 = 10 \text{ ft. and } L_3 = 10 \text{ ft.}$$

$$= \left(\frac{1/10}{1/10 \text{ ft.} + 1/10 \text{ ft.}}\right) M_3$$

$$= 0.5 \, M_3$$

(continued)

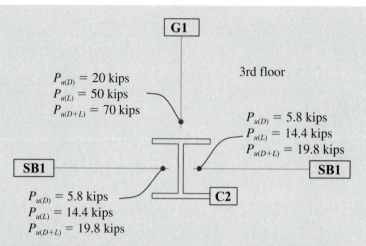

Figure 8-27 Reactions of floor beams and girders framing into column C2 at the third floor.

Load Case 1: The axial load is maximized.

$$P_3 = \underset{(\text{max } P_{roof})}{20} + (\underset{(P^R_{DL+LL} \text{ beam})}{19.8} + \underset{(P^L_{DL+LL} \text{ beam})}{19.8}) + (\underset{(P^R_{DL+LL} \text{ girder})}{70} + \underset{(P^L_{DL+LL} \text{ girder})}{0}) = 130 \text{ kips}$$

$$M_{3x-x} = (70 - 0)(0.54 \text{ ft.}) = 38 \text{ ft.-kips}$$

$$M_{3y-y} = (19.8 - 19.8)(0.25 \text{ ft.}) = 0 \text{ ft.-kips}$$

$$P_3 = 130 \text{ kips}$$

$$M_{3botx-x} = 0.5(38 \text{ ft.-kips}) = 19 \text{ ft.-kips}$$

$$M_{3boty-y} = 0.5(0 \text{ ft.-kips}) = 0 \text{ ft.-kips}$$

For this load case, design the column for axial load plus bending.

Load Case 2: The moments at this floor level are maximized.

$$P_3 = \underset{(\text{max } P_{roof})}{29} + (\underset{(P^R_{DL+LL} \text{ beam})}{5.8} + \underset{(P^L_{DL} \text{ beam})}{19.8}) + (\underset{(P^R_{DL+LL} \text{ girder})}{70} + \underset{(P^L_{DL} \text{ girder})}{0}) = 125 \text{ kips}$$

$$M_{3x-x} = (70 - 0) e_3 \textbf{ girder} = 70 \text{ kips} (0.54 \text{ ft.}) = 38 \text{ ft.-kips}$$

$$M_{3y-y} = (19.8 - 5.8) e_3 \textbf{ beam} = 14 \text{ kips} (0.25 \text{ ft.}) = 4 \text{ ft.-kips}$$

The column moments at the third-floor level is assumed to be distributed between the columns above and below that floor level in a ratio that is inversely proportional to the length of the columns. Therefore, the load and moments (for **load case 2**) on the column just below the third-floor level are calculated as

$$P_3 = 125 \text{ kips}$$

$$M_{3botx-x} \cong 0.5(38 \text{ ft.-kips}) = 19 \text{ ft.-kips}$$

$$M_{3boty-y} \cong 0.5(4 \text{ ft.-kips}) = 2 \text{ ft.-kips}$$

Design or check the column for combined axial load plus bending.
  The 0.5 term in the above equations for moment is the moment distribution factor.

Loads to Column C2 at the Second Floor: (See Figures 8-28 and 8-29)

Girder and Beam Eccentricities:

e_2 **girders**   = 2½ in. + ½ (Column depth)

                      = 2½ in. + ½(10 in.) assuming W10 column

                      = 7.5 in.

                      = **0.63 ft.**

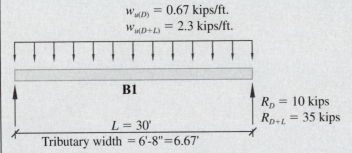

$w_{u(D)} = 0.67$ kips/ft.
$w_{u(D+L)} = 2.3$ kips/ft.

B1

$R_D = 10$ kips
$R_{D+L} = 35$ kips

$L = 30'$
Tributary width = 6'-8"=6.67'

$w_{u(D)} = (0.10 \text{ ksf})(6.67') = 0.67$ kips/ft.
$w_{u(D+L)} = (0.342 \text{ ksf})(6.67') = 2.3$ kips/ft.

$$R_D = \frac{w_{u(D)}L}{2} = \frac{(0.67)(30)}{2} = 10 \text{ kips}$$

$$R_{D+L} = \frac{w_{u(D+L)}L}{2} = \frac{(2.3)(30)}{2} = 35 \text{ kips}$$

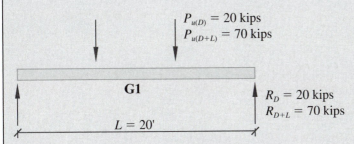

$P_{u(D)} = 20$ kips
$P_{u(D+L)} = 70$ kips

G1

$R_D = 20$ kips
$R_{D+L} = 70$ kips

$L = 20'$

$P_{u(D)} = (10 \text{ kips})(2) = 20$ kips
$P_{u(D+L)} = (35 \text{ kips})(2) = 70$ kips

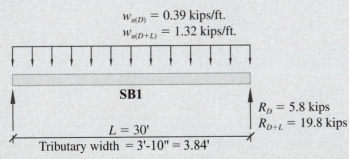

$w_{u(D)} = 0.39$ kips/ft.
$w_{u(D+L)} = 1.32$ kips/ft.

SB1

$R_D = 5.8$ kips
$R_{D+L} = 19.8$ kips

$L = 30'$
Tributary width = 3'-10" = 3.84'

$w_{u(D)} = (0.10 \text{ ksf})(3.84') = 0.39$ kips/ft.
$w_{u(D+L)} = (0.342 \text{ ksf})(3.84') = 1.32$ kips/ft.

$$R_D = \frac{w_{u(D)}L}{2} = \frac{(0.39)(30)}{2} = 5.8 \text{ kips}$$

$$R_{D+L} = \frac{w_{u(D+L)}L}{2} = \frac{(1.32)(30)}{2} = 19.8 \text{ kips}$$

Figure 8-28 Calculation of floor beam and girder reactions to C2 at the second floor.

(continued)

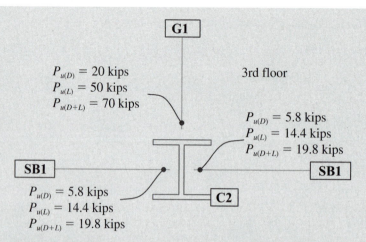

Figure 8-29 Reactions of floor beams and girders framing into column C2 at the second floor.

e_2 **beams**   = 2½ in. + ½ (Column web)

        = 2½ in. + ½(0.5 in.) assumed

        = 2.75 in. ≈ **0.25 ft.**

Factored Loads:

w_u = 0.342 ksf (6.67 ft.) = 2.3 kips/ft.

$w_{u(D)}$ = 0.10 ksf (6.67 ft.) = 0.67 kips/ft.

For the reactions in girder G1, refer to the load calculations for column C1.

For perimeter or spandrel beams, the tributary width,

TW = 6.67 ft./2 + 0.5-ft. edge distance

    = 3.84 ft.

Therefore, the loads on the spandrel beams are

w_u = 0.342 ksf (3.84 ft.) = 1.32 kips/ft.

$w_{u(D)}$ = 0.10 ksf (3.84 ft.) = 0.39 kips/ft.

Moment Split in the Column:

$$M_{2bot} = \left(\frac{1/L_1}{1/L1 \ + \ 1/L2}\right)M_2$$

$$= \left(\frac{1/10 \text{ ft.}}{1/10 \text{ ft.} \ + \ 1/10 \text{ ft.}}\right)M_2,$$

where L_1 = 10 ft. and L_2 = 10 ft.

        = $0.5M_2$

Load Case 1: The axial load is maximized.

$$P_2 = 130 + (19.8 + 19.8) + (70 + 0) = 240 \text{ kips}$$
$$(P_3\text{max})$$

$$M_{2x-x} = (70 - 0)(0.63 \text{ ft.}) = 44 \text{ ft.-kips}$$
$$M_{2y-y} = (19.8 - 19.8)(0.25 \text{ ft.}) = 0 \text{ ft.-kips}$$

$$P_2 = 240 \text{ kips}$$
$$M_{2botx-x} = 0.5(44 \text{ ft.-kips}) = 22 \text{ ft.-kips}$$
$$M_{2boty-y} = 0.5(0 \text{ ft.-kips}) = 0 \text{ ft.-kips}$$

For this load case, design the column for axial load plus bending.

Load Case 2: The moments at this floor level are maximized.

$$P_2 = 130 + (5.8 + 19.8) + (70 + 0) = 226 \text{ kips}$$
$$M_{2x-x} = (70 - 0)(0.63 \text{ ft.}) = 44 \text{ ft.-kips}$$
$$M_{2y-y} = (19.8 - 5.8)(0.25 \text{ ft.}) = 4 \text{ ft.-kips}$$

The moment at the second-floor level is assumed to be distributed between the columns above and below that floor level in a ratio that is inversely proportional to the length of the columns. Therefore, the load and moments on the column just below the second-floor level for **load case 2** are calculated as

$$P_2 = 226 \text{ kips}$$
$$M_{2botx-x} = 0.5(44 \text{ ft.-kips}) = 22 \text{ ft.-kips}$$
$$M_{2boty-y} = 0.5(4 \text{ ft.-kips}) = 2 \text{ ft.-kips}$$

Design or check column for combined axial load plus bending.

The 0.5 term in the above equations for moment is the moment distribution factor. A summary of the factored loads and moments for column C2 is shown in Table 8-5.

Table 8-5 Summary of factored loads and moments for column C2

Level	Load Case 1	Load Case 2
Roof	$P = 29$ kips $M_{roofx-x} = 10$ ft.-kips $M_{roofy-y} = 0$ ft.-kips	$P = 26$ kips $M_{roofx-x} = 10$ ft.-kips $M_{roofy-y} = 1.0$ ft.-kips
Third floor	$P = 130$ kips $M_{3botx-x} = 19$ ft.-kips $M_{3boty-y} = 0$ ft.-kips	$P = 125$ kips $M_{3botx-x} = 19$ ft.-kips $M_{3boty-y} = 2$ ft.-kips
Second floor	$P = 240$ kips $M_{2botx-x} = 22$ ft.-kips $M_{3boty-y} = 0$ ft.-kips	$P = 226$ kips $M_{2botx-x} = 22$ ft.-kips $M_{2boty-y} = 2$ ft.-kips

(continued)

Design of Column C1 (Roof to Third Floor)

Load Case 1 (with $M = 0$)	Load Case 2 or Load Case 1 with Moments
$P_u = 55$ kips: $KL \approx 1 \times 10$ ft. = 10 ft. From **Column Load Tables** (*AISCM*, Table 4-1) **Try W8 × 31 ($A = 9.12$ in.^{2})** $\phi_c P_n = 317$ kips $> P_u$ OK	**Check W8 × 31 column for** $P_u = 38$ kips $M_{rx-x} = M_{ntx} = 6$ ft.-kips $M_{ry-y} = M_{nty} = 1.4$ ft.-kips Column unbraced length, $L_b = 10$ ft. $r_x = 3.47$ in., $r_y = 2.02$ in. **LRFD Beam Design Tables (*AISCM*, Table 3-2; $F_y = 50$ ksi)** $\phi_b M_p = 114$ ft.-kips; $L_p = 7.18$ ft.; $BF = 2.37$ kips; $L_r = 24.8$ ft. $> L_b$ $\phi_b M_{nx} = \phi_b M_p - BF(L_b - L_p) = [114 - (2.37)(10 - 7.18)] = 107$ ft.-kips $\leq \phi_b M_p$ From *AISCM*: $S_x = 27.5$ in.^{3}; $Z_y = 14.1$ in.^{3}; $S_y = 9.27$ in.^{3} $\phi_b M_{nx} < 1.5 \phi_b S_x F_y / 12 = 155$ ft.-kips $\therefore \phi_b M_{nx} = \mathbf{107}$ **ft.-kips** $\phi_b M_{ny} = \phi_b Z_y F_y < 1.5 \phi_b S_y F_y$ $\phi_b M_{ny} = (0.9)(14.1)(50)/12 < 1.5(0.9)(9.27)(50)/12$ $\phi_b M_{ny} = 54$ ft.-kips $> \mathbf{52}$ **ft.-kips** Governs $\boxed{\phi_c P_n = 317 \text{ kips}; \ \phi_b M_{nx} = 107 \text{ ft.-kips}; \ \phi_b M_{ny} = 52 \text{ ft.-kips}}$ $M_{ux} = B_{1x} M_{ntx} + B_{2x} M_{ltx}; \ M_{ltx} = 0$ (for braced frames) $M_{uy} = B_{1y} M_{nty} + B_{2y} M_{lty}; \ M_{lty} = 0$ (for braced frames) *P-δ Effect for x-x Axis* $C_{mx} = 0.6 - 0.4(M_{1x}/M_{2x})$; but conservatively use $C_{mx} = 0.85$ $P_{e1x} = \dfrac{\pi^2 EA}{(KL/r_x)^2} = 2183$ kips $B_{1x} = \dfrac{C_{mx}}{(1 - P_u/P_{e1x})} = \dfrac{0.85}{[1 - (38/2183)]} = 0.87 \geq 1.0 \ \therefore \mathbf{B_{1x} = 1.0}$ $\boxed{\therefore M_{ux} = B_{1x} M_{ntx} = 1.0(6) = 6 \text{ ft.-kips}}$ *P-δ Effect for y-y Axis* $C_{my} = 0.6 - 0.4(M_{1y}/M_{2y})$; but conservatively use $C_{my} = 0.85$ $P_{e1y} = \dfrac{\pi^2 EA}{(KL/r_y)^2} = 740$ kips $B_{1y} = \dfrac{C_{my}}{(1 - P_u/P_{e1y})} = \dfrac{0.85}{[1 - (38/740)]} = 0.90 \geq 1.0 \ \therefore \mathbf{B_{1y} = 1.0}$ $\boxed{\therefore M_{uy} = B_{1y} M_{nty} = 1.0(1.4) = 1.4 \text{ ft.-kips}}$

Interaction Equation

$$P_u/\phi_c P_n = 38/317 = 0.12 < 0.2 \Rightarrow \text{Equation (8-2)}$$

$$\frac{P_u}{2\phi P_n} + \left(\frac{M_{ux}}{\phi_b M_{nx}} + \frac{M_{uy}}{\phi_b M_{ny}} \right) = \frac{0.12}{2} + \left(\frac{6}{107} + \frac{1.4}{52} \right) = 0.14 < 1.0 \text{ OK} \Rightarrow \textbf{W8 × 31 is adequate.}$$

Note: For building structures, the minimum size of column recommended is a W8 column in order to ensure sufficient flange width to accommodate the girder-to-column double-angle connections.

Design of Column C1 (Third Floor to Second Floor)

Load Case 1 (with $M = 0$)	Load Case 2 or Load Case 1 with Moments
$P_u = 248$ kips: $KL \approx 1 \times 10$ ft. $= 10$ ft. From **Column Load Tables** (*AISCM*, Table 4-1) **Try W8 $\times$ 31 ($A = 9.12$ in^2)** $\phi_c P_n = 317$ kips $> P_u$ OK	**Check W8 $\times$ 31 column for** $P_u = 173$ kips $M_{3x-x} = M_{ntx} = 14$ ft.-kips $M_{ry-y} = M_{nty} = 3.2$ ft.-kips Column unbraced length, $L_b = 10$ ft; $r_x = 3.47$ in., $r_y = 2.02$ in.

LRFD Beam Design Tables (*AISCM*, Table 3-2; $F_y = 50$ ksi)
$\phi_b M_p = 114$ ft.-kips; $L_p = 7.18$ ft.; $BF = 2.37$ kips; $L_r = 24.8$ ft. $> L_b$
$\phi_b M_{nx} = \phi_b M_p - BF(L_b - L_p) = [114 - (2.37)(10 - 7.18)] = 107$ ft.-kips $\leq \phi_b M_p$
From *AISCM*: $S_x = 27.5$ in.^{3}; $Z_y = 14.1$ in.^{3}; $S_y = 9.27$ in.^{3}
$\phi_b M_{nx} < 1.5 \phi_b S_x F_y / 12 = 155$ ft.-kips $\therefore \phi_b M_{nx} = $ **107 ft.-kips**

$\phi_b M_{ny} = \phi_b Z_y F_y < 1.5 \phi_b S_y F_y$
$\phi_b M_{ny} = (0.9)(14.1)(50)/12 < 1.5(0.9)(9.27)(50)/12$
$\phi_b M_{ny} = 54$ ft.-kips $> $ **52 ft.-kips** Governs

$\boxed{\phi_c P_n = 317 \text{ kips}; \ \phi_b M_{nx} = 107 \text{ ft.-kips}; \ \phi_b M_{ny} = 52 \text{ ft.-kips}}$

$M_{ux} = B_{1x} M_{ntx} + B_{2x} M_{ltx}; \ M_{ltx} = 0$ (for braced frames)
$M_{uy} = B_{1y} M_{nty} + B_{2y} M_{lty}; \ M_{lty} = 0$ (for braced frames)

P-δ Effect for x–x Axis
$C_{mx} = 0.6 - 0.4(M_{1x}/M_{2x})$; but conservatively use $C_{mx} = 0.85$
$$P_{e1x} = \frac{\pi^2 EA}{(KL/r_x)^2} = 2183 \text{ kips}$$
$$B_{1x} = \frac{C_{mx}}{(1 - P_u/P_{e1x})} = \frac{0.85}{[1 - (173/2183)]} = 0.92 \geq 1.0 \therefore \textbf{\textit{B}}_{\textbf{1x}} = \textbf{1.0}$$

$\boxed{\therefore M_{ux} = B_{1x} M_{ntx} = 1.0(14) = 14 \text{ ft.-kips}}$

P-δ Effect for y–y Axis
$C_{my} = 0.6 - 0.4(M_{1y}/M_{2y})$; but conservatively use $C_{my} = 0.85$
$$P_{e1y} = \frac{\pi^2 EA}{(KL/r_y)^2} = 740 \text{ kips}$$
$$B_{1y} = \frac{C_{my}}{(1 - P_u/P_{e1y})} = \frac{0.85}{[1 - (173/740)]} = 1.11 > 1.0 \therefore \textbf{\textit{B}}_{\textbf{1y}} = \textbf{1.11}$$

$\boxed{\therefore M_{uy} = B_{1y} M_{nty} = 1.11(3.2) = 4.0 \text{ ft.-kips}}$

Interaction Equation

$P_u/\phi_c P_n = 173/317 = 0.55 \geq 0.2 \Rightarrow$ Equation (8-1)

$$\frac{P_u}{\phi P_n} + \frac{8}{9}\left(\frac{M_{ux}}{\phi_b M_{nx}} + \frac{M_{uy}}{\phi_b M_{ny}}\right) = 0.55 + \frac{8}{9}\left(\frac{14}{107} + \frac{4.0}{52}\right) = 0.73 < 1.0 \text{ OK}$$

$\therefore$ **W8 $\times$ 31 is adequate.**

(continued)

Design of Column C1 (Second Floor to Ground Floor)

Load Case 1 (with $M = 0$)	Load Case 2 or Load Case 1 with Moments
$P_u = 458$ kips: $KL \approx 1 \times 10$ ft. $= 10$ ft. From **Column Load Tables** (*AISCM*, Table 4-1) **Try W8 $\times$ 48 ($A = 14.1$ in.^{2})** $\phi_c P_n = 497$ kips $> P_u$ OK	**Check W8 $\times$ 48 column for** $P_u = 383$ kips $M_{3x\text{-}x} = M_{ntx} = 16$ ft.-kips $M_{3y\text{-}y} = M_{nty} = 3.2$ ft.-kips Column unbraced length, $L_b = 10$ ft.; $r_x = 3.61$ in., $r_y = 2.08$ in.

LRFD Beam Design Tables (*AISCM*, Table 3-2; $F_y = 50$ ksi)
$\phi_b M_p = 184$ ft.-kips; $L_p = 7.35$ ft.; $BF = 2.53$ kips; $L_r = 35.2$ ft. $> L_b$
$\phi_b M_{nx} = \phi_b M_p - BF(L_b - L_p) = [184 - (2.53)(10 - 7.35)] = 177$ ft.-kips $\leq \phi_b M_p$
From *AISCM*: $S_x = 43.2$ in.^{3}; $Z_y = 22.9$ in.^{3}; $S_y = 15.0$ in.^{3}
$\phi_b M_{nx} < 1.5\phi_b S_x F_y/12 = 243$ ft.-kips $\therefore \phi_b M_{nx} = \mathbf{177}$ **ft.-kips**

$\phi_b M_{ny} = \phi_b Z_y F_y < 1.5\phi_b S_y F_y$
$\phi_b M_{ny} = (0.9)(22.9)(50)/12 < 1.5(0.9)(15.0)(50)/12$
$\phi_b M_{ny} = 86$ ft.-kips $> \mathbf{84}$ **ft.-kips** Governs

$\boxed{\phi_c P_n = 497 \text{ kips}; \ \phi_b M_{nx} = 177 \text{ ft.-kips}; \ \phi_b M_{ny} = 84 \text{ ft.-kips}}$

$M_{ux} = B_{1x}M_{ntx} + B_{2x}M_{ltx}; \ M_{ltx} = 0$ (for braced frames)
$M_{uy} = B_{1y}M_{nty} + B_{2y}M_{lty}; \ M_{lty} = 0$ (for braced frames)

P-δ Effect for x–x Axis
$C_{mx} = 0.6 - 0.4(M_{1x}/M_{2x})$; but conservatively use $C_{mx} = 0.85$
$P_{e1x} = \dfrac{\pi^2 EA}{(KL/r_x)^2} = 3652$ kips

$B_{1x} = \dfrac{C_{mx}}{(1 - P_u/P_{e1x})} = \dfrac{0.85}{[1 - (383/3652)]} = 0.95 \geq 1.0 \therefore \boldsymbol{B_{1x} = 1.0}$

$\boxed{\therefore M_{ux} = B_{1x}M_{ntx} = 1.0(16) = 6 \text{ ft.-kips}}$

P-δ Effect for y–y Axis
$C_{my} = 0.6 - 0.4(M_{1y}/M_{2y})$; but conservatively use $C_{my} = 0.85$
$P_{e1y} = \dfrac{\pi^2 EA}{(KL/r_y)^2} = 1212$ kips

$B_{1y} = \dfrac{C_{my}}{(1 - P_u/P_{e1y})} = \dfrac{0.85}{[1 - (383/1212)]} = 1.24 > 1.0 \therefore \boldsymbol{B_{1y} = 1.24}$

$\boxed{\therefore M_{uy} = B_{1y}M_{nty} = 1.24(3.2) = 4.0 \text{ ft.-kips}}$

Interaction Equation

$P_u/\phi_c P_n = 383/497 = 0.77 \geq 0.2 \Rightarrow$ Equation (8-1)

$$\frac{P_u}{\phi P_n} + \frac{8}{9}\left(\frac{M_{ux}}{\phi_b M_{nx}} + \frac{M_{uy}}{\phi_b M_{ny}}\right) = 0.77 + \frac{8}{9}\left(\frac{16}{177} + \frac{4.0}{84}\right) = 0.89 < 1.0 \text{ OK}$$

$\Rightarrow$ **W8 $\times$ 48 is adequate.**

8.9 STUDENT PRACTICE PROBLEM AND COLUMN DESIGN TEMPLATES

As an exercise, the reader should now design column C2, following the same procedure used to design column C1. To aid the reader, column design templates for W-shaped and HSS columns are presented on the following pages.

W-SHAPE COLUMN DESIGN TEMPLATE

DESIGN OF COLUMN (FLOOR to FLOOR)

Load Case 1 (with $M = 0$)	Load Case 2 or Load Case 1 with Moments

Load Case 1 (with $M = 0$)

$P_u =$ kips: $KL \approx$ =

From **Column Load Tables** (*AISCM*, Table 4-1)

Try W   ×   **(A =**   **in.^{2})**
$\phi_c P_n =$ kips $> P_u$ OK

Load Case 2 or Load Case 1 with Moments

Check W   ×   **column for**
$P_u =$ kips
$M_{x\text{-}x} = M_{ntx} =$ ft.-kips
$M_{y\text{-}y} = M_{nty} =$ ft.-kips
Column unbraced length, $L_b =$; $r_x =$, $r_y =$

LRFD Beam Design Tables (*AISCM*, Table 3-2; $F_y = 50$ ksi)
$\phi_b M_p =$ ft.-kips; $L_p =$; $BF =$ kips
$\phi_b M_{nx} = \phi_b M_p - BF(L_b - L_p) = [\ \ - (\ \ )(\ \ - \ \ )] =$ ft.-kips
From *AISCM*: $S_x =$ in.^{3}; $Z_y =$ in.^{3}; $S_y =$ in.^{3}
$\phi_b M_{nx} < 1.5 \phi_b S_x F_y / 12 =$ ft.-kips OK $\phi_b M_{nx} =$ **ft.-kips**

$\phi_b M_{ny} = \phi_b Z_y F_y / 12 < 1.5 \phi_b S_y F_y / 12$
$\phi_b M_{ny} = (0.9)(\ \ )(\ \ )/12 < 1.5(0.9)(\ \ )(\ \ )/12$
$\phi_b M_{ny} =$ ft.-kips   **ft.-kips**

$\phi_c P_n =$ kips; $\phi_b M_{nx} =$ ft.-kips; $\phi_b M_{ny} =$ ft.-kips

$M_{ux} = B_{1x} M_{ntx} + B_{2x} M_{ltx}$; $M_{ltx} = 0$ (for braced frames)
$M_{uy} = B_{1y} M_{nty} + B_{2y} M_{lty}$; $M_{lty} = 0$ (for braced frames)

P-δ Effect for x–x Axis

$C_{mx} = 0.6 - 0.4(M_{1x}/M_{2x})$; but conservatively use $C_{mx} = 0.85$

$$P_{e1x} = \frac{\pi^2 EA}{(KL/r_x)^2} = \text{ kips}$$

$$B_{1x} = \frac{C_{mx}}{(1 - P_u/P_{e1x})} = \frac{0.85}{[1 - (\ \ /\ \ )]} = \quad \geq 1.0 \therefore \boldsymbol{B_{1x}} =$$

$\therefore M_{ux} = B_{1x} M_{ntx} =$ ( ) =   ft.-kips

P-δ Effect for y–y Axis

$C_{my} = 0.6 - 0.4(M_{1y}/M_{2y})$; but conservatively use $C_{my} = 0.85$

$$P_{e1y} = \frac{\pi^2 EA}{(KL/r_y)^2} = \text{ kips}$$

$$B_{1y} = \frac{C_{my}}{(1 - P_u/P_{e1y})} = \frac{0.85}{[1 - (\ \ /\ \ )]} = \quad \geq 1.0 \therefore \boldsymbol{B_{1y}} =$$

$\therefore M_{uy} = B_{1y} M_{nty} =$ ( ) =   ft.-kips

Interaction Equation (check that the interaction equation ≤ 1.0)

$P_u / \phi_c P_n = \ /\ = \ $; If $\geq 0.2 \Rightarrow$ Use Equation (8-1), otherwise use Equation (8-2).

$$(8\text{-}1): \frac{P_u}{\phi P_n} + \frac{8}{9}\left(\frac{M_{ux}}{\phi_b M_{nx}} + \frac{M_{uy}}{\phi_b M_{ny}}\right) = \quad + \frac{8}{9}(- + -) =$$

$$(8\text{-}2): \frac{P_u}{2\phi P_n} + \left(\frac{M_{ux}}{\phi_b M_{nx}} + \frac{M_{uy}}{\phi_b M_{ny}}\right) = \quad - + (- + -) =$$

HSS COLUMN DESIGN TEMPLATE

DESIGN OF COLUMN (**FLOOR to** **FLOOR)**

Load Case 1 (with $M = 0$)

$P_u =$ kips: $KL \approx$ =

From **Column Load Tables** (*AISCM*, Table 4-3)
Try HSS × (A = in.²)
$\phi_c P_n =$ kips $> P_u$ kips OK

Load Case 2 or Load Case 1 with Moments

Check HSS × **column for**
$P_u =$ kips
$M_{x-x} = M_{ntx} =$ ft.-kips
$M_{y-y} = M_{nty} =$ ft.-kips
Column unbraced length, $L_b =$; $r_x =$, $r_y =$

From Table 1-2, F_y for HSS =
From *AISCM*: $S_x =$ in.³; $Z_y =$ in.³; $S_y =$ in.³
$\phi_b M_{nx} = \phi_b Z_x F_y / 12 < 1.5 \phi_b S_x F_y / 12 =$ ft.-kips
$\phi_b M_{nx} = (0.9)(\ )(\ )/12 < 1.5(0.9)(\ )(\ )/12$
$\phi_b M_{nx} =$ ft.-kips $<$ ft.-kips

$\phi_b M_{ny} = \phi_b Z_y F_y / 12 < 1.5 \phi_b S_y F_y / 12$
$\phi_b M_{ny} = (0.9)(\ )(\ )/12 < 1.5(0.9)(\ )(\ )/12$
$\phi_b M_{ny} =$ ft.-kips $<$ ft.-kips

$\boxed{\phi_c P_n = \text{kips}; \ \phi_b M_{nx} = \text{ft.-kips}; \ \phi_b M_{ny} = \text{ft.-kips}}$

$M_{ux} = B_{1x} M_{ntx} + B_{2x} M_{ltx}; \ M_{ltx} = 0$ (for braced frames)
$M_{uy} = B_{1y} M_{nty} + B_{2y} M_{lty}; \ M_{lty} = 0$ (for braced frames)

P-δ Effect for *x-x* Axis

$C_{mx} = 0.6 - 0.4(M_{1x}/M_{2x})$; but conservatively use $C_{mx} = 0.85$

$P_{e1x} = \dfrac{\pi^2 EA}{(KL/r_x)^2} =$ kips

$B_{1x} = \dfrac{C_{mx}}{(1 - P_u/P_{e1x})} = \dfrac{0.85}{[1 - (\ /\ )]} = \ \geq 1.0 \therefore B_{1x} =$

$\boxed{\therefore M_{ux} = B_{1x} M_{ntx} = \ (\ ) = \text{ft.-kips}}$

P-δ Effect for *y-y* Axis

$C_{my} = 0.6 - 0.4(M_{1y}/M_{2y})$; but conservatively use $C_{my} = 0.85$

$P_{e1y} = \dfrac{\pi^2 EA}{(KL/r_y)^2} =$ kips

$B_{1y} = \dfrac{C_{my}}{(1 - P_u/P_{e1y})} = \dfrac{0.85}{[1 - (\ /\ )]} = \ \geq 1.0 \therefore B_{1y} =$

$\boxed{\therefore M_{uy} = B_{1y} M_{nty} = \ (\ ) = \text{ft.-kips}}$

Interaction Equation (check that the interaction equation ≤ 1.0)

$P_u/\phi_c P_n = \ /\ = \ $; If $\geq 0.2 \Rightarrow$ Use Equation (8-1), otherwise use Equation (8-2).

$$(8\text{-}1): \frac{P_u}{\phi P_n} + \frac{8}{9}\left(\frac{M_{ux}}{\phi_b M_{nx}} + \frac{M_{uy}}{\phi_b M_{ny}}\right) = \ + \frac{8}{9}(\text{—} + \text{—}) =$$

$$(8\text{-}2): \frac{P_u}{2\phi P_n} + \left(\frac{M_{ux}}{\phi_b M_{nx}} + \frac{M_{uy}}{\phi_b M_{ny}}\right) = \text{—} + (\text{—} + \text{—}) =$$

8.10 ANALYSIS OF UNBRACED FRAMES USING THE AMPLIFIED FIRST-ORDER METHOD

In lieu of a general-purpose finite element analysis (FEA) software program that accounts for both P-Δ (frame slenderness) and P-δ (member slenderness) effects and yields directly the total factored second-order moments, M_u, acting on the columns of a moment frame, approximate methods are allowed in the AISC specification. The amplified first-order analysis is one such method that can be used. The procedure used to determine the no-translation moments, M_{nt}, and the lateral translation moments, M_{lt}, is based on the principle of superposition as follows [7]:

1. Given the original frame shown in Figure 8-30a, perform a first-order analysis of the frame for all applicable load combinations (see Chapter 2) that includes gravity and lateral loads, plus horizontal restraints (vertical rollers) added at each floor level, as shown in Figure 8-30b, to prevent any sidesway of the frame. Therefore, the moments obtained are the no-translation moments, M_{nt}; the resulting horizontal reactions, H_2, H_3, ... H_n, and H_{roof} at the vertical rollers at each floor level are also determined.

2. The horizontal restraints introduced in step 1 are removed, and for each load combination from step 1, lateral loads are applied at each level to the frame that are equal and opposite to the horizontal reactions obtained in step 1, as shown in Figure 8-30c. Note that per the AISC specification, notional loads will need to be applied at each floor level if the horizontal reactions obtained from step 1 are smaller than the notional loads (see Section 8-6). The frame is reanalyzed for the horizontal reactions and the moments obtained in this step are the lateral translation moments, M_{lt}.

The rationale behind the procedure above will now be discussed. The amplified first-order analysis uses the principle of superposition to decompose the total moment that would have been obtained if the original frame (Figure 8-30a) with all of the gravity and lateral loads was analyzed. Because we have different magnifiers for the no-translation moment, M_{nt}, and the translation moment, M_{lt}, the issue is how to separate the total moment into the two types of moments so that M_{nt} can be amplified by B_1 and M_{lt} can be amplified by B_2. The analysis in step 1, with the horizontal restraints added, represents the braced-frame portion of the original frame, so that the moments obtained from step 1—the M_{nt} moments—can then be multiplied by B_1.

For the analysis in step 2, lateral loads that are equal and opposite to the restraint reactions obtained from step 1 are applied to the frame. Using the principle of superposition, it can be seen that we have not altered the original frame (Figure 8-30a) at all; instead, we have only decomposed it into two frames (Figure 8-30b and Figure 8-30c), which when added together is equivalent to the original frame. The moments from step 2 are the M_{lt} moments that can then be multiplied by B_2. Therefore, the final second-order moments, $M_u = B_1 M_{nt}$ (from the first analysis) + $B_2 M_{lt}$ (from the second analysis). The total second-order factored moment, M_u, from the amplified first-order analysis usually compares very favorably with results from a full-blown, second-order finite element analysis that directly calculates M_u and accounts for all types of geometric nonlinearities [7].

Design of Columns in Moment, or Unbraced, Frames

The procedure for designing columns in moment frames is as follows:

1. Determine the factored axial loads, P_u, on each column in the frame from the gravity loads only.

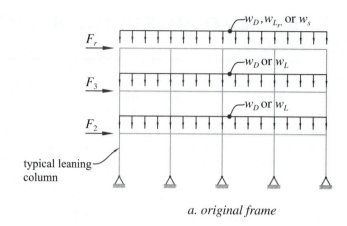

a. original frame

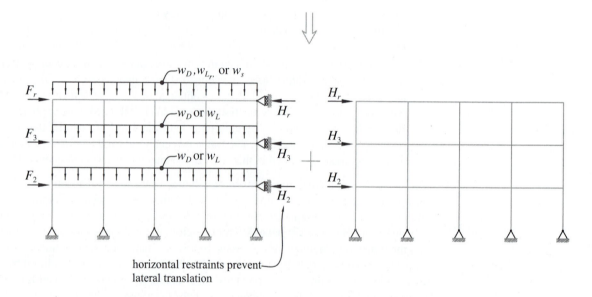

b. No translation frame: M_{nt} moments (to be amplified by β_1)

c. Lateral translation frame: M_{lt} moments (to be amplified by β_2)

Figure 8-30 Determination of M_{nt} and M_{lt} moments in moment frames.

2. Obtain the preliminary column size based on the axial load from step 1 using the column design template presented earlier. It is assumed that the preliminary sizes of the beams and girders have already been obtained for gravity loads only and assuming simple support conditions.

3. Using the preliminary column and girder sizes, perform the amplified first-order analysis as described in the previous section for all applicable load combinations. This will yield the M_{nt} and M_{lt} moments for each column, as well as the factored axial loads, P_{nt} and P_{lt}. This analysis is carried out in both orthogonal directions of the building if moment frames are used in both directions. Where braced frames are used in one direction, then the analysis has to be carried out only in a direction parallel to the moment frames.

4. Compute the moment magnification factors, B_1 and B_2, and the factored second-order moments:

$$M_{ux-x} = B_{1x-x}M_{ntx-x} + B_{2x-x}M_{ltx-x}, \text{ and}$$
$$M_{uy-y} = B_{1y-y}M_{nty-y} + B_{2y-y}M_{lty-y}$$

5. Using the column design template, design the column for the factored axial loads and the factored moments from step 4.

6. In designing columns in moment frames, distribute any direct axial loads on the *leaning* columns (i.e., columns that do not participate in resisting the lateral loads) to the moment frame columns according to the plan tributary area of the column relative to the leaning column. Thus, the moment frame columns, in providing lateral bracing to the leaning columns, will be subjected to additional axial loads from the leaning columns.

EXAMPLE 8-2

Calculation of the Sway Moment Magnifier, B_2

Calculate the sway moment magnifier for the columns in the first story (i.e., the ground floor columns) of the typical moment frame in Figure 8-31 with the service gravity and lateral wind loads shown.

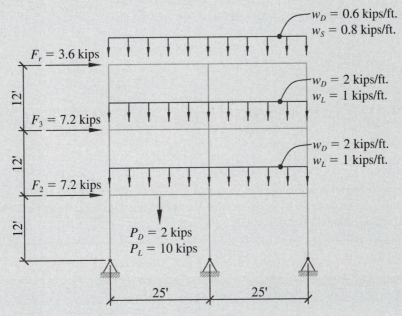

Figure 8-31 Moment frame for Example 8-2.

SOLUTION

The applicable load combinations from Chapter 2 are:

$1.2D + 1.6W + L + 0.5(L_r \text{ or } S \text{ or } R)$, and

$0.9D + 1.6W$. *(continued)*

It should be obvious that the $0.9D + 1.6W$ load combination will not be critical for calculating the sway moment magnifier, so only the first load combination needs to be checked.

Calculate the cumulative maximum factored total axial load, ΣP_u, in the ground-floor columns:

$$1.2D = 1.2[(0.6 \text{ kip/ft.})(50 \text{ ft.}) + (2 \text{ kip/ft.})(50 \text{ ft.}) + (2 \text{ kip/ft.})(50 \text{ ft.}) + (2 \text{ kips})] = 279 \text{ kips}$$
$$1.0L = 1.0[(1 \text{ kip/ft.})(50 \text{ ft.}) + (1 \text{ kip/ft.})(50 \text{ ft.}) + (10 \text{ kips})] = 110 \text{ kips}$$
$$0.5S = 0.5(0.8 \text{ kip/ft.})(50 \text{ ft.}) = 20 \text{ kips}$$

$$\Sigma P_u = 279 + 110 + 20 = 409 \text{ kips.}$$

The factored lateral shear at the ground-floor columns is

$$\Sigma H = 1.6(3.6 \text{ kips} + 7.2 \text{ kips} + 7.2 \text{ kips}) = 28.8 \text{ kips.}$$

A drift index for factored wind loads of $\dfrac{1}{312}\left(\text{i.e., } \dfrac{1}{500/1.6}\right)$ is assumed (since a drift index of $\dfrac{1}{500}$ to $\dfrac{1}{400}$ is commonly used in practice for building frames under service wind loads). Therefore,

$$\frac{\Delta_{oh}}{L} = \frac{1}{312}.$$

The sway moment magnification factor is calculated as

$$B_2 = \frac{1}{1 - \dfrac{\Delta_{oh}}{L}\left(\dfrac{\Sigma P_u}{\Sigma H}\right)} \geq 1.0$$

$$= \frac{1}{1 - \dfrac{1}{312}\left(\dfrac{409}{28.8}\right)} = 1.05 < 1.5. \text{ OK}$$

Therefore, $B_2 = \mathbf{1.05}$.

It should be reemphasized that where B_2 is greater than 1.5, this would indicate a stability-sensitive structure or frame, and the more accurate direct analysis method (see AISC specification, Appendix 7) would have to be used to determine the magnified moments in the frame.

EXAMPLE 8-3

Design of Columns in Moment Frames

The typical floor and roof plan for a two-story office building laterally braced with moment frames in both orthogonal directions is shown in Figure 8-32a. The floor-to-floor height is 12 ft. and the uniform roof and floor loads and the lateral wind load have been determined as follows:

Roof dead load = 30 psf	Snow load = 35 psf
Floor dead load = 100 psf	Floor live load = 50 psf

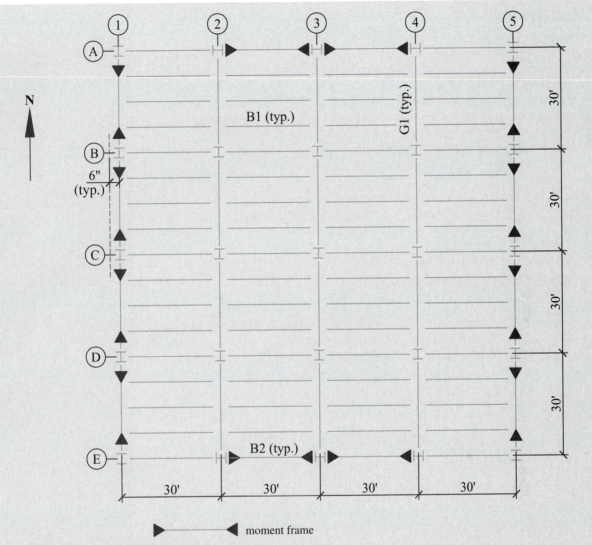

moment frame

Figure 8-32a Typical Floor and Roof Plan for Example 8-3.

Assume a uniform lateral wind load of 20 psf. The building is located in a region where seismic loads do not control.

1. Using a commercially available structural analysis software, perform an amplified first-order analysis for the East–West moment frame along line A to determine the M_{nt} and M_{lt} moments in the ground-floor columns. For this problem, consider only the load combination $1.2D + 1.6W + L + 0.5(L_r$ or S or $R)$.

2. Design column A-3 at the ground-floor level for the combined effects of axial load and moments. Consider the effect of leaning columns. To simplify the design, use K_x and K_y values from Figure 5-3 or the alignment charts.

(continued)

SOLUTION

Calculate Beam and Girder Reactions: (See Figures 8-32b and 8-32c)

Roof Level
Spandrel roof beam, RB2:

Tributary width, $TW = \left(\dfrac{7.5 \text{ ft.}}{2} + 0.5\text{-ft. edge distance}\right) = 4.25 \text{ ft.}$

$w_D = (30 \text{ psf})(4.25 \text{ ft.}) = 127.5 \text{ lb./ft.}$

$w_S = (35 \text{ psf})(4.25 \text{ ft.}) = 148.8 \text{ lb./ft.}$

$w_u = 1.2D + 0.5S = 1.2(127.5) + 0.5(148.8) = 228 \text{ lb./ft.} = 0.23 \text{ kips/ft.}$

This uniform load acts on the East–West moment frame spandrel roof beams.
Interior roof beam, RB1:

Tributary width, $TW = 7.5 \text{ ft.}$

$w_D = (30 \text{ psf})(7.5 \text{ ft.}) = 225 \text{ lb./ft.}$

$w_S = (35 \text{ psf})(7.5 \text{ ft.}) = 263 \text{ lb./ft.}$

$w_u = 1.2D + 0.5S = 1.2(225) + 0.5(263) = 402 \text{ lb./ft.} = 0.4 \text{ kips/ft.}$

Roof beam reaction, $R_u = (0.4 \text{ kips/ft.})\left(\dfrac{30 \text{ ft.}}{2}\right) = 6 \text{ kips}$

Interior roof girder, RG1:

Girder reaction, $R_u = 18 \text{ kips}$

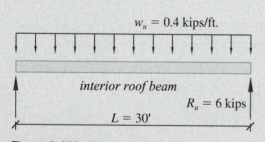

$w_u = 0.4 \text{ kips/ft.}$

interior roof beam

$R_u = 6 \text{ kips}$

$L = 30'$

$P_u = 12 \text{ kips}$ $P_u = 12 \text{ kips}$ $P_u = 12 \text{ kips}$

interior roof girder

$R_u = 18 \text{ kips}$

$L = 30'$

Figure 8-32b Beam and girder reactions.

Second-Floor Level
Spandrel floor beam, B2:

Tributary width, $TW = \left(\dfrac{7.5 \text{ ft.}}{2} + 0.5\text{-ft. edge distance}\right) = 4.25 \text{ ft.}$

$w_D = (100 \text{ psf})(4.25 \text{ ft.}) = 425 \text{ lb./ft.}$

$w_S = (50 \text{ psf})(4.25 \text{ ft.}) = 213 \text{ lb./ft.}$

$w_u = 1.2D + 1.0L = 1.2(425) + 1.0(213) = 723 \text{ lb./ft.} = 0.72 \text{ kips/ft.}$

This uniform load acts on the East–West moment frame spandrel floor beams.

Interior floor beam, B1:

Tributary width, TW $= 7.5$ ft.

$w_D = (100 \text{ psf})(7.5 \text{ ft.}) = 750 \text{ lb./ft.}$

$w_L = (50 \text{ psf})(7.5 \text{ ft.}) = 375 \text{ lb./ft.}$

$w_u = 1.2D + 1.0L = 1.2(750) + 1.0(375) = 1275 \text{ lb./ft.} = 1.28 \text{ kips/ft.}$

Roof beam reaction, $R_u = (1.28 \text{ kips/ft.})\left(\dfrac{30 \text{ ft.}}{2}\right) = 19.2 \text{ kips}$

Interior floor girder, G1:

Girder reaction, $R_u = 57.6$ kips

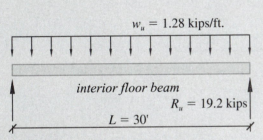
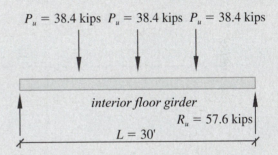

Figure 8-32c Beam and girder reactions.

Factored Lateral Wind Loads on Moment Frame along Grid Line A

$$F_{roof} = 1.6(20 \text{ psf})\left(\frac{12 \text{ ft.}}{2}\right)\left(\frac{120 \text{ ft.}}{2}\right) = 11.5 \text{ kips}$$

$$F_2 = 1.6\left[(20 \text{ psf})\left(\frac{12 \text{ ft.}}{2} + \frac{12 \text{ ft.}}{2}\right)\left(\frac{120 \text{ ft.}}{2}\right)\right] = 23 \text{ kips}$$

The amplified first-order analysis involves using the principle of superposition and replacing the original frame in Figure 8-32d with two constituent frames (Figure 8-32e and Figure 8-32f) such that when the load effects of these constituent frames are summed up, we obtain the actual load effects on the original frame. In Figure 8-32e, horizontal restraints H_{roof} and H_2 are provided at the floor levels by adding vertical rollers at these locations. The first-order analysis of this frame, using structural analysis software, yields the no-translation moments, M_{ntx-x} (frame columns are bending about their strong axis), as shown in Figure 8-32e. Next, the horizontal reactions H_{roof} and H_2 from the first analysis are applied as lateral loads to the frame in a direction opposite the direction of the reactions from the first analysis, and without any other gravity or lateral loads, and with no horizontal restraints. This second frame is then analyzed and this second analysis yields the translation moments, M_{ltx-x}. (Note that the frame columns are bending about their strong axis.) Figures 8-32e and 8-32f show the results of the computer-aided structural analysis.

(continued)

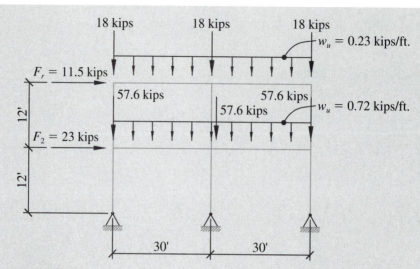

Figure 8-32d Gravity and lateral loads on East–West moment frame along gridline A.

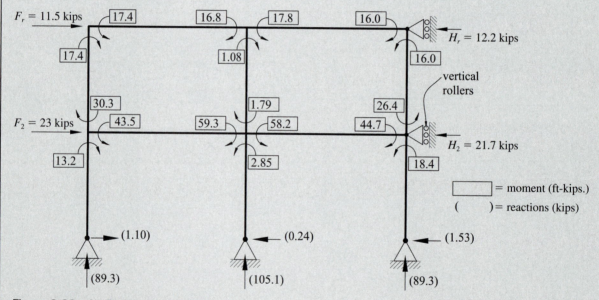

Figure 8-32e No-Translation moments, M_{ntx-x}.

The cumulative maximum total factored axial load, $\sum P_u$ (for load combination $1.2D + 1.6W + L + 0.5(L_r$ or S or $R)$), in the ground-floor columns is

$1.2D = 1.2(30 \text{ psf} + 100 \text{ psf})(120 \text{ ft.})(120 \text{ ft.}) = 2247$ kips

$1.0L = 1.0(50 \text{ psf})(120 \text{ ft.})(120 \text{ ft.}) = 720$ kips

$0.5S = 0.5(35 \text{ psf})(120 \text{ ft.})(120 \text{ ft.}) = 252$ kips

$\sum P_u = 2247 + 720 + 252 = 3219$ kips.

The factored lateral shear at the ground-floor columns is

$\sum H = 11.5 + 23 = 34.5$ kips.

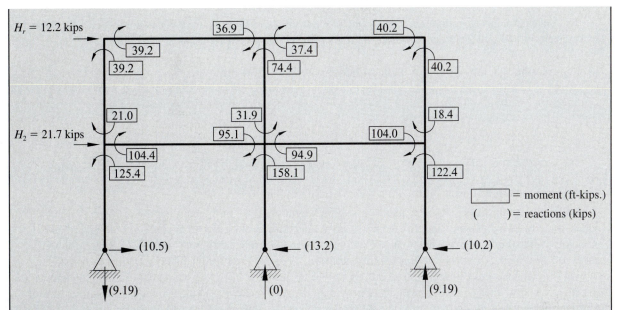

Figure 8-32f Lateral translation moments, M_{ltx-x}.

A drift index for factored wind loads of $\dfrac{1}{312}$ $\left(\text{i.e., } \dfrac{1}{500/1.6}\right)$ is assumed (because a drift index of $\dfrac{1}{500}$ to $\dfrac{1}{400}$ is commonly used in practice for frames under service wind loads). Therefore,

$$\frac{\Delta_{oh}}{L} = \frac{1}{312}.$$

The sway moment magnification factor is calculated as

$$B_2 = \cfrac{1}{1 - \cfrac{\Delta_{oh}}{L}\cfrac{\Sigma P_u}{\Sigma H}} \geq 1.0$$

$$= \cfrac{1}{1 - \cfrac{1}{312}\left(\cfrac{3219}{34.5}\right)} = 1.43 < 1.5.\ \text{OK}$$

Therefore, $B_{2x-x} = \mathbf{1.43} < 1.5$ OK

This value will be used later in the column design template for the design of Column A-3.

Weak Axis (y–y) Bending Moment in Columns A-2, A-3, and A-4

In the North–South direction, columns A-2, A-3, and A-4 will behave as **leaning columns**. They will be laterally braced at the floor levels for bending in the North–South direction (i.e., about their weak axis) by the North–South moment frames. Since these leaning columns do not participate in resisting the lateral load in the North–South direction, the sway magnifier for columns A-2, A-3, and A-4 for bending about their weak axis, $B_{2y-y} = 1.0$, and the sway moments for bending about their weak axis, $M_{lty-y} = 0$.

(continued)

The only moments in columns A-2, A-3, and A-4 acting about their weak (y–y) axis will occur due to the eccentricity of the girder or beam reactions at the floor level under consideration.

For **column A-3**, the reaction of second-floor girder G1 = 57.6 kips.

Eccentricity of girder G1 reaction, e_{y-y} = 2½ in. + ½ (column web thickness)

                                                   = 2½ in. + ½ (0.5 in.) assumed

                                                   = 2.75 in. ≈ **0.25 ft.**

M_{nty-y} due to connection eccentricity, e_{y-y} at the second-floor level = 57.6(0.25 ft.) = **14.4 ft.-kips**.

Effects of Leaning Columns on the East–West Moment Frames

In the East–West direction, the only frames providing lateral stability for the building are the frames along grid lines A and E. Since all other East–West girders are assumed to be connected to the columns along grid lines B, C, and D with simple shear connections, the columns along lines B, C, and D, as well as the corner columns at A-1, A-5, E-1, and E-5, will behave as **leaning columns**. These leaning columns are restrained or braced later-ally in the East–West direction by the six columns (A-2, A-3, A-4, E-2, E-3, and E-4) that make up the moment frames along grid lines A and E. The effect of providing this lateral restraint to the leaning columns results in ad-ditional axial load on the restraining moment frame columns [3]. The total axial load on each restraining column is the sum of the actual direct factored axial load on the restraining column, $P_{u,\,direct}$, plus the indirect axial load, $P_{u,\,indirect}$, from the leaning columns, where $P_{u,\,indirect}$ is the total axial load on all leaning columns equally divided among the six restraining columns in the East–West moment frames. Therefore, for each of the moment frame columns, the equivalent total factored axial load is

$$P_u \text{ in each restraining column} = P_{u,\,direct} + P_{u,\,indirect} = P_{u,\,direct} + \Sigma\left(\frac{P_{leaning\ columns}}{6 \text{ columns}}\right).$$

The sidesway uninhibited nomographs or Figure 5-3 will be used to determine the effective length of the **restraining columns** in the moment frame, and this column will then be designed for the total factored load, P_u, which includes the effect of the leaning columns plus the factored second-order moments, M_{ux-x} and M_{uy-y}.

It should be noted that **leaning columns** have to be designed for the factored axial load that they directly support, plus the M_{ntx-x} and M_{nty-y} moments resulting from the beam and girder connection eccentricities. For the East–West direction, the leaning columns are those columns that are not part of the East–West moment frames along grid lines A and E. Therefore, only columns A-2, A-3, A-4, E-2, E-3, and E-4 are **not** leaning columns. All other columns in Figure 8-32a are leaning columns and will be restrained by the East–West moment frame columns. The effective length factor, K, for the leaning columns is 1.0, and the sway moment magnifier, B_2, and the lateral translation moment, M_{lt}, for these leaning columns for bending in the East–West direction will be 1.0 and zero, respectively. Thus, for bending in the East–West direction, the only moments for which the leaning columns have to be designed are the moments due to the beam and girder eccentricities. In effect, the leaning columns are designed similar to columns in braced frames, as was done in Example 8-1. Table 8-6 summarizes the different types of columns in this example and the source of the bending moments in the columns, and Table 8-7 provides a summary of the factored axial loads and moments just below the second-floor level for the East–West moment frame columns along grid line A (A-2, A-3, and A-4).

Total Factored Axial Load on All Leaning Columns (for East–West Lateral Load Only)

Interior leaning columns:

$$P_{u\,roof} = [(1.2)(30 \text{ psf}) + (0.5)(35 \text{ psf})](30 \text{ ft.})(30 \text{ ft.})(9 \text{ columns}) = 433 \text{ kips}$$
$$P_{u\,floor} = [(1.2)(100 \text{ psf}) + (1.0)(50 \text{ psf})](30 \text{ ft.})(30 \text{ ft.})(9 \text{ columns}) = 1377 \text{ kips}$$

Table 8-6 Types of columns (leaning or moment frame) and the source of bending moments in Figure 8-32a

| Moment Frame Column | North–South Bending | | East–West Bending | |
	Type of Column	Source of Bending Moments	Type of Column	Source of Bending Moments
A-2, A-3, A-4	Leaning	Interior girder eccentricity	Moment frame	Moment frame
E-2, E-3, E-4	Leaning	Interior girder eccentricity	Moment frame	Moment frame
B-2, B-3, B-4, C-2, C-3, C-4, D-2, D-3, D-4	Leaning	Interior girder eccentricity	Leaning	Beam eccentricity
A-1, B-1, C-1, D-1, E-1	Moment frame	Moment frame	Leaning	Beam eccentricity
A-5, B-5, C-5, D-5, E-5	Moment frame	Moment frame	Leaning	Beam eccentricity

Corner leaning columns:

$$P_{u\,roof} = [(1.2)(30\text{ psf}) + (0.5)(35\text{ psf})](15.5\text{ ft.})(15.5\text{ ft.})(4\text{ columns}) = 51\text{ kips}$$
$$P_{u\,floor} = [(1.2)(100\text{ psf}) + (1.0)(50\text{ psf})](15.5\text{ ft.})(15.5\text{ ft.})(4\text{ columns}) = 163\text{ kips}$$

Exterior side leaning columns:

$$P_{u\,roof} = [(1.2)(30\text{ psf}) + (0.5)(35\text{ psf})](15.5\text{ ft.})(30\text{ ft.})(6\text{ columns}) = 149\text{ kips}$$
$$P_{u\,floor} = [(1.2)(100\text{ psf}) + (1.0)(50\text{ psf})](15.5\text{ ft.})(30\text{ ft.})(6\text{ columns}) = 474\text{ kips}$$

$$\Sigma P_{leaning\,columns} = 433 + 1377 + 51 + 163 + 149 + 474 = 2647\text{ kips}$$

$$P_u\text{ in each restraining column} = P_{u,\,direct} + P_{u,\,indirect} = P_{u,\,direct} + \Sigma\left(\frac{P_{leaning\,columns}}{6\text{ columns}}\right)$$

$$= P_{u,\,direct} + \left(\frac{2647\text{ kips}}{6\text{ columns}}\right)$$

$$= P_{u,\,direct} + \mathbf{441\text{ kips}}$$

The design of column A-3 at the ground-floor level, considering the effect of the leaning columns, is carried out next using a modified form of the column design template introduced earlier in this chapter. This yields a W12 × 96 column. To illustrate the impact of the leaning columns in unbraced frames, a design is also carried out, neglecting the effect of the leaning columns. This yields a W10 × 68 column. The importance of including the effect of the leaning column loads on the stability of unbraced frames is obvious.

Table 8-7 Summary of factored moments and axial loads in ground-floor columns in East–West moment frame

Moment Frame Column	M_{nt} Moments, ft.-k	M_{lt} Moments, ft.-k	$P_{u\,nt}$, kips	$P_{u\,lt}$, kips	$P_{u,\,indirect}$ (due to leaning columns), kips
A-2	−13.2	125.4	89.3	−9.19	441
A-3	2.85	158.1	105.1	0	441
A-4	18.4	122.4	89.3	−9.19	441

(continued)

Design of Column A-3 (Second Floor to Ground Floor)—EFFECT OF LEANING COLUMNS *INCLUDED*

Load Case 1 (with $M \approx 0$)	Load Case 2 or Load Case 1 with Moments

Load Case 1 (with $M \approx 0$)

$P_u = 105.1 + 441 = 546$ kips:
$KL \approx 1 \times 12$ ft. $= 12$ ft.

From Column Load Tables
(*AISCM*, Table 4-1)
Try W12 × 96 ($A = 28.2$ in.^{2})
$\phi_c P_n = 1080$ kips $> P_u$ OK

Unbraced frame column
$K_x = 2.1$ (Fig. 5-3)
$K_y = 1.0$ (leaning column in N-S direction)

Load Case 2 or Load Case 1 with Moments

Check W12 × 96 column for the following forces from Table 8-7
$P_{unt} = 105.1$ kips; $P_{ult} = 0$ kips; $P_{indirect} = 441$ kips (leaning column)
$M_{ntx-x} = 2.85$ ft.-kips; $M_{ltx-x} = 158.1$ ft.-kips (computer results)
$M_{nty-y} = 14.4$ ft.-kips; $M_{lty-y} = 0$ ft.-kips (leaning column in N–S)
Column unbraced length, $L_b = 12$ ft.; $r_x = 5.44$ in., $r_y = 3.09$ in.

LRFD Beam Design Tables (*AISCM*, Table 3-2; $F_y = 50$ ksi)
$\phi_b M_p = 551$ ft.-kips; $L_p = 10.9$ ft.; $BF = 5.81$ kips; $L_r = 46.6$ ft. $> L_b$
$\phi_b M_{nx} = \phi_b M_p - BF(L_b - L_p) = [551 - (5.81)(12 - 10.9)] = 545$ ft.-kips $\leq \phi_b M_p$
From *AISCM*: $S_x = 131$ in.^{3}; $Z_y = 67.5$ in.^{3}; $S_y = 44.4$ in.^{3}
$\phi_b M_{nx} < 1.5 \phi_b S_x F_y / 12 = 737$ ft.-kips $\therefore$ $\phi_b M_{nx} = $ **545 ft.-kips**

$\phi_b M_{ny} = \phi_b Z_y F_y < 1.5 \phi_b S_y F_y$
$\phi_b M_{ny} = (0.9)(67.5)(50)/12 < 1.5(0.9)(44.4)(50)/12$
$\phi_b M_{ny} = 253$ ft.-kips $>$ **250 ft.-kips** Governs

$\boxed{\phi_c P_n = 1080 \text{ kips}; \phi_b M_{nx} = 545 \text{ ft.-kips}; \phi_b M_{ny} = 250 \text{ ft.-kips}}$

Sway Moment Magnification Factors:
$B_{2x-x} = 1.43$; $B_{2y-y} = 1.0$ (see previous calculations/discussions)
$P_{u\ total} = P_{direct} + P_{indirect} = 105.1 + 441^* = 546$ kips (see Table 8-7)
*Learning columns included

P-δ Effect for x–x Axis
$C_{mx} = 0.6 - 0.4(M_{1x}/M_{2x})$; but conservatively use $C_{mx} = 0.85$

$P_{e1x} = \dfrac{\pi^2 EA}{(K_x L / r_x)^2} = 2612$ kips ($K_x = 2.1$ for column in E-W moment frame)

$B_{1x} = \dfrac{C_{mx}}{(1 - P_u/P_{e/x})} = \dfrac{0.85}{[1 - (546/2612)]} = 1.1 > 1.0$ $\therefore$ $B_{1x} = 1.1$

$\boxed{\therefore B_{1x} M_{ntx} = 1.1(2.85) = 3.1 \text{ ft.-kips}}$

P-δ Effect for y–y Axis
$C_{my} = 0.6 - 0.4(M_{1y}/M_{2y})$; but conservatively use $C_{my} = 0.85$

$P_{e1y} = \dfrac{\pi^2 EA}{(K_y L / r_y)^2} = 3717$ kips ($K_y = 1.0$ for leaning column in N-S direction)

$B_{1y} = \dfrac{C_{my}}{(1 - P_u/P_{e/y})} = \dfrac{0.85}{[1 - (546/3717)]} = 1.0 > 1.0$ $\therefore$ $B_{1y} = 1.0$

$\boxed{\therefore B_{1y} M_{nty} = 1.0(14) = 14 \text{ ft.-kips}}$

$M_{ux} = B_{1x} M_{ntx} + B_{2x} M_{ltx} = 3.1 + (1.43)(158.1) = 230$ ft.-kips
$M_{uy} = B_{1y} M_{nty} + B_{2y} M_{lty} = 14 + (1.0)(0) = 14$ ft.-kips
See Table 8-7 for lateral translation moments, M_{lt}.

Interaction Equation

$P_u / \phi_c P_n = 546/1080 = 0.5 \geq 0.2 \Rightarrow$ Equation (8–1)

$$\dfrac{P_u}{\phi P_n} + \dfrac{8}{9}\left(\dfrac{M_{ux}}{\phi_b M_{nx}} + \dfrac{M_{uy}}{\phi_b M_{ny}}\right) = 0.50 + \dfrac{8}{9}\left(\dfrac{230}{545} + \dfrac{14}{250}\right) = 0.93 < 1.0 \text{ OK}$$

$\therefore$ **W12 × 96 is adequate for column A-3 if the effect of leaning columns is included.**

Design of Column A-3 (Second Floor to Ground Floor)—EFFECT OF LEANING COLUMNS *EXCLUDED*

Load Case 1 (with $M \times 0$)	Load Case 2 or Load Case 1 with Moments
$P_u = 105.1$ kips: $KL \approx 1 \times 12$ ft. $= 12$ ft. From **Column Load Tables** (*AISCM*, Table 4-1) **Try W10 × 68 ($A = 20$ in.²)** $\phi_c P_n = 717$ kips $> P_u$ kips OK **Unbraced frame column** $K_x = 2.1$ (Fig. 5-3) $K_y = 1.0$ (leaning column in N-S direction)	**Check W10 × 68 column for the following forces from Table 8-7.** $P_{unt} = 105.1$ kips; $P_{ult} = 0$ kips; $P_{indirect} = 0$ kips* ***Effect of leaning columns *neglected*** $M_{ntx} = 2.85$ ft.-kips; $M_{ltx-x} = 158.1$ ft.-kips (computer results) $M_{ntyy} = 14.4$ ft.-kips; $M_{lty-y} = 0$ ft.-kips (leaning column in N-S) Column unbraced length, $L_b = 12$ ft.; $r_x = 4.44$ in., $r_y = 2.59$ in. **LRFD Beam Design Tables (*AISCM*, Table 3-2; $F_y = 50$ ksi)** $\phi_b M_p = 320$ ft.-kips; $L_p = 9.15$ ft.; $BF = 3.86$ kips; $L_r = 40.6$ ft. $> L_b$ $\phi_b M_{nx} = \phi_b M_p - BF(L_b - L_p) = [320 - (3.86)(12 - 9.15)] = 309$ ft.-kips $\leq \phi_b M_p$ From *AISCM*: $S_x = 75.7$ in.³; $Z_y = 40.1$ in.³; $S_y = 26.4$ in.³ $\phi_b M_{nx} < 1.5 \phi_b S_x F_y / 12 = 426$ ft.-kips $\therefore$ $\phi_b M_{nx} = \mathbf{309}$ **ft.-kips** $\phi_b M_{ny} = \phi_b Z_y F_y < 1.5 \phi_b S_y F_y$ $\phi_b M_{ny} = (0.9)(40.1)(50)/12 < 1.5(0.9)(26.4)(50)/12$ $\phi_b M_{ny} = 150$ ft.-kips $> $ **149 ft.-kips** Governs $\boxed{\phi_c P_n = 717 \text{ kips}; \ \phi_b M_{nx} = 309 \text{ ft.-kips}; \ \phi_b M_{ny} = 149 \text{ ft.-kips}}$ **Sway Moment Magnification Factors:** $B_{2x-x} = 1.43$; $B_{2y-y} = 1.0$ (see previous calculations/discussions) $P_{u\ total} = P_{direct} + P_{indirect} = 105.1 + 0* = 105.1$ kips (see Table 8-7) *Leaning columns neglected <u>P-δ Effect for *x–x* Axis</u> $C_{mx} = 0.6 - 0.4(M_{1x}/M_{2x})$; but conservatively use $C_{mx} = 0.85$ $P_{e1x} = \dfrac{\pi^2 EA}{(K_x L / r_x)^2} = 1234$ kips ($K_x = 2.1$ for column in E-W moment frame) $B_{1x} = \dfrac{C_{mx}}{(1 - P_u / P_{e/x})} = \dfrac{0.85}{[1 - (105.1/1234)]} = 0.93 < 1.0 \Rightarrow \mathbf{B_{1x} = 1.0}$ $\boxed{\therefore B_{1x} M_{ntx} = 1.0(2.85) = 2.9 \text{ ft.-kips}}$ <u>P-δ Effect for *y–y* Axis</u> $C_{my} = 0.6 - 0.4(M_{1y}/M_{2y})$; but conservatively use $C_{my} = 0.85$ $P_{e1y} = \dfrac{\pi^2 EA}{(K_y L / r_y)^2} = 1852$ kips ($K_y = 1.0$ for leaning column in N-S direction) $B_{1y} = \dfrac{C_{my}}{(1 - P_u / P_{e/y})} = \dfrac{0.85}{[1 - (105.1/1852)]} = 0.9 < 1.0 \Rightarrow \mathbf{B_{1y} = 1.0}$ $\boxed{\therefore B_{1y} M_{nty} = 1.0(14) = 14 \text{ ft.-kips}}$ $M_{ux} = B_{1x} M_{ntx} + B_{2x} M_{ltx} = 2.9 + (1.43)(158.1) = 230$ ft.-kips $M_{uy} = B_{1y} M_{nty} + B_{2y} M_{lty} = 14 + (1.0)(0) = 14$ ft.-kips See Table 8-7 for the lateral translation moments, M_{lt}.

Interaction Equation

$P_u / \phi_c P_n = 105.1/717 = 0.15 < 0.2 \Rightarrow$ Equation (8-2)

$$\frac{P_u}{2\phi P_n} + \left(\frac{M_{ux}}{\phi_b M_{nx}} + \frac{M_{uy}}{\phi_b M_{ny}} \right) = \frac{0.15}{2} + \left(\frac{230}{309} + \frac{14}{149} \right) = 0.92 < 1.0 \textbf{ OK}$$

$\therefore$ **W10 × 68 is adequate for column A-3 *if* the effect of leaning columns is neglected.**

8.11 ANALYSIS AND DESIGN OF BEAM–COLUMNS FOR AXIAL TENSION AND BENDING

The interaction equations used for combined axial compression plus bending are also used for the analysis and design of beam–columns under combined axial tension plus bending.

EXAMPLE 8-4

Analysis of Beam–Columns for Axial Tension and Bending

A W10 × 33 welded tension member, 15 ft. long, is subjected to a factored axial tension load of 105 kips and factored moments about the strong and weak axes of 30 ft.-k and 18 ft.-k, respectively. Assuming ASTM A572, grade 50 steel and the member braced only at the supports, check if the beam–column is adequate.

SOLUTION

Factored axial tension load, $P_u = 105$ kips

Factored moments, $M_{ux} = 30$ ft.-k and $M_{uy} = 18$ ft.-kips

Calculate the Axial Tension Capacity of the Member:

Since there are **no bolt holes**, only the **yielding** limit state is possible for the tension member:

Axial tension design strength, $\phi P_n = \phi F_y A_g$
$$= (0.9)(50 \text{ ksi})(9.71 \text{ in.}^2) = 437 \text{ kips}$$

$P_u/\phi P_n = 105/437 = 0.24 > 0.2 \therefore$ Use equation (8-1)

Calculate the Bending Moment Capacity for Both x–x and y–y Axes:

For a W10 × 33, from the LRFD beam design tables (Table 3–2 of the *AISCM*), we obtain

$\phi_b M_p = 146$ ft.-kips; $BF = 3.59$ kips; $L_p = 6.85$ ft. $< L_b$;
$\quad L_r = 21.8$ ft. $> L_b \therefore$ Zone 2

$\phi_b M_{nx} = \phi_b M_p - BF(L_b - L_p) = 146 - 3.59(15 \text{ ft.} - 6.85 \text{ ft.})$
$\quad\quad\quad = \textbf{116 ft.-kips}$

From *AISCM*, $Z_y = 14$ in.^{3}; $S_y = 9.2$ in.^{3}

$\phi_b M_{ny} = \phi Z_y F_y/12 \le 1.5 \phi S_y F_y/12$
$\phi_b M_{ny} = (0.9)(14)(50)/12 \le 1.5(0.9)(9.2)(50)/12$
$\phi_b M_{ny} = 52.5$ ft.-k $\le$ **51.8 ft.-kips**

Using equation (8-1)

$$\frac{P_u}{\phi P_n} + \frac{8}{9}\left(\frac{M_{ux}}{\phi_b M_{nx}} + \frac{M_{uy}}{\phi_b M_{ny}}\right) = \frac{105}{437} + \frac{8}{9}\left(\frac{30}{116} + \frac{18}{51.8}\right) = 0.78 < 1.0$$

$\therefore$ **W10 × 33 is adequate.**

8.12 DESIGN OF BEAM COLUMNS FOR AXIAL TENSION AND BENDING

The procedure for designing steel members for combined tension plus bending is a trial-and-error process that involves an initial guess of the size of the member, after which the member is analyzed. The design procedure is outlined as follows:

1. Determine the factored axial tension load, P_u, and the factored moments about both axes, M_{ux} and M_{uy}.

   If the connection eccentricities, e_x and e_y, about the x- and y-axes are known, the applied factored moments can be calculated as $M_{ux} = P_u e_x$ and $M_{uy} = P_u e_y$.

2. Make an **initial guess** of the member size and calculate the design tension strength, $\phi_t P_n$.

3. Calculate the design bending strength about both axes, ϕM_{nx} and ϕM_{ny}.

4. If $P_u/\phi P_n \geq 0.2 \Rightarrow$ Use the interaction equation (8-1); otherwise use equation (8-2). $\phi_t P_n$

5. If the interaction equation yields a value ≤ 1.0, the member is adequate.

   If the interaction equation yields a value > 1.0, the member is **not** adequate. Increase the member size.

8.13 COLUMN BASE PLATES

Column base plates are shop welded to the bottom of a column to provide bearing for the column usually and to help in transferring the column axial loads to the concrete pier or footing. They help prevent crushing of the concrete underneath the column and also help to provide temporary support to the column during steel erection by allowing the column (in combination with anchor bolts) to act temporarily as a vertical cantilever. The base plate is usually connected to the column with fillet welds (up to ¾ in. in size) on both sides of the web and flanges and is usually shop welded. The welds are usually sized to develop the full tension capacity of the anchor bolts or rods. Fillet welds that wrap around the tips of the flanges and the curved fillet at the intersection of the web and the flanges are not recommended because they add very little strength to the capacity of the column-to-base-plate connection and the high residual stresses in the welds may cause cracking [8]. Full-penetration groove welds are also not typically used for column base plates because of the high cost, except for column-to-base-plate connections subject to very large moments. The thickness of base plates varies from ½ in. to 6 in. and they are more commonly available in ASTM A36 steel. Steel plate availability is shown in Table 8-8 [9]. The base plate is usually larger than the column size (depending on the shape of the column) by as much as 3 to 4 in. all around to provide room for the placement of the anchor bolt holes outside of the column footprint. For W-shape columns where the anchor bolts can be located within the column footprint on either side of the web, the plan size of the base plate may only need to be just a little larger than the column size to allow for the fillet welding of the column to the base plate, but the actual plate size is still dependent on the applied load and the concrete bearing stress (see Figure 8-33).

The column base plate often bears on a layer of ¾-in. to 1½-in. nonshrink grout that provides a uniform bearing surface and, in turn, is supported by a concrete pier or directly on a concrete footing. The compressive strength of the grout should be at least equal to the compressive strength of the concrete used in the pier or footing; however, a grout compressive

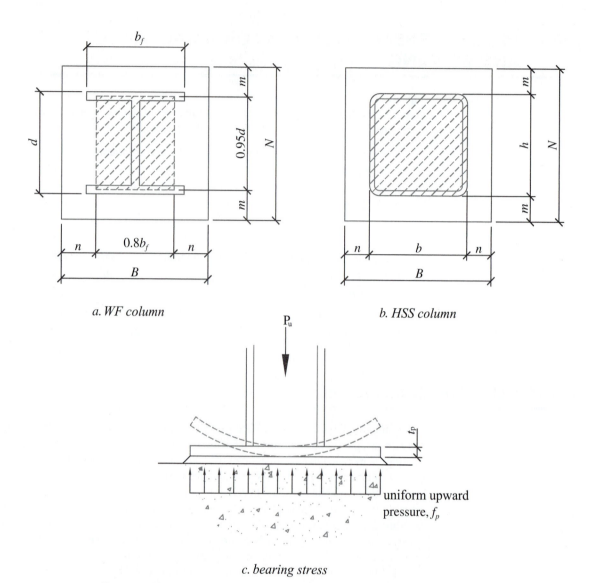

a. WF column

b. HSS column

c. bearing stress

Figure 8-33 Column base plate, bearing stresses, and critical areas.

strength of twice the concrete pier or footing compressive strength is recommended [10]. During steel erection, ¼-in.-thick leveling plates, which are slightly larger than the base plates or leveling nuts, are used at the underside of the column base plates to align or plumb the column (see Figure 8-34). For base plates larger than 24 in. in width or breadth, or for base plates supporting heavy loads, leveling nuts (see Figure 8-34b) should be used instead of leveling plates [10]. The authors recommend that high stacks of steel shims or wood shims not be used to plumb building columns during erection. The concrete piers should be larger than the base plate by at least twice the grout thickness to prevent interference between the anchor rods and the pier reinforcing, and the exterior column piers should be constructed integrally with the perimeter foundation walls. In the design of the base plate, the bearing stresses below the plate are assumed to be uniform and the base plate is assumed to bend in two directions into a bowl-shaped surface (i.e., the plate is assumed to cantilever and bend about the two orthogonal axes).

Table 8-8 Availability of base plate materials

Base Plate Thickness, in.	Plate Availability
$t_p \leq 4$ in.	ASTM A36* ASTM A572, grade 42 or 50 ASTM A588, grade 42 or 50
4 in. $< t_p \leq 6$ in.	ASTM A36* ASTM A572, grade 42 or 50 ASTM A588
$t_p > 6$ in.	ASTM A36

*ASTM A36 is the preferred material specification for steel plates.

The design strength of concrete in bearing from ACI 318 [11] is given as

$$\phi_c P_b = \phi_c(0.85f_c')A_1 \frac{A_2}{eA_1}, \tag{8-15}$$

where

A_1 = Base plate area = $B \times N$,

B = Width of base plate,

N = Length of base plate,

A_2 = Area of concrete pier concentric with the base plate area, A_1, projected at the top of the concrete pier (or at the top of the concrete footing when the column base plate is supported directly by the footing) without extending beyond the edges of the pier or footing,

f_c' = Compressive strength of the concrete pier or footing, ksi, and

ϕ_c = Strength reduction factor for concrete in bearing = 0.65 (ACI 318).

It should be noted that the strength reduction factor for concrete in bearing, ϕ_c, is 0.65, and not 0.60 as indicated in the AISC specification. As a matter of fact, reference [10], which is

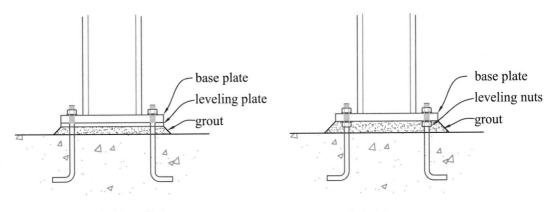

a. leveling plate   *b. leveling nuts*

Figure 8-34 Leveling plate and leveling nut.

an AISC publication, acknowledges that the discrepancy was caused by an "oversight in the AISC specification development process," and that ϕ_c should actually be 0.65 as specified in the ACI Code [11].

$$1 \le \sqrt{\frac{A_2}{A_1}} \le 2 \tag{8-16}$$

The $\sqrt{\dfrac{A_2}{A_1}}$ term accounts for the beneficial effect of confinement when the concrete pier area (or footing area when the footing directly supports the column) is greater than the base plate area.

In determining the base plate thickness, the base plate is assumed to be rigid enough to ensure a uniform bearing pressure distribution at the bottom of the base plate. The uniform bearing stress at the bottom of the base plate due to the factored column load, P_u, is

$$f_{pu} = P_u/A_1 \tag{8-17}$$

Where the base plate cantilevers beyond the critical column area by the critical distance, ℓ, the applied moment in the base plate at the edge of the critical column area due to this uniform stress is

$$M_u = \frac{f_{pu}\ell^2}{2} = \frac{P\ell^2}{2A_1}. \tag{8-18}$$

The bending strength of the plate about its weak axis is

$$\phi_b M_n = \phi_b Z_y F_y = \phi_b \left(\frac{b_p t_p^2}{4} \right) F_y, \tag{8-19}$$

where the unit width of the plate, $b_p = 1$ in.

Equating the bending strength to the applied moment (i.e., $\phi_b M_n = M_u$) yields the required plate thickness as

$$t_p = \sqrt{\frac{2P_u\ell^2}{\phi_b A_1 F_y}} = \ell \sqrt{\frac{2P_u}{\phi_b BNF_y}}. \tag{8-20}$$

The practical minimum base plate thickness used for columns in steel buildings is $\frac{3}{8}$ in., with thickness increments of $\frac{1}{8}$ in. up to $1\frac{1}{4}$ in., and $\frac{1}{4}$-in. increments beyond that. The thickness of a base plate should not be less than the column flange thickness to ensure adequate rigidity of the base plate. However, for lightly loaded columns such as posts and wind columns, $\frac{1}{4}$-in.-thick base plates may be used. The length, N, and the width, B, of the base plate should be specified in increments of 1 in. Square base plates are normally preferred and are more commonly used in practice because the directions of the length and width of the plate do not have to be specified on the drawings. The use of square base plates also helps minimize the likelihood of construction errors. The critical base plate cantilever length, ℓ, is the largest of the cantilever lengths, m, n, and $\lambda n'$, where

$$m = \frac{N - 0.95d}{2},$$

$$n = \frac{B - 0.80b_f}{2}, \text{ and}$$

$$n' = \frac{1}{4}\sqrt{db_f}.$$

The above equations are valid for W-shaped columns. For a **square HSS column,**

$$m = \frac{N - b}{2},$$

$$n = \frac{B - b}{2}, \text{ and}$$

$$n' = \frac{b}{4},$$

where

ϕ_b = 0.9 (strength reduction factor for plate bending),

F_y = Yield strength of the base plate,

d = Depth of column,

b_f = Flange width of column,

$m, n, \lambda n'$ = Cantilever lengths of the base plate beyond the edges of the critical area of the column,

ℓ = Maximum of $(m, n, \lambda n')$,

λ is conservatively taken as 1.0 [10] in this text, and

b = Width of square HSS column.

8.14 ANCHOR RODS

Anchor rods are used to safely anchor column bases and to prevent the overturning of columns during erection. They are also used to resist the base moments and the uplift forces that a column base may be subjected to due to lateral wind or earthquake forces. The most commonly used specification for anchor rods is ASTM F1554 [12]; it covers hooked, headed, and nutted anchor rods (see Figure 8-35). This specification provides for anchor rods in grades 36, 55, and 105, with grade 36 being the most commonly used in design practice. ASTM F1554, grade 36 anchor rod is weldable; the weldability of grade 55 can be enhanced by limiting the carbon content using a supplementary requirement for weldability. Grade 55 anchor rods are used to resist large tension forces due to uplift from overturning moments or moment connections at the column base plate. The weldability of anchor rods becomes a desirable property if and when field repairs that involve welding of the anchor rods are required. The use of grade 105 anchor rods is not recommended because of difficulty with weldability, so it is advisable to use a larger diameter rod size in lieu of grade 105. Prior to 1999, ASTM A36 and A307 were the commonly used material specifications for anchor

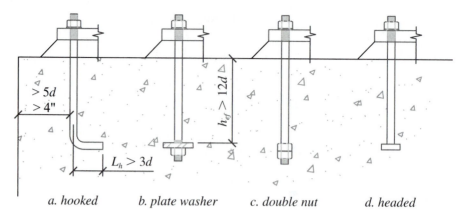

Figure 8-35 Types of anchor rods.

rods, but these have now been largely replaced by ASTM F1554. The reader should note that ASTM A325 and A490 should not be used for anchor bolts because these specifications are only valid for bolts used in steel-to-steel connections and are only available in lengths of not more than 8 in. For nutted anchor rods, the heavy-hex nut is most commonly used, and is usually tack welded at the bottom to the threaded rod to prevent the rod from turning loose from the nut when tightening the top nut above the base plate. The preferred material specification for the heavy-hex nuts is ASTM A563 where the required finish and grade of heavy-hex nuts corresponding to the various grades of anchor rods are given.

The minimum and most commonly used anchor rod size in design practice is ¾-in. diameter with a minimum recommended embedment length in the concrete of 12d and a minimum embedded edge distance of 5d or 4 in., whichever is greater, where d is the anchor rod diameter. Headed and nutted anchor rods are used to resist tension forces due to uplift from overturning or column base moments; they are typically used at braced-frame column locations or at the bases of moment frame columns. Athough a hooked anchor rod does have some nominal tension capacity, they are typically used for axially loaded columns only. In typical building columns that are not part of the lateral load resisting system, four anchor rods with 9-in. minimum embedment and 3-in. hook, are usually specified. To ensure moment restraint at the column base—thus ensuring safety during steel erection, the Occupational Safety and Health Administration (OSHA) requires that a minimum of four anchor bolts be specified at each column. The minimum moment to be resisted during the erection of a column, according to OSHA's requirements, is an eccentric 300-lb. axial load located 18 in. from the extreme outer face of the column in each orthogonal direction [10]. For a typical W12 column, this means that the anchor rods have to be capable of resisting a minimum moment about each orthogonal axis of (0.3 kip)(18 in. + 12 in./2) = 7.2 in.-kips. For a typical axially loaded W12 column with nominal ¾-in.-diameter anchor bolts, the tension pullout capacity with 9-in. minimum embedment and 3-in. hook is approximately 5.7 kips (see Section 8.16). If the anchor rods were laid out on a minimum 3-in. by 3-in. grid, the moment capacity of the four nominal anchor rods is (2 rods)(5.7 kips)(3 in.) = 34.2 in.-kips, which is much greater than the applied moment per OSHA requirements.

The accurate layout of the anchor rods is very important before the concrete piers or footings are poured. The plan location of the anchor rods should match the location of the anchor rod holes in the column base plates and must project enough distance above the top of the pier (or footing) to accommodate the nut and washer. For columns that are adjacent to intersecting foundation or basement walls, a special anchor rod

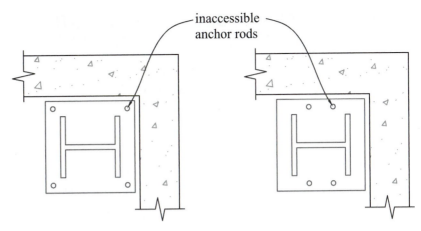

Figure 8-36 Column base plate details with inaccessible anchor rod.

layout may be required to provide accessibility to all of the anchor rods because of the presence of the wall, or a construction sequence must be specified that has the column and the anchor rods in place before the wall is poured [13]. Figure 8-36 shows some details with inaccessible anchor rods that should be avoided in practice [9]. Some contractors might request to wet-place the anchor rods immediately after the concrete is poured, but this should not be allowed (see Section 7.5 of Ref. [11]). The anchor rods must be set and tied in place within the formwork of the concrete pier before the concrete pier is poured.

Where the length of the anchor rod that projects above the top of the concrete pier is too short because of a misplacement of the anchor rod, one possible solution would be to extend the anchor rod by groove-welding a threaded anchor rod (of similar material) to the existing anchor rod and providing filler plates at the weld location as shown in Figure 8-37. The filler plates should be tack welded to each other and the washer plate should be tack welded to the filler plates. In some situations where anchor rods have been laid out incorrectly, the use of drilled-in epoxy anchors may be the only effective remedy, but care must be taken to follow the epoxy anchor manufacturer's recommendations and design criteria with regard to edge distances and minimum spacing, especially for column base plates that have to resist moments or uplift forces.

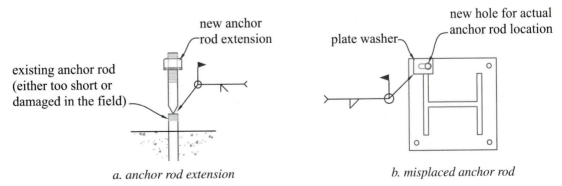

a. anchor rod extension   *b. misplaced anchor rod*

Figure 8-37 Repair of misplaced anchor rods.

8.15 UPLIFT FORCE AT COLUMN BASE PLATES

In this section, we will consider the effects of uplift forces on anchor rods.

For the three-story braced frame shown in Figure 8-38, the *unfactored* overturning moment at the base of the building is

$$OM = F_r h_r + F_3 h_3 + F_2 h_2,$$

where

F_2, F_3, and F_r are unfactored wind or seismic loads on the braced frame,

h_r, h_3, and h_2 = heights of the roof, third, and second floor levels above the ground floor

Considering only the lateral loads and neglecting the gravity loads for now and summing the moments about point B on the braced frame gives

$$\Sigma M_B = 0 \Rightarrow OM - T(b) = 0$$

∴ Unfactored uplift force, $T = OM/b$.

Vertical equilibrium yields the unfactored compression force, $C = OM/b$,

where

T = Tension force at the base of the column due to lateral loads (negative), and

C = Compression force at the base of the column due to lateral loads (positive).

b = distance between the braced frame columns.

Assuming that the unfactored cumulative dead load on column AC is P_D and using the load combinations from Chapter 2 of this text,

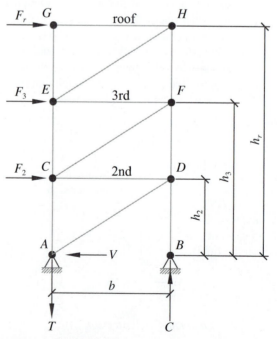

Figure 8-38 Typical braced frame.

$$\text{the factored net uplift in column AC } = 0.9\,P_D + 1.6\,T_{wind} \qquad (8\text{-}21)$$
$$\text{or } 0.9\,P_D + 1.0\,T_{earthquake}.$$

Note that in equation (8-21), the tension force, T, is entered as a negative number since the dead load acts downward, whereas the tension force acts upward in the opposite direction.

- If equation (8-21) yields a net *positive* value, then no uplift actually exists in the column and, therefore, only nominal anchor rods are required (i.e., four ¾-in.-diameter anchor rods with 9-in. minimum embedment plus 3-in. hook).
- If equation (8-21) yields a net *negative* value, then a net uplift force exists in the column, base plate, and anchor rods. Therefore, the column, column base plate, anchor rods, concrete pier, and footing have to be designed for this **net** uplift force. Headed anchor rods with adequate embedment into the concrete pier or footing are normally used to resist uplift forces.

The base plate will be subjected to upward bending due to the uplift force and the plate thickness must be checked for this upward bending moment as shown in Figure 8-39. The

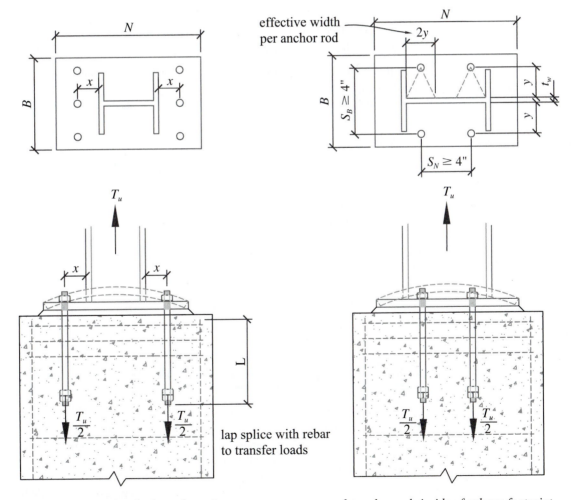

a. *anchor rods outside of column footprint*   b. *anchor rods inside of column footprint*

Figure 8-39 Bending in base plates due to uplift force.

bearing elevation of the column footing should also be embedded deep enough into the ground to provide adequate dead weight of soil to resist the net uplift force.

There are two cases of uplift that we will consider:

Case 1: Net uplift force on base plates with anchor rods outside the column footprint

For anchor rods located outside of the column footprint (see Figure 8-39a), the plate thickness is determined assuming that the base plate cantilevers from the face of the column due to the concentrated tension forces at the anchor rod locations. The applied moment per unit width in the base plate due to the uplift force is

$$M_u = \frac{\left(\dfrac{T_u}{2}\right)x}{B}, \tag{8-22}$$

where

x = Distance from the centroid of the anchor rod to the nearest face of the column,

T_u = Total net uplift force on the column, and

B = Width of base plate (usually parallel to the column flange).

The bending strength of base plate bending about its weak axis is

$$\phi_b M_n = \phi_b Z_y F_y = \phi_b \left(\frac{b_p t_p^2}{4}\right) F_y, \tag{8-23}$$

where

$b_p = 1$ in.

Equating the bending strength to the applied moment (i.e., $\phi_b M_n = M_u$) yields the required plate thickness as

$$t_p = \sqrt{\frac{2T_u x}{\phi_b B F_y}}. \tag{8-24}$$

Case 2: Net uplift force on base plates with anchor rods within the column footprint

For anchor rods located within the column footprint as shown in Figure 8-39b, the plate thickness is determined assuming that the base plate cantilevers from the face of the column web due to the concentrated tension forces at the anchor rod locations. If the distance from the centroid of the anchor rod to the face of the column web is designated as y, the effective width of the base plate for each anchor rod is assumed to be the length of the plate at and parallel to the column web and bounded by lines radiating at a 45° angle from the center of the anchor rod hole toward the column web (see Figure 8-39b). This assumption yields an effective width of $2y$ for each anchor rod. Therefore, the applied moment per unit width in the base plate due to the uplift force is

$$M_u = \frac{\left(\dfrac{T_u}{n}\right)y}{2y} = \frac{T_u}{2n}, \tag{8-25}$$

where

    y = Distance from the centroid of the anchor rod to the web of the wide flange column,

  T_u = Total net uplift force on the column, and

   n = Total number of anchor rods in tension.

The above equation assumes that the anchor bolts are spaced far enough apart that the influence lines that define the effective plate width for each anchor bolt do not overlap those of the adjacent anchor rod. The bending strength of the base plate bending about its weak axis is

$$\phi_b M_n = \phi_b Z_y F_y = \phi_b \left(\frac{b_p t_p^2}{4}\right) F_y, \tag{8-26}$$

where

  b_p = 1 in.

Equating the bending strength to the applied moment (i.e., $\phi_b M_n = M_u$) yields the required plate thickness as

$$t_p = \sqrt{\frac{2T_u}{\phi_b n F_y}}. \tag{8-27}$$

8.16 TENSION CAPACITY OF ANCHOR RODS

The failure of anchor rods embedded in plain concrete can occur either by the failure of the anchor rod in tension or by the pullout of the anchor rod from the concrete. The uplift capacity of the anchor rod is the smaller of the tension capacity of the anchor rod and the concrete pullout capacity. The calculation of the tension capacity of rods in tension was discussed in Chapter 4. The tension capacity is given as

$$\phi R_n = (0.75)(\phi) F_u A_b, \tag{8-28}$$

where

  A_b = Gross area of the anchor rod = $\pi d_b^2/4$, and

  F_u = Ultimate tensile strength of the anchor rod (58 ksi for grade 36 steel).

  ϕ = 0.75

According to Appendix D1 of ACI 318 [11], the pullout strength of a hooked anchor rod embedded in plain concrete is

$$\phi R_n = \phi \psi_4 (0.9 f'_c e_h d_b), \tag{8-29}$$

where

$\phi = 0.7,$

$\psi_4 = 1.4$ if concrete is **not** cracked at service loads (1.0 for all other cases),

e_h = Hook extension $\leq 4.5 d_b$, and

f'_c = Concrete compressive strength in psi.

For a nominal ¾-in.-diameter anchor bolts with 9-in. minimum embedment and 3-in. hook into the concrete footing or pier (ASTM F1554, grade 36 steel) in 4000-psi concrete, the uplift capacity is the smaller of

$$\phi R_n = (0.75)(\phi) F_u A_b = (0.75)(0.75)(58 \text{ ksi})(\pi)(¾)^2/4 = 14.4 \text{ kips or}$$

$$\phi R_n = \phi \psi_4 (0.9 f'_c e_h d_b) = (0.7)(1.0)(0.9)(4000)(3 \text{ in.})(¾ \text{ in.}) = \textbf{5.7 kips} \text{ Governs}$$

In Section 8-14, this anchor rod pullout capacity was used to determine the overturning moment capacity of a column base plate and was found to meet the OSHA column erection requirements.

For headed or nutted anchor rods in plain concrete, when calculating the pullout and breakout strengths of a single or a group of anchor rods, the reader should refer to Appendix D1 of ACI 318 [11]. It should be emphasized that hooked anchor rods are not recommended for resisting uplift loads or column base plates subject to moments or lateral loads. The hooked anchor rods should only be used for leaning columns and to provide temporary stability for steel columns during erection. Only headed or nutted anchor rods should be used to resist uplift forces or moments at column bases. Alternatively, the tension forces can be transferred from the anchor rods to the concrete pier through bonding, thus obviating the need to use Appendix D1 of ACI 318. For this tension force transfer mechanism, the anchor rods have to be tension lap spliced with the vertical reinforcement in the concrete pier using a tension lap splice length of 1.3 times the tension development length of the vertical reinforcement in the pier; this required lap-splice length will determine the minimum height of the concrete pier and hence the bearing elevation of the concrete footing. For anchor rods that are lap spliced with pier reinforcement, the tension capacity of the anchor rod [10] is

$$\phi R_n = \phi (0.75 A_b) F_y, \tag{8-30}$$

where

$\phi = 0.9,$

$0.75 A_b$ = Tensile stress area of the threaded anchor rod,

A_b = Nominal or unthreaded area of the anchor rod, and

F_y = Yield strength of the anchor rod.

For the base plate and anchor rod examples in this text, it is assumed that the full tension capacity of the anchor rods will be developed either by proper tension lap splice with the pier reinforcement, or by proper embedment into the concrete footing or pier, with adequate edge distance and spacing between the anchors.

EXAMPLE 8-5

Uplift at the Base of the Columns in Brace Frames

The three-story braced frame shown in Figure 8-40 is subjected to the unfactored loads shown below. Determine the maximum uplift tension force at the base of the ground-floor column.

Wind Load:

Roof level: 11.4 kips
Third-floor level: 17.9 kips
Second-floor level: 23.6 kips

Dead Load on Column:

Roof level: 10 kips
Third-floor level: 20 kips
Second-floor level: 20 kips

Live Load on Column:

Roof level (snow): 14 kips
Third-floor level: 20 kips
Second-floor level: 20 kips

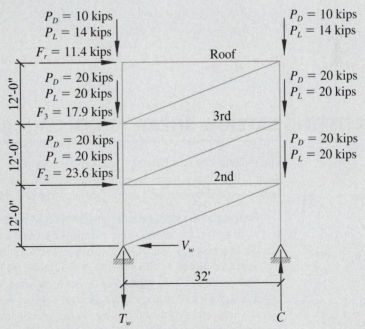

Figure 8-40 Braced frame for Example 8-5.

(continued)

SOLUTION

Maximum overturning moment at the base of the column due to wind, OM = 11.4(36 ft.) + 17.9(24 ft.) + 23.6(12 ft.) = 1123 ft.-kips

Therefore, the uplift force due to the overturning moments from wind loads is

$$T_W = OM/b = 1123 \text{ ft.-kips}/32 \text{ ft.}$$
$$= 35 \text{ kips } (-\text{ve for tension and } +\text{ve for compression}).$$

Cumulative dead load at the base of the column, $P_D = 10 + 20 + 20 = 50$ kips

Net uplift caused by the tension force due to wind is

$$0.9D + 1.6W = 0.9(50) + 1.6(-35) = -11 \text{ kips}.$$

The sign of the tension force in the load combination above is negative because the tension force acts in an opposite direction to the downward-acting dead load, D.

Therefore, the column, base plate, anchor bolts, pier, and footing will be subjected to this uplift tension force of 11 kips and must be designed for this force. The designer should also ensure that the concrete footing is embedded deep enough into the ground to engage sufficient dead load from its self-weight, plus the weight of the soil above it, to resist the net uplift force if the weight of the soil is needed to counteract the uplift. Often, the spread footings for braced frame columns may need to be placed at lower elevations compared to other columns in order to engage enough soil to provide resistance to the net uplift.

Increased Compression Load on Braced-Frame Columns

The maximum compression force on the braced-frame column should also be calculated using the load combinations from Chapter 2. The column, base plate, pier, and footing also need to be designed for this increased load.

8.17 RESISTING LATERAL SHEAR AT COLUMN BASE PLATES

The column bases in braced frames and moment frames are usually subjected to lateral shear at the column bases in addition to moments or uplift tension forces. This lateral shear can be resisted by one, or a combination, of the following mechanisms:

- **Bearing of the steel column against the concrete floor slab**
  In this case, the lateral shear is transferred into the slab by the flanges or web of the column bearing against the floor slab. The load that can be transferred by this mechanism is limited by the bending in the flange and web of the steel column due to the bearing stress. This appears to be the most commonly assumed lateral shear transfer mechanism in design practice, at least for low-rise buildings. Sometimes, horizontal steel channel struts with headed studs can be welded to the column base plate and extended for a sufficient length into the concrete slab to transfer the shear into the concrete slab, which, in turn, transfers the lateral shear into the footing through vertical dowels (see Figure 8-41a).

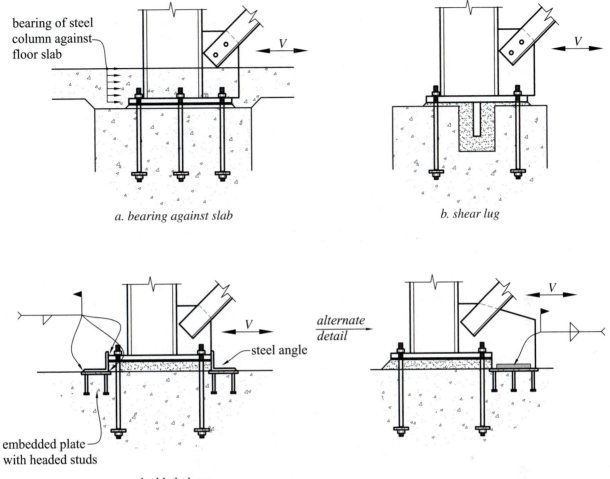

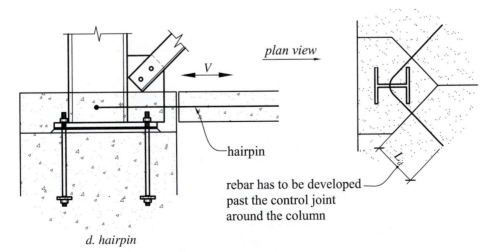

Figure 8-41 Details for resisting lateral shear at column base plates.

- **Bending of the anchor rods**

  For this mechanism, the anchor bolts are assumed to bend in double curvature in resisting the lateral shear, with a point of inflection assumed to occur in the anchor rod at the mid-depth of the grout. In addition, the anchor bolt will also resist any tension force on the column base. This bending mechanism is only possible if the anchor rod can bear directly against the column base plates; however, if oversized holes are used in the base plate, slippage of the base plate will occur until the anchor rod bears against the base plate. To avoid this slippage, the anchor rod washer and nut must be welded to the base plate. This lateral shear transfer mechanism is not recommended by the authors because of the limited bending capacity of anchor rods.

- **Shear lugs, shear stubs, or shear plates**

  For this mechanism, a steel plate (or shear key) is welded to the bottom of the column base plate; it provides the bearing surface required to transfer the lateral shear from the base plate to the concrete. The shear lug plate is subject to bending about its weak axis (see Figure 8-41b). For the design of shear lugs to resist lateral shear, the reader should refer to the AISC design guide [10].

- **Embedded plates**

  For this detail, a plate is embedded into the concrete slab in front of the column and parallel to the lateral force resisting system. A vertical gusset plate is welded to the top of the column base plate and to the top of the embedded plate; this gusset plate is used to transfer the lateral shear from the column base plate through the embedded plate into the concrete slab. An alternate detail might involve plates with headed studs embedded into the concrete on both sides of the column base plate; angles (with length parallel to the lateral shear) are field welded to the embedded plates and the base plate (see Figure 8-41c).

- **Hairpin bars or tie rods in slabs-on-grade for industrial buildings**

  Hairpin bars are U-shaped steel reinforcement that are wrapped around the steel columns in metal buildings to resist the lateral shear at the base of the column (see Figure 8-41d). The reinforcing is designed as a tension member to transfer the lateral shear in the base of the column into the concrete slab-on-grade. The reinforcing has to be developed a sufficient distance into the concrete slab to transfer this shear force, and often the reinforcing will cross a control joint in the slab-on-grade. The reinforcing must be developed past this joint.

8.18 COLUMN BASE PLATES UNDER AXIAL LOAD AND MOMENT

The base of columns in moment frames are sometimes modeled as fixed supports requiring that the base plates be designed to resist moments. Other situations with fixed supports include the bases of flag poles, light poles, handrails, and sign structures (see Figure 8-42). Two different cases will be considered here—base plates with axial load plus a small moment, and base plates with axial load plus large moments. The authors recommend that the base of columns in moment frames be modeled and designed as pinned bases (i.e., without moment restraints) because of the difficulty of achieving a fully fixed support condition in practice; however, where moment restraint is absolutely required, say, because of a need to reduce the lateral drift of the building frame, the column base plates and the anchor bolts, as well as the concrete footing, must be designed to resist the base moments while also limiting the rotation of the footing. In this case, we recommend that the spread footing be sized to ensure that the resultant load lies within the middle third of the footing dimension to

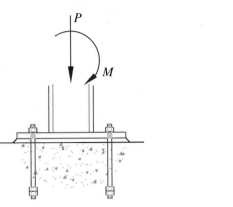

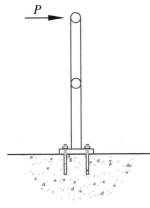

a. building column

b. handrail post

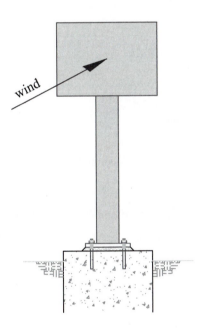

c. free-standing sign

Figure 8-42 Examples of column base subjected to axial load plus bending moment.

reduce undesirable rotations of the footing. Alternatively, the eccentric moment in the column can be resisted by strapping the column to an adjacent column footing. For a detailed design of eccentrically loaded or strap footings, the reader should refer to a reinforced concrete text (e.g., ref. [14]). It is usually much easier to achieve a fixed column base with piles or mat foundations.

There are two load cases we will consider: axial load with a small moment and axial load with a large moment.

Case 1: Axial Load Plus Small Moment with the Eccentricity of Loading, $e = M_u/P_u \leq N/6$

In this case, the moment is small enough that no tension stresses develop below the base plate. The base plate is subjected to a trapezoidally varying bearing stress that ranges from a minimum value at one edge of the base plate to a maximum value at the opposite edge, as shown in Figure 8-43.

The minimum and maximum **compression** stresses are

$$f_{u, min} = \frac{P_u}{BN} - \frac{M_u}{\left(\dfrac{BN^2}{6}\right)} \geq 0, \text{ and} \tag{8-31}$$

$$f_{u, max} = \frac{P_u}{BN} + \frac{M_u}{\left(\dfrac{BN^2}{6}\right)} \leq \phi_c f_b, \text{ respectively,} \tag{8-32}$$

where

$$\phi_c f_b = \frac{\phi_c P_b}{A_1} = 0.65(0.85f'_c),$$

P_u = Applied factored axial compression load at the base of the column,

B = Width of base plate,

N = Length of base plate,

M_u = Factored applied moment at the base of the column, and

f'_c = Concrete compressive strength.

ϕ_c = 0.65.

The procedure for the design of column base plates with small moments is as follows:

1. Determine the factored axial load, P_u, and the factored moment, M_u, for the column base.

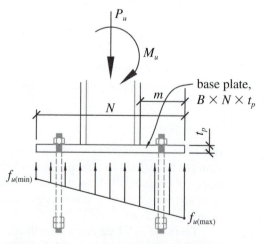

Figure 8-43 Column base plate with axial load and small moment.

2. Select a trial base plate width, B, and a base plate length, N such that

$$B \geq b_f + 4 \text{ in., and}$$
$$N \geq d + 4 \text{ in.}$$

3. Determine the load eccentricity, $e = M_u/P_u$.

    If $e \leq N/6 \Rightarrow$ Small moments. OK: This implies case 1; therefore, go to step 4.

    If $e > N/6 \Rightarrow$ Large moments. Use Case 2 (i.e., large moments).

4. Determine the plate cantilever lengths, m and n:

$$m = (N - 0.95d)/2$$
$$n = (B - 0.8b_f)/2$$

5. Determine the minimum and maximum bearing pressures using equations (8-31) and (8-32).

    If equations (8-31) and (8-32) are not satisfied, increase the base plate size until these equations are both satisfied.

6. Determine the maximum moment in the plate at the face of the column flange (i.e., at a distance m from the edge of the plate).

    The maximum bearing pressure at a distance m from the edge of the plate (see Figure 8-43) is

$$f_{u,m} = f_{u,min} + (f_{u,max} - f_{u,min})\frac{(N - m)}{N}.$$

The applied maximum bending moment per unit width in the plate at a distance m from the edge of the plate is

$$M_{u,m} = (f_{u,m})\frac{m^2}{2} + \frac{1}{2}(f_{u,max} - f_{u,m})(m)\left(\frac{2}{3}m\right).$$

The bending strength per unit width of the base plate is

$$\phi_b M_n = \phi_b Z_y F_y = \phi_b\left(\frac{b_p t_p^2}{4}\right)F_y.$$

Note that $\phi_b M_n = M_{u,m}$ yields the required minimum base plate thickness as

$$t_p = \sqrt{\frac{4M_{u,m}}{\phi_b b_p F_y}},$$

where

$M_{u,m}$ = Maximum bending moment per unit width in the base plate, in in.-kips/in. width,

b_p = Plate unit width = 1 in., and

F_y = Yield strength of the base plate.

Case 2: Axial Load Plus Large Moment with the Eccentricity of Loading, $e = M_u/P_u > N/6$

In this case, the moment at the base of the column is large enough that the base plate will lift off the grout bed on the tension side of the base plate, thus resulting in about half of the anchor rods in tension. Similar to the limit states design principles used in reinforced concrete design, the design of column base plates with large moments assumes a uniform concrete stress distribution (i.e., a rectangular concrete stress block) and a moment that is large enough for the anchor rods in the tension zone to develop their full tension capacity. The maximum stress in the compression zone is assumed to be $0.85f_c'$, acting over the full width, B, of the base plate and over a depth, a. The stress distribution is shown in Figure 8-44.

Assuming that the total area of the anchor rods in the **tension zone** is A_b, the vertical equilibrium of forces in Figure 8-44 requires that

$$0.85\phi_c f_c' Ba = P_u + T_u. \tag{8-33a}$$

Assuming that all anchor rods in the tension zone yield, then

$$T_u = \phi_t R_n = 0.75\phi_t A_b F_u.$$

If the tensile strength of the anchor rod, T_u, is limited by their pullout or breakout capacity in the concrete, then T_u will be equal to the smaller of the pullout or breakout capacity of the anchor group, and this value will have to be substituted into equation (8-33a). For a case where all anchor rods in the tension zone yield and the strength is not limited by the concrete pullout or breakout capacity, the depth of the rectangular concrete stress block, a, can be obtained from

$$0.85\phi_c f_c' Ba = P_u + 0.75\phi_t A_b F_u, \tag{8-33b}$$

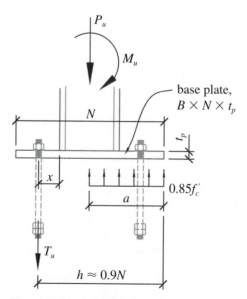

Figure 8-44 Column base plate with axial load and large moment.

where

$\phi_c = 0.65$ (ACI 318),

$\phi_t = 0.75$

$B =$ Width of base plate,

$a =$ Depth of rectangular stress block,

$P_u =$ Applied factored axial load on the column,

$F_u =$ Ultimate tensile strength of the anchor rod, and

$A_b =$ Total area of anchor bolts in the **tension zone**.

$h \;=$ distance from the anchor rods in tension to the opposite plate edge.

Summing the moments of forces about the centroid of the anchor rods in the tension zone gives

$$P_u(h - 0.5N) + M_u = 0.85\phi_c f_c' Ba\left(h - \frac{a}{2}\right). \tag{8-34}$$

Note that the above equations assumes that the full anchor rod tensile strength can be developed within the concrete embedment provided. If this is not the case, the tensile strength of the anchor bolt will be limited by the pullout strength of the anchor rod (see ref. 11). Equations (8-33) and (8-34) contain four unknowns that will be determined using the iterative approach below:

1. Determine the factored axial load, P_u, and the factored moment, M_u, for the column base, and calculate the load eccentricity, $e = M_u/P_u$.
2. Select a trial base plate width, B, and a base plate length, N, such that

   $B \geq b_f + 4$ in., and

   $N \geq d + 4$ in.

   If $e > N/6 \Rightarrow$ Case 2 (i.e., large moments). Go to step 3.

   If $e \leq N/6 \Rightarrow$ Case 1 (i.e., small moments). Stop and use the method for small moments.

3. Assume an approximate value for the effective depth, h. Assume $h = 0.9N$.
4. Assume a trial value for the area of the anchor rod in the **tension zone**, A_b.
5. Solve equation (8-33) for the depth of the rectangular concrete stress block, a, and then solve equation (8-34) for the base plate length, N.
6. Use the larger of the N values obtained from steps 2 and 5.
7. The cantilever length of the base plate in the **compression zone** is the larger of

   $m = (N - 0.95d)/2$, or

   $n = (B - 0.8b_f)/2$.

The maximum factored moment per unit width in the base plate is

$$M_u = (0.85\phi_c f_c')\frac{\ell^2}{2},$$

where

$\ell =$ Maximum of (m, n).

The bending strength of the plate per unit width is

$$\phi_b M_n = \phi_b Z_y F_y = \phi_b \left(\frac{b_p t_p^2}{4} \right) F_y.$$

Note that $\phi_b M_n = M_u$ yields the minimum required base plate thickness due to **compression stresses** in the compression zone as

$$t_p = \sqrt{\frac{4M_u}{\phi_b b_p F_y}},$$

where

b_p = Plate unit width = 1 in., and
d = Depth of the column.

8. Check the bending of the base plate in the tension zone caused by the **tension force in the anchor bolts**. The plate is assumed to cantilever from the face of the column due to the tension in the anchor rods. The applied moment per unit width in the base plate due to the tension force in the anchor bolts is

$$M_u = \frac{T_u x}{B},$$

where

x = Distance from the centroid of the anchor rod in tension to the nearest face of the column,
T_u = Total force in the anchor rods in tension = $0.75 \phi_t A_b F_u$, and
B = Width of the base plate.

The bending strength of the plate bending about its weak axis is

$$\phi_b M_n = \phi_b Z_y F_y = \phi_b \left(\frac{b_p t_p^2}{4} \right) F_y,$$

where

b_p = 1 in.

Equating the bending strength to the applied moment (i.e., $\phi_b M_n = M_u$) yields the required minimum plate thickness as

$$t_p = \sqrt{\frac{4T_u x}{\phi_b B F_y}}.$$

9. The larger of the base plate thicknesses from steps 7 and 8 governs.
10. If the designer wishes to limit the base plate thickness and also prevent excessive deformation of the base plate, vertical stiffeners may be provided for the base plate in two orthogonal directions that will prevent the plate from cantilevering the distance, m,

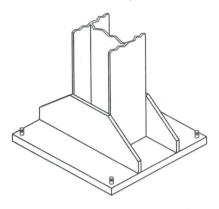

a. reinforced single direction

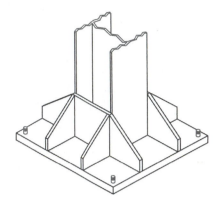

b. reinforced both directions

Figure 8-45 Column base plate with stiffeners.

indicated in step 7 and the distance, x, in step 8 (see Figure 8-45). In this case, the stiffeners bending about their strong axis will resist the maximum factored plate moments calculated in steps 7 and 8. However, in design practice, the use of stiffeners at base plates should be avoided because of the increased cost associated with the increased labor involved in welding stiffener plates to the base plate and the likely interference of the stiffener plates with the anchor rods [8]. Base plates with stiffeners should only be used for columns resisting very large base moments (e.g., columns supporting a jib crane), where the base plate thickness without stiffeners will be excessive. The presence of stiffener plates may also complicate or render impossible any field corrections that have to be made to the anchor rods in case of misplacement of the anchor rods. The authors recommend that sufficient base plate thickness be provided to obviate the need for stiffeners in base plates. However, stiffener plates on base plates may, on occasion, be required where remedial action is needed to repair underdesigned column base plates [16].

EXAMPLE 8-6

Design of Base Plate for a W-Shaped Column Subject to Axial Load Only

Design the base plate and select the minimum concrete pier size for a W12 × 58 column with a factored axial compression load of 300 kips. Assume a 1-in. grout thickness, concrete compressive strength of 4 ksi, and ASTM A36 steel for the steel plate.

SOLUTION

From Part 1 of the *AISCM*,

For W12 × 58, $d = 12.2$ in.; $b_f = 10$ in.

(continued)

- Try a base plate 4 in. larger than the column in both directions (i.e., 2 in. larger all around the column) to allow room for the placement of the anchor bolts outside the column footprint.

Try $B \approx b_f + 4$ in. $= 10 + 4 = 14$ in.
$N = d + 4$ in. $= 12.2$ in. $+ 4$ in. ≈ 17 in., say 18 in.
$A_1 = B \times N = (14$ in.$)(18$ in.$) = 252$ in.^{2}

Try a 14-in. by 18-in. base plate (see Figure 8-46).

- Select the **minimum** pier size:

$$= (B + 2h_g) \times (N + 2h_g)$$
$$= [14 \text{ in.} + (2 \times 1 \text{ in.})][(18 \text{ in.} + (2 \times 1 \text{ in.})]$$
$$\Rightarrow 16 \text{ in.} \times 20\text{-in. pier} \therefore A_2 = (16 \text{ in.})(20 \text{ in.}) = 320 \text{ in.}^2$$

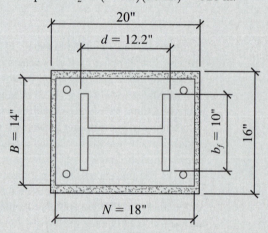

Figure 8-46 Column base plate for Example 8-6.

$$\sqrt{\frac{A_2}{A_1}} = \sqrt{\frac{320}{252}} = 1.13 < 2 \text{ and } > 1.0 \quad \text{OK}$$

$$m = \frac{N - 0.95d}{2} = \frac{18 - 0.95(12.2)}{2} = 3.21 \text{ in. (Largest value governs)}$$

$$n = \frac{B - 0.80b_f}{2} = \frac{14 - 0.80(10)}{2} = 3 \text{ in.}$$

therefore, $\ell = 3.21$ in.

$$n' = \frac{1}{4}\sqrt{db_f} = \frac{1}{4}\sqrt{(12.2)(10)} = 2.8 \text{ in.}$$

From equation (8-20), the required minimum base plate thickness is given as

$$t_p = \ell\sqrt{\frac{2P_u}{\phi_b BNF_y}} = 3.21 \text{ in.} \sqrt{\frac{(2)(300 \text{ kips})}{0.9(252 \text{ in.}^2)(36 \text{ ksi})}}$$

$$= 0.87 \text{ in.} \approx \text{⅞ in.}$$

Use ⅛-in. increments for base plate thicknesses of less than 1¼ in. and ¼-in. increments for others.

Check Bearing Capacity of Concrete Pier:

$$\phi_c P_b = \phi_c(0.85f'_c)A_1\sqrt{A_2/A_1} = (0.65)(0.85)(4 \text{ ksi})(252)(1.13)$$
$$= 628 \text{ kips} > P_u = 300 \text{ kips OK}$$

Use a 14-in. by 7/8-in. by 18-in. base plate

For this rectangular base plate, the orientation of the 14-in. and 18-in. dimensions would have to be indicated on the column schedule or on the structural plan. Where the bearing capacity of a pier, $\phi_c P_b$, is less than the factored axial load, P_u, on a column, an increase in the bearing capacity can be achieved far more efficiently by increasing the pier size (i.e., area A_2) than by increasing the base plate size (i.e., area A_1).

EXAMPLE 8-7

HSS Column Base Plate Subject to Axial Load Only

Design the base plate for an HSS 12 × 12 × 5⁄16 column with a factored axial compression load of 450 kips. Assume a 2-in. grout thickness, a concrete compressive strength of 4 ksi for the pier, and ASTM A36 steel for the base plate. Determine the minimum, pier size for this column.

SOLUTION

From Part 1 of the *AISCM*,

For HSS 12 × 12 × 5⁄16, $b = b_f = d = 12$ in.

- Try a base plate 4 in. larger than the column in both directions (i.e., 2 in. larger all around the column) to allow room for the placement of the anchor bolts outside the column footprint.

  Try $B = b_f + 4$ in. $= 12 + 4 = 16$ in.
  $N = d + 4$ in. $= 12 + 4 = 16$ in.
  $A_1 = B \times N = (16 \text{ in.})(16 \text{ in.}) = 256 \text{ in.}^2$

  Try a 16-in. by 16-in. base plate (see Figure 8-47).

- Select the **minimum** size of concrete pier
  $= (B + 2h_g) \times (N + 2h_g)$
  $= [(16 \text{ in.}) + (2)(2 \text{ in.})][(16 \text{ in.}) + (2)(2 \text{ in.})]$
  $= 20\text{-in.} \times 20\text{-in. pier} \therefore A_2 = (20 \text{ in.})(20 \text{ in.}) = 400 \text{ in.}^2$

(continued)

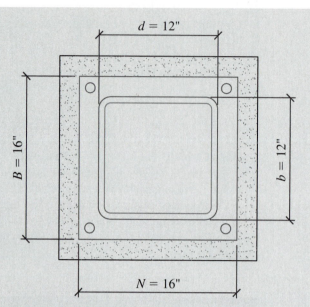

Figure 8-47 Column base plate for Example 8-7.

$$\sqrt{A_2/A_2} = \sqrt{400/256} = 1.25 < 2 \text{ and } > 1.0 \quad \text{OK}$$

$$m = \frac{N - b}{2} = \frac{16 - 12}{2} = 2 \text{ in.}$$

$$n = \frac{B - b}{2} = \frac{16 - 12}{2} = 2 \text{ in.}$$

Governs (largest value governs; therefore, $\ell = 3$ in.)

$$n' = \frac{1}{4}\sqrt{d\,b_f} = \frac{1}{4}\sqrt{(12)(12)} = 3 \text{ in.}$$

From equation (8-20), the required minimum base plate thickness is given as

$$t_p = \ell\sqrt{\frac{2P_u}{\phi_b BNF_y}} = (3 \text{ in.})\sqrt{\frac{(2)(450 \text{ kips})}{(0.9)(256 \text{ in.}^2)(36 \text{ ksi})}}$$

$$= 0.99 \text{ in.} \quad \text{Use 1-in.-thick base plate.}$$

(Use ⅛-in. increments for base plate thicknesses of less than 1¼ (in. and ¼-in. increments for others.)

Check Bearing Capacity of Concrete Pier:

$$\phi_c P_b = \phi_c(0.85f'_c)A_1\sqrt{A_2/A_1} = 0.65(0.85)(4 \text{ ksi})(256)(1.25)$$

$$= 707 \text{ kips} > P_u = 450 \text{ kips OK}$$

Use a 16-in. × 1-in. by 16-in. base plate

EXAMPLE 8-8

Design of Column Base Plate with Axial Compression Load and Moment

The base plate of a W12 × 96 column is subjected to the following service axial compression loads and moments. If the compressive strength of the concrete pier is 4000 psi, design the base plate and anchor rods for this column.

$P_D = 110$ kips

$P_L = 150$ kips

$M_D = 950$ in.-kips

$M_L = 1600$ in.-kips

SOLUTION

The section properties for a W12 × 96 column are $d = 12.7$ in. and $b_f = 12.2$ in. (*AISCM*, Part 1).

For ASTM A36 anchor rods, the ultimate strength, $F_u = 58$ ksi and the tensile strength, $F_y = 36$ ksi.

Assume, at this stage, that the moment is large and that load case 2 will govern. This will have to be verified later.

1. $P_u = 1.2(110) + 1.6(150) = 372$ kips

    $M_u = 1.2(950) + 1.6(1600) = 3700$ in.-kips

    $e = M_u/P_u = 3700/372 = 9.95$ in.

2. Select a trial base plate width, B, and base plate length, N, such that

    $B \geq b_f + 4$ in. $= 12.2 + 4 = 16.2$ in. Try $B = 20$ in., and

    $N \geq d + 4$ in. $= 12.7 + 4 = 16.7$ in. Try $N = 20$ in.

    Check if $e > N/6$:

    $e = 9.95$ in. $> N/6 = 20/6 = 3.33$ in. Therefore, use load case 2 (i.e., large moments).

3. Assume an approximate value for the effective depth, h. Assume that $h = 0.9N$

    Assume that $h = 0.9N = (0.9)(20$ in.$) = 18$ in.

4. Assume a trial value for the area of the anchor rod, A_b, in the **tension zone**.

    Assume three 1½-in.-diameter anchor rods with hex nuts in the tension zone; therefore, $A_b = 5.3$ in.^{2}

5. Solve equation (8-33b) for the depth of the rectangular concrete stress block, a, and then solve equation (8-34) for the base plate length, N.

    Solving equation (8-33b) for the depth, a, of the rectangular concrete stress block gives

    $(0.85)(0.65)(4 \text{ ksi})(20 \text{ in.}) a = (372 \text{ kips}) + (0.75)(0.75)(5.3 \text{ in.}^2)(58 \text{ ksi}).$

    Therefore, $a = 12.33$ in.

    (continued)

Solving equation (8-34) for the length, N, of the base plate gives

$$(372 \text{ k})(0.9N - 0.5N) + 3700 \text{ in.-kips}$$

$$= 0.85(0.65)(4 \text{ ksi})(20 \text{ in.})(12.33 \text{ in.})\left(0.9N - \frac{12.33 \text{ in.}}{2}\right).$$

Therefore, $N = 20.67$ in., say 21 in. Revise base plate trial size to 21 in. by 21 in.

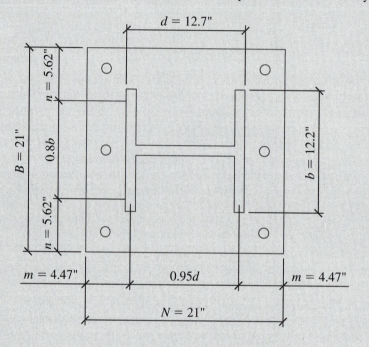

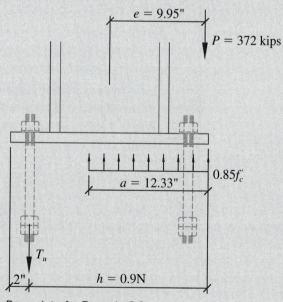

Figure 8-48 Base plate for Example 8-8.

6. Use the larger of the N values obtained from steps 2 and 5.
   Using the larger of the values from steps 2 and 5 gives $N \approx 21$ in.
7. The cantilever length of the base plate in the **compression zone** is the larger of

$$m = (N - 0.95d)/2 = [(21 \text{ in.}) - (0.95)(12.7 \text{ in.})]/2 = 4.47 \text{ in., or}$$
$$n = (B - 0.8b_f)/2 = [(21 \text{ in.}) - (0.80)(12.2 \text{ in.})]/2 = 5.62 \text{ in.}$$

$$n' = \frac{1}{4}\sqrt{db_f} = \frac{1}{4}\sqrt{(12.7 \text{ in.})(12.2 \text{ in.})} = 3.11 \text{ in.}$$

The maximum factored moment per unit width in the base plate is

$$M_u = (0.85)(0.65)(4 \text{ ksi})\frac{(5.62 \text{ in.})^2}{2} = 35 \text{ in.-kips/in. width of the base plate,}$$

where

ℓ = Maximum of (m, n).

The minimum required base plate thickness due to **compression stresses** in the compression zone is

$$t_p = \sqrt{\frac{4(35 \text{ in.-kips/in.})}{(0.9)(1 \text{ in.})(36 \text{ ksi})}} = 2.08 \text{ in.,}$$

where

b_p = Plate unit width = 1 in.

Check the bending of the base plate in the tension zone caused by the tension force, T_u, in the anchor bolts. The plate is assumed to cantilever from the face of the column due to the tension in the anchor bolts.

$$T_u = C_u - P_u = (0.85)(0.65)(4 \text{ ksi})(21 \text{ in.})(12.33 \text{ in.}) - (372 \text{ kips}) = 200 \text{ kips}$$

Assume that the edge distance of the tension zone bolts is 1.5 in. Therefore,

x = Distance from the centroid of the anchor bolts in tension to the nearest face of the column

$\quad = m - 1.5 \text{ in.} = 4.47 \text{ in.} - 1.5 \text{ in.} = 2.97 \text{ in.}$

The applied moment per unit width in the base plate due to the tension force in the anchor bolts is

$$M_u = \frac{T_u x}{B} = \frac{(200 \text{ kips})(2.97 \text{ in.})}{21 \text{ in.}} = 28.3 \text{ in.-kips/in. width of base plate.}$$

$$t_p = \sqrt{\frac{4(28.3 \text{ in.-kips/in.})}{(0.9)(1 \text{ in.})(36 \text{ ksi})}} = 1.87 \text{ in.}$$

8. The larger of the base plate thicknesses from step 7 governs. Recall that the minimum practical base plate thickness is ½ in., with increments of ¼ in. up to a 1-in. thickness, and increments of ⅛ in. for base plate thicknesses greater than 1 in.

(continued)

Comparing steps 6 and 7, and using the larger plate thickness and rounding to the nearest $\frac{1}{8}$ in. implies that

$t_p = 2\text{-}\frac{1}{8}$ in.

Therefore, use 21-in. by 2-$\frac{1}{8}$ in. by 21-in. base plate with (six) 1$\frac{1}{2}$-in. diameter threaded anchor rods with hex nut.

Note: It has been assumed in this example that the anchor rods have sufficient embedment depth into the foundation or pier and adequate edge distance and spacing between the anchors to ensure that the full tensile capacity of the anchor rods can be developed in the concrete pier or footing. If that is not the case, the tensile force, T_u, in the tension zone anchor bolts will be limited by the concrete pullout and breaking strengths, and this reduced value would need to be used in step 5. Alternatively, the anchor rods could be lap spliced with the vertical reinforcement in the pier to achieve the full tension capacity of the anchor rods, but this will increase the height of the concrete pier.

8.19 REFERENCES

1. American Institute of Steel Construction. 2006. *Steel construction manual*, 13th ed., Chicago: AISC.

2. Nair, R. Shankar. Stability analysis and the 2005 AISC specification. *Modern Steel Construction* (May 2007): 49–51.

3. Yura, J. A. The effective length of columns in unbraced frames. *Engineering Journal* 8, no. 2 (1971): 27–42.

4. Abu-Saba, Elias G. 1995. *Design of steel structures*. Chapman and Hall. New York, NY.

5. Hoffman, Edward S., Albert S. Gouwens, David P. Gustafson, and Paul F. Rice. 1996. Structural design guide to the AISC (LRFD) specification for buildings, 2nd ed. Chapman and Hall.

6. American Society of Civil Engineers. 2005. *ASCE 7: Minimum design loads for buildings and other structures*. Reston, VA: ASCE.

7. Kulak, G. L., and Grondin, G. Y. 2006. Limit states design in structural steel. Canadian Institute of Steel Construction, Willowdale, Ontario, Canada.

8. Shneur, Victor. 24 tips for simplifying braced frame connections. *Modern Steel Construction* (May 2006): 33–35.

9. Honeck, William C., and Derek Westphal. 1999. Practical design and detailing of steel column base plates. *Steel Tips*. Structural Steel Educational Council.

10. Fisher, James W., and Lawrence A. Koibler. 2006. AISC Steel Design Guide No. 1: *Base plate and anchor rod design*, 2nd ed. Chicago: American Institute for Steel Construction.

11. American Concrete Institute. 2008. *ACI 318: Building code requirements for structural concrete and commentary*, Farmington Hills, MI.

12. Carter, Charles J. Are you properly specifying materials? *Modern Steel Construction* (January 2004).

13. Swiatek, Dan, and Emily Whitbeck. Anchor rods–Can't live with 'em, can't live without 'em. *Modern Steel Construction* (December 2004): 31–33.

14. Limbrunner, George F., and Abi O. Aghayere. 2006. Reinforced concrete design, 6th ed. Upper Saddle River, NJ: Prentice Hall.

15. International Codes Council. 2006. *International building code—*. Falls Church, VA: ICC.

16. Post, Nadine M. Structural fix: VA expects disputes over delay. *Engineering News Record* (February 10, 1997): 10–11.

8.20 PROBLEMS

8-1. Determine the adequacy of a 15-ft.-long W12 × 72 column in a braced frame to resist a factored axial load of 200 kips and factored moments of 100 ft.-kips and 65 ft.-kips about the *x*- and *y*- axes, respectively. The column is assumed to be pinned at both ends. Use ASTM A992 steel.

8-2. A 15-ft.-long W12 × 96 column is part of a moment frame with column bases that are pinned. The factored axial load on the column is 250 kips and the factored moment is 120 ft.-kips about the *x*-axis. The building is assumed to be braced in the orthogonal direction. Determine whether this column is adequate to resist the applied loads. Use ASTM A992 steel.

8-3. For the mezzanine floor plan shown in Figure 8-49, assuming **HSS 7 × 7 columns**,

    a. Calculate the factored axial loads and moments (for load case 1 and load case 2) for the exterior column C1.

    b. Using the HSS beam–column design templates, design the most economical or lightest HSS 7 × 7 column size for column C1 that is adequate to resist the loads and moments.

8-4. For the floor and roof framing plan shown in Example 8-3 (Figure 8-32a), analyze the North–South moment frames using the amplified first-order analysis method and design column B1, including the effect of the leaning columns.

8-5. A W8 × 40 welded tension member, 16 ft. long, is subjected to a factored axial tension load of 150 kips and factored moments about the strong and weak axes of 30 ft.-kips and 20 ft.-kips, respectively. Assuming ASTM A572, grade 50 steel and that the member is braced only at the supports, determine whether the beam–column is adequate.

8-6. Design the base plate for an HSS 10 × 10 × 5/16 column with a factored axial compression load of 400 kips. Assume a 1-in. grout thickness, a concrete compressive strength of 4 ksi for the pier, and ASTM A36 steel for the base plate.

8-7. Design the base plate and select the concrete pier size for a W12 × 72 column with a factored axial compression load of 550 kips. Assume a 1-in. grout thickness, a concrete compressive strength of 4 ksi, and ASTM A36 steel for the base plate.

8-8. a. Design the base plate for an HSS 8 × 8 column with a factored axial load, P_u, of 500 kips.

    b. Select the most economical concrete pier size required. Assume a 1-in. nonshrink grout and a concrete 28-day strength, f'_c, of 3000 psi.

8-9. A 15-ft.-long HSS 6 × 6 × ½ hanger supports a factored axial tension load of 70 kips and factored moments, $M_{ux} = 40$ ft.-kips and $M_{uy} = 20$ ft.-kips. Assuming that the hanger is fully welded at the beam support above, is the hanger adequate? Use ASTM A500 Grade 46 steel.

8-10. Select the lightest 8-ft.-long W10 hanger to support a factored tension load of 90 kips applied with an eccentricity of 6 in. with respect to the strong (*X–X*) axis of the section and an eccentricity of 3 in. with respect to the weak (*Y–Y*) axis of the section. The member is fabricated from ASTM A36 steel, is fully welded at the connections, and is braced laterally at the supports.

8-11. For the three-story braced frame building shown in Figure 8-50, design column C2 for the axial loads and no-translation bending moments resulting from the beam-to-column and beam-to-girder reaction eccentricities. Use the W-shape column design template and present your results in a column schedule. Assume that the column will be spliced at 4 ft. above the second-floor level. Use ASTM A992 steel.

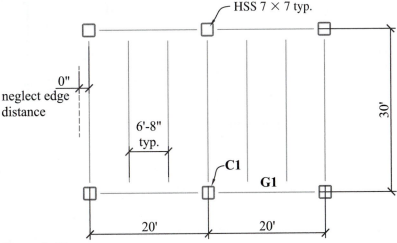

Figure 8-49

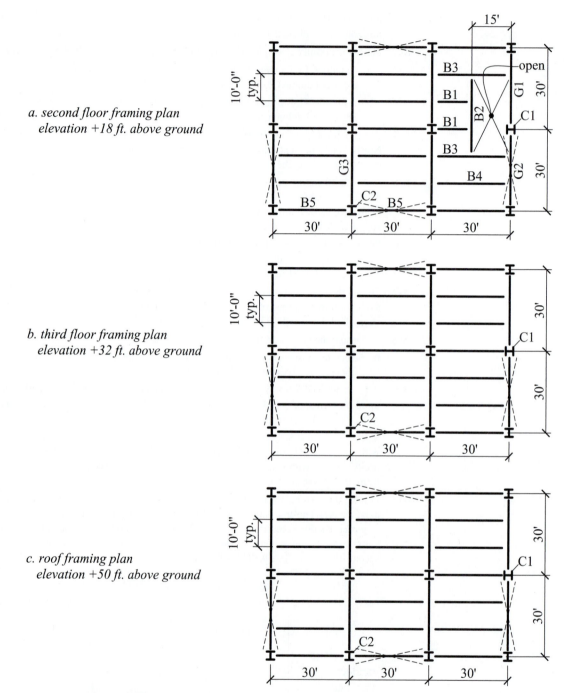

a. second floor framing plan
elevation +18 ft. above ground

b. third floor framing plan
elevation +32 ft. above ground

c. roof framing plan
elevation +50 ft. above ground

Figure 8-50 Roof and floor framing plans for problem 8-11.

Student Design Project Problem

8-12.   For the student design project building in Chapter 1, determine the factored axial loads and the no-translation moments at each level of the building for the typical interior, exterior, and corner columns. Using the column design template, select the most economical W-shape columns, presenting your results in a column schedule.

9

Bolted Connections

9.1 INTRODUCTION

In the preceding chapters, we covered the analysis and design process for basic members, such as beams, columns, and tension members. In this chapter and in the following chapter (Chapter 10, "Welded Connections"), we will cover the connections of these members to each other. In any structure, the individual components are only as strong as the connections. Consequently, designers will often specify that that some connections (such as shear connections in beams) be designed for the full capacity of the connected member to avoid creating a "weak link" at the connection.

In practice, some engineers delegate the design of the simple connections to the steel fabricator. This usually allows the fabricator to select the most economical method for fabricating and erecting the structural steel. In this scenario, the engineer will review and approve these simple connections and provide details for the more complicated connections, such as braced frames or moment connections. In other cases, engineers may delegate all of the connection designs to the steel fabricator. In this case, the engineer will provide drawings showing the member forces and reactions so that the fabricator's engineer can provide an adequate connection for the loads indicated. In practice, the delegation of connection design to the fabricator is more prevalent in the Eastern United States. In any case, the engineer of record still has to review the connection designs to ensure that they conform to the design intent.

The most common and most economical connections used are bolted connections. Riveted connections were used prior to the advent of bolted and welded connections in the 1950s. The use of riveted connections in structural steel has essentially become obsolete. Rivets required more skilled laborers for installation, as well as more inspection. They were also somewhat more dangerous in that the rivets would have to be heated and installed at a very high temperature (about 1000°F). While high-strength bolts have a greater material cost, they are installed with a greater degree of safety and with less labor.

There are two basic types of bolts—unfinished bolts (also called machine, common, or ordinary bolts) and high-strength bolts. Unfinished bolts conform to ASTM A307 and are generally used in secondary structures, such as handrails, light stairs, service platforms, and other similar structures that are not subject to cyclical loads. Unfinished bolts have a lower load-carrying capacity than high-strength bolts; therefore, their use should be limited to secondary structures that typically have lighter loads. High-strength bolts are the most common type of bolt used in steel structures and have more than twice as much tensile strength than unfinished bolts. High-strength bolts conform to either ASTM A325 or ASTM A490 and can be used in bearing, as well as slip-critical, connections (connections where slip does not occur; see Section 9-4.). There are also bolts referred to as twist-off, tension-controlled bolts, which should meet the requirements of ASTM F1852. These bolts are fabricated with a splined end and installed with a special wrench such that the splined end breaks off once the required torque is obtained. The inspection of these bolts is simplified in that only visual inspection is required. Bolts that conform to ASTM F1852 are equivalent to ASTM A325 for strength and design purposes.

9.2 BOLT INSTALLATION

There are three basic joint types that we will consider: snug tight, pretensioned, and slip-critical. The differences among these joint types are essentially the amount of clamping force that is achieved when tightening the bolts and the degree to which the connected parts can move while in service. The contact area between the connected parts is called the faying surface. In any project, the engineer must indicate the joint type and the faying surface that are to be used for any given connection.

A *snug-tight* condition occurs when the bolts are in direct bearing and the plies of a connection are in firm contact. This can be accomplished by the full effort of a worker using a spud wrench, which is an open-ended wrench approximately 16 in. long. The opposite end of the wrench is tapered to a point, which an ironworker uses to align the holes of the connecting parts. A snug-tight joint can be specified for most simple shear connections, as well as tension-only connections. Snug-tight joints are not permitted for connections supporting nonstatic loads, nor are they permitted with A490 bolts loaded in tension.

A *pretensioned joint* has a greater amount of clamping force than the snug-tight condition and therefore provides a greater degree of slip-resistance in the joint. Pretensioned joints are used for joints that are subject to cyclical loads or fatigue loads. They are also required for joints with A490 bolts in tension. Some specific examples of connections where pretensioned joints should be specified are

- Column splices in buildings with high height-to-width ratios,
- Connections within the load path of the lateral force resisting system, and
- Connections supporting impact or cyclical loads such as cranes or machinery.

It is important to note that the design strength of a pretensioned joint is equal to that of a snug-tightened joint. In a pretensioned joint, slip is prevented until the friction force is exceeded. Once the friction force is exceeded, the bolts slip into direct bearing and the pretension or clamping force is essentially zero (i.e., equivalent to a snug-tight condition). For both snug-tight and pretensioned bolts, the faying surface is permitted to be uncoated, painted, or galvanized, but must be free of dirt and other foreign material.

When pretensioned bolts are installed, they must be tightened such that a minimum clamping force is achieved between the connected parts. The AISC specification stipulates that the minimum required clamping force should be at least 70% of the nominal tensile strength, R_n, of the fastener. Table 9-1 indicates minimum tension values for various bolt types.

In order to achieve this minimum tensile force, the bolts must be installed by one of the following methods:

Table 9-1 Minimum bolt pretension (pretensioned and slip-critical joints)

Minimum Bolt Pretension, = 0.70R_n* (kips)		
Bolt Size, in.	A325 and F1852 Bolts	A490 Bolts
½	12	15
⅝	19	24
¾	28	35
⅞	39	49
1	51	64
1⅛	56	80
1¼	71	102
1⅜	85	121
1½	103	148

Adapted from Table J3.1 of the AISCM

*From equation J3-1, $R_n = F_{nt}A_b$.
F_{nt} = 90 ksi (A325 and F1852)
F_{nt} = 113 ksi (A490)
A_b = Nominal unthreaded body area of bolt

1. *Turn of the Nut*: When a nut is advanced along the length of a bolt, each turn corresponds to a certain amount of tensile force in the bolt. Therefore, there is a known relationship between the number of turns and the amount of tension in the bolt. The starting point (i.e., a point where the tensile force in the bolt is just above zero) is defined as the snug-tight condition.

2. *Calibrated Wrench Tightening*: For this method, calibrated wrenches are used so that a minimum torque is obtained, which corresponds to a specific tensile force in the bolt. On any given project, the calibration has to be done daily for each size and grade of bolt.

3. *Twist-off-type Tension-control Bolts*: As discussed in Section 9.1, these bolts conform to ASTM F1852 and are equivalent to ASTM A325 for strength and design. These bolts have a splined end that breaks off when the bolt is tightened with a special wrench (see Figure 9-1).

4. *Direct tension indicator*: Washers that conform to ASTM F959 have ribbed protrusions on the bearing surface that compress in a controlled manner such that it is

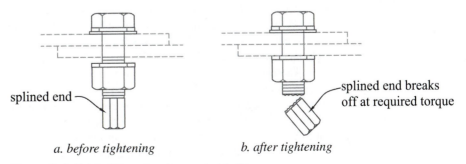

splined end —

splined end breaks off at required torque

a. before tightening   *b. after tightening*

Figure 9-1 Twist-off type tension control bolts.

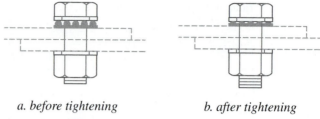

a. before tightening   *b. after tightening*

Figure 9-2 Direct tension indicator washer.

proportional to the tension in the bolt (see Figure 9-2). The deformation in the ribs is measured to determine whether the proper tension has been achieved.

One unique condition pertaining to installation is when bolted connections slip into bearing when the building is in service. When this happens, occupants might hear what sounds like a gunshot, which is naturally disturbing. However, this event is not indicative of a structural failure. This is referred to as "banging bolts." To prevent this from happening, the bolts should be snug-tight or, if it is possible, the steel erector should not tighten the bolts until the drift pins have been released and the bolts are allowed to slip into bearing before tightening (see Section 9.4 for further discussion on bolt bearing).

The final type of joint that we will consider is a *slip-critical* joint. This type of joint is similar to a pretensioned joint except that failure is assumed to occur when the applied load is greater than the friction force and thus slip does not occur between the faying surfaces. As with pretensioned joints, slip-critical joints are used for joints subjected to cyclical loads or fatigue loads. They should also be used in connections that have slotted holes parallel to the direction of the load or in connections that use a combination of welds and bolts along the same faying surface. The amount of pretension or clamping force for a slip-critical bolt is the same that was used for pretensioned joints (see Table 9-1). The design strength of a slip-critical joint is generally lower than that of a bearing-type connection since the friction resistance is usually lower than any other failure mode for a bolt (such as direct shear or bearing).

The main difference between pretensioned and slip-critical joints is the type of faying surface between the connected parts. There are three types of faying surfaces identified in the AISC specification: Class A, Class B, and Class C. Each type has a specific surface preparation and coating requirement that corresponds to a minimum coefficient of friction. A Class A faying surface has either unpainted clean mill-scale surfaces, or surfaces with Class A coatings on blast-cleaned steel. The mean slip coefficient for a Class A surface is $\mu = 0.35$. A Class A coating has a minimum mean slip coefficient of $\mu = 0.35$.

A Class B faying surface has either unpainted blast-cleaned steel surfaces or surfaces with Class B coatings on blast-cleaned steel. The mean slip coefficient for a Class B surface is $\mu = 0.50$, and a Class B coating has a minimum mean slip coefficient of $\mu = 0.50$.

A Class C faying surface has the same mean slip coefficient as a Class A surface ($\mu = 0.35$), but has roughened surfaces and a hot-dipped galvanized surface.

The mean slip coefficient for any faying surface can also be established by testing for special coatings and steel surface conditions.

Another key parameter in slip-critical connections is the probability of slip occurring at service loads or at factored loads. In some cases, slip between faying surfaces could lead to serviceability problems, but not necessarily problems at the strength level. In essence, there is slip between the faying surfaces, but that does not cause the bolt to be engaged in bearing or to cause yielding or fracture in the connected parts. In other cases, slip between the faying surfaces could cause problems at the strength level (e.g., a bolted splice in a long-span roof truss). Slip in the splice connection would cause additional deflection and increase the potential for undesirable ponding.

The AISC specification recognizes that there needs to be greater reliability in the prevention of slip for strength-sensitive connections; thus, the capacity of these connections has a reduction factor of 0.85 applied to the design strength (see ϕ in equation (9-4)). When slip is a serviceability limit state, then $\phi = 1.0$ (see equation (9-4)).

9.3 HOLE TYPES AND SPACING REQUIREMENTS

There are four basic hole types recognized in the AISC specification: standard, oversized, short-slotted, and long-slotted. Table J3.3 in the *AISCM* lists the actual hole sizes for each bolt diameter and hole type. Each hole type offers varying degrees of flexibility in the construction of the connections. Standard holes are the most common and are generally used for bolts in direct bearing. Oversized holes are only allowed for slip-critical connections. Short-slotted and long-slotted holes are used for slip-critical connections or for bearing connections where the direction of the load is normal to the length of the slot. Figure 9-2 indicates the dimensions for each bolt hole type (adapted from the *AISCM*, Table J3.3). The *AISCM*, Section B13.b indicates that when calculating the net area for shear and tension, an additional 1/16 in. should be added to the hole size to account for the roughened edges that result from the punching or drilling process. For standard holes, the hole size used for strength calculations would be the value from the *AISCM*, Table J3.3 (or Figure 9-3) plus 1/16 in.

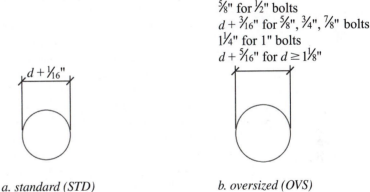

a. standard (STD)

b. oversized (OVS)

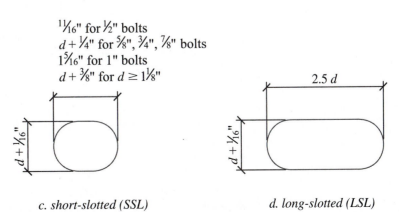

c. short-slotted (SSL)

d. long-slotted (LSL)

Figure 9-3 Bolt hole types.

Table 9-2 Recommended maximum hole sizes in column base plates

Anchor Rod Diameter, in.	Maximum Hole Diameter, in.
¾	1⁵⁄₁₆
⅞	1⁹⁄₁₆
1	1¹³⁄₁₆
1¼	2¹⁄₁₆
1½	2⁵⁄₁₆
1¾	2¾
2	3¼
2½	3¾

Adapted from Table 14-2, *AISC Manual of Steel Construction,* 13th ed.

For column bases, it has been recognized that the embedment of anchor rods into a concrete foundation generally does not occur within desirable tolerances and thus has led to numerous errors in the alignment of the columns. One way to mitigate this problem is to provide larger holes in the column base plates to allow for misaligned anchor rods. Table 9-2 indicates the recommended maximum hole sizes in column base plates.

In order to allow for standard fabrication procedures, as well as workmanship tolerances, the *AISCM,* Section J3.3 recommends that the minimum spacing between bolts be at least $3d$, with an absolute minimum of $2\frac{2}{3}\,d$, where d is the bolt diameter. The minimum distance in any direction from the center of a standard hole to an edge is given in the *AISCM,* Table J3.4. In general, the minimum edge distance is approximately $1.75d$ for bolts near a sheared edge and approximately $1.25d$ for bolts near a rolled or thermally cut edge (see Figure 9-4). One exception is that the workable gages in angle legs (*AISCM,* Table 1-7) may be used for single and double angles.

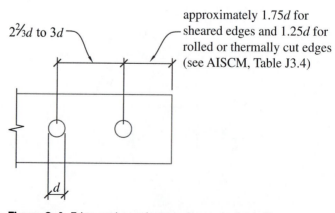

Figure 9-4 Edge and spacing requirements for bolts.

9.4 STRENGTH OF BOLTS

There are three basic failure modes of bolt strength: bearing, shear, and tension. In this section, we will consider the strength of the fasteners, as well as the connected sections.

Most basic shear connections are fastened in such a way that the bolts bear directly on the connected parts. In order for this to happen, there must be some nominal amount of displacement or slip to allow the bolts to bear directly on the connected parts. A slip-critical connection is a bolted connection where any amount of displacement is not desirable, and therefore the bolts must be pretensioned to the loads indicated in Table 9-1. For bearing connections, the bolts can be loaded either in single shear or double shear (see Figure 9-5). A lapped connection with two members has one shear plane, and therefore the bolt is considered to be loaded in single shear. For a three-member connection, there are two shear planes; therefore, the bolt is loaded in double shear. The additional shear plane reduces the amount of load to the bolts in each shear plane.

In order for the bolt to adequately transfer loads from one connected part to another, the connection material must have adequate strength in bearing. The design bearing strength for a bolt in a connection with standard, oversized, and short-slotted holes, or long-slotted hole slots parallel to the direction of the load is

$$\phi R_n = \phi 1.2 L_c t F_u \le \phi 2.4 dt F_u. \tag{9-1}$$

At connections with long-slotted holes with the slot perpendicular to the direction of the load, the design bearing strength is

$$\phi R_n = \phi 1.0 L_c t F_u \le \phi 2.0 dt F_u, \tag{9-2}$$

where

$\phi = 0.75$,

R_n = Nominal bearing strength, kips,

L_c = Clear distance between the edge of the hole and the edge of an adjacent hole, or the edge of the connected member in the direction of the load (see Figure 9-6),

t = Thickness of the connected material, in.,

d = Bolt diameter, in., and

F_u = Minimum tensile strength of the connected member.

The design shear and tension strength of a snug-tight or pretensioned bolt is

$$\phi R_n = \phi F_n A_b, \tag{9-3}$$

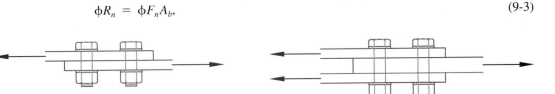

a. single shear   *b. double shear*

Figure 9-5 Bolts loaded in single and double shear.

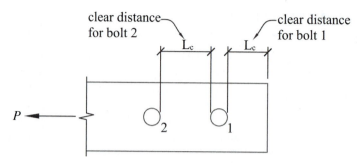

Figure 9-6 Clear distance for bolt bearing.

where

$\phi = 0.75$,

R_n = Nominal shear or tension strength, k,

F_n = Nominal shear strength (F_{nv}) or tension strength (F_{nt}) (see Table 9-3), and

A_b = Nominal unthreaded body area of bolt.

From the *AISCM,* Table 2-5, the ultimate tensile strength, F_u, is 60 ksi, for A307 grade A bolts 120 ksi for A325 and F1852 bolts (up to 1 in. in diameter), and 150 ksi for A490 bolts. The nominal tensile strength, F_{nt}, is taken as $0.75F_u$ for all cases. The nominal shear strength, F_{nv}, is a function of whether or not bolt threads are in the shear plane. When the threads are excluded from the shear plan, the nominal shear strength is taken as $0.5F_u$. When the threads are included in the shear plane, the nominal shear strength is reduced to $0.4F_u$. When bolt threads are intended to be excluded from the shear plane, they are designated as A325X or A490X. When the bolt threads are intended to be included in the shear plane, the proper designation is A325N or A490N. Table 9-3 summarizes the nominal shear and tensile strengths for various bolt types.

For slip-critical connections, the load is transmitted by friction between the connected parts. Since bearing is assumed to not occur, the strength of the fastener comes entirely from friction. However, the *AISCM,* Section J3.8 still requires that the bolts meet the strength requirements for bearing on the connected parts and shear in the bolts. For these types of connections, it is important to ensure that the connected parts are in firm contact, and that the faying surface is properly identified. The design slip resistance of a slip-critical bolted connection is

Table 9-3 Nominal shear and tensile strength of bolts

Bolt Type	Ultimate Tensile Strength, F_u, ksi	Nominal Tensile Strength, F_{nt}, ksi	Nominal Shear Strength, F_{nv}, ksi
A307, grade A	60	45	24
A325N, F1852N (up to 1" dia.)	120	90	48
A325X, F1852X (up to 1" dia.)	120	90	60
A490N	150	113	60
A490X	150	113	75

Adapted from Table J3.2, *AISC Manual of Steel Construction,* 13th ed.

$$\phi R_n = \phi \mu D_u h_{sc} T_b N_s, \tag{9-4}$$

where

> $\phi = 1.0$ if prevention of slip is a serviceability limit state
>
> $= 0.85$ if prevention of slip is at the *required strength* level,
>
> $R_n =$ Nominal shear strength, kips,
>
> $\mu =$ Mean slip coefficient
>
> $= 0.35$ for Class A surfaces
>
> $= 0.50$ for Class B surfaces
>
> $= 0.35$ for Class C surfaces,
>
> $D_u = 1.13$ (the constant value that represents the ratio between the mean installed bolt pretension and the minimum required bolt pretension; alternate values can be used if it is verified),
>
> $h_{sc} =$ Hole size factor
>
> $= 1.0$ for standard holes (STD)
>
> $= 0.85$ for oversized and short-slotted holes (OVS and SSL)
>
> $= 0.70$ for long-slotted holes (LSL),
>
> $N_s =$ Number of slip planes, and
>
> $T_b =$ Minimum bolt pretension (see Table 9-1).

When designing slip-critical connections, a Class A surface is usually assumed, which is conservative. Steel with a Class B surface would require blast cleaning, which adds labor, time, and cost. It is also generally good practice to use standard holes since oversized and slotted holes are typically not necessary. When a connection requires slip-critical bolts, they should be designated as A325SC or A490SC.

When fasteners are loaded such that there exists shear and tension components (see Figure 9-7), an interaction equation is required for design. Research has indicated that the interaction curve is

$$\left(\frac{f_t}{F_t}\right)^2 + \left(\frac{f_v}{F_v}\right)^2 \leq 1.0, \tag{9-5}$$

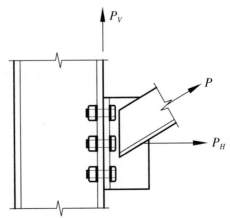

Figure 9-7 Connection subjected to combined loading.

where

f_t = Applied tensile stress,

F_t = Allowable tensile stress,

f_v = Applied shear stress, and

F_v = Allowable shear stress.

This equation is represented graphically in Figure 9-8 such that any design that falls under the curve is acceptable. The *AISCM*, Section J3.7 approximates this curve as follows:

$$F'_{nt} = 1.3F_{nt} - \frac{F_{nt}}{\phi F_{nv}}f_v \le F_{nt},$$ (9-6)

where

F'_{nt} = Nominal tension stress modified to include shear effects, ksi,

F_{nt} = Nominal tension stress, ksi (see Table 9-3),

F_{nv} = Nominal shear stress, ksi (see Table 9-3),

f_v = Applied or required shear stress, ksi, and

ϕ = 0.75.

The design tensile strength of each bolt then becomes

$$\phi R_n = \phi F'_{nt}A_b.$$ (9-7)

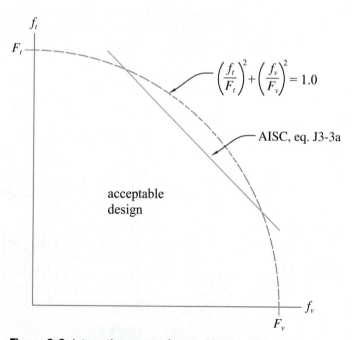

Figure 9-8 Interaction curves for combined loading.

These equations can be rewritten to determine the design shear strength. It should also be noted that when the applied shear stress is less than 20% of the available shear strength, or when the applied tension stress is less than 20% of the available tension strength, the effects of the combined stress do not have to be investigated.

When slip-critical connections are subjected to combined shear and tension loads, the tension load reduces the amount of clamping force that reduces the available slip resistance in each bolt. Therefore, the design shear strength calculated in equation (9-4) shall be reduced by the following factor:

$$k_s = 1 - \frac{T_u}{D_u T_b N_b}, \tag{9-8}$$

where

k_s = Reduction factor,

T_u = Factored tension force, kips,

D_u = 1.13 (see eq. (9-4) for discussion),

T_b = Minimum bolt pretension (see Table 9-1),

N_b = Number of bolts resisting the applied tension.

The example problems that follow will cover the design strength of bolts in common types of connections.

EXAMPLE 9-1

High-strength Bolts in Shear and Bearing

Determine whether the connection shown in Figure 9-9 is adequate to support the applied loads; consider the strength of the bolts in shear and bearing on the plate. The plates are ASTM A36 and the bolts are ¾-in.-diameter A325N in standard holes.

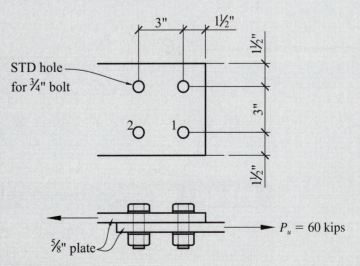

Figure 9-9 Connection details for Example 9-1.

(continued)

SOLUTION

Check bolt bearing:

$$L_{c1} = 1.5 \text{ in.} - (0.5)\left(\frac{3}{4} + \frac{1}{8}\right) = 1.06 \text{ in.}$$

$$L_{c2} = 3 \text{ in.} - \left(\frac{3}{4} + \frac{1}{8}\right) = 2.13 \text{ in.}$$

For bolt 1,

$$
\begin{aligned}
\phi R_n &= \phi 1.2 L_c t F_u \le \phi 2.4 dt F_u \\
&= (0.75)(1.2)(1.06 \text{ in.})(0.625 \text{ in.})(58 \text{ ksi}) \\
&\quad < (0.75)(2.4)(0.75)(0.625 \text{ in.})(58 \text{ ksi}) \\
&= 34.6 \text{ kips} < 48.9 \text{ kips} \\
&= 34.6 \text{ kips.}
\end{aligned}
$$

For bolt 2,

$$
\begin{aligned}
\phi R_n &= (0.75)(1.2)(2.13 \text{ in.})(0.625 \text{ in.})(58 \text{ ksi}) \\
&\quad < (0.75)(2.4)(0.75)(0.625 \text{ in.})(58 \text{ ksi}) \\
&= 69.3 \text{ kips} > 48.9 \text{ kips} \\
&= 48.9 \text{ kips.}
\end{aligned}
$$

The total strength of the connection, considering all of the bolts in bearing, is

$$
\begin{aligned}
\phi R_n &= (2)(34.6 \text{ kips}) + (2)(48.9 \text{ kips}) = \mathbf{167 \text{ kips}} \\
&> P_u = \mathbf{60 \text{ kips. OK for bearing}}
\end{aligned}
$$

Check shear on the bolts:

$$
\begin{aligned}
\phi R_n &= \phi F_n A_b \\
&= (0.75)(48 \text{ ksi})(0.442 \text{ in.}^2) \\
&= 15.9 \text{ kips/bolt (agrees with } AISCM \text{ Table 7-1)}
\end{aligned}
$$

The total shear strength of the bolt group is

$$\phi R_n = (4)(15.9 \text{ kips}) = \mathbf{63.6 \text{ kips}} > \mathbf{60 \text{ k. OK for shear}}$$

EXAMPLE 9-2

High-strength bolts in a slip-critical Connection

Repeat Example 9-1 with A325SC bolts and a Class A faying surface; the slip is a serviceability limit state.

SOLUTION

Check bolt bearing:

$$L_{c1} = 1.5 \text{ in.} - (0.5)\left(\frac{3}{4} + \frac{1}{8}\right) = 1.06 \text{ in.}$$

$$L_{c2} = 3 \text{ in.} - \left(\frac{3}{4} + \frac{1}{8}\right) = 2.13 \text{ in.}$$

For bolt 1,

$$\begin{aligned}
\phi R_n &= \phi 1.0 L_c t F_u \leq \phi 2.0 dt F_u \\
&= (0.75)(1.0)(1.06 \text{ in.})(0.625 \text{ in.})(58 \text{ ksi}) \\
&< (0.75)(2.0)(0.75)(0.625 \text{ in.})(58 \text{ ksi}) \\
&= 28.8 \text{ kips} < 40.7 \text{ kips} \\
&= 28.8 \text{ kips.}
\end{aligned}$$

For bolt 2,

$$\begin{aligned}
\phi R_n &= (0.75)(1.0)(2.13 \text{ in.})(0.625 \text{ in.})(58 \text{ ksi}) \\
&< (0.75)(2.0)(0.75)(0.625 \text{ in.})(58 \text{ ksi}) \\
&= 57.7 \text{ kips} > 40.7 \text{ kips} \\
&= 40.7 \text{ kips.}
\end{aligned}$$

The total strength of the connection, considering all of the bolts in bearing, is

$$\begin{aligned}
\phi R_n &= (2)(28.8 \text{ kips}) + (2)(40.7 \text{ kips}) = \mathbf{139} \text{ kips} \\
&> P_u = \mathbf{60 \text{ kips. OK for bearing}}
\end{aligned}$$

Check shear on the bolts:

$$\begin{aligned}
\phi R_n &= \phi \mu D_u h_{sc} T_b N_s \\
&= (1.0)(0.35)(1.13)(1.0)(28)(1) \\
&= 11.1 \text{ kips/bolt (agrees with } AISCM, \text{ Table 7-3)}
\end{aligned}$$

The total shear strength of the bolt group is

$$\phi R_n = (4)(11.1 \text{ kips}) = 44.4 \text{ kips} < 60 \text{ kips. not good for shear}$$

The connection is not adequate.

EXAMPLE 9-3

Bolted Splice Connection

For the splice connection shown in Figure 9-10, determine the maximum factored load, P_u, that can be applied; consider the strength of the bolts only. The plates are ASTM A36 and the bolts are 1-in.-diameter A325X in standard holes.

(continued)

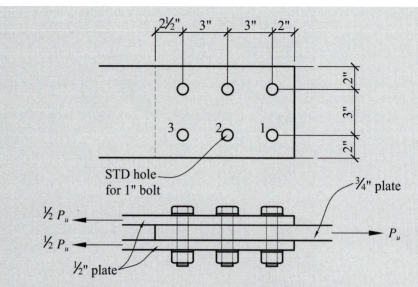

Figure 9-10 Connection details for Example 9-3.

SOLUTION

Check bolt bearing on outside plates:

$$L_{c1} = 2 \text{ in.} - (0.5)(1 + 1/8) = 1.43 \text{ in.}$$
$$L_{c2} = L_{c3} = 3 \text{ in.} - (1 + 1/8) = 1.87 \text{ in.}$$

For bolt 1,

$$\begin{aligned}
\phi R_n &= \phi 1.2 L_c t F_u \le \phi 2.4 dt F_u \\
&= (0.75)(1.2)(1.43 \text{ in.})(0.5 \text{ in.})(58 \text{ ksi}) \\
&< (0.75)(2.4)(1 \text{ in.})(0.5 \text{ in.})(58 \text{ ksi}) \\
&= 37.5 \text{ kips} < 52.2 \text{ kips} \\
&= 37.5 \text{ kips.}
\end{aligned}$$

For bolts 2 and 3,

$$\begin{aligned}
\phi R_n &= (0.75)(1.2)(1.87 \text{ in.})(0.5 \text{ in.})(58 \text{ ksi}) \\
&< (0.75)(2.4)(1 \text{ in.})(0.5 \text{ in.})(58 \text{ ksi}) \\
&= 48.8 \text{ kips} < 52.2 \text{ kips} \\
&= 48.8 \text{ kips.}
\end{aligned}$$

The total strength, considering all of the bolts in bearing on the outside plates, is

$$\phi R_n = (4)(37.5 \text{ kips}) + (8)(48.8 \text{ kips}) = 540 \text{ kips.}$$
$$P_u = 540 \text{ kips (considering bearing on outside plates only)}$$

Check bolt bearing on inside plate:

$$L_{c3} = 2.5 \text{ in.} - (0.5)(1 + \frac{1}{8}) = 1.93 \text{ in.}$$
$$L_{c2} = L_{c1} = 3 \text{ in.} - (1 + \frac{1}{8}) = 1.87 \text{ in.}$$

For bolt 1,

$$\begin{aligned}
\phi R_n &= \phi 1.2 L_c t F_u \leq \phi 2.4 d t F_u \ (\text{AISC, eq. J3.6a}) \\
&= (0.75)(1.2)(1.93 \text{ in.})(0.75 \text{ in.})(58 \text{ ksi}) \\
&\quad < (0.75)(2.4)(1 \text{ in.})(0.75 \text{ in.})(58 \text{ ksi}) \\
&= 75.5 \text{ kips} < 78.3 \text{ kips} \\
&= 75.5 \text{ kips.}
\end{aligned}$$

For bolts 2 and 3,

$$\begin{aligned}
\phi R_n &= (0.75)(1.2)(1.87 \text{ in.})(0.75 \text{ in.})(58 \text{ ksi}) \\
&\quad < (0.75)(2.4)(1 \text{ in.})(0.75 \text{ in.})(58 \text{ ksi}) \\
&= 73.2 \text{ kips} < 78.3 \text{ kips} \\
&= 73.2 \text{ kips.}
\end{aligned}$$

The total strength, considering all of the bolts in bearing on the inside plate, is

$$\phi R_n = (2)(75.5 \text{ kips}) + (4)(73.2 \text{ kips}) = 443 \text{ kips.}$$
$$P_u = 443 \text{ kips (considering bearing on inside plate only)}$$

Check shear on the bolts:

$$\begin{aligned}
\phi R_n &= \phi F_n A_b \\
&= (0.75)(60 \text{ ksi})(0.785 \text{ in.}^2) \\
&= 35.3 \text{ kips/bolt (agrees with } AISCM, \text{ Table 7-1)}
\end{aligned}$$

The shear capacity of the bolt group is

$$P_u = (2)(6)(35.3 \text{ kips}) = 424 \text{ kips (considering shear on the bolts only).}$$

Shear on the bolts controls the design, so the maximum factored load that can be applied is $P_u = 424$ kips.

EXAMPLE 9-4

Bolted Connection Loaded in Shear and Tension

For the connection shown in Figure 9-11 determine whether the bolts are adequate to support the applied load under combined loading. The bolts are ⅞-in.-diameter A490X in standard holes. Assume that each bolt takes an equal amount of shear and tension.

(continued)

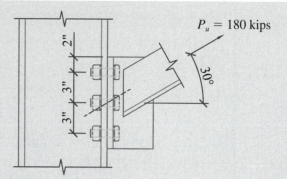

Figure 9-11 Connection details for Example 9-4.

SOLUTION

The applied load will be separated into horizontal and vertical components, respectively:

$$P_H = 180(\cos 30) = 159 \text{ kips, and}$$
$$P_V = 180(\sin 30) = 90 \text{ kips.}$$

The shear and tension stress in each bolt is

$$f_v = \frac{P_V}{A_b n} = \frac{90}{(0.601)(6)} = 25.0 \text{ ksi, and}$$

$$f_t = \frac{P_H}{A_b n} = \frac{159}{(0.601)(6)} = 44.1 \text{ ksi, respectively.}$$

The allowable shear and tension for each bolt is

$$\phi R_n = \phi F_n A_b,$$
$$\phi R_{nv} = (0.75)(75)(0.601) = 33.8 \text{ kips (agrees with } AISCM, \text{ Table 7-1)},$$
$$\phi R_{nt} = (0.75)(113)(0.601) = 51.0 \text{ kips (agrees with } AISCM, \text{ Table 7-2)},$$
$$\frac{f_v}{\phi F_{nv}} = \frac{25.0}{(0.75)(75)} = 0.44 > 0.20, \text{ and}$$
$$\frac{f_t}{\phi F_{nt}} = \frac{44.1}{(0.75)(113)} = 0.52 > 0.20.$$

Since the ratio between the actual and available strength is greater than 20% (0.20) for both shear and tension, so the combined stress effects must be considered. The interaction equation will be used to determine the modified allowable tension in each bolt:

$$F'_{nt} = 1.3F_{nt} - \frac{F_{nt}}{\phi F_{nv}}f_v \le F_{nt}$$

$$F'_{nt} = 1.3(113) - \frac{113}{(0.75)(75)}(25.0) \le 113$$

$$F'_{nt} = 96.6 \text{ ksi} < 113 \text{ ksi}$$

$$\phi F'_{nt} = (0.75)(96.6 \text{ ksi}) = 72.4 \text{ ksi}$$

$$\phi F'_{nt} = \textbf{72.4 ksi} > f_t = \textbf{44.1 ksi. OK}$$

The connection is adequate.

EXAMPLE 9-5

Bolted Connection Loaded in Shear and Tension

For the connection shown in Figure 9-11, determine the maximum factored load, P_u, that can be applied; consider the strength of the bolts only. The bolts are ⅞-in.-diameter A490SC in standard holes with a Class A faying surface; the slip is a serviceability limit state. Assume that each bolt takes an equal amount of shear and tension.

SOLUTION

The shear strength of each bolt is

$$\phi R_n = \phi \mu D_u h_{sc} T_b N_s$$
$$= (1.0)(0.35)(1.13)(1.0)(49)(1) = 19.3 \text{ kips}.$$

The shear capacity has to be reduced to account for the applied tension as follows:

$$k_s = 1 - \frac{T_u}{D_u T_b N_b}.$$

An expression can be developed to solve for the maximum load:

$$(19.3 \text{ kips})(k_s) = (P_u)(\sin 30)$$

$$19.3 \left[1 - \frac{(P_u)(\cos 30)}{D_u T_b N_b} \right] = (P_u)(\sin 30)$$

$$19.3 \left[1 - \frac{(P_u)(\cos 30)}{(1.13)(49)(6)} \right] = (P_u)(\sin 30)$$

Solving for P_u,

$$P_u = 35.1 \text{ k (per bolt)}$$
$$P_u = \textbf{(6)(35.1) = 210 kips.}$$

9.5 ECCENTRICALLY LOADED BOLTS: SHEAR

With reference to Figure 9-12, when bolts are loaded such that the load is eccentric to the bolt group in the plane of the faying surface, there are two analytical approaches that can be taken: (1) the instantaneous center (IC), of rotation method and (2) the elastic method.

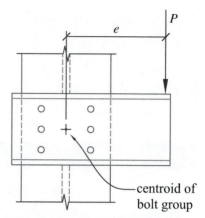

Figure 9-12 Eccentrically loaded bolt group.

In both cases, the connection is designed to resist the applied shear, P, and the additional shear generated from the moment due to the applied shear acting at an eccentricity, Pe. Both methods are relatively complex and are generally not used in practice without the aid of computers. The IC method is more accurate, but requires an iterative solution. The elastic method is less accurate and more conservative in that the ductility of the bolt group and redundancy (i.e., load distribution) are both ignored. Tables 7-7 through 7-14 in the *AISCM*, which use the IC method, are design aids for these types of connections and are more commonly used in practice and will be discussed later.

An eccentrically loaded bolt group induces both a translation and a rotation in one connected element relative to another. This combined deformation effect is equivalent to a rotation about a point called the instantaneous center (IC). The location of the IC is a function of the geometry and the direction and orientation of the load. The location of the IC (see Figure 9-13) requires an iterative solution. Note that, in the connection shown in Figure 9-13, the resulting force in each bolt will have a vertical component due to the applied vertical load and a horizontal component due to the eccentricity of the applied load.

Based on test data, the load–deformation relationship is based on the following:

$$R = R_{ult}(1 - e^{-10\Delta})^{0.55}, \tag{9-10}$$

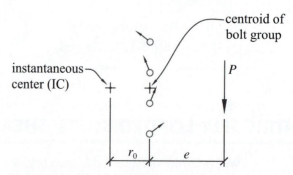

Figure 9-13 Eccentrically loaded bolt group: IC method.

where

R = Nominal shear strength of one bolt at a deformation Δ, kips,

R_{ult} = Ultimate shear strength of one bolt

  = 74 kips for ¾-in.-diameter A325 bolts, and

Δ = Total deformation, including shear, bearing, and bending in the bolt, as well as bearing deformation in the connected elements

  = 0.34 in. for ¾-in.-diameter A325 bolts.

By inspection, the bolt most remote from the IC will have the greatest load and deformation, and therefore is assumed to have the maximum load and deformation (i.e., R_{ult} = 74 kips and Δ = 0.34 in.), while the nominal shear in the other bolts in the group varies linearly with respect to the distance that the other bolts are from the IC. The constant values of R_{ult} = 74 kips and Δ = 0.34 in. are based on test data for ¾-in.-diameter A325 bolts, but can conservatively be used for bolts of other sizes and grades. Tables 7-7 through 7-14 in the *AISCM*, which use the IC method, are based on these constant values.

Using the free-body diagram shown in Figure 9-14, the forces in each bolt can be obtained from the following equilibrium equations:

$$\Sigma F_x = \sum_{n=1}^{m} (R_x)_n - P_x = 0, \tag{9-11}$$

$$\Sigma F_y = \sum_{n=1}^{m} (R_y)_n - P_y = 0, \tag{9-12}$$

$$M_{IC} = P(r_0 + e) - \sum_{n=1}^{m} r_n R_n = 0, \tag{9-13}$$

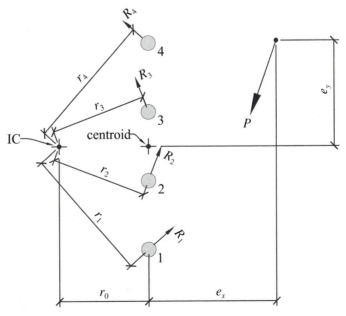

Figure 9-14 Free-body diagram of eccentrically loaded bolt group.

$$(R_x)_n = \frac{r_y}{r_n} R_n, \text{ and} \tag{9-14}$$

$$(R_y)_n = \frac{r_x}{r_n} R_n, \tag{9-15}$$

where

P = Applied load, with components P_x, P_y,

R_n = Shear in fastener, n, with components $(R_x)_n$, $(R_y)_n$,

r_n = Distance from the fastener to the IC, with components r_x, r_y,

r_0 = Distance from the IC to the centroid of the bolt group,

e = Load eccentricity; distance from the load to the centroid of the bolt group with components e_x, e_y,

n = Subscript of the individual fastener, and

m = Total number of fasteners.

In order to solve the above equilibrium equations, the location of the IC must first be assumed. If the equilibrium equations are not satisfied, then a new value must be assumed for the IC until the equations are satisfied. For most cases, the applied load, P, is vertical, and therefore equation (9-11) is eliminated by inspection.

For bolts other than ¾-in.-diameter A325N, the load capacity is also determined from a linear relationship such that the calculated nominal shear strength (R from equation (9-10)) is multiplied by the ratio between the shear strength of a ¾-in.-diameter A325 fastener and the specific fastener in question. The capacity of the connection is then the summation of the capacity of each individual fastener. Example 9-6 explains this method in further detail, as well as the use of the aforementioned design aids in the *AISCM*.

A more simplified and conservative approach is the *elastic method*. With reference to Figure 9-15, each bolt resists an equal proportion of the applied load, P, and a portion of the shear induced by the moment, Pe, proportional to its distance from the centroid of the bolt group. The shear in each bolt due to the applied load is

$$r_{px} = \frac{P_x}{n}, \text{ and} \tag{9-16}$$

$$r_{py} = \frac{P_y}{n}, \text{ respectively,} \tag{9-17}$$

where

r_p = Force in each bolt due to applied load with components r_{px}, r_{py},

P = Applied load with components P_x, P_y, and

n = Number of bolts.

The shear in the bolt most remote from the centroid of the bolt group due to the applied moment is

$$r_{mx} = \frac{Mc_y}{I_p}, \text{ and} \tag{9-18}$$

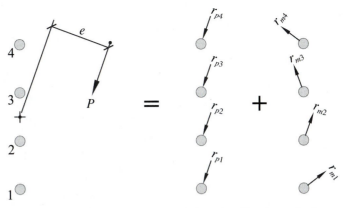

Figure 9-15 Eccentrically loaded bolt group: Elastic method.

$$r_{my} = \frac{Mc_x}{I_p},\tag{9-19}$$

where

r_m = Force in each bolt due to applied moment with components r_{mx}, r_{my},

M = Resulting moment due to eccentrically applied load

   = $P_x e_y + P_y e_x$,

c = Radial distance from the centroid of the bolt group with components c_x, c_y,

e = Load eccentricity; distance from the load to the centroid of the bolt group with components e_x, e_y, and

I_p = Polar moment of inertia of the bolt group

   = $\Sigma(I_x + I_y)$, where $I = Ad^2$

   = $\Sigma(c_x^2 + c_y^2)$ for bolts with the same cross-sectional area within a bolt group.

For this method, the critical fastener force is determined and is the basis for the connection design. The critical fastener force is usually found in the fastener located most remote to the bolt group. The critical fastener force is

$$r = \sqrt{(r_{px} + r_{mx})^2 + (r_{py} + r_{my})^2}.\tag{9-20}$$

The value of r above is used to determine the required strength of each fastener in the bolt group. See Example 9-6 for further explanation of this method.

EXAMPLE 9-6

Eccentrically Loaded Bolts in Shear

For the bracket connection shown in Figure 9-16, determine the design strength of the connection. Compare the results from the ultimate strength (or IC) method, the elastic method, and the appropriate design aid from *AISCM*, Tables 7-7 through 7-14. The bolts are ¾-in.-diameter A490N.

(continued)

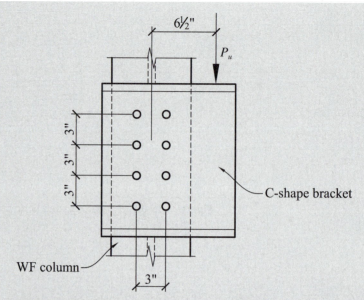

Figure 9-16 Connection details for Example 9-6.

Ultimate Strength, or IC, Method:

A location for the IC must first be assumed. By trial and error, a value of $r_0 = 1.90$ in. has been determined. A free-body diagram of the bolt group is shown in Figure 9-17.

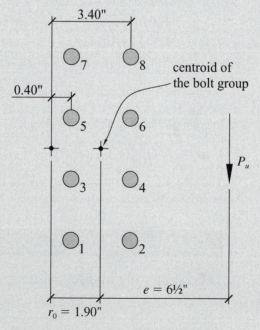

Figure 9-17 Free-Body diagram for Example 9-6. (IC method).

From the free-body diagram, the distance of each bolt from the IC can be determined:

Bolt	Distance from the IC		
	r_x, in.	r_y, in.	$r_n = \sqrt{r_x^2 + r_y^2}$, in.
1	0.40	−4.5	4.52
2	3.40	−4.5	5.64
3	0.40	−1.5	1.55
4	3.40	−1.5	3.72
5	0.40	1.5	1.55
6	3.40	1.5	3.72
7	0.40	4.5	4.52
8	3.40	4.5	5.64

For the IC method, the bolt or bolts most remote from the IC are assumed to be stressed and deformed to failure; the deformation in the remaining bolts varies linearly. The bolts at locations 2 and 8 are the most remote ($r_2 = r_8 = 5.64$ in.); therefore, the deformation of each is assumed to be $\Delta = 0.34$ in. The deformation of the remaining bolts is determined as follows:

$$\Delta_1 = \Delta_7 = \frac{4.52 \text{ in.}}{5.64 \text{ in.}}(0.34 \text{ in.}) = 0.272 \text{ in.}$$

$$\Delta_3 = \Delta_5 = \frac{1.55 \text{ in.}}{5.64 \text{ in.}}(0.34 \text{ in.}) = 0.094 \text{ in.}$$

$$\Delta_4 = \Delta_6 = \frac{3.72 \text{ in.}}{5.64 \text{ in.}}(0.34 \text{ in.}) = 0.224 \text{ in.}$$

$$\Delta_2 = \Delta_8 = 0.34 \text{ in.}$$

Using the load–deformation relationship given in equation (9-10), the shear in each bolt can be determined:

$$R = R_{ult}(1 - e^{-10\Delta})^{0.55}$$

$$R_1 = R_7 = (74)(1 - e^{-(10)(0.272)})^{0.55} = 71.3 \text{ kips}$$

$$R_3 = R_5 = (74)(1 - e^{-(10)(0.094)})^{0.55} = 56.3 \text{ kips}$$

$$R_4 = R_6 = (74)(1 - e^{-(10)(0.224)})^{0.55} = 69.6 \text{ kips}$$

$$R_2 = R_8 = (74)(1 - e^{-(10)(0.34)})^{0.55} = 72.6 \text{ kips}$$

The vertical component of each shear force, R, in the bolt is determined from equation (9-16):

$$(R_y)_n = \frac{r_x}{r_n}(R_n)$$

(continued)

$$R_{y1} = R_{y7} = \frac{0.40 \text{ in.}}{4.52 \text{ in.}}(71.3 \text{ k}) = 6.33 \text{ kips}$$

$$R_{y3} = R_{y5} = \frac{0.40 \text{ in.}}{1.55 \text{ in.}}(56.3 \text{ k}) = 14.5 \text{ kips}$$

$$R_{y4} = R_{y6} = \frac{3.40 \text{ in.}}{3.72 \text{ in.}}(69.6 \text{ k}) = 63.6 \text{ kips}$$

$$R_{y2} = R_{y8} = \frac{3.40 \text{ in.}}{5.64 \text{ in.}}(72.6 \text{ k}) = 43.8 \text{ kips}$$

From the equilibrium equation (9-12),

$$\Sigma F_y = \sum_{n=1}^{m} (R_y)_n - P_y = 0$$

$$(2)(6.33 \text{ kips} + 14.5 \text{ kips} + 63.6 \text{ kips} + 43.8 \text{ kips}) = P_y$$

$$P_y = \mathbf{256} \textbf{ kips.}$$

Since there is no horizontal component to the applied load, equation (9-11) is satisfied and $P = P_y$. To verify this value, equilibrium equation (9-13) will be used:

$$M_{IC} = P(r_0 + e) - \sum_{n=1}^{m} r_n R_n = 0$$

$$P(1.9 \text{ in.} + 6.5 \text{ in.}) = 2[(4.52 \text{ in.})(71.3 \text{ kips}) + (5.64 \text{ in.})(72.6 \text{ kips})$$
$$+ (1.55 \text{ in.})(56.3 \text{ kips}) + (3.72 \text{ in.})(69.6)]$$

$$P = \mathbf{256} \textbf{ kips.}$$

Since the equilibrium equations have been satisfied, the assumed location for the IC is correct.

For the IC method, the maximum shear and deformation values of $R_{ult} = 74$ k and $\Delta = 0.34$ in. are used as baseline values. For fasteners of other sizes and grades, linear interpolation is conservatively used. The nominal strength of the connection is then

$$P\left(\frac{R_n}{R_{ult}}\right) = 256 \text{ kips} \left[\frac{(60 \text{ ksi})(0.442 \text{ in.}^2)}{74 \text{ kips}}\right] = 92.0 \text{ kips.}$$

The design strength of the connection is

$$P_u = \phi R_n = (0.75)(92.0) = \mathbf{69} \textbf{ kips.}$$

Elastic Method:

For this method, the critical fastener force is determined from the bolt most remote from the centroid of the bolt group. For this connection, bolts at locations 1, 2, 7, and 8 are the most remote. The capacity of one of these bolts will be used to determine the strength of the connection. The shear is distributed equally to all of the bolts in the group. Since there is no horizontal component for the applied load, equation (9-16) is satisfied. From equation (9-17),

$$r_{py} = \frac{P_y}{n} = \frac{P}{8}.$$

The shear in the bolts due to the eccentricity is determined from equations (9-18) and (9-19):

$$e = 6.5 \text{ in.}$$

$$I_p = \Sigma(c_x^2 + c_y^2) = 4[(1.5)^2 + (4.5)^2] + 4[(1.5)^2 + (1.5)^2] = 108 \text{ in.}^4$$

$$r_{mx} = \frac{Mc_y}{I_p} = \frac{Pec_y}{I_p} = \frac{P(6.5)(4.5)}{108} = 0.271P$$

$$r_{my} = \frac{Mc_x}{I_p} = \frac{Pec_x}{I_p} = \frac{P(6.5)(1.5)}{108} = 0.0902P.$$

Equation (9-20) will be used to determine the capacity of the bolt most remote from the centroid. Using $r = \phi r_n$,

$$r = \sqrt{(r_{px} + r_{mx})^2 + (r_{py} + r_{my})^2}$$

$$(0.75)(60)(0.442) = \sqrt{(0 + 0.271P)^2 + \left(\frac{P}{8} + 0.0902P\right)^2}.$$

Solving for P,

$$P = P_u = 57.5 \text{ kips.}$$

The elastic method is more conservative than the IC method, so it is expected that the connection capacity is lower (57.5 kips < 69 kips).

AISC Design Aids

Based on the connection geometry, *AISCM*, Table 7-8, with Angle = 0° and $s = 3$ in. will be used:

$$e = 6.5 \text{ in., and}$$

$$n = 4.$$

By linear interpolation, $C = 3.48$.

The design strength of the connection is

$$\phi R_n = \phi C r_n,$$

where $\phi r_n = \phi F_{nv} A_b$.

$$\phi r_n = \phi F_{nv} A_b = (0.75)(60)(0.442) = 19.9 \text{ kips}$$
$$\text{(agrees with \textit{AISCM}, Table 7-1)}$$

$$\phi R_n = \phi C r_n = (3.48)(19.9)$$

$$\phi R_n = \textbf{69.2 kips}$$

This value agrees with the value of P_u calculated with the IC method.

9.6 ECCENTRICALLY LOADED BOLTS: BOLTS IN SHEAR AND TENSION

With reference to Figure 9-18, a bolt group is loaded such that the eccentricity is normal to the plane of the faying surface. This type of connection induces a shear due to the applied load (P) and a moment due to the eccentricity of the applied load (P_c) the shear in each bolt such that the applied load, P, is equally distributed to each bolt. The moment that is caused by the eccentric load, Pe, is resisted by tension in the bolts above the neutral axis of the bolt group and compression below the neutral axis of the bolt group.

For eccentrically loaded bolts, there are two possible design approaches that can be taken—Case I and Case II. In Case I, the neutral axis is not necessarily at the centroid of the bolt group and is generally below the centroid of the bolt group. Case II is more simplified and conservative in that the neutral axis is assumed to be at the centroid of the bolt group and only the bolts above the neutral axis resist the tension due to the eccentric load. In both cases, the shear is distributed equally to each bolt as follows:

$$r_v = \frac{P}{n},\tag{9-20}$$

where

r_v = Force in each bolt due to the applied load,
P = Applied load with components, and
n = Number of bolts.

For Case I, a trial position for the neutral axis has to be assumed. A value of one-sixth of the depth of the connecting element is recommended as a baseline value (see Figure 9-19). The area below the neutral axis is in compression, but only for a certain width. The width of the compression zone is defined as

$$b_{eff} = 8t_f \le b_f,\tag{9-21}$$

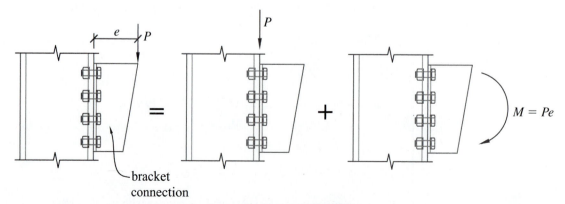

Figure 9-18 Eccentrically loaded bolts in shear in tension.

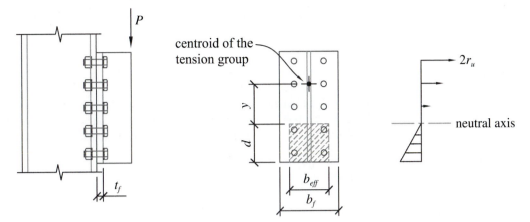

Figure 9-19 Eccentrically loaded bolts in shear in tension: Case I.

where

b_{eff} = Effective width of the compression zone,

t_f = Connecting element thickness (use the average flange thickness of the connecting element where the flange thickness is not constant), and

b_f = Width of the connecting element.

Equation (9-21) is valid for connecting elements of W-shapes, S-shapes, plates, and angles. The location assumed for the neutral axis is verified by summing moments about the neutral axis in that the moment of the bolt area above the neutral axis is compared with the moment due to the compression stress block below the neutral axis. A summation of moments yields

$$\Sigma A_b y = \frac{b_{eff}d^2}{2},$$ (9-22)

where

ΣA_b = Sum of the bolt areas above the neutral axis,

y = Distance from the centroid of the bolt group above the neutral axis to the neutral axis,

b_{eff} = Effective width of the compression zone, and

d = Depth of the compression stress block.

The location of the neutral axis is then adjusted as required until equation (9-22) is satisfied. Once the exact location of the neutral axis is determined, the tensile force in each bolt is determined as follows:

$$r_t = \frac{Mc}{I_x}A_b,$$ (9-23)

where

r_t = Force in each bolt due to the applied moment,

M = Resulting moment due to the eccentrically applied load, Pe,

A_b = Bolt area,

c = Distance from the neutral axis to the most remote bolt in the tension group,

e = Load eccentricity, and

I_x = Combined moment of inertia of the bolt group and compression block about the neutral axis.

For Case II, the neutral axis is assumed to be at the centroid of the bolt group (see Figure 9-20). Therefore, the bolts above the neutral axis are in tension and the bolts below the neutral axis are assumed to be in compression. The tensile force in each bolt is then defined as

$$r_t = \frac{M}{n' d_m},$$ (9-24)

where

r_t = Force in each bolt due to the applied moment,

M = Resulting moment due to the eccentrically applied load, Pe,

n' = Number of bolts above the neutral axis,

d_m = Moment arm between the centroid of the tension group and centroid of the compression group, and

e = load eccentricity.

Since the bolts are loaded in both shear and tension, these combined forces must be checked for conformance to interaction equation J3-3a in the *AISCM* (see equation (9-6)).

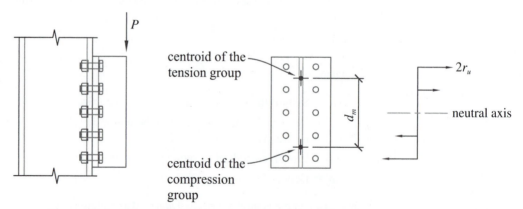

Figure 9-20 Eccentrically loaded bolts in shear in tension: Case II.

EXAMPLE 9-7

Eccentrically Loaded Bolts in Shear and Tension

Determine whether the connection shown in Figure 9-21 is adequate to support the load; consider the eccentric loading on the bolts only. Compare the results from Cases I and II. The bolts are ¾-in.-diameter A325N.

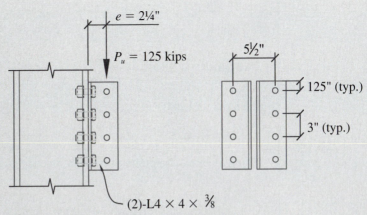

Figure 9-21 Connection details for Example 9-7.

SOLUTION

Each bolt resists an equal amount of shear for both cases:

$$r_v = \frac{P}{n} = \frac{125 \text{ kips}}{8} = 15.6 \text{ kips}$$

$$A_b = 0.442 \text{ in.}^2$$

$$f_v = \frac{r_v}{A_b} = \frac{15.6}{0.442} = 35.4 \text{ ksi}$$

Case I:

From Figure 9-22, the neutral axis is determined as follows by assuming that the neutral axis is just above the first row of bolts:

$$b_{eff} = 8t_f \le b_f$$
$$= (8)(0.375 \text{ in.}) = 3 \text{ in.} < 4 \text{ in.} + 4 \text{ in.} = 8 \text{ in.}$$

$$b_{eff} = 3 \text{ in.}$$

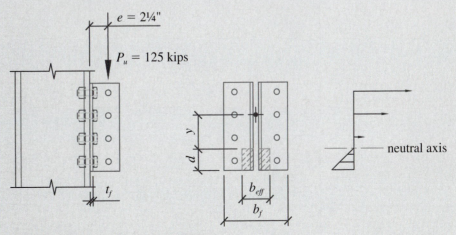

Figure 9-22 Free-body diagram: Case I.

(continued)

$$\Sigma A_b y = \frac{b_{eff} d^2}{2}$$

$$(6)(0.442)(7.25 - d) = \frac{3d^2}{2}$$

Solving for d yields,

$$d = 2.80 \text{ in.}$$

Taking a summation of moments about the neutral axis yields,

$$
\begin{aligned}
I_x &= \Sigma A d^2 \\
&= (3 \text{ in.})(2.8)(1.4 \text{ in.})^2 + (2)(0.442)(1.45)^2 \\
&\quad + (2)(0.442)(4.45)^2 + (2)(0.442)(7.45)^2 \\
I_x &= 84.9 \text{ in.}^4 \\
c &= 7.45 \text{ in.} \\
M &= Pe = (125 \text{ k})(2.25 \text{ in.}) = 281 \text{ in.-k}
\end{aligned}
$$

The tension in the bolt most remote from the neutral axis is

$$r_t = \frac{Mc}{I_x} A_b = \frac{(281)(7.45)}{84.9}(0.442) = 10.9 \text{ kips}$$

Checking combined stresses,

$$
\begin{aligned}
F_{nt} &= 90 \text{ ksi} \\
F_{nv} &= 48 \text{ ksi} \\
\phi &= 0.75 \\
F'_{nt} &= 1.3 F_{nt} - \left(\frac{F_{nt}}{\phi F_{nv}} f_v \right) \le F_{nt} \\
&= 1.3(90) - \left[\frac{90}{(0.75)(48)} 35.4 \right] \le 90 \\
&= 28.5 \text{ ksi}
\end{aligned}
$$

The design tensile strength of each bolt is

$$
\begin{aligned}
\phi R_n &= \phi F'_{nt} A_b = (0.75)(28.5)(0.442) \\
\phi R_n &= 9.45 \text{ kips} < 10.9 \text{ kips.}
\end{aligned}
$$

The connection is found to be inadequate under the Case I approach.

Case II:

As shown in Figure 9-23, only the bolts above the neutral axis resist the tension:

$$r_t = \frac{M}{n' d_m} = \frac{281}{(4)(6)} = 11.7 \text{ kips.}$$

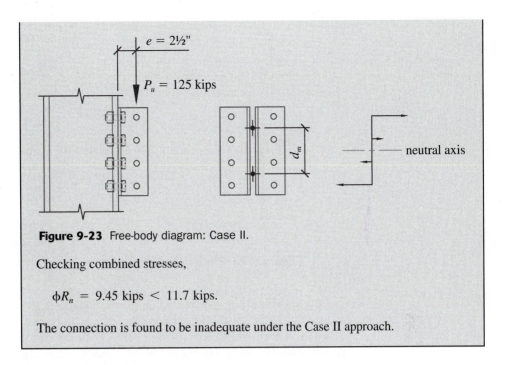

Figure 9-23 Free-body diagram: Case II.

Checking combined stresses,

$$\phi R_n = 9.45 \text{ kips} < 11.7 \text{ kips.}$$

The connection is found to be inadequate under the Case II approach.

9.7 PRYING ACTION: BOLTS IN TENSION

There are several types of connections where prying action on the connecting element results in increased tension in the bolts (see Figure 9-24). The connecting element is usually an angle or T-shape making the critical design element is the thickness of the outstanding leg that will resist the bending or prying action.

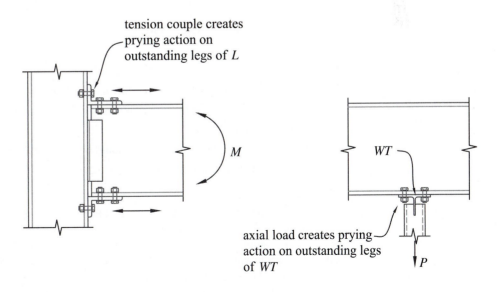

a. semi-rigid moment connection   *b. hanger connection*

Figure 9-24 Connections with prying action.

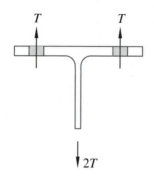
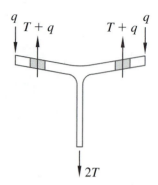

*a. hanger legs are
sufficiently rigid*

*b. hanger legs are not
sufficiently rigid*

Figure 9-25 Prying action on a hanger.

Considering the hanger connection shown in Figure 9-25, if the legs of the connecting element had sufficient stiffness, the tension force in the bolts would equal the applied load, T. This does not normally occur in these connections, due to prying action. The effect of this prying action results in a compression force near the tip of the outstanding leg, which increases the tension force in the bolt from T to $T + q$.

The derivation of the following prying action equations is discussed in references 14 and 15. Using a more simplified approach, the minimum required thickness of the connecting element to resist prying action is

$$t_{min} = \sqrt{\frac{4.44Tb'}{pF_u}}, \tag{9-26}$$

where

T = Applied tensile force in the bolt, k,

$$b' = b - \frac{d_b}{2}, \tag{9-27}$$

b = Distance from the bolt centerline to the face of the tee stem for a T-shape, in.

  = Distance from the bolt centerline to the angle-leg centerline for an L-shape, in.,

d_b = Bolt diameter, in.,

p = Tributary length of the bolts (should be less than g (see Figure 9-26)), and

F_u = Minimum tensile strength of the connecting element.

In Equation 9-26, the additional force in the bolt is nearly zero ($q \approx 0$). If the thickness of the connecting element determined from equation (9-26) is reasonable, then no further design checks need to be made.

A more complex and less conservative approach assumes a value of q that is greater than zero, thus magnifying the tensile force in the bolt. A preliminary connecting element thickness must first be selected as follows:

$$t = \sqrt{\frac{2.22Tb}{pF_u}}. \tag{9-28}$$

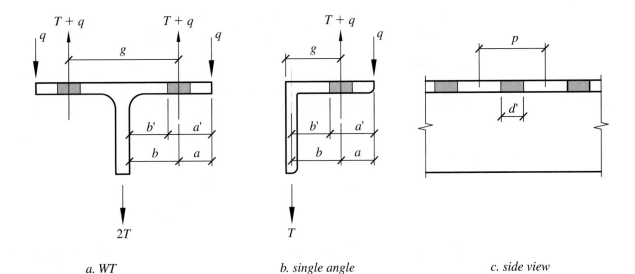

a. WT   b. single angle   c. side view

Figure 9-26 Prying action variables.

AISCM, Tables 15-1a and 15-1b, contain tabulated values based on equation (9-28). Once a preliminary thickness has been selected, then the required thickness is determined as follows:

$$t_{min} = \sqrt{\frac{4.44Tb'}{pF_u(1 + \delta\alpha')}}, \tag{9-29}$$

where

$$\delta = 1 - \frac{d'}{p}, \tag{9-30}$$

$$\alpha' = 1.0 \text{ for } \beta \geq 1,$$

$$= \frac{1}{\delta}\left(\frac{\beta}{1 - \beta}\right) < 1.0 \text{ for } \beta < 1, \tag{9-31}$$

d' = Hole width along the length of the fitting (usually the hole diameter),

$$\beta = \frac{1}{\rho}\left(\frac{B}{T} - 1\right), \tag{9-32}$$

$$\rho = \frac{b'}{a'}, \tag{9-33}$$

$$a' = \left(a + \frac{d_b}{2}\right) \leq \left(1.25b + \frac{d_b}{2}\right), \tag{9-34}$$

a = Distance from the bolt centerline to the edge of the connecting element, and

B = Available tension per bolt, ϕr_n, kips.

If t_{min} from Equation 9-29 $< t$, then the connection is adequate, otherwise the leg thickness of the connecting element will have to be increased, or the geometry of the connection will have to be changed.

EXAMPLE 9-8

Prying Action

Determine the required angle size for the moment connection shown below. The bolts are ¾-in.-diameter A490 in standard holes, and the steel is ASTM A36.

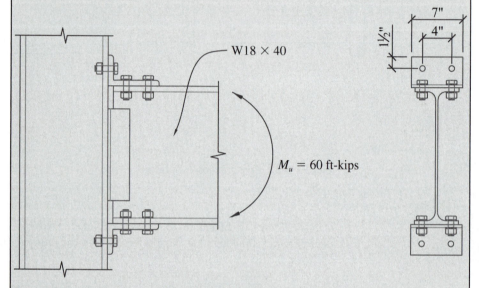

Figure 9-27 Connection details for Example 9-8.

A trial size of L7 × 4 × ¾ will be selected. The tension force due to the applied moment is found first:

$$T_u = \frac{M}{d} = \frac{(60)(12)}{17.9 \text{ in.}} = 40.2 \text{ kips.}$$

Since there are two bolts to resist the tension, the force in each bolt is 20.1 kips.

Using the geometry from Figure 9-28, the minimum angle thickness can be found using the simplified approach (eq. (9-26)):

$$a = 1.5 \text{ in.}$$

$$b = 4 \text{ in.} - 1.5 \text{ in.} - (0.5)(0.75 \text{ in.}) = 2.12 \text{ in.}$$

$$b' = b - \frac{d_b}{2} = 2.12 \text{ in.} - \frac{0.75 \text{ in.}}{2} = 1.75 \text{ in.}$$

$$p = 4 \text{ in.}$$

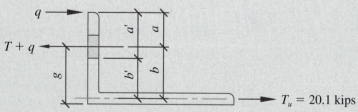

Figure 9-28 Prying action variables for Example 9-8.

$$t_{min} = \sqrt{\frac{4.44Tb'}{pF_u}} = \sqrt{\frac{(4.44)(20.1)(1.75)}{(4)(58)}} = 0.82 \text{ in.}$$

The minimum angle thickness is found to be greater plan the thickness of the selected angle, but recall that this approach is consevative. We will also consider a less conservative approacal for comparison purposes.

Using the preliminary hanger selection tables (AISC, Table 15-1a), the required strength is

$$\frac{T_u}{p} = \frac{20.1}{4} = 5.03 \text{ kips/in.}$$

With $b = 2.12$ in., we find that the required angle thickness is approximately ⅝ in. Based on this thickness, the angle is adequate. We will proceed with the in-depth analysis to compare the results.

From equation (9-28), the initial angle thickness is determined:

$$t = \sqrt{\frac{2.22Tb}{pF_u}} = \sqrt{\frac{(2.22)(20.1)(2.12)}{(4)(58)}} = 0.64 \text{ in.}$$

The minimum angle thickness, from equation (9-29), is determined as follows:

$$a' = \left(a + \frac{d_b}{2}\right) \le \left(1.25b + \frac{d_b}{2}\right)$$

$$= \left(1.5 + \frac{0.75}{2}\right) \le \left((1.25)(2.12) + \frac{0.75}{2}\right)$$

$$= 1.88 \text{ in.} < 3.03 \text{ in.}$$

$$= 1.88 \text{ in.}$$

$$\rho = \frac{b'}{a'} = \frac{1.75}{1.88} = 0.933$$

$$B = \phi r_{nt} = \phi F_{nt} A_b = (0.75)(113)(0.442) = 37.4 \text{ kips}$$
$$(\textit{agrees with AISCM}, \text{Table 7-2})$$

$$\beta = \frac{1}{\rho}\left(\frac{B}{T} - 1\right) = \frac{1}{0.933}\left(\frac{37.4}{20.1} - 1\right) = 0.922$$

$$\delta = 1 - \frac{d'}{p} = 1 - \frac{^{13}/_{16} \text{ in.}}{4 \text{ in.}} = 0.797$$

$$\alpha' = \frac{1}{\delta}\left(\frac{\beta}{1 - \beta}\right) < 1.0 \text{ for } \beta < 1$$

$$\alpha' = \frac{1}{0.797}\left(\frac{0.922}{1 - 0.922}\right) < 1.0$$

$$\alpha' = 14.8 > 1.0$$

$$\therefore \alpha' = 1.0$$

$$t_{min} = \sqrt{\frac{4.44Tb'}{pF_u(1 + \delta\alpha')}} = \sqrt{\frac{(4.44)(20.1)(1.75)}{(4)(58)[1 + (0.797)(1.0)]}}$$

$$= \mathbf{0.612} \text{ in.} < t = \mathbf{0.64} \text{ in. OK}$$

The L7 × 4 × ¾ angle size is adequate.

9.8 FRAMED BEAM CONNECTIONS

Framed beam connections are the most common connections in a steel building. The three most common variations are beam-to-beam, beam-to-column, and beam bearing. Figure 9-29 illustrates some of these connections. Any of the connecting elements can be bolted or welded; however, bolted connections are generally preferred because the labor cost and time for welding is usually greater, and bolted connections are easier to inspect. In this section, we will give a general overview of the analysis and design process for bolted connections; welded connections are covered in Chapter 10.

With the exception of welds, most of the failure modes in framed beam connections have been discussed previously. In lieu of performing each of these design checks for every type of connection, it is more practical to use the design aids in the *AISCM*, and this is what is commonly done in practice. Table 9-4 summarizes the various design and detailing aids in the AISCM.

Table 9-4 Connection design aids in the *AISCM*

AISCM Table	Connection Types
10-1	All-Bolted, Double-Angle Connections
10-2	Bolted/Welded, Double-Angle Connections
10-3	All-Welded, Double-Angle Connections
10-4	Bolted/Welded Shear End-Plate Connections
10-5	All-Bolted Unstiffened Seated Connections
10-6	All-Welded Unstiffened Seated Connections
10-7	All-Bolted Stiffened Seated Connections
10-8	Bolted/Welded Stiffened Seated Connections
10-9	Single-Plate Connections
10-10	All-Bolted Single-Angle Connections
10-11	Bolted/Welded Single-Angle Connections

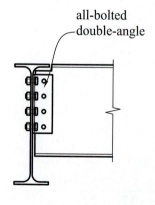

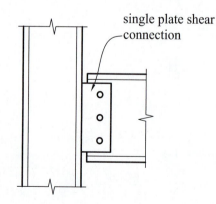

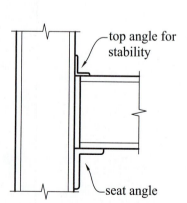

a. beam-to-beam *b. beam-to-column* *c. beam bearing*

Figure 9-29 Common types of framed beam connections.

Each table noted above is actually a compilation of several tables for nearly every common connection type. For connections with single plates or angles, the length of the plate or angle should be at least one-half the throat, or T distance, of the beam in order to provide adequate stability. The amount of cope in the connected beam will have to be checked; however, it generally does not govern the design unless the amount of cope is excessive (see Chapter 11 for further discussion of coped beams).

All-Bolted Double-Angle Connections

An all-bolted double-angle connection is shown in Figure 9-30. For this type of connection, the failure modes that must be considered are shear in the bolts, bearing of the bolts, direct shear and block shear in the angles, and direct shear and block shear in the beam. *AISCM*, Table 10-1, should be used for these types of connections.

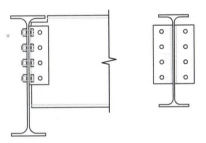

Figure 9-30 All-bolted double-angle connection.

EXAMPLE 9-9

All-Bolted Double-Angle Connection

Determine the capacity of the connection shown in Figure 9-31 using *AISCM*, Table 10-1. The bolts are F1852N in standard holes. The angles are ASTM A36 and the beam is ASTM A992, grade 50.

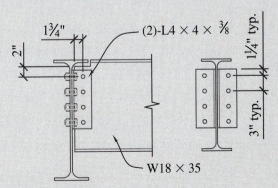

Figure 9-31 Connection details for Example 9-9.

(continued)

SOLUTION

From *AISCM*, Table 10-1, with $N = 4$ for ¾-in.-diameter bolts and an angle thickness of ⅜ in., the design strength is 127 kips. The capacity of the bolts bearing on the beam, and the block shear on the beam also must be checked.

With reference to Figure 9-32, the L_{eh} term is reduced by ¼ in. for fabrication tolerances. The capacity of the bolts bearing on the beam web is

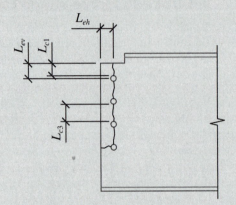

Figure 9-32 Bolt bearing and block shear in beam.

$$L_{c1} = 1.75 \text{ in.} - 0.5\left(\frac{3}{4} + \frac{1}{8}\right) = 1.31 \text{ in.}$$

$$L_{c2} = L_{c3} = L_{c4} = 3 \text{ in.} - \left(\frac{3}{4} + \frac{1}{8}\right) = 2.12 \text{ in.}$$

For bolt 1,

$$
\begin{aligned}
\phi R_n &= \phi 1.2 L_c t F_u \le \phi 2.4 dt F_u \text{ (AISC eq., J3.6a)} \\
&= (0.75)(1.2)(1.31 \text{ in.})(0.3 \text{ in.})(65 \text{ ksi}) \\
&\quad < (0.75)(2.4)(0.75 \text{ in.})(0.3 \text{ in.})(65 \text{ ksi}) \\
&= 22.9 \text{ kips} < 26.3 \text{ kips} \\
&= 22.9 \text{ kips.}
\end{aligned}
$$

For bolts 2, 3, and 4,

$$
\begin{aligned}
\phi R_n &= (0.75)(1.2)(2.12 \text{ in.})(0.3 \text{ in.})(65 \text{ ksi}) \\
&\quad < (0.75)(2.4)(0.75 \text{ in.})(0.3 \text{ in.})(65 \text{ ksi}) \\
&= 37.2 \text{ kips} > 26.3 \text{ kips} \\
&= 26.3 \text{ kips.}
\end{aligned}
$$

The total strength, considering all of the bolts in bearing, is

$$\phi R_n = 22.9 \text{ kips} + (3)(26.3 \text{ kips}) = 101 \text{ kips.}$$

Note that since the web thickness for the W24 × 55 is 0.395 in., the bolt bearing on the W18 × 35 will control.

Block shear in the W18 × 35:

$$A_{gv} = [2 \text{ in.} + (3)(3 \text{ in.})]0.3 \text{ in.} = 3.3 \text{ in.}^2$$

$$A_{nv} = \left[2 \text{ in.} + (3)(3 \text{ in.}) - 3.5\left(\frac{3}{4} + \frac{1}{8}\right)\right]0.3 \text{ in.} = 2.38 \text{ in.}^2$$

$$A_{gt} = (1.75 \text{ in.} - 0.25 \text{ in.})(0.3 \text{ in.}) = 0.45 \text{ in.}^2$$

$$A_{nt} = \left[(1.75 \text{ in.} - 0.25 \text{ in.}) - 0.5\left(\frac{3}{4} + \frac{1}{8}\right)\right]0.3 \text{ in.} = 0.318 \text{ in.}^2$$

Using *AISCM*, equation J4-5, to determine the block shear strength,

$$\begin{aligned}
\phi R_n &= \phi(0.6F_u A_{nv} + U_{bs}F_u A_{nt}) \leq \phi(0.6F_y A_{gv} + U_{bs}F_u A_{nt}) \\
&= 0.75[(0.6)(65)(2.38) + (1.0)(65)(0.318)] \\
&\leq 0.75[(0.6)(50)(3.3) + (1.0)(65)(0.31)] \\
&= 85.2 \text{ kips} \leq 89.3 \text{ kips} \\
&= 85.2 \text{ kips} < 101 \text{ k (block shear controls over bolt bearing).}
\end{aligned}$$

Alternatively, the bottom of Table 10-1 could be used to determine the bolt bearing and block shear strength of the beam web. Using $L_{eh} = 1.75$ in. and $L_{ev} = 2$ in., the design capacity for a beam web coped at the top flange is 284 kips/in. Since the web thickness of the W18 × 35 is $t_w = 0.3$ in., the capacity is

$$(0.3 \text{ in.})(284 \text{ kips/in.}) = 85.2 \text{ kips, which agrees with the value calculated above.}$$

Bolted/Welded Shear End-Plate Connections

A bolted/welded shear end-plate connection is shown in Figure 9-33. For this type of connection, the failure modes that must be considered are shear in the bolts, bearing of the bolts, direct shear and block shear in the end plate, and shear in the weld (welds are covered in Chapter 10). *AISCM*, Table 10-4, should be used for these types of connections.

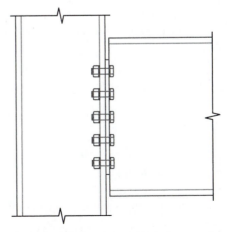

Figure 9-33 Bolted/Welded shear end-plate connections.

EXAMPLE 9-10

Bolted/Welded Shear End-Plate Connection

Determine the capacity of the connection shown in Figure 9-34 using *AISCM*, Table 10-4. The bolts are ⅞-in.-diameter A325N in standard holes. The end plate is ASTM A36 and the beam and column is ASTM A992, grade 50.

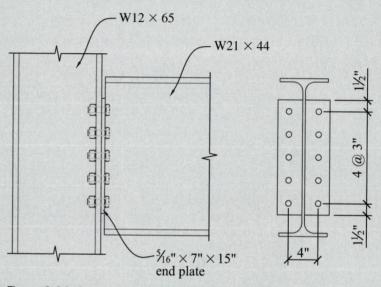

Figure 9-34 Connection details for Example 9-10.

From AISCM Table 10-4, the design strength of the connection, using ⅞-in.-diameter A325N, $n = 5$, with $t_p = 5/16$ in., is 155 k. The strength of the weld will be discussed in Chapter 10. The strength of the bolts bearing on the column flange should also be checked. From Part 1 of the *AISCM*, the flange thickness for a W12 × 65 is $t_f = 0.605$ in.

The bearing strength is the same for each bolt.

$$L_c = 3 \text{ in.} - (⅞ + ⅛) = 2.0 \text{ in.}$$

$$\phi R_n = \phi 1.2 L_c t F_u \leq \phi 2.4 d t F_u \ \text{(AISC, eq. J3.6a)}$$

$$= (0.75)(1.2)(2.0 \text{ in.})(0.605 \text{ in.})(65 \text{ ksi})$$

$$\leq (0.75)(2.4)(0.875 \text{ in.})(0.605 \text{ in.})(65 \text{ ksi})$$

$$= 70.7 \text{ k} > 61.9 \text{ k}$$

$$\phi R_n = 61.9 \text{ k (for all bolts)}$$

The total strength, considering all of the bolts in bearing, is

$$\phi R_n = (10)(61.9 \text{ k}) = 619 \text{ k}.$$

Alternatively, *AISCM*, Table 10-4, is a design aid for the available strength of the bolts bearing on the support. From *AISCM*, Table 10-4, the available strength per inch of thickness is 1020 kips/in. The strength of the bolts bearing on the flange of the W12 × 65 is

$$\phi R_n = (1020 \text{ kips/in.})(0.605 \text{ in.}) = 617 \text{ kips},$$

which is approximately equal to the calculated value of 619 k.

Single-Plate Shear Connections

A single-plate shear connection is shown in Figure 9-35. The failure modes for this connection are shear in the bolts, bearing of the bolts, direct shear and block shear in the plate, weld to the connecting elements, and direct shear and block shear in the beam. *AISCM*, Table 10-9, should be used for these types of connections. The dimensional limitations for this connection type when using *AISCM*, Table 10-9, are noted in Figure 9-32.

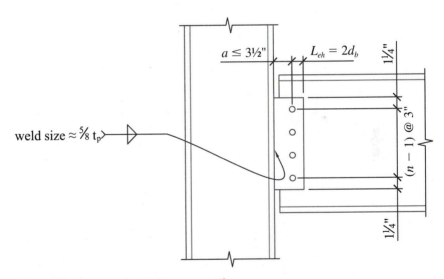

Figure 9-35 Single-plate shear connection.

EXAMPLE 9-11

Single-Plate Shear Connection

Determine the capacity of the connection shown in Figure 9-36 using the *AISCM*, Table 10-9. The bolts are ¾-in.-diameter A490N in short-slotted holes transverse to

(continued)

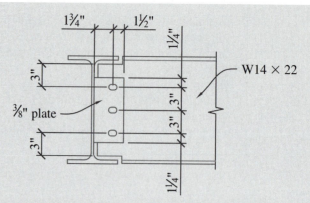

Figure 9-36 Connection details for Example 9-11.

the direction of the load (SSLT). The plate is ASTM A36 and the beam is ASTM A992, grade 50.

From *AISCM*, Table 10-9a, with $n = 3$, $L = 8\frac{1}{2}$ in. A490N SSLT, and $t_p = \frac{3}{8}$ in., the design strength of the connection is $\phi R_n = 57.5$ kips. To check the strength of the beam web for block shear, *AISCM*, Table 10-1, can be used (similar to Example 9-9).

From *AISCM*, Table 10-1, with $n = 3$, $L_{ev} = 3$ in., $L_{eh} = 1\frac{3}{4}$ in. (SSLT), the design strength is 246 kips/in. The web thickness of a W14 × 22 is $t_w = 0.23$ in. The strength of the beam in block shear is

$$\phi R_n = (246 \text{ kips/in.})(0.23 \text{ in.}) = 56.6 \text{ kips.}$$

The bolt bearing on the beam also must be checked.

$$L_{c1} = L_{c2} = L_{c3} = 3 - \left(\frac{3}{4} + \frac{1}{8}\right) = 2.12 \text{ in.}$$

For bolts 1, 2, and 3,

$$\begin{aligned}
\phi R_n &= \phi 1.2 L_c t F_u \leq \phi 2.4 dt F_u \\
&= (0.75)(1.2)(2.12 \text{ in.})(0.23 \text{ in.})(65 \text{ ksi}) \\
&\quad < (0.75)(2.4)(0.75 \text{ in.})(0.23 \text{ in.})(65 \text{ ksi}) \\
&= 28.5 \text{ kips} < 20.1 \text{ kips} \\
&= 20.1 \text{ kips.}
\end{aligned}$$

The total strength, considering all of the bolts in bearing, is

$$\phi R_n = (3)(20.1 \text{ kips}) = 60.3 \text{ kips.}$$

The strength of the single plate controls, so the capacity of the connection is $\phi R_n = 56.6$ kips.

9.9 REFERENCES

1. American Institute of Steel Construction. 2006. *Steel construction manual*, 13th ed. Chicago: AISC.

2. International Codes Council. 2006. *International building code.* Falls Church, VA: ICC.

3. American Society of Civil Engineers. 2005. *ASCE-7: Minimum design loads for buildings and other structures.* Reston, VA: ASCE.

4. American Institute of Steel Construction. 2002. *Steel design guide series 17: High-strength bolts—A primer for structural engineers.* Chicago: AISC.

5. American Institute of Steel Construction. 2006. *Steel design guide series 21: Welded connections—A primer for structural engineers.* Chicago: AISC.

6. McCormac, Jack. 1981. *Structural steel design,* 3rd ed. New York: Harper and Row.

7. Salmon, Charles G., and Johnson, John E. 1980. *Steel structures: Design and behavior*, 2nd ed. New York: Harper and Row.

8. Smith, J. C. 1988. *Structural steel design: LRFD fundamentals.* Hoboken, NJ: Wiley.

9. Blodgett, Omer. 1966. *Design of welded structures.* Cleveland: The James F. Lincoln Arc Welding Foundation.

10. Segui, William. 2006. *Steel design*, 4th ed. Toronto: Thomson Engineering.

11. Limbrunner, George F., and Leonard Spiegel. 2001. *Applied structural steel design*, 4th ed. Upper Saddle River, NJ: Prentice Hall.

12. Crawford, S. F., and G. L. Kulak. 1971. Eccentrically loaded bolted connections. *Journal of the Structural Division* (ASCE) 97, no. ST3: 765–783.

13. Chesson, Eugene, Norberto Faustino, and William Munse. 1965. *High-strength bolts subjected to tension and shear. Journal of the Structural Division* (ASCE) 91, no. ST5: 155–180.

14. Swanson, J. A. 2002. *Ultimate-strength prying models for bolted T-stub connections. Engineering Journal* (AISC) 39 (3): pp. 136–147.

15. Thornton, W. A. 1992. *Strength and servicability of hanger connections. Engineering Journal* (AISC) 29 (4): pp. 145–149.

9.10 PROBLEMS

9-1. Determine whether the connection shown in Figure 9-37 is adequate in bearing. Check the edge and spacing requirements. The steel is ASTM A36 and the bolts are in standard holes.

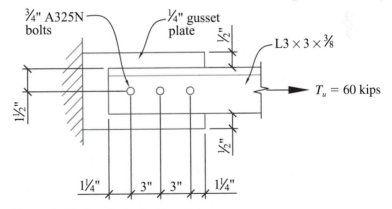

Figure 9-37 Details for Problem 9-1.

9-2. Determine the maximum tensile force, P_u, that can be applied to the connection shown in Figure 9-38 based on the bolt strength; assume ASTM A36 steel for the following bolt types in standard holes:

1. ⅞-in.-diameter A325N,
2. ⅞-in.-diameter A325SC (Class A faying surface), and
3. ¾-in.-diameter A490X.

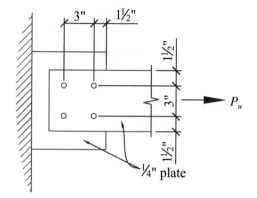

Figure 9-38 Details for Problem 9-2.

9-3. Determine the maximum tensile force, P_u, that can be applied to the connection shown in Figure 9-39 based on the bolt strength; assume ASTM A36 steel for the following bolt types in standard holes:

1. 1-in.-diameter A325N,

2. 1-in.-diameter A325SC (Class B faying surface), and

3. ⅞-in.-diameter A490X.

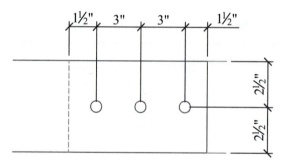

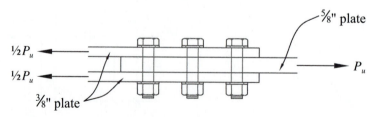

Figure 9-39 Details for Problem 9-3.

9-4. Determine the required bolt diameter for the connection shown in Figure 9-40. The steel is ASTM A36 and the bolts are in standard holes. Check the edge and spacing requirements. Use the following bolt properties:

1. A490N, and

2. A490SC.

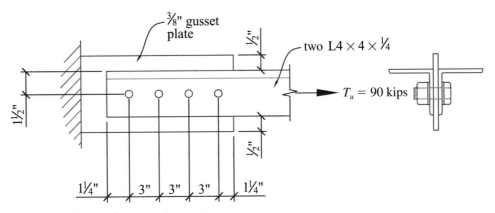

Figure 9-40 Details for Problem 9-4.

9-5. Determine whether the connection shown in Figure 9-41 is adequate to support the applied moment considering the strength of the bolts in shear and bearing. The steel is ASTM A36 and the holes are standard (STD).

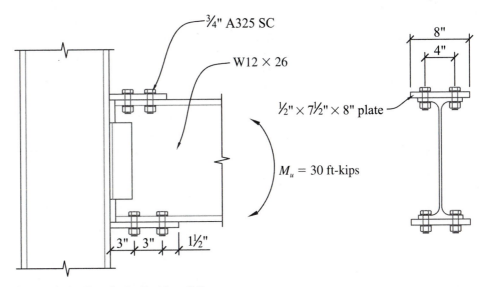

Figure 9-41 Details for Problem 9-5.

9-6. Determine the capacity of the connection shown in Figure 9-42, considering the strength of the bolts only. Assume that each bolt takes an equal amount of shear and tension. The bolts are ¾-in.-diameter A325N in standard holes.

9-7. Determine the capacity of the connection shown; in Figure 9-43, considering the strength of the bolts only assume ¾-in.-diameter A325X bolts in standard holes. Use the following methods and compare the results:

    1. Instantaneous center (IC) method,

    2. Elastic method, and

    3. AISC design aids.

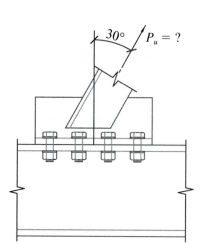

Figure 9-42 Details for Problem 9-6.

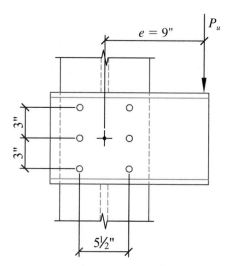

Figure 9-43 Details for Problem 9-7.

9-8. Determine the capacity of the connection shown in Figure 9-44, considering the strength of the bolts only. Assume ¾-in.-diameter A490N bolts in standard holes. Use the following methods and compare the results:

  1. Case I,
  2. Case II, and
  3. AISC design aids.

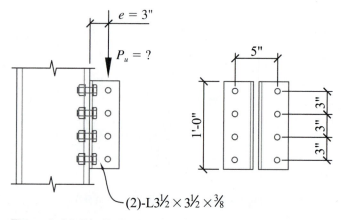

Figure 9-44 Details for Problem 9-8.

9-9. Determine whether the moment connection shown in Figure 9-45 is adequate considering the prying action on the angle, shear in the bolts, and bearing in the bolts. The steel is ASTM A36 and the holes are oversized (OVS).

Student Design Project Problems:

9-10. Select the following connections from the appropriate *AISCM* Table; check block shear and bolt bearing as required:

  1. Beam-to-girder connections at the second floor (use single-plate shear connections), and
  2. Beam-to-column and girder-to-column connections (use all-bolted, double-angle).

9-11. Check the bolted connections for the X-braces (designed in Chapter 4) for the appropriate limits states.

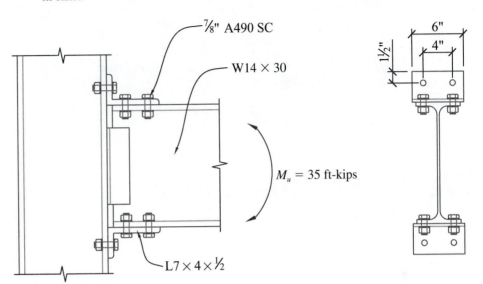

Figure 9-45 Details for Problem 9-9.

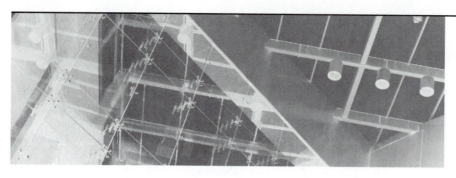

Welded Connections

10.1 INTRODUCTION

Welding is a process in which two steel members are heated and fused together with or without the use of a filler metal. In structural steel buildings, connections often incorporate both welds and bolts. The use of bolts or welds in any connection is a function of many factors, such as cost, construction sequence, constructibility, and the contractor's preference. Welded connections offer some advantages over bolted connections, although they do have some disadvantages. The following lists summarize the advantages and disadvantages of welded connections:

Advantages of welded connections

1. Welded connections can be adapted to almost any connection configuration in which bolts are used. This is especially advantageous when construction problems are encountered with a bolted connection (e.g., misaligned bolt holes) and a field-welded connection is the only reasonable solution. Connections to existing steel structures are sometimes easier with welds because greater dimension tolerances are often needed that might not be possible with bolts.

2. The full design strength of a member can be more easily developed with a welded connection with an all-around weld. For example, a beam splice would need to develop the shear and bending capacity of the member in question. Although a bolted connection would be possible for this condition (see Chapter 11), a welded connection requires less material and space to develop the design strength of the beam and would create a beam splice that is completely continuous across the joint. A beam that is fabricated too short and that needs additional length to accommodate field conditions would often be a welded splice instead of a bolted splice for just that reason.

3. Welded connections take up less material and space. Consider a bolted moment connection from a beam to a column. The beam would have a top plate, as well as bolts

protruding into space that would normally be occupied by a metal deck and concrete for a floor slab. The floor deck and slab would have to be modified and reinforced in this area to allow for the placement of the bolted connection. A welded moment connection in this case would allow the floor deck to align flush with the top of the beam.

4. Welded connections are more rigid and are subject to less deformation than bolted connections.

5. Welded connections are often preferred in exposed conditions where aesthetics are of concern because they can be modified to have a more smooth and cleaner appearance.

Disadvantages of welded connections

1. Welded connections require greater skill. A welder is often certified not just for welding in general, but also for certain welding techniques (e.g., overhead welding).

2. Welded connections often require more time to construct than equivalent bolted connections. A greater amount of time equates to more cost because of the labor, especially for welds that have to be performed in the field. In many cases, contractors prefer to avoid field-welded connections as much as possible because of the associated labor cost, even if the result is to use rather large and seemingly oversized bolted connections.

3. The inspection of welded connections is more extensive than that of bolted connections. Discontinuities and other deficiencies in welds cannot easily be found by visual inspection. A variety of inspection methods are available, such as penetrant testing, magnetic particle testing, ultrasonic testing, and radiographic testing [1]. Each of these methods requires labor and equipment, which adds to the cost of the connection. Some welds, such as full-penetration welds, require continuous inspection, which would mean that a welding inspector would have to be on the site whenever this type of welding was occurring—also adding to the cost of the connection.

4. In existing structures, welded connections may be difficult or even impossible due to the use of the structure. For example, welding in a warehouse with paper or other flammable material has an obvious element of danger associated with it. Special protective measures would then be required that might interrupt the regular business operation of the facility, in which case an alternate bolted connection might have to be used. With other structures, such as hospitals, a special permit might be required to ensure the safety of the occupants during the welding operation.

The heat energy needed to fuse two steel members together could be electrical, mechanical, or chemical, but electrical energy is normally used for welding structural steel. The most common welding processes are *shielded metal arc welding* (SMAW) and *submerged arc welding* (SAW). In each case, electric current forms an arc between the electrode and the steel that is being connected. An electrode is essentially a "stick" that acts as a conductor for the electric current. This arc melts the base metal as it moves along the welding path, leaving a bead of weld. As the steel cools, the impurities rise to the surface to form a layer of slag, which must be removed if additional passes of the electrode are required. The basic welding process is shown in Figure 10-1.

The shielded metal arc welding process, also called stick welding, is a manual process and is the most common type of weld. In this process, a coated electrode is used to heat both the base metal and the tip of the electrode, whereby part of the electrode is deposited onto the base metal. As the coating on the electrode dissolves, it forms a gaseous shield to help protect the weld from atmospheric impurities.

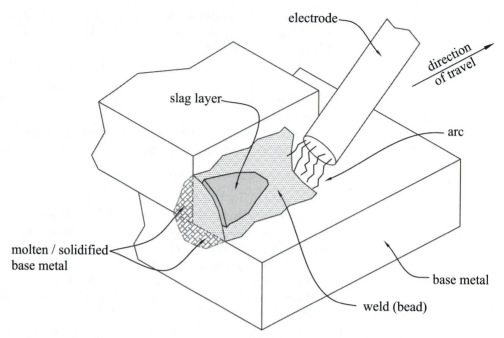

Figure 10-1 Basic welding process.

The shielded arc welding process can be either automatic or semiautomatic. This process is similar to SMAW, but an uncoated electrode is used. Granular flux is placed over the joint while submerging the electrode and the arc. This process is usually faster and results in a weld with a deeper penetration, which results in a higher weld strength.

There are several other types of welding processes, such as flux cored arc welding (FCAW), gas metal arc welding (GMAW), electroslag welding (ESW), electrogas welding (EGW), and gas tungsten arc welding (GTAW), but they are beyond the scope of this text (see ref. 2).

10.2 TYPES OF JOINTS AND WELDS

There are numerous possible joint configurations but the most common are the lap, butt, corner, and tee joint (see Figure 10-2). In Section 8 of the *AISCM*, these joints are designated as B (butt joint), C (corner joint), and T (T-joint). The following combinations are also recognized: BC (butt or corner joint), TC (T- or corner joint), and BTC (butt, T-, or corner joint). These designations are a shorthand form for specifying certain weld types.

There are several weld types shown in Figure 10-3. There are a variety of factors that determine what type of weld to use, but the most common weld is the fillet weld. Fillet welds are generally triangular in shape and join together members that are usually at right angles to each other. Fillet welds are usually the most economical since they require very little surface preparation and can be used in virtually any connection configuration.

Plug and slot welds are used to transmit shear in lap joints or to connect components of built-up members (such as web doubler plates) to prevent buckling. They are also used to conceal connections for steel that are exposed for architectural reasons. In addition, they can be used to add strength to connections with fillet welds (see Example 10-5). Neither plug

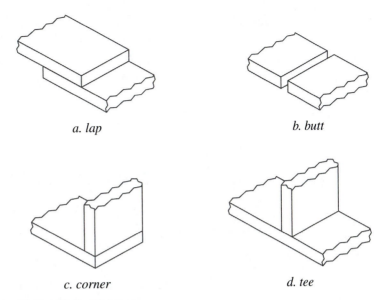

a. lap   *b. butt*

c. corner   *d. tee*

Figure 10-2 Joint types.

nor slot welds should be used where a tension force is normal to the plane of the faying surface. Nor should they be used to support cyclical loads. Plug welds are placed in round holes and slot welds are placed in elongated holes. In each case, the weld metal is placed in the hole up to a certain depth (partial penetration or full penetration). The penetration depth of a plug or slot weld is difficult to inspect visually, so such welds are not often preferred.

Groove welds are used to fill the groove between the ends of two members. Groove welds can be made in joints that are classified as square, bevel, V (or double-bevel), U, J, or flare V (or flare bevel) (see Figure 10-4). In order to contain the weld metal, a backing bar is used at the bottom of groove welds. In some cases, the backing bar should be removed so that the weld can be inspected. In other cases, the backing bar is removed so that additional weld metal can be added on the other side of the joint. In this case, any part of the weld that is incomplete is removed prior to adding additional weld. This process is called back gouging and is generally limited to connections for joints specially detailed for seismic resistance. The backing bar for groove welds can also be left in place where special seismic detailing is not required, but this decision is usually left to the engineer.

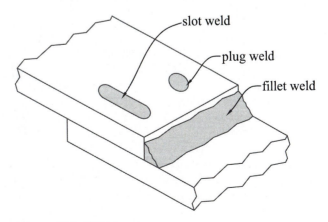

slot weld

plug weld

fillet weld

Figure 10-3 Weld types.

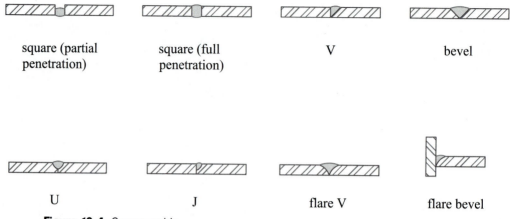

square (partial
penetration)   square (full
penetration)   V   bevel

U   J   flare V   flare bevel

Figure 10-4 Groove welds.

Groove welds can penetrate the connected member for a portion of the member thickness, or it can penetrate the full thickness of the connected member. These are called partial-joint penetration (PJP) and complete-joint penetration (CJP), respectively (see further discussion on weld strength in Section 10.4). Complete-penetration welds (also called full-penetration or "full-pen" welds) fuse the entire depth of the ends of the connected members. Partial penetration welds are more cost-effective and are used when the applied loads are such that a full-penetration weld is not required. They can also be used where access to the groove is limited to one side of the connection.

Whenever possible, welds should be performed in the shop where the quality of the weld is usually better. Shop welds are not subjected to the weather and access to the joint is fairly open. Welds can be classified as flat, horizontal, vertical, and overhead (see Figure 10-5). It can be seen that flat welds are the easiest to perform; they are the preferred method. Overhead welds, which are usually done in the field, should also be avoided where possible because they are difficult and more time-consuming, and therefore more costly.

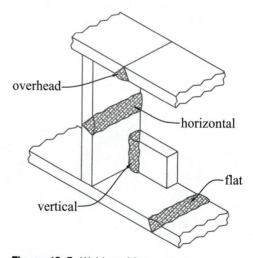

Figure 10-5 Weld positions.

10.3 WELD SYMBOLS

Weld symbols are commonly used to identify the required weld properties used in the connection design. Symbols have been standardized by the American Welding Society (see ref. 3); they are summarized in Table 8-2 of the *AISCM*.

Fillet welds are the most commonly specified welds, and will be used as the basis for the following discussion (see Figures 10-6 and 10-7). The standard symbol is an arrow pointing to the weld or joint, with a horizontal line forming the tail of the arrow. The triangular shape indicates a fillet weld, but for all welds, the vertical line of the weld symbol is always to the left. Above and below the horizontal line, information about the weld is given. If the information is below the horizontal line, then the welded joint is on the near side of the arrow. If the information is above the horizontal line, then the welded joint is on the opposite side. The size of the weld is stated first on the left side of the weld-type symbol, then the length and spacing of the weld is placed to the right of the weld-type symbol. A circle at the intersection of the horizontal line and arrow indicates that the weld is around the entire joint. A flag at this location indicates that the weld is to be made in the field. The absence of the flag indicates that the weld should be performed in the shop. At the end of the horizontal line, any special notes can be added.

Table 8-2 of the *AISCM* also lists certain pre-qualified weld symbols that can be used to identify partial-joint or complete-joint penetration weld types. Table 10-1 below summarizes the basic notation used.

As an example, in the designation BU-4a used in *AISCM*, Table 8-2, the B indicates a butt joint, and the U indicates that the thickness of the connected parts is not limited and that the weld is to have complete-joint penetration. The number 4 indicates that the joint has a

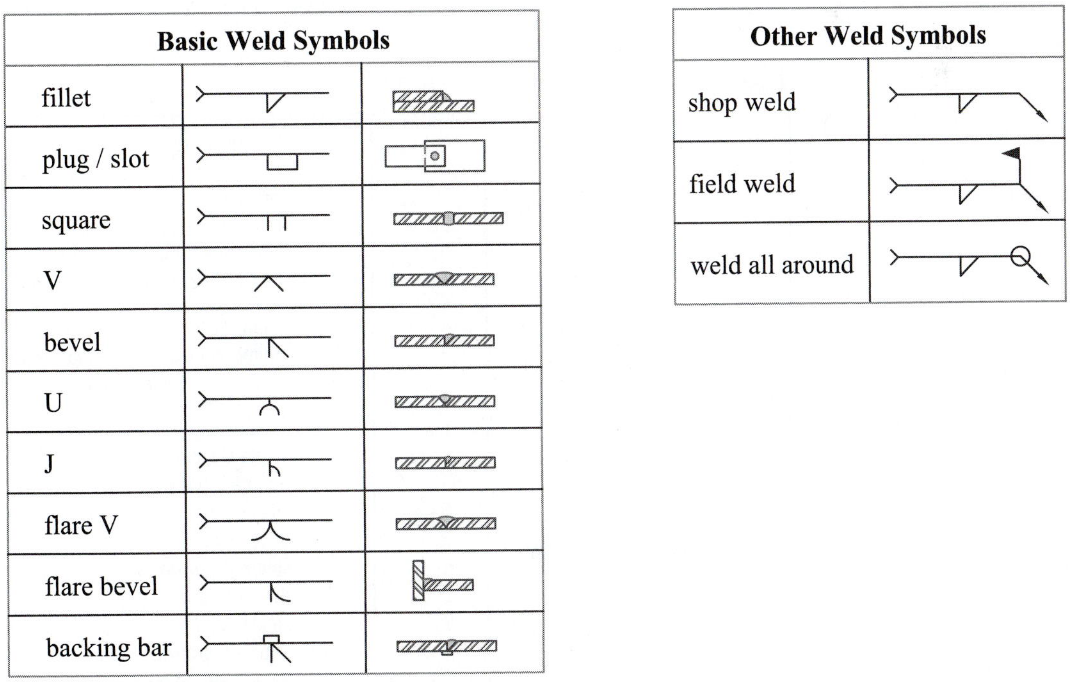

Figure 10-6 Basic weld symbols.

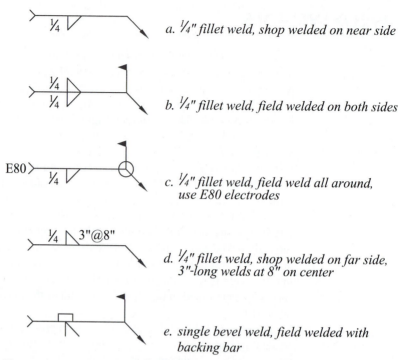

Figure 10-7 Weld symbol examples.

single-bevel groove, which means that one connected part is flat or unprepared, whereas the other connected part has a beveled edge. The letter a indicates something unique about this joint, which, in this case, means that a backing bar is used. The letter *a* also differentiates this joint type from a BU-4b. The letter b, in this case, indicates that the underside of the joint in question must be backgouged and reinforced with additional weld metal.

Table 10-1 Pre-qualified weld notation

Joint Type Symbols	B C T BC TC BTC	butt joint corner joint T-joint butt or corner joint T- or corner joint butt, T-, or corner joint
Base Metal Thickness and Penetration Symbols	L U P	limited thickness, complete-joint penetration unlimited thickness, complete-joint penetration partial-joint penetration
Weld-Type Symbols	1 2 3 4 5 6 7 8 9 10	square groove single-V groove double-V groove single-bevel groove double-bevel groove single-U groove double-U groove single-J groove double-J groove flare-bevel groove

10.4 DIMENSIONAL REQUIREMENTS FOR WELDS

There are minimum dimensional requirements for welds given in the *AISCM*. Table 10-2 gives the maximum and minimum fillet weld sizes, which are a function of the thickness of the connected parts.

The minimum total length of a fillet weld must be at least four times the nominal size of the weld, or else the size of the weld used to determine its design strength shall be assumed to be one-fourth of the weld length. With reference to Figure 10-8, when longitudinal welds are used to connect the ends of tension members, the length of each weld shall not be less that the distance between the welds. This is to prevent shear lag, which occurs when not all parts of the connected member are fully engaged in tension (see Chapter 4).

The maximum length of a fillet weld is unlimited, except for end-loaded welds. End loaded welds are longitudinal welds in axial-loaded members. End-loaded welds are permitted to have a length of 100 times the weld size. When the length-to-size ratio of an end-loaded weld exceeds 100, the design strength is adjusted by the following factor:

$$\beta = 0.60 < 1.2 - 0.002\left(\frac{L}{w}\right) \leq 1.0, \tag{10-1}$$

where

β = Weld strength adjustment factor (varies from 0.60 to 1.0),

L = Weld length, and

w = Weld size.

The minimum diameter of a plug weld or width of a slot weld is the thickness of the part containing it plus $\frac{5}{16}$ in. rounded up to the next larger odd $\frac{1}{16}$ in. (see Figure 10-9). The maximum diameter of a plug weld or width of a slot weld is $2\frac{1}{4}$ times the thickness of the weld. The length of a slot weld shall be less than or equal to 10 times the weld thickness. Plug welds should be spaced a minimum of four times the hole diameter. Slot welds should be

Table 10-2 Maximum and minimum fillet weld sizes

Maximum Fillet Weld Size	
Connected part thickness, t^1	**Maximum weld size, w**
$t < \frac{1}{4}''$	$w = t$
$t > \frac{1}{4}''$	$w = t - \frac{1}{16}''$
Minimum Fillet Weld Size^2	
Thickness of thinner connected part, t	**Minimum weld size, w**
$t \leq \frac{1}{4}''$	$\frac{1}{8}$
$\frac{1}{4}'' < t \leq \frac{1}{2}''$	$\frac{3}{16}$
$\frac{1}{2}'' < t \leq \frac{3}{4}''$	$\frac{1}{4}$
$t > \frac{3}{4}''$	$\frac{5}{16}$

[1]The term t is the thickness of the thicker connected part.

[2]Single-pass welds must be used. The maximum weld size that can be made in a single pass is 5/16 in.

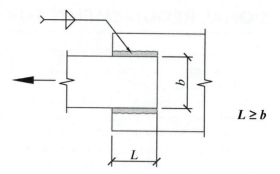

$$L \geq b$$

Figure 10-8 Longitudinal welds.

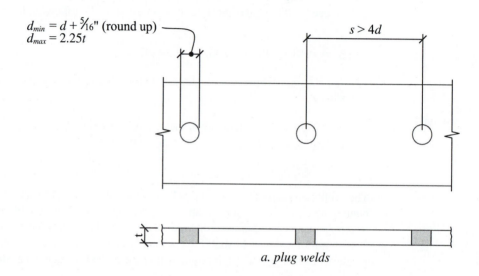

$d_{min} = d + \frac{5}{16}"$ (round up)
$d_{max} = 2.25t$

$s > 4d$

a. plug welds

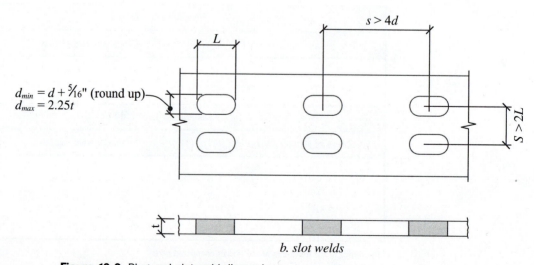

$s > 4d$

L

$d_{min} = d + \frac{5}{16}"$ (round up)
$d_{max} = 2.25t$

$S > 2L$

b. slot welds

Figure 10-9 Plug and slot weld dimensions.

spaced a minimum of four times the width of the slot in the direction transverse to their length. In the longitudinal direction, slot welds should be spaced a minimum of twice the length of the slot.

For a connected part that is ⅝ in. thick or less, the minimum thickness of plug or slot welds is the thickness of the material. For material greater than ⅝ in. thick, the thickness of plug or slot welds should be at least half the thickness of the material, but not less than ⅝ in.

Figure 10-9 shows the basic dimensional requirements for plug and slot welds.

10.5 FILLET WELD STRENGTH

The strength of a fillet weld is based on the assumption that the weld forms a right triangle with a one-to-one slope between the connected parts (see Figure 10-10). Welds can be loaded in any direction, but are weakest in shear and are therefore assumed to fail in shear. The shortest distance across the weld is called the throat, which is where the failure plane is assumed to be. Any additional weld at the throat (represented by the dashed line in Figure 10-10) is neglected in calculations of strength.

Based on the geometry shown, the strength of the weld can be calculated as

$$R_n = (\sin 45)wLF_w, \tag{10-2}$$

where

R_n = Nominal weld strength, kips,

w = Weld size, in.,

L = Weld length, in., and

F_w = Nominal weld strength, ksi.

The nominal weld strength, F_w, is a function of the weld metal or electrode used. The ultimate electrode strength can vary from 60 ksi to 120 ksi, but the most commonly used

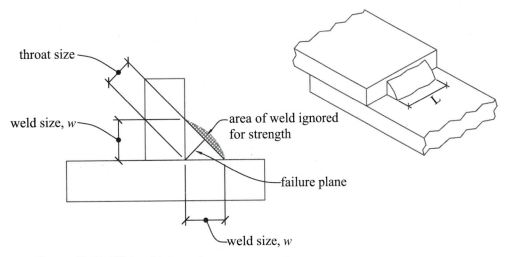

throat size

weld size, w

area of weld ignored for strength

failure plane

weld size, w

L

Figure 10-10 Fillet weld geometry.

electrode strength is 70 ksi. The designation for electrode strength is an E followed by two digits that represent the electrode strength and two additional numbers that indicate the welding process. The designation E70XX is commonly used to indicate electrodes with a nominal strength of 70 ksi. The first letter X indicates the weld position (such as overhead), and the second letter X indicates the welding current and other weld properties not directly pertinent to the weld strength. An example of a complete weld designation is E7028, where the 2 represents a horizontal and flat weld and the 8 represents an electrode with a low hydrogen coating.

From *AISCM*, Table J2.5, the nominal strength of a fillet weld in shear is

$$F_w = 0.6F_{\text{EXX}} \tag{10-3}$$

where F_{EXX} is the electrode strength. Therefore, the available strength of a fillet weld is found by combining equations (10-2) and (10-3):

$$\phi R_n = \phi 0.6F_{\text{EXX}} \frac{\sqrt{2}}{2} wL, \tag{10-4}$$

where $\phi = 0.75$.

The weld size is commonly expressed as a thickness that is a certain number of sixteenths of an inch. For example, a weld size of $w = \frac{1}{4}$ in. could be expressed as four-sixteenths or simply as $D = 4$. From *AISCM*, Table J2.5, the strength reduction factor for shear is $\phi = 0.75$. Combining these with a commonly used electrode strength of $F_{\text{EXX}} = 70$ ksi, equation (10-4) becomes

$$\phi R_n = (0.75)(0.6)(70)\left(\frac{\sqrt{2}}{2}\right)\left(\frac{D}{16}\right)L$$
$$= 1.392DL, \tag{10-5}$$

where

ϕR_n = Available weld strength, kips,

D = Weld size in sixteenths of an inch, and

L = Weld length, in.

It should be noted that the strength of the connected part must also be checked for strength in shear, tension, or shear rupture, whichever is applicable to the connection (see Chapter 4).

The above formulation for weld strength is conservative in that the load direction is not accounted for (i.e., shear failure, which has the lowest strength, was assumed). For a linear weld group loaded in the plane of the weld, the nominal weld strength is

$$F_w = 0.6F_{\text{EXX}}(1 + 0.5 \sin^{1.5}\theta), \tag{10-6}$$

where θ is given in Figure 10-11. Note that a linear weld group is one that has all of the welds in line or parallel.

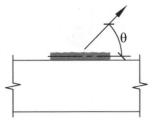

Figure 10-11 Weld loaded at an angle.

For weld groups where the welds are oriented both transversely and longitudinally to the applied load, the nominal weld strength is the greater of the following equations:

$$\phi R_n = \phi R_{wl} + \phi R_{wt} \tag{10-7}$$
$$= 0.85\phi R_{wl} + 1.5\phi R_{wt}, \tag{10-8}$$

where

R_{wl} = Nominal strength of the longitudinal fillet welds using $F_w = 0.6F_{EXX}$,

R_{wt} = Nominal strength of the transverse fillet welds using $F_w = 0.6F_{EXX}$, and

$\phi = 0.75$.

EXAMPLE 10-1

Fillet Weld Strength

Determine the capacity of the connection shown in Figure 10-12 based on weld strength alone. Electrodes are E70XX.

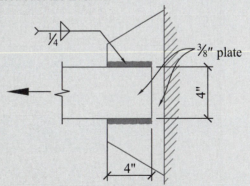

Figure 10-12 Detail for Example 10-1.

SOLUTION

The AISC requirements for weld size must be checked first.
From Table 10-2,

Minimum weld size = $\frac{3}{16}$ in. ($\frac{1}{4}$ in. provided, OK)

Maximum weld size = $t_{max} - \frac{1}{16}$ in. = $\frac{3}{8}$ in. $- \frac{1}{16}$ in. = $\frac{5}{16}$ in. ($\frac{1}{4}$ in. provided, OK)

(continued)

Spacing of longitudinal welds:

Length $= 4$ in.

Distance apart $(b) = 4$ in.

Since the weld length is not less than the distance apart, the spacing is adequate.

Weld capacity:

Since E70XX electrodes are specified, equation (10-5) can be used:

$$\phi R_n = 1.392DL$$
$$= (1.392)(4)(4 + 4) = \textbf{44.5 kips.}$$

EXAMPLE 10-2

Fillet Weld Strength

Determine the capacity of the connection shown in Figure 10-13 based on weld strength alone. Electrodes are E70XX.

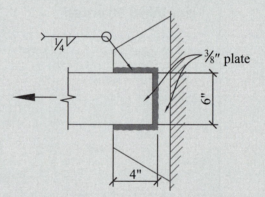

Figure 10-13 Details for Example 10-2.

From Example 10-1, the AISC requirements for weld size are satisfied.

Since both longitudinal and transverse welds are present, equations (10-7) and (10-8) are used.

From equation (10-7),

$$\phi R_n = \phi R_{wl} + \phi R_{wt}$$
$$\phi R_{wl} = 1.392DL = (1.392)(4)(4 + 4) = 44.5 \text{ kips}$$
$$\phi R_{wt} = 1.392DL = (1.392)(4)(6) = 33.4 \text{ kips}$$
$$\phi R_n = 44.5 + 33.4 = \textbf{77.9 kips.}$$

From equation (10-8),

$$\phi R_n = 0.85\phi R_{wl} + 1.5\phi R_{wt}$$
$$= (0.85)(44.5) + (1.5)(33.4) = 87.9 \text{ kips.}$$

The larger of these values controls, so the design strength of the weld is $\phi R_n = 87.9$ kips.

Thus far, only the weld strength has been considered in the analysis of connections. We will now consider the strength of the connected elements. In some cases, the connection is designed for the capacity of the connected elements in order to ensure that the connection doesn't become a weak link in the load path.

From Chapter 4, the strength of an element in tensile yielding is

$$\phi R_n = \phi F_y A_g, \tag{10-9}$$

where

$\phi = 0.90$,
$F_y = $ Yield stress, and
$A_g = $ Gross area of the connected element.

The strength of an element in tensile rupture is

$$\phi R_n = \phi F_u A_e, \tag{10-10}$$

where

$\phi = 0.75$
$F_u = $ Ultimate tensile stress,
$A_e = $ Effective area of the connected element
$\quad = A_n U$,
$A_n = $ Net area of the connected element, and
$U = $ Shear lag factor calculated from *AISCM*, Table D3.1.

The strength of an element in shear yielding is

$$\phi R_n = \phi 0.60 F_y A_{gv}, \tag{10-11}$$

where

$\phi = 1.0$, and
$A_{gv} = $ Gross area subject to shear.

The strength of an element in shear rupture is

$$\phi R_n = \phi 0.60 F_u A_{nv}, \tag{10-12}$$

where

$\phi = 0.75$, and
$A_{nv} = $ Net area subject to shear.

The block shear strength is

$$\phi R_n = \phi(0.60F_u A_{nv} + U_{bs}F_u A_{nt}) \le \phi(0.60F_y A_{gv} + U_{bs}F_u A_{nt}), \qquad (10\text{-}13)$$

where

$\phi = 0.75$,

A_{nv} = Net area subject to shear,

A_{nt} = Net area subject to tension,

A_{gv} = Gross area subject to shear, and

U_{bs} = Block shear coefficient

= 1.0 for uniform tension stress

= 0.5 for nonuniform tension stress.

For welded tension members, the second half of equation (10-13) can be ignored since the net shear area equals the gross shear area for this case. The equation then becomes

$$\phi R_n = \phi(0.60F_u A_{nv} + U_{bs}F_u A_{nt}). \qquad (10\text{-}14)$$

EXAMPLE 10-3

Design Strength of a Welded Connection

Determine the strength of the connection from Example 10-2 considering the strength of the connected elements. Assume that ASTM A36 steel is used.

From Example 10-2, the design strength of the weld was found to be $\phi R_n = 87.9$ k. The strength of the connected elements in tension, shear, and block shear will now be considered.

From the geometry of the connection,

$A_{gt} = (3/8 \text{ in.})(6 \text{ in.}) = 2.25 \text{ in.}^2$

$A_{gv} = (0.375)(4 \text{ in.} + 4 \text{ in.}) = 3.0 \text{ in.}^2$

$U = 1.0$ (from *AISCM* Table D3.1)

$A_e = A_n U = (3/8 \text{ in.})(6 \text{ in.})(1.0) = 2.25 \text{ in.}^2$

$U_{bs} = 1.0$ (stress is assumed to be uniform; see *AISCM*, Section CJ4.3)

$F_y = 36 \text{ ksi}$

$F_u = 58 \text{ ksi}$

Tensile yielding:

$\phi R_n = \phi F_y A_g$

$\quad = (0.9)(36)(2.25) = $ **72.9 kips** *Controls*

Tensile rupture:

$$\phi R_n = \phi F_u A_e$$
$$= (0.75)(58)(2.25) = \textbf{97.8 kips}$$

Shear yielding:

$$\phi R_n = \phi 0.60 F_y A_{gv}$$
$$= (1.0)(0.6)(36)(3.0) = \textbf{64.8 kips}$$

Shear rupture:

$$\phi R_n = \phi 0.60 F_u A_{nv}$$
$$= (0.75)(0.60)(58)(3.0) = \textbf{78.3 kips}$$

Note that shear yielding and shear rupture are not a valid failure mode for this connection since there is a transverse weld. The above calculations regarding shear failure are provided for reference.

Block shear:

$$\phi R_n = \phi(0.60 F_u A_{nv} + U_{bs} F_u A_{nt})$$
$$= 0.75[(0.60)(58)(3.0) + (1.0)(58)(2.25)] = \textbf{176 kips}$$

The critical failure mode is tensile yielding on the connected member where $\phi R_n = 72.9$ kips.

10.6 PLUG AND SLOT WELD STRENGTH

The strength of a plug or slot weld is a function of the size of the hole or slot. The cross-sectional area of the hole or slot in the plane of the connected parts is used to determine the strength of the weld.

The strength of plug and slot welds is as follows:

$$\phi R_n = \phi F_w A_w, \tag{10-15}$$

where

$$\phi = 0.75,$$
$$F_w = 0.6 F_{EXX}, \text{ and}$$
$$A_w = \text{Cross-sectional area of weld.}$$

The dimensional requirements for plug and slot welds are given in Section 10.4.

EXAMPLE 10-4

Plug Weld Strength

Determine the strength of the plug weld in the built-up connection shown in Figure 10-14 below considering the strength of the weld only. Electrodes are E70XX.

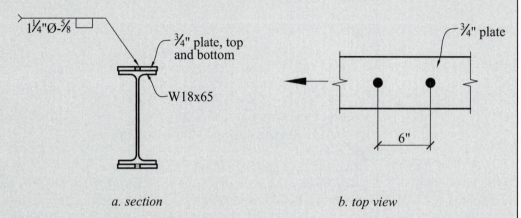

a. section   b. top view

Figure 10-14 Details for Example 10-4.

SOLUTION

Weld Dimensions

Minimum weld diameter:

t_{plate} = ¾ in.

flange thickness, t_f = ¾ in. (W18 × 65)

$d_{\min} = t + \frac{5}{16}$ in. = $1\frac{1}{16}$ in.

This should be welded up to the next odd $\frac{1}{16}$ in.; therefore, d_{min} = $1\frac{3}{16}$ in. < 1¼ in., OK

Maximum weld diameter:

d_{max} = 2.25w = (2.25) (⅝ in.) = 1.41 in. > 1¼ in. OK

Minimum thickness:

Larger of ⅝ in. or $\frac{t}{2}$

$\frac{t}{2} = \frac{\frac{3}{4} \text{ in.}}{2}$ = ⅜ in.; therefore, the minimum thickness is ⅜ in.

(⅝ in. provided)

Minimum spacing:

S_{min} = 4 × weld diameter

= (4)(1.25) = 5 in. < 6 in. OK

The dimensional requirements are all met.

Weld Strength

From equation (10-15),

$$\phi R_n = \phi F_w A_w$$

$$= (0.75)(0.6)(70 \text{ ksi})\left[\frac{\pi(1.25)^2}{4}\right] = 38.6 \text{ kips.}$$

EXAMPLE 10-5

Slot Weld Strength

Determine the strength of the connection shown in Figure 10-15 considering the strength of the weld and the steel angle. Electrodes are E70XX and the steel is ASTM A36.

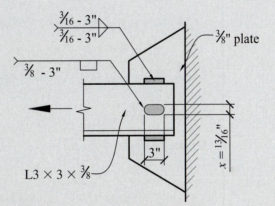

Figure 10-15 Detail for Example 10-5.

SOLUTION

Weld dimensions

Minimum weld width:

$$x_{min} = t + \tfrac{5}{16} \text{ in.} = \tfrac{11}{16} \text{ in.}$$

This should be welded up to the next odd $\tfrac{1}{16}$ in.; therefore, $x_{min} = \tfrac{13}{16}$ in. ($\tfrac{13}{16}$ in. provided). OK

Maximum weld width:

$$x_{max} = 2.25w = (2.25)(\tfrac{3}{8} \text{ in.}) = 0.85 \text{ in.} > \tfrac{13}{16} = 0.82 \text{ in. OK}$$

(continued)

Minimum thickness:

Use the thickness of the material since $t < \frac{5}{8}$ in.

Minimum thickness = $\frac{3}{8}$ in.

Maximum length = $10t = (10)(3/8) = 3.75$ in. > 3 in., OK

The dimensional requirements are all met.

Weld Strength

From equation (10-15),

$$\phi R_n = \phi F_w A_w$$
$$= (0.75)(0.6)(70\text{ ksi})(3\text{ in.})(\tfrac{13}{16}\text{ in.})$$
$$= 76.7\text{ kips (strength of the slot weld).}$$

From Equation 10-5,

$$\phi R_n = 1.392DL \ (D = 3 \text{ for } 3/16\text{-in. weld})$$
$$= (1.392)(3)(3 + 3) = 25.0\text{ kips}$$

Total weld strength: $76.7 + 25.0 = 101.7$ kips.

Angle Strength

Tensile yielding:

$$\phi R_n = \phi F_y A_g$$
$$= (0.9)(36)(2.11) = \textbf{68.3 kips}$$

Tensile rupture:

$$U = 1 - \frac{\bar{x}}{L} \text{ (from } AISCM \text{ Table D3.1)}$$

$$= 1 - \frac{0.884}{3\text{ in.}} = 0.705$$

$$A_e = A_n U$$
$$= [2.11 - (\tfrac{13}{16}\text{ in.})(\tfrac{3}{8}\text{ in.})](0.705) = 1.27\text{ in.}^2$$
$$\phi R_n = \phi F_u A_e$$
$$= (0.75)(58)(1.27) = \textbf{55.3 kips} \text{ Controls}$$

Tensile rupture controls the design of the connection. Note that the strength of the fillet welds alone would not be adequate to match the tensile rupture strength of the angle. More fillet weld could also be added, but the use if a slot need is shown here for illustrative purposes.

10.7 ECCENTRICALLY LOADED WELDS: SHEAR ONLY

The analysis of eccentrically loaded welds follows the same principles used for eccentrically loaded bolts (see Chapter 9). With reference to Figure 10-16, eccentrically loaded welds must resist the direct shear from the applied load, P, plus additional shear caused by the moment resulting from the eccentric load, Pe. There are two methods of analysis that can be used—the instantaneous center of rotation (IC), method and the elastic method.

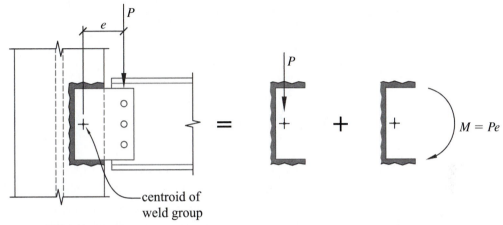

Figure 10-16 Eccentrically loaded weld.

The IC method is more accurate, but requires an iterative solution. As was the case with bolted connections, an eccentrically loaded weld will rotate about a point called the instantaneous center (see Figure 10-17). The IC has to initially be assumed for analysis. The weld must also be broken down into discrete elements of equal length for the analysis. It is recommended that at least 20 elements be selected for the longest weld in the group in order to maintain reasonable accuracy in the solution [1].

Using a load–deformation relationship, the nominal strength of the weld group is

$$R_{nx} = \Sigma F_{wix} A_{wi}, \text{ and} \qquad (10\text{-}16)$$

$$R_{ny} = \Sigma F_{wiy} A_{wi}, \qquad (10\text{-}17)$$

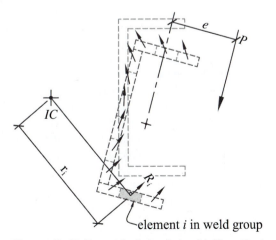

element i in weld group

Figure 10-17 Eccentrically loaded weld: IC method.

where

R_{nx} = x-component of the nominal weld strength,

R_{ny} = y-component of the nominal weld strength,

A_{wi} = Effective weld throat area of element i,

F_{wi} = Weld stress for element i

$$= 0.60F_{EXX}(1 + 0.5\ sin^{1.5}\theta)[p(1.9 - 0.9p)]^{0.3}, \tag{10-18}$$

F_{wix} = x-component of the weld stress, F_{wi},

F_{wiy} = y-component of the weld stress, F_{wi},

$$p = \frac{\Delta_i}{\Delta_m}, \tag{10-19}$$

$$\Delta_i = \frac{r_i\Delta_u}{r_{crit}}, \tag{10-20}$$

r_i = Distance from the IC to the weld element i, in.,

r_{crit} = Distance from the IC to the weld element with minimum Δ_u/r_i ratio, in.,

Δ_u = Deformation of weld element at ultimate stress, usually the weld element located the farthest from the IC (maximum value of r_i), in.

$$= 1.087w(\theta + 6)^{-0.65} \leq 0.17w, \tag{10-21}$$

Δ_m = Deformation of weld element at maximum stress, in.

$$= 0.209w(\theta + 2)^{-0.32}, \tag{10-22}$$

θ = Load angle measured from the longitudinal axis of the weld, degrees, and

w = Weld leg size, in.

The resistance of each weld element calculated above is assumed to act perpendicular to a line drawn from the IC to the weld element (see Figure 10-17). When the location of the IC is correct, then all of the equilibrium equations will be satisfied ($\Sigma F_x = 0$, $\Sigma F_y = 0$, $\Sigma M = 0$). It is evident that the above analytical process is quite tedious and is not practical for common use. Tables 8-4 through 8-11 are provided in the *AISCM* as design aids that use the IC method (see Examples 10-7 and 10-8).

In lieu of the more tedious IC method, the elastic method can be used. This method is more conservative because it ignores the ductility of the weld group and the redundancy, or load redistribution, of the weld group.

When a weld is subject to an eccentric load in the plane of the weld as shown in Figure 10-18, the weld is subjected to a direct shear component and a torsional shear component. Considering the direct shear component, the shear in the weld per linear inch is

$$r_p = \frac{P}{L}, \tag{10-23}$$

where

r_p = Shear per linear inch of weld with components r_{px} and r_{py},

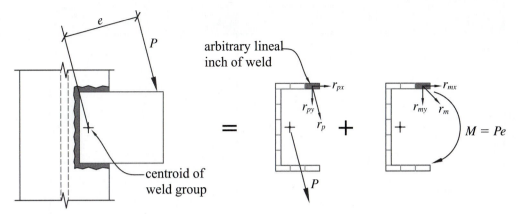

Figure 10-18 Eccentrically loaded weld: Elastic method.

$$r_{px} = \frac{P_x}{L}, \tag{10-24}$$

$$r_{py} = \frac{P_y}{L}, \tag{10-25}$$

P = Applied load with components P_x and P_y, and

L = Total weld length.

Considering the torsional component, the shear in the weld per linear inch is

$$r_m = \frac{Pec}{I_p}, \tag{10-26}$$

where

r_m = Shear per linear inch of weld with components r_{mx} and r_{my},

$$r_{mx} = \frac{Pec_y}{I_p}, \tag{10-27}$$

$$r_{my} = \frac{Pec_x}{I_p}, \tag{10-28}$$

P = Applied load,

e = Distance from the applied load to the centroid of the weld group,

c = Distance from the center of gravity of the weld group to a point most remote from the centroid of the weld group with components c_x and c_y, and

I_p = Polar moment of inertia (also commonly noted as J); see Table 10-3

  = $I_x + I_y$.

Table 10-3 Polar moment of inertia for weld groups

Weld Group	Center of Gravity, in. x, y	Section Modulus, in.^{3} S_{top}	S_{bot}	I_p (or J), in.^{4}
a.	$0, \dfrac{d}{2}$	$\dfrac{d^2}{6}$		$\dfrac{d^3}{12}$
b.	$\dfrac{b}{2}, \dfrac{d}{2}$	$\dfrac{d^2}{3}$		$\dfrac{d(3b^2 + d^2)}{6}$
c.	$\dfrac{b}{2}, \dfrac{d}{2}$	bd		$\dfrac{b^3 + 3bd^2}{6}$
d.	$\dfrac{b^2}{2(b + d)}, \dfrac{d^2}{2(b + d)}$	$\dfrac{4bd + d^2}{6}$	$\dfrac{d^2(4b + d)}{6(2b + d)}$	$\dfrac{(b + d)^4 - 6b^2d^2)}{12(b + d)}$
e.	$\dfrac{b^2}{2b + d}, \dfrac{d}{2}$	$bd + \dfrac{d^2}{6}$		$\dfrac{(2b + d)^3}{12} - \dfrac{b^2(b + d)^2}{(2b + d)}$
f.	$\dfrac{b}{2}, \dfrac{d^2}{b + 2d}$	$\dfrac{2bd + d^2}{3}$	$\dfrac{d^2(2b + d)}{3(b + d)}$	$\dfrac{(b + 2d)^3}{12} - \dfrac{d^2(b + d)^2}{(b + 2d)}$
g.	$\dfrac{b}{2}, \dfrac{d}{2}$	$bd + \dfrac{d^2}{3}$		$\dfrac{(b + d)^3}{6}$
h.	$\dfrac{b}{2}, \dfrac{d^2}{b + 2d}$	$\dfrac{2bd + d^2}{3}$	$\dfrac{d^2(2b + d)}{3(b + d)}$	$\dfrac{(b + 2d)^3}{12} - \dfrac{d^2(b + d)^2}{(b + 2d)}$
i.	$\dfrac{b}{2}, \dfrac{d^2}{2(b + d)}$	$\dfrac{4bd + d^2}{3}$	$\dfrac{4bd^2 + d^3}{6b + 3d}$	$\dfrac{d^3(4b + d)}{6(b + d)} + \dfrac{b^3}{6}$
j.	$\dfrac{b}{2}, \dfrac{d}{2}$	$bd + \dfrac{d^2}{3}$		$\dfrac{b^3 + 3bd^2 + d^3}{6}$
k.	$\dfrac{b}{2}, \dfrac{d}{2}$	$2bd + \dfrac{d^2}{3}$		$\dfrac{2b^3 + 6bd^2 + d^3}{6}$
l.	$\dfrac{d}{2}, \dfrac{d}{2}$	$\dfrac{\pi d^2}{4}$		$\dfrac{\pi d^3}{4}$

The polar moment of inertia, I_p, can be found by summing the rectangular moments of inertia ($I_x + I_y$); however, Table 10-3 is provided as a reference. In calculating the polar moment of inertia, the thickness of the weld is ignored and the weld group is treated as line elements with a unit thickness.

Once all of the shear stress components are found (r_p, r_m), they are added together to determine the point at which the shear stress is the highest, which is usually the point most remote in the weld group:

$$r = \sqrt{(r_{px} + r_{mx})^2 + (r_{py} + r_{my})^2}.$$ (10-29)

This resultant shear stress is then compared to the available strength of the weld.

EXAMPLE 10-6

Eccentrically Loaded Weld: Shear Only

Determine whether the weld for the bracket connection shown in Figure 10-19 is adequate to support the applied loads. Electrodes are E70XX.

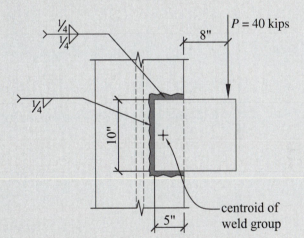

Figure 10-19 Connection detail for Example 10-6.

SOLUTION

From Table 10-3,

Center of gravity (x, y):

$$\text{CG} = \frac{b^2}{2b + d}, \frac{d}{2}$$

$$= \frac{5^2}{2(5) + (10)}, \frac{10}{2} = (1.25, 5)$$

$$e = 8 \text{ in.} + 5 \text{ in.} - 1.25 \text{ in.} = 11.75 \text{ in.}$$

(continued)

Polar moment of inertia:

$$I_p = \frac{(2b + d)^3}{12} - \frac{b^2(b + d)^2}{(2b + d)}$$

$$= \frac{[(2)(5) + (10)]^3}{12} - \frac{[(5)^2](5 + 10)^2}{[(2)(5) + 10)]} = 385 \text{ in.}^4$$

$c_x = 5 \text{ in.} - 1.25 \text{ in.} = 3.75 \text{ in.}$

$c_y = 5 \text{ in.}$

$c = \sqrt{c_x^2 + c_y^2}$

$\quad = \sqrt{3.75^2 + 5^2} = 6.25 \text{ in.}$

Weld stress due to direct shear:

From equations (10-24) and (10-25),

$$r_{px} = \frac{P_x}{L}; P_x = 0; \text{ therefore, } r_{px} = 0.$$

$$r_{py} = \frac{P_y}{L}$$

$$= \frac{40 \text{ kips}}{(5 \text{ in.} + 10 \text{ in.} + 5 \text{ in.})} = 2 \text{ kips/in.}$$

Weld stress due to torsional component:

From equations (10-27) and (10-28):

$$r_{mx} = \frac{Pec_y}{I_p}$$

$$= \frac{(40)(11.75)(5)}{385} = 6.10 \text{ kips/in.}$$

$$r_{my} = \frac{Pec_x}{I_p}$$

$$= \frac{(40)(11.75)(3.75)}{385} = 4.57 \text{ kips/in.}$$

Total weld stress:

From equation (10-29),

$$r = \sqrt{(r_{px} + r_{mx})^2 + (r_{py} + r_{my})^2}$$

$$= \sqrt{(0 + 6.10)^2 + (2 + 4.57)^2} = 8.97 \text{ kips/in.}$$

Available weld strength:

From equation (10-5),

$$\phi R_n = 1.392DL(D = 4 \text{ for a } \tfrac{1}{4} \text{ inch weld})$$
$$= (1.392)(4)(1) = 5.57 \text{ k/in.} < 8.97 \text{ k/in. Not good}$$

The applied stress is found to be greater than the available strength, so the weld size would have to be increased (by inspection, a $\tfrac{7}{16}$-in. weld would be adequate). Note that in the above calculation, the weld length, L, is 1 in., since the applied stress was calculated per unit length.

EXAMPLE 10-7

Eccentrically Loaded Weld: Shear Only, Using the *AISCM* Design Aids

Determine whether the weld in Example 10-6 is adequate using the appropriate design aid from the *AISCM*.

$$D = 4 \text{ (for } \tfrac{1}{4}\text{-in. weld)}$$

From the *AISCM*, Table 8-8, $\theta = 0°$,

$$e_x = al = 11.75$$
$$l = 10 \text{ in.}$$
$$a = \frac{11.75}{10} = 1.175$$
$$kl = 5 \text{ in.}$$
$$k = \frac{5}{10} = 0.5$$

Using $a = 1.175$ and $k = 0.5$,

$$C = 1.45 \text{ (by interpolation)}$$
$$D_{min} = \frac{P_u}{\phi CC_1 l}$$
$$(C_1 = 1.0 \text{ for E70XX, see } AISCM, \text{ Table 8-8})$$
$$D_{min} = \frac{40 \text{ k}}{(0.75)(1.45)(1.0)(10)} = \mathbf{3.68} < D = \mathbf{4 \ OK}$$

Using Table 8-8, it is found that the $\tfrac{1}{4}$-in. weld is, in fact, adequate. Example 10-6 uses the elastic method, whereas Example 10-7 and the corresponding AISC table use the instantaneous center (IC) method. As discussed previously, the elastic method is more conservative because simplifying assumptions are made.

10.8 ECCENTRICALLY LOADED WELDS: SHEAR PLUS TENSION

A load that is applied eccentrically to a weld group that is not in the plane of the weld group (see Figure 10-20) produces both shear and tension stress components in the weld group. The eccentric load is resolved into a direct shear component equal to the applied load, P, and a moment equal to Pe. Using a unit width for the weld thickness, the stress components are as follows:

Direct shear component:

$$f_p = \frac{P}{l} \tag{10-30}$$

Tension component due to moments:

$$f_m = \frac{Pe}{S}, \tag{10-31}$$

where

P = Applied load,
e = Eccentricity,
l = Total weld length in the group, and
S = Section modulus of the weld group (see Table 10-3).

The resulting weld stress is found by adding the shear and tension components as follows:

$$f_r = \sqrt{f_p^2 + f_m^2}. \tag{10-32}$$

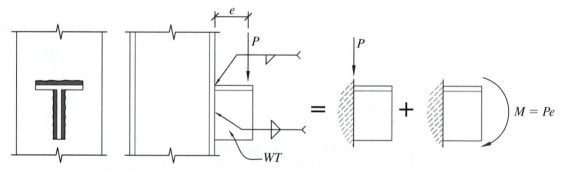

Figure 10-20 Eccentrically loaded weld in shear plus tension.

EXAMPLE 10-8

Eccentrically Loaded Weld: Shear Plus Tension

Determine the required fillet weld size for the bracket connection shown in Figure 10-21 considering the strength of the weld only. Compare the results with the appropriate table from the *AISCM*. Electrodes are E70XX.

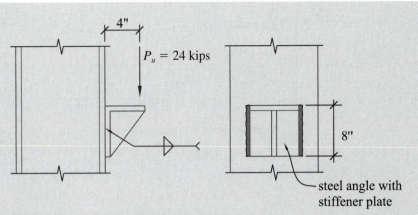

Figure 10-21 Connection details for Example 10-8.

SOLUTION

$e = 4$ in.

$l = 8$ in. $+ 8$ in. $= 12$ in.

From Table 10-3,

$$S = \frac{d^2}{3}$$

$$= \frac{(8)^2}{3} = 21.3 \text{ in.}^3$$

Weld stress due to direct shear:

From equation (10-30),

$$f_p = \frac{P}{l}$$

$$= \frac{24}{16} = 1.5 \text{ kips/in.}$$

Weld stress due to tension from applied moment:

From equation (10-31),

$$f_m = \frac{Pe}{S}$$

$$= \frac{(24)(4)}{21.3} = 4.5 \text{ kips/in.}$$

(continued)

Combined weld stress:

From equation (10-32),

$$f_r = \sqrt{f_p{}^2 + f_m{}^2}$$
$$f_r = \sqrt{(1.5)^2 + (4.5)^2} = 4.74 \text{ kips/in.}$$

Required weld thickness:

$$\phi R_n = 1.392DL$$

Therefore $\phi R_n/L = 1.392D = 474$ kips/in.

$D - 3.42 \approx$ four sixteenths $\therefore$ use ¼ in. weld

Note that the elastic method is more conservative. We will now compare these results with the design aids in the *AISCM*, which are based on the inelastic method.

From *AISCM*, Table 8-4, $\theta = 0°$,

$$e = 4 \text{ in.} = al$$
$$a = \frac{4 \text{ in.}}{8} = 0.5$$
$$k = 0 \ (\text{see } AISCM, \text{ Table 8-4})$$
$$C = 2.29$$
$$D_{\min} = \frac{P_u}{\phi C C_1 l}$$

$(C_1 = 1.0 \text{ for E70XX, see } AISCM, \text{ Table 8-3})$

$$D_{min} = \frac{24 \text{ k}}{(0.75)(2.29)(1.0)(8)} = \textbf{1.75} \approx \textbf{two-sixteenths}, \therefore \textbf{Use ⅛-in. weld.}$$

Note that while a 1/8-in. fllet weld is the minimum size reqired for strength, the weld size may have to be increased in the basis of the thickness of the connected parts (see Table 10-2).

10.9 REFERENCES

1. American Institute of Steel Construction. 2006. *Steel Construction manual*, 13th ed. Chicago: *AISC*.

2. American Institute of Steel Construction. 2006. *Steel design guide series 21: Welded connections — A primer for structural engineers*.

3. American Welding Society. 2006. *AWS D1.1 — Structural welding code*. Miami.

4. International Codes Council. 2006. *International building code*. Falls Church, VA: ICC.

5. McCormac, Jack. 1981. *Structural steel design,* 3rd Ed. Harper and Row. New York.

6. Salmon, Charles G., and John E. Johnson. 1980. *Steel structures: design and behavior*, 2nd ed. Harper and Row. New York.

7. Blodgett, Omer. 1966. *Design of welded structures*. Cleveland: The James F. Lincoln Arc Welding Foundation.

8. Segui, William. 2006. *Steel design*, 4th ed. Toronto: Thomson Engineering.

9. Limbrunner, George F., and Leonard Spiegel. 2001. *Applied structural steel design*, 4th ed. Upper Saddle River, NJ: Prentice Hall.

10. American Society of Civil Engineers. 2005. *ASCE-7: Minimum design loads for buildings and other structures*. Reston, VA: *ASCE*.

10.10 PROBLEMS

10-1. Determine the weld required for the connection shown in Figure 10-22. The steel is ASTM A36 and the weld electrodes are E70XX.

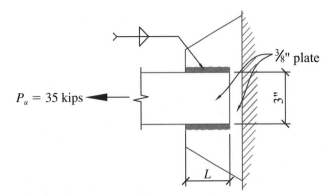

$P_u = 35$ kips

$\frac{3}{8}$" plate

3"

L

Figure 10-22 Details for Problem 10-1.

10-2. Determine the maximum tensile load that may be applied to the connection shown in Figure 10-23 based on the weld strength only. The steel is ASTM A36 and the weld electrodes are E70XX.

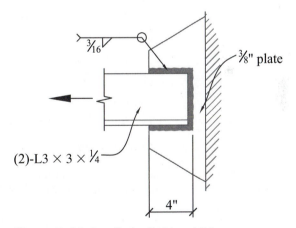

$\frac{3}{16}$

$\frac{3}{8}$" plate

(2)-L3 × 3 × ¼

4"

Figure 10-23 Details for Problem 10-2.

10-3. Determine the fillet weld required for the lap splice connection shown in Figure 10-24. The steel is ASTM A36 and the weld electrodes are E70XX.

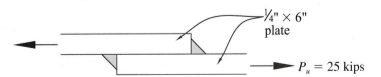

¼" × 6" plate

$P_u = 25$ kips

Figure 10-24 Details for Problem 10-3.

10-4. Determine the length of the plate required for the moment connection shown in Figure 10-25. The steel is ASTM A36 and the weld electrodes are E70XX. Ignore the strength of the W12 × 26; consider the strength of the plate in tension.

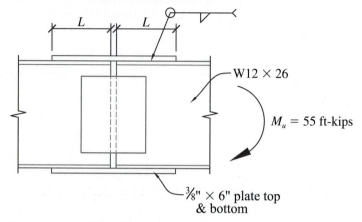

Figure 10-25 Details for Problem 10-4.

10-5. Determine the capacity of the lap splice connection shown in Figure 10-26. The steel is ASTM A36 and the weld electrodes are E70XX.

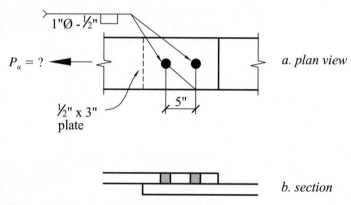

Figure 10-26 Details for Problem 10-5.

10-6. Determine the required length of the slot weld shown in the connection shown in Figure 10-27. The steel is ASTM A572 and the weld electrodes are E70XX.

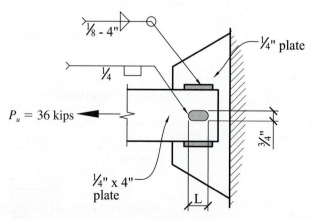

Figure 10-27 Details for Problem 10-6.

10-7. Determine the capacity of the bracket connection shown in Figure 10-28 below using the IC method and the elastic method. Compare the results. Weld electrodes are E70XX.

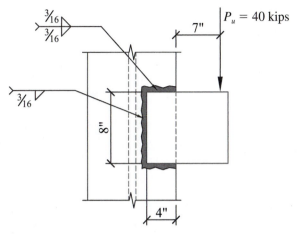

Figure 10-28 Details for Problem 10-7.

10-8. Determine the required weld size for the bracket connection shown in Figure 10-29. Weld electrodes are E70XX.

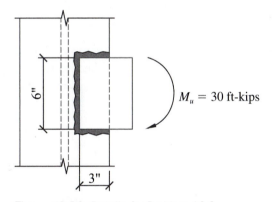

Figure 10-29 Details for Problem 10-8.

10-9. Determine the required weld length for the seat connection shown in Figure 10-30. Weld electrodes are E70XX. Ignore the strength of the bracket.

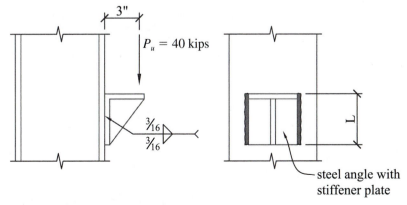

Figure 10-30 Details for Problem 10-9.

10-10. Determine the maximum eccentricity for the connection shown in Figure 10-31. Weld electrodes are E70XX.

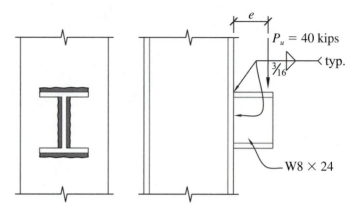

Figure 10-31 Details for Problem 10-10.

Student Design Project Problem:

10-11. Design a welded seat angle connection for the typical roof joist and joist girders, using the detail shown in Figure 10-32 as a giude.

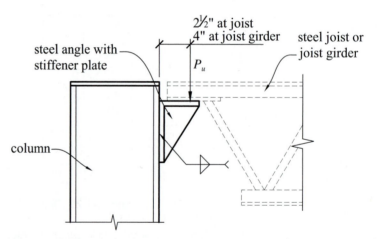

Figure 10-32 Joist seat connection detal.

10-12. Design the X-brace connections to be welded connections. Check the appropriate limit states for block shear on the gusset plates.

11

Special Connections and Details

11.1 INTRODUCTION

In Chapters 9 and 10, we considered basic bolted and welded connections, including an introduction to the use of the AISC selection tables for standard connections. In this chapter, we will consider the design of connections and details that are not specifically covered in the AISC selection tables, but are very common in practice. The connections that we will consider use a combination of bolts and welds, and many of these connections will have design components that are covered in the AISC selection tables. Below is a summary of the connections and details that will be considered in this chapter:

11.2 Coped Beams

11.3 Moment Connections: Introduction

11.4 Moment Connections: Partially Restrained and Flexible

11.5 Moment Connections: Fully Restrained

11.6 Moment Connections: Beams and Beam Splices

11.7 Column Stiffeners

11.8 Column Splices

11.9 Holes in Beams

11.10 Design of Gusset Plates in Vertical Bracing and Truss Connections

11.2 COPED BEAMS

Coped beams occur on virtually every steel-framed project, and it is therefore essential for any designer to understand the analysis and design parameters for coped connections. The geometry of a beam cope will generally be a function of the shape of the connected

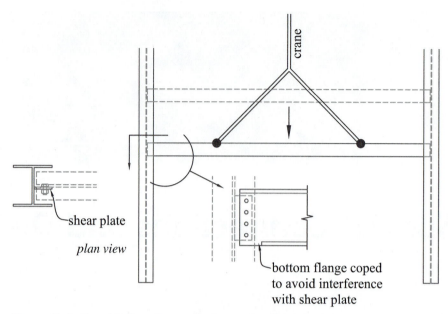

Figure 11-1 Coped bottom flange due to construction sequence.

member. In some cases, field conditions may dictate the requirement of a cope. For example, Figure 11-1 shows a beam framing into the web of a column. When this beam is erected into its final position, it is dropped down in between the column flanges. This particular beam would require a coped bottom flange in order to be placed without an obstruction. Other common coped beam connections are shown in Figure 11-2.

Additional beam modifications for connections are shown in Figure 11-3. A beam *cope* is defined as the removal of part of the beam web and flange. A *block* is the removal of the flange only, and a *cut* is the removal of one side of a flange. In each of these cases, it is common to refer to any of them as copes since the analysis and design procedures are similar.

When beams are coped, the load path of the end reaction must pass through a reduced section of the connected beam (see Figure 11-4). The strength of the beam in block shear was covered in Chapter 9, and we will now consider the effect of bending stresses at the critical section due to the eccentric load. The eccentricity, e, is the distance from the face of the cope to the point of inflection. In some connections, the point of inflection may not be at the face of the support, but it is conservative to assume that the point of inflection is at the

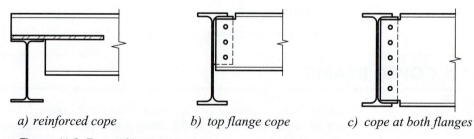

a) reinforced cope   *b) top flange cope*   *c) cope at both flanges*

Figure 11-2 Types of beam copes.

plan

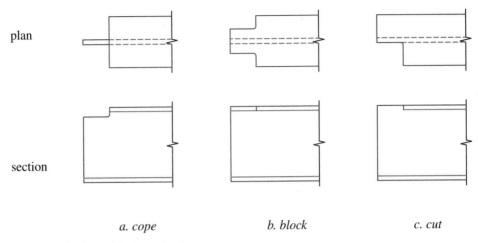

section

| *a. cope* | *b. block* | *c. cut* |

Figure 11-3 Cope, block, and cut.

face of the support or at the line of action of the bolt group. A lower value of e can be used if justified by analysis.

When the geometry of the beam cope is such that the localized stresses exceed the design strength, the beam can be reinforced at the cope. In Figure 11-5a, a web doubler plate is shown; in Figure 11-5b, a longitudinal stiffener is shown. In these two cases, the reinforcement should be extended a distance of d_c past the critical section, where d_c is the depth of the beam cope. In Figure 11-5c, both transverse and longitudinal stiffeners are shown. The given geometry of a connection will dictate the type of reinforcement to use where reinforcement is required.

For beams coped at either the top or bottom flange, or both flanges, the design strength for flexural rupture is

$$\phi_{br}M_n = \phi_{br}F_uS_{net}, \tag{11-1}$$

where

$\phi_{br} = 0.75$,

F_u = Tensile rupture strength, and

S_{net} = Net section modulus at the critical section.

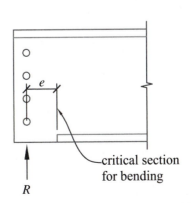

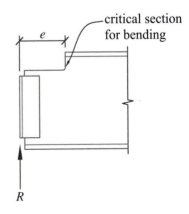

Figure 11-4 Reaction at a coped beam.

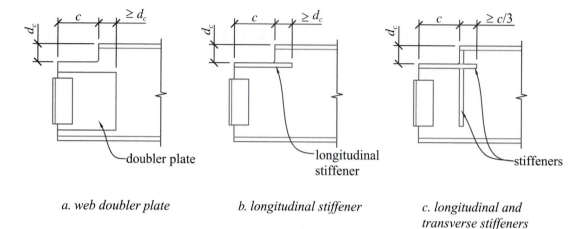

a. web doubler plate   b. longitudinal stiffener   c. longitudinal and
                                                     transverse stiffeners

Figure 11-5 Reinforcement of coped beams.

The net section modulus, S_{net}, for various beams and cope depths is provided in *AISCM*, Table 9-2. In this table, the cope depth is limited to the smaller of $0.5d$ and 10 in.

For beams coped at either the top or bottom flange, or both flanges, the design strength for flexural local buckling is

$$\phi_b M_n = \phi_b F_{cr} S_{net}, \tag{11-2}$$

where

$\phi_b = 0.9$, and

F_{cr} = Available buckling stress.

For beams coped at one of the flanges only, the available buckling stress when $c \le 2d$ and $d_c \le 0.5d$ is

$$F_{cr} = 26{,}210 \left(\frac{t_w}{h_1} \right)^2 fk, \tag{11-3}$$

where

f = Plate buckling model adjustment factor

$$= \frac{2c}{d} \quad \text{when } \frac{c}{d} \le 1.0 \tag{11-4}$$

$$= 1 + \frac{c}{d} \quad \text{when } \frac{c}{d} > 1.0, \tag{11-5}$$

k = Plate buckling coefficient

$$= 2.2 \left(\frac{h_1}{c} \right)^{1.65} \quad \text{when } \frac{c}{h_1} \le 1.0 \tag{11-6}$$

$$= \frac{2.2 h_1}{c} \quad \text{when } \frac{c}{h_1} > 1.0, \tag{11-7}$$

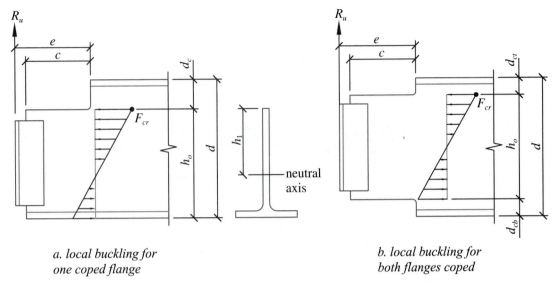

Figure 11-6 Dimensional parameters for local buckling in a coped beam.

t_w = Beam web thickness,

h_1 = Distance from the horizontal cope edge to the neutral axis (Note: The AISC specification also allows the more conservative value of $h_0 = d - d_c$ to be used in lieu of h_1 for beams coped at one flange. For beams coped at both flanges, the reduced beam depth, h_0, shown in Figure 11-6b, is used.)

c = Cope length (see Figure 11-6),

d = Beam depth, and

d_c = Cope depth (see Figure 11-6).

When beams are coped at both flanges, the available buckling stress when $c \leq 2d$ and $d_c \leq 0.2d$ is

$$F_{cr} = 0.62\pi E\left(\frac{t_w^2}{ch_0}\right)f_d, \tag{11-8}$$

where

$$f_d = 3.5 - 7.5\left(\frac{d_c}{d}\right), \text{ and} \tag{11-9}$$

$$E = 29 \times 10^6 \text{ psi}$$

When beams are coped at both flanges, the available buckling stress when $d_c > 0.2d$ is

$$F_{cr} = F_y Q, \tag{11-10}$$

where

F_y = Yield stress,

Q = 1.0 for $\lambda \leq 0.7$

   = $1.34 - 0.486\lambda$ for $0.7 < \lambda \leq 1.41$

$$= \frac{1.30}{\lambda^2} \text{ for } \lambda > 1.41, \text{ and}$$

$$\lambda = \frac{h_0 \sqrt{F_y}}{10t_w \sqrt{475 + 280 \left(\dfrac{h_0}{c}\right)^2}}. \tag{11-11}$$

EXAMPLE 11-1

Beam Coped at the Top Flange

For the beam shown in Figure 11-7, determine whether the beam is adequate for flexural rupture and flexural yielding at the beam cope. The steel is ASTM A992, grade 50.

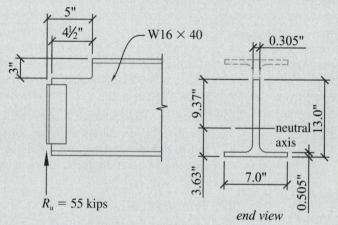

Figure 11-7 Details for Example 11-1.

From *AISCM*, Table 1-1,

$$d = 16.0 \text{ in.} \qquad t_w = 0.305 \text{ in.}$$
$$b_f = 7.0 \text{ in.} \qquad t_f = 0.505 \text{ in.}$$
$$h_o = d - d_c = 16.0 - 3 = 13.0 \text{ in.}$$

Section properties at the critical section:

Web component is $(13.0 \text{ in.} - 0.505 \text{ in.})$ by 0.305 in.

Flange component is 7.0 in. by 0.505 in.

Table 11-1 Section properties at the beam cope

Element	A	y	Ay	I	$d = y - \bar{y}$	$I + Ad^2$
Web	3.81	6.75	25.73	49.6	3.13	86.9
Flange	3.53	0.252	0.89	0.0751	−3.37	40.2
$\Sigma =$	7.34		26.63			$I_{net} = 127.1$

$$\bar{y} = \frac{\Sigma Ay}{\Sigma A} = \frac{26.6}{7.34} = 3.63 \text{ in.}$$

$$h_1 = h_o - \bar{y} = 13.0 - 3.63 = 9.37 \text{ in.}$$

$$S_{net} = \frac{I_{net}}{h_1} = \frac{127.1}{9.37} = 13.6 \text{ in.}^3$$

Alternatively, using *AISCM*, Table 9-2 with $d_c = 3$ in., confirm that $S_{net} = 13.6$ in.^{3}.

$$M_u = R_u e = (55 \text{ kips})(5 \text{ in.}) = 275 \text{ in.-kips}$$

Flexural rupture strength:

$$\phi_{br} M_n = \phi_{br} F_u S_{net}$$

$$= (0.75)(65)(13.6) = 663 \text{ in.-kips} > 275 \text{ in.-kips OK}$$

$$c \le 2d \rightarrow 4.5 \text{ in.} < (2)(16.0 \text{ in.}) = 32 \text{ in. OK}$$

$$d_c \le 0.5d \rightarrow 3 \text{ in.} < (0.5)(16.0 \text{ in.}) = 8 \text{ in. OK}$$

$$\frac{c}{d} = \frac{4.5}{16.0} = 0.281 \le 1.0 \quad \rightarrow \quad f = \frac{2c}{d} = \frac{(2)(4.5)}{16.0} = 0.563 \text{ in.}$$

$$\frac{c}{h_1} = \frac{4.5}{9.37} = 0.480 \le 1.0 \quad \rightarrow \quad k = 2.2\left(\frac{h_1}{c}\right)^{1.65} = 2.2\left(\frac{9.37}{4.5}\right)^{1.65} = 7.38$$

$$F_{cr} = 26{,}210\left(\frac{t_w}{h_1}\right)^2 fk = 26{,}210\left(\frac{0.305}{9.37}\right)^2 (0.563)(7.38) = 115.4 \text{ ksi}$$

$$\phi_b M_n = \phi_b F_{cr} S_{net} = (0.9)(115.4)(13.6) = 1412 \text{ in.-kips} > 275 \text{ in.-kips, OK}$$

Shear strength (from Chapter 6):

$$\phi_v V_n = \phi_v 0.6 F_y A_w C_v$$

$$= (0.9)(0.6)(50)[(13.0)(0.305)](1.0) = 107 \text{ kips} > V_u = 55 \text{ kips OK}$$

The coped connection is adequate in shear and in flexural rupture and flexural yielding.

EXAMPLE 11-2

Beam Cope with Reinforcing

For the beam shown in Figure 11-8, determine whether the beam is adequate for flexural rupture and flexural yielding at the beam cope. The steel has a yield strength of 50 ksi.

This is a common connection condition when W-shaped beams are used in the same framing plan as open-web steel joists that have a seat depth of 2.5 in. at the bearing ends. For the connection shown, the depth of the cope, dc, is 5.49 in. (greater than $0.5d = 4.0$ in.) and therefore must be reinforced as shown with the ⅜-in. by 2-in. plates. Note that the plates extend beyond the face of the cope at least 5.49 in. (the depth of the cope; see Figure 11-5).

(continued)

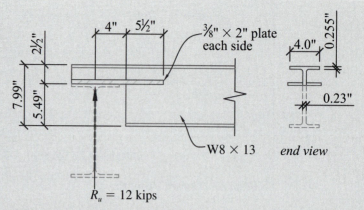

Figure 11-8 Details for Example 11-2.

The capacity of this connection must be checked for flexural rupture. The section properties of the critical section are summarized as follows.

From *AISCM*, Table 1-1,

$$d = 7.99 \text{ in.} \qquad t_w = 0.23 \text{ in.}$$
$$b_f = 4.0 \text{ in.} \qquad t_f = 0.255 \text{ in.}$$

Section properties at the critical section:

Web component is $(2.5 \text{ in.} - 0.255 \text{ in.})$ by 0.23 in.

Flange component is 4.0 in. by 0.255 in.

Plate component is 4.0 in. by 0.375 in.

Table 11-2 Section properties of reinforced beam cope

Element	A	y	Ay	I	$d = y - \bar{y}$	$I + Ad^2$
Top flange	1.02	2.373	2.42	0.006	1.292	1.706
Web	0.516	1.123	0.58	0.217	0.042	0.218
Plates	1.5	0.188	0.281	0.018	−0.893	1.22
$\Sigma =$	**3.036**		**3.281**			$I_{net} = \textbf{3.14}$

$$\bar{y} = \frac{\Sigma Ay}{\Sigma A} = \frac{3.281}{3.036} = 1.08 \text{ in.}$$

$$S_{net} = \frac{I_{net}}{2.5 - \bar{y}} = \frac{3.14}{2.5 - 1.08} = 2.21 \text{ in.}^3$$

Flexural rupture strength:

$$\phi_{br}M_n = \phi_{br}F_u S_{net} = (0.75)(65)(2.21) = 107 \text{ in.-kips}$$
$$M_u = R_u e = (12 \text{ kips})(4 \text{ in.}) = 48 \text{ in.-kips} < 107 \text{ in.-kips OK}$$

Shear strength (from Chapter 6):

$$\phi_v V_n = \phi_v 0.6 F_y A_w C_v$$
$$= (0.9)(0.6)(50)[(2.5)(0.23)](1.0) = 15.5 \text{ kips} > V_u = 12 \text{ kips OK}$$

EXAMPLE 11-3

Beam Cope at Both Flanges

For the beam shown in Figure 11-9, determine the maximum reaction that could occur based on flexural rupture and flexural yielding at the beam cope. The steel is ASTM A992, grade 50.

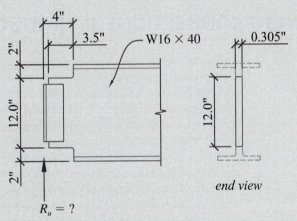

Figure 11-9 Details for Example 11-3.

From *AISCM*, Table 1-1,

$$d = 16.0 \text{ in.} \qquad t_w = 0.305 \text{ in.}$$
$$b_f = 7.0 \text{ in.} \qquad t_f = 0.505 \text{ in.}$$

$$h_o = d - d_{ct} - d_{cb} = 16.0 - 2 \text{ in.} - 2 \text{ in.} = 12.0 \text{ in.}$$
$$S_{net} = \frac{t_w h_o^2}{6} = \frac{(0.305)(12.0)^2}{6} = 7.32 \text{ in.}^3$$

Flexural rupture strength:

$$\phi_{br} M_n = \phi_{br} F_u S_{net} = (0.75)(65)(7.32) = 356 \text{ in.-kips}$$
$$R_{u,max} = \frac{\phi_{br} M_n}{e} = \frac{356}{4} = 89.2 \text{ kips}$$
$$c \leq 2d \rightarrow 3.5 \text{ in.} < (2)(16.0 \text{ in.}) = 32 \text{ in. OK}$$
$$d_c \leq 0.2d \rightarrow 2 \text{ in.} < (0.2)(16.0 \text{ in.}) = 3.2 \text{ in. OK}$$
$$f_d = 3.5 - 7.5\left(\frac{d_c}{d}\right) = 3.5 - 7.5\left(\frac{2}{16.0}\right) = 2.56$$

(continued)

$$F_{cr} = 0.62\pi E \left(\frac{t_w^2}{ch_0} \right) f_d = (0.62)(\pi)(29,000) \left[\frac{0.305^2}{(3.5)(12.0)} \right] 2.56 = 320 \text{ ksi}$$

$$\phi_b M_n = \phi_b F_{cr} S_{net} = (0.9)(320)(7.32) = 2112 \text{ in.-kips OK}$$

$$R_{u,max} = \frac{\phi_{br} M_n}{e} = \frac{2112}{4} = 528 \text{ kips}$$

Shear strength (from Chapter 6):

$$\phi_v V_n = \phi_v 0.6 F_y A_w C_v = (0.9)(0.6)(50)[(12.0)(0.305)](1.0) = 98.8 \text{ kips}$$

Flexural rupture controls, so $R_{u,max} = 89.2$ kips.

11.3 MOMENT CONNECTIONS: INTRODUCTION

A moment connection is capable of transferring a moment couple or reaction across a joint or to a support. Beam support conditions that are assumed to be fixed are considered moment connections; however, most steel moment connections that occur in practice do not have absolute fixity in that some rotation will occur at the connection. Keep in mind that a purely fixed end condition is one in which the end has zero rotation. Support conditions that are assumed to be pinned have connected ends that are completely free to rotate. Virtually all steel connections have some degree of fixity such that they are neither perfectly fixed nor pinned; they have a degree of fixity somewhere between these two extremes. For simplicity, connections designed to transfer shear only are considered to be pinned even though some degree of rotation is limited. Connections designed to resist some moment are considered to be moment connections with a relatively smaller degree of rotation occurring at the joint (see Figure 11-10).

The AISC specification in Section B3.6 identifies three basic connection types: *simple connections*, *fully restrained moment connections* (FR), and *partially restrained moment connections* (PR).

Simple connections are assumed to allow complete rotation at a joint and will therefore transmit a negligible moment across the connection. Standard shear connections that connect to the webs of beams are the most common type of simple connection.

Fully restrained (FR) moment connections transfer a moment across a joint with a negligible rotation. FR connections are designed to maintain the angle between the connected members at the factored load level.

Partially restrained (FR) connections will transfer some moment across a connection, but there will also exist a corresponding rotation at the connection as well. The use of PR connections is only permitted when the force versus deformation characteristics of the connection are known either by documented research or analysis. However, the deformation characteristics of a PR connection cannot be easily determined because such behavior is a function of the sequence in which the loads are applied. Since the actual load sequence for any structure cannot truly be known, engineering judgment must be used to identify the possible load sequences in order to properly design such a connection in accordance with ASCE 7.

A more simplified and conservative approach to PR connection design is to use a flexible moment connection (FMC). In the past, FMCs have been refered to as Type 2 with wind, semi-rigid, or flexible wind connections [2].

With FMC design, the gravity loads are taken in the shear connection only and any end restraint provided by the moment connection is neglected for gravity loads. For lateral loads, the FMC is assumed to have the same behavior as an FR connection and all of the lateral loads are taken

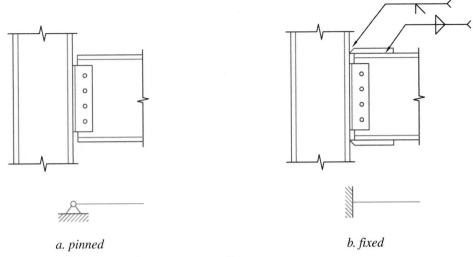

a. pinned   *b. fixed*

Figure 11-10 Pinned and moment connections.

by the moment connection. It has been shown that this type of connection is adequate for resisting lateral loads provided that the plastic moment capacity of the connection is not exceeded [2]. Therefore, the full plastic moment capacity of the beam is available to resist lateral loads.

The general moment versus rotation curve for each of the above-mentioned connection types is shown in Figure 11-11. Each of the three curves (1, 2, and 3) represents a different connection type.

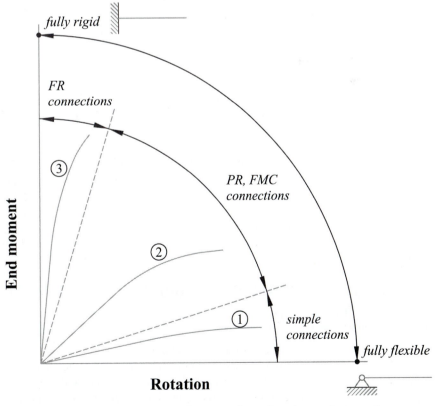

Figure 11-11 Moment versus rotation curve for various types of connection.

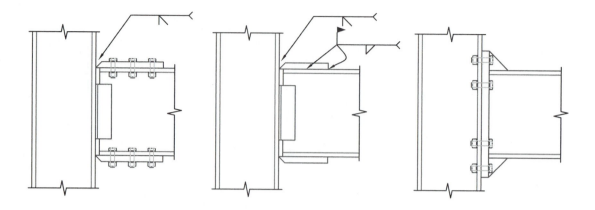

a. FR (fully restrained) connections

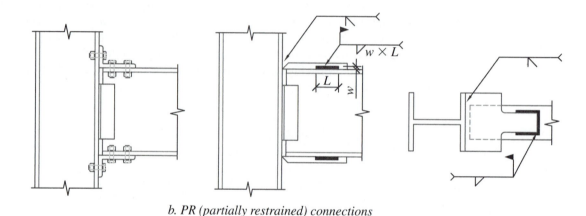

b. PR (partially restrained) connections

Figure 11-12 FR and PR connections.

It is also important to note that some moment connections might provide little or no rotation even though an assumption has been made that an FMC has been used when, in fact, a more rigid connection exists. For this reason, the designer should select a known flexible connection when one is required. Figure 11-12 shows common FR and PR connection types. FR connection behavior would resemble curve 3 in Figure 11-11 and PR connection behavior would resemble curve 2 in Figure 11-11.

11.4 MOMENT CONNECTIONS: PARTIALLY RESTRAINED AND FLEXIBLE

As discussed in the previous section, the use of PR connections requires knowledge of the moment versus rotation curve, as well as the load sequence. Since very little data is available for PR connections, most designers will use an FMC, which allows conservative and simplifying assumptions to be made. It is important to note that the use of a PR connection

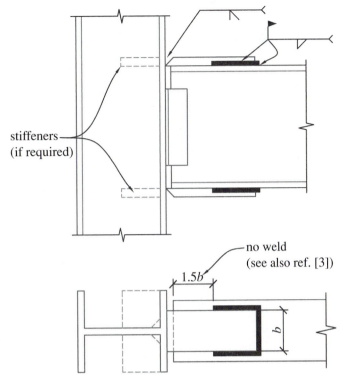

Figure 11-13 Flange-plated FMC.

or an FMC requires that the seismic response modification factor, R, be taken as less than or equal to 3.0. When R is greater than 3.0, the moment connections must be designed as an FR connection and must include the gravity load effects.

Common FMCs are shown in Figure 11-12, and the reader is referred to Example 9-8 for an analysis of a flange-angle FMC. For flange-plated FMCs, the flange plate has an un-welded length equal to 1.5 times the width of the plate in order to allow for the elongation of the plate, thus creating flexible behavior (see Figure 11-13).

EXAMPLE 11-4

Determine whether the FMC shown in Figure 11-14 is adequate to support the factored moment due to wind loads. The beam and column are ASTM A992, the steel plate is ASTM A36, and the weld electrodes are E70XX.

From *AISCM*, Table 1-1,

$d = 12.2$ in., and

$t_f = 0.38$ in.

Flange force:

$$P_{uf} = \frac{M}{d} = \frac{50}{12.2/12} = 49.2 \text{ kips}$$

(continued)

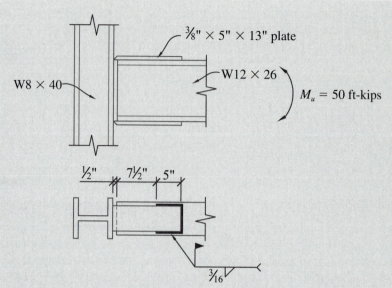

Figure 11-14 Details for Example 11-4.

Tension on gross plate area:

$$\phi P_n = \phi A_g F_y = (0.9)(0.375)(5)(36) = 60.7 \text{ kips} > 49.2 \text{ kips OK}$$

Compression:

$$K = 0.65 \text{ (Figure 5-3)}$$
$$L = 0.5 \text{ in.} + 7.5 \text{ in.} = 8.0 \text{ in.}$$
$$r_y = \frac{t}{\sqrt{12}} = \frac{0.375}{\sqrt{12}} = 0.108 \text{ in.}$$
$$\frac{KL}{r} = \frac{(0.65)(8.0)}{0.108} = 48.0$$

From *AISCM*, Table 4-22, $\phi_c F_{cr} = 28.7$ ksi.

$$\phi P_n = \phi F_{cr} A_g = (28.7)(0.375)(5) = 53.8 \text{ kips} > 49.2 \text{ kips OK}$$

Weld strength:

$$b = 5 \text{ in.} \left(\tfrac{3}{8}\text{-in. by 5-in. plate}\right)(L \geq b) \text{ OK}$$

From Table 10-2,

Minimum weld size $= \tfrac{3}{16}$ in. $\therefore D_{min} = 3$

Maximum weld size $= t - \tfrac{1}{16} = \tfrac{5}{16}'' \therefore D_{max} = 5 \left(\tfrac{3}{16}\text{-in. weld OK}\right)$

From equation (10-7),

$$\phi R_n = \phi R_{wl} + \phi R_{wt}$$
$$= 1.392DL_l + 1.392DL_t$$
$$= [(1.392)(3)(5+5)] + [(1.392)(3)(5)] = 62.6 \text{ kips} > 49.2 \text{ kips. OK}$$

Alternatively, from equation (10-8),

$$\phi R_n = 0.85\phi R_{wl} + 1.5\phi R_{wt}$$
$$= (0.85)1.392DL_l + (1.5)1.392DL_t$$
$$= [(0.85)(1.392)(3)(5+5)] + [(1.5)(1.392)(3)(5)]$$
$$= 66.8 \text{ kips} > 49.2 \text{ kips. OK}$$

11.5 MOMENT CONNECTIONS: FULLY RESTRAINED

As discussed in Section 11.3, fully restrained (FR) connections are sufficiently rigid to maintain the angle between the connected members. FR connections are designed to carry both gravity and lateral loads. For seismic loads, when the seismic response modification factor, R, is less than or equal to 3.0, the design approach is the same as for FMCs and the connection type must be one that provides adequate rigidity (see curve 3 in Figure 11-11 and the details in Figure 11-12a).

When the seismic response modification factor is greater than 3.0, additional design requirements must be met for supporting seismic loads [4]. These additional requirements are a combination of strength and stability design parameters, depending on the type of moment connections used. The AISC seismic provisions identify three basic moment frame types with $R > 3.0$: ordinary moment frames (OMF), intermediate moment frames (IMF), and special moment frames (SMF). The seismic response modification factors are 4.0, 6.0, and 8.0, respectively. Each of these moment frame types requires varying degrees of additional strength and stability requirements; the OMF has the least stringent requirments and the SMF has the most stringent requirments. Generally speaking, each of these connections types is designed for a certain moment and rotation. These connections are generally designed around the concept of creating a plastic hinge away from the beam–column joint and a strong column/weak beam scenario, which is the preferred failure mode [5]. Figure 11-15 illustrates this concept.

There are several connections that have published analysis and testing data that can be used for OMF, IMF, and SMF connections (see refs. 4 and 5). Using these pre-qualified connections is generally preferred since a rigorous analysis would be required for other connections that have not been tested. Figure 11-16 illustrates some of the basic types of connections that can be used for OMF, IMF, and SMF frames. Note that only Figures 11-16d and 11-16e are recognized in the AISC seismic provisions (see ref. 4), the other connection types are found in reference 5.

The prescriptive requirements for the above connection types are found in their respective standards (see refs. 4 and 5) and are beyond the scope of this text. It can be observed that using a seismic response modification factor equal to or less than 3.0 is highly desirable in that the analysis and design procedure is more simplified than a procedure that uses the OMF, IMF, or SMF requirements. In general, buildings with a Seismic Design Category of A, B, or C can usually be economically designed with $R \le 3.0$. This approach is recommended where possible.

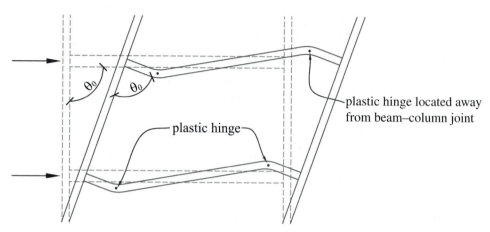

Figure 11-15 Plastic hinge formation in FR connections.

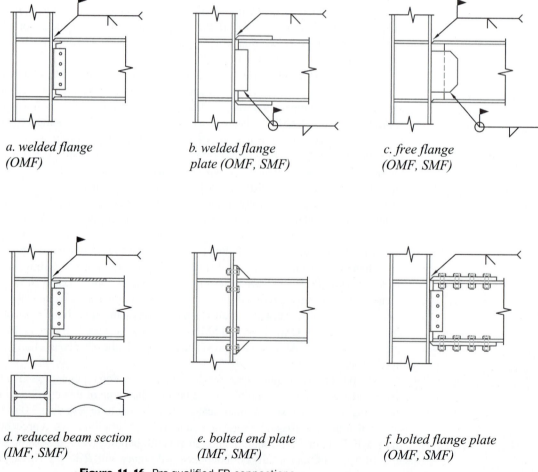

*a. welded flange
(OMF)*

*b. welded flange
plate (OMF, SMF)*

*c. free flange
(OMF, SMF)*

*d. reduced beam section
(IMF, SMF)*

*e. bolted end plate
(IMF, SMF)*

*f. bolted flange plate
(OMF, SMF)*

Figure 11-16 Pre-qualified FR connections.

EXAMPLE 11-5

Bolted FR Moment Connection

Determine whether the FR connection shown below is adequate to support the factored moment due to wind loads. The steel plate is ASTM A36, the beam is ASTM A992, and the bolts are ¾-in. A325N in standard (STD) holes.

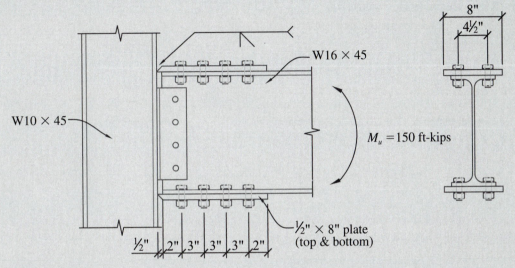

Figure 11-17 Details for Example 11-5.

From *AISCM*, Table 1-1,

W16 × 45
d = 16.1 in.
b_f = 7.04 in.
t_f = 0.565 in.
S_x = 72.7 in.^{3}

Flange force:

$$P_{uf} = \frac{M}{d} = \frac{150}{16.1/12} = 112 \text{ kips}$$

When the flanges of moment connections are bolted, the flexural strength of the beam is reduced due to the presence of the holes at the connection. The following provisions apply (AISC specification, Section F13):

When $F_u A_{fn} \geq Y_t F_y A_{fg}$, (11-12)

the reduced flexural strength does not need to be checked. When $F_u A_{fn} < Y_t F_y A_{fg}$, the design flexural strength at the moment connection is

$$\phi_b M_n = \phi_b F_u S_x \left(\frac{A_{fn}}{A_{fg}} \right), \tag{11-13}$$

(continued)

where

$\phi_b = 0.9,$

F_u = Tensile rupture strength,

S_x = Section modulus,

A_{fg} = Gross area of tension flange,

A_{fn} = Net area of tension flange, and

Y_t = 1.0 for $F_y/F_u \le 0.8$

    = 1.1 for all other cases.

Check the reduced flexural strength of the W16 × 45:

$$F_y/F_u = 50/65 = 0.77 < 0.80 \therefore Y_t = 1.0$$
$$A_{fg} = (7.04)(0.565) = 3.98 \text{ in.}^2$$
$$A_{fn} = A_{fg} - \Sigma A_{holes}$$
$$= 3.98 - \left[(2)\left(\frac{3}{4} + \frac{1}{8}\right)(0.565)\right] = 2.99 \text{ in.}^2$$
$$F_u A_{fn} = (65)(2.99) = 194 \text{ kips}$$
$$Y_t F_y A_{fg} = (1.0)(50)(3.98) = 199 \text{ kips}$$

194 k < 199 k → Reduced flexural strength must be checked.

$$\phi_b M_n = \phi_b F_u S_x\left(\frac{A_{fn}}{A_{fg}}\right) = (0.9)(65)(72.7)\left(\frac{2.99}{3.98}\right) = 3195 \text{ in.-kips} = 266 \text{ ft.-kips} > 150 \text{ ft.-kips OK}$$

Check shear on bolts:

From *AISCM*, Table 7-1, $\phi_v r_n = 15.9$ kips/bolt.

$$N_{b,required} = \frac{P_{uf}}{\phi_v r_n} = \frac{112}{15.9} = 7.05 < 8 \text{ bolts provided OK}$$

Check bolt bearing:

Since the thickness of the plate is less than the flange thickness, and since F_u for the plate is less than F_u for the beam, the bolt bearing on the plate will control. Figure 11-18 illustrates the failure modes for bearing, tension, and block shear.

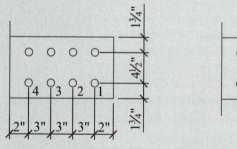
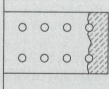

a. bolt bearing   b. tension on net area   c. block shear

Figure 11-18 Bolt bearing and block shear.

$$L_{c1} = 2 \text{ in.} - 0.5\left(\frac{3}{4} + \frac{1}{8}\right) = 1.56 \text{ in.}$$

$$L_{c2} = L_{c3} = L_{c4} = 3 \text{ in.} - \left(\frac{3}{4} + \frac{1}{8}\right) = 2.13 \text{ in.}$$

For bolt 1, $\phi R_n = \phi 1.2 L_c t F_u \leq \phi 2.4 dt F_u$

$$= (0.75)(1.2)(1.56 \text{ in.})(0.5 \text{ in.})(58 \text{ ksi}) < (0.75)(2.4)(0.75)(0.5 \text{ in.})(58 \text{ ksi})$$

$$= 40.7 \text{ kips} > 39.2 \text{ kips}$$

$$\phi R_n = 39.2 \text{ k.}$$

For bolts 2, 3, and 4, $\phi R_n = (0.75)(1.2)(2.13 \text{ in.})(0.5 \text{ in.})(58 \text{ ksi}) < (0.75)(2.4)(0.75)(0.5 \text{ in.})(58 \text{ ksi})$

$$= 55.4 \text{ kips} > 39.2 \text{ kips}$$

$$= 39.2 \text{ kips}$$

$$\phi R_n = [(2)(39.2 \text{ kips})] + [(6)(39.2 \text{ kips})] = \textbf{313 kips} > \textbf{P}_{uf} = \textbf{112 kips OK for bearing}$$

Check tension on gross and net area of the flange plate:

$$A_g = (0.5 \text{ in.})(8 \text{ in.}) = 4 \text{ in.}^2$$

$$A_n = A_g - \Sigma A_{holes} = 4 - \left[(2)\left(\frac{3}{4} + \frac{1}{8}\right)(0.5)\right] = 3.13 \text{ in.}^2$$

$$A_n \leq 0.85 A_g = (0.85)(4) = 3.4 \text{ in.}^2 \text{ OK}$$

$$A_e = A_n U = (3.13 \text{ in.}^2)(1.0) = 3.13 \text{ in.}^2$$

Strength based on gross area is

$$\phi P_n = \phi F_y A_g = (0.9)(36)(4 \text{ in.}^2) = 129 \text{ kips} > P_{uf} = 112 \text{ kips OK}$$

Strength based on effective area is

$$\phi P_n = \phi F_u A_e = (0.75)(58)(3.13 \text{ in.}^2) = 136 \text{ kips} > P_{uf} = 112 \text{ kips OK}$$

Block shear:

$$A_{gv} = (2)(3 \text{ in.} + 3 \text{ in.} + 3 \text{ in.} + 2 \text{ in.})(0.5 \text{ in.}) = 11 \text{ in.}^2$$

$$A_{gt} = (1.75 \text{ in.} + 1.75 \text{ in.})(0.5 \text{ in.}) = 1.75 \text{ in.}^2$$

$$A_{nv} = A_{gv} - \Sigma A_{holes} = 11 - \left[(3.5)(2)\left(\frac{3}{4} + \frac{1}{8}\right)(0.5)\right] = 7.93 \text{ in.}^2$$

$$A_{nt} = A_{gt} - \Sigma A_{holes} = 1.75 - \left[\left(\frac{3}{4} + \frac{1}{8}\right)(0.5)\right] = 1.31 \text{ in.}^2$$

The available block shear strength is found from equation (4-11) ($U_{bs} = 1.0$):

$$\phi P_n = \phi(0.60 F_u A_{nv} + U_{bs} F_u A_{nt}) \leq \phi(0.60 F_y A_{gv} + U_{bs} F_u A_{nt})$$

$$= 0.75[(0.60)(58)(7.93) + (1.0)(58)(1.31)]$$

$$\leq 0.75[(0.60)(36)(11) + (1.0)(58)(1.31)]$$

(continued)

263 kips $>$ 235 kips

ϕP_n = 235 kips $> P_{uf}$ = 112 kips OK

Compression:

K = 0.65 (Figure 5-3)

L = 3 in.

$$r_y = \frac{t}{\sqrt{12}} = \frac{0.5}{\sqrt{12}} = 0.144 \text{ in.}$$

$$\frac{KL}{r} = \frac{(0.65)(3)}{0.144} = 13.5$$

Since $KL/r < 25$, $F_{cr} = F_y$ (AISC specification, Section J4.4).

$$\phi P_n = \phi F_{cr} A_g = (0.9)(36)(0.5)(8) = 129 \text{ kips} > P_{uf} = 112 \text{ kips OK}$$

The connection is adequate for the applied moment. Note that the column should also be checked for the need of stiffeners to support the concentrated flange force (see Example 11-8).

11.6 MOMENT CONNECTIONS: BEAMS AND BEAM SPLICES

In the previous sections, we considered moment connections as they related to connections used in moment frames. In this section, we will consider moment connections for beam elements supporting mainly gravity loads.

The simplest type of moment connection is one in which welds are used to connect one beam to another beam or other element. Figure 11-19a shows a small beam cantilevered from the face of a column. The flanges and web have a groove weld to the column. Note that the flanges are welded to develop the flexural strength and the web is welded to develop shear capacity. Figure 11-19b indicates the welds used at a beam splice. In each of these cases, the welds could be partial or full penetration, or fillet welds, depending on the shear and moment loads at the connection.

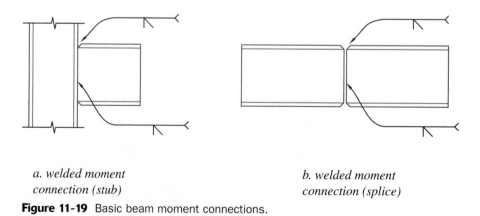

*a. welded moment
connection (stub)*

*b. welded moment
connection (splice)*

Figure 11-19 Basic beam moment connections.

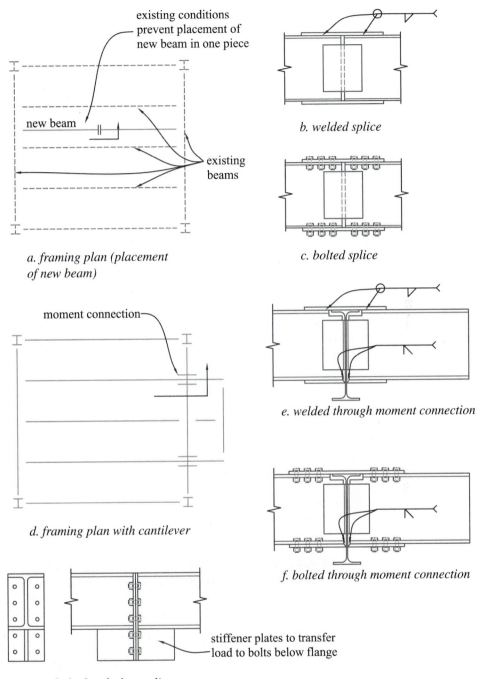

a. framing plan (placement of new beam)

b. welded splice

c. bolted splice

d. framing plan with cantilever

e. welded through moment connection

f. bolted through moment connection

g. bolted end-plate splice

Figure 11-20 Beam splices.

There are various types of connections where moment transfer occurs in a beam. Figure 11-20 indicates two types of moment connections that occur in beams. Figure 11-20a shows a new beam added to an existing floor framing plan. The existing conditions would usually make it impossible to place this beam in one section, and so one solution is to cut the beam at midspan and place the beam in two sections. The splice that is created would then

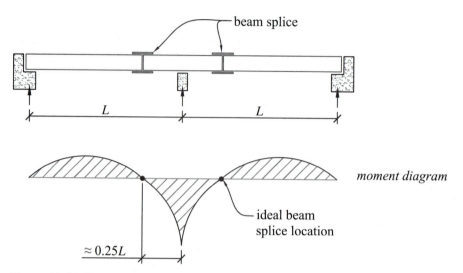

Figure 11-21 Beam splice location.

be designed for the shear and moment that is required at the splice. The top and bottom plates would be designed for the moment, and the web plate would be design for the shear (see Figures 11-20b and 11-20c). Figure 11-20d shows a cantilever condition where one beam frames through another. In a similar manner, the connection is designd to transfer shear and moment across the intermediate beam (see Figures 11-20e and 11-20f).

In each of these cases, the connection can be made with either bolts or welds, but not with bolts and welds in the same plane of loading. In many cases, the splice or moment connection is designed for the full moment capacity and full shear capacity for simplicity in the design. The location of the beam splice is also a design consideration. When a splice is located at midspan of a simply supported beam, the design moment is generally at a maximum, but the shear is generally at a minimum. The location of the splice may also have a practical significance. Figure 11-21 shows a two-span beam, which is common in bridge construction. The basic moment diagram for this beam is such that the moment reaches zero at about the one-quarter point of each span measured from the center support. It is ideal to place a beam splice at this location since localized bending stresses are minimized.

EXAMPLE 11-6

Welded Beam Splice

A simply supported W16 × 36 beam requires a beam splice at midspan. Design the splice for the full moment capacity of the beam using welded plates. The beam is ASTM A992, the plates are ASTM A36, and the welds are E70XX.

SOLUTION

From *AISCM*, Table 1-1,

$$d = 15.9 \text{ in.} \qquad T = 13\text{-}\tfrac{5}{8} \text{ in.}$$
$$t_f = 0.43 \text{ in.} \qquad b_f = 6.99 \text{ in.}$$
$$t_w = 0.295 \text{ in.}$$

From *AISCM*, Table 3-6, $\phi_b M_p = 240$ ft kips

Flange force:

$$P_{uf} = \frac{M}{d} = \frac{240}{15.9/12} = 182 \text{ kips}$$

Tension on gross plate area:

$$\phi P_n = \phi A_g F_y = (0.9)(A_g)(36) = 182 \text{ k} \rightarrow \text{Solving for } A_g, \text{ yields } 5.59 \text{ in.}^2.$$

Size	A_g, in.^{2}	
½″ × 11¼″	5.62	
⅝″ × 9″	5.62	← Select
¾″ × 7½″	5.62	

Weld strength:

$$b = 6.99 \text{ in.} (b_f = 6.99 \text{ in.})(L > b)$$

From Table 10-2,

Minimum weld size $= \frac{1}{4}$ in. $\therefore D_{min} = 4$

Maximum weld size $= t - \frac{1}{16} = 0.43 - \frac{1}{16} = 0.367$ in.

$$\therefore D_{max} = 5(\frac{1}{4}\text{-in. weld}) \text{ OK}$$

From equation (10-7),

$$\phi R_n = \phi R_{wl} + \phi R_{wt}$$
$$182 = 1.392 D L_l + 1.392 D L_t$$
$$= (1.392)(4)(2L_l) + (1.392)(4)(6.99) \rightarrow L_l = 12.84 \text{ in.}$$

Alternatively, from equation (10-8),

$$\phi R_n = 0.85 \phi R_{wl} + 1.5 \phi R_{wt}$$
$$182 = 0.85(1.392 D L_l) + 1.5(1.392 D L_t)$$
$$= (0.85)(1.392)(4)(2L_l) + (1.5)(1.392)(4)(6.99) \rightarrow L_l = 13.05 \text{ in.}$$

Use $L_l = 13$ in.

Shear strength:

The amount of shear at midspan of a simply supported beam is usually close to zero; however, heavy concentrated loads could be present on any given beam in a building, so engineering judgment will be needed to select the necessary amount of shear capacity. In this case, we will assume that 50 kips is the required shear capacity.

Since the throat length, T, is 13.625 in., this is the maximum depth of the shear plate.

(continued)

Solving for the plate thickness, assuming that the height of the plate is 13 in.,

$$\phi_v V_n = \phi_v 0.6 F_y A_w$$
$$50 = (1.0)(0.6)(36)(13)(t_p) \therefore t_p = 0.178 \text{ in.}$$

For practical reasons, a plate that has a thickness equal to or greater than the web thickness should be used. Therefore, use $t_p = \frac{3}{8}$ in.

From Table 10-2,

Minimum weld size $= \frac{3}{16}$ in. (Table 10-2) $\therefore D_{min} = 3$

Maximum weld size $= t - \frac{1}{16} = 0.295 - \frac{1}{16} = 0.232$ in. $\therefore D_{max} = 3$

Assume that $b = 3$ in. (see Figure 11-22):

$$CG = \frac{b^2}{2b + d} \text{ (see Table 10-3)}$$

$$= \frac{3^2}{(2)(3) + 13} = 0.473 \text{ in.}$$

$$e = (3 \text{ in.} - 0.473 \text{ in.}) + \frac{1}{2} \text{ in.} + (3 - 0.473 \text{ in.}) = 5.55 \text{ in.}$$

From *AISCM*, Table 8-8,

$$a = \frac{e}{l} = \frac{5.55}{13} = 0.427$$

$$k = \frac{b}{l} = \frac{3}{13} = 0.230$$

$$C = 2.10$$

$$D_{min} = \frac{P_u}{\phi C C_l l} = \frac{50}{(0.75)(2.10)(1.0)(13)} = 2.46 < 3 \text{ OK}$$

Note that the final design detail includes a ½-in. gap between the ends of the beam added for construction tolerances.

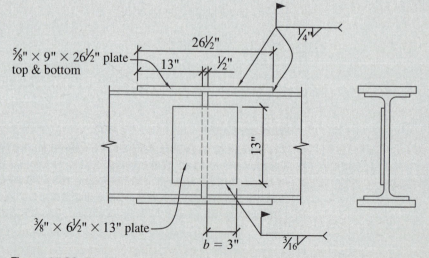

Figure 11-22 Details for Example 10-6.

EXAMPLE 11-7

Bolted Beam Splice

Repeat Example 11-6 for a bolted moment connection assuming $M_u = 195$ ft·k and $V_u = 50$ k. Use ⅞-in.-diameter A490 SC bolts in STD holes and a Class B faying surface.

SOLUTION

From *AISCM*, Table 1-1,

$$d = 15.9 \text{ in.} \qquad T = 13\text{-}\tfrac{5}{8} \text{ in.}$$
$$t_f = 0.43 \text{ in.} \qquad b_f = 6.99 \text{ in.}$$
$$t_w = 0.295 \text{ in.} \qquad S_x = 56.5 \text{ in.}^3$$

From *AISCM*, Table 3-6, $\phi_b M_p = 240$ ft-kips.

Flange force:

$$P_{uf} = \frac{M}{d} = \frac{195}{15.9/12} = 148 \text{ kips}$$

Bolts in Top and Bottom Flanges:

In this case, slip is a strength limit state since slip at the joint would cause additional beam deflection (see discussion in Chapter 9).

From *AISCM*, Table 7-4, $\phi_v R_n = (1.43)(16.5) = 23.6$ kips/bolt.

$$N = \frac{P_{uf}}{\phi_v R_n} = \frac{148}{23.6} = 6.24 \ \therefore \ \text{Use 8 bolts.}$$

Check the reduced flexural strength of the W16 × 36:

$$F_y/F_u = 50/65 = 0.77 < 0.80 \rightarrow Y_t = 1.0$$
$$A_{fg} = (6.99)(0.43) = 3.00 \text{ in.}^2$$
$$A_{fn} = A_{fg} - \Sigma A_{holes}$$
$$= 3.00 - \left[(2)\left(\frac{7}{8} + \frac{1}{8} \right)(0.43) \right] = 2.14 \text{ in.}^2$$
$$F_u A_{fn} = (65)(2.14) = 139 \text{ kips}$$
$$Y_t F_y A_{fg} = (1.0)(50)(3.00) = 150 \text{ kips}$$
$$\phi_b M_n = \phi_b F_u S_x \left(\frac{A_{f_n}}{A_{f_g}} \right) = (0.9)(65)(56.5)\left(\frac{2.14}{3.0} \right) = 2357 \text{ in·k}$$
$$= 196 \text{ ft.k} > 195 \text{ ft.k, OK}$$

Plates

Assume that the top and bottom plates are still ⅝ in. by 9 in. Tension on the gross area was checked in the previous example; now we will check tension on the net area (see Figure 11-23).

Tension on the net plate area:

$$A_g = (0.625 \text{ in.})(9 \text{ in.}) = 5.62 \text{ in.}^2$$
$$A_n = A_g - \Sigma A_{holes} = 5.62 - \left[(2)\left(\frac{7}{8} + \frac{1}{8} \right)(0.625) \right] = 4.38 \text{ in.}^2$$

(continued)

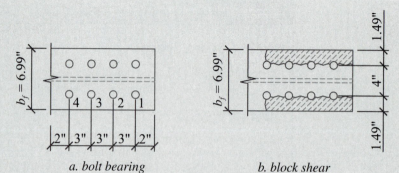

a. bolt bearing   b. block shear

Figure 11-23 Bolt bearing and block shear for Example 11-7.

$A_n \leq 0.85A_g = (0.85)(5.62) = 4.78$ in.^{2}. Use $A_n = 4.38$ in.^{2}.
$A_e = A_n U = (4.38$ in.$^2)(1.0) = 4.38$ in.^{2}

Strength based on the effective area is

$$\phi P_n = \phi F_u A_e = (0.75)(58)(4.38 \text{ in.}^2) = 190 \text{ kips} > P_{uf} = 148 \text{ kips OK}$$

Check bolt bearing (see Figure 11-23):

Since $t_f F_{ub} = (0.43)(65) = 27.9$ kips/in. $< t_p F_{up} = (0.625)(58) = 36.2$ kips/in., bearing on the beam flange will control.

$$L_{c1} = 2 \text{ in.} - \left[0.5\left(\frac{7}{8} + \frac{1}{8}\right)\right] = 1.50 \text{ in.}$$

$$L_{c2} = L_{c3} = L_{c4} = 3 \text{ in.} - \left[\left(\frac{7}{8} + \frac{1}{8}\right)\right] = 2.0 \text{ in.}$$

For bolt 1, $\phi R_n = \phi 1.2 L_c t F_u \leq \phi 2.4 d t F_u$
$$= (0.75)(1.2)(1.50 \text{ in.})(0.43 \text{ in.})(65 \text{ ksi})$$
$$< (0.75)(2.4)(0.875)(0.43 \text{ in.})(65 \text{ ksi})$$
$$= 37.7 \text{ kips} < 44.0 \text{ kips}$$
$$\phi R_n = 37.7 \text{ kips}$$

For bolts 2, 3, and 4, $\phi R_n = (0.75)(1.2)(2.0 \text{ in.})(0.43 \text{ in.})(65 \text{ ksi})$
$$< (0.75)(2.4)(0.875)(0.43 \text{ in.})(65 \text{ ksi})$$
$$= 50.3 \text{ kips} > 44.0 \text{ kips.}$$
$$\phi R_n = 44.0 \text{ kips.}$$

$$\phi R_n = (2)(37.1 \text{ k}) + (6)(44.0 \text{ k}) = \textbf{338 kips} > P_{uf} = \textbf{148 kips OK for bearing}$$

Block shear:

$$A_{gv} = (2)(3 \text{ in.} + 3 \text{ in.} + 3 \text{ in.} + 2 \text{ in.})(0.43 \text{ in.}) = 9.46 \text{ in.}^2$$
$$A_{gt} = (1.49 \text{ in.} + 1.49 \text{ in.})(0.43 \text{ in.}) = 1.28 \text{ in.}^2$$

$$A_{nv} = A_{gv} - \Sigma A_{holes} = 9.46 - \left[(3.5)(2)\left(\frac{7}{8} + \frac{1}{8}\right)(0.43) \right] = 6.45 \text{ in.}^2$$

$$A_{nt} = A_{gt} - \Sigma A_{holes} = 1.28 - \left[\left(\frac{7}{8} + \frac{1}{8}\right)(0.43) \right] = 0.85 \text{ in.}^2$$

The available block shear strength is found from Chapter 4 ($U_{bs} = 1.0$):

$$\phi P_n = \phi(0.60 F_u A_{nv} + U_{bs} F_u A_{nt}) \leq \phi(0.60 F_y A_{gv} + U_{bs} F_u A_{nt})$$
$$= 0.75[(0.60)(65)(6.45) + (1.0)(65)(0.85)]$$
$$\leq 0.75[(0.60)(50)(9.46) + (1.0)(65)(0.85)]$$

230 kips < 254 kips

$\phi P_n = 230 \text{ kips} > P_{uf} = 148 \text{ kips OK}$

Compression:

$$K = 0.65 \text{ (Figure 5-3)}$$
$$L = 2 \text{ in.} + 0.5 \text{ in.} + 2 \text{ in.} = 4.5 \text{ in.}$$
$$r_y = \frac{t}{\sqrt{12}} = \frac{0.625}{\sqrt{12}} = 0.180 \text{ in.}$$
$$\frac{KL}{r} = \frac{(0.65)(4.5)}{0.180} = 16.2$$

Since $KL/r < 25$, $F_{cr} = F_y$ (AISC, Section J4.4).

$$\phi P_n = \phi F_{cr} A_g = (0.9)(36)(0.625)(9) = 182 \text{ kips} > P_{uf} = 148 \text{ kips OK}$$

Shear strength:

A ⅜-in. by 13-in. plate will be used (same size used for previous example). Assuming that a shear strength of 50 kips is needed, *AISCM*, Table 7-7 can be used to check the capacity of the bolt group (see Figure 11-24):

$$e = 1.5 \text{ in.} + 0.5 \text{ in.} + 1.5 \text{ in.} = 3.5 \text{ in.}$$

Using $N = 4$ and $s = 3$ in., $C = 2.58$.

Assuming (4)- ⅞-in.-diameter A490 bolts, $\phi_v R_n = 27.1$ kips/bolt.
The strength of the bolt group is $(2.58)(27.1) = 69.9$ kips > 50 kips. OK

Check bearing on beam web:

$$L_c(\text{all bolts}) = 3 \text{ in.} - \left(\frac{7}{8} + \frac{1}{8}\right) = 2.0 \text{ in.}$$

$$\phi R_n = (0.75)(1.2)(2.0 \text{ in.})(0.295 \text{ in.})(65 \text{ ksi})$$
$$< (0.75)(2.4)(0.875)(0.295 \text{ in.})(65 \text{ ksi})$$

(continued)

$$= 34.5 \text{ kips} > 30.2 \text{ kips}$$

$$= 30.2 \text{ kips}$$

$$\phi R_n = (4)(30.2) = \textbf{120 kips} > \textbf{50 kips} \text{ OK for bearing}$$

Block shear in plate:
From *AISCM*, Table 10-1, $L_{ev} = 2$ in. and $L_{eh} = 1.5$ in.; the available strength, considering block shear and bearing, is 256 kips/in. Multiplying by the plate thickness yields

$$(256)(0.375) = 96 \text{ kips} > 50 \text{ kips OK}$$

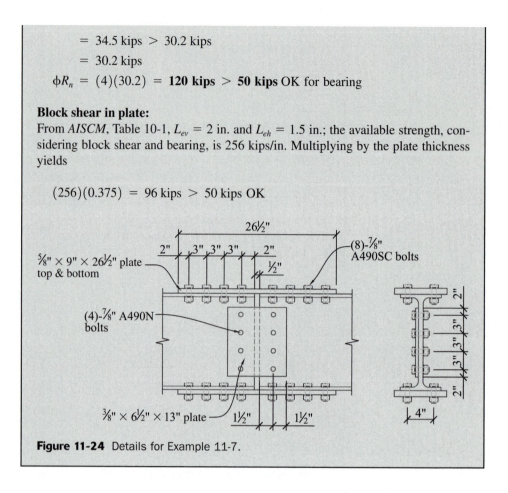

Figure 11-24 Details for Example 11-7.

11.7 COLUMN STIFFENERS

In Chapter 6, we considered the design of beams for concentrated forces. In this section, we will expand on this topic as it applies to concentrated forces in columns. In the previous sections, we considered the design of moment connections at the ends of beams. When these connections are made to column flanges, there are several localized failure modes that need to be investigated. When an end moment from a beam is applied to the flanges of a column as shown in Figure 11-25, the flange force due to the moment is transferred to the column through the flanges and to the web. This force could be either tension or compression and could cause localized bending of the flange and localized buckling of the web. To prevent such behavior, stiffener plates can be added to the flange and the web. In some cases, these stiffeners might be added even if the column were adequate to support the concentrated forces as a means of providing redundancy to the connection. However, the addition of these plates can create constructibility issues in that the stiffeners might conflict with a beam framing into the web of the column. The stiffeners could be field applied, but this adds cost to the connection since field welds are more expensive than welds applied in the shop. Figure 11-26 shows several possible column stiffening details.

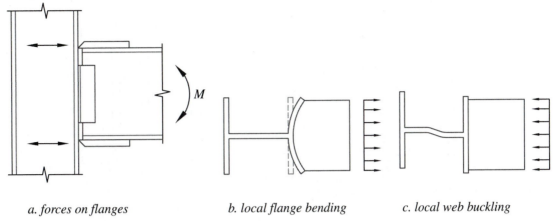

a. forces on flanges   b. local flange bending   c. local web buckling

Figure 11-25 Concentrated forces on columns.

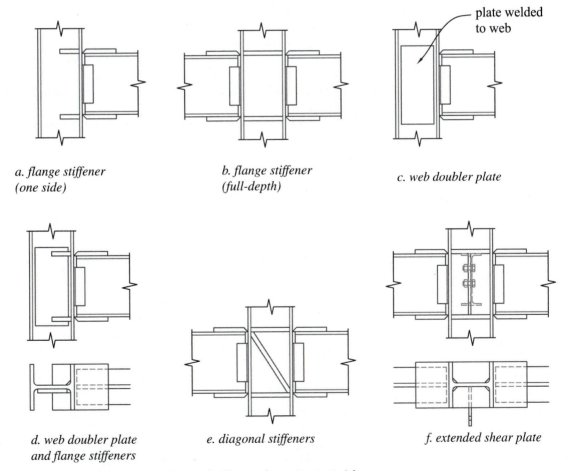

a. flange stiffener
(one side)

b. flange stiffener
(full-depth)

c. web doubler plate

plate welded
to web

d. web doubler plate
and flange stiffeners

e. diagonal stiffeners

f. extended shear plate

Figure 11-26 Column stiffeners for concentrated forces.

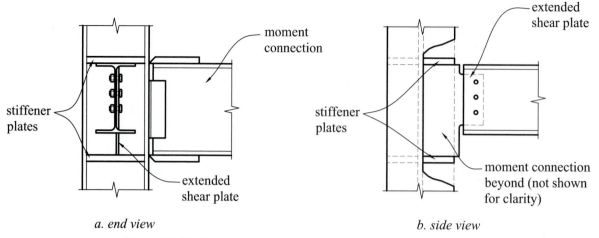

a. end view   *b. side view*

Figure 11-27 Extended shear plate.

Another possible solution is to use an extended shear plate connection for the beam framing into the web of the column as shown in Figure 11-27. This connection increases the bending moment to the column due to the increased eccentricity of the shear connection of the beam to the column web, and therefore the column would have to be designed accordingly. A simpler solution might also be to increase the size of the column such that it has adequate capacity to support these localized concentrated forces without the use of stiffener plates.

The various failure modes for concentrated forces on column flanges will now be described in greater detail.

Flange local bending occurs when a concentrated tension force is applied to the flange. The design strength for flange local bending when the concentrated force is applied at a distance greater than $10t_f$ from the end of the member is

$$\phi R_n = \phi 6.25 t_f^2 F_y \text{ for } y \geq 10 b_f. \tag{11-14}$$

When the location of the concentrated force occurs at a distance less than $10t_f$ from the end of the member, the design strength is

$$\phi_{fb} R_n = \phi_{fb} 3.125 t_f^2 F_y \text{ for } y < 10 b_f, \tag{11-15}$$

where

$\phi_{fb} = 0.9$ for flange local bending,

R_n = Nominal design strength,

t_f = Flange thickness,

b_f = Flange width, and

F_y = Yield strength of flange.

When the concentrated force exceeds the design strength for flange local bending, transverse stiffeners are required (stiffener design will be covered later). When the loading across the flange is less than $0.15 b_f$, then flange local bending does not need to be checked.

Web local yielding applies to concentrated compression forces. The design equations for this case were discussed in Chapter 6 (beam bearing), but will be repeated here for clarity.

The design strength for web local yielding is

$$\phi_{wy} R_n = \phi_{wy}(5k + N)F_y t_w \text{ for } y > d, \text{ and} \tag{11-16}$$

$$\phi_{wy} R_n = \phi_{wy}(2.5k + N)F_y t_w \text{ for } y \leq d, \tag{11-17}$$

where

ϕ_{wy} = 1.0 for web local yielding,

R_n = Nominal design strength,

t_w = Column web thickness,

N = Bearing length,

d = Column depth,

y = Distance from the end of the column to the concentrated load (see Figure 11-28),

k = Section property from *AISCM*, Part 1, and

F_y = Yield strength, ksi.

When the concentrated force exceeds the design strength for web local yielding, either transverse stiffeners or a web doubler plate is required.

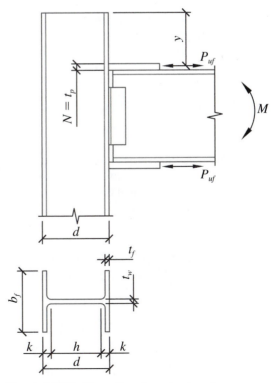

Figure 11-28 Dimensional parameters for concentrated forces.

Web crippling occurs when the concentrated load causes a local buckling of the web. The design strength for web crippling is

$$\phi_{wc}R_n = \phi_{wc}\,0.8t_w^{\,2}\left[1 + 3\left(\frac{N}{d}\right)\left(\frac{t_w}{t_f}\right)^{1.5}\right]\sqrt{\frac{EF_y t_f}{t_w}}\ \text{for } y \geq \frac{d}{2}, \tag{11-18}$$

$$\phi_{wc}R_n = \phi_{wc}\,0.4t_w^{\,2}\left[1 + 3\left(\frac{N}{d}\right)\left(\frac{t_w}{t_f}\right)^{1.5}\right]\sqrt{\frac{EF_y t_f}{t_w}}\ \text{for } y < \frac{d}{2}\ \text{and } \frac{N}{d} \leq 0.2, \text{and} \tag{11-19}$$

$$\phi_{wc}R_n = \phi_{wc}\,0.4t_w^{\,2}\left[1 + \left(\frac{4N}{d} - 0.2\right)\left(\frac{t_w}{t_f}\right)^{1.5}\right]\sqrt{\frac{EF_y t_f}{t_w}}\ \text{for } y < \frac{d}{2}\ \text{and } \frac{N}{d} > 0.2, \tag{11-20}$$

where

ϕ_{wc} = 0.75 for web crippling, and
$E = 29 \times 10^6$ psi.

When the concentrated force exceeds the design strength for web crippling, either transverse stiffeners or a web doubler plate extending at least $d/2$ is required.

Web compression buckling can occur when a concentrated compression force is applied to both sides of a member at the same location. This occurs at a column with moment connections on each side. The design strength for web compression buckling is

$$\phi_{wb}R_n = \frac{24t_w^{\,3}\sqrt{EF_y}}{h}\ \text{for } y \geq \frac{d}{2}, \text{and} \tag{11-21}$$

$$\phi_{wb}R_n = \frac{12t_w^{\,3}\sqrt{EF_y}}{h}\ \text{for } y < \frac{d}{2}, \tag{11-22}$$

where

ϕ_{wb} = 0.90 for web buckling, and
h = Clear distance between the flanges, excluding the fillets (see Figure 11-28).

When the concentrated force exceeds the design strength for web buckling, a single transverse stiffener, a pair of transverse stiffeners, or a full-depth web doubler plate is required.

The concentrated force on a column flange could also cause large shear forces across the column web. The region in which these forces occur is called the panel zone. The shear in the panel zone is the sum of the shear in the web and the shear due to the flange force, P_{uf}.

The design strength for *web panel zone shear* assumes that the effects of panel zone deformation on frame stability are not considered:

When $P_r \leq 0.4P_c$,

$$\phi_{cw}R_n = \phi_{cw}\,0.6F_y dt_w; \tag{11-23}$$

when $P_r > 0.4P_c$,

$$\phi_{cw}R_n = 0.6F_y dt_w\left(1.4 - \frac{P_r}{P_c}\right), \tag{11-24}$$

where

$\phi_{cw} = 0.9$,

P_r = Factored axial load in column, P_u,

P_c = Yield strength of the column, $P_y = F_y A$, and

A = Area of the column.

When the effects of panel zone deformation on frame stability is considered, the reader is referred to the AISC specification, Section J10.6.

When the shear strength in the web panel zone is exceeded, a full-depth web doubler plate or a pair of diagonal stiffeners are required.

For all of the previous design checks for concentrated forces, stiffener plates are required when the applied forces are greater than the design strength for each failure mode. When transverse stiffeners are required, the force is distributed to the web or flange and the stiffener plate based on their relative stiffnesses. However, the AISC specification allows a more simplified approach where the size of the plate is based on the difference between the required strength and the available strength of the failure mode in question, which is expressed as follows:

$$R_{u\,st} = P_{uf} - \phi R_{n,min},\qquad(11\text{-}25)$$

where

$R_{u\,st}$ = Required strength of the stiffener (tension or compression),

P_{uf} = Flange force, and

$\phi R_{n,\,min}$ = Lesser design strength of flange local bending, web local yielding, web crippling, and compression buckling.

The transverse stiffeners are then designed to provide adequate cross-sectional area as follows:

$$R_{u\,st} < \phi F_{y\,st} A_{st}.\qquad(11\text{-}26)$$

Solving for the plate area yields

$$A_{st} > \frac{R_{u\,st}}{\phi F_{y\,st}},\qquad(11\text{-}27)$$

where

ϕ = 0.9 (yielding),

A_{st} = Area of the transverse stiffeners, and

$F_{y\,st}$ = yield stress of the transverse stiffeners.

Web doubler plates are required when the shear in the column exceeds the web panel zone shear strength. The required design strength of the web doubler plate or plates is expressed as follows:

$$V_{u\,dp} = V_u - \phi_{cw} R_n,\qquad(11\text{-}28)$$

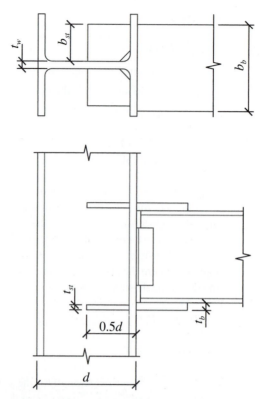

Figure 11-29 Dimensional requirements for column stiffeners.

where

$V_{u\,dp}$ = Required strength of the web doubler plate or plates,

V_u = Factored shear in the column web at the concentrated force, and

$\phi_{cw}R_n$ = Design shear strength of the web panel zone (eq. (11-23) or (11-24)).

With reference to Figure 11-29, column stiffeners are proportioned to meet the following requirements per the AISC specification.

The minimum width of a transverse stiffener is

$$b_{st} \geq \frac{b_b}{3} - \frac{t_w}{2}, \tag{11-29}$$

where

b_{st} = Width of the transverse stiffener,

b_b = Width of the beam flange or moment connection plate, and

t_w = Column web thickness.

The minimum thickness of the stiffener is the larger of the following:

$$t_{st} \geq \frac{t_b}{2} \tag{11-30}$$

or

$$t_{st} \geq \frac{b_{st}}{15},$$ (11-31)

where

b_{st} = Width of the transverse stiffener,

t_{st} = Thickness of the transverse stiffener, and

t_b = Thickness of the beam flange or moment connection plate.

Transverse stiffeners are required to extend the full depth of the column when there are applied forces on both sides of the column. For concentrated forces, the length of the transverse stiffener should extend to half of the column depth.

Transverse stiffeners are welded to both the web and the loaded flange. The weld to the flange is designed for the difference between the required strength and the design strength of the controlling limit state (eq. (11-25)).

When web doubler plates are required, they are designed for the shear in the column that exceeds the web panel zone shear strength (eq. (11-27)). This force in the doubler plate could be compression, tension, or shear, and therefore the doubler plate must be designed for these limit states. The web doubler plate is welded to the column web based on the force in the doubler plate (eq. (11-27)).

EXAMPLE 11-8

Column with Concentrated Flange Forces

Determine whether the W8 × 40 column in Example 11-4 is adequate for the applied concentrated flange forces. Assume that the beam connection occurs at a location remote from the ends (i.e., $y > d$ and $y > 10b_f$) and that $P_r \leq 0.4P_c$.

From Example 11-4,

P_{uf} = 49.2 kips, and

N = $\frac{3}{8}$ in. (flange plate thickness, t_b).

From *AISCM*, Table 1-1, for a W8 × 40,

d = 8.25 in. t_w = 0.36 in.

t_f = 0.56 in. k = 0.954 in.

Flange local bending:

$$\phi R_n = \phi 6.25 t_f^2 F_y = (0.9)(6.25)(0.56)^2(50) = 88.2 \text{ kips} > P_{uf}$$
$$= 49.2 \text{ kips OK}$$

Web local yielding:

$$\phi_{wy} R_n = \phi_{wy}(5k + N)F_y t_w = 1.0[(5)(0.954) + 0.375](50)(0.36)$$
$$= 92.6 \text{ kips} > P_{uf} = 49.2 \text{ kips OK}$$

(continued)

Web crippling:

$$\phi_{wc}R_n = \phi_{wc}0.8t_w^2\left[1 + 3\left(\frac{N}{d}\right)\left(\frac{t_w}{t_f}\right)^{1.5}\right]\sqrt{\frac{EF_yt_f}{t_w}}$$

$$= (0.75)(0.8)(0.36)^2\left[1 + 3\left(\frac{0.375}{8.25}\right)\left(\frac{0.36}{0.56}\right)^{1.5}\right]\sqrt{\frac{(29,000)(50)(0.56)}{0.36}}$$

$$= 125 \text{ kips} > P_{uf} = 49.2 \text{ kips OK}$$

Web panel zone shear:

$$\phi_{cw}R_n = \phi_{cw}0.6F_ydt_w = (0.9)(0.6)(50)(8.25)(0.36)$$

$$= 80.2 \text{ kips} > P_{uf} = 49.2 \text{ kips OK}$$

The W8 × 40 column is adequate for concentrated forces. Note that web compression buckling does not need to be checked since the concentrated forces are applied to one side of the column only.

11.8 COLUMN SPLICES

In buildings less than four stories in height, it may be advantageous from a constructibility standpoint to use a single column for all of the stories instead of using smaller column sizes for the upper levels, even though it is more economical from a design standpoint to use smaller columns for the upper levels. In multistory buildings, columns could be spliced every two, three, or four floor levels depending on the design and construction parameters. The OSHA (Occupational Safety and Health Administration) safety regulations for column erection require that a steel erector have adequate protection from fall hazards of more than two stories or 30 ft. above a lower level, whichever is less [6]. It is therefore recommended to avoid column splices at every third level and allow column splices at every second or fourth level in order to meet safety requirements.

An additional safety requirement related to column splices is that perimeter columns should have holes or other attachment devices sufficient to support a safety cable or other similar rail system. The holes or attachment devices are located 42 to 45 in. above the finished floor and at the midpoint between the top cable or rail and the finished floor. The column splice is therefore required to be a minimum of 48 in. above the finished floor or at a higher distance in order to avoid interference with the safety attachments. The safety regulations do allow for exceptions to the above requirement where constructibility does not allow such a distance, but overall safety compliance would be left to the steel erector to resolve [6].

The simplest column splice is one in which only compression forces are transferred between columns of the same nominal depth (Figure 11-30). The design strength in bearing between the area of contact is

$$\phi R_n = \phi 1.8F_yA_{pb},\tag{11-32}$$

where

$$\phi = 0.75,$$
$$F_y = \text{Yield strength, and}$$
$$A_{pb} = \text{Contact area.}$$

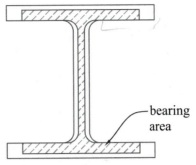

Figure 11-30 Column splice (bearing).

By inspection, it can be seen that the bearing strength for a column with full-contact area will not be more critical than the design strength of the column in compression.

The *AISCM* provides column splice details for several framing conditions in Table 14-3. This table, which is actually a series of tables for various column splice configurations, is used mainly for splices that support compression load only. Load conditions that have tension, shear, bending, or some combination of the three would have to be designed accordingly using the procedures previously discussed (see Chapter 4 for tension members and this chapter for shear and bending).

EXAMPLE 11-9

Column Splice

A column splice design between a W12 × 65 from a lower level and a W12 × 53 above will be investigated. The steel is ASTM A992, grade 50 and the floor-to-floor height is 15 ft.

Compression:
The depth of the W12 × 65 is equal to the depth of the W12 × 53, so there will be full contact for transferring the compression load. The design strength in bearing is

$$\phi R_n = \phi 1.8 F_y A_{pb} = (0.75)(1.8)(50)(15.6) = 1053 \text{ kips.}$$

This value far exceeds the design strength in compression, even at a very small unbraced length, so bearing is not a concern.

Tension:
The design strength in tension for a W12 × 53 is

$$\phi P_n = \phi F_y A_g = (0.9)(50)(15.6) = 702 \text{ kips.}$$

While it is likely that the column does not experience that magnitude of tension under an actual design condition, the design strength in tension would likely have to be developed with a welded splice plate instead of bolts. Assuming that a plate on either side

(continued)

of the column web is provided with a cross-sectional area equal to the area of the column, a ⅞-in.-thick plate would be required on each side with a width less than or equal to the T distance of the column ($T = 9⅛$ in. for the W12 × 65)

$$A_p = (2)(7/8)(9) = 15.75 \text{ in.}^2 > 15.6 \text{ in.}^2. \text{ OK}$$

Assuming a 5/16-in. fillet weld and a 42-in.-long plate (21 in. on each side of the splice), the weld strength is

$$\phi R_n = (1.392 \text{ k/in})(5)(2)(21 \text{ in.} + 9 \text{ in.} + 21 \text{ in.})$$
$$= 709 \text{ kips} > 702 \text{ kips. OK}$$

Shear and Bending in the W12 × 53:
From *AISCM*, Table 3-6, $\phi V_n = 125$ kips.
    From *AISCM*, Table 3-10, $\phi_b M_{nx} = 258$ ft-kips ($L_b = 15$ ft.).

$$M_{ny} = M_{py} = F_y Z_y \leq 1.6 F_y S_y$$
$$= (50)(29.1) < (1.6)(50)(19.2) = 1455 \text{ in.-kips} < 1536 \text{ in.-kips}$$
$$\phi_b M_{ny} = \frac{(0.9)(1455)}{12} = 109 \text{ ft.-kips}$$

The splice detail for shear and for the bending moments in the strong and weak axes will not be developed here; however, Examples 11-6 and 11-7 provide the procedure for designing these splice plates (welded or bolted).

11.9 HOLES IN BEAMS

Holes in any steel section are generally not desirable, especially if the holes are made after construction and were not part of the original design. In many connections, bolt holes are intentional and necessary, and are always considered in the original design. In other cases, field conditions might dictate the need for an opening in a steel beam, usually for the passage of mechanical, electrical, or plumbing ducts that conflict with a steel member. Careful attention to coordination among all of the construction trades could avoid such a conflict, but it is sometime unavoidable. The use of open-web steel joists or castellated beams would provide a framing system that allows for the passage of moderately sized ducts through the framing members (see Figure 11-31).

Other references provide more detailed coverage of beams with web openings [7], but the approach taken here will be similar to a common approach taken in practice. When an opening needs to be made through a steel member, the following general guidelines apply:

1. Provide reinforcement above and below the opening so that all of the original section properties are maintained. For example, if an opening is required in a W-shaped beam, provide a steel angle or plate so that the addition of the angles or plates will yield a composite steel section that has a cross-sectional area, section modulus, and moment of inertia that is equal to or greater than the original section. This will help to ensure the same behavior of the beam with respect to the applied loads. For beams that are to be left unreinforced, refer to reference 8.

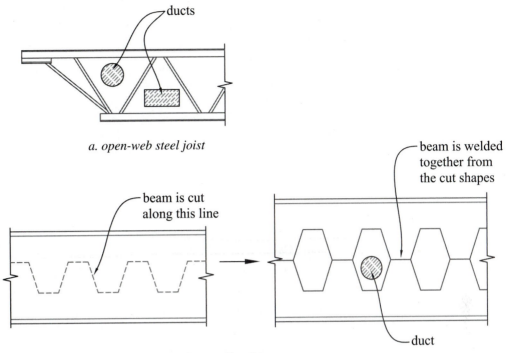

a. open-web steel joist

b. castellated beam

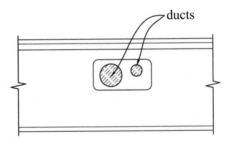

c. wide flange beam

Figure 11-31 Beams with web openings.

2. Concentrated loads should not be permitted above the opening, nor should the opening be within a distance d from the bearing end of the beam.

3. Circular openings are preferred because they are less susceptible to stress concentrations. Square openings should have a minimum radius at the corners of $2t_w$ or ⅝ in., whichever is greater.

4. Openings should be spaced as follows:

    a. *Rectangular openings:*

    $$S \geq h_o \tag{11-33}$$

    $$S \geq a_o \left(\frac{V_u/\phi V_p}{1 - V_u/\phi V_p} \right) \tag{11-34}$$

b. *Circular openings:*

$$S \geq 1.5 D_o \qquad\qquad (11\text{-}35)$$

$$S \geq D_o \left(\frac{V_u/\phi V_p}{1 - V_u/\phi V_p} \right) \qquad\qquad (11\text{-}36)$$

c. *Openings in composite beams:*

$$S \geq a_o \qquad\qquad (11\text{-}37)$$
$$S \geq 2d, \qquad\qquad (11\text{-}38)$$

where

S = Clear space between openings,
h_o = Opening depth,
a_o = Opening width,
D_o = Opening diameter,
d = Beam depth,
V_u = Factored shear,
ϕ = 0.9 for noncomposite beams
    = 0.85 for composite beams,
V_p = Plastic shear capacity

$$= \frac{F_y t_w s}{\sqrt{3}}, \qquad\qquad (11\text{-}39)$$

F_y = Yield stress,
t_w = Web thickness, and
s = Depth of the remaining section.

5. The weld strength within the length of the opening should be as follows:

$$R_{wr} = \phi 2 P_r, \qquad\qquad (11\text{-}40)$$

where

R_{wr} = Required weld strength,
ϕ = 0.9 for noncomposite beams
    = 0.85 for composite beams,

$$P_r = F_y A_r \leq \frac{F_y t_w a_o}{2\sqrt{3}}, \text{ and} \qquad\qquad (11\text{-}41)$$

A_r = Cross-sectional area of the reinforcement above or below the opening.

6. The length of the extension beyond the opening should be as follows:

$$L_1 = \frac{a_o}{4} \geq \frac{A_r \sqrt{3}}{2t_w}.$$ (11-42)

7. The weld strength within the length of the extension should be as follows:

$$R_{wr} = \phi F_y A_r.$$ (11-43)

The design parameters indicated above are illustrated in Figure 11-32.

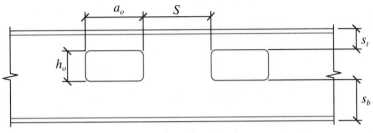

a. rectangular opening

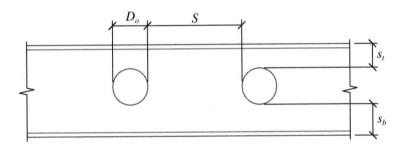

b. circular opening

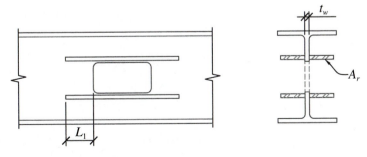

c. reinforcement

Figure 11-32 Web opening design parameters.

EXAMPLE 11-10

Beam with Web Opening

Determine the required reinforcement for the noncomposite beam shown in Figure 11-33. The 6-in. opening is centered in the beam depth and is at midspan of a simple-span beam away from any significant concentrated loads. The steel is ASTM A992, grade 50.

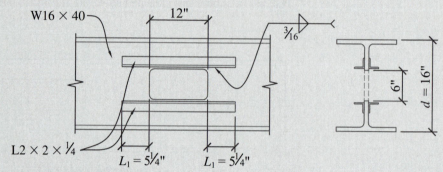

Figure 11-33 Details for Example 11-10.

From *AISCM*, Table 1-1,

$$d = 16.0 \text{ in.} \qquad t_w = 0.305 \text{ in.}$$
$$A = 11.8 \text{ in.}^2 \qquad I_x = 518 \text{ in.}^4$$
$$S_x = 64.7 \text{ in.}^3$$

The area of the web that has been removed is (6 in.)(0.305 in.) = 1.83 in.^{2}

Assuming that a pair of angles are provided on the top and bottom of the opening, the required area for each vertical leg is

$$1.83 \text{ in.}^2/4 = 0.46 \text{ in.}^2.$$

An L2 × 2 × ¼ angle is selected that has an area of (2)(0.25) = 0.50 in.^{2} per vertical leg. This ensures that an equivalent amount of shear strength is provided to replace the portion of the web that has been removed. By inspection, the amount of area provided is greater than the area removed and the location of the angles is such that the composite section properties will be greater than the original section properties.

Required weld length within the opening:

$$P_r = F_y A_r \le \frac{F_y t_w a_o}{2\sqrt{3}} = (50)(0.938) \le \frac{(50)(0.305)(12)}{2\sqrt{3}}$$
$$= 46.9 \text{ kips} < 52.8 \text{ kips}$$
$$R_{wr} = \phi 2 P_r = (0.9)(2)(46.9) = 84.4 \text{ kips}$$

Assuming a ³⁄₁₆-in.-long weld, the weld strength is

$$(1.392)(3)(2)(12) = 100 \text{ kips} > 84.4 \text{ kips} \quad \text{OK}$$

Required extension length:

$$L_1 = \frac{a_o}{4} \geq \frac{A_r\sqrt{3}}{2t_w} = \frac{12}{4} \geq \frac{0.938\sqrt{3}}{(2)(0.305)} = 3 > 2.66$$

Required weld strength within the length of the extension:

$$R_{wr} = \phi F_y A_r = (0.9)(50)(0.938) = 42.2 \text{ kips}$$

Assuming a ³⁄₁₆-in.-long weld, the weld strength is

$$(1.392)(3)(2)(3) = 25.1 \text{ kips} < 42.2 \text{ kips. Not good}$$

The weld size could be increased, but a simpler solution would be to increase the extension length so that a ³⁄₁₆-in.-long weld could be used throughout:

$$L_1 = \left(\frac{42.2}{25.2}\right)(3 \text{ in.}) = 5.05 \rightarrow \text{Use } L_1 = 5¼ \text{ in.}$$

11.10 DESIGN OF GUSSET PLATES IN VERTICAL BRACING AND TRUSS CONNECTIONS

Gusset plates are flat structural elements that are used to connect adjacent members meeting at truss panel joints and at diagonal brace connections as shown in Figure 11-34 and Figures 11-35a through 11-35d. Gusset plates help transmit loads from one member to

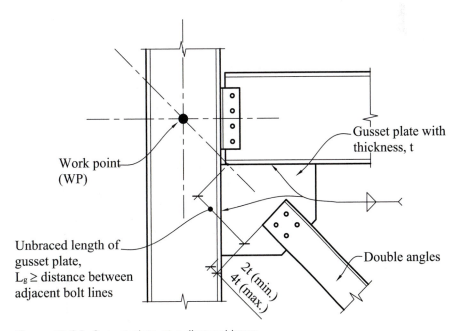

Figure 11-34 Gusset plate at a diagonal brace.

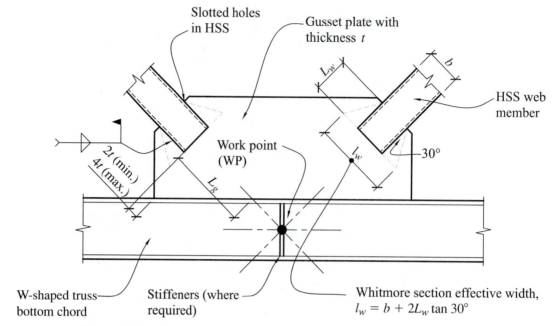

Figure 11-35a Gusset plate at a truss panel point (bottom chord).

another. Their design is covered in Part 9 of the *AISCM*, where the gusset plate is part of a seismic force-resisting system with a seismic response modification factor, R, greater than 3. The requirements of the *AISC* seismic provisions for steel buildings must also be satisfied [4]. The gusset plates may be bolted or welded to the members meeting at the joint, and the practical minimum thickness of gusset plates used in design practice is usually ⅜ in. For diagonal brace connections with gusset plates, the gusset plate helps to transfer the diaphragm lateral forces from the beams or girders to the diagonal brace and the adjoining

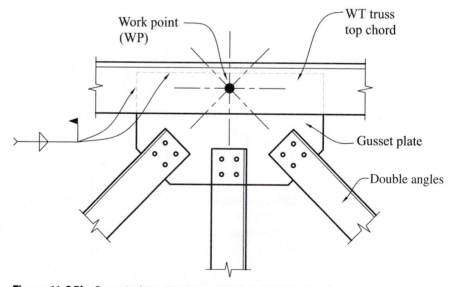

Figure 11-35b Gusset plate at a truss panel point (top chord).

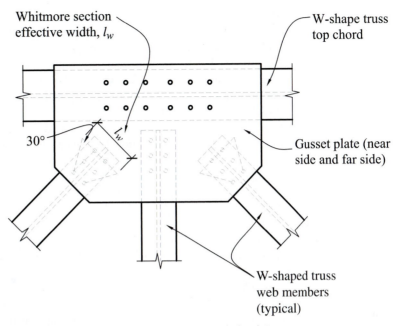

Figure 11-35c Gusset plate at a truss panel point.

column. Several connection interfaces must be designed: the diagonal brace-to-gusset connection, the gusset-to-beam connection, the gusset-to-column connection, and the beam-to-column connection. At truss joints, the gusset plates connect the web members to the chord members and at the diagonal brace connections. The gusset plates are connected to the adjoining members with welds or bolts. At truss panel joints or diagonal brace connections, it is common practice to choose the geometry of the joint such that the centroidal axes of the members meeting at the joint coincide at one point, called the work point (WP), in

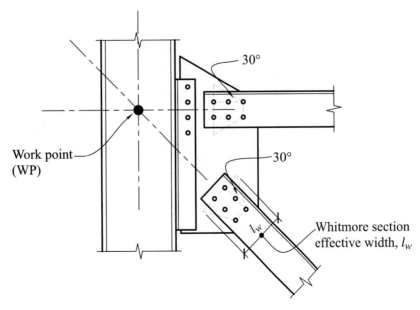

Figure 11-35d Gusset plate at a truss support.

order to minimize bending moments in the gusset plate and the adjoining members. Where it is not feasible to have a common work point for all the members meeting at a joint, there will be moments induced in the gusset plate and the connecting members, in addition to other stresses. In addition to bending moments, gusset plates are usually subjected to shear and axial tension or compression stresses. For direct axial stresses (i.e., tension and compression), the effective cross-section is defined by the Whitmore effective width, l_w (typical throughout), at the end of the connection, and this is obtained by projecting lines at an angle of spread of 30° on both sides of the connection starting from the first row of bolts to the last row of bolts in the connection. For welded connections, the lines are projected on both sides of the longitudinal weld and at the spread-out angle of 30° starting from the edge of the longitudinal weld to the end of the weld. (see Figures 11-35a, 11-35c, and 11-35d). The effective gross area of the plate is the Whitmore effective width, l_w, times the plate thickness, t.

To ensure adequate out-of-plane rotation of the gusset plate when the bracing or truss web member is subjected to out-of-plane buckling under cyclic loading (e.g., seismic loads), Astaneh-Asl recommends that the end of the bracing member or truss web member be terminated at least a distance of $2t$ away from the re-entrant corner of the gusset plate at the gusset-to-beam and gusset-to-column interfaces [9] (see Figures 11-34 and 11-35a). This requirement can be relaxed for connections subject to monotonic or static loading. Although gusset plates may appear to be small and insignificant structural elements, it is important that they be adequately designed, detailed, and protected against corrosion to avoid connection failures that could lead to the collapse of an entire structure.

In this section, only the design of the gusset plate itself is covered. The determination of the forces acting on gusset plate interface connections is discussed, but the design of the welds and bolts at these connection interfaces is not covered in this section because this matter has already been covered in previous chapters. The observed failure modes of gusset plates include the following [9]:

1. **Out-of-plane buckling** of the gusset plate due to the axial compression force
   The unbraced length, L_g, of a gusset plate is taken as the larger of the length of the plate between adjacent lines of bolts parallel to the direction of the axial compression force, or the length of the plate along the centroidal axis of the diagonal brace or truss web member between the end of the brace or truss web member and the connected edge of the gusset plate (see Figures 11-34 and 11-35a). The gusset plate will buckle out of plane about its weaker axis; the buckling is assumed to occur over a plate width equal to the Whitmore effective width, l_w. To determine the design compression load, determine the slenderness ratio, KL_g/r, where r is approximately $0.3t$ and the effective length factor, K, is conservatively taken as 1.2. The critical buckling stress, ϕF_{cr}, is obtained from *AISCM*, Table 4-22, and the design compressive strength of the gusset plate is calculated as

$$\phi P_{cr} = \phi F_{cr} l_w t, \tag{11-44}$$

   where

   l_w = Whitmore effective width,
   t = Thickness of the gusset plate, and
   ϕ = 0.9.

2. **Buckling of the free or unsupported edge** of the gusset plate

To prevent buckling at the unsupported edges of gusset plates, the minimum gusset plate thickness (without edge stiffeners) required in the American Association of State and Highway Transportation Officials (AASHTO) code [10] and commonly used in design practice for monotonic or static loading is

$$t \geq 0.5L_{fg}\sqrt{\frac{F_y}{E}},$$ (11-45a)

For $F_y = 36$ ksi, $t \geq \dfrac{L_{fg}}{56}$, and

For $F_y = 50$ ksi, $t \geq \dfrac{L_{fg}}{48}$,

where

L_{fg} = Length of the free or unsupported edge of the gusset plate (see Figure 11-36),

F_y = Yield strength of the gusset plate, and

E = Modulus of elasticity of the gusset plate.

For gusset plates subjected to cyclic (or seismic) loading, the minimum required gusset plate thickness is

$$t \geq 1.33L_{fg}\sqrt{\frac{F_y}{E}}$$ (11-45b)

3. Tension or compression failure of the gusset plate due to **yielding** within the Whitmore effective area

Tension yielding is the most desirable form of failure because of the ductility associated with this failure mode.

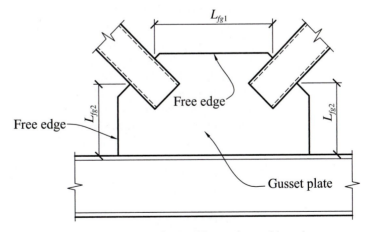

Use the largest free unbraced length,
L_{fg} = larger of L_{fg1} or L_{fg2}.

Figure 11-36 Free or unsupported edge length of gusset plates.

The design tension or compression yield strength is

$$\phi P_n = \text{Gross area of the Whitmore section} \times \phi F_y$$
$$= \phi F_y l_w t,$$ (11-46)

where

$\phi = 0.9,$

$l_w =$ Whitmore effective width (see Figure 11-35a), and

$t =$ Thickness of the gusset plate.

4. Tension failure of the gusset plate due to **fracture** at a bolt line within the Whitmore effective area

This is the least desirable form of failure and should be avoided for structures subjected to cyclic loading (e.g., seismic loads, traffic loads, and crane loads) because of the sudden and brittle nature of this failure mode.

The design tension strength, $\phi P_n = \text{Net area of the Whitmore section} \times \phi F_u$
$$= \phi F_u (W - n d_{hole})t,$$ (11-47)

where

$F_u =$ Tensile strength of the gusset plate,

$\phi = 0.75,$

$n =$ Number of bolt holes perpendicular to the applied axial force for each line of bolt, and

$d_{hole} =$ Diameter of the bolt hole $= d_{bolt} + \frac{1}{8}$ in. (see Chapter 4).

5. Tension failure of the gusset plate due to **block shear**
The calculation of the tensile strength of a plate for this failure mode has been discussed in Chapter 4.

6. **Fracture** of the connecting welds and bolts
The design of bolts and welds has been discussed in Chapters 9 and 10.

7. Yielding failure of the gusset plate from **combined axial tension or compression load, bending moment, and shear**
The applied loads on a diagonal brace or truss connection may result in a combination of shear, V_u, bending moment, M_u, and tension or compression force, P_u, acting on a critical section of the gusset plate. These forces are determined from a free-body diagram of the gusset plate using equilibrium and statics principles. The following interaction equation from plasticity theory is recommended for the design of gusset plates under combined loads [9, 11, 12]:

$$\frac{M_u}{\phi M_p} + \left(\frac{P_u}{\phi P_y}\right)^2 + \left(\frac{V_u}{\phi V_y}\right)^4 \leq 1.0,$$ (11-48)

where

$\phi = 0.9,$

$\phi M_p =$ Plastic moment capacity of the gusset plate at the critical section

$$= \phi \frac{t L_{g,cr}^2}{4} F_y,$$

$$t = \text{Thickness of gusset plate,}$$

$$L_{g,\,cr} = \text{Length of the gusset plate at the critical section,}$$

$$F_y = \text{Yield strength of the gusset plate,}$$

$$\phi P_y = \text{Axial yielding capacity} = A_{g,\,cr}\,F_y,$$

$$A_{g,\,cr} = t L_{g,\,cr},\ \text{and}$$

$$\phi V_y = \text{Shear yielding capacity of the gusset plate} = \phi(0.6 A_{g,\,cr}\,F_y).$$

EXAMPLE 11-11

Gusset Plate at a Symmetrical Truss Joint

For the truss joint shown in Figure 11-37, determine the following, assuming a ⅝-in. gusset plate, ¾-in.-diameter bolts and grade 50 steel:

a. Whitmore effective width for the gusset plate on diagonal web members A and B,

b. Compression buckling capacity of the gusset plate on diagonal member A, and

c. Tension capacity of the gusset plate on diagonal member B (assume that block shear does not govern).

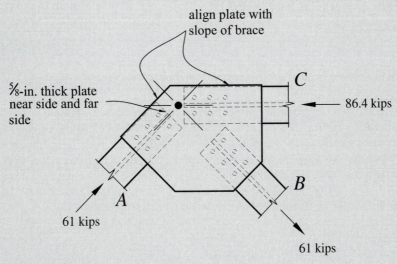

Figure 11-37 Gusset plate details for Example 11-11.

SOLUTION

1. From Figure 11-38, the effective width of the gusset plate on diagonal web members A and B is

$$l_{wA} = 1.5\ \text{in.} + b_A + L_{wA}\tan\theta$$
$$= 1.5\ \text{in.} + 3.08\ \text{in.} + (6)(\tan 30)$$

$$l_{wA} = 8.04\ \text{in. (the part of the Whitmore section that falls outside of the}$$
$$\qquad \text{gusset plate is ignored), and}$$
$$l_{wB} = b_B + 2L_{wB}\tan\theta = 3.08\ \text{in.} + (2)(6)(\tan 30)$$
$$l_{wB} = 10\ \text{in.}$$

(continued)

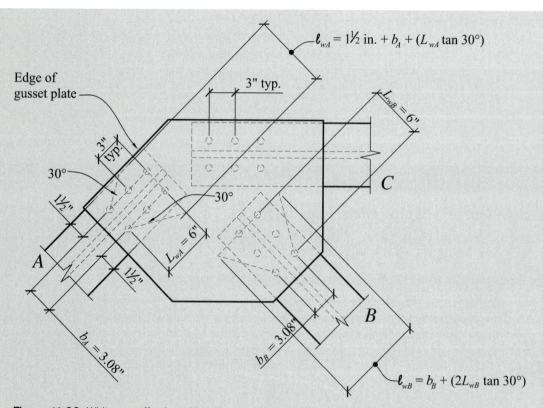

Figure 11-38 Whitmore effective width of gusset plate in Example 11-11.

2. The maximum unbraced length of the gusset plate is the larger of the distance between bolt lines (3 in.) or the maximum unsupported distance of the gusset plate measured along the centroid of the brace or diagonal member from the end of the diagonal member to the connected edge of the gusset plate (0 inches in this case since the W6 diagonal members have been extended to the point where they abut other W6 members at the joint; see Figure 11-38).

Therefore, the maximum unbraced length of gusset plate, $L_g = 3$ in.

$$\frac{KL_g}{r} = \frac{1.2(3 \text{ in.})}{0.3(0.625 \text{ in.})} = 19.2 < 25$$

Therefore, according to *AISCM*, Section J4.4, buckling can be neglected. Note that for cases where $KL/r > 25$, the buckling capacity, ϕP_n, is determined from $\phi F_{cr} A_g$, where ϕF_{cr} is obtained from *AISCM*, Table 4-22.

a. The tension yielding capacity of the gusset plate on diagonal member B is

$$\phi P_n = \phi F_y l_{wB} t = (0.9)(50 \text{ ksi})(10 \text{ in.})(0.625 \text{ in.}) = 281 \text{ kips.}$$

b. The tension capacity of the gusset plate on diagonal member B due to fracture of the gusset plate is

$$\phi P_n = \phi F_u (l_{wB} - nd_{hole})t$$
$$= (0.75)(65 \text{ ksi})[10 \text{ in.} - (2)(\tfrac{7}{8} \text{ in.})](0.625 \text{ in.})$$
$$= 251 \text{ kips,}$$

where

$$l_{wB} = 10 \text{ in.,}$$
$$n = 2 \text{ bolts per line,}$$
$$d_{bolt} = \tfrac{3}{4} \text{ in.,}$$
$$d_{hole} = (\tfrac{3}{4} + \tfrac{1}{8}) = \tfrac{7}{8} \text{ in., and}$$
$$t = \tfrac{5}{8} \text{ in.} = 0.625 \text{ in.}$$

The smaller of the above two values will govern for the tension capacity of the gusset plate on diagonal member B. Therefore, the design tension strength, $\phi P_n = 251$ kips.

EXAMPLE 11-12

Gusset Plate at a Nonsymmetrical Truss Joint

The gusset plate for a truss bridge is subjected to the factored loads shown in Figure 11-39, assuming grade 50 steel:

1. Determine the combined moment, M_u, shear, V_u, and axial load, P_u, acting on the gusset plate at point 'c' along the critical section C-C just below the work point.

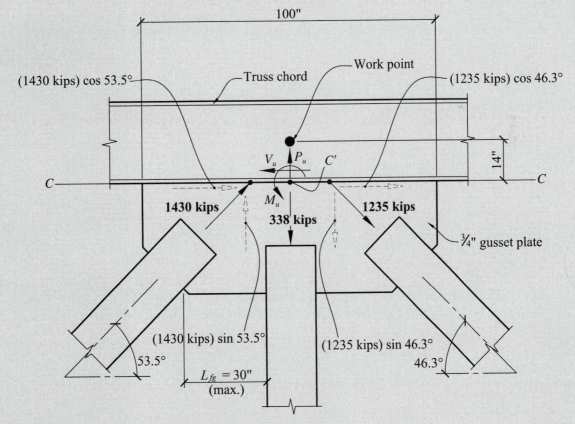

Figure 11-39 Gusset plate details for Example 11-12.

(continued)

2. Using the plasticity theory interaction formula, determine whether the gusset plate is adequate to resist the combined loads.
3. Determine whether the gusset plate is adequate for free-edge buckling.

SOLUTION

1. Summing the forces in the horizontal direction (i.e., $\Sigma F_x = 0$) yields

$$1430 \cos 53.5° + 1235 \cos 46.3° - V_u = 0.$$

Therefore, $V_u = 1704$ kips.

$$\phi V_y = \phi(0.6A_{g,cr}F_y) = (0.9)(0.6)(50 \text{ ksi})(100 \text{ in.})(0.75 \text{ in.}) = 2025 \text{ kips} > V_u \text{ OK}$$

Summing the forces in the vertical direction (i.e., $\Sigma F_y = 0$) yields

$$-338 + 1430 \sin 53.5° - 1235 \sin 46.3° + P_u = 0.$$

Therefore, $P_u = 81$ kips.

$$\begin{aligned}\phi P_y &= \phi A_{g,cr}F_y = (0.9)(100 \text{ in.})(0.75 \text{ in.})(50 \text{ ksi}) \\ &= 3375 \text{ kips} > P_u \text{ OK}\end{aligned}$$

Summing the moments about a point C' on the critical section C-C just below the work point (i.e., $\Sigma M = 0$) yields

$$[(1430 \sin 53.5°)(14/\tan 53.5°)] + [(1235 \sin 46.3°)(14/\tan 46.3°)] - M_u = 0.$$

Therefore, $M_u = 23{,}854$ in.-kips.

$$\begin{aligned}\phi M_p &= \phi\frac{tL_{g,cr}^2}{4}F_y = (0.9)\left[\frac{(0.75 \text{ in.})(100 \text{ in.})^2}{4}\right](50 \text{ ksi}) \\ &= 84{,}375 \text{ in.-kips} > M_u \text{ OK}\end{aligned}$$

From equation (11-48), the interaction equation for a gusset plate under combined loading is

$$\frac{M_u}{\phi M_p} + \left(\frac{P_u}{\phi P_y}\right)^2 + \left(\frac{V_u}{\phi V_y}\right)^4 = \frac{23{,}854}{84{,}375} + \left(\frac{81}{3375}\right)^2 + \left(\frac{1704}{2025}\right)^4 = 0.78 < 1.0. \text{ OK}$$

2. The maximum unbraced length of the free (unsupported) edge of the gusset plate, $L_{fg} = 30$ in. From equation (11-45a), the required minimum gusset plate thickness to avoid unsupported edge buckling is

$$\begin{aligned}t \geq 0.5L_{fg}\sqrt{\frac{F_y}{E}} &= (0.5)(30 \text{ in.})\left(\sqrt{\frac{50 \text{ ksi}}{29{,}000 \text{ ksi}}}\right) \\ &= 0.62 \text{ in.} < \frac{3}{4} \text{ in. provided. OK}\end{aligned}$$

If this gusset plate were subjected to cyclic (i.e., seismic) loading, the minimum required thickness of the gusset plate, without edge stiffeners, would be 1.66 in. from equation 11-45b.

Gusset Plate Connection Interface Forces

The determination of the gusset plate connection interface forces is a complex indeterminate problem and the most efficient connection design method for calculating the shear, axial tension or compression forces, and moments acting on gusset plate connections is the so-called uniform force method (UFM) illustrated in Figure 11-40 [9, 13–17]. For economic reasons, it is desirable to select a connection geometry that will avoid, or at least minimize, the moments acting on the gusset-to-beam, gusset-to-column, and beam-to-column connection interfaces, but in order to successfully avoid these moments and thus have only axial and shear forces on the gusset connection interfaces, the UFM requires that the gusset plate connection satisfy the following conditions:

$$\alpha - \beta \tan \theta = 0.5d_b \tan \theta - 0.5d_c, \tag{11-49}$$

$$r = \sqrt{(\alpha + 0.5d_c)^2 + (\beta + 0.5d_b)^2}, \tag{11-50}$$

$$V_B = \frac{0.5d_b}{r} P_{brace}, \tag{11-51}$$

$$H_B = \frac{\alpha}{r} P_{brace}, \tag{11-52}$$

$$V_C = \frac{\beta}{r} P_{brace}, \text{ and} \tag{11-53}$$

$$H_C = \frac{0.5d_c}{r} P_{brace}, \tag{11-54}$$

where

$\theta = $ Angle between the diagonal brace and the vertical plane,

$\alpha = $ *Ideal* distance from the face of the column flange or web to the centroid of the gusset-to-beam connection (Note: The setback between the gusset and the face of the column flange can typically be assumed to be approximately 0.5 in.),

$\beta = $ *Ideal* distance from the face of the beam flange to the centroid of the gusset-to-column connection (where the beam is connected to the column web, set $\beta = 0$),

$d_b = $ Depth of beam,

$d_c = $ Depth of column (where the diagonal brace is not connected to a column flange, set $0.5d_c \approx 0$ and $H_C = 0$),

$V_B, H_B = $ Vertical and horizontal forces on the gusset–beam connection interface,

$V_C, H_C = $ Vertical and horizontal forces on the gusset–column connection interface,

$X = $ Horizontal length of the gusset plate $= 2(\alpha - 0.5\text{-in. setback})$, and

$Y = $ Vertical length of the gusset plate $= 2\beta$.

The *design* procedure for a *new* gusset plate connection, with the beam and column sizes already determined, is as follows:

1. Knowing the beam and column sizes and the brace geometry, determine $0.5d_b$, $0.5d_c$, and θ.

   Note: If the brace is connected to a column web, set $d_c = 0$ and $H_C = 0$.

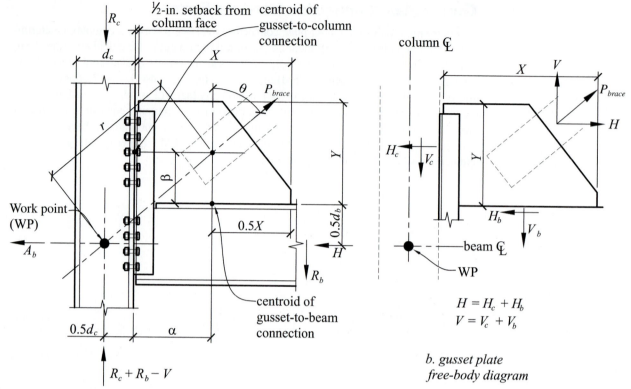

a. diagonal bracing connection and applied forces

b. gusset plate free-body diagram

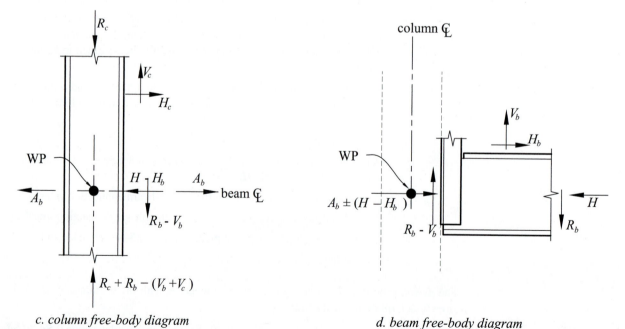

c. column free-body diagram

d. beam free-body diagram

Figure 11-40 Gusset connection interface forces using the uniform force method. Adapted from reference [1].

Table 11-3 Gusset connection interface design forces

Connection	Shear Force	Axial Force	Moment	Remarks
Gusset-to-beam	H_B	V_B	0	Assuming that α and β satisfy equation (11-49)
Gusset-to-column	V_C	H_C	0	
Beam-to-column	$R - V_B$	$A_B \pm (H - H_B)$	0	

R = Factored or required end reaction in the beam,

A_B = Factored or required horizontal axial force from the adjacent bay (due to drag strut action), and

H = Horizontal component of the factored or required diagonal brace force = $P_{brace} \sin \theta$.

If the gusset is connected only to the beam and not to a column, set $d_C = 0$, $\beta = 0$, $V_C = 0$, and $H_C = 0$.

2. Select a value for Y, the vertical dimension of the gusset plate, and determine $\beta = 0.5Y$.

3. Substitute β, θ, $0.5d_b$, and $0.5d_c$ into equation (11-49) to determine α.

4. Knowing α, with the centroid of the gusset-to-beam connection interface assumed to be at the midpoint of this interface, the horizontal length of the gusset plate can be determined from the relationship $\alpha = \dfrac{X}{2} + 0.5$ in. where 0.5 in. is the gusset setback from the column face.

5. Determine the horizontal and vertical forces on the gusset plate connection interfaces, V_B, H_B, V_C, and H_C, using Equations (11-50) to (11-54).

The forces on the gusset connection interfaces are given in Table 11-3.

For new connection designs, it is relatively easy to select a gusset geometry with values of X and Y or α and β that satisfy equation (11-49) and thus ensure that there are no moments on the gusset connection interfaces. However, for existing gusset plate connections or where constraints have been placed on the gusset plate dimensions, it may not be possible to satisfy equation (11-49), and therefore, moments may exist on one or both gusset connection interfaces or on the beam-to-column connection interface. It is usual design practice to assume that the more rigid gusset connection interface will resist all of the moment required to satisfy equilibrium [20].

The procedure for the *analysis* of *existing* gusset plate connections is as follows:

1. Determine $0.5d_b$, $0.5d_c$, X, Y, and θ from the geometry of the connection.

Note that where the brace is not connected to a column flange, set $0.5d_c \approx 0$ and $H_C = 0$.

2. Using the known horizontal and vertical lengths of the gusset plate, determine the actual α and β values as follows:

$$\overline{\alpha} = \frac{X}{2} + 0.5 \text{ in. } (0.5 \text{ in. is the assumed setback between the gusset plate and}$$

the face of the column flange),

$$\overline{\beta} = \frac{Y}{2},$$

$\overline{\alpha}$ = *Actual* distance between the face of the column flange or web and the centroid of the gusset-to-beam connection, and

$\overline{\beta}$ = *Actual* distance between the face of the beam flange and the centroid of the gusset-to-column connection.

3. If the gusset-to-beam connection is more rigid than the gusset-to-column connection (e.g., if a welded connection is used for the gusset-to-beam connection and a bolted connection is used for the gusset-to-column connection), set the ideal β equal to the actual β (i.e., $\overline{\beta}$ from step 2) and calculate the ideal α using equation (11-49). These ideal values of α and β are used in equation (11-50) to calculate the parameter, r.

   If the ideal α calculated in step 3 equals the actual α (i.e., $\overline{\alpha}$ from step 2), then no moment exists on the gusset-to-beam connection interface. If not, the forces and moment on the gusset-to-beam connection interface are calculated as follows:

$$H_B = \frac{\alpha_{step3}}{r_{step3}}P_{brace},\tag{11-55}$$

$$V_B = \frac{0.5d_b}{r_{step3}}P_{brace}, \text{ and}\tag{11-56}$$

$$M_{u,g-b} = V_B|\alpha - \overline{\alpha}|,\tag{11-57}$$

where

$$r = \sqrt{(\alpha_{step3} + 0.5d_c)^2 + (\overline{\beta}_{step2} + 0.5d_b)^2}.\tag{11-58}$$

4. If the gusset-to-column connection is more rigid than the gusset-to-beam connection (e.g., if a welded connection is used for the gusset-to-column connection and a bolted connection is used for the gusset-to-beam connection), set the ideal α equal to the actual α (i.e., $\overline{\alpha}$ from step 2) and calculate the ideal β using equation (11-49). These ideal values of α and β are used in equation (11-50) to calculate the parameter, r.

   If the ideal β, calculated in step 4, equals the actual β (i.e., $\overline{\beta}$ from step 2), then no moment exists on the gusset-to-column connection interface. If not, the forces and moment on the gusset-to-column connection interface are calculated as follows:

$$H_C = \frac{0.5d_c}{r_{step4}}P_{brace},\tag{11-59}$$

$$V_C = \frac{\beta_{step4}}{r_{step4}}P_{brace}, \text{ and}\tag{11-60}$$

$$M_{u,g-c} = H_C|\beta - \overline{\beta}|,\tag{11-61}$$

where

$$r = \sqrt{(\overline{\alpha}_{step2} + 0.5d_c)^2 + (\beta_{step4} + 0.5d_b)^2}.\tag{11-62}$$

EXAMPLE 11-13

Design of gusset plate connection

Determine the dimensions of the gusset plate and the gusset connection interface forces for the diagonal brace connection shown in Figure 11-41.

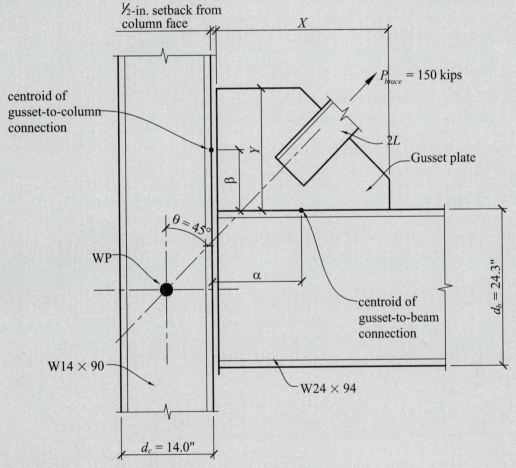

Figure 11-41 Diagonal brace connection design example.

SOLUTION

1. d_c(W14 × 90) = 14.0 in.; d_b(W24 × 94) = 24.30 in.

   $0.5d_c$(W14 × 90) = 7.0 in.; $0.5d_b$(W24 × 94) = 12.15 in.

   $\theta = 45°$ from the geometry of the connection

2. Assume $Y = 18$ in.

   $$\beta = \frac{Y}{2} = \frac{18}{2} = 9 \text{ in.}$$

(continued)

3. Using equation (11-49), we have

$\alpha - \beta \tan\theta = e_b \tan\theta - e_c$. Then,

$\alpha - (9\text{ in.})(\tan 45°) = (12.15\text{ in.})(\tan 45°) - 7.0\text{ in.}$

Therefore, $\alpha = 14.15$ in.

4. Hence, the horizontal length of the gusset plate, X, is obtained from

$$\alpha = \frac{X}{2} + 0.5\text{ in.}$$

That is, $14.15\text{ in.} = \dfrac{X}{2} + 0.5\text{ in.}$

Thus, $X = (14.15 - 0.5)(2) = 27.3\text{ in.} = 2\text{ ft. }4\text{ in.}$

From step 3, $Y = 18\text{ in.} = $ **1 ft. 6 in.**

5. Use equation (11-50) to determine

$$r = \sqrt{(\alpha + 0.5d_c)^2 + (\beta + 0.5d_b)^2} = \sqrt{(14.15\text{ in.} + 7.0\text{ in.})^2 + (9\text{ in.} + 12.15\text{ in.})^2}$$
$$= 29.92\text{ in.}$$

Using equations (11-51) to (11-54) the gusset connection interface forces are determined as follows:

$$V_B = \frac{0.5d_b}{r}P_{brace} = \frac{12.15\text{ in.}}{29.92\text{ in.}}(150\text{ k}) = 61.0\text{ kips}$$

$$H_B = \frac{\alpha}{r}P_{brace} = \frac{14.15\text{ in.}}{29.92\text{ in.}}(150\text{ k}) = 71.0\text{ kips}$$

$$V_C = \frac{\beta}{r}P_{brace} = \frac{9\text{ in.}}{29.92\text{ in.}}(150\text{ k}) = 45.0\text{ kips}$$

$$H_C = \frac{0.5d_c}{r}P_{brace} = \frac{7.0\text{ in.}}{29.92\text{ in.}}(150\text{ k}) = 35.1\text{ kips}$$

EXAMPLE 11-14

Analysis of Existing Diagonal Brace Connection

Determine the force distribution at the gusset plate connection interfaces for the existing diagonal brace connection shown in Figure 11-42, assuming that

a. The gusset-to-beam connection is more rigid than the gusset-to-column connection, and

b. The gusset-to-column connection is more rigid than the gusset-to-beam connection.

Assume a ½-in. setback between the gusset plate and the face of the column.

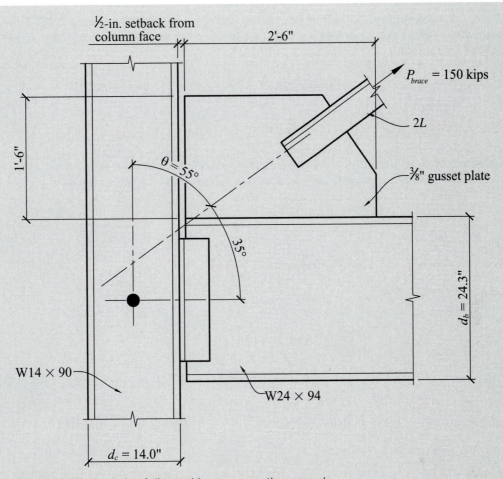

Figure 11-42 Analysis of diagonal brace connection example.

SOLUTION

1. d_c(W14 × 90) = 14.0 in.; d_b(W24 × 94) = 24.3 in.

   $0.5d_c$(W14 × 90) = 7.0 in.; $0.5d_b$(W24 × 94) = 12.15 in.

   θ = 55° from the geometry of the connection

   X = 30 in. and Y = 18 in. (see Figure 11-42)

2. The actual centroidal distances are calculated as follows:

$$\overline{\alpha} = \frac{X}{2} + 0.5 \text{ in.} = \frac{30 \text{ in.}}{2} + 0.5 \text{ in.} = 15.5 \text{ in.}$$

$$\overline{\beta} = \frac{Y}{2} = \frac{18 \text{ in.}}{2} = 9 \text{ in.}$$

(continued)

3. If the gusset-to-beam connection is more rigid than the gusset-to-column connection (e.g., if a welded connection is used for the gusset-to-beam connection and a bolted connection is used for the gusset-to-column connection),

$$\beta = \bar{\beta} \text{ (from step 2)} = 9 \text{ in. Substituting this in equation (11-49) yields}$$

$$\alpha - \beta \tan \theta = e_b \tan \theta - e_c$$

$$\alpha - (9 \text{ in.})(\tan 55°) = (12.15 \text{ in.})(\tan 55°) - 7.0 \text{ in.}$$

Therefore, set $\alpha = 23.2 \text{ in.} = \bar{\alpha}$ from step 2. Thus, we find that a moment exists on the gusset-to-beam connection interface; the moment and forces on this interface are calculated as follows:

$$r = \sqrt{(\alpha_{step3} + 0.5d_c)^2 + (\bar{\beta}_{step2} + 0.5d_b)^2}$$

$$= \sqrt{(23.2 \text{ in.} + 7 \text{ in.})^2 + (9 \text{ in.} + 12.15 \text{ in.})^2} = 36.9 \text{ in.}$$

The forces on the gusset-to-beam interface are

$$H_B = \frac{\alpha_{step3}}{r_{step3}} P_{brace} = \frac{23.2 \text{ in.}}{36.9 \text{ in.}}(150 \text{ k}) = 94.3 \text{ kips}$$

$$V_B = \frac{0.5d_b}{r_{step3}} P_{brace} = \frac{12.15 \text{ in.}}{36.9 \text{ in.}}(150 \text{ k}) = 49.4 \text{ kips}$$

$$M_{u,g-b} = V_B|\alpha - \bar{\alpha}| = (49.4 \text{ k})(23.2 \text{ in.} - 15.5 \text{ in.})$$

$$= 380.4 \text{ in.-kips} = \mathbf{31.7 \text{ ft.-kips}}$$

The forces on the gusset-to-column interface are

$$H_C = \frac{0.5d_c}{r} P_{brace} = \frac{7.0 \text{ in.}}{36.9 \text{ in.}}(150 \text{ kips}) = 28.5 \text{ kips}$$

$$V_C = \frac{\beta}{r} P_{brace} = \frac{9 \text{ in.}}{36.9 \text{ in.}}(150 \text{ kips}) = 36.6 \text{ kips}$$

$$M_{u,g-c} = 0 \text{ ft.-kips}$$

4. If the gusset-to-column connection is more rigid than the gusset-to-beam connection (e.g., if a welded connection is used for the gusset-to-column connection and a bolted connection is used for the gusset-to-beam connection), set $\alpha = \bar{\alpha}$(from step 2) = 15.5 in. Substituting this in equation (11-49) yields

$$15.5 \text{ in.} - \beta(\tan 55°) = (12.15 \text{ in.})(\tan 55°) - 7.0 \text{ in.}$$

Therefore, $\beta = 3.6 \text{ in.} \neq \bar{\beta}$ from step 2. Thus, we find that a moment exists on the gusset-to-column connection interface; the moment and forces on this interface are calculated as follows:

$$r = \sqrt{(\bar{\alpha}_{step2} + 0.5d_c)^2 + (\beta_{step4} + 0.5d_b)^2}$$

$$= \sqrt{(15.5 \text{ in.} + 7.0 \text{ in.})^2 + (3.6 \text{ in.} + 12.15 \text{ in.})^2} = 27.5 \text{ in.}$$

The forces on the gusset-to-column interface are

$$H_C = \frac{0.5d_c}{r_{step4}} P_{brace} = \frac{7.0 \text{ in.}}{27.5 \text{ in.}}(150 \text{ k}) = 38.2 \text{ kips}$$

$$V_C = \frac{\beta_{step4}}{r_{step4}} P_{brace} = \frac{3.6 \text{ in.}}{27.5 \text{ in.}}(150 \text{ k}) = 19.6 \text{ kips}$$

$$M_{u,g-c} = H_C|\beta - \overline{\beta}| = (38.2 \text{ k})(3.6 \text{ in.} - 9 \text{ in.})$$
$$= 206.3 \text{ in.-kips} = \textbf{17.2 ft.-kips}$$

The forces on the gusset-to-beam interface are

$$H_B = \frac{\alpha}{r} P_{brace} = \frac{15.5 \text{ in.}}{27.5 \text{ in.}}(150 \text{ kips}) = 84.5 \text{ kips}$$

$$V_B = \frac{0.5d_b}{r} P_{brace} = \frac{12.15 \text{ in.}}{27.5 \text{ in.}}(150 \text{ kips}) = 66.3 \text{ kips}$$

$$M_{u,g-b} = 0 \text{ ft.-kips}$$

The available strength of the gusset plate connection is determined using the methods presented in Chapters 9 and 10 for calculating bolt and weld capacities.

For more information and detailed design examples of gusset-plated connections, the reader should refer to Chapter 2 of reference 18.

11.11   REFERENCES

1. American Institute of Steel Construction. 2006. *Steel construction manual*, 13th ed. Chicago: AISC.

2. Geschwinder, L. F., and R. O. Disque. 2005. Flexible moment connections for unbraced frames—A return to simplicity. *Engineering Journal* 42, no. 2 (2nd quarter): 99–112.

3. Blodgett, Omer. 1966. *Design of welded structures*. Cleveland: The James F. Lincoln Arc Welding Foundation.

4. American Institute of Steel Construction. 2006. ANSI/AISC 341-05 and ANSI/AISC 358-05. *Seismic design manual*. Chicago: AISC.

5. Federal Emergency Management Agency. 2000. *FEMA-350: Recommended seismic design criteria for new steel moment frame buildings*. Washington, D.C.: FEMA.

6. Occupational Safety and Health Administration. *Part 1926: Safety and health regulations for construction*. Washington, D.C.: OSHA.

7. Holt, Reggie, and Joseph Hartmann. 2008. *Adequacy of the U10 and L11 gusset plate designs for Minnesota bridge no. 9340 (I-35W over the Mississippi River)*, interim report. Washington, D.C.: U.S. Government Printing Office.

8. Darwin, David. American Institute of Steel Construction. 1990. *Steel design guide series 2: Steel and composite beams with web openings*. Chicago.

9. Astaneh-Asl, A. 1998. *Seismic behavior and design of gusset plates, steel tips*. Moraga, CA: Structural Steel Educational Council.

10. American Association of State Highway Transportation Officials. 2004. *Standard specification for highway bridges*, Washington.

11. Thornton, W. A. 2000. *Combined stresses in gusset plates*. Roswell, GA: CIVES Engineering Corporation.

12. Neal, B. G., 1977. *The plastic methods of structural analysis*. New York: Halsted Press, Wiley.

13. Richard, Ralph M. 1986. Analysis of large bracing connection designs for heavy construction. In *Proceedings of the AISC National Engineering Conference*. Nashville, TN: American Institute of Steel Construction.

14. Thornton, W. A. 1991. On the analysis and design of bracing connections. In *Proceedings of the AISC National Steel Construction Conference*. Washington, D.C.: American Institute of Steel Construction.

15. Thornton, W. A. 1995. Connections—Art, science and information in the quest for economy and safety. *AISC Engineering Journal* 32, no. 4 (4th quarter): 132–144.

16. American Institute of Steel Construction. 1994. Manual of steel construction—Load and resistance factor design, 2nd ed. 2 vols. Chicago: AISC.

17. Muir, Larry S. 2008. Designing compact gussets with the uniform force method. *AISC Engineering Journal*, (1st quarter): 13–19.

18. Tamboli, Akbar R. 1999. Handbook of structural steel connection design and details, 35. New York: McGraw-Hill.

19. Miller, Duane. American Institute of Steel Construction. 2006. *Steel design guide series 21: Welded connections—A primer for structural engineers*. Chicago.

20. McCormac, Jack. 1981. *Structural steel design*, 3rd ed. New York: Harper and Row.

21. Salmon, Charles G., and John E. Johnson. 1980. *Steel structures: Design and behavior*, 2nd ed. New York: Harper and Row.

22. Smith, J. C. 1988. *Structural steel design: LRFD fundamentals*. Hoboken, NJ: Wiley.

23. Segui, William. 2006. *Steel design*, 4th ed. Toronto: Thomson Engineering.

24. Limbrunner, George F., and Leonard Spiegel. 2001. *Applied structural steel design*, 4th ed. Upper Saddle River, NJ: Prentice Hall.

25. Carter, Charles. American Institute of Steel Construction. 1999. *Steel design guide series 13: Wide-flange column stiffening at moment connections*. Chicago.

26. Brockenbrough, Roger L. and Merrit, Frederick S. 2005. *Structural Steel Designer's Handbook*. Fourth Edition, McGraw Hill, New York.

11.12 PROBLEMS

11-1. Determine whether the W14 × 34 beam shown in Figure 11-43 is adequate for flexural rupture and flexural yielding at the cope. The steel is ASTM A992, grade 50.

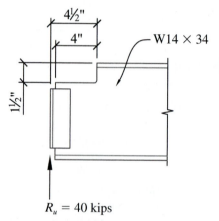

Figure 11-43 Details for Problem 11-1.

11-2. Determine the maximum reaction, R_u, that could occur, considering flexural rupture and flexural yielding at the cope for the beam shown in Figure 11-44. The steel is grade 50. What is the required plate length for the transverse stiffener?

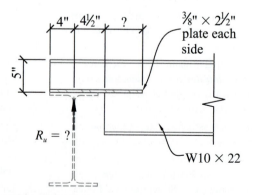

Figure 11-44 Details for Problem 11-2.

11-3. Determine whether the W21 × 44 beam shown in Figure 11-45 is adequate for flexural rupture and flexural yielding at the cope. The steel is ASTM A36.

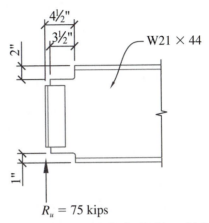

$R_u = 75$ kips

Figure 11-45 Details for Problem 11-3.

11-4. Design the top and bottom plates for the moment connection shown in Figure 11-46, including the welds to the beam. The beam is ASTM A992 and the welds are E70xx.

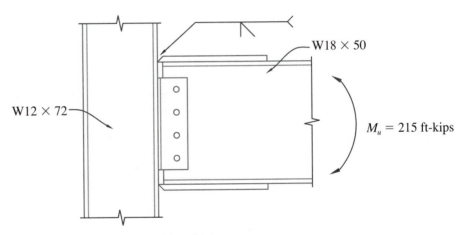

$M_u = 215$ ft-kips

Figure 11-46 Details for Problem 11-4.

11-5. Repeat Problem 11-4 assuming that the plates are bolted to the beam flange. Use ¾-in.-diameter ASTM A490N bolts.

11-6. Design a beam splice for a W21 × 44 beam assuming that the top and bottom plates and the shear plate are welded. The steel is ASTM A36 and the welds are E70XX. Assume that the required design bending strength is the full plastic moment capacity ($\phi_b M_p$) and that the required shear strength is one-third of the design shear strength ($\phi_v V_n$).

11-7. Determine whether the column in Problem 11-4 is adequate for the concentrated forces due to the applied moment. Assume that the beam connection occurs at a location remote from the ends.

11-8. For a W16 × 50 beam with the top and bottom flanges fully welded to the flanges of a W10 × 54 supporting column, determine the maximum end moment that could be applied without the use of stiffeners, taking into consideration the strength of the column in supporting concentrated forces. Assume that the beam connection occurs at a location remote from the ends. The steel is ASTM A992.

11-9. A diagonal brace in Chevron vertical bracing is connected to a W12 × 72 column and a W21 × 44 beam with a gusset plate in a connection similar to that shown in Figure 11-40. The factored brace force is 100 kips and the diagonal brace is inclined 60° to the horizontal. Determine the dimensions of the gusset plate and the forces on the gusset–column and gusset–beam interfaces. Assume a ½-in. setback between the gusset plate and the face of the column.

11-10. A diagonal brace in Chevron vertical bracing is connected to a W12 × 72 column and a W21 × 44 beam with a gusset plate in a connection similar to that shown in Figure 11-41, but with a gusset horizontal length, X-2 ft. 4 in. and a vertical length, Y-1 ft. 2 in. The factored brace force is 90 kips and the diagonal brace is inclined 35° to the horizontal. Determine the force distribution in the gusset plate connection interfaces assuming that the gusset-to-beam connection is more rigid than the gusset-to-column connection. Assume a ½-in. setback between the gusset plate and the face of the column.

11-11. Repeat Problem 11-10 assuming that the gusset-to-column connection is more rigid than the gusset-to-beam connection.

Student Design Project Problems

11-12. For the student design project, see Figure 1-22:

    a. Perform a lateral analysis of the building assuming that the X-brace locations are partially restrained moment frames where $R = 3.0$.

    b. Design a bolted moment connection to the column assuming a single top and bottom plate. Check the appropriate limit states for the beam and moment plates.

    c. Check the column for concentrated loads on the flange and web. Design stiffeners as required.

11-13. For the shear connections designed for the design project problems in Chapter 9, determine which of the connections require a coped connection and determine whether the beam is adequate for the appropriate limit states for coped beams.

11-14. For the vertical X-brace option in the design project problem, design the gusset plate connections between the diagonal brace, the ground-floor column, and the second-floor beam. Assume a ½-in. setback between the gusset plate and the face of the column.

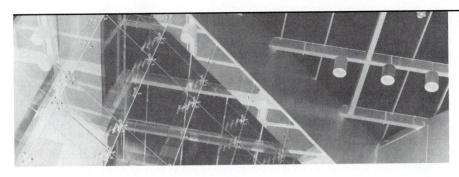

Floor Vibrations

12.1 INTRODUCTION

Steel-framed floors are generally designed to satisfy strength and serviceability requirements, as discussed in previous chapters. However, the topic of floor vibrations, as it relates to serviceability, deserves special attention. The regular activity of human occupancy can be annoying to other occupants, and thus a vibration analysis should be part of the design process. Steel-framed floors have traditionally been designed to satisfy a live load deflection limit of $L/360$, as well as a span-to-depth ratio, L/d, not to exceed 24. These rules of thumb are still in use today and are usually sufficient for shorter spans. Floors framed with longer spans, or floors framed with open-web steel joists, are the most susceptible to vibration problems. In this chapter, we will adopt the vibration analysis procedures presented in reference [1].

There are two basic types of floor vibrations: *steady state* and *transient*. Steady-state vibrations are due to continuous harmonic dynamic forces, such as vibration due to equipment or rotating machinery. These vibrations are best controlled by isolating the equipment from the structure.

Transient vibrations are due to low-impact activities such walking and dancing, and high-impact activities such as aerobics, jumping, concerts, and athletics. This type of vibration decays or fades out eventually due to damping (i.e., friction or viscous forces). Transient vibrations are best controlled by relocating the activity, increasing the damping, increasing the mass of the structure, stiffening the structure, or a combination of these options. In this chapter, we will discuss the basic design process for controlling transient vibrations in a floor structure.

12.2 VIBRATION TERMINOLOGY

Several basic terms that will be referred to later need to be defined:

Damping (β): Damping in a structural system is usually expressed as a percentage of critical damping. Critical damping is that which is required to bring the system to rest in one-half cycle. See Table 12-2 for typical damping values.

Period (*T*): Time, in seconds, for one complete cycle of oscillation.

Frequency (*f*): Number of oscillations per second in hertz (Hz) or cycles per second, $f = 1/T$. A person walking at a pace of two steps per second is said to walk at a frequency of 2 Hz or two cycles per second.

Forcing Frequency (f_f): The frequency, in hertz, of the applied dynamic force.

Harmonic: An integer multiple of a forcing frequency, f_f. Any forcing frequency can have an infinite number of harmonics, but human activities are generally limited to a maximum of three harmonics. For example, for an applied forcing frequency of 2 Hz, the first harmonic is 2 Hz, the second harmonic is 4 Hz, and the third harmonic is 6 Hz.

Natural Frequency (f_n): The frequency at which a structure vibrates when it is displaced and then suddenly released from an at-rest state. This is also called free vibration since no other external forces are applied. The natural frequency of a structure is proportional to its stiffness.

Resonance: A phenomenon where the forcing frequency, f_f (or a harmonic multiple of the forcing frequency), of the dynamic activity coincides with one of the natural frequencies, f_n, of the structure. This causes very large displacements, velocity, acceleration, and stresses. For a person walking at a pace of two steps per second, the floor will have to be checked for the first three harmonics of this forcing frequency (i.e., 2 Hz, 4 Hz, and 6 Hz).

Mode Shape: The deflected shape of a structural system that is subjected to free vibration. Each natural frequency of a structure has a corresponding deflected or mode shape.

Modal Analysis: Analytical method for calculating the natural frequencies, mode shapes, and responses of individual modes of a structure to a given dynamic force. The total structural system response is the sum of all individual mode responses.

12.3 NATURAL FREQUENCY OF FLOOR SYSTEMS

The most critical parameter in a vibration analysis of a floor system is the natural frequency of the floor system. There are several factors that impact the natural frequency of a floor system, so a simplified approach will be used here. The floor system will be assumed to be a concrete slab with or without a metal deck supported by steel beams or joists that are supported by some combination of steel girders, walls, or columns.

The natural frequency of a beam or joist can be estimated as follows, assuming a uniformly loaded, simple-span beam:

$$f_n = \frac{\pi}{2}\sqrt{\frac{gEI}{wL^4}}, \tag{12-1}$$

where

f_n = Natural frequency, Hz,

g = Acceleration due to gravity (386 in./s^2),

E = Modulus of elasticity of steel (29×10^6 psi),

I = Moment of inertia, in.^{4},

$\quad$ = Transformed moment of inertia, I_t, for composite floors,

w = Uniformly distributed load, lb./in. (see discussion below), and

L = Beam, joist, or girder span, in.

Equation (12-1) can be simplified as follows:

$$f_n = 0.18\sqrt{\frac{g}{\Delta}}, \tag{12-2}$$

where

Δ = Maximum deflection at midspan

$\quad = \dfrac{5wL^4}{384EI}.$

For a combined beam and girder system, the natural frequency can be estimated as follows:

$$\frac{1}{f_n^2} = \frac{1}{f_j^2} + \frac{1}{f_g^2}, \tag{12-3}$$

where

f_j = Frequency of the beam or joist, and

f_g = Frequency of the girder.

Combining equations (12-2) and (12-3) yields the natural frequency for the combined system:

$$f_n = 0.18\sqrt{\frac{g}{\Delta_j + \Delta_g}}, \tag{12-4}$$

where

Δ_j = Beam or joist deflection, and

Δ_g = Girder deflection.

The effect of column deformation must be considered for taller buildings with rhythmic activities. For this case, the natural frequency of the floor system becomes

$$f_n = 0.18\sqrt{\frac{g}{\Delta_j + \Delta_g + \Delta_c}}, \tag{12-5}$$

where

Δ_c = Column deformation.

Table 12-1 Recommended sustained live load values

Occupancy	Sustained Live Load
Office floors	11 psf
Residential floors	6 psf
Footbridges, gymnasiums, shopping center floors	0 psf

The uniform load, w, in the above equations represents the actual dead load plus the sustained live load that is present. Table 12-1 lists recommended values for the sustained live load, but the actual value should be used if it is known. The above equations are also based on the assumption that the beams and girders are uniformly loaded, which is typically the case. The exception to this assumption are the girders that support a single beam or joist at midspan, in which case the deflection should be increased by a factor of $4/\pi$.

EXAMPLE 12-1

Frequency of a Floor System

For the floor system shown in Figure 12-1, calculate the natural frequency of the floor system. The dead load of the floor is 40 psf. The moment of inertia of the beam is $I_b = 371$ in.^{4} and the moment of inertia of the girder is $I_g = 500$ in.^{4}. Neglect the column deformation.

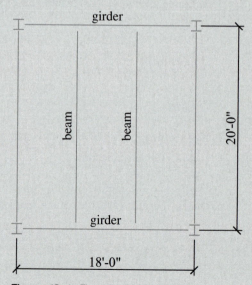

Figure 12-1 Floor framing plan for Example 12-1.

SOLUTION

The free-body diagram of the beam is shown in Figure 12-2.

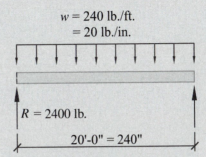

$w = 240$ lb./ft.
$= 20$ lb./in.

$R = 2400$ lb.

$20'\text{-}0'' = 240''$

Figure 12-2 Free-body diagram of the beam.

where

$$w = (40 \text{ psf})(6 \text{ ft.}) = 240 \text{ lb./ft.}(20 \text{ lb./in.})$$

$$R = (240 \text{ lb./ft.})\left(\frac{20 \text{ ft.}}{2}\right) = 2400 \text{ lb.}$$

The beam deflection is

$$\Delta = \frac{5wL^4}{384EI}$$

$$= \frac{(5)(20)(240)^4}{(384)(29 \times 10^6)(371)} = 0.0803 \text{ in.}$$

The natural frequency of the beam is

$$f_b = 0.18\sqrt{\frac{g}{\Delta}}$$

$$= 0.18\sqrt{\frac{386}{0.0803}} = 12.48 \text{ Hz.}$$

Using the beam reactions calculated previously, the free-body diagram of the girder is as shown in Figure 12-3.

From *AISCM*, Table 3-23, the maximum deflection in the girder is

$$\Delta_g = \frac{Pa}{24EI}(3L^2 - 4a^2)$$

$$= \frac{(2400)(72)}{(24)(29 \times 10^6)(500)}[(3)(216)^2 - (4)(72)^2] = 0.0592 \text{ in.}$$

(continued)

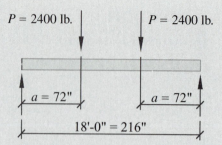

Figure 12-3 Free-body diagram of the girder.

The natural frequency of the girder is

$$f_g = 0.18 \sqrt{\frac{g}{\Delta}}$$

$$= 0.18 \sqrt{\frac{386}{0.0592}} = 14.53 \text{ Hz.}$$

The natural frequency of the combined system is

$$\frac{1}{f_n{}^2} = \frac{1}{f_b{}^2} + \frac{1}{f_g{}^2}$$

$$= \frac{1}{(12.48)^2} + \frac{1}{(14.53)^2}$$

$$f_n = 9.47 \text{ Hz.}$$

Alternatively, the natural frequency can be calculated as follows:

$$f_n = 0.18 \sqrt{\frac{g}{\Delta_b + \Delta_g}}$$

$$= 0.18 \sqrt{\frac{386}{0.0803 + 0.0592}} = 9.47 \text{ Hz.}$$

It should be noted that the above calculation illustrates the application of the natural frequency equations to an isolated floor. The effects of continuity and other factors will be considered later.

12.4 FLOOR SYSTEMS WITH OPEN-WEB STEEL JOISTS

For wide-flange beams with solid webs, shear deformation is usually small enough to be neglected. For open-web steel joists, shear deformation occurs due to the eccentricity at the joints, which occurs as a result of the fabrication process (see Figure 12-4). For this reason, shear deformation must be considered for open web steel joists for serviceability.

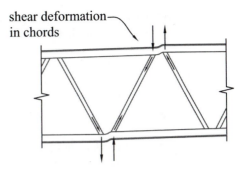

Figure 12-4 Shear deformation in an open-web steel joist.

For the purposes of vibration analysis, the effective moment of inertia of a simply supported joist is calculated as follows:

$$I_{jeff} = \frac{I_{comp}}{1 + \dfrac{0.15I_{comp}}{I_{chords}}}, \tag{12-6}$$

where

I_{jeff} = Effective moment of inertia of the joist, in.[4] (accounts for shear deformation),

I_{comp} = Composite moment of the joist, in.[4], and

I_{chords} = Moment of inertia of the joist chords, in.[4].

Equation (12-6) is only valid for span-to-depth ratios greater than or equal to 12.

For the case when open-web steel joists are supported by wide flange girders or joist girders, the girders do not act as a fully composite section because they are physically separated from the floor slab by the joist seats (see Figure 12-5).

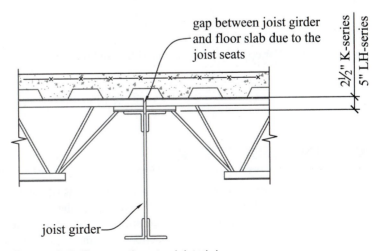

Figure 12-5 Floor section at a joist girder.

For the purposes of vibration analysis, the effective moment of inertia of the girders supporting open-web steel joists is calculated as follows:

$$I_{geff} = I_{nc} + \frac{(I_c - I_{nc})}{4},$$ (12-7)

where

I_{geff} = Effective moment of inertia of the girder, in.⁴ (accounts for shear deformation),
I_{nc} = Noncomposite moment of inertia, in.⁴, and
I_c = Composite moment of inertia, in.⁴.

The effective moment of inertia of the joists is modified as follows for this case:

$$I_{jeff} = \frac{1}{\dfrac{\gamma}{I_{chords}} + \dfrac{1}{I_{comp}}},$$ (12-8)

where

I_{jeff} = Effective moment of inertia of the joist, in.⁴,
I_{comp} = Composite moment of the joist, in.⁴,
I_{chords} = Moment of inertia of the joist chords, in.⁴, and

$$\gamma = \frac{1}{C_r} - 1,$$ (12-9)

where C_r is a parameter that is determined as follows:

For open-web steel joists with single- or double-angle web members,

$$C_r = 0.9\left(1 - e^{-0.28\left(\frac{L_j}{D}\right)}\right)^{2.8} \text{ for } 6 \leq \frac{L_j}{D} \leq 24.$$ (12-10)

For open-web steel joists with round web members,

$$C_r = 0.721 + 0.00725\left(\frac{L_j}{D}\right) \text{ for } 10 \leq \frac{L_j}{D} \leq 24.$$ (12-11)

12.5 WALKING VIBRATIONS

Most common structures support live loads where the dynamic force is in the form of walking. This motion could be annoying to other stationary occupants if the structure in question does not have adequate stiffness or damping. The perception of the floor vibrations can be subjective and is a function of the individual sensitivities of the building occupants. For example, individuals with a hearing impairment might have a higher degree of sensitivity to floor vibrations. Some common occupancy types that should be analyzed for walking vibrations are office, residential, and retail space; footbridges; and hospitals. It should be noted that hospitals typically have sensitive equipment, which is subject to more stringent vibration criteria and will be covered later in this chapter.

For a floor to satisfy the criteria for walking vibrations, the peak floor acceleration due to dynamic forces from walking should not exceed a specified limit, as shown in the following equation:

$$\frac{a_p}{g} = \frac{P_o e^{(-0.35 f_n)}}{\beta W} \leq \frac{a_o}{g}, \tag{12-12}$$

where

$\dfrac{a_p}{g}$ = Peak floor acceleration as a fraction of gravity,

$\dfrac{a_o}{g}$ = Human acceleration limit as a fraction of gravity (see Table 12-2),

P_o = Constant that represents the magnitude of the walking force (see Table 12-2),

f_n = Natural frequency of the floor system,

β = Modal damping ratio (see Table 12-2), and

W = Weighted average mass of the floor system (see equation (12-13)).

The weight of the floor system, W, is equal to the total weight of the floor system for simply supported floor systems. For other structures, the weight is a function of the beam or joist panel combined with the girder panel. The weight of the floor system is as follows:

$$W = \left(\frac{\Delta_j}{\Delta_j + \Delta_g}\right) W_b + \left(\frac{\Delta_g}{\Delta_j + \Delta_g}\right) W_g, \tag{12-13}$$

where

Δ_j = Midspan deflection of the beam or joist

$\quad = \dfrac{5wL_j{}^4}{384EI_j}$, and

Table 12-2 Recommended vibration values for variables in equation (12-12)

Occupancy	Constant Force, P_o	Damping Ratio, β	Acceleration Limit, $\dfrac{a_o}{g} \times 100\%$
Offices, residences, churches	65 lb.	0.02–0.05[1]	0.5%
Shopping malls	65 lb.	0.02	1.5%
Footbridges—Indoor	92 lb.	0.01	1.5%
Footbridges—Outdoor	92 lb.	0.01	5.0%

[1]β = 0.02 for floors with few nonstructural components (e.g., paperless or electronic office)
$\quad$ = 0.03 for floors with nonstructural components and furnishings and small demountable partitions
$\quad$ = 0.05 for fixed full-height partitions between floors
Adapted from Table 4.1, reference 1.

Δ_g = Midspan deflection of the girder

$$= \frac{5wL_g^{\,4}}{384EI_g}.$$

Note that the uniformly distributed load, w, is the dead load plus the sustained live load, which should not be confused with the occupancy live load used in the design of the floor system for strength.

The weighted average mass of the beam panel, W_b (or W_j for a joist panel), is as follows:

$$W_j = \gamma w_t B_j L_j, \text{ lb.,} \tag{12-14}$$

where

γ = 1.0 for all joists or beams, except

    = 1.5 for rolled steel beams that are connected to girders at both ends with simple shear connectors (e.g., clip angles and shear plates) and the adjacent beam span is greater than 70% of the span of the beam being considered,

w_t = Dead plus sustained live load per unit area, psf,

L_j = Length of the beam or joist,

B_j = Effective width of the joist panel

$$= C_j \left(\frac{D_s}{D_j} \right)^{0.25} L_j \leq \tfrac{2}{3} \text{ floor width (see Figure 12-7),} \tag{12-15}$$

C_j = 2.0 for most beams and joists

    = 1.0 for beams or joists parallel to an interior edge,

D_s = Transformed moment of inertia of slab per unit width

$$= \frac{d_e^{\,3}}{n}, \text{ in.}^4/\text{ft.,} \tag{12-16}$$

d_e = Average depth of concrete slab on metal deck (see Figure 12-6)

$$= t_c + \frac{h_r}{2} \tag{12-17}$$

where t_c = Concrete slab thickness above the deck ribs,

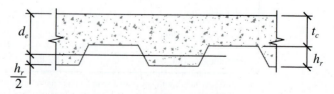

Figure 12-6 Floor slab section.

n = Dynamic modular ratio

$$= \frac{E_s}{1.35E_c},$$ (12-18)

E_s = Modulus of elasticity of steel
     = 29,000,000 psi,

E_c = Modulus of elasticity of concrete

$$= 33w_c^{1.5}\sqrt{f'_c}, \text{ psi,}$$ (12-19)

w_c = Unit weight of the concrete, lb./ft.^{3},

f'_c = 28-day compressive strength of the concrete, psi,

D_j = Transformed moment of inertia of beam or joist per unit width

$$= \frac{I_j}{S}, \text{ in.}^4/\text{ft.,}$$ (12-20)

I_j = Transformed moment of inertia of the beam or joist, and

S = Beam or joist spacing, ft.

The weighted average mass of the girder panel, W_g, is as follows:

$$W_g = w_t B_g L_g, \text{ lb.,}$$ (12-21)

where

B_g = Effective width of the girder panel

$$= C_g\left(\frac{D_j}{D_g}\right)^{0.25} L_g \leq \text{⅔ floor length (see Figure 12-7)}$$ (12-22)

C_g = 1.8 for girders supporting rolled steel beams connected to the web

   = 1.8 for girders supporting joists with extended chords

   = 1.6 for girders supporting joists without extended chords (joist seats on the girder flange)

D_g = Transformed moment of inertia of girder per unit width

$$= \frac{I_g}{L_j} \text{ for all but edge girders}$$ (12-23)

$$= \frac{2I_g}{L_j} \text{ for edge girders}$$ (12-24)

L_g = Girder span

With reference to Figure 12-7, the plan aspect ratio of the floor needs to be checked. If the girder span is more than twice the beam or joist span, $L_g > (2)(L_j \text{ or } L_b)$, then the beam or joist panel mode, as well as the combined mode, should be checked separately. It should be

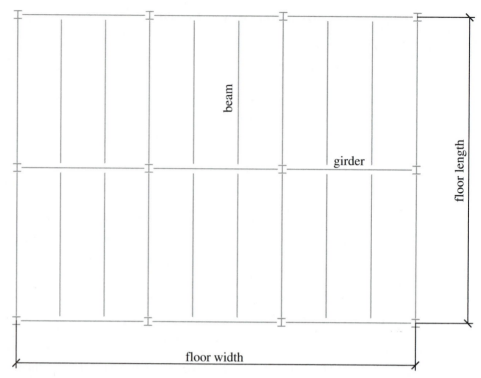

Figure 12-7 Floor width and length.

noted that this is not likely to be the case since it is generally more economical to have beam or joist spans that are longer than the girder span in a typical steel-framed floor.

When the girder span is less than the beam or joist panel width, $L_g < B_j$, the combined mode exhibits greater stiffness. To account for this, the girder deflection, Δ_g, used in equation (12-13) is modified as follows:

$$\Delta'_g = \frac{L_g}{B_j}\Delta_g \geq 0.5\Delta_g. \tag{12-25}$$

Damping, β, in a floor system is a function of the amount of nonstructural components present, such as walls, furniture, and occupants (for rhythmic vibrations, see Section 12.6). Table 12-2 lists recommended damping values. A value of $\beta = 0.02$ is used for floors with large, open areas with very few nonstructural components, such as a mall or an open office. A value of $\beta = 0.03$ is used for floors with some nonstructural components, such as an office area with demountable partitions (i.e., cubicles). For floors with full-height partitions (i.e., the partitions span from floor to floor), $\beta = 0.05$.

For floor systems with a natural frequency, f_n, greater than 9 Hz, the floor stiffness also needs to be checked to ensure that the floor has a stiffness greater than 5700 lb./in. The floor stiffness is calculated as follows:

$$K_s = \frac{1}{\Delta_p} \geq 5700 \text{ lb./in.,} \tag{12-26}$$

where

K_s = Floor stiffness,

Δ_p = Total floor deflection under a unit concentrated load

$$= \Delta_{jp} + \frac{\Delta_{gP}}{2}, \tag{12-27}$$

Δ_{gp} = Deflection of the more flexible girder under a unit concentrated load at midspan

$$= \frac{L_g^{\,3}}{48EI_g}, \tag{12-28}$$

Δ_{jp} = Joist panel deflection under a unit concentrated load

$$= \frac{\Delta_{oj}}{N_{eff}}, \tag{12-29}$$

Δ_{oj} = Deflection of the joist or beam under a unit concentrated load at midspan

$$= \frac{L^3}{48EI_j}, \text{ and} \tag{12-30}$$

N_{eff} = Number of effective beams or joists

$$= 0.49 + 34.2\frac{d_e}{S} + (9.0 \times 10^{-9})\frac{L_j^{\,4}}{I_t} - 0.00059\left(\frac{L_j}{S}\right)^2 \geq 1.0. \tag{12-31}$$

Equation (12-31) is only valid when the following conditions are met:

$$0.018 \leq \frac{d_e}{S} \leq 0.208, \tag{12-32}$$

$$4.5 \times 10^6 \leq \frac{L_j^{\,4}}{I_t} \leq 257 \times 10^6, \text{ and} \tag{12-33}$$

$$2 \leq \frac{L_j}{S} \leq 30. \tag{12-34}$$

The joist and girder deflections under a concentrated load at midspan, given in equations (12-28) and (12-30), assume a simply supported condition, hence the 1/48 coefficient. To account for the rotational restraint provided by typical beam connections, this coefficient may be reduced to 1/96. The 1/96 term is the average of the 1/48 coefficient for a simply supported beam and the 1/192 coefficient for a beam with fixed ends (see diagrams 7 and 16 in Figure 3-23 of the *AISCM*). This reduction would not apply to open web steel joist connections.

12.6 ANALYSIS PROCEDURE FOR WALKING VIBRATIONS

The analysis procedure for walking vibrations is outlined as follows:

1. Determine the effective slab width, b_{eff}, and use this to calculate the composite moment of inertia of the joists or beams and girders. Use the composite moment of inertia unless the upper flange of the beam, joist, or girder is separated from the concrete slab, or where the deck frames into the web of the beam or girder, in which case, use the noncomposite moment of inertia is used. (see Example 12-2).

$$b_{eff} \leq 0.4 \times \text{Length of the member}$$
$$\leq \text{Tributary width of the member}$$

Note: For footbridges that are supported by beams only, with no girders, calculate only the beam properties and the beam panel mode.

2. Calculate the dead loads and sustained live loads supported by the joist, beam, or girder.
3. Calculate the joist or beam and girder deflections and the natural frequency of the floor using equation (12-4). Ensure that the natural frequency of the floor is greater than 3 Hz; otherwise, stiffen the joist or beams and girders.
4. Calculate the effective weight, W, using equation (12-13).
5. Obtain the constant, P_o, and the damping ratio, β, from Table 12-2 and use equation (12-12) to calculate the acceleration of the floor system, a_p/g.
6. Ensure that the acceleration of the floor system, a_p/g, is less than or equal to the human acceleration limit, a_p/g, given in Table 12-2; otherwise, increase the floor stiffness and/or the floor dead load, and/or increase the floor damping (see Section 12.9 for remedial measures).

EXAMPLE 12-2

Walking Vibrations with Wide-flange Beam Framing

The floor framing system shown in Figure 12-8 is to be used in an office building. The building has light, demountable partitions about 5 ft. in height. Check if the floor is adequate for walking vibrations. Assume normal weight concrete with a density of 145 pcf and a 28-day compressive strength of 4000 psi. The floor loads are as follows:

Floor dead load = 55 psf (includes weight of slab, deck, finishes, mechanical and electrical fixtures, and partitions)
Sustained live load = 11 psf (see Table 12-1)

SOLUTION

Beam Section Properties and Deflection
From Part 1 of the *AISCM* for a W16 × 26:

$d = 15.7$ in.
$A = 7.68$ in.^{2}
$I = 301$ in.^{4}

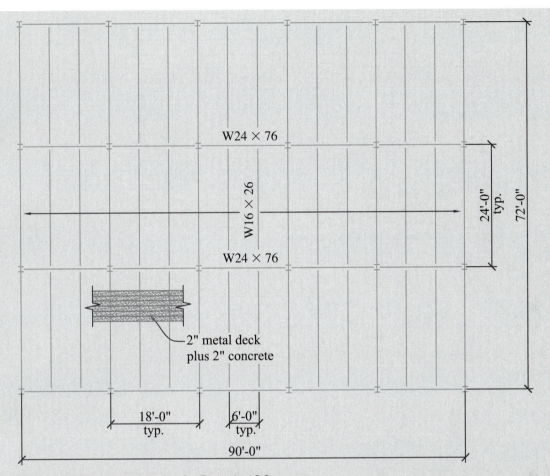

Figure 12-8 Floor framing plan for Example 12-2.

Effective slab width, b_{eff} $\leq 0.4 \times$ Beam span $= (0.4)(24 \text{ ft.}) = 9.6 \text{ ft.}$

$\leq$ Beam tributary width $= 6 \text{ ft.} = \textbf{72 in.} \rightarrow \textbf{Governs}$

t_c = Concrete slab thickness above deck ribs = 2 in.

h_r = Depth of metal deck ribs = 2 in.

Note: Only concrete above the deck ribs is used to compute the I_{comp} for the beam

$E_s = 29,000 \text{ ksi}$

$E_c = 33w_c^{1.5}\sqrt{f'_c}$

$\quad = (33)(145)^{1.5}\sqrt{4000} = 3644 \text{ ksi}$

$n = \dfrac{E_s}{1.35E_c}$

$\quad = \dfrac{29,000}{(1.35)(3644)} = 5.9$

(continued)

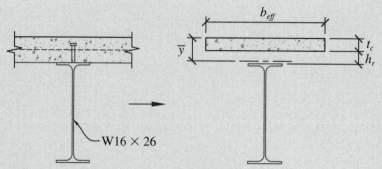

Figure 12-9 Section through composite beam.

b_e = Transformed effective width of the concrete slab

$$= \frac{b_{eff}}{n}$$

$$= \frac{72 \text{ in.}}{5.9} = 12.2 \text{ in.}$$

The centroid (see Figure 12-9) is found by summing the moments of areas about the top of the slab. From statics,

$$\bar{y} = \frac{\Sigma Ay}{\Sigma A}$$

$$= \frac{\left[(2 \text{ in.})(12.2 \text{ in.})\left(\dfrac{2 \text{ in.}}{2}\right)\right] + \left[(7.68 \text{ in.}^2)\left(\dfrac{15.7 \text{ in.}}{2} + 4 \text{ in.}\right)\right]}{[(2 \text{ in.})(12.2 \text{ in.})] + 7.68 \text{ in.}^2} = 3.6 \text{ in.}$$

The composite moment of inertia is

$$I_{comp} = \Sigma I + Ad^2$$

$$= (12.2 \text{ in.})\left(\frac{(2)^3}{12}\right) + (12.2 \text{ in.})(2 \text{ in.})\left(3.6 \text{ in.} - \frac{2 \text{ in.}}{2}\right)^2$$

$$+ 301 \text{ in.}^4 + (7.68 \text{ in.}^2)\left(\frac{15.7 \text{ in.}}{2} + 4 \text{ in.} - 3.6 \text{ in.}\right)^2$$

$$= 996 \text{ in.}^4$$

$$w = (55 \text{ psf} + 11 \text{ psf})(6 \text{ ft.}) = 396 \text{ lb./ft.}$$

$$= 33 \text{ lb./in.}$$

$$\Delta_j = \frac{5wL_j^4}{384EI_j}$$

$$= \frac{(5)(33)(24 \text{ ft.} \times 12)^4}{(384)(29 \times 10^6)(996)} = 0.102 \text{ in.}$$

Girder Section Properties and Deflection

From Part 1 of the *AISCM* for a W24 × 76:

$d = 23.9$ in.

$A = 22.4$ in.^{2}

$I = 2100$ in.^{4}

Effective slab width, $b_{eff} \leq 0.4 \times$ Girder span $= 0.4 \times 18$ ft. $= 7.2$ ft. $= \textbf{86.4 in.} \rightarrow \textbf{Governs}$
$\leq$ Girder tributary width $= 24$ ft.

$t_c =$ Concrete slab thickness above deck ribs $= 2$ in.

$h_r =$ Depth of metal deck ribs $= 2$ in.

$$d_e = t_c + \frac{1}{2}h_r$$

$$= (2 \text{ in.}) + \frac{1}{2}(2 \text{ in.}) = 3 \text{ in.}$$

Note: Concrete within and above the deck ribs is used to compute the I_{comp} for the girder.

$b_e =$ Transformed effective width of the slab

$$= \frac{b_{eff}}{n}$$

$$= \frac{86.4 \text{ in.}}{5.9} = 14.6 \text{ in.}$$

The centroid (see Figure 12-10) is found by taking a summation of the moments of the areas about the top of the slab. From statics,

$$\bar{y} = \frac{\Sigma Ay}{\Sigma A}$$

$$= \frac{\left[(3 \text{ in.})(14.6 \text{ in.})\left(\dfrac{3 \text{ in.}}{2}\right)\right] + \left[(22.4 \text{ in.}^2)\left(\dfrac{23.9 \text{ in.}}{2} + 4 \text{ in.}\right)\right]}{[(3 \text{ in.})(14.6 \text{ in.})] + 22.4 \text{ in.}^2} = 6.4 \text{ in.}$$

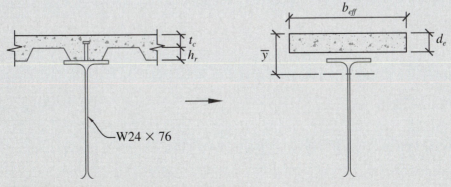

Figure 12-10 Section through composite girder.

(continued)

The composite moment of inertia is

$$I_{comp} = \Sigma I + Ad^2$$

$$= (14.6 \text{ in.})\left(\frac{(3)^3}{12}\right) + (14.6 \text{ in.})(3 \text{ in.})\left(6.4 \text{ in.} - \frac{3 \text{ in.}}{2}\right)^2$$

$$+ 2100 \text{ in.}^4 + (22.4 \text{ in.}^2)\left(\frac{23.9 \text{ in.}}{2} + 4 \text{ in.} - 6.4 \text{ in.}\right)^2$$

$$= 5230 \text{ in.}^4$$

$$w = (55 \text{ psf} + 11 \text{ psf})(24 \text{ ft.}) = 1584 \text{ plf}$$

$$= 132 \text{ lb./in.}$$

$$\Delta_g = \frac{5wL_g{}^4}{384EI_g}$$

$$= \frac{(5)(132)(18 \text{ ft.} \times 12)^4}{(384)(29 \times 10^6)(5230)} = 0.025 \text{ in.}$$

Beam effective weight, W_j:

$$W_j = \gamma w_t B_j L_j$$

The beams are likely to be connected to girders at both ends with simple shear connectors (e.g., clip angles and shear plates) and the adjacent beam span is greater than 70% of the span of the beam being considered (both spans are equal). Therefore, $\gamma = 1.5$.

$$D_s = \frac{d_e{}^3}{n}$$

$$= \frac{(3 \text{ in.})^3}{5.9} = 4.6 \text{ in.}^4/\text{ft.}$$

$$D_j = \frac{I_j}{S}$$

$$= \frac{996}{6 \text{ ft.}} = 166 \text{ in.}^4/\text{ft.}$$

$C_j = 2.0$ (interior beam is being considered)

$$B_j = C_j\left(\frac{D_s}{D_j}\right)^{0.25} L_j \leq \tfrac{2}{3} \text{ floor width}$$

$$= 2.0\left(\frac{4.6}{166}\right)^{0.25} (24 \text{ ft.}) \leq (\tfrac{2}{3})(90 \text{ ft.})$$

$$= 19.6 \text{ ft.} < 60 \text{ ft.}$$

$$B_j = 19.6 \text{ ft.}$$

$$W_j = (1.5)(55 \text{ psf} + 11 \text{ psf})(19.6 \text{ ft.})(24 \text{ ft.}) = \mathbf{46{,}570 \text{ lb.}}$$

Girder effective weight, W_g:

$$W_g = w_t B_g L_g$$

Since the span of the girder is less than the width of the beam panel ($L_g = 18$ ft. $< B_j = 19.6$ ft.), the girder deflection has to be modified using equation (12-25):

$$\Delta_g' = \frac{L_g}{B_j}\Delta_g \geq 0.5\Delta_g$$

$$= \left(\frac{18 \text{ ft.}}{19.6 \text{ ft.}}\right)(0.025 \text{ in.}) \geq (0.5)(0.025 \text{ in.})$$

$$= 0.023 \text{ in.} > 0.013 \text{ in.}$$

$$\Delta_g' = 0.023 \text{ in.}$$

$$D_g = \frac{I_g}{L_j}$$

$$= \frac{5230}{24 \text{ ft.}} = 218 \text{ in.}^4/\text{ft.}$$

$C_g = 1.8$ (girders supporting rolled steel beams)

$$B_g = C_g\left(\frac{D_j}{D_g}\right)^{0.25} L_g \leq \tfrac{2}{3} \text{ floor length}$$

$$= 1.8\left(\frac{166}{218}\right)^{0.25} 18 \text{ ft.} \leq (\tfrac{2}{3})(72 \text{ ft.})$$

$$= 30.3 \text{ ft.} < 48 \text{ ft.}$$

$$B_g = 30.3 \text{ ft.}$$

$$W_g = (55 \text{ psf} + 11 \text{ psf})(30.3 \text{ ft.})(18 \text{ ft.}) = \textbf{36,000 lb.}$$

Effective weight of the floor system, W:
From equation (12-13),

$$W = \left(\frac{\Delta_j}{\Delta_j + \Delta_g}\right)W_b + \left(\frac{\Delta_g}{\Delta_j + \Delta_g}\right)W_g$$

$$= \left(\frac{0.102 \text{ in.}}{0.102 \text{ in.} + 0.023 \text{ in.}}\right)(46,570 \text{ lb.}) + \left(\frac{0.023 \text{ in.}}{0.102 \text{ in.} + 0.023 \text{ in.}}\right)(36,000 \text{ lb.})$$

$$= 44,625 \text{ lb.}$$

The natural frequency of the floor system is determined from equation (12-4):

$$f_n = 0.18\sqrt{\frac{g}{\Delta_j + \Delta_g}}$$

$$= 0.18\sqrt{\frac{386}{0.102 \text{ in.} + 0.023 \text{ in.}}} = 10.0 \text{ Hz} (> 3 \text{ Hz, initially OK}).$$

From Table 12-2,

$P_o = 65$ lb., and

$\beta = 0.03$ (light, demountable partitions, less than 5 ft. in height).

(continued)

The acceleration of the floor system due to walking vibrations is calculated using equation (12-12):

$$\frac{a_p}{g} = \frac{P_o e^{(-0.35f_n)}}{\beta W} \leq \frac{a_o}{g}$$

$$= \frac{65e^{(-0.35)(10.0)}}{(0.03)(44,625)} = 0.00146, \text{ or } \mathbf{0.146\%}.$$

From Table 12-2, the recommended acceleration limit is $a_o/g = 0.5\%$, so this floor is adequate for walking vibrations. Since the natural frequency of the floor is greater than 9 Hz, the floor stiffness is required to be greater than 5.7 kips/in. (see equation (12-26)). This condition will now be checked.

The number of effective floor beams is determined from equation (12-31), but we must first determine whether the aforementioned limitations are met (see eqs. (12-32) through (12-34)):

$$0.018 \leq \frac{d_e}{S} \leq 0.208$$

$$0.018 \leq \frac{3 \text{ in.}}{72 \text{ in.}} = 0.042 \leq 0.208 \text{ OK}$$

$$4.5 \times 10^6 \leq \frac{L_j^4}{I_j} \leq 257 \times 10^6$$

$$4.5 \times 10^6 \leq \frac{(24 \times 12)^4}{996} = 6.9 \times 10^6 \leq 257 \times 10^6 \text{ OK}$$

$$2 \leq \frac{L_j}{S} \leq 30$$

$$2 \leq \frac{24 \text{ ft.}}{6 \text{ ft.}} = 4 \leq 30 \text{ OK}$$

If the above parameters calculated from equations (12-32) through (12-34) were not satisfied, the variables in these equations would have to be changed until the conditions were met.

The number of effective beams is

$$N_{eff} = 0.49 + 34.2\frac{d_e}{S} + (9.0 \times 10^{-9})\frac{L_j^4}{I_j} - 0.00059\left(\frac{L_j}{S}\right)^2 \geq 1.0$$

$$= 0.49 + 34.2\left(\frac{3 \text{ in.}}{72 \text{ in.}}\right) + (9.0 \times 10^{-9})\frac{(24 \times 12)^4}{996} - 0.00059\left(\frac{24}{6}\right)^2 \geq 1.0$$

$$= 1.97.$$

The individual joist or beam deflection under a unit concentrated load is

$$\Delta_{oj} = \frac{L^3}{48EI_j} \text{ (a factor of } {}^{1}\!/_{48} \text{ is conservatively assumed)}$$

$$= \frac{(24 \times 12)^3}{48(29 \times 10^6)(996)} = 17.2 \times 10^{-6} \text{ in./lb. (deflection under a unit load of 1 lb.)}$$

Beam panel deflection under a unit concentrated load:

$$\Delta_{jp} = \frac{\Delta_{oj}}{N_{eff}}$$

$$= \frac{0.0000172}{1.97} = 8.73 \times 10^{-6} \text{ in./lb.}$$

The girder panel deflection under a unit concentrated load is

$$\Delta_{gp} = \frac{L_g{}^3}{48EI_g} \text{ (a factor of } {}^1\!/_{48} \text{ is conservatively assumed)}$$

$$= \frac{(18 \times 12)^3}{48(29 \times 10^6)(5230)} = 1.38 \times 10^{-6} \text{ in./lb. (deflection under a unit load of 1 lb.)}$$

Total floor deflection under a unit concentrated load:

$$\Delta_p = \Delta_{jp} + \frac{\Delta_{gp}}{2}$$

$$= (8.73 \times 10^{-6}) + \frac{1.38 \times 10^{-6}}{2} = 9.42 \times 10^{-6} \text{ in./lb.}$$

The floor stiffness is then

$$K_s = \frac{1}{\Delta_p} \geq 5700 \text{ lb./in.}$$

$$= \frac{1}{9.42 \times 10^{-6}} = 106,000 \text{ lb./in.} \geq 5700 \text{ lb./in. Floor stiffness is OK}$$

EXAMPLE 12-3

Walking Vibrations with Open-web Steel Joist Framing

The floor framing system shown in Figure 12-11 is to be used for an office building. The building has full-height partitions. Determine whether the floor is adequate for walking vibrations. Assume lightweight concrete with a density of 110 pcf and a 28-day compressive strength of 3500 psi. The floor loads are as follows:

Floor dead load = 55 psf (includes weight of slab, deck, finishes, mechanical and electrical fixtures, and partitions)
Sustained live load = 11 psf (see Table 12-1)

SOLUTION

Joist Section Properties and Deflection
For a 24K5 open-web steel joist,

(continued)

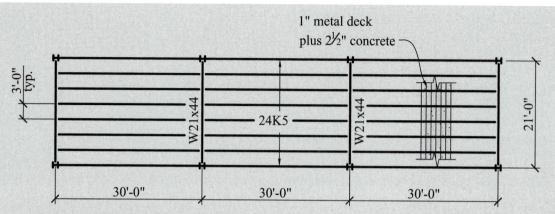

Figure 12-11 Floor framing plan for Example 12-3.

$d = 24$ in.

$A = 2.0$ in.^{2}(1.0 in.^{2} per chord)

$I = 210$ in.^{4} (Note: Exact section properties for the joist would be provided by the joist supplier)

Effective slab width, $b_{eff} \le 0.4 \times$ Beam span $= (0.4)(30 \text{ ft.}) = 12$ ft.

$\qquad\qquad\qquad\qquad \le$ Beam trib width $= 3$ ft. $= 36$ in. $\rightarrow$ Governs

$t_c =$ Concrete slab thickness above deck ribs $= 2.5$ in.

$h_r =$ Depth of metal deck ribs $= 1$ in.

Note: Only concrete above the deck ribs is used to compute the I_{comp} for the joist (see Figure 12-12)

$E_s = 29{,}000$ ksi

$E_c = 33w_c^{1.5}\sqrt{f'_c}$

$\quad = (33)(110)^{1.5}\sqrt{3500} = 2252$ ksi

$n = \dfrac{E_s}{1.35E_c}$

$\quad = \dfrac{29{,}000}{(1.35)(2252)} = 9.5$

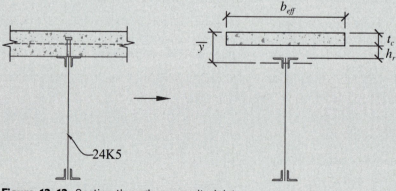

Figure 12-12 Section through composite joist.

b_e = Transformed effective width of the slab

$$= \frac{b_{eff}}{n}$$

$$= \frac{36 \text{ in.}}{9.5} = 3.8 \text{ in.}$$

The centroid (see Figure 12-9) is found by summing the moments of areas about the top of the slab. From statics,

$$\bar{y} = \frac{\Sigma Ay}{\Sigma A}$$

$$= \frac{\left[(2.5 \text{ in.})(3.8 \text{ in.})\left(\frac{2.5 \text{ in.}}{2} \right) \right] + \left[(2.0 \text{ in.}^2)\left(\frac{24 \text{ in.}}{2} + 3.5 \text{ in.} \right) \right]}{[(2.5 \text{ in.})(3.8 \text{ in.})] + 2.0 \text{ in.}^2} = 3.73 \text{ in.}$$

The composite moment of inertia is

$$I_{comp} = \Sigma(I + Ad^2)$$

$$= \left[(3.8 \text{ in.})\left(\frac{(2.5)^3}{12} \right) \right] + \left[(3.8 \text{ in.})(2.5 \text{ in.})\left(3.73 \text{ in.} - \frac{2.5 \text{ in.}}{2} \right)^2 \right]$$

$$+ 210 \text{ in.}^4 + \left[(2.0 \text{ in.}^2)\left(\frac{24 \text{ in.}}{2} + 3.5 \text{ in.} - 3.73 \text{ in.} \right)^2 \right]$$

$$= 550 \text{ in.}^4.$$

Since $6 < L_j/D = (30 \times 12)/24 \text{ in.} = 15 < 24$, equation (12-10) is used to calculate C_r:

$$C_r = 0.9\left(1 - e^{-0.28\left(\frac{L_j}{D} \right)} \right)^{2.8}$$

$$= 0.9\left(1 - e^{-0.28\left(\frac{30 \times 12}{24} \right)} \right)^{2.8} = 0.862$$

$$\gamma = \frac{1}{C_r} - 1$$

$$= \frac{1}{0.862} - 1 = 0.159$$

The effective moment of inertia of the joists is modified as follows for this case:

$$I_{jeff} = \frac{1}{\dfrac{\gamma}{I_{chords}} + \dfrac{1}{I_{comp}}}$$

$$= \frac{1}{\dfrac{0.159}{210} + \dfrac{1}{550}} = 388 \text{ in.}^4$$

$$w = (55 \text{ psf} + 11 \text{ psf})(3 \text{ ft.}) = 198 \text{ lb./ft.}$$

$$= 16.5 \text{ lb./in.}$$

(continued)

$$\Delta_j = \frac{5wL_j{}^4}{384EI_j}$$

$$= \frac{(5)(16.5)(30 \text{ ft.} \times 12)^4}{(384)(29 \times 10^6)(388)} = 0.321 \text{ in.}$$

Girder Section Properties and Deflection

From Part 1 of the *AISCM* for a W21 × 44,

d = 20.7 in.

A = 13.0 in.^{2}

I = 843 in.^{4}

Effective slab width, $b_{eff} \leq 0.4 \times$ Beam span = 0.4×21 ft. = 8.4 ft. = 100.8 in. → Governs
$\qquad\qquad\qquad\qquad \leq$ Girder tributary width = 30 ft.

t_c = Concrete slab thickness above deck ribs = 2.5 in.

h_r = Depth of metal deck ribs = 1 in.

$$d_e = t_c + \frac{h_r}{2}$$

$$= 2.5 \text{ in.} + \frac{1 \text{ in.}}{2} = 3 \text{ in.}$$

Note: Concrete within and above the deck ribs is used to compute the I_{comp} for the girder.

b_e = Transformed effective width of the slab

$\quad= \dfrac{b_{eff}}{n}$

$\quad= \dfrac{100.8 \text{ in.}}{9.5} = 10.6 \text{ in. (see Figure 12-13)}$

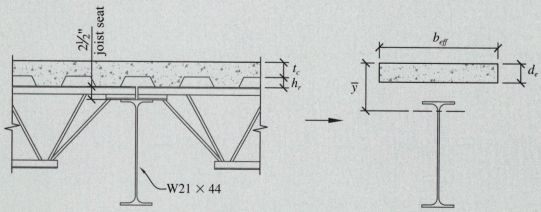

Figure 12-13 Section through composite girder.

The centroid (see Figure 12-13) is found by summing the moments of areas about the top of the slab. From statics,

$$\bar{y} = \frac{\Sigma Ay}{\Sigma A}$$

$$= \frac{\left[(3 \text{ in.})(10.6 \text{ in.})\left(\dfrac{3 \text{ in.}}{2} \right) \right] + \left[(13.0 \text{ in.}^2)\left(\dfrac{20.7 \text{ in.}}{2} + 2.5 \text{ in.} + 3.5 \text{ in.} \right) \right]}{[(3 \text{ in.})(10.6 \text{ in.})] + 13.0 \text{ in.}^2} = 5.78 \text{ in.}$$

The composite moment of inertia is

$$\begin{aligned}
I_{comp} &= \Sigma(I + Ad^2) \\
&= \left[(10.6 \text{ in.})\left(\frac{(3)^3}{12} \right) \right] + \left[(10.6 \text{ in.})(3 \text{ in.})\left(5.78 \text{ in.} - \frac{3 \text{ in.}}{2} \right)^2 \right] \\
&\quad + 843 \text{ in.}^4 + \left[(13.0 \text{ in.}^2)\left(\frac{20.7 \text{ in.}}{2} + 2.5 \text{ in.} + 3.5 \text{ in.} - 5.78 \text{ in.} \right)^2 \right] \\
&= 2901 \text{ in.}^4
\end{aligned}$$

The composite moment of inertia is reduced since the joist seats prevent the slab from acting fully compositely with the girder:

$$\begin{aligned}
I_{geff} &= I_{nc} + \frac{(I_c - I_{nc})}{4} \\
&= 843 + \frac{(2901 - 843)}{4} = 1357 \text{ in.}^4
\end{aligned}$$

Girder deflection:

$$\begin{aligned}
w &= (55 \text{ psf} + 11 \text{ psf})(30 \text{ ft.}) = 1980 \text{ lb./ft.} \\
&= 165 \text{ lb./in.}
\end{aligned}$$

$$\begin{aligned}
\Delta_g &= \frac{5wL_g^{\,4}}{384EI_g} \\
&= \frac{(5)(165)(21 \text{ ft.} \times 12)^4}{(384)(29 \times 10^6)(1356)} = 0.22 \text{ in.}
\end{aligned}$$

Joist effective weight, W_j:

$$W_j = \gamma w_r B_j L_j$$

Since joists are used, $\gamma = 1.0$.

$$\begin{aligned}
D_s &= \frac{d_e^{\,3}}{n} \\
&= \frac{(3 \text{ in.})^3}{9.5} = 2.84 \text{ in.}^4/\text{ft.}
\end{aligned}$$

(continued)

$$D_j = \frac{I_j}{S}$$

$$= \frac{388}{3 \text{ ft.}} = 129 \text{ in.}^4/\text{ft.}$$

$C_j = 2.0$ (interior joist is being considered)

$$B_j = C_j \left(\frac{D_s}{D_j} \right)^{0.25} L_j \leq \tfrac{2}{3} \text{ floor width}$$

$$= (2.0) \left(\frac{2.84}{129} \right)^{0.25} (30 \text{ ft.}) \leq \tfrac{2}{3} (90 \text{ ft.}) = 60 \text{ ft.}$$

$$= 23.1 \text{ ft.}$$

$$W_j = (1.0)(55 \text{ psf} + 11 \text{ psf})(23.1 \text{ ft.})(30 \text{ ft.}) = 45,738 \text{ lb.}$$

Girder effective weight, W_g:

$$W_g = w_t B_g L_g$$

Since the span of the girder is less than the width of the beam panel ($L_g = 21$ ft. $< B_j = 23.1$ ft.), the girder deflection has to be modified using equation (12-25):

$$\Delta_g' = \frac{L_g}{B_j} \Delta_g \geq 0.5 \Delta_g$$

$$= \left(\frac{21 \text{ ft.}}{23.1 \text{ ft.}} \right) (0.22 \text{ in.}) \geq (0.5)(0.22 \text{ in.})$$

$$= 0.20 \text{ in.} > 0.11 \text{ in.}$$

$$\Delta_g' = 0.20 \text{ in.}$$

$$D_g = \frac{I_g}{L_j}$$

$$= \frac{1357}{30 \text{ ft.}} = 45.2 \text{ in.}^4/\text{ft.}$$

$C_g = 1.8$ (girders supporting rolled steel beams)

$$B_g = C_g \left(\frac{D_j}{D_g} \right)^{0.25} L_g \leq \tfrac{2}{3} \text{ floor length} \rightarrow \text{Ignore since floor length is not given}$$

$$= (1.8) \left(\frac{129}{45.2} \right)^{0.25} (21 \text{ ft.})$$

$$= 49.1 \text{ ft.}$$

$$W_g = (55 \text{ psf} + 11 \text{ psf})(49.1 \text{ ft.})(21 \text{ ft.}) = 68,095 \text{ lb.}$$

Effective weight of the floor system, W:

From equation (12-13),

$$W = \left(\frac{\Delta_j}{\Delta_j + \Delta_g}\right)W_b + \left(\frac{\Delta_g}{\Delta_j + \Delta_g}\right)W_g$$

$$= \left(\frac{0.321 \text{ in.}}{0.321 \text{ in.} + 0.20 \text{ in.}}\right)(45{,}738 \text{ lb.}) + \left(\frac{0.20 \text{ in.}}{0.321 \text{ in.} + 0.20 \text{ in.}}\right)(68{,}095 \text{ lb.})$$

$$= 54320 \text{ lb.}$$

The natural frequency of the floor system is determined from equation (12-4):

$$f_n = 0.18\sqrt{\frac{g}{\Delta_j + \Delta_g}}$$

$$= 0.18\sqrt{\frac{386}{0.321 \text{ in.} + 0.20 \text{ in.}}} = 4.90 \text{ Hz. Since} > 3 \text{ Hz, initially OK}$$

From Table 12-2,

$$P_o = 65 \text{ lb., and}$$

$$\beta = 0.05 \text{ (floor with full-height partitions)}$$

The acceleration of the floor system due to walking vibrations is found by using equation (12-12):

$$\frac{a_p}{g} = \frac{P_o e^{(-0.35 f_n)}}{\beta W} \leq \frac{a_o}{g}$$

$$= \frac{(65)(e^{(-0.35)(4.90)})}{(0.05)(54320)} = 0.0043, \text{ or } \mathbf{0.43\%}.$$

From Table 12-2, the recommended acceleration limit is $a_o/g = 0.5\%$, so the floor is adequate for walking vibrations. Note that if partial-height or no partitions were used, then it can be seen by inspection that the floor would not be adequate for walking vibrations ($a_p/g = 0.72\%$ and 1.07%, respectively).

12.7 RHYTHMIC VIBRATION CRITERIA

Rhythmic activities include dancing, jumping exercises, aerobics, concerts, and sporting events. To avoid floor vibration problems during rhythmic activities, the natural frequency of the floor system should be greater than a specified minimum value. For any given activity, there can be several harmonics of vibration. For example, the participants in a lively concert might induce a forcing frequency of 2.0 Hz for the first harmonic and 4.0 Hz for the second harmonic. The floor would have to be checked for both harmonics. For most rhythmic activities, not more than three harmonics would need to be checked.

The minimum required natural frequency of a floor supporting rhythmic activities is as follows:

$$f_{n \text{ required}} \geq f_f \sqrt{1 + \left(\frac{k}{a_o/g}\right)\left(\frac{\alpha_i w_p}{w_t}\right)}, \qquad (12\text{-}35)$$

Table 12-3 Recommended rhythmic activity variables

Activity	Forcing Frequency, f_f, Hz	Weight of Participants, w_p, psf[1]	Dynamic Coefficient, α_i	Dynamic Constant, k
Dancing:				
First Harmonic ($i = 1$)	1.5–3.0	12.5	0.5	1.3
Lively concert, sporting event:		31.0	0.25	
First Harmonic ($i = 1$)	1.5–3.0	31.0	0.05	1.7
Second Harmonic ($i = 2$)	3.0–5.0			
Jumping Exercises:				
First Harmonic ($i = 1$)	2.0–2.75	4.2	1.5	
Second Harmonic ($i = 2$)	4.0–5.5	4.2	0.6	2.0
Third Harmonic ($i = 3$)	6.0–8.25	4.2	0.1	

[1]Based on maximum density of participants on the occupied area of the floor for commonly encountered conditions. For special events, the density of participants can be greater.

Adapted from Table 5.2, reference 1.

where

$f_{n\ required}$ = Minimum required natural frequency of the floor system,

f_f = Forcing frequency (frequency of the rhythmic activity; see Table 12-3),

i = Subscript indicating the harmonic number (see Table 12-3),

k = Activity constant (see Table 12-3),

α_i = Dynamic coefficient (see Table 12-3),

a_o/g = Acceleration limit for rhythmic activity (see Table 12-4),

w_p = Weight of the participants, psf (i.e., sustained live load; see Table 12-3), and

w_t = Dead plus sustained live load per unit area, psf.

If equation (12-35) is not satisfied, then a more accurate analysis can be performed. The peak floor acceleration for each harmonic of the specific rhythmic activity, using the more accurate analysis, is given as

$$\frac{a_{pi}}{g} = \frac{1.3\alpha_i w_p}{w_t \sqrt{\left[\left(\frac{f_n}{f_f}\right)^2 - 1\right]^2 + \left[\frac{2\beta f_n}{f_f}\right]^2}}, \tag{12-36}$$

Table 12-4 Recommended floor acceleration limits for rhythmic activity

Affected Occupancies (Rhythmic activities adjacent to …)	a_o/g
Office, residential	0.4%–0.7%
Dining, weightlifting	1.5%–2.5%
Rhythmic activity	4%–7%

Adapted from Table 5.1, reference 1 and reference 9.

where

$$\frac{a_{pi}}{g} = \text{Peak floor acceleration for each harmonic, and}$$

$$\beta = \text{Damping ratio}$$

$$= 0.06 \text{ for rhythmic activity.}$$

It should be noted here that the damping ratio, β, is higher than the values indicated in Table 12-2 since the participants contribute to the damping of the system.

The effective maximum acceleration, taking all of the harmonics into consideration is given as

$$\frac{a_{pm}}{g} = \left(\Sigma \frac{a_{pi}}{g}^{1.5} \right)^{\frac{2}{3}} \leq \frac{a_o}{g} \tag{12-37}$$

$$= \left[\left(\frac{a_{p1}^{1.5}}{g} \right) + \left(\frac{a_{p2}^{1.5}}{g} \right) + \left(\frac{a_{p3}^{1.5}}{g} \right) \right]^{\frac{2}{3}} \leq \frac{a_o}{g}.$$

For rhythmic vibrations, the natural frequency is calculated from either equation (12-4) or (12-5). The decision as to whether or not to include the effects of axial column deformation is left to the designer. Buildings fewer than five stories generally do not have a significant contribution from column deformation.

EXAMPLE 12-4

Rhythmic Vibrations

Determine the adequacy of the floor framing shown in Figures 12-14 and 12-15 for jumping exercises. The slab properties are as follows:

Dead load $= 120$ psf

$I = 1400$ in.^{4}/ft.

$f'_c = 5000$ psi

$\gamma_c = 145$ pcf

SOLUTION

Slab Deflection:

From Table 12-3, $w_p = 4.2$ psf

$$E_c = 33w_c^{1.5}\sqrt{f'_c}$$
$$= 33(145)^{1.5}\sqrt{5000} = 4074 \text{ ksi}$$

$$w = (120 \text{ psf} + 4.2 \text{ psf}) = 124.2 \text{ lb./ft.}$$
$$= 10.35 \text{ lb./in.}$$

(continued)

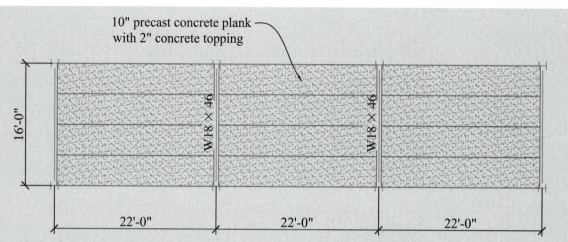

Figure 12-14 Floor framing plan for Example 12-4.

$$\Delta_s = \frac{5wL^4}{384EI}$$

$$= \frac{(5)(10.35)(22 \times 12)^4}{(384)(4,074,000)(1400)} = 0.111 \text{ in.}$$

Girder Deflection:

$$w = (120 \text{ psf} + 4.2 \text{ psf})(22 \text{ ft.}) = 2732 \text{ lb./ft.}$$
$$= 228 \text{ lb./in.}$$

$$\Delta_g = \frac{5wL^4}{384EI}$$

$$= \frac{(5)(228)(16 \times 12)^4}{(384)(29,000,000)(712)} = 0.195 \text{ in.}$$

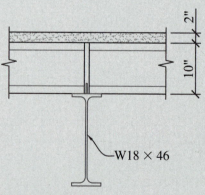

Figure 12-15 Floor section for Example 12-4.

Natural Frequency of the System:

$$f_n = 0.18\sqrt{\frac{g}{\Delta_s + \Delta_g}}$$

$$= 0.18\sqrt{\frac{386}{0.111 + 0.195}} = 6.39 \text{ Hz}$$

Minimum Required Natural Frequency for Each Harmonic (eq. (12-35)):

$$f_{n\,required} \geq f_f\sqrt{1 + \left(\frac{k}{a_o/g}\right)\left(\frac{\alpha_i w_p}{w_t}\right)}$$

First harmonic:

$$2.5\sqrt{1 + \left[\frac{2.0}{0.055}\right]\left[\frac{(1.5)(4.2)}{124.2}\right]} = 4.22 \text{ Hz} < f_n = 6.39 \text{ Hz OK}$$

Second harmonic:

$$5.0\sqrt{1 + \left[\frac{2.0}{0.055}\right]\left[\frac{(0.6)(4.2)}{124.2}\right]} = 6.59 \text{ Hz} > f_n = 6.39 \text{ Hz Not Good}$$

Third harmonic:

$$7.5\sqrt{1 + \left[\frac{2.0}{0.055}\right]\left[\frac{(0.1)(4.2)}{124.2}\right]} = 7.95 \text{ Hz} > f_n = 6.39 \text{ Hz Not Good}$$

The floor system does not meet the criteria given in equation (12-35), so the more accurate analysis needs to be performed:

$$\frac{a_{pi}}{g} = \frac{1.3\alpha_i w_p}{w_t\sqrt{\left[\left(\frac{f_n}{f_f}\right)^2 - 1\right]^2 + \left(\frac{2\beta f_n}{f_f}\right)^2}}.$$

First harmonic:

$$\frac{a_{p1}}{g} = \frac{(1.3)(1.5)(4.2)}{124.2\sqrt{\left[\left(\frac{6.39}{2.5}\right)^2 - 1\right]^2 + \left[\frac{(2)(0.06)(6.39)}{2.5}\right]^2}} = 0.0119$$

Second harmonic:

$$\frac{a_{p2}}{g} = \frac{(1.3)(0.6)(4.2)}{124.2\sqrt{\left[\left(\frac{6.39}{5.0}\right)^2 - 1\right]^2 + \left[\frac{(2)(0.06)(6.39)}{5.0}\right]^2}} = 0.0405$$

Third harmonic:

$$\frac{a_{p3}}{g} = \frac{(1.3)(0.1)(4.2)}{124.2\sqrt{\left[\left(\frac{6.39}{7.5}\right)^2 - 1\right]^2 + \left[\frac{(2)(0.06)(6.39)}{7.5}\right]^2}} = 0.0150$$

(continued)

The effective maximum acceleration, considering all of the harmonics, is

$$\frac{a_{pm}}{g} = \left(\frac{a_{pi}^{1.5}}{g}\right)^{2/3} \leq \frac{a_o}{g}$$

$$= (0.0119^{1.5} + 0.0405^{1.5} + 0.0150^{1.5})^{2/3} = 0.503 < \frac{a_o}{g} = 0.055.$$

From Table 12-4, the recommended acceleration limit is a range between 4% and 7% in this case, the average between these two values (5.5%) is taken as the acceleration limit.

The floor acceleration is less than the specified limit, so this floor is adequate for jumping exercises. If this floor were to be used for jumping exercises adjacent to a dining area, the average acceleration limit, from Table 12-4, would be 2%. Therefore, this floor would not be adequate in that case.

12.8 SENSITIVE EQUIPMENT VIBRATION CRITERIA

In the previous sections, we considered transient floor vibrations in terms of floor acceleration (i.e., a_o/g). For floor structures that support sensitive equipment, the floor motion is expressed in terms of velocity, because the design criteria for equipment often corresponds to a constant velocity over the frequency range considered. To convert from acceleration to velocity, the following expression is used:

$$\frac{a}{g} = \frac{2\pi f V}{g} \tag{12-38}$$

where

V = velocity

Any given floor structure that supports sensitive equipment will likely have several types of equipment in use, so the floor structure is usually designed for the equipment with the most stringent criteria. Table 12-5 lists several types of sensitive equipment, with corresponding vibrational velocity limits, that can be used as a guide for preliminary design. The limits listed are compared with the calculated floor velocities due to walking or footsteps. The required floor structure parameters for any given facility supporting sensitive equipment should always be obtained from the equipment supplier or manufacturer, to ensure that the correct parameters are used and that an economical design is produced.

With reference to Table 12-5, the most sensitive equipment can be found in a Class E microelectronics manufacturing facility, where the maximum allowable vibrational velocity is about 130 μ-in./sec. This would require a floor structure that has a natural frequency close to 50 Hz [2]. By contrast, a typical floor supporting an office occupancy that has approximate bay sizes of 30 ft. by 30 ft. would have a vibrational velocity of about 16,000 μ-in./sec. and a floor frequency between 5 and 8 Hz. Floor structures that support microelectronics manufacturing are often designed with concrete waffle slabs, but a steel design that uses heavy composite slab (3-inch metal deck plus 3 inches of concrete) with W21 beams spaced 8 feet apart in 16 ft.-by-16-ft. column bays also has been found to provide adequate stiffness for this type of facility [2].

The maximum floor velocity is determined as follows:

$$V = 2\pi f_n X_{max}, \tag{12-39}$$

Table 12-5 Vibration criteria for sensitive equipment

Equipment or Use	Vibrational Velocity, V, µ-in./s
Computer systems, operating rooms, bench microscopes up to $100 \times$ magnification	8000
Laboratory robots	4000
Bench microscopes up to $400 \times$ magnification, optical and other precision balances, coordinate measuring machines, metrology laboratories, optical comparators, microelectronics manufacturing equipment—*Class A*	2000
Microsurgery, eye surgery, neurosurgery; bench microscopes greater than $400 \times$ magnification; optical equipment on isolation tables; microelectronics manufacturing equipment—*Class B*	1000
Electron microscopes up to $30,000 \times$ magnification, microtomes, magnetic resonance imagers, microelectronics manufacturing equipment—*Class C*	500
Electron microscopes greater than $30,000 \times$ magnification, mass spectrometers, cell implant equipment, microelectronics manufacturing equipment—Class D	250
Microelectronics manufacturing equipment—Class E, unisolated laser and optical research systems	130

Class A: Inspection, probe test, and other manufacturing support equipment.
Class B: Aligners, steppers, and other critical equipment for photolithography with line widths of 3 microns or more.
Class C: Aligners, steppers, and other critical equipment for photolithography with line widths of 1 micron.
Class D: Aligners, steppers, and other critical equipment for photolithography with line widths of ½ micron, includes electron–beam systems.
Class E: Aligners, steppers, and other critical equipment for photolithography with line widths of ¼ micron, includes electron–beam systems.

Adapted from Table 6.1, reference 1.

where

f_n = Natural frequency of the floor system,

X_{max} = Maximum dynamic displacement

$$= \frac{F_m \Delta_p f_o^2}{2 f_n^2} \text{ for } f_n > 5 \text{ Hz} \tag{12-40}$$

$$= A_m F_m \Delta_p \text{ for } f_n \leq 5 \text{ Hz}, \tag{12-41}$$

F_m = Maximum footstep force (see Table 12-6),

Δ_p = Total floor deflection under a unit concentrated load (see eq. (12-27)),

f_o = Inverse of the footstep pulse rise or decay time (see Table 12-6)

$$= \frac{1}{t_o}, \text{ and}$$

Table 12-6 Values of footfall impulse parameters

Walking Pace (steps per minute)	Dynamic Load Factor (DLF)	F_m = DLF × 185 lb.	$f_o = \dfrac{1}{t_o}$, Hz
100 (fast)	1.7	315	5.0
75 (moderate)	1.5	280	2.5
50 (slow)	1.3	240	1.4

Adapted from Table 6.2, reference 1.

A_m = Maximum dynamic amplitude

$$= 2.0 \text{ for } \frac{f_n}{f_o} \le 0.5.$$

$$= \frac{f_o^{\,2}}{2f_n^{\,2}} \text{ for } \frac{f_n}{f_o} > 0.5. \tag{12-42}$$

The exact value of the maximum dynamic amplitude, A_m, is shown on the solid curve in Figure 12-16. The dashed line is an approximation of the solid curve (see eq. (12-41)). The maximum footstep force, F_m, is the weight of a person (assumed to be 185 lb.) multiplied by a dynamic load factor that is dependent on the walking speed (see Table 12-6).

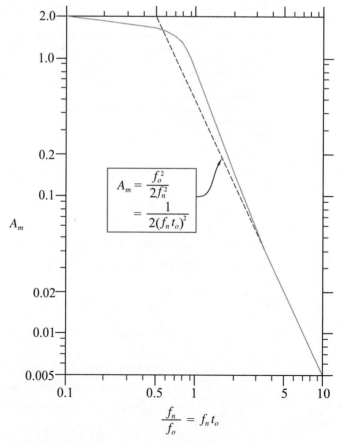

$$A_m = \frac{f_o^2}{2f_n^2}$$
$$= \frac{1}{2(f_n t_o)^2}$$

$$\frac{f_n}{f_o} = f_n t_o$$

Figure 12-16 Maximum dynamic amplitude.
Adapted from Figure 6.5 reference 1

EXAMPLE 12-5

Sensitive Equipment Vibrations for $f_n > 5$ Hz

For the floor system in Example 12-2, investigate the adequacy of the floor to support sensitive equipment.

SOLUTION

From Example 12-2,

$$f_n = 10.0 \text{ Hz, and}$$
$$\Delta_p = 9.42 \times 10^{-6} \text{ in./lb.}$$

From Table 12-6,

$$F_m = 315 \text{ lb.}, f_o = 5.0 \text{ Hz (fast walking)}$$
$$= 280 \text{ lb.}, f_o = 2.5 \text{ Hz (moderate walking)}$$
$$= 240 \text{ lb.}, f_o = 1.4 \text{ Hz (slow walking).}$$

Maximum Dynamic Displacement:

Since $f_n = 10.0 \text{ Hz} > 5.0 \text{ Hz}$, use equation (12-40).

$$X_{max} = \frac{F_m \Delta_p f_o^2}{2 f_n^2}$$

$$= \frac{(315)(9.42 \times 10^{-6})(5.0)^2}{(2)(10.0)^2} = 371 \text{ }\mu\text{-in. (fast walking)}$$

$$= \frac{(280)(9.42 \times 10^{-6})(2.5)^2}{(2)(10.0)^2} = 82.4 \text{ }\mu\text{-in. (moderate walking)}$$

$$= \frac{(240)(9.42 \times 10^{-6})(1.4)^2}{(2)(10.0)^2} = 22.2 \text{ }\mu\text{-in. (slow walking)}$$

The maximum floor velocity is then determined from equation (12-39):

$$V = 2\pi f_n X_{max}$$
$$= (2)(\pi)(10.0)(371) = 23{,}300 \text{ }\mu\text{-in./s (fast walking)}$$
$$= (2)(\pi)(10.0)(82.4) = 5170 \text{ }\mu\text{-in./s (moderate walking)}$$
$$= (2)(\pi)(10.0)(22.2) = 1390 \text{ }\mu\text{-in./s (slow walking)}$$

We see from Table 12-5 that this floor system could not support any of the equipment listed for fast walking. Reducing the criteria to moderate walking would allow the floor to support any of the equipment listed under 8000 μ-in./s and less. Further reducing the criteria to slow walking would allow equipment listed under 2000 μ-in./s and less.

Sensitive equipment located adjacent to a long, straight corridor would have a greater likelihood of being impacted by "fast walking." If "slow walking" is desired in order to reduce the vibrational velocity, then short-length corridors with turns or bends should be used to reduce the walking speed, or the equipment should be relocated away from areas with heavy foot traffic.

EXAMPLE 12-6

Sensitive Equipment Vibrations for $f_n < 5$ Hz

Assuming that the natural frequency of the floor system in Example 12-5 was 3.5 Hz, investigate the adequacy of the floor to support sensitive equipment.

SOLUTION

From Example 12-5,

$$f_n = 3.5 \text{ Hz, and}$$
$$\Delta_p = 9.42 \times 10^{-6} \text{ in./lb.}$$

From Table 12-6,

$$F_m = 315 \text{ lb.}, f_o = 5.0 \text{ Hz (fast walking)}$$
$$= 280 \text{ lb.}, f_o = 2.5 \text{ Hz (moderate walking)}$$
$$= 240 \text{ lb.}, f_o = 1.4 \text{ Hz (slow walking)}$$

Maximum Dynamic Displacement:

Since $f_n = 3.5$ Hz < 5.0 Hz, use equation (12-41).

$$X_{max} = A_m F_m \Delta_p$$
$$\frac{f_n}{f_o} = \frac{3.5}{5.0} = 0.70 \text{ (fast walking)}$$
$$= \frac{3.5}{2.5} = 1.4 \text{ (moderate walking)}$$
$$= \frac{3.5}{1.4} = 2.5 \text{ (slow walking)}$$

The maximum dynamic amplitude, A_m, is determined from Figure 12-16:

$$A_m = 1.4 \text{ (fast walking)}$$
$$= \frac{f_o^2}{2f_n^2} = \frac{2.5^2}{(2)(3.5)^2} = 0.255 \text{ (moderate walking)}$$
$$= \frac{f_o^2}{2f_n^2} = \frac{1.4^2}{(2)(3.5)^2} = 0.08 \text{ (slow walking)}$$

The maximum dynamic displacement is then

$$X_{max} = A_m F_m \Delta_p$$
$$= (1.4)(315)(9.42 \times 10^{-6}) = 4154 \text{ }\mu\text{-in. (fast walking)}$$
$$= (0.255)(280)(9.42 \times 10^{-6}) = 673 \text{ }\mu\text{-in. (moderate walking)}$$
$$= (0.08)(240)(9.42 \times 10^{-6}) = 180 \text{ }\mu\text{-in. (slow walking)}$$

The maximum floor velocity is then determined from equation (12-38):

$$V = 2\pi f_n X_{max}$$
$$= (2)(\pi)(3.5)(4154) = 91{,}350 \ \mu\text{-in./s (fast walking)}$$
$$= (2)(\pi)(3.5)(673) = 14{,}800 \ \mu\text{-in./s (moderate walking)}$$
$$= (2)(\pi)(3.5)(180) = 3960 \ \mu\text{-in./-s (slow walking).}$$

We see from Table 12-5 that this floor system could not support any of the equipment listed for fast or moderate walking. Reducing the criteria to slow walking would allow the floor to support any of the equipment listed at less than 4000 μ-in./s.

12.9 VIBRATION CONTROL MEASURES

For small dynamic forces such as those caused by walking vibrations, the vibration effects can be more effectively controlled by increasing the mass of the structure, increasing the stiffness of the structure, increasing the damping, or a combination of these.

For large dynamic forces such as those caused by aerobics, the vibration effects are most effectively controlled by keeping the natural frequency of any mode of vibration most affected by the dynamic force away from the forcing frequency causing the vibrations. To achieve this, the natural frequency of the structure must be much greater than the forcing frequency of the highest harmonic dynamic force causing the vibration. This can be achieved by stiffening the structure (e.g., by adding beam depth, columns, or posts).

Floors with a natural frequency greater than 10 Hz and a stiffness greater than 5.7 kips/in. do not generally have vibration problems due to human activities. Occupants may experience discomfort in floor systems with natural frequencies in the 5- to 8-Hz range because this frequency range coincides with the natural frequencies of many internal human organs.

Use of a "floating floor" completely separated from the surrounding slabs is effective mostly for controlling vibrations due to aerobics and other rhythmic activities (see Figure 12-17 for examples). The floating floor concept is similar to that used in vibration isolation of equipment. The floor is supported on very soft springs (e.g., neoprene pads) attached to the structural floor. The combined natural frequency of the floating floor and

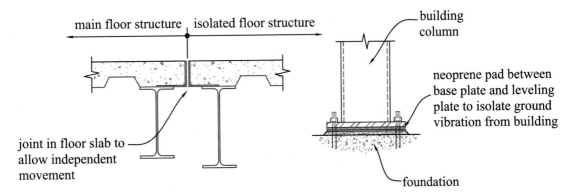

Figure 12-17 Floating or isolated structures.

the springs should be very small—less than 2 to 3 Hz. This can be achieved by using a thick slab (4 in. to 8 in. thick). The space between the floating floor and the structural slab must be properly vented to prevent the change in pressure due to the movement of the floating floor from causing the structural floor to move.

Rhythmic activities, such as aerobics, can be located on the ground floor of a building or an isolated framing system can be used. Weightlifting activity can be accommodated on framed floors; however, one must consider the dynamic effect of weights being dropped on the floor, which is a common occurrence with this activity. This dynamic force can be mitigated by using an appropriate mat to absorb some of the energy of the falling weight, but it is generally advisable to locate this activity on the ground floor of a building.

Floors with a natural frequency of less than 3 Hz should be avoided. Walking speed in an office is usually between 1.25 to 1.5 steps per second (or 1.25 to 1.5 Hz). Floors with a natural frequency of 3 Hz may experience resonance at the second harmonic frequency (i.e., between 2.5 and 3 Hz).

Damping is a critical component of the vibration analysis, but the designer typically has very little control over the amount of damping that is present in a floor system. For example, full-height partitions provide the best form of damping, but are usually specified by someone other than the structural engineer. Furthermore, such partitions may not be present for the life of the structure. The same holds true for other damping components, such as furniture and ductwork. The point is that increasing damping to reduce vibration is generally not feasible, since damping is a parameter that is difficult to quantity and control, so the designer has to make a reasonable assumption as to the amount of damping that will be present in a structure.

12.10 REFERENCES

1. American Institute of Steel Construction. 2003. Murray, Thomas, David Allen, and Eric Ungar. *Steel design guide series 11: Floor vibrations due to human activity.* Chicago: AISC.

2. Charlton, Nathan. "Framing systems for microelectronic facilities. *Modern Steel Construction*, May 1997.

3. American Institute of Steel Construction. 2006. *Steel construction manual*, 13th ed. Chicago: AISC.

4. American Concrete Institute. 2005. *ACI 318: Building code requirements for structural concrete and commentary.* Farmington Hills, MI.

5. Vulcraft. 2001. *Steel roof and floor deck.* Florence, SC: Vulcraft/Nucor.

6. Steel Joist Institute. 2005. Standard specifications—Load tables and weight tables for steel joists and joist girders, 42nd ed. Myrtle Beach, SC: SJI.

7. Spancrete Manufacturers Association. *Span limitations: Floor vibrations—Rhythmic activity.* Research Notes #1021. Waukesha, WI: SMA. 2005

8. Spancrete Manufacturers Association. *Span limitations: Floor vibrations—Flexible supports.* Research Notes #1023. Waukesha, WI: SMA. 2005

9. National Research Council of Canada, 1990 National Building Code of Canada, Supplement-Commentary A, Serviceability Criteria for Deflection and Vibration, Ottawa, Canada.

12.11 PROBLEMS

12-1. Calculate the natural frequency of a W18 × 35 beam spanning 30 ft. with an applied load (dead plus sustained live loads) of 500 plf.

12-2. Given the following, based on the floor plan and floor section shown in Figure 12-18,

- Dead load = 60 psf,
- Concrete is normal weight with f'_c = 3500 psi, and
- Assume loads to the joist and girder are uniformly distributed,

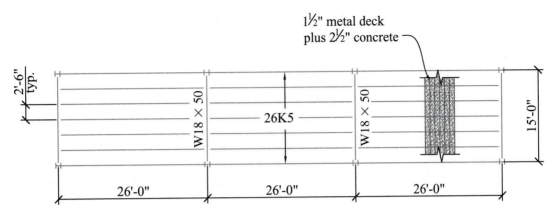

Figure 12-18 Floor framing for Problem 12-2.

determine the following:

1. Natural frequency of the floor system, and
2. Adequacy for walking vibrations in a shopping mall.

12-3. Given the following, based on the floor plan shown in Figure 12-19,

- Dead load = 70 psf, Sustained live load = 11 psf,
- Floor is an open office area with few nonstructural components,
- Transformed moment of inertia, I_{BEAM} = 2100 in.^{4}, I_{GIRDER} = 3100 in.^{4}, and
- Assume loads to the beam and girder are uniformly distributed,

determine the following:

1. Natural frequency of the floor system, and
2. Adequacy for walking vibrations in an office.

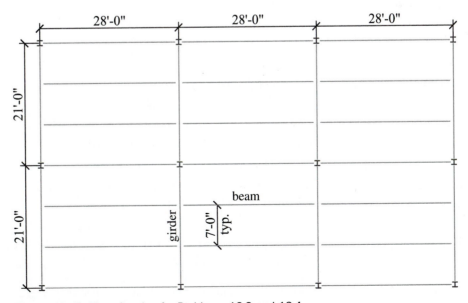

Figure 12-19 Floor framing for Problems 12-3 and 12-4.

12-4. Given the following, based on the floor plan shown in Figure 12-19

- Total slab depth is 5.5 in. (2 in. of metal deck plus 3.5 in. of concrete), and
- Moderate walking pace (75 steps per minute),

determine the following:

1. Vibrational velocity of the floor system in μ-in./s, and
2. Type of sensitive equipment that could be placed on the floor system.

12-5. Given the following, based on the floor plan shown in Figure 12-20

- Dead load = 75 psf, Sustained live load = 11 psf, and
- Transformed moment of inertia, I_{BEAM} = 2100 in.⁴, I_{GIRDER} = 3400 in.⁴,

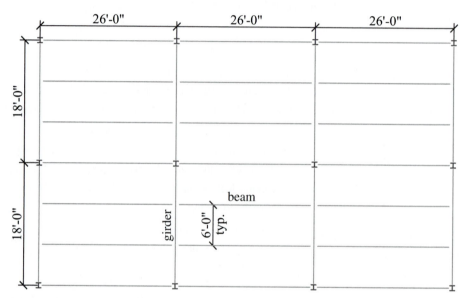

Figure 12-20 Floor framing for Problems 12-5 and 12-6.

determine the following:

1. Natural frequency of the floor system, and
2. Adequacy for jumping exercises in a facility with rhythmic activities.

12-6. Given the following, based on the floor plan shown in Figure 12-20

- Total slab depth is 6.5 in. (3 in. metal deck plus 3.5 in. of concrete), and
- Slow walking pace (50 steps per minute),

determine the following:

1. Vibrational velocity of the floor system in μ-in./s, and
2. Type of equipment that could be placed on the floor system.

12-7. Given the floor plan shown in Figure 12-21 with a dead load of 55 psf and a sustained live load of 11 psf, determine whether the floor system is adequate for walking vibrations in an office area with some nonstructural components. The concrete is normal weight with f'_c = 4000 psi.

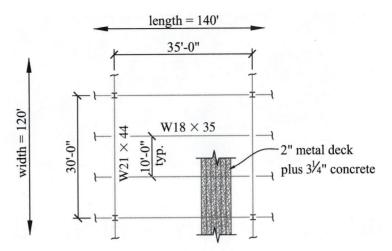

Figure 12-21 Floor framing for Problem 12-7.

Student Design Project Problems:

12-8. For the floor framing in the student design project (see Figure 1-22), analyze the floor structure for walking vibrations assuming β = 0.03.

Built-up Sections—Welded Plate Girders

13.1 INTRODUCTION TO WELDED PLATE GIRDERS

Welded plate girders, which are the most common form of plate girders, are built-up structural steel members that consists of flange plates welded to a web plate with fillet welds [1]. They are used to support loads over long spans (60 ft. to 200 ft.) [2] and to support structural loads that are too large to be supported by the rolled steel shapes shown in the *AISCM*. Plate girders are rarely used in building structures, but are commonly used in bridge structures [3]. They are used as transfer girders in building structures to support columns above large column-free areas, such as atriums, auditoriums, and assembly areas as shown in Figure 13-1. Plate girders may also be used in the retrofitting of existing building structures where column-free areas are needed and existing columns have to be cut off or removed below a certain floor level. One such detail is shown in Figure 13-2. Plate girders are also used as crane support girders in heavy industrial structures with long spans. One disadvantage of plate girders when used in building structures is that mechanical ducts may have to be placed below the girder to avoid cutting holes in the web of the girder that will reduce the strength of the transfer girder. Furthermore, locating mechanical ducts below a deep transfer plate girder increases the floor-to-floor height of the building, with a resulting increase in construction costs. In any case, it is more common to use transfer trusses in place of transfer girders in building structures because it allows for passage of mechanical ducts between the web members of the truss without adversely affecting the strength of the truss.

The term "plate girder" no longer exists in the most recent AISC specification [4]. Instead, this term has been replaced by the term "built-up sections," and the design requirements for flexure and shear for these sections are found in Sections F5 and G of the AISC specification, respectively. In this text, we classify plate girders as built-up sections with

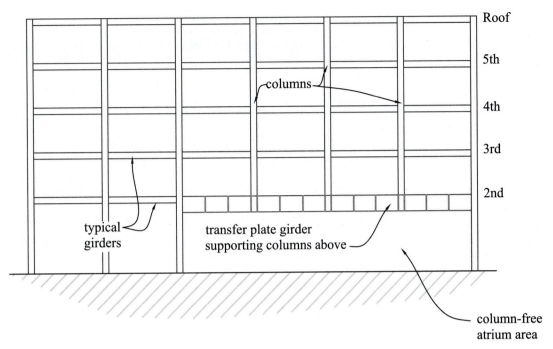

Figure 13-1 Transfer plate girder.

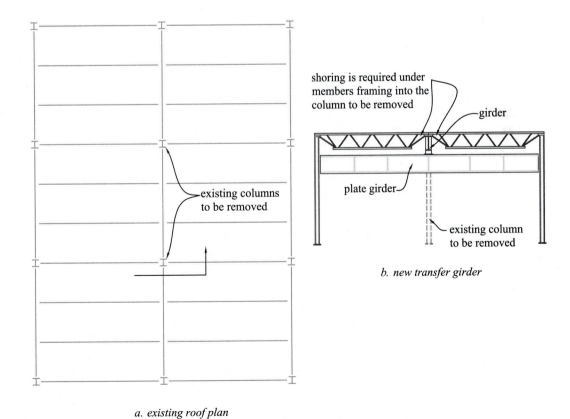

a. existing roof plan

Figure 13-2 Transfer plate girder details in retrofitting an existing building.

Table 13-1 Web depth-to-thickness ratios for noncompact and slender webs

	Doubly Symmetric I-Shaped Section	Singly-Symmetric I-Shaped Section
Noncompact web	$3.76\sqrt{\dfrac{E}{F_y}} < \dfrac{h}{t_w} \leq 5.70\sqrt{\dfrac{E}{F_y}}$	$\dfrac{\dfrac{h_c}{h_p}\sqrt{\dfrac{E}{F_y}}}{\left(0.54\dfrac{M_p}{M_y} - 0.09\right)^2} < \dfrac{h_c}{t_w} \leq 5.70\sqrt{\dfrac{E}{F_y}}$
Slender web	$\dfrac{h}{t_w} > 5.70\sqrt{\dfrac{E}{F_y}}$	$\dfrac{h_c}{t_w} > 5.70\sqrt{\dfrac{E}{F_y}}$

noncompact or slender webs, and the design for flexure is carried out using Section F5 of the AISC specification. The web depth-to-thickness ratios that define the limits of non-compactness or slenderness of the web of doubly symmetric and singly symmetric I-shaped built-up sections are given in Table 13-1.

To prevent web buckling, web stiffeners can be added for stability (see Figure 13-3c). The AISC specification prescribes mandatory limits for the web depth-to-thickness ratios as a function of the clear spacing between the transverse stiffeners (see Table 13-2) [4].

where

a = Clear horizontal distance between transverse stiffeners, if any,

h = Depth of the web = Clear distance between the flanges of a plate girder minus the fillet or corner radius for rolled sections; for welded built-up sections, the clear distance between the flanges (see Figure 13-3),

h_c = Twice the distance between the *elastic* neutral axis and the inside face of the compression flange for **nonsymmetric** welded built-up sections

= clear distance, h, between inside faces of the flanges for welded built-up sections with equal flange areas,

h_p = Twice the distance between the *plastic* neutral axis and the inside face of the compression flange for **nonsymmetric** welded built-up sections

= clear distance, h, between inside faces of the flanges for welded built-up sections with equal flange areas,

t_w = Web thickness,

d = Overall depth of the plate girder,

M_p = Plastic moment of the section = $Z_x F_y$,

M_y = Yield moment of the section = $S_x F_y$,

A_w = Area of the web = $h t_w$,

A_{fc} = Area of the compression flange = $b_{fc} t_{fc}$,

I_y = Moment of inertia of the built-up cross section about the weak axis (y–y) or the vertical axis in the plane of the web, and

I_{yc} = Moment of inertia of the compression flange of the built-up cross section about the weak axis (y–y) or the vertical axis in the plane of the web.

In a typical welded plate girder, the top and bottom flange plates resist the bending moments through a tension–compression couple in the top and bottom flanges, and the web plate primarily resists the shear. Transverse vertical stiffeners are used to increase the web buckling

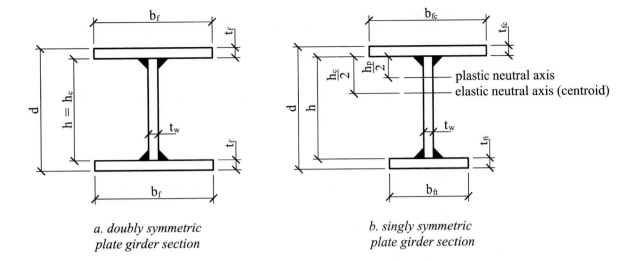

a. doubly symmetric
plate girder section

b. singly symmetric
plate girder section

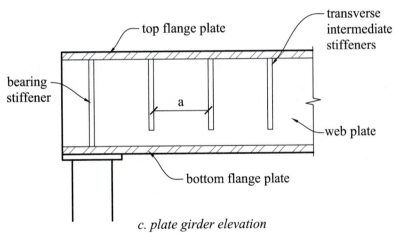

c. plate girder elevation

Figure 13-3 Welded plate girder.

Table 13-2 Size limitations for I-shaped built-up sections (*AISCM*, Section F13.2)

Without Transverse Web Stiffeners (i.e., unstiffened girders)	With Transverse Web Stiffeners	
$\dfrac{h}{t_w} \leq 260$ and $\dfrac{A_w}{A_{fc}} \leq 10$	$\dfrac{a}{h} \leq 1.5$	$\dfrac{a}{h} > 1.5$
	$\dfrac{h}{t_w} \leq 11.7\sqrt{\dfrac{E}{F_y}}$	$\dfrac{h}{t_w} \leq \dfrac{0.42E}{F_y}$
For **singly** symmetric sections, $0.1 \leq \dfrac{I_{yc}}{I_y} \leq 0.9$		

capacity of the built-up section. Longitudinal web stiffener plates are rarely used except for very deep webs [2]. The web plate is fillet welded to the top and bottom flange plates and these welds resist the horizontal interface shear between the flanges and the web. The width and thickness of the compression and tension flanges may be varied along the span in proportion to the bending moment, but the web thickness is generally kept constant along the span of the girder. The depth of the girder may also be varied for long-span girders. When higher yield strength steel is used for the flange plates and conventional steel is used for the web plates, the built-up section is referred to as a *hybrid* girder.

13.2 DESIGN OF PLATE GIRDERS

In the design of plate girders, there are two possible options:

1. *Unstiffened Plate Girder:* The plate girder is proportioned with adequate flange and web thicknesses to avoid the need for web stiffeners. The unstiffened plate girder option results in a thicker web and flanges, and therefore a heavier plate girder self-weight, but the complexity of fabrication is minimized. Since fabrication and erection costs make up more than 60% of the construction costs of structural steel buildings, this may be a more economical option in some cases.

2. *Stiffened Plate Girder:* A stiffened plate girder is designed with web stiffeners. This stiffened plate girder option results in a lighter weight plate girder, but the fabrication costs increase.

It is possible to find optimum thicknesses for the web and flanges, and an optimum size and number of stiffeners that will yield a plate girder with self-weight and fabrication and erections costs such that the stiffened plate girder is less costly than an equivalent unstiffened girder. The presence of vertical web stiffeners improves the buckling resistance of the web. After initial buckling of the web, the plate girder is still capable of resisting additional loads because of the "diagonal tension fields" that develop in the web of the plate girder between the stiffeners in the post-web buckling range [3]. This diagonal tension field makes the plate girder behave like a Pratt truss as shown in Figure 13-4.

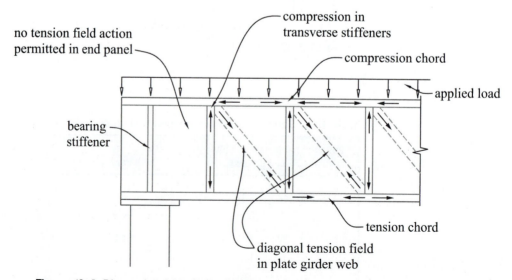

Figure 13-4 Diagonal tension field and truss action in plate girders.

13.3 BENDING STRENGTH OF WELDED PLATE GIRDERS

When a plate girder section has a noncompact or slender web, the nominal moment capacity, M_n, will be less than the plastic moment capacity, M_p, of the section because of several limit states that are attained before the section can reach its full plastic moment capacity. The following are the possible limit states that may occur in built-up sections in bending:

- Compression flange yielding,
- Lateral torsional buckling,
- Compression flange local buckling, and
- Tension flange yielding.

The design moment capacity for the built-up section depends on the compactness, noncompactness, or slenderness of the flanges, and will be the smallest strength obtained for the following four limit states.

1. *Compression flange yielding*

   Design moment capacity, $\phi M_n = \phi R_{pg} F_y S_{xc}$,   (13-1)

   where

   $\phi = 0.9$,

   R_{pg} is as defined in equation (13-7),

   F_y = Yield strength of the compression flange,

   S_{xc} = Elastic section modulus about the strong axis $(x–x)$ relative to the outermost face of the compression flange of the built-up section

   $= I_{x–x}/y_c$, and

   y_c = Distance from the neutral axis to the outer most face of the compression flange.

   $I_{x–x}$ = moment of inertia of the plate girder about the strong axis $(x–x)$

2. *Lateral torsional buckling*

   Lateral torsional buckling is a function of the lateral unbraced length, L_b, of the compression flange of the plate girder. This limit state only applies when the unbraced length of the compression flange, L_b, is greater than L_p. The design bending strength for the case is

   $$\phi M_n = \phi R_{pg} F_{cr} S_{xc}$$ (13-2)

   where

   The critical bending stress, F_{cr}, is obtained from Table 13-3,

   $\phi = 0.9$,

   R_{pg} is as defined in equation (13-7),

   S_{xc} is as defined previously for the compression flange yielding limit state,

   The lateral support length parameters are

   $$L_p = 1.1 r_t \sqrt{\frac{E}{F_y}},$$ (13-5)

Table 13-3 Critical bending stress as a function of the unbraced length

Unbraced Length, L_b	Critical Bending Stress, F_{cr}	
$L_b \leq L_p$	Lateral torsional buckling limit states **does not** apply.	
$L_p \leq L_b \leq L_r$	$F_{cr} = C_b \left[F_y - (0.3F_y) \left(\dfrac{L_b - L_p}{L_r - L_p} \right) \right] \leq F_y$	(13-3)
$L_b > L_r$	$F_{cr} = \dfrac{C_b \pi^2 E}{\left(\dfrac{L_b}{r_t} \right)^2} \leq F_y$	(13-4)

$$L_r = \pi r_t \sqrt{\frac{E}{0.7F_y}}, \text{ and} \tag{13-6}$$

C_b = Bending moment coefficient (see Chapter 6). It is conservative to assume a C_b value of 1.0.

The bending strength reduction factor, R_{pg}, is given as

$$R_{pg} = 1 - \left(\frac{a_w}{1200 + 300a_w} \right) \left(\frac{h_c}{t_w} - 5.7\sqrt{\frac{E}{F_y}} \right) \leq 1.0, \tag{13-7}$$

where

a_w = Ratio of two times the web area in compression due to the application of major axis bending moment alone to the area of the compression flange components. Mathematically,

$$a_w = \frac{h_c t_w}{b_{fc} t_{fc}} \leq 10, \tag{13-8}$$

b_{fc} = Width of the compression flange,

t_{fc} = Thickness of the compression flange,

t_w = Thickness of the web,

h_c = Twice the distance between the elastic neutral axis and the inside face of the compression flange for **nonsymmetric** welded built-up sections, and

= clear distance, h, between inside faces of the flanges for welded built-up sections with equal flange areas.

The parameter r_t in equations 13-4 and 13-6 is the radius of gyration of the flange components in flexural compression plus one-third of the web area in compression due to the application of major axis bending moment. Mathematically, r_t can be approximated for I-shaped sections as

$$r_t \approx \frac{b_{fc}}{\sqrt{12 \left(1 + \dfrac{1}{6} a_w \right)}} \text{ in.} \tag{13-9}$$

Table 13-4 Critical bending stress, F_{cr}

Flange Compactness	Controlling Failure Mode	Critical Bending Stress, F_{cr}
Compact flange $\dfrac{b_{fc}}{2t_{fc}} \leq 0.38\sqrt{\dfrac{E}{F_y}}$	Flange yielding	F_y
Noncompact flanges $0.38\sqrt{\dfrac{E}{F_y}} < \dfrac{b_{fc}}{2t_{fc}} \leq 0.95\sqrt{\dfrac{k_c E}{F_L}}$	Inelastic flange local buckling	$F_{cr} = \left[F_y - (0.3F_y)\left(\dfrac{\lambda - \lambda_{pf}}{\lambda_{rf} - \lambda_{pf}}\right) \right]$
Slender flanges $\dfrac{b_{fc}}{2t_{fc}} > 0.95\sqrt{\dfrac{k_c E}{F_L}}$	Elastic flange local buckling	$F_{cr} = \dfrac{0.9Ek_c}{\left(\dfrac{b_f}{2t_f}\right)^2}$

F_y = Yield strength of the compression flange,

$\lambda = \dfrac{b_{fc}}{2t_{fc}}$,

λ_{pf} = Limiting slenderness for the **compact** compression flange obtained from *AISCM*, Table B4.1

$\quad = 0.38\sqrt{\dfrac{E}{F_y}}$ for the compression flange,

λ_{rf} = Limiting slenderness for **noncompact** compression flange obtained from *AISCM*, Table B4.1

$\quad = 0.95\sqrt{\dfrac{k_c E}{F_L}}$ for the compression flange,

b_{fc} = Width of the compression flange,

t_{fc} = Thickness of the compression flange,

$k_c = \dfrac{4}{\sqrt{\dfrac{h}{t_w}}}$ $(0.35 \leq k_c \leq 0.76)$.

t_w = Web thickness, and

h = Clear distance between the inside faces of the flanges for welded built-up sections.

3. *Compression flange local buckling*
   This limit state is not applicable to built-up sections with **compact** flanges. For all other sections, the design moment capacity is given as

   $$\phi M_n = \phi R_{pg} F_{cr} S_{xc}, \tag{13-10}$$

   where the critical bending stress, F_{cr}, is obtained as shown in Table 13-4 and R_{pg} is as defined in equation 13-7.
   S_{xc} is as defined previously for the compression flange yielding limit state, and

   $$\phi = 0.9 \text{ and } F_L \text{ is obtained from Table 13-5,}$$

4. *Tension flange yielding*
   This limit state **does not** apply to built-up sections when the section modulus of the built-up section with respect to the tension face, S_{xt}, is greater than or equal to the section modulus with respect to the compression face, S_{xc} (i.e., when $S_{xt} \geq S_{xc}$).

   For all other sections, the design moment capacity is given as

   $$\phi M_n = \phi F_y S_{xt}. \tag{13-11}$$

Table 13-5 Parameter, F_L, for major axis bending (from *AISCM*, Table B4.1)

Description	F_L
I-shaped built-up section with **noncompact** web *and* $\dfrac{S_{xt}}{S_{xc}} \geq 0.7$	$0.7F_y$
I-shaped built-up section with **noncompact** web *and* $\dfrac{S_{xt}}{S_{xc}} < 0.7$	$F_y \dfrac{S_{xt}}{S_{xc}} \geq 0.5F_y$
I-shaped built-up section with **slender** web	$0.7F_y$

h_c = Twice the distance between the elastic neutral axis and the inside face of the compression flange for **nonsymmetric** welded built-up sections

= Clear distance, h, between the inside faces of the flanges for **symmetric** welded built-up sections,

S_{xc} = Elastic section modulus about the strong axis (x-x) with respect to the compression flange of the built-up section, and

S_{xt} = Elastic section modulus about the strong axis (x-x) with respect to the tension flange of the built-up section.

13.4 DESIGN FOR SHEAR IN PLATE GIRDERS WITHOUT DIAGONAL TENSION FIELD ACTION (AISCM, SECTION G2)

Due to the relatively thin webs used in plate girders, the design for shear is more complicated than for rolled sections. In fact, there are two approaches available for designing for shear in plate girders. One method accounts only for the buckling strength of the web, while the second method accounts for the post-buckling strength of the web panels between stiffeners as a result of diagonal tension field action. Therefore, unless diagonal tension field action is to be relied on, it is recommended that sufficient web thickness be used to avoid the need for stiffeners.

For unstiffened and stiffened webs of doubly symmetric and singly symmetric shapes subject to shear in the plane of the web, the design shear strength is given as

$$\phi V_n = \phi 0.6 F_{yw} A_w C_v,\tag{13-12}$$

where

$\phi = 0.9$,

A_w = Overall depth of built-up section times the web thickness = dt_w,

F_{yw} = Yield strength of the web material, and

Web shear coefficient, C_v, and shear parameter, k_v, are obtained from Table 13-6 and Table 13-7, respectively, as a function of the web depth-to-thickness ratio.

Stiffened Webs Without Diagonal Tension Field Action

Without diagonal tension field action, the transverse stiffeners, if provided, are there to prevent web buckling. Transverse stiffeners are *not* required if one of the following conditions is satisfied:

- $\dfrac{h}{t_w} \leq 2.46\sqrt{\dfrac{E}{F_{yw}}}$

Table 13-6 Web shear coefficient, C_v

Web Depth-to-Thickness Ratio, h/t_w	Web Shear Coefficient, C_v
$\dfrac{h}{t_w} \leq 1.10\sqrt{\dfrac{k_v E}{F_{yw}}}$	1.0
$1.10\sqrt{\dfrac{k_v E}{F_{yw}}} < \dfrac{h}{t_w} \leq 1.37\sqrt{\dfrac{k_v E}{F_{yw}}}$	$\dfrac{1.10\sqrt{\dfrac{k_v E}{F_{yw}}}}{h/t_w}$
$\dfrac{h}{t_w} > 1.37\sqrt{\dfrac{k_v E}{F_{yw}}}$	$\dfrac{1.51 E k_v}{(h/t_w)^2 F_{yw}}$

or

- $V_u \leq \phi V_n$ for $k_v = 5$.

For all other conditions, transverse stiffeners are required and the spacing, a, and thickness, t_{st}, of the stiffener must be selected to satisfy the following conditions:

$$I_{st,\, z-z} \geq a t_w^3 j,$$

where

$I_{st,\, z-z}$ = Moment of inertia for **a pair of stiffeners** (i.e., stiffeners on both sides of the web) about a horizontal axis at the centerline of the web (see Figure 13-5a)

$$= \frac{t_{st}(2b_{st} + t_w)^3}{12},$$

$I_{s,\, z-z}$ = Moment of inertia for a **single stiffener** about a horizontal axis at the face of the web in contact with the stiffener (see Figure 13-5b)

Table 13-7 Shear parameter, k_v

Web Depth-to-Thickness Ratio, h/t_w	Shear Parameter, k_v
Unstiffened web of I-shaped members with $\dfrac{h}{t_w} < 260$	5
Stiffened web with $\dfrac{a}{h} \leq 3$ or $\dfrac{a}{h} \leq \left(\dfrac{260}{h/t_w}\right)^2$	$5 + \dfrac{5}{(a/h)^2}$
Stiffened web with $\dfrac{a}{h} > 3$ or $\dfrac{a}{h} > \left(\dfrac{260}{h/t_w}\right)^2$	5

a = Horizontal clear distance between transverse stiffeners, and

h = Clear distance between the flanges for welded plate girders.

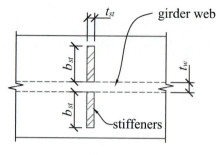
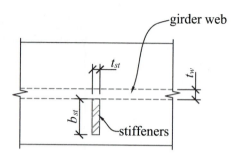

a. stiffener pairs

b. single stiffener

Figure 13-5 Plan view of web *Stiffeners*.

$$= \frac{t_{st}(b_{st})^3}{12} + t_{st}b_{st}\left(\frac{b_{st}}{2}\right)^2,$$

$$j = \frac{2.5}{(a/h)^2} - 2 \geq 0.5,$$

t_{st} = Thickness of transverse stiffener, and

b_{st} = Width of transverse stiffener perpendicular to the longitudinal axis of the girder.

13.5 DIAGONAL TENSION FIELD ACTION IN PLATE GIRDERS (*AISCM*, SECTION G3)

When a stiffened plate girder is loaded, it continues to support loads even after initial buckling of the web panels between the transverse stiffeners [3]. After this initial web buckling, diagonal tension fields develop in the webs of the plate girder between the stiffeners during the post-buckling phase, leading to an increase in the strength of the plate girder. The formation of these diagonal tension bands, or fields, during the post-buckling phase are balanced by vertical compression bands in the transverse stiffeners (which are assumed to support no loads prior to web buckling), thus creating a Pratt truss-like load-carrying mechanism [3] as shown in Figure 13-4. The consideration of diagonal tension field action, which is only permitted if certain conditions are met and if the web plate is supported on all four sides by the flanges and the stiffeners, can increase the load capacity of a plate girder by up to two to three times the initial web-buckling capacity [3]. It should be noted, however, that in the AISC specification, it is not mandatory to consider tension field action in the design of plate girders [4].

The AISC specification does not permit diagonal tension field action to be considered for the following situations:

- End panels in stiffened plate girders cannot fully develop tension field action because the end panels are typically narrow and thus the vertical shear in the plate girder is resisted by beam action in this panel.

- Built-up sections with web members for which $\dfrac{a}{h} > 3$ or $\dfrac{a}{h} > \left(\dfrac{260}{h/t_w}\right)^2$ because for these a/h ratios, the diagonal tension bands become too flat to function effectively as truss members.

- Built-up sections with $\dfrac{2A_w}{A_{fc} + A_{ft}} > 2.5$, or $\dfrac{h}{b_{fc}}$ or $\dfrac{h}{b_{ft}} > 6.0$.

The design shear strength, when the diagonal tension field is considered and the tension field is assumed to yield, is given in Table 13-8.

where

$\phi = 0.9$,

k_v and C_v have been defined previously in Table 13-6 and Table 13-7, respectively,

A_{fc} = Compression flange area = $b_{fc}\, t_{fc}$, in.^{2},
A_{ft} = Tension flange area = $b_{ft}\, t_{ft}$, in.^{2},
b_{fc}, t_{fc} = Compression flange width and thickness, respectively, in.,
b_{ft}, t_{ft} = Tension flange width and thickness, respectively, in.,
F_{yw} = Yield strength of the web material,
A_w = Overall depth of the built-up section times the web thickness = dt_w, and

a and h are as previously defined in Section 13.4.

Stiffened Webs with Diagonal Tension Field Action Considered

When diagonal tension field action is considered, the transverse stiffeners must satisfy the requirements indicated below. Transverse stiffeners *do not* have to be used if one of the following conditions is satisfied:

- $\dfrac{h}{t_w} \le 2.46\sqrt{\dfrac{E}{F_{yw}}}$

  or

- $V_u \le \phi V_n$ for $k_v = 5$.

Table 13-8 Design shear strength with diagonal tension field action (i.e., post-buckling strength considered)

Web Depth-to-Thickness Ratio, h/t_w	Design Shear Capacity, ϕV_n
$\dfrac{h}{t_w} \le 1.10\sqrt{\dfrac{k_v E}{F_{yw}}}$	$\phi 0.6 F_{yw} A_w$
$\dfrac{h}{t_w} > 1.10\sqrt{\dfrac{k_v E}{F_{yw}}}$	$\phi 0.6 F_{yw} A_w \left(C_v + \dfrac{1 - C_v}{1.15\sqrt{1 + (a/h)^2}} \right)$

For all other conditions, transverse stiffeners *must* be provided and the spacing, a, and thickness, t_{st}, of the stiffener must be selected to satisfy the following conditions:

$$\frac{b_{st}}{t_{st}} \leq 0.56\sqrt{\frac{E}{F_{y,st}}},$$

$$A_{st} > \frac{F_{yw}}{F_{y,st}}\left[0.15D_s h t_w(1 - C_v)\frac{V_u}{\phi V_n} - 18t_w^2\right] \geq 0, \text{ and}$$

$$I_{st,z-z} \geq a t_w^3 j,$$

where

$C_v =$ As defined earlier in Table 13-6,

$I_{st,z-z} =$ Moment of inertia for **a pair of stiffeners** (i.e., stiffeners on both sides of the web) about a horizontal axis at the centerline of the web (see Figure 13-5a)

$$= \frac{t_{st}(2b_{st} + t_w)^3}{12},$$

$I_{s,z-z} =$ Moment of inertia for a **single stiffener** about a horizontal axis at the face of the web in contact with the stiffener (see Figure 13-5b)

$$= \frac{t_{st}(b_{st})^3}{12} + t_{st}b_{st}\left(\frac{b_{st}}{2}\right)^2,$$

$$j = \frac{2.5}{(a/h)^2} - 2 \geq 0.5,$$

$t_{st} =$ Thickness of the transverse stiffener,

$b_{st} =$ Width of the transverse stiffener perpendicular to the longitudinal axis of the girder,

$\dfrac{b_{st}}{t_{st}} =$ Stiffener width-to-thickness ratio,

$F_{y,st} =$ Yield strength of the stiffener material,

$F_{yw} =$ Yield strength of the web material,

$D_s = 1.0$ for stiffeners in pairs

$\quad = 1.8$ for single-angle stiffeners

$\quad = 2.4$ for single-plate stiffeners,

$V_u =$ Factored shear force or required shear strength, and

$\phi V_n =$ Design shear strength.

13.6 CONNECTION OF WELDED PLATE GIRDER COMPONENTS

The connections of the components of a welded plate girder include the fillet welds at the flange-to-web interface, the fillet welds at the transverse stiffener-to-web interface, and the fillet welds at the bearing stiffener-to-web interface. For very long span girders, it might also include the welds at the flange and web splice locations.

Connection of Plate Girder Flange and Web

The connection between the flange plate and the web plate of the built-up section can be achieved by using intermittent or continuous fillet welds, with the latter more commonly preferred in design practice. However, care should be taken with continuous fillet welds, especially for thin plates, as this could lead to a distortion in these plates, creating undesirable residual stresses. The horizontal shear force per unit length or shear flow at the interface between the web plate and the flange plate is calculated as follows [1]:

$$v_h = \frac{V_u Q_f}{I_{x-x}},$$
(13-13)

where

V_u = Factored vertical shear force, kips,

Q_f = Statical moment of the flange area about the neutral axis of the built-up section, in.^{3}, and

I_{x-x} = Moment of inertia of the built-up section about the strong axis $(x–x)$, in.^{4}.

Although the vertical shear force, V_u, may vary along the length of the girder depending on the loading, and thus the required weld size will also vary, it is practical to use the maximum factored shear force in the girder in equation (13-13) and to provide a constant size of fillet weld to resist the horizontal shear flow, v_h.

Connection of Intermediate Stiffeners to Girder Web (*AISCM*, Section G2.2)

The intermediate stiffeners are connected to the web with intermittent fillet welds, and the end distance between the intermittent stiffener-to-web fillet weld and the nearest toe of the web-to-flange fillet weld shall not be less than $4t_w$ nor more than $6t_w$ (see Figure 13-6).

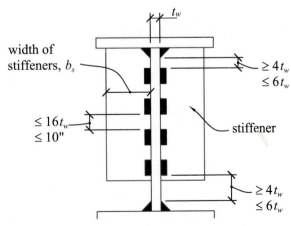

Figure 13-6 Spacing requirement for the stiffener-to-web weld.

The clear spacing between the web-to-stiffener fillet welds shall not be greater than $16t_w$, or 10 in. Pairs of stiffeners do not have to be connected to the compression flange, but to the web only. Single stiffeners should be connected to both the web and the compression flange.

Connection of Bearing Stiffeners to Girder Web

The bearing stiffeners are connected to the girder web with *continuous* fillet welds on both sides of the stiffener plate because bearing stiffeners are direct load-carrying elements that have to transfer the support reactions into the girder web. The design of bearing stiffeners has already been discussed in Chapter 6.

13.7 PLATE GIRDER PRELIMINARY DESIGN

In the preliminary design of plate girders, it can be assumed that the web resists all of the vertical shear force and the flanges resist all of the bending moments. The design procedure is as follows:

1. Calculate the maximum factored moment, M_u, and maximum factored shear, V_u, initially neglecting the self-weight of the plate girder.
2. Assume that the plate girder overall depth, d, is between $L/8$ and $L/12$ [1]. For $d \leq$ 48 in., it is recommended that a sufficient web thickness be provided to avoid the use of transverse stiffeners. Even for $d > 48$ in., as few stiffeners as possible should be used to minimize the cost of fabrication [2].
3. Assume a flange width, $b_f > d/6$, but preferably greater than 12 in. [2].
4. Assume that the couple of forces in the bottom and tension flange, $T_u = C_u \approx M_u/0.95d$.
5. Assume that the initial flange thickness, $t_f = T_u/0.9F_y b_f > \frac{3}{4}$ in.
6. The depth of the web, $h = d - 2t_f$.
7. Select a web thickness, t_w, that meets the proportion limits given in Table 13-2, but it should be greater than or equal to $\frac{1}{2}$ in. [2].

13.8 PLATE GIRDER FINAL DESIGN

Using the size obtained from the preliminary design above, the procedure for the final design is as follow:

1. Calculate the self-weight of the plate girder and recalculate the total factored moment, M_u, and the total factored shear, V_u, on the plate girder.
2. Decide whether or not to take into account diagonal tension field action. Note that it is **not** mandatory to account for the post-buckling strength of stiffened webs resulting from diagonal tension field action.

   a. If the diagonal tension field action **is not** considered, stiffeners should not be used and a sufficient web thickness should be provided.

   b. If diagonal tension field action **is** to be used, then select a stiffener spacing such that the a/h ratio is between 1.0 and 2.0.

3. With the trial section selected in the preliminary design phase, determine the design moment capacity, ϕM_n, considering all four possible limit states (see Section 13.3). The smallest ϕM_n is the design moment capacity of the built-up section.

4. Determine k_v from Table 13-7, and C_v from Table 13-6.

5. Based on the choice made in step 2 regarding diagonal tension field action, determine the design shear strength, ϕV_n.

6. Check deflections and vibrations as discussed previously in Chapters 6 and 12, respectively.

7. Design the bearing stiffeners at the support reactions or at concentrated load locations as discussed in Chapter 6.

8. Design the fillet welds required at the flange-to-web interface and the stiffener-to-web interface.

EXAMPLE 13-1

Design of a Welded Transfer Plate Girder

A five-story building has the elevation shown in Figure 13-7. The building has typical column grids at 30 ft. on center in both orthogonal directions. The floor dead load is 100 psf and the live load is 50 psf. The roof dead load is 30 psf with a flat roof snow load of 40 psf. In order to span over a large atrium area at the ground floor, a series

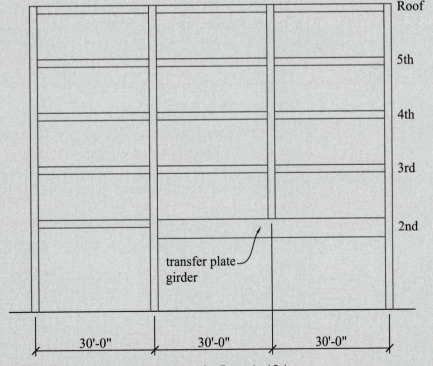

Figure 13-7 Interior building section for Example 13-1.

(continued)

of transfer plate girders spanning 60 ft. and spaced at 30 ft. on center are to be used at the second-floor level. Design a typical interior second-floor transfer plate girder to support the column loads from above, as well as the second-floor load. Neglect any live load reduction and ignore diagonal tension field action. Assume grade 50 steel and assume the compression flange is fully braced.

SOLUTION

Second-Floor Load on Transfer Girder

At this preliminary stage, we will initially neglect the self-weight of the girder, but this will be calculated later in the final design stage.

Tributary width of girder = 30 ft.

Assume that the infill beams are to be spaced such that a uniform floor load can be assumed on the girder. Therefore, the uniformly distributed loads on the girder are

$$w_D = (100\,\text{psf})(30\,\text{ft.}) = 3000\,\text{lb./ft.} = 3\,\text{kips/ft., and}$$
$$w_L = (50\,\text{psf})(30\,\text{ft.}) = 1500\,\text{lb./ft.} = 1.5\,\text{kips/ft.}$$

The column loads are calculated as

$$P_D = [(30\,\text{psf} + (3\,\text{floors})(100\,\text{psf})](30\,\text{ft.} \times 30\,\text{ft.})$$
$$= 297,000\,\text{lb.} = 297\,\text{kips,}$$
$$P_S = (40\,\text{psf})(30\,\text{ft.})(30\,\text{ft.}) = 36,000\,\text{lb.} = 36\,\text{kips, and}$$
$$P_L = (3\,\text{floors})(100\,\text{psf})(30\,\text{ft.})(30\,\text{ft.}) = 270,000\,\text{lb.} = 270\,\text{kips.}$$

Note that the self-weight of the column has been neglected since this is usually negligible compared to the total dead load on the girder. Using the load combinations in Chapter 2, the factored uniformly distributed load and concentrated loads on the transfer girder are calculated as

$$w_u = 1.2(3\,\text{kips/ft.}) + 1.6(1.5\,\text{kips/ft.}) = 6\,\text{kips/ft., and}$$
$$P_u = 1.2(297\,\text{kips}) + 1.6(270\,\text{kips}) + 0.5(36\,\text{kips}) = 806.4\,\text{kips.}$$

A free-body diagram of the transfer girder is shown in Figure 13-8.

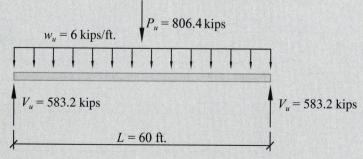

Figure 13-8 Free-body diagram of simply supported transfer girder.

The factored or required shear and bending moment are calculated as

$$V_u = 6 \text{ kips/ft.}\left(\frac{60 \text{ ft.}}{2}\right) + \frac{806.4 \text{ kips}}{2} = 583.2 \text{ kips, and}$$

$$M_u = 6 \text{ kips/ft.}\frac{(60 \text{ ft.})^2}{8} + \frac{806.4 \text{ kips}(60 \text{ ft.})}{4}$$
$$= 14{,}796 \text{ ft.-kips, respectively.}$$

Preliminary Design

1. $M_u = 14{,}796$ ft.-kips and $V_u = 583$ kips
2. Assume a plate girder overall depth, $d = L/10 = (60 \text{ ft.}) (12)/10 = 72$ in.
3. Assume a flange width, $b_f = d/6 (\geq 12 \text{ in.}) = 72 \text{ in.}/6 = 12$ in. Try $b_f = 15$ in.
4. The couple of forces in the bottom and tension flange is

$$T_u = C_u \approx \frac{M_u}{0.95d} = \frac{14{,}796 \text{ ft.-kips}}{0.95(72 \text{ in.}/12)} = 2596 \text{ kips.}$$

5. Assume an initial flange thickness, $t_f = \dfrac{T_u}{0.9F_y b_f} = \dfrac{2596 \text{ kips}}{0.9(50 \text{ ksi})(15 \text{ in.})}$
$$= 3.84 \text{ in.} \geq \tfrac{3}{4} \text{ in.}$$

   Try $t_f = 4$ in.

6. Depth of web, $h = d - 2t_f = 72 \text{ in.} - (2)(4 \text{ in.}) = 64$ in.

7. From Table 13-2, assuming an unstiffened web, the minimum web thickness is

$$t_w = h/260 \geq \tfrac{1}{2} \text{ in.}$$
$$= 64/260 = 0.25 \text{ in.} < \tfrac{1}{2} \text{ in.}$$

Based on shear strength and web compactness, which will be checked later in this example, a ½-in. web will not be adequate.

Therefore, try $t_w = 0.75$ in.

The trial plate girder cross section resulting from the preliminary considerations above is shown in Figure 13-9.

After the preliminary sizing of the plate girder, the final design is now carried out as follows:

1. Using the preliminary size selected, the self-weight of the plate girder is now calculated.

   Total area of the trial plate girder = (15 in.)(4 in.)(2 flanges) + (64 in.)(0.75 in.)
   = 168 in.^{2}

   With the density of steel of 490 lb./ft.^{3}, the self-weight of the plate girder is

$$\left(\frac{168 \text{ in.}^2}{144}\right)490 \text{ lb./ft.}^3 = 572 \text{ lb./ft.} \approx 0.6 \text{ kips/ft.}$$

(continued)

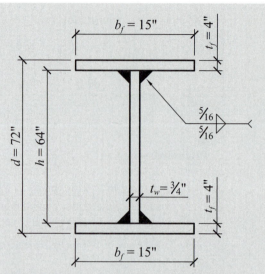

Figure 13-9 Trial plate girder for Example 13-1.

The revised factored shear and bending moment are

$$V_u = 6.6 \text{ kips/ft.}\left(\frac{60 \text{ ft.}}{2}\right) + \frac{806.4 \text{ kips}}{2} = 601 \text{ kips, and}$$

$$M_u = 6.6 \text{ kips/ft.}\frac{(60 \text{ ft.})^2}{8} + \frac{806.4 \text{ kips}(60 \text{ ft.})}{4}$$
$$= 15,066 \text{ ft.-kips, respectively.}$$

2. To avoid using stiffeners, we will ignore the diagonal tension field action and thus ignore the post-buckling strength of the plate girder as well. Consequently, sufficient web thickness will be provided to avoid the use of intermediate stiffeners. The upper limit of the web depth-to-thickness ratio for a compact web is given in Table 13-1 as

$$3.76\sqrt{\frac{E}{F_y}} = 3.76\sqrt{\frac{29,000 \text{ ksi}}{50 \text{ ksi}}} = 90.55$$

The web thickness from step 7 of the preliminary design, $t_w = 0.75$ in.

Therefore, the web-depth-to-thickness ratio is
$h/t_w = 64 \text{ in.}/0.75 \text{ in.} = 85.3 < 90.55$ OK

Therefore, using Table 13-1, we conclude that this built-up section has a compact web and therefore the plastic moment capacity of the section, M_p, is attainable. However, for illustrative purposes, we will still check the four limit states of flexure in Step 3.

$$A_w = (64 \text{ in.})(0.75 \text{ in.}) = 48 \text{ in.}^2$$
$$A_{fc} = (15 \text{ in.})(4 \text{ in.}) = 60 \text{ in.}^2$$

$$\frac{A_w}{A_{fc}} = \frac{48}{60} = 0.80 < 10$$

3. The four possible limit states for bending will now be considered (see Section 13.3).

Unbraced length of the compression flange, $L_b = 0$ (i.e., fully braced)

a. **Compression flange yielding**

$F_y = 50$ ksi

$h_c = h = 64$ in. (welded plate girder with equal flange areas)

$\phi = 0.9$

From equation (13-8),

$$a_w = \frac{h_c t_w}{b_{fc} t_{fc}} = \frac{(64)(0.75)}{(15)(4)} = 0.8 < 10. \text{ OK}$$

From equation (13-7),

$$R_{pg} = 1 - \left(\frac{a_w}{1200 + 300a_w}\right)\left(\frac{h_c}{t_w} - 5.7\sqrt{\frac{E}{F_y}}\right) \leq 1.0$$

$$= 1 - \left(\frac{0.8}{1200 + 300(0.8)}\right)\left(\frac{64}{0.75} - 5.7\sqrt{\frac{29,000}{50}}\right)$$

$$= 1.029 > 1.0.$$

Therefore, use $R_{pg} = 1.0$.

The moment of inertia of the plate girder about the strong axis is

$$I_{x-x} = \left[\frac{(15 \text{ in.})(4 \text{ in.})^3}{12} + (15 \text{ in.})(4 \text{ in.})\left(\frac{64 \text{ in.}}{2} + \frac{4 \text{ in.}}{2}\right)^2\right]$$

$$\times (2 \text{ flanges}) + \frac{(0.75 \text{ in.})(64 \text{ in.})^3}{12}$$

$$= 155,264 \text{ in.}^4.$$

The distance from the elastic neutral axis to the outermost face of the compression flange is

$$y_c = \frac{72 \text{ in.}}{2} = 36 \text{ in.}$$

The elastic section modulus relative to the compression face is

$$S_{xc} = \frac{I_{x-x}}{y_c} = \frac{155,264}{36} = 4312 \text{ in.}^3.$$

(continued)

The design moment capacity for this limit state is

$$\phi M_n = \phi R_{pg} F_y S_{xc} = (0.9)(1.0)(50 \text{ ksi})(4312 \text{ in.}^3)/12$$
$$= 16{,}173 \text{ ft.-kips.}$$

b. **Lateral torsional buckling**
From equation (13-9),

$$r_t \approx \frac{b_{fc}}{\sqrt{12\left(1 + \frac{1}{6}a_w\right)}} = \frac{15 \text{ in.}}{\sqrt{12\left(1 + \frac{1}{6}(0.8)\right)}} = 4.07 \text{ in.}$$

From equation (13-5),

$$L_p = 1.1 r_t \sqrt{\frac{E}{F_y}} = 1.1(4.07 \text{ in.})\sqrt{\frac{29{,}000}{50}} = 107.8 \text{ in.}$$

The floor deck is assumed to be welded to the top of the plate girder; therefore, the lateral unbraced length, $L_b = 0$.
Since $L_b = 0 < L_p$, the lateral torsional buckling limit state is not applicable.

c. **Compression flange buckling**

$$\frac{b_{fc}}{2t_{fc}} = \frac{15 \text{ in.}}{2(4 \text{ in.})} = 1.88 < 0.38\sqrt{\frac{29{,}000 \text{ ksi}}{50 \text{ ksi}}} = 9.15.$$

Therefore, from Table 13-4, this is a compact flange, and $F_{cr} = F_y$.
From equation (13-10),

$$\phi M_n = \phi R_{pg} F_y S_{xc} = (0.9)(1.0)(50 \text{ ksi})(4312 \text{ in.}^3)/12$$
$$= 16{,}173 \text{ ft.-kips.}$$

d. **Tension flange yielding**
Since the plate girder section selected is doubly symmetric, $S_{xt} = S_{xc}$; therefore, the tension flange limit state is not critical. In fact, since this is a compact section, the design moment capacity for tension flange yielding is exactly the same as the design moment capacity for compression flange yielding.

Thus, the design moment capacity for the plate girder, which is the smallest strength of all of the four limit states, is

$$\phi M_n = \mathbf{16{,}173 \text{ ft.-kips}} > M_u = 15{,}066 \text{ ft.-kips.}   \text{OK}$$

4. From Table 13-7, considering an unstiffened web with $\dfrac{h}{t_w} = \dfrac{64 \text{ in.}}{0.75 \text{ in.}}$

$= 85.33 < 260$, therefore, $k_v = 5$

The web depth-to-thickness ratio limits, from Table 13-6, are

$$1.10\sqrt{\frac{k_v E}{F_y}} = 1.10\sqrt{\frac{(5)(29,000\,\text{ksi})}{50\,\text{ksi}}} = 59.23$$

$$1.37\sqrt{\frac{k_v E}{F_y}} = 1.37\sqrt{\frac{(5)(29,000\,\text{ksi})}{50\,\text{ksi}}} = 73.8$$

$$\frac{h}{t_w} = \frac{64\,\text{in.}}{0.75\,\text{in.}} = 85.33 > 73.8$$

Therefore, the web shear coefficient can be calculated from Table 13-6 as

$$C_V = \frac{1.51 E k_v}{(h/t_w)^2 F_y} = \frac{1.51(29,000\,\text{ksi})(5)}{(85.33)^2(50\,\text{ksi})} = 0.6014.$$

5. Based on the choice made in step 2 to ignore diagonal tension field action, determine the design shear strength, ϕV_n.

From equation (13-12), the design shear strength is

$$\phi V_n = \phi 0.6 F_{yw} A_w C_v$$
$$= (0.9)(0.6)(50\,\text{ksi})(64\,\text{in.})(0.75\,\text{in.})(0.6)$$
$$= 779\,\text{kips} > V_u = 601\,\text{kips. OK}$$

Steps 6 and 7 have not been checked in this case, but these can be done using the procedures used previously in Chapter 6.

6. Check the deflections and vibrations as discussed previously in Chapters 6 and 12, respectively.
7. Design the bearing stiffeners at the support reactions or at concentrated load locations as discussed in Chapter 6.
8. Design the fillet welds required at the flange-to-web interface and the stiffener-to-web interface.

Since there are no intermediate stiffeners, only the weld at the web–flange interface will be designed for this plate girder.

The moment of inertia of the plate girder about the strong axis is

$$I_{x-x} = \left[\frac{(15\,\text{in.})(4\,\text{in.})^3}{12} + (15\,\text{in.})(4\,\text{in.})\left(\frac{64\,\text{in.}}{2} + \frac{4\,\text{in.}}{2}\right)^2\right]$$
$$\times (2\,\text{flanges}) + \frac{(0.75\,\text{in.})(64\,\text{in.})^3}{12}$$
$$= 155,264\,\text{in.}^4.$$

The statical moment of the flange area about the elastic neutral axis of the built-up section is

$$Q_f = (15\,\text{in.})(4\,\text{in.})\left(\frac{64\,\text{in.}}{2} + \frac{4\,\text{in.}}{2}\right) = 2040\,\text{in.}^3.$$

(continued)

Maximum factored vertical shear, $V_u = 601$ kips

From equation (13-13), the maximum horizontal shear force per unit length at the web-to-flange fillet weld is

$$v_h = \frac{V_u Q_f}{I_{x-x}} = \frac{(601 \text{ kips})(2040 \text{ in.}^3)}{155,264 \text{ in.}^4}$$

$$= 7.90 \text{ kips/in. for welds on both sides of the web.}$$

Therefore, the shear flow on the fillet weld on each side of the web is $7.90/2 = 3.95$ kips/in.

A $\frac{3}{16}$-in. continuous fillet weld that has a design shear strength of $(1.392 \text{ kips/in.}) \times (3) = 4.18$ kips/in. is adequate to resist the shear flow of 3.95 kips/in. However, the minimum weld size from *AISCM*, Table J2.4, is $\frac{5}{16}$ in. for material $\frac{3}{4}$ in. thick and greater.

With respect to bending and shear, the 72-in.-deep unstiffened plate girder consisting of two 15-in. by 4-in.-thick flange plates and a 64-in. by $\frac{3}{4}$-in.-thick web plate as shown in Figure 13-9 is adequate to resist the applied loads and will suffice as the final girder selection. The reader should check deflections and bearing as was done in Chapter 6 to ensure that they are also within allowable limits.

13.9 REFERENCES

1. Spiegel, Leonard, and George F. Limbrunner. 2002. *Applied structural steel design*, 4th ed. Upper Saddle River, NJ: Prentice Hall.

2. New York State Department of Transportation. 2006. *Bridge Manual*. Albany, NY: NYSDOT.

3. Taly, Narendra. 1998. *Design of Modern Highway Bridges.* New York: McGraw-Hill.

4. American Institute of Steel Construction. 2006. *Steel construction manual*, 13th ed. Chicago: AISC.

13.10 PROBLEMS

13-1. What are plate girders? List some conditions that might warrant their use in building structures.

13-2. List some of the disadvantages of using plate girders in building structures and list one alternative to plate girders in building structures.

13-3. Determine the moment capacity of the welded built-up section shown in Figure 13-10, which has the following dimensions:

$$d = 82 \text{ in.}, \ t_f = 2 \text{ in.}, \ b_f = 13 \text{ in.}, \ t_w = \frac{3}{4} \text{ in.}$$

13-4. What is diagonal tension field action and when does it occur in plate girders? List the conditions under which a diagonal tension field may be considered in the design for shear in plate girders.

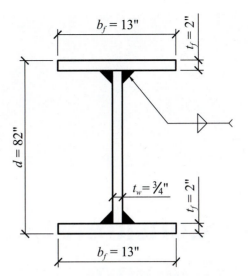

Figure 13-10 Plate girder for Problem 13-3.

13-5. Repeat the design exercise in Example 13-1, taking into account diagonal tension field action.

Student Design Project Problem

13-6. In the student design project building in Figure 1-22, it is desired to create a column-free atrium area at the ground floor level by terminating Columns B-3 and C-3 at the second floor. Design a transfer girder spanning between Column A-3 and Column D-3 to support the second floor and the loads from Columns B-3 and C-3. Neglect diagonal tension field action.

Practical Considerations in the Design of Steel Buildings

14.1 RULES OF THUMB AND PRACTICAL CONSIDERATIONS FOR STRUCTURAL STEEL DESIGN

Some design rules of thumb available for the preliminary sizing of structural steel members, are presented in Table 14-1. In selecting steel members for buildings, the designer should endeavor to use simple details and to use as few different members and connections as possible in order to keep the labor cost down since fabrication and erection accounts for more than 60% of the cost of structural steel buildings, while materials cost accounts for only about 30% [1].

In designing the lateral force resisting system for steel structures in seismic design category (SDC) A, B, or C, an R-value of 3.0, corresponding to "Structural steel systems not specifically detailed for seismic resistance," is recommended in order to reduce the complexity of the member and connection details and hence avoid the increased cost that is associated with using higher R-values. In addition, the use of braced frames for the lateral force resisting system should be given first consideration in the absence of other requirements because of its relative economy when compared to moment frames. For more ductile buildings or structures in SDC D, E, or F, where higher R-values are required, the lateral force resisting systems and other building components must be detailed to satisfy the special seismic detailing requirements in the *AISC Seismic Provisions for Structural Steel Buildings* (AISC 341-05) [2].

If moment frames cannot be avoided, wide-flange sections bending about their strong axis should be used as the columns and girders in these frames, and it is recommended that no one column in the building be part of any two orthogonal moment frames to avoid columns bending about their weak (y-y) axis.

Table 14-1 Design rules of thumb [1], [3], [4]

Structural Steel Components	Design Rules of Thumb
Joists	Joist depth $\geq \dfrac{L}{24}$ (Use $L/20$ for floor joists; deeper joists may be needed to control floor vibrations.)
Beams and girders	Depth $\geq \dfrac{L}{24}$ (Use $L/20$ for floor beams), and $I_{required} \geq \dfrac{wL^3}{64}$ (for uniform loads) or $I_{required} \geq \dfrac{ML}{8}$ (for nonuniform loads), where w = Unfactored uniform load, kips/ft., L = Beam or girder span, ft., M = Unfactored maximum moment, ft.-kips, and $I_{required}$ = Required moment of inertia, in.^{4}
Trusses	Depth $\geq \dfrac{L}{12}$
Plate girders	Depth $\geq \dfrac{L}{8}$ to $\dfrac{L}{12}$
Continuous girders	Locate hinge or splice locations between $0.15L$ and $0.25L$, where L = Length of the back span.
Cantilevered roof system	Make cantilever length between $0.15L$ and $0.25L$, where L = Length of the back span.
Braced frames and moment frames subject to wind loads	Limit lateral drift due to wind to between $H/400$ and $H/500$, where H = Total height of building.
Columns	Practical minimum size to accommodate beam-to-column and girder-to-column connections is a W8 (some engineers use a minimum of W10 for columns). The *approximate* axial load capacity for W-shapes (grade 50 steel) not listed in the column load tables is $\phi P_n \approx 1.5\left(30 - 0.15\dfrac{KL}{r}\right)A_g,$ where L = Unbraced length, R = Minimum radius of gyration, and A_g = Gross area of column.
Optimum bay sizes (roof and floor)	Most economical bay size: • Rectangular bay size with length-to-width ratio between 1.25 and 1.50 • Bay area ≈ 1000 ft.^{2}

14.2 LAYOUT OF STRUCTURAL SYSTEMS IN STEEL BUILDINGS

The practical and economic considerations necessary to achieve optimum and functional layout for the horizontal (roof and floor) systems and vertical lateral force resisting structural systems in steel buildings are discussed in this section.

Layout of Floor and Roof Framing Systems

The design process for a steel building begins with a review of the architectural drawings and the laying out of the roof and floor framing and lateral force resisting systems, taking into account architectural, mechanical, and electrical considerations. Issues such as headroom limitations, roof or floor openings, and door and window openings must be kept in mind when laying out the structural system. The different types of floor and roof framing schemes used commonly in design practice are presented in this section. The following design considerations should be taken into account when selecting floor and roof framing systems.

1. For rectangular bays, the beams and joists should span in the longer direction, while the girders should span in the shorter direction in order to produce optimum sizes for the structural framing.

2. The roof and floor deck should be framed in a direction perpendicular to the beams and joists.

3. The framing around stair and floor or roof openings should consist of wide-flange sections, except for very short spans and lightly loaded members where channel or angle sections may suffice. It should be noted that at stair openings, the edge framing also supports the stair stringer reactions and partition wall loads.

4. Use wide-flange steel beams and girders along column grid lines, along the grid lines with vertical lateral force resisting systems, and along the perimeter of the building. Open-web steel joists should not be used along the perimeter of a building since they cannot provide adequate lateral support for exterior cladding perpendicular to their span; they should also not be used along grid lines that have lateral force resisting systems, unless the joists are specifically designed for these loads.

5. Heavy partition walls should be directly supported directly by a floor beam or girder located under and running parallel to the wall.

6. Roof or floor openings should be located offset from the column grid lines in order to minimize interruption of the beam and girder framing.

Several examples of different roof and floor framing layouts are shown in Figures 14-1 through Figure 14-5 [5].

Layout of Lateral Force Resisting Systems

The simple shear connections between horizontal members (i.e., beams or girders) and the vertical linear members (i.e., columns) in steel buildings behave as pinned joints, and therefore do not provide inherent lateral stability to the structure. Consequently, steel buildings must be adequately braced for lateral loads by using braced frames, moment frames, shear walls, or a combination of these lateral force resisting systems (LFRS). Moment frames provide the most flexibility with regard to architectural considerations since door or window

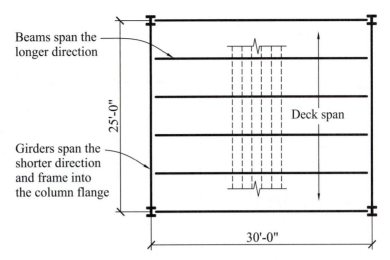

Figure 14-1 Simple roof/floor framing layout.

layouts do not have any adverse effect on the layout of moment frames. On the other hand, since the main lateral force resisting members in braced frames are the diagonal brace members, care has to be taken to ensure that the locations of these members do not conflict with the locations of doors or windows. Careful coordination with the architect is necessary at the preliminary design stage to avoid the unpleasant situation of having to cut or modify the diagonal member in a braced frame because a door or window opening is interrupted by the brace member.

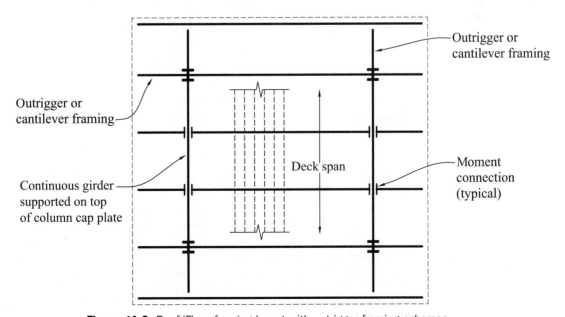

Figure 14-2 Roof/Floor framing layout with outrigger framing schemes.

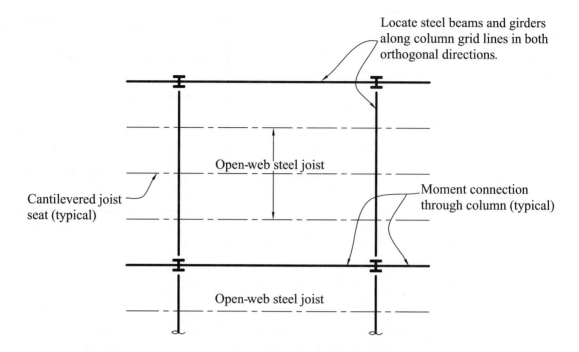

Locate steel beams and girders along column grid lines in both orthogonal directions.

Open-web steel joist

Cantilevered joist seat (typical)

Moment connection through column (typical)

Open-web steel joist

a. Framing plan

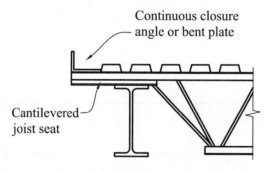

Continuous closure angle or bent plate

Cantilevered joist seat

b. Joist seat extension

Figure 14-3 Roof framing with cantilevered a joist seat.

Braced frames should preferably be located along column lines at partition wall locations to help conceal the diagonal brace within the wall, unless the architect intends to expose the diagonal braces for aesthetic purposes. The size of the diagonal brace member should be selected to fit within the partition wall thickness. X-braced frames with single- or double-angle diagonal braces may require the use of gusset plates at the intersection of the diagonal members to ensure that the intersecting diagonal X-brace members lie on the same vertical plane, thus minimizing the required wall thickness necessary to accommodate the diagonal brace members. A typical layout of diagonal bracing in a floor or roof framing plan is shown in Figure 14-6. The vertical bracing may be located on the interior or the perimeter of the building.

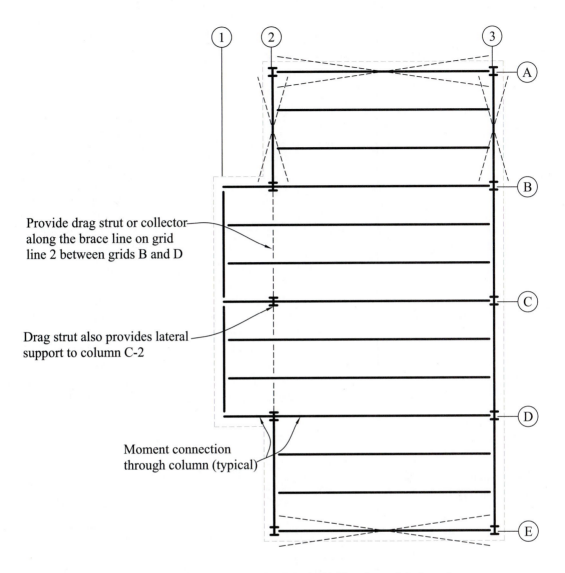

Provide drag strut or collector along the brace line on grid line 2 between grids B and D

Drag strut also provides lateral support to column C-2

Moment connection through column (typical)

Note: this framing minimizes the number of moment connections

Figure 14-4 Roof/Floor framing layout with interruption of drag strut and chords.

Masonry shear walls are frequently used to resist lateral forces in steel buildings, in addition to serving as exterior cladding for wind loads perpendicular to the face of the wall. In this system, the steel framing resists only gravity loads, and the connection between the steel beams or girders and the masonry shear wall is detailed so as to prevent the gravity load from being transferred from the steel roof or floor framing into the masonry shear wall below. This is achieved by using a connection detail with vertically slotted bolt holes as shown in Figure 14-7. This allows the lateral load to be transferred from the roof or floor diaphragm to the masonry shear wall without transmitting the gravity loads to the shear wall. The structural steel is usually erected

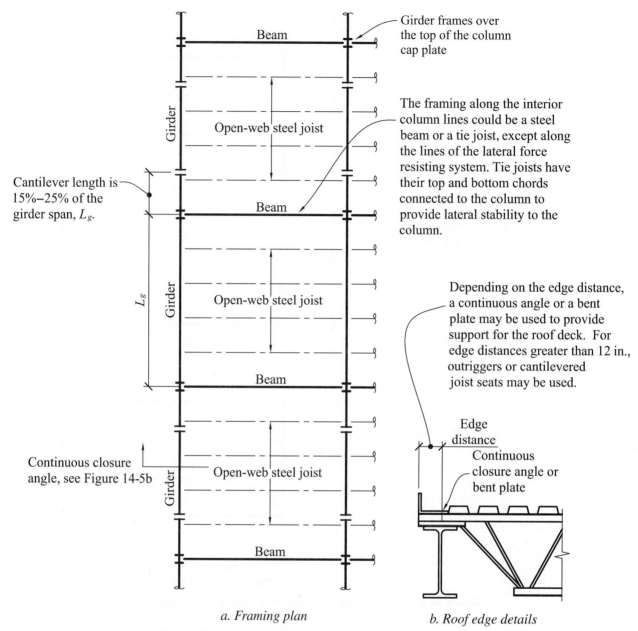

Girder frames over the top of the column cap plate

The framing along the interior column lines could be a steel beam or a tie joist, except along the lines of the lateral force resisting system. Tie joists have their top and bottom chords connected to the column to provide lateral stability to the column.

Cantilever length is 15%–25% of the girder span, L_g.

Open-web steel joist

Beam

Girder

L_g

Open-web steel joist

Beam

Depending on the edge distance, a continuous angle or a bent plate may be used to provide support for the roof deck. For edge distances greater than 12 in., outriggers or cantilevered joist seats may be used.

Continuous closure angle, see Figure 14-5b

Girder

Open-web steel joist

Beam

Edge distance

Continuous closure angle or bent plate

a. Framing plan

b. Roof edge details

Figure 14-5 Cantilevered roof framing.

before the masonry shear walls are built; therefore, the building must be temporarily braced (usually with guy wires) during construction until the masonry shear walls are built.

Masonry shear walls may also be used in infill frames where the block walls are built tightly to the columns and to the underside of the steel beams and girders. In this system, both the gravity loads and the lateral loads are resisted by the steel beam/girder and masonry wall combination.

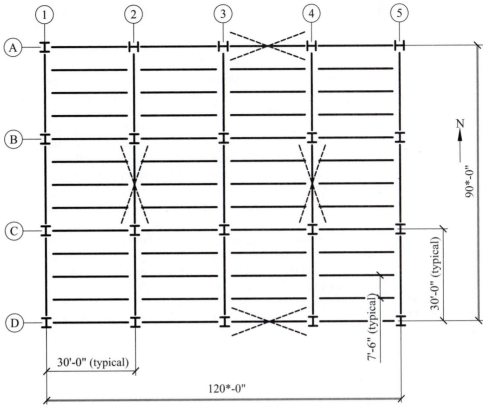

Figure 14-6 Diagonal bracing layout.

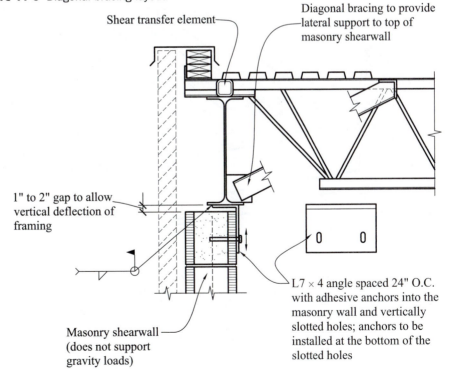

Figure 14-7 Lateral force transfer details.

14.3 DIAPHRAGM ACTION OF ROOF AND FLOOR DECKS

In addition to supporting gravity loads, the roof and floors of steel buildings also act as horizontal diaphragms in resisting lateral loads such as wind or earthquake loads and help to transfer the lateral load to the lateral force resisting system. The seismic force for the design of diaphragms, chords, and drag struts is determined using the larger of the F_x seismic lateral force from equations 3-5, 3-7, or 3-14 and the F_{px} diaphragm seismic lateral force from equation 12.10-1 of ASCE 7 [25]. The horizontal diaphragm (see Figure 14-8) consists of the roof or floor deck, the diaphragm chord members perpendicular to the lateral force, and the drag struts or collectors that are parallel to and lie along the same line as the lateral force resisting system [6]. The function of drag struts or collectors is to transfer or "collect" (or drag) the lateral force from the roof or floor diaphragm into the lateral force resisting system while ensuring that the diaphragm shear is uniform. Drag struts, which resist axial tension and compression forces due to wind and seismic lateral forces acting in both directions, should be provided along the drag lines (i.e., lines of lateral force resisting systems) for the entire length and width of the roof and floor diaphragms [7]. The drag beams and their connections have to be designed for these axial forces acting in combination with the gravity loads. The beams and girders along the drag lines should be designed as noncomposite members for gravity loads, even for floors with shear studs, because of the possibility of stress reversals that

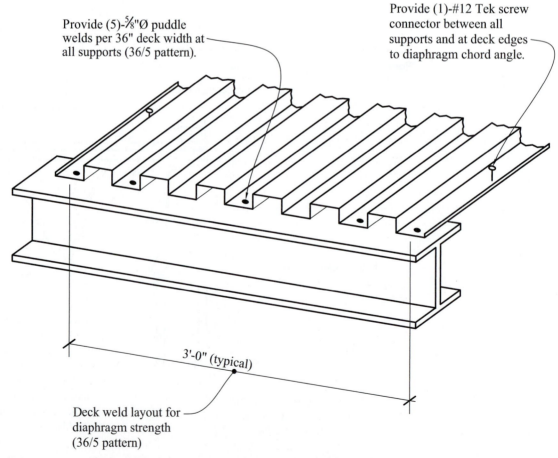

Provide (5)-$\frac{5}{8}$"Ø puddle welds per 36" deck width at all supports (36/5 pattern).

Provide (1)-#12 Tek screw connector between all supports and at deck edges to diaphragm chord angle.

3'-0" (typical)

Deck weld layout for diaphragm strength (36/5 pattern)

Figure 14-8 Deck diaphragm weld layout details.

may occur due to upward bending caused by the axial compression forces in the drag beams. In designing the drag beams for the axial compression forces, the unbraced length should be taken as the smaller of the length of the beam between column supports and the distance between adjacent infill framing members that are supported by the drag beam. The use of drag struts reduces the diaphragm shear compared to cases where the lateral force is transferred from the deck directly to the lateral force resisting system (LFRS) only within the length of the LFRS. To illustrate these scenarios, if there were no drag struts between points B and F in Figure 14-9a, the diaphragm shear would be 50 kips/ 20 ft. = 2.5 kips/ft. since the lateral shear would be transferred only between points A and B (i.e., distributed over the length of the lateral force resisting system only), instead of the 0.5 kip/ft. (50 kips/100 ft.) diaphragm shear obtained when drag struts are provided between points B and F, making the full 100-ft. length of the diaphragm effective. It is recommended that roof or floor openings should not be located adjacent to the drag struts, as this increases the shear in the diaphragm and the forces in the drag struts. The diaphragm action of the roof or floor deck results in the in-plane bending of the diaphragm, with the metal deck subjected to shear forces (or stresses) and the chord members and drag struts subjected to axial tension and compression forces.

The rigidity of a horizontal diaphragm affects the distribution of lateral forces to the lateral force resisting systems in a building as already discussed in Chapter 3. In general, for steel buildings, concrete-on-metal deck floors can usually be classified as a rigid diaphragm if the maximum span-to-width ratio of the diaphragm between adjacent and parallel lateral force resisting systems does not exceed 3:1; roof decks can be classified as rigid diaphragms if the maximum span-to-width ratio does not exceed 2:1 [8]. As discussed in Chapter 3 (see Figure 3-10), the typical diaphragm behaves like a horizontal beam in which the metal deck or concrete-on-metal deck web member, acting as the web member, resists the lateral shear. Chord members resist the tension–compression couple caused by the in-plane moments due to bending of the diaphragm between the adjacent and parallel lateral force resisting systems, and the drag struts resist tension and compression forces that result from the difference between the uniform deck diaphragm shear and the lateral shear in the lateral force resisting system.

The focus in this section is on the design of the horizontal diaphragms for lateral loads. The chord and drag strut forces are usually shown on the structural plans or on elevation views of the lateral force resisting systems and these forces are used for the design of the connections. The design of the chords and drag struts, after the compression and tension forces are determined, follows the same approach as presented in Chapters 4 and 5. The common standard width for metal, roof or floor decks is usually 36 in. center to center of the end flutes; they are usually available in lengths up to 40 ft. The common practice is for the deck to be fastened to the steel framing members using 5/8-in.-diameter puddle welds in the flutes. Sometimes screws or powder-actuated fasteners are also used. The sidelap connections or the attachment of adjacent deck panels to each other, and the attachment of deck panels to the diaphragm chords, are made using self-drilling screws, button punching, or welds with a spacing not exceeding 2 ft. 6 in [9]. The number of puddle welds for each standard 36 in. width of deck can vary from a minimum of three (i.e., 36/3 pattern) to a maximum of nine (i.e., 36/9 pattern) (see Figure 14-8). The diaphragm shear strength of a roof or floor deck is a function of the number of puddle welds used within a standard deck width and the type and spacing of sidelap connections. A typical diaphragm lateral shear strength table, from the Steel Deck Institute (SDI) *Diaphragm Design Manual* (SDI *DDM03*) [9], is shown in Table 14-2. When the required lateral shear strength cannot be achieved using puddle welds and sidelap connections, different and stronger connection materials may be used or a horizontal bracing system in the plane of the roof or floor may be provided on the underside of the deck or, alternatively, more lateral force resisting systems could be provided to reduce

Table 14-2 Nominal shear strength for selected diaphragm assemblies

1.5 in. (WR, IR, NR) $t = 0.0358$ in. (20 gage)		Support fasteners: 5/8-in. puddle welds (or equivalent) Sidelap fasteners: #10 screws									
		Nominal shear strength, plf									
		Span, ft.									
Fastener layout	**Sidelaps per span**	**4.0**	**4.5**	**5.0**	**5.5**	**6.0**	**6.5**	**7.0**	**7.5**	**8.0**	**K1**
36/7	0	1035	915	820	740	675	620	575	530	495	0.535
	1	1220	1090	975	880	805	–	–	–	–	0.415
	2	1385	1245	1130	1020	930	855	795	735	690	0.340
	3	1540	1390	1265	1160	1060	975	900	840	785	0.287
	4	1695	1530	1395	1280	1185	1095	1010	940	880	0.249
	5	1835	1665	1525	1400	1295	1205	1120	1045	(a)	0.219
	6	1970	1795	1645	1515	1405	1305	(b)	(c)	(d)	0.196
36/5	0	945	845	760	685	625	575	530	490	455	0.642
	1	1100	995	905	825	750	–	–	–	–	0.477
	2	1245	1130	1030	950	880	810	750	695	650	0.380
	3	1375	1255	1150	1065	985	920	860	800	745	0.315
	4	1495	1370	1265	1170	1090	1015	950	895	840	0.270
	5	1605	1480	1370	1270	1185	1110	1040	980	925	0.236
	6	1705	1580	1465	1365	1280	1200	1130	(e)	(f)	0.209
36/4	0	725	640	575	515	470	430	395	365	340	0.803
	1	875	795	725	655	600	–	–	–	–	0.561
	2	1015	925	845	780	725	665	615	570	535	0.431
	3	1135	1040	960	890	825	770	725	675	630	0.350
	4	1240	1145	1060	990	925	865	810	765	725	0.294
	5	1335	1240	1155	1080	1010	950	895	845	800	0.254
	6	1415	1325	1240	1165	1095	1030	975	920	875	0.224

Notes: Courtesy of Steel Deck Institute, Fox River Grove, IL

1. Shaded values do not comply with the minimum spacing requirements for sidelap connections and shall not be used except with properly spaced sidelap connections.
2. Strength factors: $\phi = 0.55$ (EQ) $\Omega = 3.00$ (EQ)
                                     $= 0.70$ (Wind) $= 2.35$ (Wind)
                                     $= 0.60$ (Other) $= 2.65$ (Other)
3. Adjusted shear strength due to panel buckling:
   a. Narrow Rib (NR): 815, Intermediate Rib (IR): 865, Wide Rib (WR): 975
   b. NR: 1065, IR: 1130, WR: 1220
   c. NR: 925, IR: 985, WR: 1145
   d. NR: 815, IR: 865, WR: 1070
   e. NR: 925, IR: 985, WR: 1065
   f. NR: 815, IR: 865, WR: 1005

the span of the diaphragm and hence the diaphragm shear. If a diaphragm system is not capable of transferring lateral loads, as in the case of diaphragms with glass skylights, a horizontal bracing system in the plane of the roof must be provided on the underside of the skylight to transfer the lateral loads to the vertical lateral force resisting system.

EXAMPLE 14-1

Design of Horizontal Diaphragms

The roof framing for a five-story building with a 15-ft. floor-to-floor height is similar to that shown in Figure 14-6. The roof deck is 1½-in. by 20-gage metal deck and the floor deck is 2½-in. concrete on 3-in. by 20-gage galvanized composite metal deck. The building is subjected to a uniform 25-psf wind load in the East–West direction. Determine the required size and spacing of the puddle weld and sidelap connection, the chord forces along grid lines 1 and 5, and the drag strut forces along lines A and D at the roof level. Assume a standard 36-in. deck width.

SOLUTION

The horizontal diaphragm is modeled as a simply supported horizontal beam where the supports are the LFRS at both ends of the diaphragm.

For the East–West wind, the diaphragm spans between the X-brace frames along lines A and D in Figure 14–6. Therefore,

Diaphragm span, l = 90 ft. (for East–West wind), and

Diaphragm width (parallel to lateral force), b = 120 ft.

The load tables from the SDI *Diaphragm Design Manual* will be used to determine the diaphragm shear strength that will be compared to the applied factored unit shear [9].

Factored uniform horizontal load on the roof diaphragm, $= (1.6)(25 \text{ psf})\left(\dfrac{15 \text{ ft.}}{2}\right)$

$= 300 \text{ lb./ft.}$

Maximum factored shear in the diaphragm, $V = 300 \text{ lb./ft.} \left(\dfrac{90 \text{ ft.}}{2}\right) =$

13,500 lb.

Maximum factored moment in the diaphragm, $M_{du} = 300 \text{ lb./ft.} \dfrac{(90 \text{ ft.})^2}{8} =$

303,750 ft.-lb.

Applied factored unit shear in the diaphragm,

$$v_d = \left(\frac{\text{Lateral shear, } V}{\begin{array}{c}\text{Length of diaphragm} \\ \text{parallel to LFRS}\end{array} - \begin{array}{c}\text{Cumulative length of diaphragm openings} \\ \text{adjacent to drag struts}\end{array}}\right).$$

$$= \frac{13,500 \text{ lb.}}{(120 \text{ ft.} - 0 \text{ ft.})}$$

$$= 112.5 \text{ lb./ft.}$$

(continued)

Determine the diaphragm unit shear strength from the SDI *Diaphragm Design Manual*:

Thickness of deck (20 gage) = 0.0358 in.

Span of deck (between adjacent steel beams or joists) = 7.5 ft.

Assume four 5/8-in.-diameter puddle welds on a 36-in. deck section (i.e., 36/4 pattern; this is the minimum weld configuration presented in the SDI *Diaphragm Design Manual* for the 36-in. standard deck width). Assume #10 TEK screws sidelap connection spaced at no more than 30 in. on center (o.c.) (7.5 ft./2.5 ft. = 3 spaces, which implies two (i.e., 3 spaces − 1) sidelap connections within the span.

From Table 14-2 (reproduced from page AV-14 of the SDI *Diaphragm Design Manual*), the nominal diaphragm shear strength is v_n = 570 lb./ft.

The design diaphragm shear strength for a 20-gage deck with a span of 7.5 ft. = ϕv_n = (0.70)(570 lb./ft.) = 399 lb./ft. > 112.5 lb./ft., (where ϕ = 0.70 for wind) OK

Chord Forces:

$$\text{Maximum factored chord force, } T_u = C_u = \frac{M_{du}}{b}$$

$$= \frac{303{,}750 \text{ ft.-lb.}}{120 \text{ ft.}} = 2531 \text{ lb.} = 2.53 \text{ kips}$$

The chord member for this roof framing would typically be a continuous deck closure angle.

Assuming a continuous L3 × 3 × ¼ (ASTM A36 steel), the tensile capacity of the angle is

$$\phi P_n = (0.9)(1.44 \text{ in.}^2)(36 \text{ ksi}) = 46.7 \text{ kips} > 2.53 \text{ kips. OK}$$

EXAMPLE 14-2

Drag Strut Forces

The building in Figures 14-9a through 14-9f is subjected to a factored 100-kips lateral force at the roof level in the North–South direction. For each X-brace configuration and layout, calculate the drag strut forces and determine which X-brace configuration results in the lowest drag strut forces.

SOLUTION

The lateral shear force to be transferred from the diaphragm to the drag struts along the vertical brace lines is

$$V = \frac{100 \text{ kips}}{2} = 50 \text{ kips}$$

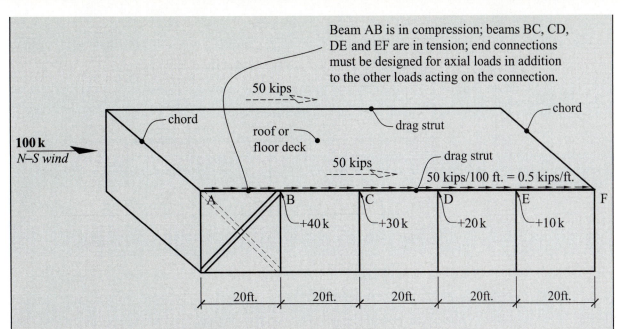

Beam AB is in compression; beams BC, CD, DE and EF are in tension; end connections must be designed for axial loads in addition to the other loads acting on the connection.

Drag strut beam AB is in compression for N–S wind:

$$P_{uAB} = (0.5 \text{ kips/ft.})(20 \text{ ft.}) = 10 \text{ kips}$$

Drag strut beams BC, CD, DE and EF are in tension:

$$T_{uBC} = (0.5 \text{ kips/ft.})(20 \text{ ft.} + 20 \text{ ft.} + 20 \text{ ft.} + 20 \text{ ft.}) = 40 \text{ kips}$$
$$T_{uCD} = (0.5 \text{ kips/ft.})(20 \text{ ft.} + 20 \text{ ft.} + 20 \text{ ft.}) = 30 \text{ kips}$$
$$T_{uDE} = (0.5 \text{ kips/ft.})(20 \text{ ft.} + 20 \text{ ft.}) = 20 \text{ kips}$$
$$T_{uEF} = (0.5 \text{ kips/ft.})(20 \text{ ft.}) = 10 \text{ kips}$$

Figure 14-9a

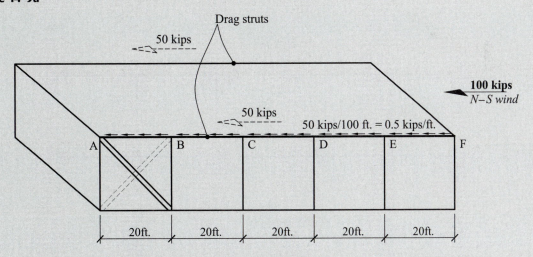

Drag strut beams AB, BC, CD, DE and EF are in compression:

$$P_{uAB} = (0.5 \text{ kips/ft.})(20 \text{ ft.} + 20 \text{ ft.} + 20 \text{ ft.} + 20 \text{ ft.} + 20 \text{ ft.}) = 50 \text{ kips}$$
$$P_{uBC} = (0.5 \text{ kips/ft.})(20 \text{ ft.} + 20 \text{ ft.} + 20 \text{ ft.} + 20 \text{ ft.}) = 40 \text{ kips}$$
$$P_{uCD} = (0.5 \text{ kips/ft.})(20 \text{ ft.} + 20 \text{ ft.} + 20 \text{ ft.}) = 30 \text{ kips}$$
$$P_{uDE} = (0.5 \text{ kips/ft.})(20 \text{ ft.} + 20 \text{ ft.}) = 20 \text{ kips}$$
$$P_{uEF} = (0.5 \text{ kips/ft.})(20 \text{ ft.}) = 10 \text{ kips}$$

Figure 14-9b

(continued)

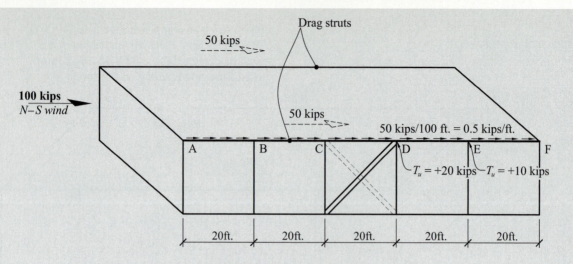

Drag strut beams AB, BC and CD are in compression:

$$P_{uAB} = (0.5 \text{ kips/ft.})(20 \text{ ft.}) = 10 \text{ kips}$$
$$P_{uBC} = (0.5 \text{ kips/ft.})(20 \text{ ft.} + 20 \text{ ft.}) = 20 \text{ kips}$$
$$P_{uCD} = (0.5 \text{ kips/ft.})(20 \text{ ft.} + 20 \text{ ft.} + 20 \text{ ft.}) = 30 \text{ kips}$$

Drag strut beams DE and EF are in tension:

$$T_{uDE} = (0.5 \text{ kips/ft.})(20 \text{ ft.} + 20 \text{ ft.}) = 20 \text{ kips}$$
$$T_{uEF} = (0.5 \text{ kips/ft.})(20 \text{ ft.}) = 10 \text{ kips}$$

Figure 14-9c

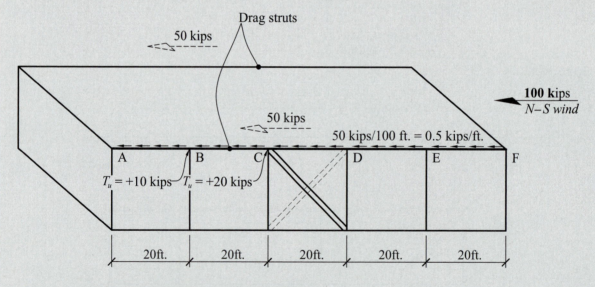

Drag strut beams CD, DE and EF are in compression:

$$P_{uCD} = (0.5 \text{ kips/ft.})(20 \text{ ft.} + 20 \text{ ft.} + 20 \text{ ft.}) = 30 \text{ kips}$$
$$P_{uDE} = (0.5 \text{ kips/ft.})(20 \text{ ft.} + 20 \text{ ft.}) = 20 \text{ kips}$$
$$P_{uEF} = (0.5 \text{ kips/ft.})(20 \text{ ft.}) = 10 \text{ kips}$$

Drag strut beams AB and BC are in tension:

$$T_{uAB} = (0.5 \text{ kips/ft.})(20 \text{ ft.}) = 10 \text{ kips}$$
$$T_{uBC} = (0.5 \text{ kips/ft.})(20 \text{ ft.} + 20 \text{ ft.}) = 20 \text{ kips}$$

Figure 14-9d

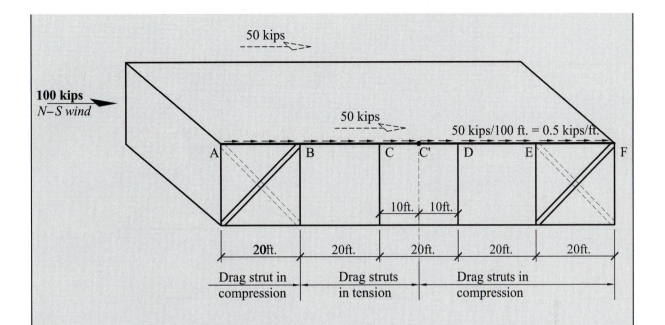

Drag strut beam AB is in compression:

$$P_{uAB} = (0.5 \text{ kips/ft.})(20 \text{ ft.}) = 10 \text{ kips}$$

Drag strut beams DE, EF, and ½C'D are in compression:

$$P_{uC'D} = (0.5 \text{ kips/ft.})(0.5)(20 \text{ ft.}) = 5 \text{ kips}$$
$$P_{uDE} = (0.5 \text{ kips/ft.})[(0.5)(20 \text{ ft.}) + 20 \text{ ft.}] = 15 \text{ kips}$$
$$P_{uEF} = (0.5 \text{ kips/ft.})[(0.5)(20 \text{ ft.}) + 20 \text{ ft.} + 20 \text{ ft.}] = 25 \text{ kips}$$

Drag strut beams BC and ½CC' are in tension:

$$T_{uC'C} = (0.5 \text{ kips/ft.})(0.5)(20 \text{ ft.}) = 5 \text{ kips}$$
$$T_{uBC} = (0.5 \text{ kips/ft.})[(0.5)(20 \text{ ft.}) + 20 \text{ ft.}] = 15 \text{ kips}$$

Figure 14-9e

Assuming that the drag struts extend through the full length of the building and since there are no openings in the diaphragm adjacent to the drag struts, the uniform unit factored shear in the diaphragm

$$= \frac{\text{Lateral shear, } V}{\begin{array}{c}\text{Length of diaphragm} \\ \text{parallel to LFRS}\end{array} - \begin{array}{c}\text{Cumulative length of diaphragm openings} \\ \text{adjacent to drag struts}\end{array}}$$

$$= \frac{50 \text{ kips}(1000)}{100 \text{ ft.} - 0 \text{ ft.}} = 500 \text{ lb/ft.} = 0.5 \text{ kips/ft.}$$

The reader should refer to Figures 14-9a to 14-9f for the drag strut forces. For the single X-brace frame the configuration with a symmetrically located X-braced frame (i.e., X-braced frame located within bay CD), yields

(continued)

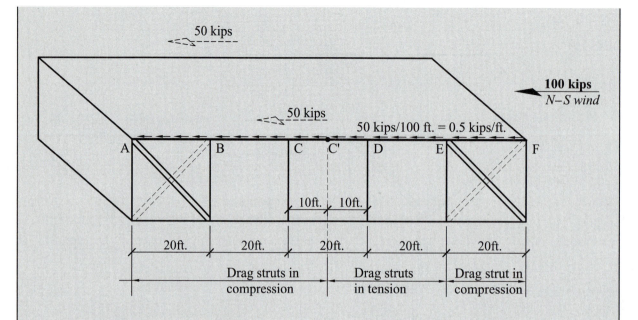

Figure 14-9f

Drag strut beam EF is in compression:

$P_{uEF} = (0.5 \text{ kips/ft.})(20 \text{ ft.}) = 10 \text{ kips}$

Drag strut beams AB, BC, and ½C'C are in compression:

$P_{uC'C} = (0.5 \text{ kips/ft.})(0.5)(20 \text{ ft.}) = 5 \text{ kips}$
$P_{uBC} = (0.5 \text{ kips/ft.})[(0.5)(20 \text{ ft.}) + 20 \text{ ft.}] = 15 \text{ kips}$
$P_{uAB} = (0.5 \text{ kips/ft.})[(0.5)(20 \text{ ft.}) + 20 \text{ ft.} + 20 \text{ ft.}] = 25 \text{ kips}$

Drag strut beams DE and ½C'D are in tension:

$T_{uC'D} = (0.5 \text{ kips/ft.})(0.5)(20 \text{ ft.}) = 5 \text{ kips}$
$T_{uDE} = (0.5 \text{ kips/ft.})[(0.5)(20 \text{ ft.}) + 20 \text{ ft.}] = 15 \text{ kips}$

the smallest drag strut force. If the lateral load on the building is caused by wind, the drag strut force calculated in this example is substituted, as W, directly into the load combination equations in Section 2.3 to determine the design loads for the drag beams. However, if the lateral load is caused by earthquake, or seismic effects, a special seismic force has to be calculated for the drag struts as discussed in Section 2.3 before substituting into the load combination equations to determine the design loads. The special seismic force is

$$E_m = \Omega_0 Q_E \pm 0.2 S_{DS}D,$$

where

$\Omega_0 = $ Overstrength factor from ASCE-7, Table 12.2-1, (Ensures that structure remains elastic during a seismic event.)

$Q_E = $ Drag strut force due to horizontal earthquake

= Drag strut forces calculated in Figures 14-9a through 14-9f,

D = Dead load supported by the drag beams,

S_{DS} = Design spectral response acceleration at short period (see Section 2.3 and Chapter 3),

$\Omega_0\, Q_E$ = horizontal component of the earthquake force on the drag strut, and

$0.2\, S_{DS}D$ = vertical component of the earthquake force on the drag strut.

To illustrate the calculation of this special seismic force for the drag beam BC in Figures 14-9a and 14-9b, assume $S_{DS} = 0.27$, and that the building is located in SDC C, and can be classified as "Steel systems not specifically detailed for seismic resistance" (i.e., System H in ASCE-7, Table 12.2-1); therefore, the overstrength factor is $\Omega_0 = 3.0$.

The horizontal seismic forces in drag beam BC is $\Omega_0\, Q_E = (3.0)(\pm 40 \text{ kips}) = 120$ kips (tension and compression). The vertical seismic force in drag beam BC $= 0.2\, S_{DS}D = (0.2)(0.27)D = 0.054D$. Therefore, the special seismic force in the drag beam, $E_m = (40 \pm 0.054D)$ kips. To complete the design load calculations for the drag beams, this special seismic force is substituted, as the seismic force, E, into the load combination equations in Section 2.3.

Thus, the drag beam-to-column connections will be subjected to combined shear and axial loads. The design of simple connections for shear and axial loads has been covered in Chapters 9 and 10. There are several possible connection details that could be used to transfer the drag strut force and gravity loads, but the most economical method is to use a simple shear connection. In Figures 14-10a and 14-10b, the connection will be subjected to vertical and horizontal loads. For the single plate connection (Figure 14-10a), the bolts will subjected to shear only and the weld to the column will be subjected to both shear and tension. In Figure 14-10b, the weld to the supported beam (or drag strut) will be subjected to shear and tension and the bolts connecting the double angles to the column will also be subjected to both shear and tension. The double angles will be subjected as well to prying action (see Chapter 9). An alternative detail would be to provide a shear connection to resist the gravity loads only and then provide an additional member to transfer the drag strut force from the beam to the column (see Figure 14-10c).

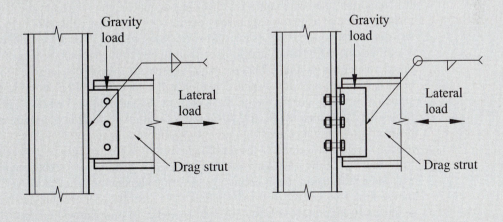

a. single plate

b. double angle

Figure 14-10 Drag strut-to-column connections.

(continued)

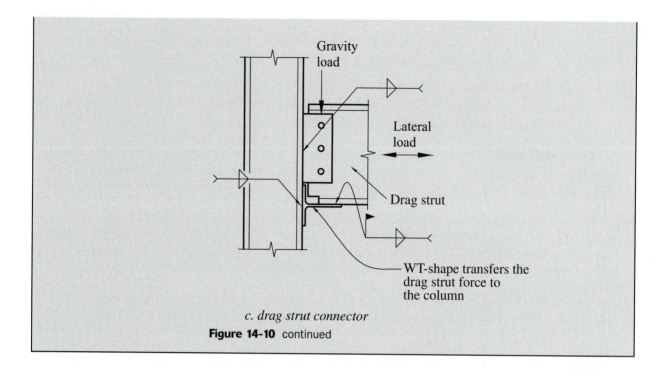

c. drag strut connector

Figure 14-10 continued

14.4 TRANSFER OF LATERAL LOADS FROM ROOF AND FLOOR DIAPHRAGMS TO LATERAL FORCE RESISTING SYSTEMS

It is important that the diaphragm shear be adequately transferred to the lateral force resisting system; as previously discussed, the drag struts or collectors help make this transfer of lateral shear forces possible. Where the roof or floor diaphragm is attached directly to the infill beams and girders, the transfer of the lateral shear is direct, since the perimeter beams and girders act as the chords and drag struts, and the tops of these beams and girders are at the same elevation as the underside of the deck. However, one situation in which proper lateral shear transfer is commonly overlooked is shown in Figure 14-11. In this detail, the joists are framed in a direction perpendicular to the beam or girder framing, and without a drag strut or collector element located between the joist seats and welded to the top flange of the roof beams along the vertical brace lines, the web of the joist seats will be subjected to bending or racking about its weak axis when the lateral force is acting in a direction perpendicular to the joists. To prevent this racking effect of the joist seats, a drag strut shear transfer element can be provided as shown in Figures 14-10a and 14-10b. However, testing has indicated that standard 2½ inch joist seats have a nominal rollover strength in excess of 8,000 lb. Using an appropriate factor of safety yields a safe working load of 1,920 lb at each joist seat, as shown in Figure 14-10c for joist girders with 7½ inch seats, the safe working load is 4,000 lb [10]. The size and spacing of the puddle weld connection between the deck and the drag strut element are dependent on the lateral shear transferred from the deck diaphragm to the shear transfer element.

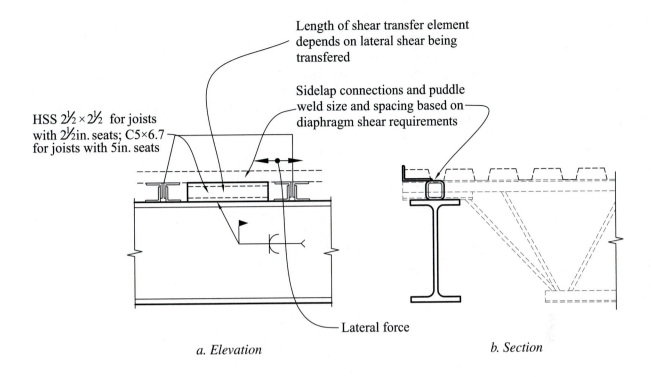

Length of shear transfer element depends on lateral shear being transfered

Sidelap connections and puddle weld size and spacing based on diaphragm shear requirements

HSS $2\frac{1}{2} \times 2\frac{1}{2}$ for joists with $2\frac{1}{2}$ in. seats; C5×6.7 for joists with 5in. seats

Lateral force

a. Elevation

b. Section

$P = 1920$ lb ($2\frac{1}{2}$" seats)
$P = 4000$ lb ($7\frac{1}{2}$" seats)

P   P

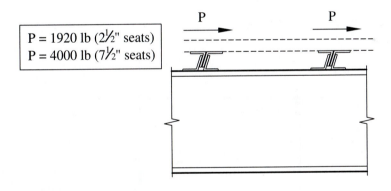

c. Joist seat rollover strength

Figure 14-11 Drag strut shear transfer elements.

14.5 GIRTS AND WIND COLUMNS

For exterior wall systems in steel buildings, the maximum vertical distance that the metal wall cladding can span is usually limited, and girts or horizontal bending members may need to be used to reduce the vertical span of the cladding. The girts, which are usually C-shaped rolled sections or light-gage C- or Z-shaped sections, span between the building columns and the wind columns, and are usually oriented with the channel toes pointing downwards to prevent moisture or debris from accumulating. The channels support their self-weight through bending about their weak axis, and for this reason, hanger or sag rods are typically used in these situations to limit the vertical deflection of the channel girt. The sag rods also provide lateral bracing for the girts for bending about their strong axis in resisting wind loads from the exterior wall cladding (see Figure 14-12a).

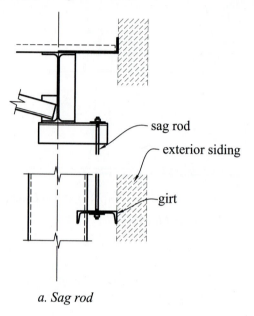

a. Sag rod

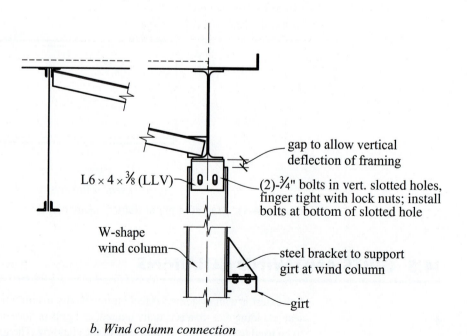

b. Wind column connection

Figure 14-12 Sag rod and wind column details.

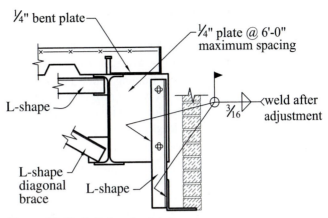

Figure 14-13 Relief angle details.

Wind columns are vertical structural members that resist wind loads, but do not support axial loads. They span from floor to floor and resist bending due to wind load applied perpendicular to the face of the exterior wall. Wind columns are used when the exterior wall cladding system is unable to span from floor to floor and therefore girts, which span between the wind columns and the building columns, have to be used. Since the wind columns are not usually gravity-load-supporting elements, the connection details for the beams or girders at the top of the column must allow for a vertical deflection gap to enable the floor or roof above to deflect without imposing any gravity load on the wind column. The deflection gap is achieved by using a vertically slotted connection as shown in Figure 14-12b. Note that the wind columns are laterally braced by the girts.

14.6 RELIEF ANGLES FOR BRICK VENEER

In multistory buildings, horizontal angles supported by the floor framing are normally used to support the brick cladding and to provide horizontal expansion joints for the brick veneer at each floor level [11]. However, for buildings three stories or fewer and with the brick cladding height from top of foundation wall to top of brick cladding not greater than 30 ft., the brick cladding is allowed to span the full 30-ft. height without any relief angles. In this case, the full height of the brick wall will be supported at the ground-floor level on a foundation wall; however, adequate provisions should be made in the lateral tie system to account for the differential vertical movement between the brick veneer and the building frame. Typical relief angle details are shown in Figure 14-13.

14.7 ACHIEVING ADEQUATE DRAINAGE IN STEEL-FRAMED ROOFS

Most steel buildings have relatively flat roofs. These types of roofs may be highly susceptible to ponding—the tendency for flat roofs to be subjected to increased loads due to additional accumulation of rain or melted snow in the deflected areas of a flat roof. To avoid drainage problems, the roof slope in steel-framed roofs is achieved in one of two ways:

1. Maintain an essentially flat structural roof framing and use tapered insulation that is placed underneath the roofing membrane and on top of the metal roof deck to

achieve the roof slope. Since a minimum roof slope of ¼-in. per 1 foot is recommended for proper drainage, this option could result in large insulation thicknesses, especially at the high points on the roof.

2. The more frequently used method for achieving proper roof drainage is to slope the structural roof framing with the high and low points (i.e., drain location) located near the columns. A typical roof framing plan layout showing the top of steel elevations (TOS) with allowance for proper drainage (i.e., minimum slope of ¼-in. rise per 1-ft. horizontal length) is shown in Figure 14-14.

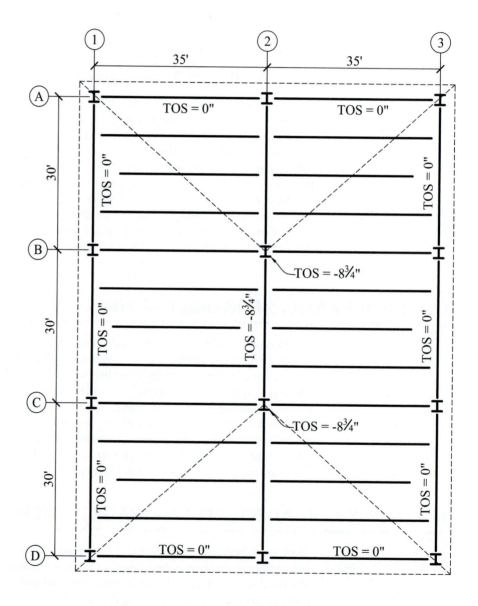

TOS = Top of steel (deck bearing)

Figure 14-14 Typical roof framing plan with top of steel elevations for proper drainage.

In Figure 14-14, the tops of steel elevations at columns B-2 and C-2 (i.e., the low points) to accommodate a minimum slope of ¼-in. rise per 1 ft. horizontal length are the larger of 8¾ in. (i.e., 35 ft. × ¼ in./ft.) or 7½ in. (i.e., 30 ft. × ¼ in./ft.); therefore, we will assume that the top of steel elevation at the low points (i.e., drain locations) is −8¾ in. relative to the top of steel elevation at the high points of the roof. The roof drains will be located near columns B-2 and C-2. The roof framing around the roof drain opening usually consists of steel angles spanning between the roof joists or roof beams.

If the infill members in the roof framing in Figure 14-14 were open-web steel joists instead of steel beams, the top of steel elevations for the girders along grid lines 1, 2, and 3 will be lowered further by an amount equal to the depth of the joist seat. Typical joist seat depths are 2½ in. and 5 in., depending on the span of the steel joist. The joist seat depth that is appropriate for a given span is usually tabulated in joist manufacturers' catalogs. For example, if a joist with a seat depth of 2½ in. were used, the top of steel elevation for the girders along lines 1 and 3 would be −2½ in.

14.8 PONDING IN STEEL FRAMED ROOF SYSTEMS

Ponding occurs when rainwater accumulates on a roof due to progressive deflection. In general, roofs are constructed to be flat, but could have some initial deflection due to the dead weight of the roofing and structure. When rainwater accumulates (see Chapter 2 for rain load calculations), the roof framing will deflect under the weight of the water; this deflection will allow more water to accumulate, thus creating a situation in which instability and eventual collapse could occur.

Section 1611 of the International Building Code [12] (IBC) requires that roofs with a slope of less than ¼ in./ft. be investigated for adequate stiffness so that instability does not occur. The IBC also requires that roofs have a primary and secondary drainage system to accommodate the possibility that the primary system may become blocked. The height of the rainwater used for design is the distance from the roof to the inlet elevation of the secondary drainage system. Careful attention to the detailing of the roof drains will allow rainwater to properly drain from the roof and mitigate the effects of ponding. This detailing, however, is usually the responsibility of the project architect.

The AISC specification provides two methods for determining whether or not a roof structure has adequate stiffness to resist ponding: a simplified method and an improved method.

The simplified method states that when the following criteria are met, the roof structure is considered adequate with regard to ponding stability:

$$C_p + 0.9C_s \leq 0.25, \text{ and} \tag{14-1}$$
$$I_d \geq 25(S^4)10^{-6}, \tag{14-2}$$

where

$$C_p = \frac{32L_sL_p^{\,4}}{10^7I_p}, \tag{14-3}$$

$$C_s = \frac{32SL_s^{\,4}}{10^7I_s}, \tag{14-4}$$

L_p = Primary member length between columns (length of the girders), ft.,

L_s = Secondary member length between columns (length of the members perpendicular to the girders), ft.,

S = Spacing of the secondary members, ft.,

I_p = Moment of inertia of the primary members, in.^{4} (reduce by 15% for trusses),

I_s = Moment of inertia of the secondary members, in.^{4} (reduce by 15% for trusses), and

I_d = Moment of inertia of the steel deck supported by the secondary members, in.^{4}/ft.

When steel trusses and joists are used, the moment of inertia shall be decreased by 15%. When the steel deck is supported by primary members only, then the deck shall be considered the secondary member. For a more in-depth analysis (the improved method), the reader is referred to Appendix 2 of the AISC specification.

EXAMPLE 14-3

Design for Ponding in Roof Framing

Determine whether the framing below is stable for ponding.

SOLUTION

From Figure 14-15,

L_p = 24 ft.
L_s = 35 ft.
S = 6 ft.

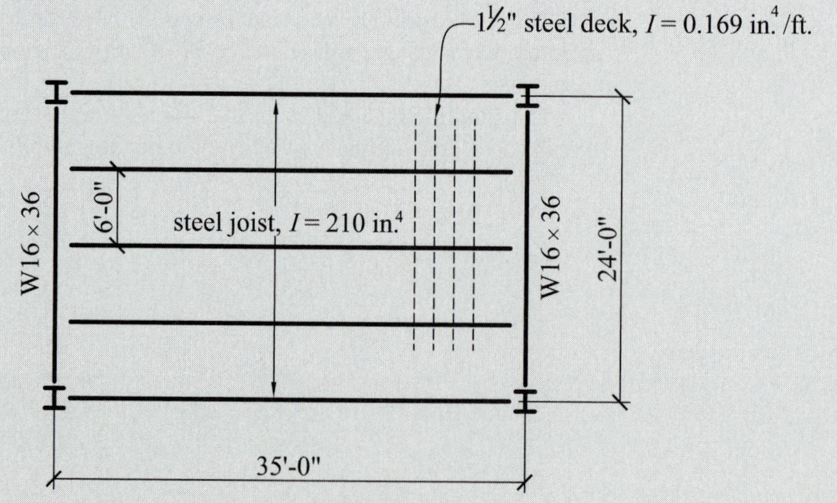

Figure 14-15 Details for Example 14-3.

$I_p = 448$ in.^{4}

$I_s = 210/1.15 = 182$ in.^{4} (15% reduction for a steel joist)

$I_d = 0.169$ in.^{4}/ft.

$$C_p = \frac{32L_sL_p{}^4}{10^7I_p} = \frac{(32)(35)(24)^4}{10^7(448)} = 0.0829$$

$$C_s = \frac{32SL_s{}^4}{10^7I_s} = \frac{(32)(6)(35)^4}{10^7(182)} = 0.1583$$

$C_p + 0.9C_s^s \ 0.25 = 0.0829 + (0.9)(0.1583) = 0.225 < 0.25$ OK

$I_d \geq 25(S^4)10^{-6}$

$0.169 > (25)(6^4)10^{-6} = 0.0324$ OK

The roof framing has adequate stability with regard to ponding.

14.9 STABILITY BRACING FOR BEAMS AND COLUMNS

Beam and columns require bracing for stability at their support points, as well as at points where lateral stability is required for the limit state of buckling. We will consider both cases in this section.

When beams are supported by means of direct bearing on their bottom flange, the top flange needs to be restrained against rotation. The AISC specification, Section J10-7 requires that a full-depth stiffener plate be added when no other restraint is provided. This is often the case when a steel beam is bearing on concrete. When the bearing occurs at a steel column, for example, a stabilizing angle could be added at the top of the beam (see Figure 14-16).

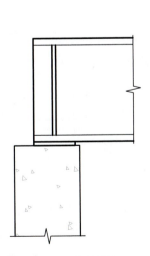

Angle at top to stabilize beam

a. Bearing on concrete   *b. Bearing on steel*

Figure 14-16 Stiffener plate for end bearing.

In roof framing with open-web steel joists, there are two basic framing conditions at the columns that are commonly used: girders framing into the columns and girders framing over the columns. When the girders frame into the column, the column is adequately stabilized and no further detailing needs to be considered. The steel joists, however, will require a stabilizer plate at the bottom chord to meet OSHA regulations. This plate must be at least 6 in. by 6 in., and it should extend at least 3 in. below the bottom chord of the joist and have a 13/16-in. hole for attaching guy wires (see Figure 14-17a).

When the girder frames over the column, the top of the column needs lateral support perpendicular to the length of the girder to prevent lateral buckling. This can be accomplished by either adding a diagonal brace or by extending the bottom chords of the joists to brace the column (see Figure 14-17b).

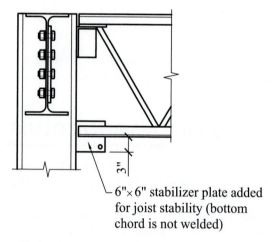

3"

6"×6" stabilizer plate added
for joist stability (bottom
chord is not welded)

a. Girder frames into column

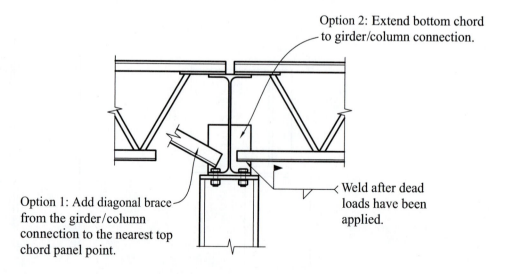

Option 2: Extend bottom chord
to girder/column connection.

Option 1: Add diagonal brace
from the girder/column
connection to the nearest top
chord panel point.

Weld after dead
loads have been
applied.

b. Girder frames over column

Figure 14-17 Column bracing with steel joists.

When beams or columns require lateral bracing to prevent the limit state of buckling, the strength and stiffness of the brace have to be checked. When bracing is provided, the axial force in the bracing is assumed to act perpendicular to the member being braced. When bracing is oriented diagonally, the axial force has to be adjusted in accordance with the geometry. Additionally, the strength and stiffness of the bracing end connections should also be considered. Stability bracing should always be located between the neutral or centroidal axis of the beam or column and the compression flange (see discussion on lateral torsional buckling in Chapter 6).

There are also two types of bracing mechanisms that are recognized in the AISC specification: relative and nodal bracing. Relative bracing is a line of bracing connected to more than one brace point. The movement at a relative brace is controlled with respect to adjacent brace points. A nodal brace is independent and controls movement at only one brace point without regard to adjacent brace points.

Appendix 6 of the AISC specification provides the design strength and stiffness of lateral bracing as follows:

Columns—Relative Bracing
    Required strength:

$$P_{br} = 0.004P_u \tag{14-5}$$

    Required stiffness:

$$\beta_{br} = \frac{2P_u}{\phi L_b} \tag{14-6}$$

Columns—Nodal Bracing
    Required strength:

$$P_{br} = 0.01P_u \tag{14-7}$$

    Required stiffness:

$$\beta_{br} = \frac{8P_u}{\phi L_b} \tag{14-8}$$

Beams—Relative Bracing
    Required strength:

$$P_{br} = \frac{0.008M_u C_d}{h_o} \tag{14-9}$$

    Required stiffness:

$$\beta_{br} = \frac{4M_u C_d}{\phi L_b h_o} \tag{14-10}$$

Beams—Nodal Bracing
    Required strength:

$$P_{br} = \frac{0.02M_u C_d}{h_o} \tag{14-11}$$

Required stiffness:

$$\beta_{br} = \frac{10M_u C_d}{\phi L_b h_o} \tag{14-12}$$

where

P_u = Factored axial load,

M_u = Factored bending moment,

L_b = Unbraced length (distance between braces),

C_d = 1.0 for single-curvature bending

  = 2.0 for double-curvature bending,

h_o = Distance between flange centroids, and

ϕ = 0.75.

EXAMPLE 14-4

Lateral Bracing for a Column

A 32-ft.-long W8 × 31 column requires a lateral brace at midheight of the weak axis. Determine whether an L3 × 3 × ¼ nodal brace provides adequate lateral bracing if the factored axial load in the column is P_u = 200 kips. The L3 × 3 × ¼ is ASTM A36 and has a length of 6 ft. from the column web to an isolated support.

SOLUTION

From equation 14-7, the required strength of the brace is

$$P_{br} = 0.01P_u = (0.01)(200) = 2 \text{ kips}$$

From *AISCM*, Table 4-11, the design strength of the L3 × 3 × ¼ at an unbraced length of 6 ft. is ϕP_n = 20.9 kips > 2 kips, OK.

From equation 14-8, the required stiffness of the brace is

$$\beta_{br} = \frac{8P_u}{\phi L_b} = \frac{(8)(200)}{(0.75)(16)} = 134 \text{ kips/ft.} = 11.1 \text{ kips/in.},$$

where L_b = 32 ft./2 = 16 ft. (midspan brace).

From *AISCM*, Table 1-7 for an L3 × 3 × ¼,

$$A = 1.44 \text{ in.}^2.$$

The stiffness of the brace is

$$\beta_{br} = \frac{EA}{L} = \frac{(29,000)(1.44)}{(6)(12)} = 580 \text{ kips/in.} > 11.1 \text{ kips/in.}$$

The L3 × 3 × ¼ provides adequate strength and stiffness as lateral bracing.

EXAMPLE 14-5

Lateral Bracing for a Beam

A 32-ft.-long W18 × 50 beam requires a lateral brace at midspan of a simple span. Determine whether an L3 × 3 × ¼ nodal brace provides adequate lateral bracing if the factored bending moment in the beam is M_u = 240 ft.-kips. The L3 × 3 × ¼ is ASTM A36 and has a length of 8 ft. from the compression flange of the beam to an isolated support.

SOLUTION

From *AISCM*, Table 1-1,

$$h_o = 17.4 \text{ in. for a W18} \times 50$$

From equation 14-11, the required strength of the brace is

$$P_{br} = \frac{0.02 M_u C_d}{h_o} = \frac{(0.02)[(200)(12)](1.0)}{(17.4)} = 2.76 \text{ kips.}$$

From *AISCM*, Table 4-11, the design strength of the L3 × 3 × ¼ at an unbraced length of 8 ft. is ϕP_n = 12.0 kips > 2.76 kips, OK.

From equation 14-12, the required stiffness of the brace is

$$\beta_{br} = \frac{10 M_u C_d}{\phi L_b h_o} = \frac{(10)[(200)(12)](1.0)}{(0.75)[(16)(12)](17.4)} = 9.58 \text{ kips/in.,}$$

where L_b = 32 ft./2 = 16 ft. (midspan brace) and C_d = 1.0 (simple span, therefore single curvature bending).

From *AISCM*, Table 1-7 for an L3 × 3 × ¼,

$$A = 1.44 \text{ in.}^2.$$

The stiffness of the brace is

$$\beta_{br} = \frac{EA}{L} = \frac{(29,000)(1.44)}{(8)(12)} = 435 \text{ kips/in.} > 9.58 \text{ kips/in.}$$

The L3 × 3 × ¼ provides adequate strength and stiffness as lateral bracing.

14.10 STEEL PREPARATIONS, FINISHES, AND FIREPROOFING

The preparation and finishing of structural steel is generally beyond the scope of responsibilities of the structural engineer; however, it is important for any designer to have a basic understanding of steel coatings because they often impact the detailing and selection of steel members. This section will provide a general overview of steel coatings and fireproofing, but given the wide variety of paint and fireproofing products available, it is recommended that the designer reviews the required steel preparation and coating options for any given project.

We will first consider the preparation of steel for the application of coatings. The Society for Protective Coatings identifies several preparation methods based on the final coating to be applied [13]. When a coating system is selected, the required surface preparation is usually dictated by the manufacturer of the coating. Table 14-3 below summarizes the various surface preparations that can be specified for steel sections.

A more detailed description of each method is given in reference 13, but most are self-explanatory. When structural steel is intended to be left uncoated, the AISC Code of Standard Practice states that SSPC-SP 1 (Solvent Cleaning) is required. Solvent cleaning uses a commercial cleaner to remove visible grease and soil. When shop painting is required, then the minimum surface preparation is SSPC-SP 2 (Hand Tool Cleaning). Hand tool cleaning removes loose mill scale, rust, and other deleterious matter using nonpowered hand tools. The other surface preparations indicated are more labor-intensive and should be specified in accordance with the coating manufacturer's recommendations.

The use of coatings is a function of several factors, such as aesthetics and the intended use of the structure. For most structural steel applications, the use of a shop coat of paint is sufficient in that the shop coat provides moderate resistance to environmental conditions during the construction process when the steel is directly exposed to the weather. Once the steel is in place, adequate protection for weather is usually assured for covered structures. It should be noted that structural steel can be left uncoated for this typical condition, and it is recommended that the steel be left uncoated when spray-applied fireproofing is used because the presence of primer often reduces the adhesion quality of the fireproofing. When the term "shop coat" is used to describe a coating, it is assumed that the paint has a minimum dry film thickness of 1 mil, which is 1/1000th of an inch.

When structural steel is intended for use in an exterior environment, or in a corrosive environment such as an indoor swimming pool, the coating system should be carefully selected and should be applied in consultation with the coating manufacturer. The coating

Table 14-3 Description of various steel surface preparations [13]

SSPC-SP 1: Solvent Cleaning
SSPC-SP 2: Hand Tool Cleaning
SSPC-SP 3: Power Tool Cleaning
SSPC-SP 5: White Metal Blast Cleaning
SSPC-SP 6: Commercial Blast Cleaning
SSPC-SP 7: Brush-Off Blast Cleaning
SSPC-SP 8: Pickling
SSPC-SP 10: Near-White Blast Cleaning
SSPC-SP 11: Power Tool Cleaning to Bare Metal
SSPC-SP 12: Surface Preparation and Cleaning of Metals by Waterjetting Prior to Recoating

system for these types of service conditions usually consists of a more extensive surface preparation (often SSPC-SP 6, Commercial Blast Cleaning) in conjunction with an epoxy primer and a finish coat. In addition to the preparation and painting of the structural steel members, there are often field welds at the connections also need a proper coating system. The surface preparation for field welds is typically SSPC-SP 11, Power Tool Cleaning to Bare Metal, and the coating system would match the system for the other members. Again, it is stressed that the variety of products available requires that the coating system for any project be examined on a case-by-case basis to determine the best system for a given exposure condition.

In lieu of the coating systems mentioned above for exterior exposed or corrosive environments, the process of hot-dip galvanizing can be specified. Hot-dip galvanizing is the process of coating structural steel with a layer of zinc by dipping it into a tank of molten zinc. The zinc layer will react with oxygen to create zinc oxide while in service; therefore, the layer will decrease in thickness over time. The minimum thickness of this zinc layer is given in ASTM A123; it varies, depending on the thickness of the steel section, from 1.4 to 3.9 mils. These minimum values have proven to provide a service life that varies from 20 to 50 years, depending on the zinc thickness and the exposure conditions.

When hot-dip galvanizing is specified with bolted connections, the fabricator may prefer to use larger bolt holes to accommodate the added layers of zinc. The standard hole size is 1/16 in. or 62.5 mils larger than the bolt, but the addition of zinc layers on the bolts and at least two connected parts (clip angles or shear tabs) could reduce the ease of installing the bolts given that the hole diameter is 62.5 mils larger than the bolt in each connected part and that a hole size reduction of as much as 3.9 mils could be present in each connected part. For field welds, cold galvanizing repair paint in accordance with ASTM A780 is used to repair previously galvanized areas affected by welding.

Structural steel loses strength at high temperatures (e.g., in a fire), so steel members often are required to be protected from fire depending on the occupancy category of the building and the member type. Steel can be left exposed in some buildings, but in cases where the aesthetics require an exposed steel member, as well as adequate protection from fire, a product called intumescent paint could be applied. During a fire event, intumescent paint will expand up to 50 times its original thickness, creating a foamlike layer of insulation that will help reduce the rate of heat transfer, and therefore the loss of strength, to the steel. Intumescent paint is usually applied in several layers, varying from 1/8 in. to 5/8 in., depending on the product and the required fire rating for the member. Another method used for meeting the fire protection requirements for a building with exposed steel is the use of a sprinkler system. The use and applicability of this system would be coordinated between the architect and plumbing engineer.

One of the more common methods of protecting steel from fire is to use spray-applied fire-resistant materials (SFRM). SFRM can be either fibrous or cementitious, both of which have noncombustible properties. The use of primer prior to the application of SFRM is not recommended because the primer could reduce the adhesion quality of the SFRM.

Another common method used for protecting steel members from fire is to provide an enclosure, usually composed of gypsum wall board (GWB). This type of protection can be specified regardless of the coating used on the structural steel. The steel could also be encased in concrete or masonry, which was a common method of fire projection in the past, but is not as economical today compared with other fire protection methods.

The selection of a fire protection system is often the responsibility of someone other than the structural engineer and it rarely has a major impact on the final design of the main structural system. For more information and details on fire-rated assemblies, the reader is referred to references 14 and 15.

14.11 STRENGTHENING AND REHABILITATION OF EXISTING STEEL STRUCTURES

Additions and renovations to existing buildings are common during the life of a structure. These modifications often occur to accommodate the changing use or occupancy category of a building. For example, a building addition in which the new building is taller than an adjacent existing building would create a condition in which snow can accumulate on the lower roof of the existing building. The original roof of the existing building would not likely have been designed for this additional snow load and would have to be investigated for structural adequacy. Another common change in the use of a roof structure occurs due to the addition or replacement of new equipment, that is supported by the roof framing. This would also require an investigation into the existing roof framing for structural adequacy. Occupancy changes to floor structures are common as well, such as the addition of a high-density file storage system to a portion of a floor. In any occupancy change, both strength and serviceability must be checked, and in some cases, serviceability can be the primary concern (e.g., when a structure must support deflection-sensitive elements or if the structure is intended to support dynamic loads from equipment or human occupants).

Chapter 34 of the IBC [12] provides direction for repairs, additions, and occupancy changes to existing structures. In general, modifications to existing structures must comply with the design requirements for new construction and any modification should not cause any part of the existing structure to be in violation of the requirements for new construction. Unaffected portions of the existing structure are not required to comply with the current code requirements unless an existing member is found to be structurally inadequate, in which case it must be modified or replaced to conform to the current code. Another key provision is that modifications to an existing structure shall not increase the stress to an existing member by more than 5% unless the member can adequately support the increased stress, and the strength of any existing member shall not be decreased by more than what is required for new construction. For seismic loads, the existing structure does not have to be investigated for such loads if the addition is seismically independent of the existing structure. If this is not the case, then the existing lateral force resisting system (LFRS) must conform to the requirements for new structures unless the following are met: The addition must conform to the requirements for new construction; the addition does not increase the stress to any member in the LFRS by more than 10% unless the member is adequate to support the additional stress; and the addition shall not decrease the strength of any member in the LFRS by more than 10% unless the member is adequate to support the additional stress.

Once a new occupancy or proposed structural modification is defined, an investigation of the existing structural condition is required. If the existing construction drawings are available, the investigation is simpler; however, drawings for older structures are often not available and a more extensive field investigation is required. For any structural modification, a field investigation is usually necessary to help assess practical issues, such as the constructibility of an addition or alteration. There are often physical barriers to structural modifications, such as mechanical equipment, lights, ducts, and other similar elements.

Once a structure is examined for adequacy, the nature of the reinforcement can be selected. For floor or roof systems that require a relatively large increase in load-carrying capacity, the addition of new intermediate beams between existing beams may be required as shown in Figure 14-18. This scheme also decreases the span of the deck system and therefore increases the load-carrying capacity of the slab system, which may be desirable in some cases. Note that the constructibility of this scheme must be carefully considered. There will

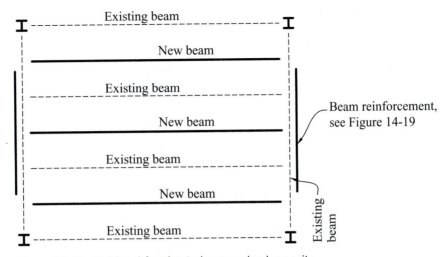

Figure 14-18 Additional framing to increase load capacity.

likely be lights, ducts, and other physical obstructions that might make placing new beams difficult or even impossible. Additionally, it may not be possible to install the new beams in one piece; a beam splice may be needed (see Figure 11-20).

The girders in Figure 14-18 would probably require reinforcement to support the new design loads. Figure 14-19 shows several possible beam reinforcement details to increase the flexural strength of a beam section. In each case, the intent is to increase the section properties by adding more steel near the flanges or by increasing the depth of the section. Noncomposite beams usually need reinforcement at both flanges; however, it is not usually possible to add reinforcement at the top flange due to the likely presence of a deck system. Composite beams usually only require reinforcement at the bottom flange since additional compression stresses can often be resisted by the concrete slab. The position of the field weld should be considered when selecting a reinforcement detail. Note that in the details in Figures 14-19a, 14-19c, and 14-19d, the reinforcing steel is wider than the flange of the existing beam, which allows for the desirable downward field weld. The use of a wider steel section rather than a smaller steel section would likely be preferred as reinforcement in these cases regardless of material cost, to avoid an overhead welding.

Noncomposite and partially composite floor beams can also be reinforced to increase their degree of composite action, and therefore their load-carrying capacity, as shown in Figure 14-20. To accomplish this for noncomposite beams and girders, round holes are cut into the slab along the length of the beam to allow for the placement of a shear stud. The size of the hole must be such that the code-required clearances are provided around the shear stud (see Chapter 7). A new shear stud is then placed and the hole is filled with shrinkage-compensating cementitious grout. It is possible to increase the degree of composite action of an existing partially composite beam by adding new shear studs, but it is difficult to accomplish given that the location of the existing shear studs may not be easy to discern.

Open-web steel joists can also be reinforced to increase their load-carrying capacity. The investigation of these members is usually more difficult because the individual web and chord shapes can usually only be determined by a field investigation. Furthermore, the geometry of the truss members likely has joint eccentricities that must be accounted for in the analysis.

Figure 14-21 shows common joist reinforcement methods. In Figure 14-21a, round bars are added near the top and bottom chords to increase the flexural strength and stiffness

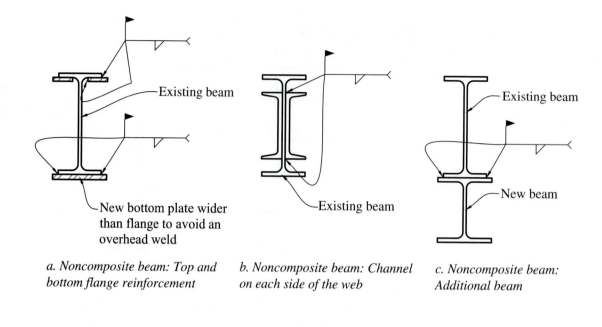

a. Noncomposite beam: Top and bottom flange reinforcement

b. Noncomposite beam: Channel on each side of the web

c. Noncomposite beam: Additional beam

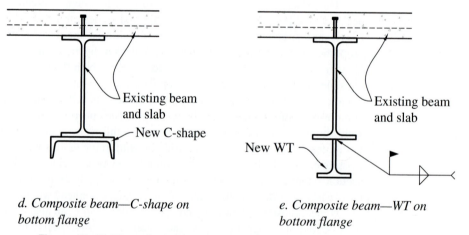

d. Composite beam—C-shape on bottom flange

e. Composite beam—WT on bottom flange

Figure 14-19 Flexural reinforcement of beams.

of the overall open-web steel joist. The web members may also require reinforcement to support additional loads. In Figure 14-21b, new web members are added adjacent to existing web members to increase the axial load capacity of the existing web member. Figure 14-21b also shows the reinforcement of a joist when a concentrated load is applied away from a panel point. When loads are applied away from panel points, the section properties and therefore the flexural strength of the chords are such that they cannot support concentrated loads of any significance (200 lb. is a common upper limit on concentrated loads away from a panel point). In this detail, a new web member is added from the concentrated load to the nearest panel point to ensure that the concentrated load is resisted by direct tension in a web member rather than by bending of a chord member.

When beam members are reinforced to support additional loads, it is also important to consider the strength of the connections and any other elements along the load path. Two

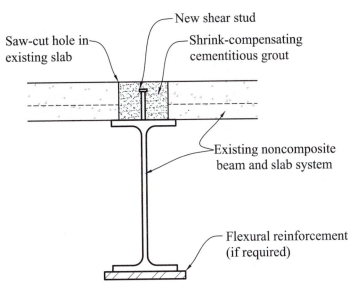

Figure 14-20 Reinforcement of a noncomposite or partially composite beam.

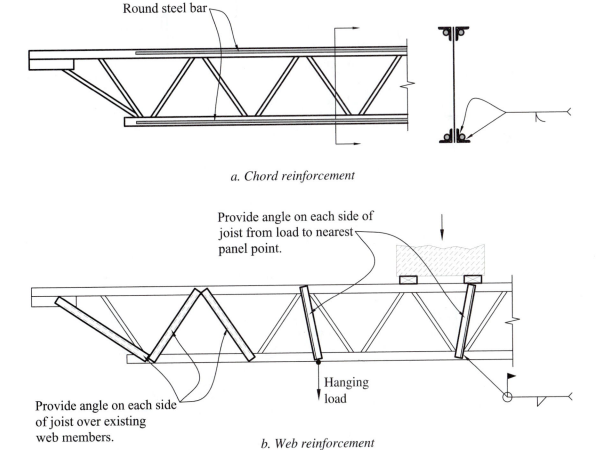

a. Chord reinforcement

b. Web reinforcement

Figure 14-21 Flexural reinforcement of steel joists.

possible shear reinforcement details are shown in Figures 14-22a and 14-22b. In these details, welds are added to existing bolted connections, which is a common method of reinforcing existing connections. The AISC specification does not give much guidance on the design of connections that combine both welds and bolts; however, reference 16 provides some background on the use of such details. One important parameter in the use of a combined welded and bolted connection is the use of welds that are oriented longitudinally with respect to the load direction. Test results have shown that when bolts are loaded in combination with transverse welds, the contribution of bolt shear strength is negligible, but when

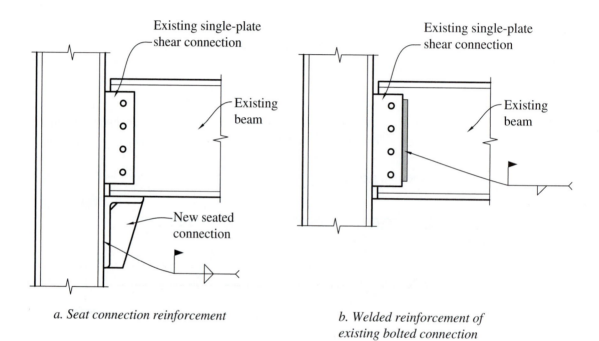

a. Seat connection reinforcement

b. Welded reinforcement of existing bolted connection

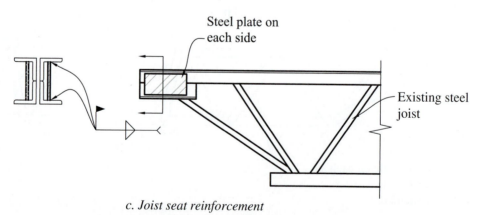

c. Joist seat reinforcement

Figure 14-22 Reinforcement of connections.

bolts are loaded in combination with longitudinal welds, the contribution of bolt shear is about 75 percent of the ultimate strength of the bolts [16].

In lieu of adding additional steel components to an existing connection, it may be possible to completely remove and replace the existing connection, or some selected components of the existing connection, such as the replacement of A325 bolts with A490 bolts. An existing welded connection could also be reinforced by the placement of additional weld length or by increasing the weld size.

The end connections of open-web steel joists can be reinforced by the addition of vertical stiffeners at the bearing points as shown in Figure 14-22c.

The load-carrying capacity of columns can be increased by the addition of steel shapes, similar to the reinforcement for beams. Figure 14-23 shows several possible column reinforcement details. The addition of steel shapes to an existing column will increase the column design strength because of the additional area of steel and because the location of the steel sections could increase the radius of gyration, thus decreasing the column slenderness ratio, leading to an increase in the design compressive strength.

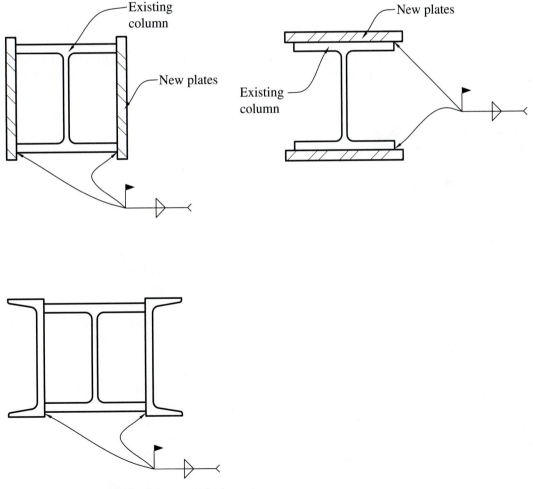

Figure 14-23 Column reinforcement.

One common issue with older steel structures is the weldability of the existing material. Steel replaced wrought iron as a building material after about 1900, but the chemical and mechanical properties were not initially standardized and thus some of the steel was weldable, but some was not. One could quickly determine the weldability of an unknown steel section by looking for existing welds on the member which would indicate that the steel is weldable. From about 1910 to 1960, most steel used in buildings was ASTM A7 or A9, which is a medium carbon steel and is generally weldable; however, it has a relatively wide range of acceptable carbon content and should therefore be tested prior to welding (see ref. 17 for the test method). After about 1960, ASTM A36 and A572 were primarily used in building construction, and these steels are weldable.

Another important consideration in the reinforcement of existing members is the use of shoring to relieve existing stresses. When existing members are shored, the strength of the reinforced section is maximized, but it may not be possible in some cases to use shoring. Columns are usually not shored prior to reinforcement because it has been shown that the addition of shoring does not significantly increase the load-carrying capacity of a reinforced section [18].

The design of the shoring and reinforcement for existing composite beams must consider several factors. If the existing structure was constructed with shoring prior to the placement of the concrete slab, then shoring can be used to reinforce the existing member since the dead and live loads on the member were applied after the concrete hardened. If the existing structure was not constructed without shoring prior to the placement of the concrete slab, then it is not recommended to use shoring while adding the reinforcement. When reinforcement is required for an unshored floor system, then any unnecessary dead and live loads should be removed from the floor area while the reinforcement is being installed.

14.12 REFERENCES

1. Ricker, David T. 2000. Value engineering for steel construction. *Modern Steel Construction.* (April).

2. AISC. 2005. *Seismic provisions for structural steel building,* Chicago: AISC.

3. Ioannides, Socrates A., and John L. Ruddy. 2000. Rules of thumb for steel design. *Modern Steel Construction.* (February).

4. Carter, Charles J., and Steven M. Ashton. 2004. Look before you leap: Practical advice for laying out steel framing. *Modern Steel Construction.* (March).

5. Ambrose, James. 1997. *Simplified design of steel structures,* 7th ed. Hoboken, NJ: John Wiley and Sons.

6. Naeim, Favzad. 2001. *The Seismic design handbook.* Country Club Hills, IL: International Conference of Building Officials (ICBO).

7. Lochrane, Michael L., and William C. Honeck. 2004. *Design of special concentric braced frames (with comments on ordinary concentric braced frames).* Moraga, CA: Structure Steel Education Council.

8. Matthes, John, ed. 1993. *Masonry designers' guide.* Boulder, CO: The Masonry Society.

9. Luttrel, Larry. 2004. *Diaphragm design manual,* 3rd ed. Fox River Grove, IL: Steel Deck Institute.

10. Fisher, James M, Michael A. West, and Julius P. Van de Pas. 2002. *Designing with Vulcraft steel joists, joist girders and steel deck.* Milwaukee, WI: Nucor Co.

11. Krogstad, Nobert V. 2003. Using steel relief angles. *Masonry Construction.* (May)

12. International Codes Council. 2006. *International building code.* Falls Church, VA: ICC.

13. The Society for Protective Coatings. 2006. *Surface preparation specification and practices.* Chicago: SPC.

14. American Institute of Steel Construction. 2003. *Fire resistance of structural steel framing.* Chicago: AISC.

15. Underwriters Laboratories, Inc. 2003. *Fire resistance directory,* vol. I. Northbrook, IL.

16. Kulak, Geoffrey L., and Gilbert Y. Grondin. 2003. Strength of joints that combine bolts and welds. *Engineering Journal,* (2nd Quarter).

17. Ricker, David T. 1988. Field welding to existing structures. *Engineering Journal.* (1st Quarter).

18. Tall, Lambert. 1989. The reinforcement of steel columns. *Engineering Journal.* (1st Quarter).

19. Al Nageim, Hassan, and T. J. MacGinley. 2005. *Steel structures—Practical design studies,* 3rd ed. Taylor and Francis.

20. American Institute of Steel Construction. 2006. *Steel construction manual,* 13th ed. Chicago: AISC.

21. American Institute of Steel Construction. 2003. *Steel design guide series 3: Serviceability design considerations for steel buildings.* Chicago: ASTM.

22. ASTM International. 2002. *Standard specification for Zinc (hot-dip galvanized) coatings on iron and steel products— ASTM A123.* West Conshohocken, PA: ASTM.

23. American Institute of Steel Construction. 2002. *Steel design guide series 15: AISC rehabilitation and retrofit guide.* Chicago: ASTM.

24. Miller, John P. 1996. Strengthening of existing composite beams using LRFD procedures. *Engineering Journal.* (2nd Quarter).

25. American Society of Civil Engineers, 2005. ASCE 7, Minimum design loads for buildings and other structures, Reston, VA.

APPENDIX A

OPEN–WEB STEEL JOIST AND JOIST GIRDER TABLES

STANDARD LRFD LOAD TABLE
OPEN WEB STEEL JOISTS, K-SERIES

Based on a 50 ksi Maximum Yield Strength
Adopted by the Steel Joist Institute May 1, 2000
Revised to November 10, 2003 – Effective March 01, 2005

The black figures in the following table give the TOTAL safe factored uniformly distributed load-carrying capacities, in pounds per linear foot, of **LRFD K-Series** Steel Joists. The weight of factored DEAD loads, including the joists, must be deducted to determine the factored LIVE load-carrying capacities of the joists. Sloped parallel-chord joists shall use span as defined by the length along the slope.

The figures shown in **RED** in this load table are the unfactored nominal LIVE loads per linear foot of joist which will produce an approximate deflection of 1/360 of the span. LIVE loads which will produce a deflection of 1/240 of the span may be obtained by multiplying the figures in **RED** by 1.5. In no case shall the TOTAL load capacity of the joists be exceeded.

The approximate joist weights per linear foot shown in these tables do not include accessories.

The approximate moment of inertia of the joist, in inches[4] is;

$$I_j = 26.767(W_{LL})(L^3)(10^{-6}),$$ where W_{LL} = **RED** figure in the Load Table and L = (Span - 0.33) in feet.

For the proper handling of concentrated and/or varying loads, see Section 6.1 in the Code of Standard Practice for Steel Joists and Joist Girders.

Where the joist span exceeds the unshaded area of the Load Table, the row of bridging nearest the mid span shall be diagonal bridging with bolted connections at the chords and intersections.

LRFD

STANDARD LOAD TABLE FOR OPEN WEB STEEL JOISTS, K-SERIES
Based on a 50 ksi Maximum Yield Strength - Loads Shown in Pounds per Linear Foot (plf)

Joist Designation	8K1	10K1	12K1	12K3	12K5	14K1	14K3	14K4	14K6	16K2	16K3	16K4	16K5	16K6	16K7	16K9
Depth (in.)	8	10	12	12	12	14	14	14	14	16	16	16	16	16	16	16
Approx. Wt (lbs./ft.)	5.1	5.0	5.0	5.7	7.1	5.2	6.0	6.7	7.7	5.5	6.3	7.0	7.5	8.1	8.6	10.0
Span (ft.)																
8	825/550															
9	825/550															
10	825/480	825/550														
11	798/377	825/542														
12	666/288	825/455	825/550	825/550	825/550											
13	565/225	718/363	825/510	825/510	825/510											
14	486/179	618/289	750/425	825/463	825/463	825/550	825/550	825/550	825/550							
15	421/145	537/234	651/344	814/428	825/434	766/475	825/507	825/507	825/507							
16	369/119	469/192	570/282	714/351	825/396	672/390	825/467	825/467	825/467	825/550	825/550	825/550	825/550	825/550	825/550	825/550
17		415/159	504/234	630/291	825/366	592/324	742/404	825/443	825/443	768/488	825/526	825/526	825/526	825/526	825/526	825/526
18		369/134	448/197	561/245	760/317	528/272	661/339	795/397	825/408	684/409	762/456	825/490	825/490	825/490	825/490	825/490
19		331/113	402/167	502/207	681/269	472/230	592/287	712/336	825/383	612/347	682/386	820/452	825/455	825/455	825/455	825/455
20		298/97	361/142	453/177	613/230	426/197	534/246	642/287	787/347	552/297	615/330	739/386	825/426	825/426	825/426	825/426
21			327/123	409/153	555/198	385/170	483/212	582/248	712/299	499/255	556/285	670/333	754/373	822/405	825/406	825/406
22			298/106	373/132	505/172	351/147	439/184	529/215	648/259	454/222	505/247	609/289	687/323	747/351	825/385	825/385
23			271/93	340/116	462/150	321/128	402/160	483/188	592/226	415/194	462/216	556/252	627/282	682/307	760/339	825/363
24			249/81	312/101	423/132	294/113	367/141	442/165	543/199	381/170	424/189	510/221	576/248	627/269	697/298	825/346
25						270/100	339/124	408/145	501/175	351/150	390/167	469/195	529/219	576/238	642/263	771/311
26						249/88	313/110	376/129	462/156	324/133	360/148	433/173	489/194	532/211	592/233	711/276
27						231/79	289/98	349/115	427/139	300/119	334/132	402/155	453/173	493/188	549/208	658/246
28						214/70	270/88	324/103	397/124	279/106	310/118	373/138	421/155	459/168	510/186	612/220
29										259/95	289/106	348/124	391/139	427/151	475/167	570/198
30										241/86	270/96	324/112	366/126	399/137	444/151	532/178
31										226/78	252/87	304/101	342/114	373/124	415/137	498/161
32										213/71	237/79	285/92	321/103	349/112	388/124	466/147

LRFD

STANDARD LOAD TABLE FOR OPEN WEB STEEL JOISTS, K-SERIES
Based on a 50 ksi Maximum Yield Strength - Loads Shown in Pounds per Linear Foot (plf)

Joist Designation	18K3	18K4	18K5	18K6	18K7	18K9	18K10	20K3	20K4	20K5	20K6	20K7	20K9	20K10	22K4	22K5	22K6	22K7	22K9	22K10	22K11
Depth (In.)	18	18	18	18	18	18	18	20	20	20	20	20	20	20	22	22	22	22	22	22	22
Approx. Wt. (lbs./ft.)	6.6	7.2	7.7	8.5	9	10.2	11.7	6.7	7.6	8.2	8.9	9.3	10.8	12.2	8	8.8	9.2	9.7	11.3	12.6	13.8
Span (ft.)																					
18	825	825	825	825	825	825	825														
	550	550	550	550	550	550	550														
19	771	825	825	825	825	825	825														
	494	523	523	523	523	523	523														
20	694	825	825	825	825	825	825	775	825	825	825	825	825	825							
	423	490	490	490	490	490	490	517	550	550	550	550	550	550							
21	630	759	825	825	825	825	825	702	825	825	825	825	825	825							
	364	426	460	460	460	460	460	453	520	520	520	520	520	520							
22	573	690	777	825	825	825	825	639	771	825	825	825	825	825	825	825	825	825	825	825	825
	316	370	414	438	438	438	438	393	461	490	490	490	490	490	548	548	548	548	548	548	548
23	523	630	709	774	825	825	825	583	703	793	825	825	825	825	825	825	825	825	825	825	825
	276	323	362	393	418	418	418	344	402	451	468	468	468	468	491	518	518	518	518	518	518
24	480	577	651	709	789	825	825	535	645	727	792	825	825	825	712	804	825	825	825	825	825
	242	284	318	345	382	396	396	302	353	396	430	448	448	448	431	483	495	495	495	495	495
25	441	532	600	652	727	825	825	493	594	669	729	811	825	825	657	739	805	825	825	825	825
	214	250	281	305	337	377	377	266	312	350	380	421	426	426	381	427	464	474	474	474	474
26	408	492	553	603	672	807	825	456	549	618	673	750	825	825	606	682	744	825	825	825	825
	190	222	249	271	299	354	361	236	277	310	337	373	405	405	338	379	411	454	454	454	454
27	378	454	513	558	622	747	825	421	508	573	624	694	825	825	561	633	688	768	825	825	825
	169	198	222	241	267	315	347	211	247	277	301	333	389	389	301	337	367	406	432	432	432
28	351	423	477	519	577	694	822	391	472	532	579	645	775	825	522	588	640	712	825	825	825
	151	177	199	216	239	282	331	189	221	248	269	298	353	375	270	302	328	364	413	413	413
29	327	394	444	483	538	646	766	364	439	495	540	601	723	825	486	547	597	664	798	825	825
	136	159	179	194	215	254	298	170	199	223	242	268	317	359	242	272	295	327	387	399	399
30	304	367	414	451	502	603	715	340	411	462	504	561	675	799	453	511	556	619	745	825	825
	123	144	161	175	194	229	269	153	179	201	218	242	286	336	219	245	266	295	349	385	385
31	285	343	387	421	469	564	669	318	384	433	471	525	631	748	424	478	520	580	697	825	825
	111	130	146	158	175	207	243	138	162	182	198	219	259	304	198	222	241	267	316	369	369
32	267	322	363	396	441	529	627	298	360	406	442	492	592	702	397	448	489	544	654	775	823
	101	118	132	144	159	188	221	126	147	165	179	199	235	276	180	201	219	242	287	337	355
33	252	303	342	372	414	498	589	280	339	381	415	463	556	660	373	421	459	511	615	729	798
	92	108	121	131	145	171	201	114	134	150	163	181	214	251	164	183	199	221	261	307	334
34	237	285	321	349	390	468	555	264	318	358	391	435	523	621	352	397	432	481	579	687	774
	84	98	110	120	132	156	184	105	122	137	149	165	195	229	149	167	182	202	239	280	314
35	223	268	303	330	367	441	523	249	300	339	369	411	493	585	331	373	408	454	546	648	741
	77	90	101	110	121	143	168	96	112	126	137	151	179	210	137	153	167	185	219	257	292
36	211	253	286	312	348	417	495	235	283	319	348	388	466	553	313	354	385	429	516	612	700
	70	82	92	101	111	132	154	88	103	115	125	139	164	193	126	141	153	169	201	236	269
37								222	268	303	330	367	441	523	297	334	364	406	487	579	663
								81	95	106	115	128	151	178	116	130	141	156	185	217	247
38								211	255	286	312	348	418	496	280	316	345	384	462	549	628
								74	87	98	106	118	139	164	107	119	130	144	170	200	228
39								199	241	271	297	330	397	471	267	300	327	364	438	520	595
								69	81	90	98	109	129	151	98	110	120	133	157	185	211
40								190	229	258	282	313	376	447	253	285	310	346	417	495	565
								64	75	84	91	101	119	140	91	102	111	123	146	171	195
41															241	271	295	330	396	471	538
															85	95	103	114	135	159	181
42															229	259	282	313	378	448	513
															79	88	96	106	126	148	168
43															219	247	268	300	360	427	489
															73	82	89	99	117	138	157
44															208	235	256	286	343	408	466
															68	76	83	92	109	128	146

LRFD

STANDARD LOAD TABLE FOR OPEN WEB STEEL JOISTS, K-SERIES Based on a 50 ksi Maximum Yield Strength - Loads Shown in Pounds per Linear Foot (plf)															
Joist Designation	24K4	24K5	24K6	24K7	24K8	24K9	24K10	24K12	26K5	26K6	26K7	26K8	26K9	26K10	26K12
Depth (In.)	24	24	24	24	24	24	24	24	26	26	26	26	26	26	26
Approx. Wt. (lbs./ft.)	8.4	9.3	9.7	10.1	11.5	12.0	13.1	16.0	9.8	10.6	10.9	12.1	12.2	13.8	16.6
Span (ft.)															
24	780 516	825 544	825 544	825 544	825 544	825 544	825 544	825 544							
25	718 456	810 511	825 520	825 520	825 520	825 520	825 520	825 520							
26	663 405	748 453	814 493	825 499	825 499	825 499	825 499	825 499	813 535	825 541	825 541	825 541	825 541	825 541	825 541
27	615 361	693 404	754 439	825 479	825 479	825 479	825 479	825 479	753 477	820 519	825 522	825 522	825 522	825 522	825 522
28	571 323	643 362	700 393	781 436	825 456	825 456	825 456	825 456	699 427	762 464	825 501	825 501	825 501	825 501	825 501
29	531 290	600 325	652 354	727 392	804 429	825 436	825 436	825 436	651 384	709 417	790 463	825 479	825 479	825 479	825 479
30	496 262	559 293	609 319	679 353	750 387	816 419	825 422	825 422	607 346	661 377	738 417	816 457	825 459	825 459	825 459
31	465 237	523 266	570 289	636 320	702 350	765 379	825 410	825 410	568 314	619 341	690 378	763 413	825 444	825 444	825 444
32	435 215	490 241	535 262	595 290	658 318	717 344	823 393	823 393	534 285	580 309	648 343	715 375	778 407	823 431	823 431
33	409 196	462 220	502 239	559 265	619 289	673 313	798 368	798 368	501 259	546 282	609 312	672 342	732 370	798 404	798 404
34	385 179	435 201	472 218	526 242	582 264	634 286	753 337	774 344	472 237	514 257	573 285	633 312	688 338	774 378	774 378
35	363 164	409 184	445 200	496 221	549 242	598 262	709 308	751 324	445 217	484 236	540 261	597 286	649 310	751 356	751 356
36	343 150	387 169	421 183	469 203	519 222	565 241	670 283	730 306	420 199	457 216	510 240	564 263	613 284	729 334	730 334
37	324 138	366 155	399 169	444 187	490 205	534 222	634 260	711 290	397 183	433 199	483 221	534 242	580 262	690 308	711 315
38	307 128	346 143	378 156	421 172	465 189	507 204	601 240	691 275	376 169	411 184	457 204	505 223	550 241	654 284	691 299
39	292 118	328 132	358 144	399 159	441 174	480 189	570 222	673 261	357 156	390 170	433 188	480 206	522 223	619 262	673 283
40	277 109	312 122	340 133	379 148	420 161	456 175	541 206	657 247	340 145	370 157	412 174	456 191	496 207	589 243	657 269
41	264 101	297 114	324 124	361 137	399 150	435 162	516 191	640 235	322 134	352 146	393 162	433 177	472 192	561 225	640 256
42	252 94	283 106	309 115	343 127	379 139	414 151	490 177	625 224	307 125	336 136	373 150	412 164	450 178	534 210	625 244
43	240 88	270 98	294 107	328 116	363 130	394 140	468 165	609 213	294 116	319 126	357 140	394 153	429 166	508 195	610 232
44	229 82	258 92	280 100	313 110	346 121	376 131	447 154	580 199	280 108	306 118	340 131	376 143	409 155	486 182	597 222
45	219 76	246 86	268 93	298 103	330 113	360 122	427 144	555 185	268 101	291 110	325 122	360 133	391 145	465 170	583 212
46	208 71	235 80	256 87	286 97	316 106	345 114	408 135	531 174	256 95	279 103	310 114	343 125	375 135	444 159	570 203
47	199 67	225 75	246 82	274 90	303 99	330 107	391 126	508 163	246 89	267 96	298 107	328 117	358 127	426 149	553 192
48	192 63	216 70	235 77	262 85	291 93	316 101	375 118	487 153	235 83	256 90	285 100	315 110	343 119	408 140	529 180
49									225 78	246 85	274 94	303 103	330 112	391 131	508 169
50									216 73	235 80	262 89	291 97	316 105	375 124	487 159
51									208 69	226 75	252 83	279 91	304 99	361 116	469 150
52									199 65	217 71	243 79	268 86	292 93	346 110	451 142

LRFD

STANDARD LOAD TABLE FOR OPEN WEB STEEL JOISTS, K-SERIES
Based on a 50 ksi Maximum Yield Strength - Loads Shown in Pounds per Linear Foot (plf)

Joist Designation	28K6	28K7	28K8	28K9	28K10	28K12	30K7	30K8	30K9	30K10	30K11	30K12
Depth (In.)	28	28	28	28	28	28	30	30	30	30	30	30
Approx. Wt. (lbs./ft.)	11.4	11.8	12.7	13.0	14.3	17.1	12.3	13.2	13.4	15.0	16.4	17.6
Span (ft.)												
28	822 / 541	825 / 543	825 / 543	825 / 543	825 / 543	825 / 543						
29	766 / 486	825 / 522	825 / 522	825 / 522	825 / 522	825 / 522						
30	715 / 439	796 / 486	825 / 500	825 / 500	825 / 500	825 / 500	825 / 543	825 / 543	825 / 543	825 / 543	825 / 543	825 / 543
31	669 / 397	745 / 440	825 / 480	825 / 480	825 / 480	825 / 480	801 / 508	825 / 520	825 / 520	825 / 520	825 / 520	825 / 520
32	627 / 361	699 / 400	772 / 438	823 / 463	823 / 463	823 / 463	751 / 461	823 / 500	823 / 500	823 / 500	823 / 500	823 / 500
33	589 / 329	657 / 364	726 / 399	790 / 432	798 / 435	798 / 435	706 / 420	780 / 460	798 / 468	798 / 468	798 / 468	798 / 468
34	555 / 300	618 / 333	684 / 364	744 / 395	774 / 410	774 / 410	664 / 384	735 / 420	774 / 441	774 / 441	774 / 441	774 / 441
35	523 / 275	583 / 305	645 / 333	702 / 361	751 / 389	751 / 389	627 / 351	693 / 384	751 / 415	751 / 415	751 / 415	751 / 415
36	495 / 252	550 / 280	609 / 306	663 / 332	730 / 366	730 / 366	592 / 323	654 / 353	712 / 383	730 / 392	730 / 392	730 / 392
37	468 / 232	522 / 257	576 / 282	627 / 305	711 / 344	711 / 344	559 / 297	619 / 325	673 / 352	711 / 374	711 / 374	711 / 374
38	444 / 214	493 / 237	546 / 260	594 / 282	691 / 325	691 / 325	531 / 274	586 / 300	639 / 325	691 / 353	691 / 353	691 / 353
39	420 / 198	469 / 219	519 / 240	564 / 260	670 / 306	673 / 308	504 / 253	556 / 277	606 / 300	673 / 333	673 / 333	673 / 333
40	399 / 183	445 / 203	492 / 222	535 / 241	636 / 284	657 / 291	478 / 234	529 / 256	576 / 278	657 / 315	657 / 315	657 / 315
41	379 / 170	424 / 189	468 / 206	510 / 224	606 / 263	640 / 277	454 / 217	502 / 238	547 / 258	640 / 300	640 / 300	640 / 300
42	361 / 158	403 / 175	445 / 192	486 / 208	576 / 245	625 / 264	433 / 202	480 / 221	522 / 240	619 / 282	625 / 284	625 / 284
43	345 / 147	385 / 163	426 / 179	463 / 194	550 / 228	610 / 252	414 / 188	457 / 206	498 / 223	591 / 263	610 / 270	610 / 270
44	330 / 137	367 / 152	406 / 167	442 / 181	525 / 212	597 / 240	394 / 176	436 / 192	475 / 208	564 / 245	597 / 258	597 / 258
45	315 / 128	351 / 142	388 / 156	423 / 169	501 / 198	583 / 229	376 / 164	417 / 179	454 / 195	538 / 229	583 / 246	583 / 246
46	301 / 120	336 / 133	372 / 146	405 / 158	480 / 186	570 / 219	361 / 153	399 / 168	435 / 182	516 / 214	570 / 236	570 / 236
47	288 / 112	321 / 125	355 / 136	387 / 148	459 / 174	558 / 210	345 / 144	382 / 157	415 / 171	493 / 201	558 / 226	558 / 226
48	276 / 105	309 / 117	340 / 128	370 / 139	441 / 163	547 / 201	331 / 135	366 / 148	399 / 160	472 / 188	543 / 215	547 / 216
49	265 / 99	295 / 110	327 / 120	355 / 130	423 / 153	535 / 193	318 / 127	351 / 139	382 / 150	454 / 177	520 / 202	535 / 207
50	255 / 93	283 / 103	313 / 113	342 / 123	405 / 144	525 / 185	304 / 119	337 / 130	367 / 141	436 / 166	499 / 190	525 / 199
51	244 / 88	273 / 97	301 / 106	328 / 115	390 / 136	507 / 175	292 / 112	324 / 123	352 / 133	418 / 157	480 / 179	514 / 192
52	235 / 83	262 / 92	289 / 100	315 / 109	375 / 128	487 / 165	282 / 106	312 / 116	339 / 126	402 / 148	462 / 169	504 / 184
53	226 / 78	252 / 87	279 / 95	304 / 103	360 / 121	469 / 156	271 / 100	300 / 109	327 / 119	387 / 140	444 / 159	495 / 177
54	217 / 74	243 / 82	268 / 89	292 / 97	348 / 114	451 / 147	261 / 94	288 / 103	313 / 112	373 / 132	427 / 150	486 / 170
55	210 / 70	234 / 77	259 / 85	282 / 92	334 / 108	435 / 139	252 / 89	277 / 98	303 / 106	360 / 125	412 / 142	468 / 161
56	202 / 66	226 / 73	249 / 80	271 / 87	322 / 102	420 / 132	243 / 84	268 / 92	292 / 100	346 / 118	397 / 135	451 / 153
57							234 / 80	259 / 88	282 / 95	334 / 112	384 / 128	435 / 145
58							226 / 76	250 / 83	271 / 90	322 / 106	370 / 121	420 / 137
59							219 / 72	241 / 79	262 / 86	312 / 101	358 / 115	406 / 130
60							211 / 69	234 / 75	253 / 81	301 / 96	346 / 109	393 / 124

LRFD

JOIST DESIGNATION	DEPTH (inches)	MOMENT CAPACITY (inch-kips)	SHEAR CAPACITY* (lbs)	APPROX. WEIGHT** (lbs/ft)	GROSS MOMENT OF INERTIA (in.⁴)	BRIDGING TABLE SECTION NUMBER
			STANDARD LOAD TABLE FOR KCS OPEN WEB STEEL JOISTS			
			Based on a 50 ksi Maximum Yield Strength			
10KCS1	10	258	3000	6.0	29	1
10KCS2	10	337	3750	7.5	37	1
10KCS3	10	444	4500	10.0	47	1
12KCS1	12	313	3600	6.0	43	3
12KCS2	12	411	4500	8.0	55	5
12KCS3	12	543	5250	10.0	71	5
14KCS1	14	370	4350	6.5	59	4
14KCS2	14	486	5100	8.0	77	6
14KCS3	14	642	5850	10.0	99	6
16KCS2	16	523	6000	8.5	99	6
16KCS3	16	705	7200	10.5	128	9
16KCS4	16	1080	7950	14.5	192	9
16KCS5	16	1401	8700	18.0	245	9
18KCS2	18	592	7050	9.0	127	6
18KCS3	18	798	7800	11.0	164	9
18KCS4	18	1225	8550	15.0	247	10
18KCS5	18	1593	9300	18.5	316	10
20KCS2	20	663	7800	9.5	159	6
20KCS3	20	892	9000	11.5	205	9
20KCS4	20	1371	11850	16.5	308	10
20KCS5	20	1786	12600	20.0	396	10
22KCS2	22	732	8850	10.0	194	6
22KCS3	22	987	9900	12.5	251	9
22KCS4	22	1518	11850	16.5	377	11
22KCS5	22	1978	12900	20.5	485	11
24KCS2	24	801	9450	10.0	232	6
24KCS3	24	1080	10800	12.5	301	9
24KCS4	24	1662	12600	16.5	453	12
24KCS5	24	2172	13350	20.5	584	12
26KCS2	26	870	9900	10.0	274	6
26KCS3	26	1174	11700	12.5	355	9
26KCS4	26	1809	12750	16.5	536	12
26KCS5	26	2364	13800	20.5	691	12
28KCS2	28	939	10350	10.5	320	6
28KCS3	28	1269	12000	12.5	414	9
28KCS4	28	1954	12750	16.5	626	12
28KCS5	28	2556	13800	20.5	808	12
30KCS3	30	1362	12000	13.0	478	9
30KCS4	30	2100	12750	16.5	722	12
30KCS5	30	2749	13800	21.0	934	12

*MAXIMUM UNIFORMLY DISTRIBUTED LOAD CAPACITY IS 825 PLF AND SINGLE CONCENTRATED LOAD CANNOT EXCEED SHEAR CAPACITY
**DOES NOT INCLUDE ACCESSORIES

STANDARD LRFD LOAD TABLE
LONGSPAN STEEL JOISTS, LH-SERIES

Based on a 50 ksi Maximum Yield Strength
Adopted by the Steel Joist Institute May 1, 2000
Revised to November 10, 2003 - Effective March 01, 2005

The black figures in the following table give the TOTAL safe factored uniformly distributed load-carrying capacities, in pounds per linear foot, of **LRFD LH-Series** Steel Joists. The weight of factored DEAD loads, including the joists, must in all cases be deducted to determine the factored LIVE load-carrying capacities of the joists. The approximate DEAD load of the joists may be determined from the weights per linear foot shown in the tables.

The **RED** figures in this load table are the unfactored, nominal LIVE loads per linear foot of joist which will produce an approximate deflection of 1/360 of the span. LIVE loads which will produce a deflection of 1/240 of the span may be obtained by multiplying the **RED** figures by 1.5. In no case shall the TOTAL load capacity of the joists be exceeded.

This load table applies to joists with either parallel chords or standard pitched top chords. When top chords are pitched, the carrying capacities are determined by the nominal depth of the joists at the center of the span. Standard top chord pitch is 1/8 inch per foot. If pitch exceeds this standard, the load table does <u>not</u> apply. Sloped parallel-chord joists shall use span as defined by the length along the slope.

Where the joist span is in the <u>RED SHADED</u> area of the load table, the row of bridging nearest the midspan shall be diagonal bridging with bolted connections at chords and intersection. <u>Hoisting cables shall not be released until this row of bolted diagonal bridging is completely installed.</u>

Where the joist span is in the **BLUE SHADED** area of the load table, all rows of bridging shall be diagonal bridging with bolted connections at chords and intersection. <u>Hoisting cables shall not be released until the two rows of bridging nearest the third points are completely installed.</u>

The approximate moment of inertia of the joist, in inches[4] is; $I_j = 26.767(W_{LL})(L^3)(10^{-6})$, where W_{LL} = RED figure in the Load Table, and L = (clear span + 0.67) in feet.

When holes are required in top or bottom chords, the carrying capacities must be reduced in proportion to the reduction of chord areas.

The top chords are considered as being stayed laterally by floor slab or roof deck.

The approximate joist weights per linear foot shown in these tables do <u>not</u> include accessories.

LRFD

STANDARD LOAD TABLE FOR LONGSPAN STEEL JOISTS, LH-SERIES
Based on a 50 ksi Maximum Yield Strength - Loads Shown in Pounds per Linear Foot (plf)

Joist Designation	Approx. Wt in Lbs. Per Linear Ft (Joists only)	Depth in inches	SAFE LOAD in Lbs. Between	CLEAR SPAN IN FEET															
			21-24	**25**	**26**	**27**	**28**	**29**	**30**	**31**	**32**	**33**	**34**	**35**	**36**				
18LH02	10	18	12000	702	663	627	586	550	517	486	459	433	409	388	367				
				313	284	259	234	212	193	175	160	147	135	124	114				
18LH03	11	18	13300	781	739	700	657	613	573	538	505	475	448	424	400				
				348	317	289	262	236	213	194	177	161	148	136	124				
18LH04	12	18	15500	906	856	802	750	703	660	619	582	547	516	487	462				
				403	367	329	296	266	242	219	200	182	167	153	141				
18LH05	15	18	17500	1026	972	921	871	814	762	714	672	631	595	562	532				
				454	414	378	345	311	282	256	233	212	195	179	164				
18LH06	15	18	20700	1213	1123	1044	972	907	849	796	748	705	664	627	594				
				526	469	419	377	340	307	280	254	232	212	195	180				
18LH07	17	18	21500	1260	1213	1170	1089	1017	952	892	838	789	744	703	666				
				553	513	476	428	386	349	317	288	264	241	222	204				
18LH08	19	18	22400	1314	1264	1218	1176	1137	1075	1020	961	906	856	810	768				
				577	534	496	462	427	387	351	320	292	267	246	226				
18LH09	21	18	24000	1404	1351	1302	1257	1215	1174	1138	1069	1006	949	897	849				
				616	571	527	491	458	418	380	346	316	289	266	245				
			22-24	**25**	**26**	**27**	**28**	**29**	**30**	**31**	**32**	**33**	**34**	**35**	**36**	**37**	**38**	**39**	**40**
20LH02	10	20	11300	663	655	646	615	582	547	516	487	460	436	412	393	373	355	337	322
				306	303	298	274	250	228	208	190	174	160	147	136	126	117	108	101
20LH03	11	20	12000	703	694	687	678	651	621	592	558	528	499	474	448	424	403	382	364
				337	333	317	302	280	258	238	218	200	184	169	156	143	133	123	114
20LH04	12	20	14700	861	849	837	792	744	700	660	624	589	558	529	502	477	454	433	412
				428	406	386	352	320	291	265	243	223	205	189	174	161	149	139	129
20LH05	14	20	15800	924	913	903	892	856	816	769	726	687	651	616	585	556	529	504	481
				459	437	416	395	366	337	308	281	258	238	219	202	187	173	161	150
20LH06	15	20	21100	1233	1186	1144	1084	1018	952	894	840	790	745	703	666	631	598	568	541
				606	561	521	477	427	386	351	320	292	267	246	226	209	192	178	165
20LH07	17	20	22500	1317	1267	1221	1179	1140	1066	1000	940	885	834	789	745	706	670	637	606
				647	599	556	518	484	438	398	362	331	303	278	256	236	218	202	187
20LH08	19	20	23200	1362	1309	1263	1219	1177	1140	1083	1030	981	931	882	837	795	754	718	685
				669	619	575	536	500	468	428	395	365	336	309	285	262	242	225	209
20LH09	21	20	25400	1485	1429	1377	1329	1284	1242	1203	1167	1132	1068	1009	954	904	858	816	775
				729	675	626	581	542	507	475	437	399	366	336	309	285	264	244	227
20LH10	23	20	27400	1602	1542	1486	1434	1386	1341	1297	1258	1221	1186	1122	1060	1005	954	906	862
				786	724	673	626	585	545	510	479	448	411	377	346	320	296	274	254

LRFD

STANDARD LOAD TABLE FOR LONGSPAN STEEL JOISTS, LH-SERIES
Based on a 50 ksi Maximum Yield Strength - Loads Shown in Pounds per Linear Foot (plf)

24LH Series — CLEAR SPAN IN FEET

Joist Designation	Approx. Wt (lbs/ft)	Depth (in)	SAFELOAD 28-32	33	34	35	36	37	38	39	40	41	42	43	44	45	46	47	48
24LH03	11	24	17250	513	508	504	484	460	439	418	400	382	366	351	336	322	310	298	286
				235	226	218	204	188	175	162	152	141	132	124	116	109	102	96	90
24LH04	12	24	21150	628	597	568	540	514	490	468	447	427	409	393	376	361	346	333	321
				288	265	246	227	210	195	182	169	158	148	138	130	122	114	107	101
24LH05	13	24	22650	673	669	660	628	598	570	544	520	496	475	456	436	420	403	387	372
				308	297	285	264	244	226	210	196	182	171	160	150	141	132	124	117
24LH06	16	24	30450	906	868	832	795	756	720	685	655	625	598	571	546	522	501	480	460
				411	382	356	331	306	284	263	245	228	211	197	184	172	161	152	142
24LH07	17	24	33450	997	957	919	882	847	811	774	736	702	669	639	610	583	559	535	514
				452	421	393	367	343	320	297	276	257	239	223	208	195	182	171	161
24LH08	18	24	35700	1060	1015	973	933	895	858	817	780	745	712	682	652	625	600	576	553
				480	447	416	388	362	338	314	292	272	254	238	222	208	196	184	173
24LH09	21	24	42000	1248	1212	1177	1146	1096	1044	994	948	903	861	822	786	751	720	690	661
				562	530	501	460	424	393	363	337	313	292	272	254	238	223	209	196
24LH10	23	24	44400	1323	1284	1248	1213	1182	1152	1105	1053	1002	955	912	873	834	799	766	735
				596	559	528	500	474	439	406	378	351	326	304	285	266	249	234	220
24LH11	25	24	46800	1390	1350	1312	1276	1243	1210	1180	1152	1101	1051	1006	963	924	885	850	816
				624	588	555	525	498	472	449	418	388	361	337	315	294	276	259	243

28LH Series — CLEAR SPAN IN FEET

Joist Designation	Approx. Wt (lbs/ft)	Depth (in)	SAFELOAD 33-40	41	42	43	44	45	46	47	48	49	50	51	52	53	54	55	56
28LH05	13	28	21000	505	484	465	445	429	412	397	382	367	355	342	330	319	309	298	289
				219	205	192	180	169	159	150	142	133	126	119	113	107	102	97	92
28LH06	16	28	27900	672	643	618	592	568	546	525	505	486	469	451	436	421	406	393	379
				289	270	253	238	223	209	197	186	175	166	156	148	140	133	126	120
28LH07	17	28	31500	757	726	696	667	640	615	591	568	547	528	508	490	474	457	442	427
				326	305	285	267	251	236	222	209	196	186	176	166	158	150	142	135
28LH08	18	28	33750	810	775	744	712	684	657	630	604	580	556	535	516	496	478	462	445
				348	325	305	285	268	252	236	222	209	196	185	175	165	156	148	140
28LH09	21	28	41550	1000	958	918	879	844	810	778	748	721	694	669	645	622	601	580	561
				428	400	375	351	329	309	291	274	258	243	228	216	204	193	183	173
28LH10	23	28	45450	1093	1056	1018	976	937	900	864	831	799	769	742	715	690	666	643	622
				466	439	414	388	364	342	322	303	285	269	253	238	225	215	204	193
28LH11	25	28	48750	1170	1143	1104	1066	1023	982	943	907	873	841	810	781	753	727	702	679
				498	475	448	423	397	373	351	331	312	294	278	263	249	236	223	212
28LH12	27	28	53550	1285	1255	1227	1200	1173	1149	1105	1053	1023	984	948	913	880	849	819	790
				545	520	496	476	454	435	408	383	361	342	321	303	285	270	256	243
28LH13	30	28	55800	1342	1311	1281	1252	1224	1198	1173	1149		1083	1041	1002	964	930	897	865
				569	543	518	495	472	452	433	415	396	373	352	332	314	297	281	266

32LH Series — CLEAR SPAN IN FEET

Joist Designation	Approx. Wt (lbs/ft)	Depth (in)	SAFELOAD 38-46	SAFELOAD 47-48	49	50	51	52	53	54	55	56	57	58	59	60	61	62	63	64
32LH06	14	32	25050	25050	507	489	472	456	441	426	412	399	385	373	363	351	340	330	321	312
					211	199	189	179	169	161	153	145	138	131	125	119	114	108	104	99
32LH07	16	32	28200	28200	568	549	529	511	493	477	462	447	432	418	406	393	381	370	360	349
					235	223	211	200	189	179	170	162	154	146	140	133	127	121	116	111
32LH08	17	32	30600	30600	616	595	574	553	535	517	499	483	468	453	439	426	412	400	388	378
					255	242	229	216	205	194	184	175	167	159	151	144	137	131	125	120
32LH09	21	32	38400	38400	774	747	720	694	670	648	627	606	586	568	550	534	517	502	487	472
					319	302	285	270	256	243	230	219	208	198	189	180	172	164	157	149
32LH10	21	32	42450	42450	856	825	796	768	742	717	693	667	645	624	603	583	564	546	529	513
					352	332	315	297	282	267	254	240	228	217	206	196	186	178	169	162
32LH11	24	32	46500	46500	937	903	870	840	811	783	757	732	709	687	664	643	624	604	585	567
					385	363	343	325	308	292	277	263	251	239	227	216	206	196	187	179
32LH12	27	32	54600	54600	1101	1068	1032	996	961	928	897	867	838	811	786	762	738	715	694	673
					450	428	406	384	364	345	327	311	295	281	267	255	243	232	221	211
32LH13	30	32	60900	60900	1225	1201	1177	1156	1113	1072	1035	999	964	931	900	871	843	816	790	766
					500	480	461	444	420	397	376	354	336	319	304	288	275	262	249	238
32LH14	33	32	62700	62700	1264	1239	1215	1192	1170	1149	1107	1069	1032	997	964	933	903	874	846	820
					515	495	476	458	440	417	395	374	355	337	321	304	290	276	264	251
32LH15	35	32	64800	64800	1305	1279	1255	1231	1207	1186	1164	1144	1125	1087	1051	1017	984	952	924	895
					532	511	492	473	454	438	422	407	393	374	355	338	322	306	292	279

36LH Series — CLEAR SPAN IN FEET

Joist Designation	Approx. Wt (lbs/ft)	Depth (in)	SAFELOAD 42-46	SAFELOAD 47-56	57	58	59	60	61	62	63	64	65	66	67	68	69	70	71	72
36LH07	16	36	25200	25200	438	424	411	399	387	376	366	355	345	336	327	318	310	301	294	286
					177	168	160	153	146	140	134	128	122	117	112	107	103	99	95	91
36LH08	18	36	27750	27750	481	466	453	439	426	414	402	390	379	369	358	349	340	331	322	313
					194	185	176	168	160	153	146	140	133	128	123	118	113	109	104	100
36LH09	21	36	35550	35550	616	597	579	561	544	528	513	499	484	471	459	445	433	423	412	400
					247	235	224	214	204	195	186	179	171	163	157	150	144	138	133	127
36LH10	21	36	39150	39150	681	660	639	619	601	583	567	550	535	520	507	492	480	466	454	442
					273	260	248	236	225	215	206	197	188	180	173	165	159	152	146	140
36LH11	23	36	42750	42750	742	720	697	676	657	637	618	601	583	567	552	537	522	508	495	483
					297	283	269	257	246	234	224	214	205	196	188	180	173	166	159	153
36LH12	25	36	51150	51150	889	862	835	810	784	762	739	717	696	675	655	636	618	600	583	567
					354	338	322	307	292	279	267	255	243	232	222	213	204	195	187	179
36LH13	30	36	60150	60150	1045	1012	981	951	922	894	868	843	819	796	774	753	732	712	694	676
					415	395	376	359	342	327	312	298	285	273	262	251	240	231	222	213
36LH14	36	36	66300	66300	1152	1132	1093	1059	1024	991	961	931	903	876	850	826	802	780	757	738
					456	434	412	392	373	356	339	323	309	295	283	270	259	247	237	228
36LH15	36	36	69900	69900	1213	1192	1171	1153	1116	1081	1047	1015	984	955	927	900	874	850	826	804
					480	464	448	434	418	394	375	358	342	327	312	299	286	274	263	252

LRFD

STANDARD LOAD TABLE FOR LONGSPAN STEEL JOISTS, LH-SERIES
Based on a 50 ksi Maximum Yield Strength - Loads Shown in Pounds per Linear Foot (plf)

Joist Designation	Approx. Wt in Lbs. Per Linear Ft. (Joists Only)	Depth in inches	SAFELOAD* in Lbs. Between 47-59	60-64	65	66	67	68	69	70	71	72	73	74	75	76	77	78	79	80
40LH08	16	40	24900	24900	381	370	361	351	342	333	325	316	309	301	294	288	280	274	267	261
					150	144	138	132	127	122	117	112	108	104	100	97	93	90	86	83
40LH09	21	40	32700	32700	498	484	472	459	447	436	424	414	403	394	384	375	366	358	349	342
					196	188	180	173	166	160	153	147	141	136	131	126	122	118	113	109
40LH10	21	40	36000	36000	550	535	520	507	493	481	469	457	445	435	424	414	403	393	382	373
					216	207	198	190	183	176	169	162	156	150	144	139	134	129	124	119
40LH11	22	40	39300	39300	598	582	567	552	537	523	510	498	484	472	462	450	439	429	418	409
					234	224	215	207	198	190	183	176	169	163	157	151	145	140	135	130
40LH12	25	40	47850	47850	729	708	688	670	652	636	619	603	588	573	559	546	532	519	507	495
					285	273	261	251	241	231	222	213	205	197	189	182	176	169	163	157
40LH13	30	40	56400	56400	859	835	813	792	771	750	730	712	694	676	660	643	628	613	598	585
					334	320	307	295	283	271	260	250	241	231	223	214	207	199	192	185
40LH14	35	40	64500	64500	984	957	930	904	880	856	834	813	792	772	753	735	717	699	682	666
					383	367	351	336	323	309	297	285	273	263	252	243	233	225	216	209
40LH15	36	40	72150	72150	1101	1068	1036	1006	978	949	924	898	874	850	828	807	786	766	747	729
					427	408	390	373	357	342	328	315	302	290	279	268	258	248	239	230
40LH16	42	40	79500	79500	1212	1194	1176	1158	1141	1126	1095	1065	1036	1009	982	957	933	909	886	864
					469	455	441	428	416	404	387	371	356	342	329	316	304	292	282	271

Joist Designation	Wt	Depth	52-59	60-72	73	74	75	76	77	78	79	80	81	82	83	84	85	86	87	88
44LH09	19	44	30000	30000	408	397	388	379	370	363	354	346	339	331	324	316	310	303	297	291
					158	152	146	141	136	131	127	122	118	114	110	106	103	99	96	93
44LH10	21	44	33150	33150	450	439	429	418	408	399	390	381	373	364	357	349	342	334	327	321
					174	168	162	155	150	144	139	134	130	125	121	117	113	110	106	103
44LH11	22	44	35850	35850	487	475	465	453	442	433	423	414	403	396	387	378	370	363	354	348
					188	181	175	168	162	157	151	146	140	136	131	127	123	119	115	111
44LH12	25	44	44400	44400	603	589	574	561	547	534	520	508	496	484	472	462	450	439	430	420
					232	224	215	207	200	192	185	179	172	166	160	155	149	144	139	134
44LH13	30	44	52650	52650	715	699	681	666	649	634	619	606	592	579	565	553	541	529	519	507
					275	265	254	246	236	228	220	212	205	198	191	185	179	173	167	161
44LH14	31	44	60600	60600	823	801	780	759	739	721	703	685	669	654	637	622	609	594	580	568
					315	302	291	279	268	259	249	240	231	223	215	207	200	193	187	181
44LH15	36	44	70500	70500	958	934	912	889	868	847	826	805	786	768	750	732	714	699	682	667
					366	352	339	326	314	303	292	281	271	261	252	243	234	227	219	211
44LH16	42	44	81300	81300	1105	1078	1051	1026	1002	978	955	933	912	891	870	852	832	814	796	780
					421	405	390	375	362	348	336	324	313	302	291	282	272	263	255	246
44LH17	47	44	87300	87300	1185	1170	1153	1138	1125	1098	1072	1048	1024	1000	978	957	936	915	895	876
					450	438	426	415	405	390	376	363	351	338	327	316	305	295	285	276

Joist Designation	Wt	Depth	56-59	60-80	81	82	83	84	85	86	87	88	89	90	91	92	93	94	95	96
48LH10	21	48	30000	30000	369	361	354	346	339	331	325	318	312	306	300	294	288	282	277	271
					141	136	132	127	123	119	116	112	108	105	102	99	96	93	90	87
48LH11	22	48	32550	32550	399	390	382	373	366	358	351	343	337	330	324	318	312	306	300	294
					152	147	142	137	133	129	125	120	117	113	110	106	103	100	97	94
48LH12	25	48	41100	41100	504	493	483	472	462	451	442	433	424	415	408	399	391	384	376	369
					191	185	179	173	167	161	156	151	147	142	138	133	129	126	122	118
48LH13	29	48	49200	49200	603	589	576	564	552	540	529	517	507	498	487	477	468	459	450	441
					228	221	213	206	199	193	187	180	175	170	164	159	154	150	145	141
48LH14	32	48	58050	58050	712	696	681	666	651	637	624	610	598	585	574	562	550	540	529	519
					269	260	251	243	234	227	220	212	206	199	193	187	181	176	171	165
48LH15	36	48	66750	66750	817	799	781	765	748	732	717	702	687	672	658	645	633	619	607	595
					308	298	287	278	269	260	252	244	236	228	221	214	208	201	195	189
48LH16	42	48	76950	76950	943	922	901	882	864	844	826	810	792	777	760	745	730	715	702	688
					355	343	331	320	310	299	289	280	271	263	255	247	239	232	225	218
48LH17	47	48	86400	86400	1059	1035	1012	990	969	948	928	909	889	871	853	837	820	804	787	772
					397	383	371	358	346	335	324	314	304	294	285	276	268	260	252	245

* The safe factored uniform load for the clear spans shown in the Safe Load Column is equal to (Safe Load) / (Clear span + 0.67). (The added 0.67 feet (8 inches) is required to obtain the proper length on which the Load Tables were developed).

In no case shall the safe factored uniform load, for clear spans less than the minimum clear span shown in the Safe Load Column, exceed the uniform load calculated for the minimum clear span listed in the Safe Load Column.

To solve for *live* loads for clear spans shown in the Safe Load Column (or lesser clear spans), multiply the live load of the shortest clear span shown in the Load Table by the (the shortest clear span shown in the Load Table + 0.67 feet)^2 and divide by (the actual clear span + 0.67 feet)^2. The live load shall *not* exceed the safe uniform load.

DESIGN GUIDE LRFD WEIGHT TABLE
FOR JOIST GIRDERS

Based on a 50 ksi Maximum Yield Strength

JOIST GIRDER WEIGHT – POUNDS PER LINEAR FOOT
FACTORED LOAD ON EACH PANEL POINT – KIPS

GIRDER SPAN (ft.)	JOIST SPACES (ft.)	GIRDER DEPTH (in.)	6.0	9.0	12.0	15.0	18.0	21.0	24.0	27.0	30.0	36.0	42.0	48.0	54.0	60.0	66.0	72.0	78.0	84.0
20	2N@10.00	20	16	19	19	19	19	20	24	24	25	30	37	41	46	50	56	62	70	75
		24	16	19	19	19	19	20	21	21	25	28	32	36	41	42	49	52	53	66
		28	16	19	19	19	19	20	20	21	23	26	28	32	39	40	42	46	48	49
	3N@6.67	20	15	15	19	19	20	23	24	27	31	36	44	48	54	74	75	81	84	89
		24	15	16	16	16	19	20	23	26	27	33	36	45	47	53	56	68	79	82
		28	15	16	16	16	17	20	24	24	26	31	36	44	46	49	53	57	68	80
	4N@5.00	20	15	15	19	21	25	29	33	38	41	50	57	65	71	88	97	100	107	120
		24	15	16	17	20	23	26	29	32	35	44	50	55	62	71	85	90	100	102
		28	16	16	17	19	22	25	28	30	34	39	49	50	59	63	72	86	91	91
	5N@4.00	20	15	17	21	26	31	36	39	48	51	62	71	82	99	99	109	120	141	142
		24	16	16	20	23	26	30	35	39	43	53	60	68	80	91	101	103	110	120
		28	16	16	18	22	27	28	33	37	39	48	55	64	68	77	93	95	107	111
	6N@3.33	20	16	19	25	29	36	41	50	57	58	72	82	99	107	118	138	141		
		24	16	18	22	28	31	37	43	46	53	61	70	85	102	102	111	123	144	147
		28	17	18	22	26	30	33	40	42	47	58	68	76	83	96	109	112	119	130
	8N@2.50	20	19	25	32	41	51	58	65	72	82	99	118	139	142					
		24	17	22	29	36	42	50	54	61	69	86	103	107	128	149	153			
		28	18	22	29	34	40	47	54	61	67	76	88	107	112	124	135	155	166	
22	2N@11.00	20	21	21	21	22	22	23	24	24	25	34	39	43	49	55	62	69	76	78
		24	18	21	21	22	22	22	23	24	30	33	41	41	45	51	55	61		73
		28	18	21	21	21	22	22	22	23	24	37	30	33	41	42	46	48	51	58
	3N@7.33	20	15	18	18	19	22	24	26	29	33	42	45	53	68	70	76	84	88	94
		24	15	15	19	19	20	23	24	26	30	35	40	45	48	55	61	74	81	84
		28	15	16	16	16	19	20	23	24	27	32	36	45	47	52	54	59	74	82
	4N@5.50	20	15	16	19	23	26	30	36	39	44	55	62	71	82	95	96	106	119	134
		24	15	15	17	20	25	27	29	34	38	48	52	58	71	79	89	98	101	107
		28	16	16	16	19	22	25	28	32	35	40	49	54	60	72	79	87	90	97
	5N@4.40	20	15	17	24	27	34	38	42	49	55	65	75	96	98	111	126	137		
		24	16	16	20	24	28	33	38	40	48	56	62	73	85	100	101	110	116	133
		28	16	16	18	22	26	30	32	38	41	51	57	65	73	86	92	102	105	111
	6N@3.67	20	16	21	27	33	39	49	56	57	65	79	97	106	118	137				
		24	16	19	23	28	32	39	45	51	58	66	82	98	101	109	120	142	144	
		28	16	18	22	26	30	34	39	44	50	61	70	76	89	102	104	113	127	148
	8N@2.75	20	19	27	36	43	56	64	71	80	96	106	135	138						
		24	18	24	31	38	46	53	60	68	75	101	105	125	145	149				
		28	18	22	28	34	40	47	54	62	69	79	87	106	118	131	152	164		
25	3N@8.33	20	18	18	19	22	26	27	30	37	41	49	59	66	70	76	86	89	97	102
		24	15	18	19	20	22	25	26	28	32	39	43	51	59	67	71	81	84	89
		28	15	15	19	19	20	23	24	27	29	34	39	45	47	55	59	67	81	82
		32	15	16	16	16	20	21	23	24	27	32	36	44	46	52	54	58	74	81
		36	16	16	16	17	17	20	24	24	26	32	36	40	45	48	53	54	68	79
	4N@6.25	20	15	18	20	25	29	35	39	42	49	55	70	78	93	99	109	119	134	135
		24	15	16	19	21	26	29	33	37	40	50	57	64	72	88	97	100	106	120
		28	15	15	17	20	24	25	29	34	37	43	51	58	66	72	89	90	101	102
		32	16	16	17	19	21	25	28	32	35	40	49	54	60	69	79	86	91	96
		36	16	16	17	19	21	26	26	29	34	38	49	50	56	63	73	85	88	92
	5N@5.00	20	15	18	25	31	38	43	51	55	58	73	93	100	109	125	134			
		24	15	17	23	26	32	36	42	47	53	61	75	81	98	102	112	129	140	
		28	16	16	20	24	28	31	37	41	47	56	62	72	79	93	101	106	117	125
		32	16	16	19	23	26	30	33	38	41	51	57	65	73	83	93	102	105	111
		36	16	17	18	22	26	28	31	36	39	48	54	64	69	75	88	96	101	108
	6N@4.17	20	16	24	29	38	45	55	58	69	78	94	104	116	134					
		24	16	20	25	31	37	44	50	56	64	75	97	99	107	118	138			
		28	16	18	23	28	32	38	44	51	55	67	73	87	101	104	120	134	143	145
		32	16	18	22	26	30	34	39	44	50	61	69	77	89	102	105	113	127	148
		36	16	18	24	25	30	36	39	43	49	58	67	74	84	98	108	116	117	129
	8N@3.12	20	21	29	39	48	58	70	78	94	99	115	134							
		24	19	26	33	41	50	57	65	75	81	99	118	138						
		28	18	23	30	38	44	53	60	67	75	86	103	116	127	147				
		32	18	24	28	34	39	47	54	65	71	78	87	105	117	129	152	154		
		36	18	22	29	34	40	46	52	61	63	76	87	101	114	121	136	148	166	167
	10N@2.50	20	26	38	49	63	78	94	100	115	134									
		24	23	33	42	54	65	75	89	99	104	130								
		28	21	30	38	48	56	64	74	84	101	109	134	147						
		32	21	28	36	43	52	62	69	76	87	107	118	130	153					
		36	22	28	37	44	52	64	71	77	85	100	116	130	151	157				

LRFD

GIRDER SPAN (ft.)	JOIST SPACES (ft.)	GIRDER DEPTH (in.)	JOIST GIRDER WEIGHT – POUNDS PER LINEAR FOOT / FACTORED LOAD ON EACH PANEL POINT – KIPS																	
			6.0	9.0	12.0	15.0	18.0	21.0	24.0	27.0	30.0	36.0	42.0	48.0	54.0	60.0	66.0	72.0	78.0	84.0
28	3N@ 9.33	24	18	18	19	22	24	27	29	36	39	43	53	62	70	71	78	85	89	98
		28	18	18	19	20	22	25	26	28	31	39	43	46	55	61	66	76	83	86
		32	15	18	19	21	23	24	27	28	34	39	45	48	53	58	66	80	81	
	4N@ 7.00	24	15	16	20	24	27	32	38	40	48	55	62	71	82	95	104	106	120	135
		28	15	15	18	21	25	28	32	36	39	49	56	64	71	79	96	97	106	107
		32	15	15	17	20	23	25	29	33	37	43	50	58	62	70	85	90	99	102
	5N@ 5.60	24	15	18	24	29	34	39	46	52	58	66	78	96	102	111	126	136		
		28	15	17	21	26	30	35	39	46	50	61	68	77	90	99	107	114	130	142
		32	16	17	20	24	32	37	41	44	56	62	70	80	93	102	107	112	119	
	6N@ 4.67	24	16	21	28	35	41	49	55	63	70	79	96	106	134	137				
		28	15	20	24	30	36	42	50	54	58	71	82	99	107	118	138	142		
		32	16	19	23	28	32	37	43	49	53	64	74	84	101	102	111	123	144	146
	7N@ 4.00	24	18	24	32	41	49	56	64	74	79	96	110	135						
		28	17	22	27	35	43	51	57	62	69	82	99	108	129	140				
		32	16	21	27	31	38	44	52	55	63	74	85	102	108	123	143	146		
	8N@ 3.50	24	20	28	37	48	55	64	74	79	95	105	134							
		28	18	25	32	39	50	58	65	72	81	99	108	129	141					
		32	17	24	29	38	43	53	60	64	70	86	103	113	127	147	149			
	10N@ 2.80	24	24	36	46	57	70	79	96	102	117	137								
		28	23	30	41	50	60	69	82	99	100	120	141							
		32	21	30	38	46	55	66	71	80	93	109	126	147						
30	3N@ 10.00	24	18	18	21	24	27	31	35	38	40	48	58	66	71	80	92	98	117	119
		28	18	18	19	22	25	27	30	35	37	42	49	56	63	70	79	82	93	99
		32	15	18	19	20	22	26	28	31	32	39	46	51	57	64	71	73	83	84
		36	16	19	19	19	21	23	26	28	31	35	39	46	52	57	64	65	73	75
	4N@ 7.50	24	16	18	23	29	33	37	42	49	53	64	76	85	101	104	126	127	149	150
		28	15	16	21	25	30	33	37	42	45	53	61	73	81	86	103	104	126	128
		32	15	16	18	22	26	30	34	37	43	51	55	62	70	77	87	103	105	116
		36	16	16	17	22	24	27	31	34	36	46	52	59	64	74	78	88	91	105
	5N@ 6.00	24	15	19	25	30	37	43	51	55	58	73	86	109	125	134				
		28	15	17	23	27	32	37	44	47	53	61	75	88	97	102	112	128	138	
		32	16	17	21	24	29	35	39	43	48	56	63	77	90	100	101	107	117	133
		36	16	17	20	24	27	31	36	40	43	51	60	70	80	86	94	103	110	118
	6N@ 5.00	24	16	24	29	37	45	52	58	66	73	94	104	116	134					
		28	16	20	27	32	38	44	50	57	65	75	97	99	107	137	140			
		32	16	19	24	29	34	40	45	51	58	65	82	98	100	109	121	142	144	
		36	16	18	23	26	31	37	41	46	52	61	70	84	101	102	111	123	126	148
	8N@ 3.75	24	21	32	40	51	63	73	83	99	111	124	146							
		28	20	30	37	44	53	61	73	80	86	114	126	149						
		32	18	26	34	42	55	63	71	79	104	117	130	154	161					
		36	17	23	32	39	46	54	61	69	76	89	108	121	134	154	169			
	10N@ 3.00	24	25	38	51	66	78	99	111	123	134									
		28	24	36	47	57	69	80	94	113	116	138								
		32	22	31	39	52	58	74	82	95	105	129	142							
		36	22	30	39	48	54	68	79	84	91	119	132	151						
32	3N@ 10.67	24	18	19	21	26	27	34	38	40	42	54	61	70	75	84	88	102	102	113
		28	16	17	18	24	26	28	31	34	37	43	55	60	69	70	76	85	89	93
		32	17	17	18	21	25	26	28	32	34	39	44	54	61	62	67	77	80	86
		36	15	17	17	19	20	23	25	26	28	38	40	45	51	53	58	67	81	77
	4N@ 8.00	24	18	19	23	26	32	37	40	47	55	61	72	86	94	103	114	133	134	
		28	15	18	20	24	28	32	37	40	45	55	62	70	78	94	96	105	121	135
		32	15	15	20	22	25	29	32	36	39	49	56	64	71	83	82	97	102	107
		36	15	16	17	21	24	26	30	34	36	43	50	58	65	70	85	90	99	102
	5N@ 6.40	24	15	20	27	33	39	44	51	57	65	77	93	100	123	133				
		28	15	18	24	28	34	39	46	52	58	66	74	96	101	110	126	137		
		32	15	17	22	26	32	35	41	46	53	61	68	77	90	99	105	114	130	142
		36	16	17	21	24	27	33	37	42	47	56	62	70	79	93	102	106	117	120
	6N@ 5.33	24	17	24	31	39	47	55	61	69	76	94	103	133	134					
		28	16	21	27	35	40	48	55	60	67	79	96	105	117	137				
		32	16	20	25	30	36	42	50	54	58	71	82	99	103	118	139	142		
		36	16	19	24	28	34	38	44	49	55	66	73	84	101	102	111	123	144	146
	8N@ 4.00	24	22	32	40	54	61	72	86	93	103	133								
		28	19	27	35	45	55	63	70	80	95	105	134	137						
		32	18	25	32	39	50	58	65	71	81	99	109	120	141					
		36	18	24	31	38	43	53	59	67	71	86	103	113	127	147				

LRFD

GIRDER SPAN (ft.)	JOIST SPACES (ft.)	GIRDER DEPTH (in.)	JOIST GIRDER WEIGHT – POUNDS PER LINEAR FOOT — FACTORED LOAD ON EACH PANEL POINT – KIPS																
			6.0	9.0	12.0	15.0	18.0	21.0	24.0	27.0	30.0	36.0	42.0	48.0	54.0	60.0	66.0	72.0	78.0	84.0
35	4N@ 8.75	28	16	19	23	27	31	36	41	46	52	60	74	79	94	100	111	117	137	138
		32	15	18	21	24	28	33	37	39	45	53	60	73	80	92	100	106	112	127
		36	15	16	20	23	27	30	33	37	41	561	55	62	74	83	94	97	107	113
		40	15	16	17	21	26	27	30	37	38	46	52	61	64	75	90	95	96	108
	5N@ 7.00	28	15	20	26	32	37	43	52	57	59	73	86	100	109	126	136			
		32	15	18	24	29	34	37	45	50	53	66	75	88	100	102	112	128	138	
		36	16	17	23	27	29	35	40	46	48	62	68	77	90	100	104	115	131	133
		40	16	17	22	25	27	33	37	43	47	56	63	70	80	95	102	107	115	125
	6N@ 5.83	28	17	24	30	37	44	52	58	65	73	93	103	115	134					
		32	16	21	27	33	38	46	53	57	65	79	96	100	117	139	140			
		36	16	20	25	31	36	41	48	54	58	70	81	99	102	113	121	142	144	
		40	16	20	24	28	34	38	44	49	55	64	77	84	101	104	115	123	145	146
	7N@ 5.00	28	19	27	34	43	52	59	66	74	86	101	115	135						
		32	17	24	30	39	47	53	61	67	75	97	103	118	137					
		36	17	23	28	35	42	48	55	62	69	82	99	105	120	141	144			
		40	17	22	27	32	39	44	50	55	63	73	86	102	107	118	133	147		
	8N@ 4.38	28	21	30	39	48	59	69	78	94	98	115	136							
		32	20	27	36	42	53	61	69	79	88	101	118	138						
		36	19	26	32	39	48	55	62	71	77	99	109	121	141					
		40	18	24	30	37	44	54	60	65	73	86	102	113	127	147	149			
38	4N@ 9.50	32	16	19	21	26	31	34	39	43	48	58	67	74	87	100	101	111	127	138
		36	15	17	21	24	28	33	35	39	44	53	60	74	75	93	97	106	112	123
		40	15	16	20	23	27	30	34	37	41	51	55	62	74	83	94	98	107	109
		44	16	16	20	22	26	28	30	35	38	46	52	58	65	75	90	95	95	108
	5N@ 7.60	32	15	20	25	31	36	42	46	52	59	70	86	96	101	111	126	137		
		36	16	20	24	28	33	38	45	47	53	64	74	89	98	103	112	129	138	
		40	16	20	23	26	31	35	40	46	48	59	70	78	91	101	105	113	117	134
		44	17	20	22	25	30	33	39	41	48	56	63	75	80	93	102	107	111	118
	6N@ 6.33	32	17	24	30	35	41	49	55	62	70	86	98	105	125	136				
		36	16	21	27	33	39	47	50	57	61	75	89	100	107	118	141	142		
		40	16	21	25	31	36	40	48	55	59	71	82	99	102	109	121	143	142	
		44	17	20	24	29	33	38	44	49	55	64	77	84	102	104	115	123	145	147
	8N@ 4.75	32	20	29	38	47	56	64	74	86	95	105	135							
		36	19	28	35	42	50	57	65	76	81	101	113	138	140					
		40	19	26	32	40	48	55	62	67	78	100	103	121	142	144				
		44	20	24	30	39	47	51	57	64	71	86	102	113	127	147	149			
40	4N@ 10.00	32	17	20	23	29	37	40	47	50	56	64	73	86	103	114	126	128	149	151
		36	17	19	22	31	37	40	44	51	57	65	74	87	103	104	125	127	128	
		40	17	18	22	25	29	33	37	40	47	52	62	73	77	87	96	104	117	127
		44	16	17	20	24	29	31	36	38	41	49	59	66	74	78	84	96	106	106
		48	17	17	20	24	25	30	32	37	39	48	53	59	67	78	78	85	99	106
	5N@ 8.00	32	15	21	26	32	38	43	52	55	62	73	86	101	109	124	134			
		36	16	20	24	30	34	39	45	53	55	66	74	88	102	102	112	128	138	
		40	16	20	24	27	32	37	41	46	51	62	68	77	90	100	105	115	130	142
		44	17	20	23	29	32	37	41	49	50	58	70	82	84	99	116	118	130	141
		48	17	20	23	26	31	34	40	41	50	57	68	75	85	95	100	119	120	132
	6N@ 6.67	32	16	24	30	38	44	52	58	65	72	93	100	115	133					
		36	17	22	27	34	39	47	53	60	67	79	97	102	117	137	141			
		40	16	21	26	30	36	43	48	54	62	71	82	99	103	114	130	142		
		44	17	21	24	28	36	40	47	51	55	66	78	91	102	107	116	134	142	146
		48	17	21	24	31	36	42	46	53	57	69	79	86	100	109	132	133	135	164
	7N@ 5.71	32	18	26	33	43	52	58	66	74	86	101	115	135						
		36	17	24	31	39	47	53	61	67	75	97	103	117	136					
		40	17	24	29	35	43	49	55	62	69	82	99	105	119	140				
		44	20	22	28	33	39	48	55	59	64	78	92	102	111	122	143			
		48	20	23	28	36	41	48	54	61	66	80	86	108	122	134	136	164	167	
	8N@ 5.00	32	21	29	38	48	58	67	78	94	96	115	135							
		36	19	27	36	46	53	60	68	80	88	102	118	137						
		40	19	25	34	39	49	58	65	72	82	99	109	120	141					
		44	21	27	33	39	47	56	63	70	75	93	103	120	136	147				
		48	20	25	32	42	47	55	62	69	80	90	104	122	136	155	170			
	10N@ 4.00	32	29	39	51	64	79	92	112	123	125	149								
		36	25	36	47	60	69	81	94	103	125	150								
		40	24	36	45	56	66	75	82	96	115	129	152							
		44	23	32	41	51	60	71	82	84	99	119	143	161						
		48	23	32	41	52	58	68	76	85	94	121	134	152						

LRFD

GIRDER SPAN (ft.)	JOIST SPACES (ft.)	GIRDER DEPTH (in.)	JOIST GIRDER WEIGHT – POUNDS PER LINEAR FOOT / FACTORED LOAD ON EACH PANEL POINT – KIPS																	
			6.0	9.0	12.0	15.0	18.0	21.0	24.0	27.0	30.0	36.0	42.0	48.0	54.0	60.0	66.0	72.0	78.0	84.0
42	4N@ 10.50	32	16	21	25	29	34	38	43	49	53	67	74	86	99	101	112	125	134	138
		36	16	19	22	26	32	35	39	44	47	58	67	73	87	95	101	112	118	129
		40	16	19	21	24	28	34	36	41	45	53	61	73	76	93	97	97	113	122
		44	16	19	20	23	27	31	34	38	42	51	55	62	74	84	94	97	108	109
		48	16	19	21	24	26	29	32	36	39	47	54	62	65	75	90	95	97	108
	5N@ 8.40	32	16	22	28	35	41	45	52	57	66	74	88	100	110	125				
		36	15	21	25	31	36	42	46	52	59	70	85	96	102	111	126	137		
		40	16	21	24	28	33	39	44	51	54	64	74	89	98	103	113	129	130	
		44	16	20	24	27	31	37	40	46	52	59	69	78	91	101	105	113	126	134
		48	17	20	23	27	30	35	39	42	48	57	63	75	81	95	102	107	115	118
	6N@ 7.00	32	18	25	32	39	45	55	61	69	77	93	103	124	135					
		36	17	23	30	35	41	49	56	60	67	79	96	105	117	137				
		40	17	21	26	33	39	46	54	57	61	75	89	100	108	119	141	142		
		44	16	21	24	31	35	41	48	54	59	71	81	100	102	109	121	143	142	
		48	20	20	25	29	33	39	44	49	56	64	77	85	102	104	115	124	145	147
	7N@ 6.00	32	20	28	36	45	52	65	72	85	93	102	125							
		36	19	26	34	40	49	56	67	74	79	90	110	127	138					
		40	18	24	31	38	46	54	61	68	75	90	101	113	129	142				
		44	20	23	29	35	41	49	55	63	70	78	100	106	116	132	145			
		48	18	23	28	34	39	44	50	56	64	73	92	102	108	118	136	149		
	8N@ 5.25	32	22	32	40	51	62	72	78	94	100	124	135							
		36	20	27	38	56	64	74	79	96	105	126	138							
		40	20	26	35	42	51	57	65	76	81	101	113	138	141					
		44	20	25	32	39	49	55	63	70	78	99	107	121	142	147				
		48	21	26	32	41	48	56	63	67	74	93	103	112	128	148				
	10N@ 4.20	32	27	38	52	62	77	94	101	114	134									
		36	25	36	46	60	70	86	97	102	112	140								
		40	24	34	45	54	64	75	89	99	104	129								
		44	23	31	41	52	61	70	79	91	100	114	143							
		48	23	30	39	49	56	66	72	80	93	107	125	146						
45	4N@ 11.25	36	18	21	25	28	33	38	42	46	52	62	72	79	95	100	112	117	128	138
		40	19	21	22	27	31	35	39	44	47	55	64	75	87	95	101	112	113	128
		44	19	21	22	24	29	33	37	39	45	53	61	74	76	89	95	102	108	114
		48	18	21	22	24	28	31	34	38	40	51	55	63	75	83	94	95	107	109
		52	18	22	23	24	27	29	33	37	39	47	52	60	66	76	91	95	96	109
	5N@ 9.00	36	16	22	27	33	38	44	52	55	63	74	86	101	109	125	136			
		40	16	21	25	30	36	42	45	53	56	68	75	88	102	111	122	128		
		44	16	21	24	29	34	38	44	46	54	65	74	85	90	103	110	123	130	142
		48	20	21	24	27	32	36	41	45	52	59	67	75	91	95	106	112	118	134
		52	20	21	24	27	30	35	39	42	48	57	64	75	81	94	98	107	117	119
	6N@ 7.50	36	19	24	31	38	45	52	58	66	74	93	100	115	134					
		40	19	23	28	34	40	47	53	60	67	79	97	103	117	137	140			
		44	19	21	27	32	38	46	50	54	62	76	90	100	107	118	139	142		
		48	20	21	26	30	36	42	48	55	59	69	78	92	102	110	122	143	143	
		52	20	21	25	29	34	39	44	50	56	64	77	85	102	102	116	124	136	148
	7N@ 6.43	36	20	27	35	44	52	58	66	74	86	101	115	135						
		40	20	26	33	40	47	54	61	67	75	97	105	127	138					
		44	20	24	30	39	46	54	61	62	69	90	100	113	129	143				
		48	20	23	29	36	41	49	55	63	70	79	92	107	117	133	145			
		52	18	23	28	34	39	45	50	56	65	73	93	102	109	118	136	149		
	8N@ 5.62	36	21	30	38	48	58	67	78	94	98	114	135							
		40	20	28	36	46	53	61	68	80	89	105	118	137						
		44	20	27	34	41	51	58	66	73	81	99	109	130	141					
		48	21	26	32	39	47	55	63	68	74	92	104	116	142	146				
		52	22	28	33	42	48	54	59	67	71	94	102	112	127	148				
	9N@ 5.00	36	24	34	45	55	66	74	88	98	104	135								
		40	22	31	39	49	61	69	80	89	100	113	138							
		44	23	31	39	48	58	66	76	89	99	108	132							
		48	23	29	37	47	55	63	70	79	91	106	117	133						
		52	23	28	36	46	55	60	70	73	84	102	112	135	148					
	10N@ 4.50	36	26	38	49	60	73	86	98	105	116	137								
		40	25	35	47	60	66	76	90	102	112	140								
		44	24	33	46	54	64	72	89	99	104	130	142							
		48	24	31	40	49	62	71	78	91	100	114	134							
		52	23	31	39	50	56	67	72	80	93	107	123	147						

LRFD

GIRDER SPAN (ft.)	JOIST SPACES (ft.)	GIRDER DEPTH (in.)	JOIST GIRDER WEIGHT – POUNDS PER LINEAR FOOT / FACTORED LOAD ON EACH PANEL POINT – KIPS																	
			6.0	9.0	12.0	15.0	18.0	21.0	24.0	27.0	30.0	36.0	42.0	48.0	54.0	60.0	66.0	72.0	78.0	84.0
48	5N@ 9.60	36	19	26	31	37	45	52	59	66	71	87	111	113	135	136				
		40	19	23	29	35	41	46	52	59	68	77	92	112	114	136	138			
		44	19	22	27	32	37	44	48	54	61	69	80	93	113	116	126	139	150	
		48	19	21	25	30	36	40	48	48	55	69	78	90	96	115	116	128	140	142
		52	20	21	25	29	33	39	42	50	54	62	71	82	92	99	117	130	141	
		56	20	21	24	29	33	38	40	46	50	59	71	79	85	100	100	119	120	133
	6N@ 8.00	36	20	28	35	42	51	62	70	78	83	100	122	134	147					
		40	19	25	33	39	47	56	64	71	79	93	112	124	137	148				
		44	19	24	31	36	45	50	57	65	73	81	102	115	127	138	151			
		48	19	23	30	35	40	48	52	59	67	78	95	105	116	129	141	160		
		52	20	23	27	32	38	46	51	59	60	75	83	97	107	130	131	144	162	
		56	20	22	27	31	37	42	48	54	61	69	80	91	107	120	132	134	153	165
	8N@ 6.00	36	30	36	45	56	64	78	91	100	122	134								
		40	28	33	42	51	59	70	80	92	101	124	148							
		44	27	32	39	49	55	65	74	82	95	114	127	150						
		48	26	30	37	47	53	60	68	76	84	105	129	131	154					
		52	26	30	36	44	51	59	65	71	80	99	119	132	146	164				
		56	25	28	36	43	49	57	63	69	78	90	109	123	136	155				
	9N@ 5.33	36	35	44	55	70	79	91	99	121	122	146								
		40	34	42	52	63	74	88	93	101	113	136								
		44	33	39	50	59	69	83	91	94	103	126	150							
		48	33	37	46	56	66	76	85	94	97	118	130							
		52	31	36	46	54	63	72	80	95	101	108	132	152						
		56	31	35	44	53	62	69	80	89	98	103	123	137	165					
	12N@ 4.00	36	35	52	71	84	100	123	135	148										
		40	34	48	65	76	93	113	125	137	149									
		44	31	44	57	73	82	102	115	126	139									
		48	30	41	53	67	76	88	104	117	130	153								
		52	30	39	52	61	76	84	97	107	131	144								
		56	27	38	49	61	70	81	91	108	122	135	165							
50	5N@ 10.00	40	18	23	30	38	44	47	56	60	68	79	93	113	124	136	138			
		44	17	22	29	34	40	46	51	56	61	76	89	94	113	126	137	139		
		48	19	22	28	31	38	42	48	55	61	69	78	94	96	115	127	139	141	
		52	20	22	25	31	35	40	45	49	55	62	74	82	96	116	117	129	141	142
		56	20	22	25	30	32	40	43	50	51	63	71	83	92	99	117	119	131	142
		60	20	20	24	30	33	36	42	46	51	58	65	76	86	96	101	120	121	133
	6N@ 8.33	40	20	28	34	42	48	56	64	71	80	100	112	124	147					
		44	19	24	31	38	47	50	57	65	73	85	102	124	127	149				
		48	19	23	30	37	40	49	57	65	67	82	95	115	127	129	151			
		52	20	23	30	36	40	46	52	59	67	75	84	105	117	129	131	153	162	
		56	20	23	26	33	39	42	51	54	60	72	84	98	107	120	132	144	163	164
		60	21	23	27	33	38	43	49	53	61	70	80	87	102	110	123	134	154	165
	8N@ 6.25	40	22	31	39	51	59	67	78	86	96	110	135							
		44	21	29	37	47	53	61	70	80	96	103	118	139						
		48	21	27	35	42	51	58	69	76	81	99	114	130	142					
		52	21	25	33	40	49	55	63	70	78	99	107	121	141					
		56	24	29	36	42	47	56	64	68	78	94	108	118	137	148				
		60	24	27	35	40	47	55	61	69	74	83	103	110	123	139	149			
	9N@ 5.56	40	24	34	44	55	66	74	86	96	104	134								
		44	23	32	40	53	61	69	80	88	98	113	138							
		48	24	32	42	52	58	69	77	90	99	111	133							
		52	24	31	40	47	58	66	74	79	92	106	126	143						
		56	24	30	38	46	55	60	68	77	89	102	116	135						
		60	24	32	38	49	53	61	70	75	83	97	111	125	141					
	10N@ 5.00	40	26	38	49	60	74	87	96	104	116	136								
		44	25	36	47	60	68	84	96	102	112	140								
		48	24	34	46	54	65	76	89	99	103	130								
		52	24	34	45	52	62	70	79	91	100	114	134							
		56	23	32	41	48	60	70	76	87	93	107	134	146						
		60	24	31	40	49	57	66	73	81	94	109	119	138						
	12N@ 4.17	40	34	49	65	80	100	112	125	147										
		44	31	44	57	73	86	102	126	127	149									
		48	30	41	58	67	82	96	115	127	130	154								
		52	30	39	53	68	76	84	105	118	130	154								
		56	27	40	52	61	70	85	99	108	122	135	164							
		60	27	39	49	61	70	82	88	104	112	135	166							

LRFD

JOIST GIRDER WEIGHT – POUNDS PER LINEAR FOOT

FACTORED LOAD ON EACH PANEL POINT – KIPS

GIRDER SPAN (ft.)	JOIST SPACES (ft.)	GIRDER DEPTH (in.)	6.0	7.5	9.0	10.5	12.0	13.5	15.0	16.5	18.0	19.5	21.0	24.0	27.0	30.0	33.0	36.0	39.0	42.0
55	5N@ 11.00	44	21	21	24	25	29	32	35	38	41	43	47	53	59	63	71	82	83	86
		48	21	21	23	24	28	30	32	35	38	41	43	49	56	60	64	71	73	83
		52	20	22	23	25	27	29	32	33	36	39	42	44	52	57	65	66	74	74
		56	20	21	24	24	26	28	31	33	36	37	39	44	51	53	58	66	66	74
		60	23	24	24	24	27	27	31	33	35	38	38	45	47	52	60	61	67	68
		66	24	24	24	25	26	28	28	33	34	37	37	42	47	48	55	56	62	69
	6N@ 9.17	44	19	22	26	29	33	36	38	43	45	51	52	59	66	75	86	86	98	101
		48	20	22	24	28	31	33	36	40	44	46	50	56	64	68	75	87	89	98
		52	20	22	24	26	29	33	35	37	41	59	59	66	74	86	93	99	109	110
		56	18	21	24	25	28	31	35	36	39	42	47	52	55	63	70	71	78	91
		60	20	21	24	25	29	30	33	35	38	39	43	48	55	60	64	71	75	80
		66	19	20	22	24	28	30	31	33	36	39	40	47	50	56	62	65	73	73
	7N@ 7.86	44	21	24	28	33	36	39	44	50	53	59	59	70	75	87	97	102	111	120
		48	21	24	27	31	34	38	43	45	51	54	56	65	72	76	89	98	103	110
		52	21	23	26	29	33	36	39	44	46	52	55	62	69	74	86	91	100	105
		56	20	22	25	28	31	35	38	40	46	48	53	55	64	70	79	87	92	101
		60	21	22	24	27	30	33	36	39	41	47	49	56	64	68	72	81	93	94
		66	22	22	24	26	30	32	36	37	40	43	48	52	58	65	70	74	83	84
	9N@ 6.11	44	24	29	34	39	46	52	55	60	67	74	74	87	98	105	116	135	137	
		48	24	28	32	38	40	47	53	57	61	68	69	81	97	103	107	118	129	139
		52	25	30	33	39	43	47	52	57	65	73	77	90	104	105	114	125	133	
		56	24	29	32	38	43	46	51	53	59	66	67	75	87	92	105	107	117	128
		60	24	27	32	36	40	45	47	52	56	60	67	71	80	93	95	108	109	118
		66	24	27	31	35	39	42	46	49	54	58	61	71	78	83	91	97	111	113
	11N@ 5.00	44	30	36	43	49	55	63	67	74	87	88	97	106	126	137				
		48	28	33	39	45	54	61	65	69	76	87	89	103	112	128	139			
		52	27	34	37	44	52	55	62	66	73	77	88	99	105	115	131	142		
		56	27	33	39	42	48	54	60	64	68	77	80	93	102	107	118	134	146	
		60	26	31	37	40	47	49	58	64	67	72	77	82	95	108	110	121	137	148
		66	26	31	36	39	45	50	54	60	65	68	74	82	97	98	113	117	126	141
60	5N@ 12.00	48	21	23	27	29	33	35	39	43	44	49	51	57	63	69	76	87	89	94
		52	21	22	27	28	31	33	36	40	44	45	47	52	60	65	69	77	85	90
		56	22	23	24	28	30	31	34	36	41	44	45	52	59	63	69	74	78	87
		60	22	23	24	28	29	32	34	35	40	42	45	49	53	60	66	70	75	80
		66	24	24	24	24	30	30	33	35	36	38	42	47	51	56	61	67	72	73
		72	25	25	25	25	29	30	31	34	37	39	45	48	56	56	64	63	69	70
	6N@ 10.00	48	20	24	29	32	36	38	41	47	49	56	60	67	72	80	93	93	112	113
		52	20	23	28	30	33	37	39	46	50	57	62	69	78	80	94	94	113	
		56	19	24	25	30	33	38	39	42	48	49	51	58	66	69	79	83	95	96
		60	19	23	24	29	32	34	39	40	43	49	50	57	63	70	75	83	83	96
		66	19	23	24	27	32	32	34	40	42	44	50	52	61	65	69	77	84	85
		72	22	22	24	27	28	33	34	36	41	43	44	52	54	63	68	71	75	87
	8N@ 7.50	48	24	29	34	39	43	49	56	57	64	72	72	80	93	112	123	125	136	148
		52	23	29	31	37	40	48	50	57	58	66	72	81	94	103	114	125	127	139
		56	23	26	31	36	38	44	49	51	58	60	66	75	83	96	104	116	127	129
		60	23	26	32	33	39	42	47	50	53	59	61	69	77	85	98	106	118	129
		66	28	30	33	34	41	43	46	48	53	57	62	70	78	82	90	100	108	120
		72	29	30	31	34	36	41	46	47	52	58	59	66	73	80	90	92	104	110
	10N@ 6.00	48	26	32	37	44	49	55	60	67	74	79	87	97	105	118	137	138		
		52	28	34	38	44	50	56	64	65	71	75	88	97	103	113	130	138		
		56	27	33	37	43	46	51	58	66	65	72	76	90	104	105	123	131	143	
		60	25	31	37	39	45	51	57	60	66	70	73	86	93	104	111	126	134	
		66	27	32	37	42	49	51	56	62	65	72	74	85	95	102	120	122	134	145
		72	26	32	33	38	42	47	50	55	59	66	69	74	83	96	98	111	111	121
	12N@ 5.00	48	33	39	46	53	59	68	75	86	87	97	102	111	135					
		52	31	37	45	51	57	65	69	76	88	89	98	104	118	139				
		56	29	36	41	48	55	62	66	72	77	89	91	102	106	116	133	140		
		60	30	35	39	47	54	56	64	73	74	79	80	87	100	110	122	134	147	164
		66	32	35	41	48	53	61	62	70	77	77	80	86	100	110	114	127	142	
		72	29	33	38	42	50	52	60	61	69	72	77	86	100	110	114	127	142	151
	15N@ 4.00	48	40	49	64	72	80	93	102	113	124	126	136							
		52	39	48	57	66	74	81	94	103	114	126	127	150						
		56	38	46	53	67	71	80	83	96	104	116	127	140	153					
		60	38	42	51	60	68	76	83	89	98	106	118	132	144					
		66	35	41	49	55	62	70	81	87	87	103	110	123	136	153	167			
		72	35	44	46	55	64	66	77	85	90	93	106	125	139	142	160	171		

LRFD

GIRDER SPAN (ft.)	JOIST SPACES (ft.)	GIRDER DEPTH (in.)	JOIST GIRDER WEIGHT – POUNDS PER LINEAR FOOT — FACTORED LOAD ON EACH PANEL POINT – KIPS																	
			6.0	7.5	9.0	10.5	12.0	13.5	15.0	16.5	18.0	19.5	21.0	24.0	27.0	30.0	33.0	36.0	39.0	42.0
65	6N@10.83	52	22	28	30	33	39	41	45	49	54	58	61	69	78	83	95	97	115	116
		56	21	25	29	33	35	40	42	48	49	55	58	63	70	80	84	97	97	117
		60	23	24	29	32	34	39	41	44	50	50	56	64	71	76	82	92	98	99
		66	22	24	26	31	33	35	40	42	45	51	51	58	65	73	78	83	87	100
		72	24	25	27	31	32	35	37	42	43	47	49	54	60	68	76	80	87	89
	8N@8.12	52	25	31	38	40	44	51	58	62	66	74	74	83	97	115	127	129	141	153
		56	24	30	34	39	43	50	52	59	63	68	74	83	97	105	118	129	131	143
		60	23	28	33	39	41	47	51	53	60	68	69	77	85	99	108	119	130	133
		66	24	28	33	35	42	44	49	52	56	63	63	75	80	89	101	110	122	124
		72	38	39	39	39	42	45	47	52	56	58	65	73	78	89	92	104	113	125
	9N@7.22	52	30	32	38	44	49	58	62	67	74	79	83	97	116	128	129	142	153	
		56	26	32	39	42	48	53	59	68	68	76	81	98	106	118	130	142	144	155
		60	25	32	38	40	47	51	58	60	69	70	78	86	100	109	120	132	145	146
		66	28	32	37	41	44	50	53	60	64	71	72	81	89	103	112	124	136	138
		72	29	30	35	38	44	46	52	57	62	66	71	79	91	108	115	127	140	
	10N@6.50	52	31	36	41	49	58	62	67	75	82	89	97	116	128	131	154	155		
		56	31	36	40	46	52	60	68	69	77	85	91	107	119	132	144			
		60	29	34	40	44	51	57	61	70	74	78	87	100	109	122	134	146		
		66	27	34	39	43	50	54	60	65	72	74	82	90	103	113	125	138	140	163
		72	27	33	37	44	47	52	56	62	67	75	76	87	93	110	127	129	141	143
	11N@5.91	52	33	39	45	52	59	67	75	83	89	98	106	118	131	153				
		56	32	39	44	51	60	64	69	77	85	91	99	119	132	144	156			
		60	33	38	44	49	55	63	70	74	79	86	92	109	122	134	147			
		66	30	37	42	46	54	57	64	72	73	81	90	104	113	125	139	147	164	
		72	30	36	41	47	51	57	62	67	77	77	88	93	110	118	131	144	156	173
	13N@5.00	52	37	45	55	64	72	79	89	98	106	117	130	142						
		56	37	43	53	61	69	77	86	91	99	108	120	133	146					
		60	35	41	50	58	64	71	77	85	93	100	108	131	134	158				
		66	34	41	49	53	62	70	75	80	87	93	102	122	134	137	161			
		72	34	41	46	58	64	72	78	85	90	90	113	127	138	141	170			
70	7N@10.00	56	24	25	30	35	39	43	46	51	56	57	64	71	83	88	102	102	110	121
		60	23	26	30	33	37	43	44	50	52	57	61	66	73	85	90	102	105	
		66	24	27	30	32	35	39	44	46	51	53	58	67	73	75	87	93	104	106
		72	24	25	29	32	34	38	42	46	47	53	54	60	69	76	78	89	94	102
		78	25	26	28	31	34	37	40	43	47	49	50	58	63	71	78	83	90	96
		84	24	27	29	31	35	37	39	42	44	49	51	57	65	69	72	80	85	94
	9N@7.78	56	26	31	37	40	45	53	56	61	67	72	75	88	102	110	122	128		
		60	25	30	35	39	45	47	54	61	65	70	73	89	99	105	114	129	131	
		66	31	34	38	43	48	51	56	63	67	70	74	86	92	106	112	122	127	
		72	32	33	37	43	45	51	56	58	64	67	69	77	89	100	108	114	124	131
		78	32	34	36	39	45	48	53	59	60	66	66	76	87	93	102	110	116	118
		84	33	34	35	38	45	47	50	55	59	63	67	72	81	94	95	103	113	118
	10N@7.00	56	27	34	38	45	53	57	60	68	75	80	88	100	106	118	137			
		60	30	36	41	48	55	60	65	69	71	84	88	102	109	122	130			
		66	29	35	42	44	51	55	62	66	70	73	85	91	105	109	123	132		
		72	30	34	38	43	47	52	59	63	66	69	78	88	94	106	112	127	133	
		78	30	33	37	40	46	51	55	61	65	71	71	79	94	96	108	115	130	137
		84	31	33	36	40	47	49	55	57	63	70	72	80	92	98	109	112	121	133
	11N@6.36	56	32	41	45	51	60	64	71	83	87	89	102	108	127	138				
		60	30	39	44	50	57	65	66	73	85	89	90	104	114	129				
		66	31	38	43	46	53	59	67	67	76	86	88	105	106	117	132			
		72	32	37	42	48	55	57	62	70	70	78	82	94	108	109	119	136	148	
		78	29	35	40	47	50	55	61	65	73	72	80	92	110	118	124	140	141	
		84	30	36	39	45	49	52	59	66	68	73	78	84	97	102	116	124	129	144
	12N@5.83	56	34	41	50	56	63	68	76	87	88	102	103	113	129					
		60	33	39	46	55	58	65	74	76	89	90	103	112	128	139				
		66	32	37	45	48	55	63	67	76	78	90	92	105	115	130	143			
		72	32	37	42	48	55	61	65	69	77	80	89	102	107	119	135	148		
		78	30	36	42	48	51	56	64	70	72	80	84	97	106	113	123	141	151	
		84	30	36	40	45	51	53	61	68	73	77	83	89	102	115	118	128	144	151
	14N@5.00	56	36	44	53	63	71	75	87	96	102	111	120	137						
		60	37	43	54	61	69	75	88	89	99	103	112	128						
		66	35	42	48	55	64	70	77	90	92	102	106	115	132					
		72	34	40	49	55	61	69	73	81	91	95	103	110	120	138				
		78	33	39	44	52	58	67	72	76	84	92	97	111	120	138	141			
		84	33	40	44	51	58	62	69	78	79	86	97	106	116	127	143	155		

LRFD

GIRDER SPAN (ft.)	JOIST SPACES (ft.)	GIRDER DEPTH (in.)	JOIST GIRDER WEIGHT – POUNDS PER LINEAR FOOT																	
			FACTORED LOAD ON EACH PANEL POINT – KIPS																	
			6.0	7.5	9.0	10.5	12.0	13.5	15.0	16.5	18.0	19.5	21.0	24.0	27.0	30.0	33.0	36.0	39.0	42.0
75	8N@ 9.38	56	29	33	40	43	49	55	61	65	73	79	82	95	115	116	128	140	152	
		60	26	32	38	42	48	51	58	63	70	75	80	92	97	116	118	130	142	153
		66	27	32	35	41	44	51	53	60	64	69	72	82	98	99	118	120	132	144
		72	26	32	34	41	43	46	52	58	61	66	71	79	87	100	101	121	122	134
		78	27	29	34	37	43	45	54	54	61	64	69	77	81	89	103	105	123	125
	10N@ 7.50	60	32	39	42	50	59	67	69	76	83	89	98	117	129	131	154			
		66	32	37	42	49	55	62	69	70	78	86	87	100	119	132	134			
		72	30	36	42	45	54	57	63	72	73	81	86	101	111	123	136	138		
		78	31	35	39	46	48	56	63	66	74	75	82	91	105	114	127	139	152	
		84	31	36	39	45	49	55	59	65	69	77	78	94	95	110	128	131	143	156
	12N@ 6.25	60	38	43	51	59	68	76	84	90	98	106	118	131	144					
		66	35	42	50	55	62	70	79	87	90	100	110	122	135	148				
		72	36	41	46	54	63	65	73	81	90	91	104	124	126	141	154			
		78	35	42	47	54	61	68	76	78	86	90	98	105	126	139	152	163		
		84	34	39	46	52	56	64	70	78	79	90	92	106	126	139	141	164	171	
	14N@ 5.36	66	41	48	56	63	72	80	89	102	111	122	125	137						
		72	41	46	52	61	70	75	84	95	101	110	121	134	148					
		78	37	44	53	61	68	76	80	89	98	103	107	125	139	151				
		84	38	44	52	57	64	71	79	86	92	100	108	127	130	153	171			
		90	37	42	50	58	66	73	77	87	94	94	110	119	142	144	173	176		
	15N@ 5.00	66	41	52	60	69	77	85	98	106	118	120	132	146						
		72	42	52	59	67	74	84	87	99	110	121	123	146	160					
		78	41	47	54	65	73	77	88	91	104	112	124	139	152	169				
		84	39	46	55	63	67	76	86	92	93	109	116	131	143	171	174			
		90	38	46	52	60	69	74	81	90	95	103	118	133	145	146	177			
80	8N@ 10.00	60	28	31	37	42	45	51	56	63	64	72	75	88	97	103	112	127	137	
		66	30	31	35	38	45	47	52	57	62	65	70	77	90	103	105	113	129	131
		72	29	32	33	38	41	46	48	53	59	63	68	76	87	92	106	108	116	126
		78	30	31	33	37	41	42	47	53	56	60	64	73	81	88	94	109	111	118
		84	30	32	35	37	39	43	48	52	56	59	63	71	79	83	96	98	112	114
		90	53	54	56	56	57	57	58	60	63	67	70	79	79	90	95	103	105	118
	10N@ 8.00	60	31	35	41	47	53	60	68	75	76	88	97	103	112	129	139			
		66	31	35	39	46	52	55	62	70	75	78	90	100	107	115	132	142		
		72	33	37	43	50	55	62	63	70	74	83	87	97	106	120	127			
		78	32	36	42	46	51	56	63	68	71	76	86	90	100	112	122	130		
		84	33	37	42	45	51	57	61	65	70	77	78	91	100	109	115	125	131	
		90	34	36	40	44	49	53	60	65	68	72	77	87	92	102	111	118	132	136
	12N@ 6.67	66	36	44	50	57	65	70	73	86	90	103	103	115	130					
		72	34	42	47	54	59	67	72	77	86	92	101	107	125	133				
		78	33	39	46	53	60	65	69	79	80	88	94	108	114	129	136			
		84	34	38	47	49	56	63	70	72	79	83	92	99	111	121	138	140		
		90	36	39	44	50	56	59	66	72	74	82	86	101	113	116	125	143	149	
		96	34	37	43	50	54	60	68	71	75	79	85	98	104	117	120	130	147	156
	14N@ 5.71	66	39	47	57	64	73	77	89	98	103	109	113	129						
		72	38	46	54	59	67	76	79	91	101	106	106	125	143					
		78	36	43	50	58	66	70	78	90	95	96	109	118	136	149				
		84	36	42	50	56	64	71	74	80	92	98	99	112	124	143				
		90	36	41	48	53	61	68	74	82	86	95	100	115	121	136	146			
		96	37	40	47	53	61	67	74	79	84	88	100	108	118	127	145	152		
	16N@ 5.00	66	42	53	62	70	78	90	101	105	113	129	130							
		72	41	50	57	69	76	81	93	102	109	116	118	145						
		78	41	49	58	66	73	83	91	96	104	112	120	137	149					
		84	39	45	54	61	69	76	84	97	100	109	115	126	143					
		90	39	46	54	62	70	74	80	86	101	102	114	119	144	155				
		96	40	46	55	58	68	73	81	88	94	106	110	121	133	155	164			

LRFD

GIRDER SPAN (ft.)	JOIST SPACES (ft.)	GIRDER DEPTH (in.)	JOIST GIRDER WEIGHT – POUNDS PER LINEAR FOOT																
			FACTORED LOAD ON EACH PANEL POINT – KIPS																
			6.0	7.5	9.0	10.5	12.0	13.5	15.0	16.5	18.0	19.5	21.0	24.0	27.0	30.0	33.0	36.0	39.0	42.0
90	9N@ 10.00	72	40	42	46	49	55	60	64	72	81	82	92	98	117	119	141	143		
		84	41	44	48	48	50	54	60	67	75	76	84	88	102	121	124	135	148	149
		90	54	55	56	56	57	59	62	65	72	77	85	88	99	105	125	128	138	
		96	55	56	57	57	58	59	64	65	69	74	80	91	98	107	110	128	131	142
		102	55	57	57	58	59	60	62	65	69	74	75	87	95	105	112	130	133	134
	10N@ 9.00	72	42	46	48	52	61	64	72	78	85	93	99	118	130	142	155			
		84	42	45	49	51	58	62	69	73	81	94	97	115	117	137	148			
		90	42	46	50	51	56	60	66	71	79	81	89	100	107	126	129	141		
		96	43	46	48	53	56	59	66	70	74	82	87	95	108	113	129	133	153	
		102	43	45	48	53	57	60	65	69	76	77	84	97	105	115	124	131	137	155
	11N@ 8.18	72	43	47	51	59	65	73	78	86	99	100	119	120	143					
		84	43	49	50	55	62	67	74	78	87	91	100	113	126	138	150			
		90	45	48	51	53	59	66	72	77	85	90	93	107	128	129	142			
		96	47	48	53	56	60	63	71	75	81	87	95	105	113	132	134	148		
		102	48	49	57	58	61	64	70	73	82	86	94	101	116	124	138	150	163	
	12N@ 7.50	78	44	49	53	60	68	72	79	88	102	103	111	124	149					
		84	45	49	52	56	65	75	79	84	91	103	105	125	137	149				
		90	46	50	52	60	68	75	79	88	89	100	106	126	128	151	152			
		96	46	48	52	58	63	72	76	82	90	93	103	110	129	132	153	156		
		108	45	49	55	56	64	66	76	81	85	92	97	107	115	135	137	160	168	
	15N@ 6.00	78	47	54	66	75	82	94	99	120	121	133	145	148						
		84	49	54	62	68	76	86	97	103	122	124	125	149						
		90	50	52	60	69	78	82	90	99	106	125	127	140	153					
		96	48	53	58	66	72	80	93	95	108	112	129	131	154	173				
		108	51	57	59	64	72	78	87	99	101	109	115	136	139	168	172			
	18N@ 5.00	78	51	62	74	84	99	102	120	133	145	148	159							
		84	51	61	73	80	89	104	113	124	137	150	151							
		90	52	58	70	79	90	93	106	126	129	142	153	166						
		96	53	58	68	78	87	95	108	113	131	133	144	158						
		108	57	59	64	76	85	95	103	113	120	127	139	151	172					
100	10N@ 10.00	78	45	49	52	55	58	62	68	75	79	91	92	106	115	131	140			
		84	47	50	53	55	58	61	69	72	77	81	93	102	109	118	133	143		
		96	55	56	56	57	62	64	68	74	84	86	87	102	116	125	126			
		102	55	56	57	58	61	64	66	73	77	86	89	100	106	121	127	133		
		108	56	57	58	59	61	64	67	70	76	80	87	92	106	107	127	130		
	12N@ 8.33	78	48	53	56	62	70	74	86	92	97	105	112	124						
		84	48	52	55	63	68	72	84	88	98	99	107	126	133					
		96	47	51	55	58	66	67	75	81	91	93	102	111	116	131				
		102	48	52	55	58	62	69	73	79	90	94	95	113	118	133	141			
		108	48	51	55	59	62	70	72	76	85	92	97	106	117	123	139	149		
	15N@ 6.67	78	53	56	67	75	86	91	104	106	115	125	133							
		84	53	56	61	69	78	88	94	107	113	118	128							
		96	52	56	61	68	72	82	93	99	105	114	118	133						
		102	53	56	60	66	74	83	85	97	102	116	117	125	144					
		108	53	56	59	65	73	77	87	99	103	104	118	123	140	149				
	16N@ 6.25	84	53	58	69	72	80	92	106	107	117	127	133							
		96	53	57	63	71	75	85	98	100	115	115	124	140						
		102	53	57	62	66	74	84	97	102	111	117	118	136	154					
		108	54	58	62	67	76	82	87	100	104	117	118	129	148					
		120	56	61	64	70	76	83	86	93	104	109	116	128	140	161				
	17N@ 5.88	84	55	61	70	77	88	94	107	114	127	133	145							
		96	54	59	65	72	80	93	99	113	115	121	135	151						
		102	55	59	66	73	79	87	98	102	118	118	127	144						
		108	55	60	65	69	78	87	91	105	107	119	120	140	160					
		120	56	62	67	71	78	87	93	100	110	112	125	133	149	168				
	18N@ 5.56	84	55	61	70	81	94	102	109	118	134	144								
		96	55	60	65	72	84	97	100	114	120	124	140							
		102	56	61	66	73	84	89	102	112	118	125	137	154						
		108	57	60	68	73	82	91	104	106	119	121	130	148						
		120	59	64	69	75	84	88	98	108	113	122	129	142	163					
	20N@ 5.00	84	58	66	77	94	103	109	118	134	146									
		96	60	65	73	83	99	108	115	123	125	144	153							
		102	59	65	71	80	89	103	114	121	129	147	147							
		108	60	67	71	80	89	106	110	123	126	134	149	164						
		120	68	73	90	101	108	113	123	133	152	155	166	182	200					

LRFD

| GIRDER SPAN (ft.) | JOIST SPACES (ft.) | GIRDER DEPTH (in.) | JOIST GIRDER WEIGHT – POUNDS PER LINEAR FOOT | | | | | | | | | | | | | | | | | |
| | | | FACTORED LOAD ON EACH PANEL POINT – KIPS | | | | | | | | | | | | | | | | | |
			6.0	7.5	9.0	10.5	12.0	13.5	15.0	16.5	18.0	19.5	21.0	24.0	27.0	30.0	33.0	36.0	39.0	42.0
110	10N@ 11.00	84	54	58	61	65	69	73	82	83	94	99	100	120	143	144				
		96	62	62	63	65	69	72	81	82	91	97	98	107	125					
		108	63	63	64	67	69	72	75	82	86	91	95	105	113	131	133			
		114	63	64	67	68	72	73	76	79	86	88	88	108	115	133	136			
		120	64	64	66	69	72	74	76	81	83	88	90	100	111	128	137	140		
	12N@ 9.17	84	58	62	66	70	74	84	88	101	109	120	122	144						
		96	57	62	66	70	74	79	88	92	101	107	125	127	151					
		108	58	64	68	72	75	79	84	90	95	106	111	132	136	158				
		114	59	65	66	71	75	79	84	89	102	106	107	126	134	156	158			
		120	59	62	67	72	74	79	82	91	96	107	109	126	135	158	161			
	14N@ 7.86	84	60	66	71	76	84	97	102	122	123	134	147							
		96	60	65	69	74	83	95	100	105	124	125	136	150						
		108	60	64	69	72	78	87	99	103	108	120	128	142	155					
		114	61	65	69	74	79	84	93	103	105	111	124	133	157					
		120	60	66	69	74	80	82	90	96	106	109	126	135	158	160				
	16N@ 6.88	96	62	68	72	79	89	104	106	125	126	147	149							
		102	63	67	74	80	89	103	108	125	127	128	152	156						
		108	64	68	73	81	83	95	104	110	127	130	142	158						
		114	65	70	74	80	86	95	105	111	114	132	135	161	162					
		120	66	69	75	81	88	97	99	109	117	135	138	152	165					
	18N@ 6.11	96	64	71	77	87	99	106	125	127	148	151								
		102	66	70	80	89	101	109	127	128	139	152	153							
		108	66	71	77	83	94	106	111	129	131	144	157							
		114	67	73	79	85	97	107	113	132	134	137	159	163						
		120	68	74	79	88	91	101	110	118	136	139	152	166						
	20N@ 5.50	96	68	77	82	99	106	125	139	152	154									
		102	69	75	81	94	109	129	130	142	154	155								
		108	69	77	83	94	106	114	132	133	145	157	169							
		114	69	77	86	91	101	115	134	135	147	160	161							
		120	66	72	77	83	93	106	113	126	128	137	154	167						
120	10N@ 12.00	96	63	66	69	72	76	78	82	86	89	89	94	108	115	129				
		102	64	67	69	71	75	79	83	83	86	91	92	110	117	131	137			
		108	78	79	82	83	83	83	86	91	95	94	100	108	126					
		114	78	79	82	83	83	84	86	91	90	95	95	109	127	128				
		120	79	81	83	84	84	85	86	88	92	92	97	102	113	133				
	12N@ 10.00	96	68	69	71	77	82	86	90	99	100	113	125	130						
		102	68	69	72	78	80	85	88	96	101	102	116	130						
		108	69	70	72	75	81	86	90	91	99	103	105	128	134					
		114	70	70	71	75	82	86	87	92	95	100	130	121	135					
		120	70	71	72	76	80	84	88	92	93	102	107	123	133	138				
	15N@ 8.00	96	69	74	77	82	90	96	109	115	125	129	134							
		102	70	73	78	84	88	93	103	113	118	129	132							
		108	70	73	80	85	90	95	101	106	115	119	133							
		114	70	73	78	83	88	93	98	107	117	121	122	137						
		120	72	74	78	84	89	94	99	100	110	118	124	140						
	16N@ 7.50	96	70	76	80	85	90	100	109	114	128	134								
		102	70	74	78	86	92	97	110	112	120	131	137							
		108	70	74	80	85	90	95	100	114	120	124	133							
		114	70	73	81	86	91	96	101	107	117	122	135	145						
		120	70	75	79	85	90	94	99	103	118	119	126	147						
	18N@ 6.67	96	71	77	85	89	95	109	116	129	136									
		102	72	78	83	87	97	111	113	121	138	138								
		108	72	79	84	88	94	101	115	121	156	157								
		114	72	76	85	90	96	102	116	117	123	136	143							
		120	73	77	84	89	95	99	105	118	125	129	140							
	20N@ 6.00	96	76	82	89	94	110	116	130	136										
		102	75	83	87	92	105	114	123	140	150									
		108	75	81	88	94	101	115	121	135	142	152								
		114	77	82	87	93	103	113	119	128	138	146								
		120	77	84	90	96	102	107	121	124	133	148	150							
	24N@ 5.00	96	83	90	96	111	121	136												
		102	81	88	99	108	118	140	151											
		108	83	91	96	103	119	129	147	157										
		114	86	96	109	121	141	143	152	160										
		120	86	97	107	117	143	146	152	163	165									

APPENDIX B

B.1 PLASTIC ANALYSIS AND DESIGN OF CONTINUOUS BEAMS AND GIRDERS

Plastic design is an economical method for the design of statically indeterminate beams, girders, and frames; it is covered in Appendix 1 of the AISC specification [1]. In plastic design, the structure is assumed to undergo inelastic deformations under the factored loads from the formation of plastic hinges [2]. The only difference between the plastic design (PD) method and the LRFD method is in the analysis procedure used to determine the load effects. In fact, the load combinations introduced in Chapter 2 apply to both methods. In plastic design, the load effect (e.g., required plastic moment strength) is determined using *plastic* structural analysis, whereas the LRFD method uses *elastic* structural analysis to determine the required moment strength, M_u. For both design methods, the material or design strength, ϕM_n, of the member is determined in exactly the same way. In this text, the focus will be on the plastic design of continuous beams and girders. More extensive coverage of plastic design of statically indeterminate frames can be found in reference 3.

The process of plastic hinge formation is as follows. When a steel member is subjected to a bending moment, M, the stress in the critical section increases from zero, as the load is increased, until the yield stress is reached at the critical section when the moment reaches M_y, the yield moment. As the load is further increased, the stress at this critical section can no longer increase beyond the yield stress, F_y, at the extreme fibers, but the yielding spreads further into the section between the neutral axis and the extreme fibers (see Figure B-1). As the load or moment is increased further, the yielding at the critical section continues to

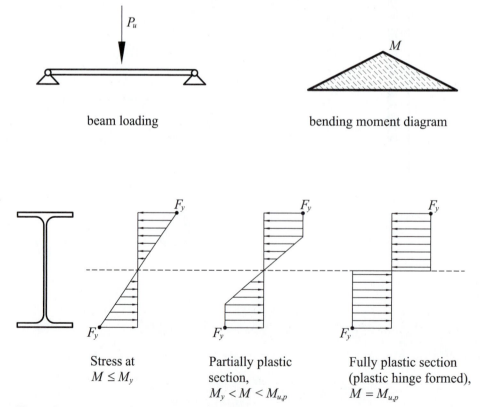

beam loading   bending moment diagram

Stress at $M \leq M_y$

Partially plastic section, $M_y < M < M_{u,p}$

Fully plastic section (plastic hinge formed), $M = M_{u,p}$

Figure B-1 Process of plastic hinge formation.

spread until the whole section reaches the yield stress, F_y. At this stage, the section is now in the plastic state and a plastic hinge forms at that section. Beyond this point, this plastified section can no longer participate in resisting any increase in load or moment, and any further increase in load now has to be resisted, through a process of redistribution of moments and stresses, by other yet unplastified sections, although there will be increased strain or rotations at the already plastified sections.

This process of plastification is continued until an adequate number of plastic hinges are formed at all critical sections within the span to create a collapse mechanism. The principle of equilibrium or the principle of virtual work can then be used to determine the plastic moment strength or capacity of the structural member. It should be noted that in order for the plastic moment capacity to be reached, the ductility of the member must be ensured and lateral torsional buckling must be prevented by adequately bracing the member. If plastic design is used, the maximum permitted unbraced length for the compression flange of W-shaped members is obtained from Section 1.7 of the appendix of the AISC specification as

$$L_{pd} = \left[0.12 + 0.076\left(\frac{M_1}{M_2}\right) \right]\left(\frac{E}{F_y}\right)r_y,$$

where

M_1 = Smaller moment at the end of the unbraced length of the beam, in.-kips,

M_2 = Larger moment at the end of the unbraced length of the beam, kip-in. = $M_{u,p}$,

r_y = Radius of gyration about the weak axis,

M_1/M_2 = Positive for double-curvature bending in the beam, and

M_1/M_2 = Negative for single-curvature bending in the beam.

If the maximum permitted unbraced length above cannot be satisfied, the load effects on the member must be obtained using elastic analysis and not plastic analysis. The principle of virtual work is used in this text for plastic analysis and it involves the following steps:

1. Based on the elastic moment distribution, determine where plastic hinges (i.e., apart from the naturally occurring hinges at rollers or pinned supports) need to form within a span to create a mechanism in that span. Note that each span is considered separately and independently of the other spans.

2. Assume a virtual vertical displacement at the plastic hinge within the span; also assume that rotations can only occur at the plastic hinges and at roller or pinned supports. Therefore, the member segments between the plastic hinges and roller or pinned supports deflect rigidly.

3. Determine the virtual rotations at the hinges resulting from the applied virtual displacement.

4. At each formed hinge, excluding pinned or roller supports at the ends of a member, calculate the internal work done by the plastic moment, $M_{u,p}$.

5. Determine the external work done by the applied loads acting through the applied virtual displacement. For a concentrated load, the external work done is the product of the factored concentrated load and the virtual vertical displacement at that load. For a uniform load, the external work done is the product of the factored uniform load and the area of the virtual displacement diagram over the extent of the uniform load.

6. Sum up the external work done by all of the applied factored loads within the span.

7. Sum up all of the internal work done by the plastic moment, $M_{u,p}$, at the formed hinges (i.e., excluding pinned or roller supports at the exterior end of the span).

8. Using the principle of virtual work, equate the total internal work done to the total external work done, and solve for the required plastic moment strength, $M_{u,p}$, as a function of the applied factored loads. $M_{u,p}$ is calculated for each span separately and independently of the other spans, and the largest value is the required plastic moment strength for the continuous beam or girder.

9. Determine the required plastic section modulus, Z_x, using the limit states design relationship that the required plastic moment strength, $M_{u,p}$, shall not be greater than the design strength, ϕM_n (i.e., $M_{u,p} \leq \phi M_n = \phi Z_x F_y$).

We will now illustrate the plastic design of continuous beams and girders with several examples.

EXAMPLE B-1

Plastic Load Capacity for Statically Indeterminate Beam

Determine the plastic load capacity for the statically indeterminate beams shown in Figure B-2.

(continued)

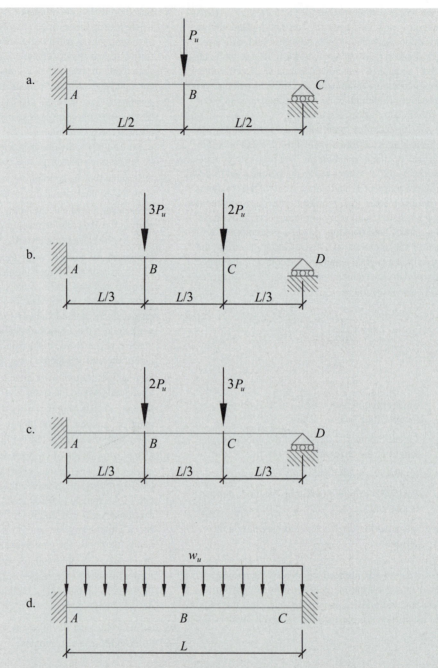

Figure B-2 Statically indeterminate beams for Example B-1.

SOLUTION

(a) For this problem, two plastic hinges are necessary to create a collapse mechanism (i.e., an unstable structure that will continue to undergo deflection without any increase in load). The plastic hinges will form at points *A* and *B*. Because point *C* is already a hinge, these hinges will result in a mechanism or an unstable structure.

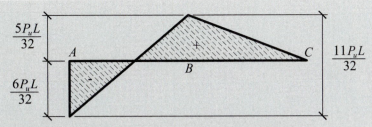

Figure B-3 Elastic moment distribution for Figure B-2a.

From the elastic moment distribution shown in Figure B-3, the plastic hinge will first form at point A because point A has the higher elasic moment compared with point B. Figure B-4 shows the beam in its deflected shape with plastic hinges formed at points A and B. The member segments between the hinges are assumed to deflect rigidly.

To determine whether there are sufficient hinges to create a mechanism or an unstable structure, the reader should recall from structural analysis that Figure B-4 is an unstable structure and thus a mechanism. Assuming that point A rotates through an angle, θ, as shown, the rotation at point C, by geometry, will also be θ, and the rotation at point B will therefore be 2θ (i.e., $\theta + \theta$).

The total internal work done by the plastic moments at the plastic hinges is the sum of the product of the plastic moment and the corresponding rotations. Thus, the total internal work done is $M_{u,p}\,\theta + M_{u,p}\,(2\theta) = 3M_{u,p}\,\theta$.

The total external work done by the applied factored loads is Σ(Factored concentrated load $\times$ Vertical displacement of the mechanism at the load) $= P_u\,(L/2)\theta = P_u L\theta/2$.

Using the principle of virtual work, the internal work done must be equal to the external work done; thus,

$$3M_{u,p}\theta = P_u L\theta/2.$$

Therefore, the required plastic moment strength, $M_{u,p} = P_u L/6$.

A steel section can be selected that will provide this required plastic moment.

(b) For this problem, the plastic hinges will form at point A and at point B or C. Looking at the applied loads, it is more likely that the the plastic hinge will form at point B rather than at point C.

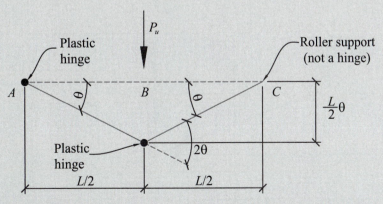

Figure B-4 Virtual displacement of the beam due to plastic hinges for Figure B-2a.

(continued)

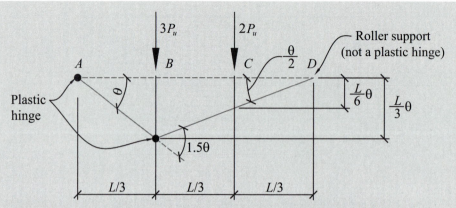

Figure B-5 Virtual displacement of the beam due to plastic hinges for Figure B-2b.

From Figure B-5, the total internal work done at the plastic hinges is $M_{u,p}\,\theta + M_{u,p}(1.5\theta) = 2.5\,M_{u,p}\,\theta$.

The total external work done by the factored loads is Σ(Factored concentrated load $\times$ Vertical displacement of the mechanism at the load) $= 3P_u(L/3)\theta + 2P_u(L/6)\theta = 1.33P_uL\theta$.

Using the principle of virtual work, the internal work done must be equal to the external work done; thus,

$$2.5M_{u,p}\theta = 1.33P_uL.$$

Therefore, the required plastic moment capacity is $M_{u,p} = 0.533P_uL$.

A section can be selected that will provide this required plastic moment.

(c) For this problem, plastic hinges will form at point A and at point B or C, depending on the elastic moment distribution. Given the higher load applied at point C, it is more likely that a plastic hinge will form there first before forming at point B.

From Figure B-6a, the total internal work done at the plastic hinges is $M_{u,p}\theta + M_{u,p}(3\theta) = 4M_{u,p}\theta$.

The total external work done by the factored loads is Σ(Factored concentrated load $\times$ Vertical displacement of the mechanism at the load) $= 2P_u(L/3)\theta + 3P_u(2/3L\theta) = 2.67P_uL\theta$.

Using the principle of virtual work, the internal work done must be equal to the external work done; thus,

$$4M_{u,p}\theta = 2.67P_uL\theta.$$

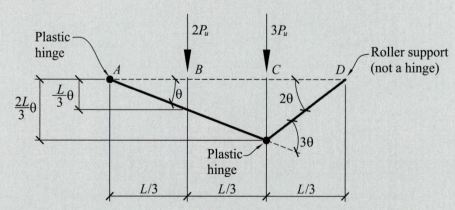

Figure B-6a Virtual displacement of the beam due to plastic hinges at points A and C in Figure B-2c.

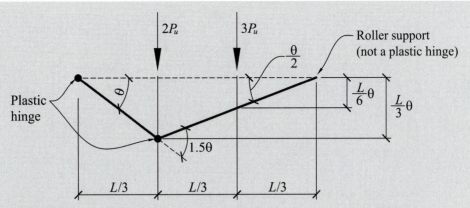

Figure B-6b Virtual displacement of the beam due to plastic hinges at points A and B in Figure B-2c.

Therefore, the required plastic moment capacity, $M_{u,p} = 0.667 P_u L$.

As a check, it can be verified that the plastic hinge does occur at point C and not at point B. If a plastic hinge were assumed to form at point B instead of point C, the deflected shape of the mechanism would be as shown in Figure B-6b.

From Figure B-6b, the total internal work done at the plastic hinges is $M_{u,p}\theta + M_{u,p}(1.5\theta) = 2.5M_{u,p}\theta$.

The total external work done by the factored loads is Σ(Factored concentrated load $\times$ Vertical displacement at the load) $= 2P_u(L/3)\theta + 3P_u(L/6)\theta = 1.17 P_u L\theta$.

Using the principle of virtual work, the internal work done must be equal to the external work done; thus,

$$2.5M_{u,p}\theta = 1.17 P_u L\theta.$$

Therefore, the required plastic moment capacity, $M_p = 0.467\ P_u L$, which is less than the plastic moment capacity of $0.667 P_u L$ required for the plastic hinge forming at point C. Therefore, the condition with the plastic hinge at point C requires a higher capacity and thus governs the design, as expected.

(d) For this problem, based on the elastic moment distribution (see Figure B-7), the plastic hinges will form at points A, B, and C to create a mechanism.

From Figure B-8, the total internal work done at the plastic hinges is $M_{u,p}\theta + M_{u,p}(2\theta) + M_{u,p}\theta = 4M_p\theta$.

The total external work done by the factored uniformly distributed loads is Σ(Factored uniform load $\times$ Area of the deflected shape of the mechanism under the load) $= w_u\,(\frac{1}{2}L)(L/2)\theta = w_u L^2\theta/4$.

Using the principle of virtual work, the internal work done must be equal to the external work done; thus,

$$4M_{u,p}\theta = w_u L^2\theta/4.$$

Therefore, the required plastic moment capacity, $M_{u,p} = w_u L^2/16$.

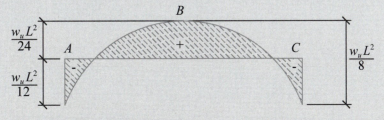

Figure B-7 Elastic moment distribution for Figure B-2d.

(continued)

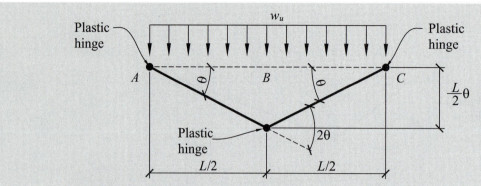

Figure B-8 Virtual displacement of the beam due to plastic hinges for Figure B-2d.

EXAMPLE B-2

Plastic Moment Capacity for a grider with Uniform Loads

Determine the required plastic moment, $M_{u,p}$, for the statically indeterminate three-span girder shown in Figure B-9 with a typical span of 40 ft. The girder is part of a floor framing with a uniformly distributed dead load of 3.0 kips/ft. and a uniformly distributed live load of 3.0 kips/ft. Select a W-shaped section to support this load. Assume ASTM A992 steel.

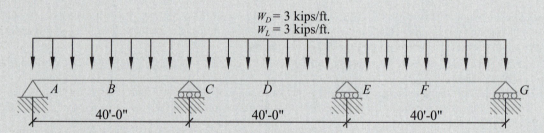

Figure B-9 Statically indeterminate beams for Example B-2.

SOLUTION

Based on the elastic moment distribution (see Figure B-10), the collapse mechanism will form first in the end spans with plastic hinges at points B (not at midspan), and C or at points E, and F (not at midspan). If plastic hinges were to form

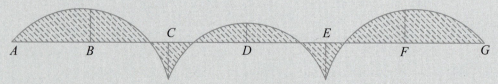

Figure B-10 Elastic moment distribution for Example B-2.

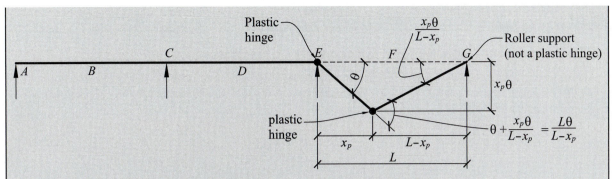

Figure B-11 Virtual displacement of the beam due to plastic hinges in span *EFG*.

in the middle span, based on the symmetrical loading, they would form at points C and E and at the midspan between points C and E. Considering span EFG, assume that the distance from the plastic hinge at point F to the first interior support is x_p.

From Figure B-11, the total internal work done at the plastic hinges is $M_{u,p}\theta + M_{u,p}\left(\dfrac{L}{L - x_p}\right)\theta$.

The total external work done by the factored uniformly distributed loads is Σ(Factored uniform load $\times$ Area of the deflected shape of the mechanism under the load) $= \dfrac{1}{2}(L)(x_p\theta)w_u$.

Using the principle of virtual work, the internal work done must be equal to the external work done; therefore,

$$\frac{1}{2}Lx_pw_u\theta = M_{u,p}\left(\frac{2L - x_p}{L - x_p}\right)\theta.$$

Thus, the plastic uniform load capacity, $w_u = \dfrac{2M_{u,p}}{L}\left(\dfrac{2L - x_p}{Lx_p - x_p^2}\right)$.

The distance x_p required to obtain the maximum load capacity can be determined using calculus principles by requiring that $\dfrac{dw_u}{dx_p} = 0$, which yields $x_p = 0.586L$.

Therefore, the required plastic moment capacity is obtained by solving the equation above:

$$\frac{1}{2}L(0.586L)w_u = M_{u,p}\left(\frac{2L - 0.586L}{L - 0.586L}\right).$$

Thus, $M_{u,p} = 0.0858w_uL^2$.

The factored load, $w_u = 1.2w_D + 1.6w_L = 1.2(3.0 \text{ kips/ft.}) + 1.6(3.0 \text{ kips/ft.}) = 8.4 \text{ kips/ft.}$

Therefore, $M_{u,p} = 0.0858(8.4 \text{ kips/ft.})(40 \text{ ft.})^2 = 1153 \text{ ft.-kips} \le \phi M_n = \phi Z_x F_y.$

Hence, the section modulus required, $Z_x = \dfrac{M_{u,p}}{\phi F_y} = \dfrac{1153 \text{ ft.-kips}(12 \text{ in./ft.})}{(0.9)(50 \text{ ksi})} = 307.5 \text{ in}^3.$

Therefore, use W30 $\times$ 99 ($Z_x = 312 \text{ in.}^3$) (see AISCM, Table 3-2).

EXAMPLE B-3

Plastic Moment Capacity for a Girder with Concentrated Loads

Determine the required plastic moment capacity M_p, for the statically indeterminate three-span girder shown in Figure B-12 with a typical span of 35 ft. and the service loads shown. Select a W-shaped section to support this load. Assume ASTM A992 steel.

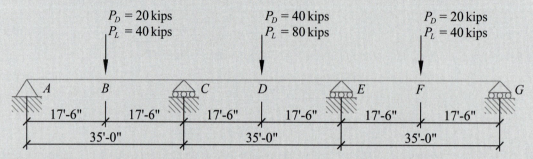

Figure B-12 Statically indeterminate beams for Example B-3.

SOLUTION

We will investigate the maximum load required for mechanisms to develop in spans ABC, CDE, and EFG. The mechanism that yields the highest required plastic moment will be the governing mechanism.

Span ABC:

For this span, the factored concentrated load at midspan, $P_u = 1.2P_D + 1.6P_L = 1.2(20 \text{ kips}) + 1.6(40 \text{ kips}) = 88$ kips.

The location of the hinges in this span will be at the concentrated load at point B and at point C as shown in Figure B-13.

From Figure B-13, the total internal work done at the plastic hinges is $M_{u,p}\theta + M_{u,p}(2\theta) = 3M_{u,p}\theta$.

The total external work done by the factored uniformly distributed loads is Σ(Factored concentrated load $\times$ Vertical displacement of the mechanism at the load) $= P_{u,ABC}\left(\dfrac{L}{2}\theta\right)$.

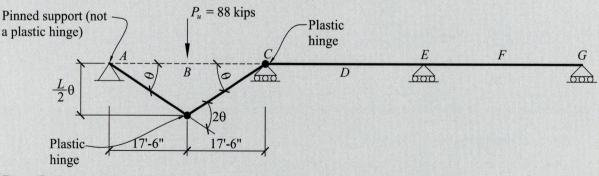

Figure B-13 Virtual displacement in span ABC.

Using the principle of virtual work, the internal work done must be equal to the external work done; thus,

$$3M_{u,p}\theta = P_{u,ABC}\left(\frac{L}{2}\theta\right) = (88 \text{ kips})\left(\frac{35 \text{ ft.}}{2}\right)\theta.$$

Therefore, $M_{u,p} = 513.3$ ft.-kips.

Span *CDE*:

For this span, the total factored concentrated load at midspan, $P_u = 1.2P_D + 1.6P_L = 1.2(40 \text{ kips}) + 1.6(80 \text{ kips}) = 176$ kips.

The location of the hinges in this span will be at the concentrated load at point *D* and at points *C* and *E* as shown in Figure B-14.

From Figure B-14, the total internal work done at the plastic hinges is $M_{u,p}\theta + M_{u,p}(2\theta) + M_{u,p}\theta = 4M_{u,p}\theta.$

The total external work done by the factored uniformly distributed loads is Σ(Factored concentrated load $\times$ Vertical displacement of the mechanism at the load) $= P_{u,CDE}\left(\frac{L}{2}\theta\right)$. Using the principle of virtual work, the internal work done must be equal to the external work done; thus,

$$4M_{u,p}\theta = P_{u,CDE}\left(\frac{L}{2}\theta\right) = (176 \text{ kips})\left(\frac{35 \text{ ft.}}{2}\right)\theta.$$

Therefore, $M_{u,p} = 770$ ft.-kips. Governs

Span *EFG*:

The location of hinges for this span will be similar to that for span *ABC* and thus the required plastic moment will also be $M_{u,p} = 513.3$ ft.-kips.

The highest required plastic moment for all three spans will govern the capacity of the girder. Using the limit states design equation ($M_u \leq \phi M_n$) yields

$$770 \text{ ft-kips} \leq \phi M_n = \phi Z_x F_y.$$

The required section modulus, $Z_x = \dfrac{M_{u,p}}{\phi F_y} = \dfrac{770 \text{ ft.-kips}(12 \text{ in./ft.})}{0.9(50 \text{ ksi})} = 205.3 \text{ in.}^3.$

Therefore, use W24 $\times$ 84 ($Z_x = 224 \text{ in.}^3$).

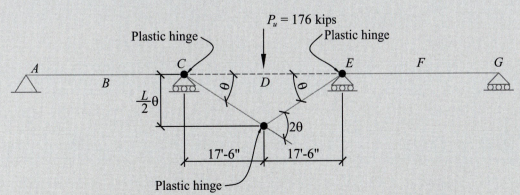

Figure B-14 Deflected shape of the beam after formation of plastic hinges in span *CDE*.

B.2 REFERENCES

1. American Institute of Steel Construction. 2006. *Steel construction manual*, 13th ed. Chicago: Author.

2. Bhatt, P, and H. M. Nelson. 1990. *Marshall and Nelson's structures*, 3rd ed. London: Longman.

3. Disque, R. O. 1971. *Applied design in steel*. New York: Van Nostrand Reinhold.

B.3 PROBLEMS

B-1. Determine the required plastic moment capacity, $M_{u,p}$, for the statically indeterminate two-span girder shown in Figure B-15. Select a W-shaped section to support this load. Assume ASTM A992 steel.

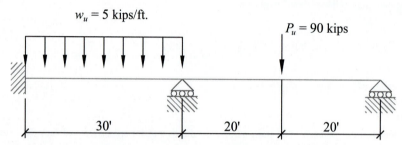

Figure B-15 Statically indeterminate beams for Problem B-1.

INDEX